VOLUME 2

Molecular Cloning

A LABORATORY MANUAL

FOURTH EDITION

OTHER TITLES FROM CSHL PRESS

LABORATORY MANUALS

Antibodies: A Laboratory Manual
Imaging: A Laboratory Manual
Live Cell Imaging: A Laboratory Manual, 2nd Edition
Manipulating the Mouse Embryo: A Laboratory Manual, 3rd Edition
RNA: A Laboratory Manual

HANDBOOKS

Lab Math: A Handbook of Measurements, Calculations, and Other Quantitative Skills for Use at the Bench
Lab Ref, Volume 1: A Handbook of Recipes, Reagents, and Other Reference Tools for Use at the Bench
Lab Ref, Volume 2: A Handbook of Recipes, Reagents, and Other Reference Tools for Use at the Bench
Statistics at the Bench: A Step-by-Step Handbook for Biologists

WEBSITES

Molecular Cloning, A Laboratory Manual, 4th Edition, *www.molecularcloning.org*
Cold Spring Harbor Protocols, www.cshprotocols.org

Molecular Cloning
A LABORATORY MANUAL

FOURTH EDITION

Michael R. Green

Howard Hughes Medical Institute
Programs in Gene Function and Expression and in Molecular Medicine
University of Massachusetts Medical School

Joseph Sambrook

Peter MacCallum Cancer Centre and the
Peter MacCallum Department of Oncology
The University of Melbourne, Australia

COLD SPRING HARBOR LABORATORY PRESS
Cold Spring Harbor, New York • www.cshlpress.org

Molecular Cloning
A LABORATORY MANUAL

FOURTH EDITION

Publisher	John Inglis
Acquisition Editors	John Inglis, Ann Boyle, Alexander Gann
Director of Development, Marketing, & Sales	Jan Argentine
Developmental Editors	Judy Cuddihy, Kaaren Janssen, Michael Zierler
Project Manager	Maryliz Dickerson
Permissions Coordinator	Carol Brown
Production Editor	Kathleen Bubbeo
Production Manager	Denise Weiss
Sales Account Manager	Elizabeth Powers

Cover art direction and design: Pete Jeffs, 2012

Library of Congress Cataloging-in-Publication Data

Green, Michael R. (Michael Richard), 1954-
  Molecular cloning : a laboratory manual / Michael R. Green,
Joseph Sambrook. – 4th ed.
     p. cm.
  Rev. ed. of: Molecular cloning : a laboratory manual / Joseph
Sambrook, David W. Russell. 2001.
  Includes bibliographical references and index.
  ISBN 978-1-936113-41-5 (cloth) – ISBN 978-1-936113-42-2
(pbk.)
  1. Molecular cloning–Laboratory manuals. I. Sambrook, Joseph.
II. Sambrook, Joseph. Molecular cloning. III. Title.

QH442.2.S26 2012
572.8–dc23
                          2012002613

10 9 8 7 6

Contents

VOLUME 1

PART 1: ESSENTIALS

CHAPTER 1

Isolation and Quantification of DNA   1

CHAPTER 2
Analysis of DNA   81

CHAPTER 3
Cloning and Transformation with Plasmid Vectors   157

CHAPTER 4

Gateway Recombinational Cloning   261

John S. Reece-Hoyes and Albertha J.M. Walhout

CHAPTER 5

Working with Bacterial Artificial Chromosomes and Other High-Capacity Vectors   281

Nathaniel Heintz and Shiaoching Gong

CHAPTER 6

Extraction, Purification, and Analysis of RNA from Eukaryotic Cells   345

CHAPTER 7

Polymerase Chain Reaction   455

VOLUME 2

PART 2: ANALYSIS AND MANIPULATION OF DNA AND RNA

CHAPTER 10

Nucleic Acid Platform Technologies   683

Oliver Rando

CHAPTER 11

DNA Sequencing   735

Elaine Mardis and W. Richard McCombie

CHAPTER 12

Analysis of DNA Methylation in Mammalian Cells 893

Paul M. Lizardi, Qin Yan, and Narendra Wajapeyee

INTRODUCTION

CHAPTER 13
Preparation of Labeled DNA, RNA, and Oligonucleotide Probes   943

INFORMATION PANELS

CHAPTER 14

Methods for In Vitro Mutagenesis 1059

Matteo Forloni, Alex Y. Liu, and Narendra Wajapeyee

INTRODUCTION

PROTOCOLS

PART 3: INTRODUCING GENES INTO CELLS

CHAPTER 15

Introducing Genes into Cultured Mammalian Cells   1131

Priti Kumar, Arvindhan Nagarajan, and Pradeep D. Uchil

CHAPTER 16

Introducing Genes into Mammalian Cells: Viral Vectors 1209

Guangping Gao and Miguel Sena-Esteves

VOLUME 3

PART 4: GENE EXPRESSION

CHAPTER 17

Analysis of Gene Regulation Using Reporter Systems   1335

Pradeep D. Uchil, Arvindhan Nagarajan, and Priti Kumar

PART 5: INTERACTION ANALYSIS

CHAPTER 20

Cross-Linking Technologies for Analysis of Chromatin Structure and Function 1637

Tae Hoon Kim and Job Dekker

CHAPTER 21

Mapping of In Vivo RNA-Binding Sites by UV-Cross-Linking Immunoprecipitation (CLIP)   1703

Jennifer C. Darnell, Aldo Mele, Ka Ying Sharon Hung, and Robert B. Darnell

APPENDICES

APPENDIX 2
Commonly Used Techniques   1843

APPENDIX 3
Detection Systems   1855

List of Tables

CHAPTER 15

Introducing Genes into Cultured Mammalian Cells

CHAPTER 16

Introducing Genes into Mammalian Cells: Viral Vectors

CHAPTER 17

Analysis of Gene Regulation Using Reporter Systems

CHAPTER 18

RNA Interference and Small RNA Analysis

Preface

THE READY AVAILABILITY OF THE COMPLETE GENOME SEQUENCES for humans and model organisms has profoundly affected the way in which biologists of every discipline now practice science. Exploration of the vast genomic landscape required the development of a great variety of new experimental techniques and approaches. Inevitably, venerable cloning manuals became outdated and established methods were rendered obsolete. These events provided the major impetus for a thorough revision of *Molecular Cloning*.

In the first stages of *Molecular Cloning 4* (MC4), we undertook an extensive review process to determine what old material should be kept, what new material should be added, and, most difficult of all, what material should be deleted. The names of the many scientists who contributed invaluable advice during the review process are listed on the following acknowledgments page. We are grateful to them all.

It is of course impossible to include in any single laboratory manual all experimental methods used in molecular biology, and so choices, sometimes tough ones, had to be made. Individuals may disagree with some of our choices (and we suspect some will). However, our two guiding principles are as follows. First, *Molecular Cloning* is a "nucleic acid–centric" laboratory manual, and therefore in general we have not included methods that do not directly involve DNA or RNA. So, although MC4 contains a chapter on yeast two-hybrid procedures for analyzing protein–protein interactions, we did not include the many other approaches for studying protein–protein interactions that do not directly involve nucleic acids. Second, we have attempted to include as many as possible of those nucleic acid–based methods that are widely used in molecular and cellular biology laboratories with the hope in the John Lockean sense that we "do the greatest good for the most people." The harder task for us was to decide which material to delete. However, our task was made somewhat easier by the agreement of our publisher, Cold Spring Harbor Laboratory Press, to make these older methods freely available on the *Cold Spring Harbor Protocols* website (www.cshprotocols.org).

The explosion of new experimental approaches has made it impractical, if not impossible, for any single person (or even two) to write with authority on all the relevant experimental methods. As a result, a major departure from previous editions of *Molecular Cloning* was the commissioning of experts in the field to write specific chapters and to contribute specific protocols. Without the enthusiastic participation of these scientists, MC4 simply would not exist.

Since the last edition of *Molecular Cloning*, there has been a relentless and continuing proliferation of commercial "kits," which is both a blessing and a curse. On the one hand, kits offer tremendous convenience, particularly for procedures that are not routinely used in an individual laboratory. On the other hand, kits can often be too convenient, enabling users to perform procedures without understanding the underlying principles of the method. Where possible, we have attempted to deal with this dilemma by providing lists of commercially available kits and also describing how they work.

There are many people who played essential roles in the production of MC4 and whom we gratefully acknowledge. Ann Boyle was instrumental in getting MC4 off the ground and played a critical organizational role during the early stages of the project. Subsequently, her responsibilities were taken over by the able assistance of Alex Gann. Sara Deibler contributed at all stages of

MC4 in many ways, but in particular in assisting with writing, editing, and proofreading. Monica Salani made substantial contributions to the content and writing of Chapter 9.

We are especially grateful for the enthusiastic support, extraordinary cooperation, and tolerance of the staff of the Cold Spring Harbor Laboratory Press—in particular Jan Argentine, who managed the entire project and kept a close eye on its finances; Maryliz Dickerson, our project manager; our developmental editors Kaaren Janssen, Judy Cuddihy, and Michael Zierler; Denise Weiss, the Production Manager; our production editor, Kathleen Bubbeo; and, of course, John Inglis, the éminence grise of Cold Spring Harbor Laboratory Press.

MICHAEL R. GREEN
JOSEPH SAMBROOK

A Note from the Publisher

Readers are encouraged to visit the website www.molecularcloning.org to obtain up-to-date information about all aspects of this book and its contents.

Acknowledgments

The authors wish to thank the following colleagues for their valuable assistance:

H. Efsun Arda

Michael F. Carey

Darryl Conte

Job Dekker

Claude Gazin

Paul Kaufman

Nathan Lawson

Chengjian Li

Ling Lin

Donald Rio

Sarah Sheppard

Stephen Smale

Narendra Wajapeyee

Marian Walhout

Phillip Zamore

Maria Zapp

The Publisher wishes to thank the following:

Paula Bubulya

Tom Bubulya

Nicole Nichols

Sathees Raghavan

Barton Slatko

CHAPTER 9

Quantification of DNA and RNA by Real-Time Polymerase Chain Reaction

POLYMERASE CHAIN REACTION (PCR)-BASED METHODS that simply measure the amount of amplified product that has accumulated over the course of the reaction (so-called end-point methods)
cannot be reliably used to quantify the amount of nucleic acid present in a sample. The primary
reason lies in the fact that as the reaction progresses, primers and nucleotides are consumed and
eventually become limiting, which reduces the efficiency of amplification. Because PCR is an exponential process, small differences in the efficiency at each cycle of the reaction can lead to large differences in the yield of the amplified product. Thus, in traditional PCR, there is a nonlinear
relationship between the starting copy number and the final yield of the amplified product.

Rudimentary methods for quantitative PCR typically rely on the addition of known amounts of
an external reference nucleic acid, which is amplified using the same set of primers as those used to
amplify the target sequence. Quantification is achieved by comparing the amount of amplified

products generated by the reference and target sequences and determining the ratio of reference to target sequence in the reaction mixture at cycle zero. Commonly, this approach involves quantifying the amounts of reference and target sequences in consecutive cycles within the exponential phase of the PCR. A variety of different methods can be used to detect and quantify amplified products, including measurement of radioactivity incorporated during amplification or computer analysis of gels stained with ethidium bromide.

In recent years, the ability to quantify nucleic acid concentrations has been revolutionized by the development of sophisticated instruments that amplify specific nucleic acid sequences and simultaneously measure their concentration, allowing the kinetics of PCR amplification to be monitored in "real time" as the reaction progresses. In real-time PCR, also called quantitative real-time PCR [or simply quantitative PCR (qPCR)] or kinetic PCR, the amplification of DNA is monitored by the detection and quantitation of a fluorescent reporter signal, which increases in direct proportion to the amount of PCR product in the reaction. The fluorescent reporter is excited by light from the real-time PCR machine, a fluorescence-detecting thermocycler. By recording the amount of fluorescence emission at each cycle, the PCR can be monitored during the exponential phase when the first significant increase in the amount of PCR product correlates with the initial amount of target template (see PCR in Theory in Chapter 7). The ability to quantify the amplified DNA during the exponential phase of the PCR, when none of the components of the reaction is limiting, has resulted in dramatically improved precision in the quantitation of target sequences. In addition, because of the high sensitivity of fluorometric detection, real-time PCR is capable of measuring the initial concentration of target DNA over a vast dynamic range (up to eight or nine orders of magnitude) and with a high degree of sensitivity (as little as one copy of template DNA).

Real-time PCR is used in a wide variety of applications in both basic research and clinical settings (Klein 2002). In research laboratories, real-time PCR, combined with reverse transcription (and called real-time RT-PCR or quantitative RT-PCR), can be used to quantify microRNA (miRNA) levels (Benes and Castoldi 2010) and has become the preferred method to quantify messenger RNA (mRNA) levels and to validate gene expression data generated by microarray analysis and other genomics techniques (Nolan et al. 2006). Real-time PCR can also be used to detect changes in gene copy number (D'Haene et al. 2010) or quantify the abundance of a particular DNA sequence in a sample, such as that derived from chromatin immunoprecipitation (Taneyhill and Adams 2008). In the clinical setting, real-time PCR has facilitated the development of new biomedical diagnostics for detecting pathogenic microorganisms (Gupta et al. 2008), quantifying viral load (Niesters 2001), and monitoring residual disease in cancer patients (Martinelli et al. 2006). Real-time PCR is also used for allelic discrimination and identification, as well as single nucleotide polymorphism (SNP) genotyping (please see the information panel SNP Genotyping).

The chief advantages of real-time PCR are its ability to measure the concentrations of nucleic acids over a vast dynamic range, its high sensitivity, and its capacity to process many samples simultaneously. Although it is a powerful technique, researchers often face challenges in reliability and reproducibility because of the lack of assay standardization. Therefore, it is critical to optimize the reagents and reaction conditions, include proper internal and external controls, and perform rigorous data analysis in order to generate accurate and reproducible results in real-time PCR experiments.

REAL-TIME PCR CHEMISTRIES

There are several chemistries available that allow detection of real-time PCR products via the generation of a fluorescent signal (Mackay and Landt 2007). They can be classified into three basic categories: DNA-binding dyes, probe-based chemistries, and quenched dye primers (Table 1).

DNA-Binding Dyes

Fluorogenic DNA-binding dyes bind to double-stranded DNA in a non-sequence-specific manner. Following excitation, these fluorogenic dyes show little fluorescence when in solution but emit a

TABLE 1. Summary of real-time PCR chemistries

	DNA-binding dyes	Probe-based chemistries	Quenched dye primers
Basic principle	Uses a fluorogenic, nonspecific DNA-binding dye to detect a PCR product as it accumulates during PCR	Uses one or more fluorogenic oligonucleotide probes to detect a PCR product as it accumulates during PCR; relies on FRET	Uses a fluorogenic primer that becomes incorporated into the PCR product; relies on FRET
Specificity	Detects all amplified double-stranded DNA, including nonspecific reaction products, such as primer dimers	Detects specific amplification products only	Detects amplification products and nonspecific reaction products, such as primer dimers
Applications	DNA and RNA quantitation; validation of gene expression	DNA and RNA quantitation; validation of gene expression; allelic discrimination; SNP genotyping; pathogen and viral detection; multiplex PCR	DNA and RNA quantitation; validation of gene expression; allelic discrimination; SNP genotyping; pathogen and viral detection; multiplex PCR
Advantages	Allows quantification of any double-stranded DNA sequence; does not require probes, thereby reducing assay setup and running costs; useful for analyzing a large number of genes; simple to use	Specific hybridization between the probe and target is required to generate a fluorescent signal, thereby reducing background and false positives; probes can be labeled with different, distinguishable reporter dyes, allowing multiplex reactions	Specific hybridization between the probe and target is required to generate a fluorescent signal, thereby reducing background and false positives; probes can be labeled with different, distinguishable reporter dyes, allowing multiplex reactions
Disadvantages	Because the dye detects both specific and nonspecific PCR products, may generate false-positive signals; requires a post-PCR processing step	A different probe must be synthesized for each unique target sequence, increasing material costs	A different probe must be synthesized for each unique target sequence, increasing material costs
Examples	SYBR Green I	TaqMan, molecular beacon, Scorpion, and hybridization probes	Amplifluor and LUX fluorogenic primers

strong fluorescent signal after binding to DNA (Fig. 1). Because many dye molecules can bind to each DNA product, the intensity of signal generated is high and proportional to the total mass of DNA generated during the PCR.

The most commonly used fluorogenic DNA-binding dye is SYBR Green I. In fact, SYBR Green I is currently the most popular real-time PCR chemistry overall, because of its relatively high sensitivity and reliability, low cost, and simplicity of use. Unlike probe-based chemistries and quenched dye primers, which require the synthesis of expensive fluorescently labeled oligonucleotides for each target to be analyzed (explained in more detail below), SYBR Green I is an economical option, especially for assays involving a large number of genes. The absorbance and emission maxima for SYBR Green I are 494 nm and 521 nm, respectively, and following binding to double-stranded DNA, the fluorescence of the dye is enhanced ~1000-fold. SYBR Green I has been reported to have reduced sensitivity compared with probe-based chemistries; however, as with conventional PCR,

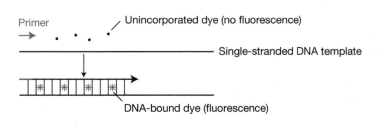

FIGURE 1. **The fluorescence intensity of fluorogenic dyes increases following binding to double-stranded DNA.** Fluorogenic dyes show little fluorescence when in solution, but, following excitation, emit a strong fluorescent signal when bound to double-stranded DNA. Fluorogenic DNA-binding dyes bind to double-stranded DNA in a sequence-independent manner.

this sensitivity is largely dependent on the primers used in the PCR. With appropriate design, SYBR Green I offers comparable or better sensitivity than probe-based applications (Schmittgen et al. 2000; Newby et al. 2003). Other fluorogenic DNA-binding dyes for use in real-time PCR include LC Green I (Wittwer et al. 2003), SYTO 9 (Monis et al. 2005), and a family of asymmetric cyanine minor groove binders: BEBO (Bengtsson et al. 2003; Karlsson et al. 2003b), BETO (Karlsson et al. 2003a), BETIBO (Ahmad and Ghasemi 2007), BOXTO (Karlsson et al. 2003a), BOXTO-PRO (Eriksson et al. 2006), and BOXTO-MEE (Eriksson et al. 2006), which bind to both single- and double-stranded DNA.

The disadvantage of fluorogenic DNA-binding dyes is that they bind to any double-stranded DNA in the reaction, including primer dimers (which generate much of the background noise in real-time PCR) and other nonspecific reaction products. Because the fluorescence intensity of the dye depends on the total amount of double-stranded DNA, the lack of specificity can result in an overestimation of the target concentration. In the worst cases, the emitted fluorescence may bear little or no relationship to the amount of starting target DNA or to the amount of full-length product produced. For this reason, detection by fluorogenic DNA-binding dyes requires extensive optimization and follow-up analyses to validate the results. Often, at the end of the PCR, a melting curve (also called a thermal denaturation curve or dissociation curve) of the amplified DNA is generated to measure the melting temperature (T_m) of the PCR product(s). The shape of the melting curve indicates whether the amplified products are homogeneous, and the T_m provides reassurance that the correct product has been specifically amplified (for more detailed information, see Protocol 1). Primer dimers, which are short in length, generally denature at much lower temperatures and can easily be distinguished from the amplified target DNA. However, for single PCR product reactions with well-designed primers (i.e., those aimed at generating highly specific amplification), SYBR Green I can work extremely well, with spurious nonspecific background only showing up in very late cycles.

In combination with a well-designed assay that generates highly specific amplicons, DNA-binding dyes can be used for nucleic acid quantification and validation of gene expression. However, for diagnostic applications, it is generally advisable to use probe-based assays to guarantee specificity of the amplified product.

Probe-Based Chemistries

Probe-based chemistries involve one or more fluorescently labeled oligonucleotides that hybridize specifically to an internal sequence of the amplified product. The strength of the fluorescent signal is therefore proportional to the amount of the target amplicon and is not influenced by the accumulation of nonspecific products such as primer dimers. In general, probe-based chemistries use a pair of fluorescent dyes, which may be present on the same oligonucleotide or two different oligonucleotides, and depend on fluorescence (or Förster) resonance energy transfer (FRET) from one fluorophore to the other to generate the fluorescence signal. For FRET to occur, the emission spectrum of one dye must overlap with the excitation spectrum of the other dye, such that a transfer of energy occurs when the two dyes are spatially close to one another.

Probe-based chemistries can be used for all real-time PCR applications, including nucleic acid quantification, gene expression analysis, verification of microarray data, SNP genotyping, and diagnostic tests. In addition, probe-based chemistries allow distinct DNA species to be amplified and measured in the same sample ("multiplex PCR"), because each probe can be designed with a spectrally unique fluorescent dye pair (see the information panel Multiplex PCR). These probes afford a level of discrimination that is not possible with DNA-binding dyes because they will only hybridize to true targets in a PCR and not to primer dimers or other nonspecific products. Therefore, in contrast to DNA-binding dyes, post-PCR processing is not required for probe-based chemistries. However, the main disadvantage of probe-based chemistries is that they are relatively expensive to synthesize; because an individual probe must be synthesized for each target to be analyzed, probe-based chemistries are not an economical option for applications involving analysis of large numbers of genes.

Probe-based chemistries include TaqMan probes, molecular beacons, Scorpion probes, and hybridization probes. Of these, TaqMan probes and molecular beacons are the most popular options.

TaqMan Probes

TaqMan probes (also known as hydrolysis probes or dual-labeled probes) are linear molecules, usually 18–24 bases in length, that contain a fluorescent "reporter" dye covalently attached to the 5′ end and a "quencher" dye coupled to the 3′ end (Heid et al. 1996). In the nonhybridized state, the reporter and quencher dyes are in close proximity. When irradiated, the excited reporter dye transfers energy to the nearby quencher through FRET, thereby suppressing (or quenching) emission of a fluorescent signal from the reporter. During PCR, the TaqMan probe specifically anneals to an internal region of the PCR product between the forward and reverse primers. When *Taq* polymerase extends the primer and replicates the template on which the TaqMan probe is bound, the 5′–3′ nuclease activity of the *Taq* polymerase degrades the probe, thereby separating the reporter and quencher dyes, interrupting FRET, and allowing the reporter to fluoresce (see Fig. 2). Thus, the intensity of fluorescence increases in direct proportion to the amount of target DNA synthesized during the course of the PCR. Because an increase in the fluorescence signal is detected only if the target sequence is complementary to the probe, nonspecific amplification is not detected.

A variety of reporter and quencher dyes are available for TaqMan probes as well as other probe-based chemistries, such as molecular beacons and Scorpion probes (see below) (Table 2).

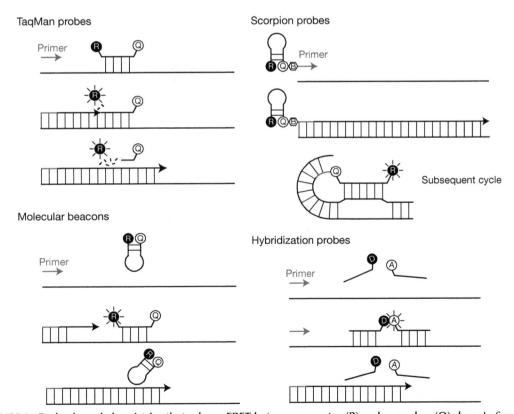

FIGURE 2. **Probe-based chemistries that rely on FRET between reporter (R) and quencher (Q) dyes.** In Scorpion probes, the probe element is separated from the PCR primer sequence by a nonamplifiable "blocker" (B). Hybridization probes are used in pairs. The donor (D) and the acceptor (A) moieties have to be close enough to one another for FRET to occur.

TABLE 2. Commonly used reporter and quencher dyes for real-time PCR

	Maximum absorbance (nm)	Maximum emission (nm)
Reporters		
6-FAM (6-carboxyfluorescein)	495	517
HEX (hexachlorofluorescein)	537	553
TET (tetrachlorofluorescein)	521	538
Cy3 (cyanine 3)	550	570
Cy5 (cyanine 5)	650	667
JOE	520	548
ROX	581	607
TAMRA (tetramethylrhodamine)	550	576
Texas Red	589	610
Quenchers		
BHQ-1 (black-hole quencher)	535	None
BHQ-2	579	None
DABCYL [4-((4-(dimethylamino)phenyl)azo)benzoic acid]	453	None
TAMRA (tetramethylrhodamine)	550	576

Commonly used reporters include the fluorescein derivatives 6-carboxyfluorescein (6-FAM), hexachlorofluorescein (HEX), and tetrachlorofluorescein (TET); cyanine dye family members Cy3 and Cy5; and rhodamine-based dyes tetramethylrhodamine (TAMRA) and Texas Red.

Quenchers must show a spectral overlap with the reporter fluorophore and, thus, should be selected accordingly. Popular quenchers include tetramethylrhodamine (TAMRA) and 4-((4-(dimethylamino)phenyl)azo)benzoic acid (DABCYL), although they each have certain limitations: TAMRA has an inherent fluorescence, resulting in a relatively poor signal-to-noise ratio, and DABCYL has relatively poor spectral overlap with fluorophores emitting above 480 nm, which limits its ability to quench via FRET. To overcome these limitations, a new class of high-efficiency "dark" quenchers has been developed called the "black-hole quencher" dyes (BHQ-1, -2, and -3). BHQ dyes are nonfluorescent, which results in lower background fluorescence and higher signal-to-noise ratios, thereby providing higher sensitivity and a more precise measure of the reporter dye fluorescence (for review, see Marras et al. 2002). In addition, they have been designed to provide maximal spectral overlap, thereby increasing quenching efficiency.

The most common reporter–quencher combination for TaqMan probes is 6-FAM or TET as the 5' reporter dye and TAMRA as the 3' quencher; other commonly used reporter–quencher pairs for TaqMan probes are listed in Table 3.

TABLE 3. Commonly used reporter–quencher pairs for TaqMan probes

5' Reporter dye	3' Quencher dye
HEX	TAMRA, BHQ-1, or BHQ-2
TET	TAMRA or BHQ-1
6-FAM	TAMRA or BHQ-1
JOE	BHQ-1
Cy3	BHQ-2
Cy5	BHQ-2
ROX	BHQ-2
TAMRA	BHQ-2
Texas Red	BHQ-2

TABLE 4. Commonly used reporter–quencher pairs for molecular beacon probes

5′ Reporter dye	3′ Quencher dye
HEX	DABCYL
TET	DABCYL
6-FAM	DABCYL
Cy3	DABCYL
Cy5	DABCYL
TAMRA	DABCYL
ROX	DABCYL
Texas Red	DABCYL

Molecular Beacon Probes

Molecular beacons are hairpin-shaped molecules that, like TaqMan probes, contain a fluorescent reporter dye coupled to the 5′ end and a nonfluorescent quencher moiety attached to the 3′ end (Tyagi and Kramer 1996). Typically, a molecular beacon probe is 28–44 nucleotides in length; the 5–7-nucleotide arms on each end are complementary to each other and form the stem, whereas the central 18–30 nucleotides that form the loop are complementary to the target DNA. When free in solution, the stem portion of the molecule keeps the reporter and quencher moieties in close proximity, causing the fluorescence of the fluorophore to be quenched by FRET (see Fig. 2). During PCR, the central portion of the probe hybridizes to the target, causing separation of the reporter and quencher, which interrupts FRET and results in a fluorescence signal. Although DABCYL is a relatively weak quencher (see above), it is well suited for use in molecular beacons because the hairpin places the quencher in close proximity to the fluorescent dye (Table 4). Unlike TaqMan probes, molecular beacons are designed to remain intact during PCR amplification and must rebind to the target sequence in every cycle for signal measurement.

Scorpion Probes

Scorpion probes are bifunctional molecules that contain a fluorescent probe covalently linked to a PCR primer (Whitcombe et al. 1999). The probe is a hairpin-shaped molecule that has a fluorescent reporter attached to the 5′ end and an internal quencher dye that is directly linked to the 5′ end of the primer (see Fig. 2). The probe element is separated from the PCR primer sequence by a non-amplifiable "blocker" (e.g., hexethylene glycol [HEG]) that prevents amplification of the probe sequence during PCR. The sequence of the hairpin loop is complementary to the extension product of the PCR primer. In the unhybridized state, the Scorpion probe maintains a hairpin configuration that keeps the reporter and quencher in close proximity and fluorescence quenched. During PCR, *Taq* polymerase extends the PCR primer; in the following cycle, the hairpin unfolds, and the loop region of the probe hybridizes intramolecularly to its complementary sequence in the newly synthesized target amplicon (the probe effectively curls back on itself, hence the "scorpion" moniker). In this new arrangement, the reporter and quencher dyes are no longer in close proximity, and a fluorescence signal can be detected. Some reporter–quencher pairs found in Scorpion probes are listed in Table 5.

The chief benefit of Scorpion probes derives from the fact that the probe element is physically coupled to the primer, such that the fluorescent signal is generated by a unimolecular rearrangement. In contrast, other probe chemistries such as TaqMan or Molecular Beacons require bimolecular collisions. The benefits of a unimolecular rearrangement are significant: The reaction is effectively instantaneous and therefore occurs before any competing or side reactions such as target amplicon reannealing or inappropriate target folding. This leads to stronger signals, more reliable probe design, shorter reaction times, and better discrimination.

TABLE 5. Commonly used reporter–quencher pairs for Scorpion probes

5' Reporter dye	Internal quencher dye
HEX	DABCYL, BHQ-1, or BHQ-2
TET	DABCYL or BHQ-1
6-FAM	DABCYL or BHQ-1
Cy3	DABCYL or BHQ-2
Cy5	DABCYL or BHQ-2
TAMRA	DABCYL or BHQ-2
ROX	DABCYL or BHQ-2
Texas Red	DABCYL or BHQ-2
JOE	BHQ-1

Hybridization Probes

The hybridization probes system uses a pair of probes that are designed to hybridize to adjacent regions of the template DNA (separated by 1–5 bases) (Wittwer et al. 1997). In contrast to probe-based chemistries, hybridization probes are designed using a pair of fluorophores—called a "donor" and an "acceptor"—that result in fluorescence only if the two fluorophores are in close proximity (see Fig. 2). Typically, the upstream probe is labeled at the 3' end with the donor dye, such as fluorescein, whereas the downstream probe is labeled at the 5' end with the acceptor dye, such as LC Red 640 or LC Red 705. To prevent extension by the DNA polymerase, the upstream probe is also modified with a nonamplifiable blocker such as a 3'-phosphate group or a carbon-based group (Cradic et al. 2004). The fluorescent signal is measured at the end of the annealing step of the PCR, when the probes are hybridized to their specific target regions; the energy emitted from the donor excites the acceptor dye, which emits a fluorescent signal. The probes subsequently melt off during the high temperature (72°C) elongation step; probes that are still annealed to their target sequence are displaced by the polymerase. (Note that this type of chemistry forms the basis for LightCycler probes, which are designed for use with the LightCycler thermocycler from Roche.) Hybridization probes provide a high level of specificity and sensitivity in signal detection and are suitable for most real-time PCR applications. However, hybridization probes are not suitable for multiplex assays because the presence of additional oligonucleotide hybridization probe(s) in addition to two PCR primers increases the complexity of the PCR system and limits the capability of multiplex detection.

Quenched Dye Primers

Quenched dye primers combine both primer and probe into one molecule: They are fluorescently labeled primers that rely on FRET to provide a fluorescence signal following extension of the primer during PCR. Similar to dye-based methods, quenched dye primers can provide a fluorescent signal even after the formation of nonspecific reaction products, and therefore assays using these primers require the same careful validation as dye-based methods. However, in contrast to dye-based methods, quenched dye primers can be used in multiplex applications. In fact, quenched dye primers are often used for SNP/mutation detection studies, as primers labeled with different fluorophores can be used with one common reverse primer to identify sequences differing by a single nucleotide.

Amplifluor Primers

Like Scorpion probes, Amplifluor primers consist of a primer sequence linked to an intramolecular hairpin structure that normally keeps the fluorescence signal quenched (Fig. 3). In the first cycle of the PCR, the Amplifluor primer anneals to the template and is extended. In subsequent cycles, extension of the opposite strand displaces and opens the hairpin, thereby separating the reporter and quencher fluorophores, leading to a fluorescence signal.

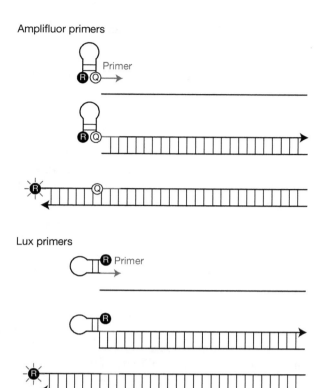

Amplifluor primers

Lux primers

FIGURE 3. **Quenched dye primers combine both primer and probe into one molecule.** Like probe-based chemistries, Amplifluor primers rely on FRET between the reporter (R) and quencher (Q) dyes. LUX primers are labeled with a single fluorophore in the absence of a separate quenching moiety.

LUX Primers

LUX (Light Upon Extension) primers are oligonucleotides, typically 20–30 bases in length, labeled with a single fluorophore in the absence of a separate quenching moiety. This technology is based on the observation that some fluorescent dyes are naturally quenched by nearby guanosine residues and/or the secondary structure of the DNA (Crockett and Wittwer 2001). Accordingly, LUX primers are designed—using proprietary software (D-LUX Designer)—to form a hairpin conformation that places the fluorophore in proximity to a guanosine residue. When the primer is incorporated into the double-stranded PCR product, the hairpin structure is lost and the fluorophore is dequenched, resulting in a significant increase in fluorescent signal (see Fig. 3). Most LUX primers are labeled with either FAM or JOE, but other fluorophores, such as HEX, TET, ROX, and TAMRA, which change their fluorescence depending on the sequence and the structure of the oligonucleotide, can also be used.

INSTRUMENTS FOR REAL-TIME PCR

Real-time PCR systems consist of a thermal cycler, optics for fluorescence excitation and emission collection, and a computer, as well as accompanying software for data acquisition, management, and analysis. The technology has advanced rapidly, and numerous systems are on the market, making the choice of a suitable instrument potentially very challenging. There are several parameters to consider when purchasing a real-time PCR system, including the following:

- *Sample capacity.* Some are 96-well standard format; others process fewer samples or require specialized glass capillary tubes.

- *Method of excitation.* Some use lasers; others use broad-spectrum light sources, such as light emitting diodes (LEDs) or tungsten-halogen lamps, with tunable filters.

- *Optical detection method.* Some use photodiodes; others use a CCD camera.

- *Overall sensitivity.* Some machines can detect down to one copy of the target sequence.

- *Dynamic range.* Typically this is 4 to 9 logs of linear dynamic range.

- *Capacity for multiplex assays*

- *Supported chemistries.* For example, SYBR Green dye, TaqMan probes, etc.

Real-time PCR systems are relatively expensive; at time of writing, they are in the range of $25,000 to $95,000.

One of the more important parameters to consider when purchasing a real-time PCR system is the method of excitation or, more specifically, the *range* of excitation. Instruments relying on single excitation lasers tend to be the least flexible because their laser excitation range (488–514 nm) is too narrow to efficiently excite the wide range of fluorophores available today. On the other hand, lasers offer high spectral brightness and sensitivity for fluorophores at their central wavelength. In contrast, tungsten-halogen lamps provide uniform excitation over a much broader range of wavelengths, which is particularly important when considering multiplex reactions that require a choice of spectrally well-resolved fluorophores to minimize cross talk.

The following is a brief list of real-time PCR systems that are currently available. Detailed information can be found on each manufacturer's website.

- The third-generation real-time cyclers from Applied Biosystems are the 7500 Real-Time PCR System and the 7500 Fast Real-Time PCR System. The thermal cycling system is a Peltier-based, 96-well block for use with either 96-well optical plates or 0.2-mL tubes. The optical system consists of a tungsten-halogen lamp and a set of five calibrated filter sets that can be recalibrated to new dyes without requiring additional filter sets. Compared with the previous generation of Applied Biosystems instruments, these systems offer more reliable and complete software for data analysis. Impressively, the 7500 Fast Real Time PCR System purportedly offers high-quality results in as little as 30 min. These systems use a passive internal reference dye, ROX, whose fluorescent output is measured during the denaturing step of each cycle of PCR, thus providing a baseline against which non-PCR-related, well-to-well variations can be normalized.

- The Bio-Rad iCycler iQ Real Time PCR System uses a tungsten-halogen lamp, which with appropriate filters permits fluorophore excitation from 400 to 700 nm, allowing up to five different reporter fluorophores to be multiplexed per sample tube. It comes with easy, intuitive software that allows access to raw data. It can amplify 96 samples at one time, and, with a recently launched module, extend its capacity to 384 samples.

- Roche Applied Science offers two different LightCycler Systems for real-time PCR. The (relatively) low price of the original Roche Lightcycler 2.0 Real-Time PCR System makes it a very attractive instrument for individual laboratories. Instead of loading PCR mixtures into tubes or plates, they are loaded into disposable glass capillary tubes arranged in a carousel for thermocycling. The glass capillaries are heated and cooled in an airstream, which greatly reduces the time required for each cycle of PCR. In contrast, the LightCycler 480 Real-Time PCR System is a high-throughput gene quantification or genotyping real-time PCR platform with exchangeable blocks for 96 or 384 samples in multiwell plates.

- The Rotor-Gene Q Cycler from QIAGEN uses a unique centrifugal rotary design that is radically different from conventional block cyclers: The reactions are carried out in standard microcentrifuge tubes inside a 36- or 72-well rotor that spins at 500 rpm, which is designed to keep all the samples at precisely the same temperature during thermal cycling. High speed is maintained throughout the run for both data collection and heating/cooling phases; all the tubes pass through the optical detector in 0.15 sec, allowing data to be captured very quickly. The centrifugal force of the spinning rotor pulls down any air bubbles introduced by pipetting and any condensation droplets. This eliminates any need to wait for temperature equilibration and creates thermal uniformity among the samples; accordingly, a sample-to-sample variation of <0.01°C is claimed, which is approximately 20 times less than that of block cyclers. The system has a wide optical range consisting of up to six channels spanning UV-to-infrared wavelengths.

- The Stratagene Mx4000 multiplex quantitative PCR system uses a tungsten-halogen lamp, with an excitation range from 350 to 750 nm, and four photomultiplier tubes with a detection range from 350 to 830 nm. Uniquely, the computer, which provides the interface between the user and the Mx4000 system, operates independently from the instruments' embedded microprocessor. In the event of a computer power loss or communication error, data are not lost, because once communication is reestablished, data are transferred automatically from the instrument's embedded software to the software on the external computer. Thus, the data are saved, and the experiment can be completed. This instrument has been designed very much with multiplexing in mind, and each of the four scanning fiber-optic heads independently excites and detects dyes, reading up to four dyes in a single tube. Optimized interference filters precisely match the excitation and emission wavelengths for each fluorophore to block out unwanted cross talk from spectrally adjacent fluorophores, thereby minimizing background and cross talk while enhancing dye discrimination.

EXTRACTING DATA FROM A REAL-TIME PCR EXPERIMENT: DATA ANALYSIS AND NORMALIZATION METHODS

The output from a real-time PCR is in the form of a graph—known as an amplification plot—showing the number of PCR cycles against increasing fluorescence (shown schematically in Fig. 4) (see the box Internal Passive Reference Dyes). The amplification plot is a sigmoidal curve. For the first 10 to 20 cycles, the curve is flat and at a baseline level, as the amount of amplified product has not yet accumulated to the point at which the fluorescent signal is detectable. The curve then increases sharply for several cycles. This steep increase, which is fairly narrow, represents the exponential phase of the reaction; it is only in this phase that you can assume there is a direct relationship between the intensity of the fluorescent signal and the amount of accumulated product in the reaction. Finally, as reagents become limiting, the curve plateaus and becomes flat again.

The greater the initial concentration of target sequence in the reaction (i.e., the higher the starting copy number), the earlier in the PCR a significant increase in fluorescence is observed and the fewer are the number of cycles required to achieve a particular yield of amplified product. The initial concentration of target nucleic acids can therefore be expressed as the cycle number required to achieve a preset threshold of amplification; this value is referred to as the threshold cycle, or C_T. The threshold represents a statistically significant level of fluorescence over the background and is associated with the exponential phase of the PCR (see Fig. 4).

To convert a C_T value to a meaningful number, the data are analyzed using either of two methodologies: absolute quantification or relative quantification. The choice depends on the application and will influence the design of the experiment.

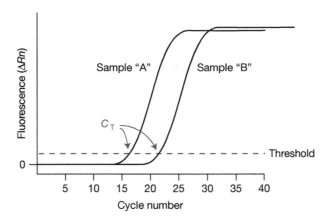

FIGURE 4. **Schematic representation of an amplification plot showing the number of PCR cycles against increasing fluorescence (ΔR_n).** The initial concentration of target nucleic acids can therefore be expressed as the cycle number required to achieve a preset threshold of amplification; this value is referred to as the threshold cycle, or C_T. In this example, sample A contains a higher initial concentration of nucleic acid than sample B and therefore has a lower C_T.

Absolute Quantification

In absolute quantification, the C_T value of the target nucleic acid in the test sample is compared with the C_T values of standards of known quantity plotted on a standard curve (C_T vs. log quantity); the quantity of the target nucleic acid is then interpolated from the standard curve. This method is used to determine the exact number of target molecules present in a sample; for example, determining the number of virus particles (DNA or RNA) in a blood sample or the chromosome or gene copy number in a cell. Accordingly, the quantity of target nucleic acid obtained from the standard curve is normalized to a unit amount of sample, such as number of cells, volume, or total amount of nucleic acid. The end result is a quantitative description of a single sample and does not depend on the properties of any other sample. This method of quantification is conceptually simple, and the mathematics of the analysis is easy to perform.

To use this method, you must have a reliable source of template of known concentration (which has been determined by some independent means), and the standards must be amplified in parallel with the test samples every time the experiment is performed. Absolute quantification also requires equivalent amplification efficiencies between the target and the external standard. It is important to note that the standard curve may only be used to interpolate—not extrapolate—the quantity of the unknown sample, because the assay may not be linear outside the range covered by the standards tested.

Relative Quantification

In relative quantification, the C_T value of the target nucleic acid in the test sample is compared with that in a control or reference sample (generally called a calibrator). The results are therefore expressed as a ratio (or fold difference) of the amount of a target nucleic acid in the test sample relative to the calibrator. This method is used to determine the difference in target levels between two different samples, for example, how much the expression level of a gene changes with a particular treatment. This method is most commonly used for quantifying mRNA levels by real-time RT-PCR and is adequate for most purposes to investigate physiological changes in gene expression levels. In these cases, the calibrator can be, for example, an untreated control, a sample at time zero in a time-course study, or a normal tissue sample.

In relative quantification, it is important to ensure that the target levels in the test sample and calibrator are compared from equivalent amounts of starting material (because finding that there is twice as much target mRNA in the test sample relative to the calibrator means nothing if there was also twice as much starting material). Several normalization methods have been described, including normalizing against either the starting number of cells or total RNA (Huggett et al. 2005). The most common normalization method, however, is to normalize the expression of the target gene to that of an endogenous reference gene (traditionally called a housekeeping gene), whose expression is constant in all samples tested; the endogenous reference gene acts as a proxy for the total amount of mRNA in the sample. In relative quantification, the target gene in the test sample and calibrator is first normalized to the endogenous reference gene, and the normalized values are then compared with each other to obtain a fold difference; typically, the normalized target gene expression in the calibrator is set to a value of 1, and the normalized target gene expression in the test sample is expressed as some n-fold increase or decrease relative to the calibrator. The advantage of this technique over absolute quantification is that the use of an internal standard (the endogenous reference gene) can minimize potential variations in sample preparation and handling and circumvents the need for accurate quantification and loading of the starting material.

Relative quantification can be performed using one of two methods: the standard curve method or the comparative C_T method.

Standard Curve Method

In the standard curve method, the amounts of the target and endogenous reference genes in both the test sample and calibrator are first determined using a standard curve, and the target gene is then normalized to the endogenous reference gene in both samples. Unlike the standard curve that is prepared for absolute quantification, the standards here do not need to be of known quantity because ultimately the normalized target gene expression in the test sample will be divided by the calibrator, and the units from the standard curve will drop out.

The accuracy of this method depends on the appropriate choice of a reference template for the standards. For each target gene analyzed, a separate standard curve must be constructed, along with the standard curve for the endogenous reference gene, and run in parallel with the test and calibrator sample(s). This method can therefore be somewhat laborious and time- and reagent-consuming when a large number of genes are to be analyzed. In addition, the inclusion of standard curves in every experiment limits the number of samples that can be analyzed at one time. However, this method is suitable for low-throughput experiments requiring the analysis of one or only a few genes.

Comparative C_T Method

The comparative C_T method, also referred to as the $\Delta\Delta C_T$ method or the Livak method, uses an arithmetic formula to compare the C_T value of the target gene with that of the endogenous reference gene. This method does not require the use of standard curves and in this regard is ostensibly less tedious than the standard curve method. However, to use the comparative C_T method, the amplification efficiencies of the target and endogenous reference genes must be approximately equal (at or near 100% and within 5% of each other)—a parameter that can require a great deal of optimization. Nonetheless, this method is preferred for experiments requiring high throughput.

DESIGNING PRIMERS AND PROBES AND OPTIMIZING CONDITIONS FOR REAL-TIME PCR

The next several sections of this introduction provide details on the steps required to perform a real-time PCR or real-time RT-PCR experiment: designing primers and probes and optimizing their concentrations; constructing a standard curve; and performing the assay.

The accuracy of a real-time PCR (or RT-PCR) experiment depends on properly optimizing the conditions of the reaction to yield maximal efficiency, specificity, and sensitivity. Achieving high specificity is particularly important for real-time PCR applications such as viral quantification and SNP genotyping; high sensitivity is important for detection of pathogens and rare mRNAs. Proper optimization is crucial for ensuring successful quantitation and the acquisition of valid, reproducible results.

A key component of optimizing a real-time PCR assay is maximizing the efficiency of DNA amplification. Ideally, the efficiency of DNA amplification in a real-time PCR experiment should be very high—in the range of 85%–110% (an efficiency of 100% means that the amount of amplified product exactly doubles with each cycle). As in traditional PCR, several technical variables can affect real-time PCR efficiency, including the length of the amplicon, the quality of the template, the presence of secondary structure, and the quality of the primers (i.e., hybridization efficiency, specificity, and the ability to form primer dimers). PCR efficiency can also be influenced by the concentration of the primers, which affects the thermodynamic stability of a duplexed primer–target structure.

Optimizing a real-time PCR involves a series of steps: selecting a suitable target sequence for amplification; designing the primers (and probe); optimizing the concentration of primers (and probe) in a real-time PCR assay; and analyzing the amplification plot (the latter two steps are detailed in Protocol 1). In addition, if SYBR Green I chemistry is used, it is also necessary to perform a postamplification melting curve analysis (see Protocol 1) to ascertain if nonspecific products (such as primer dimers) have been amplified, which will reduce both the efficiency and the sensitivity of the reaction. Once the optimal concentration of primers (and probe) has been determined, they are then tested over a wide range of template concentrations to determine assay efficiency, sensitivity, and reproducibility through the construction and analysis of a standard curve (Protocol 2). These optimization steps should be performed each time a new assay (i.e., new primer pair) is designed.

Selecting A Target Sequence

The first step in designing primers and probes for use in real-time PCR is to examine the target sequence. There are several parameters to consider when selecting a target region to amplify.

- *Length.* To promote high-efficiency amplification, the length of the amplicon should be relatively short—ideally in the range of 50–150 bp (and not exceeding 400 bp). Shorter amplicons yield higher PCR efficiencies and require shorter polymerization times for replication, making amplification of contaminating genomic DNA less likely.

- *Sequence.* Using a BLASTn search (http://blast.ncbi.nlm.nih.gov/Blast.cgi), the target sequence should be analyzed for the presence of polymorphisms or possible sequencing errors, which can affect primer binding; regions containing polymorphisms or errors should therefore be avoided for primer or probe design. The target sequence should also lack repetitive sequences and identity/homology to other regions of the genome, which result in nonproductive primer binding, leading to reduced efficiency of DNA amplification and decreased assay sensitivity.

- *G+C content.* The amplicon G+C content must be ≤60% to ensure efficient denaturation during thermal cycling, which leads to a more efficient reaction. Moreover, sequences that are GC-rich are susceptible to nonspecific interactions that result in a nonspecific signal in assays using DNA-binding dyes like SYBR Green.

- *Secondary structure.* The amplicon should be devoid of inverted repeat sequences because the formation of highly structured regions could promote inefficient hybridization of primers or probes.

- *Number of introns.* For real-time RT-PCR, it is important to select an amplicon that, in the genomic locus, includes two or more introns to avoid the coamplification of contaminating genomic sequence, pseudogenes, and other related genes (for more information, see Designing Primers and Probes). A BLAST search of the target cDNA sequence against the genomic DNA database can be used to determine intron positions. It is also important to identify possible splice variants and ensure that mRNA species that will be detected in the real-time PCR assay are, indeed, expressed in the cell type being analyzed. If using oligo(dT) priming in the reverse transcription reaction, the amplicon should be designed toward the 3' region of the template.

Quality of the Template

Real-time PCR can only be used reliably when the template can be amplified and when the amplification method is without error. PCR will only amplify DNA with an intact phosphodiester backbone between the priming sites. In addition, DNA containing lesions that affect the efficiency of amplification, such as abasic sites and thymine dimers, will be either underrepresented or may not be represented at all in real-time PCR. Finally, some additives (e.g., DMSO) and contaminants (e.g., SDS, which can be residual from extraction procedures) can inhibit the DNA polymerase, thus affecting the results of the real-time PCR. If long oligonucleotides are used as the amplicon target molecule, they should be PAGE-purified.

Designing Primers and Probes

Below we discuss the general principles involved in designing primers and probes for real-time PCR or RT-PCR. However, before starting the time-consuming process of designing primers and probes for your gene or target of interest, it is worth searching both published literature and online public databases to ascertain whether a validated real-time PCR assay has already been developed for your gene/target, and whether information about efficiency, specificity, sensitivity, and primer-dimer formation is available. However, even if a validated assay is available and provides all the relevant information, it is still necessary to validate the performance of the assay in your own hands, particularly with respect to its amplification efficiency and sensitivity limits. There are several primer and probe public databases available; the following three databases are excellent sources of information.

- *RTPrimerDB* (http://medgen.ugent.be/rtprimerdb/) (Pattyn et al. 2003, 2006; Lefever et al. 2009). This database houses validated primer and probe sequences for approximately 6000 genes; the sequences have been submitted by researchers, and the majority of the primer and probe sets (∼70%) have been published. The database can be queried using a variety of parameters (e.g., official gene name or symbol, Entrez or Ensemble gene identifier, or SNP identifier). Furthermore, it is possible to restrict a query to a particular application (e.g., gene expression quantification/detection, DNA copy number quantification/detection, SNP detection, mutation analysis, fusion gene quantification/detection, or chromatin immunoprecipitation), organism (e.g., human, mouse, zebrafish, or *Drosophila*), or detection chemistry.

- *PrimerBank* (http://pga.mgh.harvard.edu/primerbank/index.html) (Wang and Seed 2003; Spandidos et al. 2008). This database is a public resource for more than 300,000 PCR primers that have been designed using an algorithm that "has been extensively tested by real-time PCR experiments for PCR specificity and efficiency." The database houses primers against both human and mouse genes. To date, approximately 27,000 of the mouse primer pairs have been validated, with a purported ∼83% success rate; all experimental validation data are available from the PrimerBank website. The database can be queried using a variety of parameters (e.g., GenBank Accession number, NCBI gene ID or symbol, or protein accession); it is also possible to BLAST your gene sequence against the database.

- *Real Time PCR Primer Sets* (http://www.realtimeprimers.org/). This database lists primer sets and probes that have been synthesized, tested, and optimized. The database is first queried according to primer/probe type (e.g., SYBR Green primers, hybridization probes, hydrolysis probes [TaqMan], or molecular beacons), then by organism (human, mouse, rat, or other), and finally by gene name.

It should also be noted that there are a variety of commercial software programs and free web tools available to assist researchers in primer and probe design. Perhaps the most comprehensive commercially available program is Beacon Designer (from Premier Biosoft International; http://www.premierbiosoft.com), which can be used to design primers for SYBR Green PCR assays and primer/probe sets for a variety of probe-based chemistries. Beacon Designer designs highly specific primers and probes by automatically interpreting BLAST results to avoid regions of significant homology and by screening primers and probes for their thermodynamic properties and secondary structure. Another excellent commercially available software program is Primer Express (from Applied Biosystems), which allows you to design your own primers and probes using SYBR Green I dye and TaqMan chemistries for real-time PCR applications. An excellent, free option is the primer/probe design program called RealTimeDesign, which is available on the Biosearch Technologies website (http://www.biosearchtech.com/realtimedesign). To design primers for a SYBR Green I assay, a web-based free software like Primer3Plus (http://www.bioinformatics.nl/cgi-bin/primer3-plus/primer3plus.cgi) often yields very good results.

Primers

Primers pairs for real-time PCR may be designed using the same standard primer design algorithms that are used for conventional PCR (for more detailed information, see the Introduction to Chapter 7).

- *Length.* To maximize binding specificity, primers should have a length of 18–30 nucleotides.

- *Melting temperature.* Primers should have a melting temperature (T_m) of 55°C–60°C. The T_m of both primers in a primer pair should be within 2°C–3°C.

- *Sequence.* Each primer should have one—but no more than two or three—G or C nucleotide in the last five bases at the 3′ end. The presence of a single G or C nucleotide at the 3′ end can reduce nonspecific priming during the PCR. However, too many G or C nucleotides at the 3′ end can cause ambiguous binding of primers to the target site, resulting in misprimed elongation (known as the "slippage effect"). In addition, each probe should not have a long run (i.e., more than 3 or 4 nucleotides) of a single nucleotide (especially G or C) because homopolymeric runs can also cause the slippage effect. Finally, the primer pair should not have any sequence complementarity at the 3′ ends that could result in the formation of primer dimers. Because primer dimers have a negative ΔG value, primers should be chosen with a ΔG value no less than -10 kcal/mol.

- *G+C content.* Primers should have a G+C content of ∼50% (ideally 40%–60%). For primers binding to very AT-rich sequences (>70% AT-rich), it is advantageous to substitute one or more of the bases with a locked nucleic acid (LNA) analog to reduce the overall primer length while maintaining the T_m (see below for a discussion of LNA analogs).

- *Secondary structure.* Primers should lack any inverted repeat sequences, which can form a stable hairpin and result in inefficient (or complete lack of) primer binding.

- *Genomic DNA avoidance.* In real-time RT-PCR assays, false positives can be obtained because of amplification of genomic DNA. Accordingly, the primers should be designed to flank a long intron or multiple short introns or to span an exon–exon junction (Fig. 5).

- Primers should be ordered with desalt purification.

It should be noted that despite all attempts to optimize the reaction conditions, many primer sets fail to amplify the desired template, and, accordingly, a new set of primers must be designed and tested.

TaqMan Probes

Well-designed TaqMan probes require very little optimization. TaqMan probes should be designed with the following considerations.

- *Length.* TaqMan probes should be optimally 20 nucleotides in length (and no more than 30 nucleotides) to maximize quenching. Lengths greater than 30 nucleotides are possible, but in these cases, an internal quencher should be positioned ∼18–25 bases from the 5′ end.

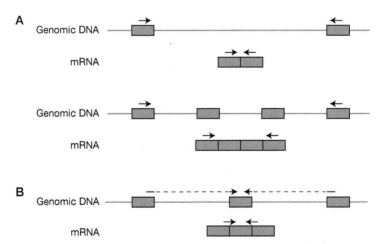

FIGURE 5. **Schematic of primer design for RT-PCR experiments.** (*A*) Intron-flanking primers. The primers are designed to flank either a long intron (*top*) or multiple short introns (*bottom*). (*B*) Exon–exon junction-spanning primers. The primers are only partly complementary to the genomic DNA and will not hybridize or generate a PCR product unless the annealing temperature is extremely low. (Rectangles) Exons; (gray line) introns; (arrows) primers.

- *Melting temperature.* TaqMan probes should have a T_m of $\sim$10°C higher than the T_m of the primers, usually in the 65°C–70°C range. The T_m difference is necessary to ensure that the probe binds to the template before hybridization of the primers. With a T_m of 8°C–10°C higher than that of the primers, the probe will anneal to the target sequence before the primers to ensure detection as the nearby primer is extended.

- *Sequence.* Long runs of a single base, especially G, should be avoided because these can affect the secondary structure of the probe and reduce hybridization efficiency. In cases in which an alternative sequence cannot be selected, disruption of a series of guanines by the substitution of an inosine can significantly improve probe performance. In addition, the probe should not contain a G nucleotide at the 5′-most base because it can quench fluorescence. In general, mismatches between the probe and target should be avoided (unless the probes are used for SNP genotyping) (see the information panel SNP Genotyping), and the probe should not have complementarity to either of the primers.

- *G+C content.* TaqMan probes should be designed to have a G/C content of $\sim$50% (ideally 30%–80%). If the target sequence is AT-rich, incorporate analogs such as locked nucleic acids (LNA; Roche) or minor groove binders (MGB; Applied Biosystems) (for details on LNAs and MGBs, see the next section).

- The 5′ end of the probe should be as close as possible to the 3′ end of one of the primers, without overlap, to ensure rapid cleavage by the *Taq* polymerase.

- TaqMan probes should be HPLC-purified.

Locked Nucleic Acid (LNA) Bases and Minor Groove Binders (MGBs). Newer adaptations of TaqMan probes incorporate the use of LNA bases or conjugated MGB groups to greatly increase the thermal stability of the probe–target duplex (Letertre et al. 2003). LNA bases are nucleic acid analogs with a chemical structure that restricts the flexibility of the ribofuranose ring and "locks" it into a rigid conformation that is constrained in the ideal conformation for Watson–Crick binding. Depending on sequence context, insertion of an LNA base into a DNA oligonucleotide can increase the T_m by 3°C–6°C. MGB groups consist of a tripeptide moiety (e.g., dihydrocyclopyrroloindole tripeptide, DPI3) that binds to the minor groove of DNA and forms highly specific and strong hybrids with complementary single-stranded DNA. The increase in probe–target duplex stability has two advantages. First, it increases the hybridization specificity of the probe (Kutyavin et al. 2000). As such, background fluorescence from nonspecific binding is reduced, and the signal-to-noise ratio is increased, thereby increasing overall assay sensitivity. Second, it allows shorter probe sequences to be used, which can overcome many of the limitations associated with probe design. For example, AT-rich probes often need to be 30–40 nucleotides in length to satisfy design guidelines; however, the substitution of an LNA base or MGB group facilitates the optimal design of highly specific, shorter probes that perform well, even at lengths of 13 to 20 nucleotides.

The addition of an LNA base or MGB group into a TaqMan probe is most commonly used for SNP genotyping applications (Johnson et al. 2004); the presence of a single base mismatch has a greater destabilizing effect on the duplex formation between an LNA/MGB-containing probe and its target nucleic acid than with a conventional DNA probe (see the information panel SNP Genotyping). Incorporating LNA bases and MGB groups also makes it possible to adjust the T_m values of primers and probes, which is important in multiplex assays in which the annealing temperatures of all primers and all probes must be the same (see the information panel Multiplex PCR). Note that so-called TaqMan MGB probes contain, in addition to the MGB group, a nonfluorescent quencher, which permits a more precise measure of the reporter dye fluorescence.

Storage of Primers and Probes

Stock solutions of primers and probes should be prepared using DNase/RNase-free water and aliquoted to avoid repeated freezing/thawing and whole-batch contamination. Primers should be stored at −20°C, at a working concentration of 10–100 μM. Probes should be protected from light and stored at −70°C, either as a lyophilized salt or as a 2–10 μM solution. Long-term storage of stock solutions is variable, ranging from 6 months to several years.

Optimizing the Concentration of Primers and Probes

Optimal primer concentrations for real-time PCR assays are determined empirically. Primer combinations that will yield the most sensitive and reproducible assays are those that, following analysis of the amplification plot, produce the lowest C_T (the cycle number where the fluorescence signal crosses the threshold) (see the section Extracting Data from a Real-Time PCR Experiment) and the highest ΔR_n (an indicator of the magnitude of the fluorescence signal generated by the PCR, and a measure of sensitivity). To facilitate the identification of a primer concentration that meets these criteria, a primer optimization matrix is prepared in which the concentration of the forward and reverse primers are independently varied and combinatorially tested (Protocol 1).

For assays using TaqMan probe chemistry, a primer optimization matrix is prepared while maintaining a constant probe concentration. Because the TaqMan probe is destroyed during the reaction, it is important to maintain the probe in vast excess; for this reason, a standard probe concentration of 250 nM (final concentration in the reaction volume) is recommended for primer optimization experiments—and real-time PCR experiments in general—to avoid probe limitation and ensure maximum sensitivity. However, if maximum sensitivity is not required (e.g., if the target sequence is abundant), then lower levels of the probe may suffice, which will have the added benefit of reducing assay cost. A probe optimization experiment can be performed using a constant, optimized primer concentration and testing several probe concentrations, typically in the range of 50–250 nM. Again, an analysis of the resulting amplification plots will identify the probe concentration that ensures the lowest C_T and the highest ΔR_n. As a practical note, even if the real-time PCR is designed to use TaqMan chemistry, perform the primer optimization experiments in the presence of SYBR Green I, which will allow detection of primer dimers and other nonspecific products that reduce efficiency and specificity.

Although primer optimization experiments are recommended for ensuring maximal assay specificity, they are somewhat tedious and time-consuming to set up and analyze. Instead, in practice most researchers simply start with a standard primer concentration: typically 500 nM for TaqMan assays (and a probe concentration of 250 nM) and 200–400 nM for SYBR Green I assays (final concentrations in the reaction). The primers are then tested and evaluated in a standard curve experiment (see Protocol 2). If detection is linear and >85% efficient over the range of standards tested, then further optimization of primer and probe concentrations is unnecessary.

To perform optimization experiments, the template sequence can be an artificial amplicon (which is best for reliable quantification), a linearized plasmid, a PCR product, or a cDNA that contains the PCR target of interest. For primer/probe optimization experiments, it is not necessary to know the concentration of the template per se; however, a template concentration should be chosen that results in C_T values between 20 and 30. The concentration of template is therefore determined empirically.

The protocols in this chapter use a PCR volume of 20 μL, although most manufacturers recommend 50-μL reactions. However, if the PCR target is not very abundant (i.e., present at one to 10 copies per sample), a larger volume may yield better reproducibility between samples.

CONSTRUCTING A STANDARD CURVE

Preparing a standard curve is a critical component of every real-time PCR experiment. When designing the assay and optimizing primer concentrations, a standard curve is used to determine the efficiency, sensitivity, reproducibility, and working range of the assay. Subsequently, during data analysis, a standard curve is used for absolute quantification or for the standard curve method of relative quantification. Standard curves are also required when using the comparative C_T method to show that the amplification efficiencies of the target and reference genes are equivalent.

To construct a standard curve (Protocol 2), a dilution series of a reference template must be carefully prepared and run under the same conditions that will be used to amplify the test sample. The standard curve should have a minimum of three replicates (ideally five or more) and a minimum of five logs of template concentration; this level of rigor is required for accurate quantification or calculation of efficiency. Furthermore, the dynamic range of the standard curve should extend past both the lowest and highest C_T values expected in the test samples. For absolute quantification, samples of unknown copy number can only be quantified if they fall within the range of the dilution series. For optimization, it is important to confirm that the assay is efficient, sensitive, and reproducible within the range of concentrations tested.

When constructing a standard curve for quantification, the plot of the standard curve will be C_T versus log of known quantity, and should yield a straight line. If the standard curve is being constructed for determining assay efficiency, sensitivity, and reproducibility, it is not necessary to know the template concentration; the plot of the standard curve will be C_T versus log of the dilution factor (in arbitrary units). Viewing the plot of the C_T obtained for the different concentrations yields important information about the efficiency of amplification, the sensitivity of the assay, and the consistency across replicates (i.e., reproducibility).

Efficiency

Slope

In a standard curve plot of log quantity or log dilution factor (x-axis) versus C_T (y-axis), the amplification efficiency of the reaction can be calculated from the slope of the line. The software for most real-time PCR instruments will prepare a standard curve and calculate efficiency. If this feature is not available, prepare a plot of C_T versus the log of nucleic acid input level (or log of the dilution factor) and perform linear regression. Calculate the slope of the line using the following equation:

$$E = 10^{(-1/\text{slope})}.$$

If the efficiency were 100%, then the amount of product would exactly double with each cycle and the slope of the graph would be −3.32. (Note that some instruments may plot C_T on the x-axis and log quantity or log dilution factor on the y-axis; in this case, if the efficiency were 100%, the slope of the line would be −0.301.)

$$\text{Efficiency} = 10^{(-1/\text{slope})}$$
$$= 10^{(-1/-3.32)}$$
$$= 10^{(0.3012)}$$
$$= 2.00.$$

To convert the E value to a percentage, use

$$\text{Percent efficiency} = (\text{Efficiency} - 1) \times 100.$$

R^2 Value

Another critical parameter for evaluating PCR efficiency is the coefficient of determination (or correlation coefficient), R^2, which is a measure of how well the experimental data fit the regression line (or, in other words, a measure of the linearity of the standard curve). If all data points lie perfectly along the line, the R^2 value would be equal to 1. In practice, an R^2 value of >0.98 is considered sufficient to provide reliable results.

Note that although the standard curve method is the most frequently used method to calculate amplification efficiency, several methods have been developed to calculate efficiency of PCR assays based on the kinetics of a single reaction (see, e.g., Tichopad et al. 2003; Zhao and Fernald 2005). These alternative methods avoid the need to generate a standard curve, saving money and time.

Sensitivity

Any assay capable of effectively amplifying and detecting one copy of starting template has achieved the ultimate level of sensitivity, regardless of the absolute value of the C_T. A real-time PCR with an efficiency of <100% will have lower sensitivity.

Reproducibility

Generally, each sample in a real-time PCR assay is tested and analyzed in triplicate. The degree to which repetitions of a sample produce similar quantitative values is referred to as "reproducibility." The most common measure of reproducibility is the standard deviation (i.e., the square root of the variance).

PERFORMING REAL-TIME PCR

There are few differences between the experimental steps necessary for amplifying template DNA in a real-time thermocycler (Protocol 3) and a standard PCR (see Chapter 7). In real-time PCR it is necessary, as noted, to optimize the concentration of primers and probe and to perform a standard curve. It is also important to consider the data analysis method that will be used (see above).

PERFORMING REAL-TIME RT-PCR

Real-time RT-PCR (also frequently called quantitative RT-PCR [abbreviated qRT-PCR or sometimes RT-qPCR]) is the most sensitive technique for RNA detection and quantification currently available; in fact, real-time RT-PCR is sensitive enough to enable quantification of RNA from a single cell (Ståhlberg and Bengtsson 2010). In recent years, real-time RT-PCR has become the most commonly used method for applications involving analysis of mRNA levels, such as verifying results from microarray analysis and analyzing changes in gene expression in response to, for example, pharmacological treatment. Compared with other mRNA analysis techniques, such as northern blotting, RNase protection, and in situ hybridization, a real-time RT-PCR experiment can be performed more quickly and does not require the use of toxic chemicals or radioactive probes. Real-time RT-PCR can also be used to quantify absolute mRNA copy number in the cell (Bustin 2000) and detect and quantify viral load (see, for example, Le Guillou-Guillemette and Lunel-Fabiani 2009; Piqueur et al. 2009).

Like conventional RT-PCR (see Chapter 7, Protocol 8), real-time RT-PCR involves two steps. In the first step, the RNA is converted into complementary DNA (cDNA) using an RNA-dependent DNA polymerase (reverse transcriptase); in the second step, the cDNA is amplified by a thermostable DNA polymerase. Variability and lack of reproducibility can be observed in quantitative real-time RT-PCR experiments. Consequently, it is crucial to assess the quality of every step of the real-time RT-PCR assay. Sample acquisition and purification of RNA represent the initial steps of every real-time RT-PCR assay, and the quality of the template is a critically important determinant of the reproducibility of real-time RT-PCR results. For gene expression studies, the reverse transcription step must be carried out with high-purity reagents and in multiple replicates because this step can introduce variability in template replication. However, whereas the reverse transcription step is highly variable, the PCR segment of the assay is highly reproducible when run under optimal conditions.

Preparing High-Quality RNA

Purity

One important consideration when choosing an RNA extraction method is the possibility of carrying over an inhibiting agent in the template RNA preparation. The presence of culture media, components of the RNA extraction reagent (e.g. phenol), or copurified components from the biological sample can result in a significant reduction of the sensitivity and kinetics of real-time RT-PCR (Guy et al. 2003; Perch-Nielsen et al. 2003; Lefevre et al. 2004; Rådström et al. 2004; Sunén et al. 2004; Jiang et al. 2005). At best, inhibitors can generate inaccurate quantitative results; at worst, a high degree of inhibition will create false-negative results.

RNA extracted using monophasic lysis reagents is the template of choice for real-time RT-PCR. (Protocols for isolating total RNA using monophasic lysis reagents are given in Chapter 6, Protocols 1–4.) However, although monophasic lysis reagents typically give a high yield, there is a strong probability of copurifying phenolic compounds that can inhibit subsequent PCR. Thus, care should be taken to remove all traces of phenol in the RNA sample.

Another troublesome contaminant in an RNA preparation is genomic DNA, which may be coamplified with the target mRNA and therefore interfere with accurate quantification. If the target mRNA is relatively abundant (i.e., hundreds or thousands of copies per cell), then genomic DNA amplification will be negligible compared with the target mRNA products. However, if the target mRNA abundance is relatively low (i.e., less than 100 copies per cell), then genomic DNA amplification can lead to erroneously high estimates of mRNA levels. To avoid genomic DNA amplification during RT-PCR, select an amplicon that includes one or more introns, and design primers to flank or span the intron(s), in order to avoid coamplification of genomic sequence (see Fig. 5). If the target gene does not contain any introns or if the primers to be used in the real-time RT-PCR span a single exon or a short intron, it may be necessary to treat the RNA sample with RNase-free or amplification-grade DNase I to remove any contaminating genomic DNA that may be present (see Chapter 6, Protocol 8). Genomic DNA amplification is easily detected by performing a control reaction lacking reverse transcriptase (commonly called a "no-RT" control). If genomic DNA amplification has occurred, a signal will be observed in the no-RT control.

Integrity

Moderately degraded RNA samples can be reliably analyzed and quantified as long as the expression of the mRNA target is normalized against an internal reference and the amplicons are kept short (<250 bp) (Fleige and Pfaffl 2006). To assess the integrity of total RNA, run an aliquot of the purified RNA sample on a denaturing agarose gel stained with ethidium bromide (see Checking the Quality of Preparations of RNA at the end of Chapter 6, Protocol 10). Alternatively, the RNA sample can be analyzed using an Agilent 2100 Bioanalyzer (Agilent Technologies), which is capable of simultaneously analyzing the concentration, integrity, and purity of the RNA in a single sample. In the absence of a reliable measurement of mRNA integrity, it is possible to analyze a specific target sequence as a representative of the integrity of all mRNAs in a given RNA sample (Nolan et al. 2006). In this assay, the integrity of the ubiquitously expressed mRNA (e.g., *GAPDH*) is measured by designing a multiplex PCR assay to quantify the levels of three distinct target amplicons inside the *GAPDH* sequence. The ratio of the three amplicons reflects the relative success of the RT-PCR to proceed along the entire length of the transcript. However, because different mRNAs degrade at different rates, it may be necessary to test multiple targets using similar assays.

Choosing a Priming Method for the Reverse Transcription Reaction

The step in which RNA is converted into a cDNA template represents an important potential contributor to the variability and lack of reproducibility that can be observed in real-time RT-PCR experiments. If the mRNA abundance is low, the efficiency of RNA-to-cDNA conversion, which is dependent on template abundance, will also be low (Karrer et al. 1995). Furthermore, each of the different priming approaches used to synthesize cDNA—gene-specific priming, oligo(dT) priming, random priming, or a combination of oligo(dT) and random priming (see Chapter 7, Protocol 8)— differs significantly in its specificity and cDNA yield. Consequently, real-time RT-PCR results are comparable only when the same priming strategy and reaction conditions have been used (Ståhlberg et al. 2004).

The most commonly used priming method in real-time RT-PCR assays is oligo(dT) priming. More specific than random priming, this method is the best approach to use when it is necessary to amplify several target mRNAs from a limited RNA sample; an oligo(dT) primer can be used to first generate a pool of cDNA, followed by separate PCR assays for each target using aliquots from the cDNA pool. Owing to amplicon length limitations, oligo(dT) is a good choice when all of the target amplicons are located near the 3′ end of the polyadenylated mRNA. However, the reverse transcriptase may fail to reach the upstream primer-binding site if secondary structures exist or if the target mRNA contains a very long untranslated 3′ region (Sanderson et al. 2004). Also, because the oligo(dT) priming reaction requires the annealing of an oligo(dT) to the 3′-poly(A) tail, it is not an effective choice for transcribing RNA that is likely to be fragmented.

The next most commonly used priming method for real-time RT-PCR assays is random priming (Bustin et al. 2005). Because random primers (hexamers, octamers, nonamers, decamers, or dodecamers) consist of every possible combination of bases, they produce more than one cDNA target per original mRNA target. Moreover, the majority of cDNA synthesized from total RNA is derived from ribosomal RNA (rRNA); therefore, if the mRNA target of interest is present at low levels, it may not be primed proportionately, and its subsequent amplification may not be quantitative. Indeed, it has been shown that using random hexamers can result in the overestimation of mRNA copy numbers by up to 19-fold compared with a sequence-specific primer (Zhang and Byrne 1999). Another disadvantage is that a reaction primed by random primers is linear over a narrower range than a similar reaction primed by target-specific primers (Bustin and Nolan 2004). Nonetheless, if the PCR targets are more than a few kilobases from the 3′ end or if the RNA is not polyadenylated, random primers will give better detection than oligo(dT) primers. If the location of PCR targets or the polyadenylation level of RNAs varies, a mixture of oligo(dT) and random oligomers will give the best results.

The least popular priming method for real-time RT-PCR is gene-specific priming. Although it may provide the greatest sensitivity for quantitative assays (Lekanne Deprez et al. 2002), this method requires separate priming reactions for each target RNA to be analyzed; therefore, it is not possible to return to the same preparation and amplify other targets at a later stage. Although it is possible to amplify more than one target in a single reaction tube (multiplex RT-PCR) (Wittwer et al. 2001), this approach requires careful experimental design and optimization to obtain accurate quantitative data. One advantage of gene-specific primers is that all of the reverse transcription product will encode the gene of interest, which may allow quantitation of very-low-abundance mRNAs that may not be detectable using nonspecific reverse transcription primers.

Selecting the Enzyme(s)

The real-time RT-PCR assay can be performed using either a single enzyme that functions as both a reverse transcriptase and thermophilic DNA polymerase, or separate reverse transcriptase and DNA polymerase enzymes.

Single Enzyme for Reverse Transcription and PCR

Tth polymerase is able to function as both a reverse transcriptase and a DNA polymerase for PCR. In this method, all reagents are added to a single tube at the beginning of the reaction (Cusi et al. 1994; Juhasz et al. 1996), thereby reducing hands-on time and the potential for contamination. However, it is not possible to optimize the two reactions separately, and, because of the less efficient reverse transcriptase activity of Tth polymerase, the assay may be less sensitive (Easton et al. 1994). Moreover, this approach can only be carried out using gene-specific primers, and the reaction is often characterized by extensive accumulation of primer dimers, which may obscure the true results in a quantitative assay (Vandesompele et al. 2002). For these reasons, the single-enzyme approach is rarely used.

Separate Enzymes for Reverse Transcription and PCR

The two-enzyme method provides increased flexibility, sensitivity, and potential for optimization, making it the preferred choice over the single-enzyme procedure. In the two-enzyme approach, the reaction can be performed either in one tube or two tubes.

In the one-step (or one-tube) reaction, the reverse transcription and PCR are performed in a single buffer system. The reverse transcriptase synthesizes cDNA in the presence of high concentrations of dNTPs with a reverse gene-specific primer, and, subsequently, a thermostable DNA polymerase, PCR buffer (without Mg^{2+}), and gene-specific primers are added, and the PCR is performed in the same tube. One-step RT-PCR simplifies high-throughput applications and helps minimize carryover contamination because the tubes remain closed between cDNA synthesis and amplification. There are two primary disadvantages of this approach: First, a template switching activity of viral reverse transcriptases can generate artifacts during transcription (Mader et al. 2001). Second, the reverse transcriptase enzyme can inhibit the PCR assay even after inactivation, resulting in an overestimation of amplification efficiency and target quantification (Suslov et al. 2005).

In the two-step (or two-tube) reaction (Protocol 4), reverse transcription and PCR are performed in separate tubes. In the first tube the reverse transcriptase synthesizes cDNA under optimal conditions, using an oligo(dT), random, or reverse gene-specific primer. An aliquot of the reverse transcriptase reaction is then transferred to a second tube containing the thermostable DNA polymerase, buffer, and PCR primers, and the reaction is carried out under conditions that are optimal for the DNA polymerase. This method is the most adaptable—allowing flexibility in the choice of primers for cDNA synthesis and in the ability to detect multiple messages from a single RNA sample—and the most frequently used.

Selecting an Endogenous Reference Gene

Using an endogenous reference gene is a simple and popular method for normalization in real-time RT-PCR experiments and is currently the preferred option (Huggett et al. 2005). To serve as a suitable normalization factor, the endogenous reference gene must be expressed at a constant level in all samples tested, and its expression must not be altered by the treatment under study. Historically, the most commonly used reference genes have been those that are expressed at relatively high levels in all cell types: β-actin, glyceraldehyde-3-phosphate dehydrogenase (*GAPDH*), hypoxanthine-guanine phosphoribosyl transferase (*HPRT*), and 18S ribosomal RNA. However, even these classic reference genes have been shown to have varying levels in certain cell types or under certain conditions. For example, 18S rRNA levels increase following cytomegalovirus infection (Tanaka et al. 1975); *HPRT* is constitutively expressed at low levels in most human tissues but at high levels in certain parts of the central nervous system (Stout et al. 1985); β-actin mRNA appears to be differentially expressed in different leukemia tumor samples (Blomberg et al. 1987); and *GAPDH* expression is elevated in certain aggressive cancers (Goidin et al. 2001). Therefore, in any experiment, it is important to verify that the expression of the endogenous reference gene is constant in all samples being analyzed; this can be validated in a real-time RT-PCR experiment in which the expression of the reference gene is normalized to the amount of total RNA in the sample (Huggett et al. 2005). If a small variability in the C_T values is observed between the samples, the reference gene may still be usable if the differences in the test samples are much greater than the reference gene variation.

In some cases, it may be necessary to use multiple reference genes, rather than a single gene, to achieve accurate quantification and normalization (Vandesompele et al. 2002). Normalization using multiple genes is a robust method for providing accurate results and is preferable if fine measurements are to be made. However, this approach requires a large quantity of samples, which increases the cost of the experiments, and is therefore not always feasible. Several programs are available that allow the assessment of multiple reference genes. Nevertheless, reference gene expression must be carefully analyzed under the experimental conditions used, and the variability must be determined and reported.

- The freely available software geNorm identifies the most appropriate reference gene by using the geometric mean of the expression of the candidate cDNA (Vandesompele et al. 2002).

- BestKeeper, like geNorm, selects the gene that is the least variable using the geometric mean, but unlike geNorm, BestKeeper uses raw data (Pfaffl et al. 2004).

- NormFinder not only measures the variation but also groups the potential reference genes by how much the experimental conditions can affect their expression (Andersen et al. 2004).

When designing a real-time RT-PCR experiment, it is important to consider the type of quantification (data analysis) method that will be used (see Introduction to this chapter), as the choice will reflect the experimental setup. Chiefly, when performing either absolute quantification or the standard curve method of relative quantification, serial dilutions of a standard must be run in parallel with the test samples (see Protocol 1). When preparing standard curves for multiple plates, it is essential to use the same stock template to be able to compare the relative quantities across different plates. Finally, it is important to keep in mind that it is generally not possible to use DNA as a standard for absolute quantitation of RNA because there is no control for the efficiency of the reverse transcription step.

MIQE GUIDELINES

Although its conceptual and practical simplicity makes real-time PCR the most popular method for nucleic acid quantification, a lack of consensus exists on how best to perform and interpret real-time PCR experiments. As a result, the quality of real-time PCR data can be variable and therefore difficult to evaluate, and the results may be impossible to repeat. Compounding the problem is the frequent lack of sufficient experimental details in publications, which further prevents critical evaluation of the data. To address these issues, a set of guidelines, called The Minimum Information for Publication of Quantitative Real Time PCR Experiments (MIQE) guidelines, has been developed in an attempt to provide a common publication standard that describes real-time PCR experiments in a comprehensive manner (Bustin et al. 2009). The aim of these guidelines is "to provide authors, reviewers and editors specifications for the minimum information that must be reported for a qPCR experiment in order to ensure its relevance, accuracy, correct interpretation and repeatability." These guidelines have been designed to help promote consistency between laboratories and to increase experimental transparency. Currently, there is an effort to have journals adopt the MIQE standards, which will mandate the inclusion of a completed checklist along with a paper, much like the MIAME (Minimum Information about a Microarray Experiment) standards for published microarray data.

REAL-TIME PCR PROTOCOLS

The remainder of this chapter provides protocols for carrying out real-time PCR and RT-PCR assays using the two most common real-time PCR chemistries: SYBR Green I and TaqMan probes. For other chemistries, refer to the references mentioned above for information on primer/probe design and assay development. In general, the basic experimental practices for real-time PCR are similar to those of standard PCRs (see Chapter 7); the same issues apply with respect to producing clean templates, designing primers, and optimizing the reaction conditions. The sequential series of protocols

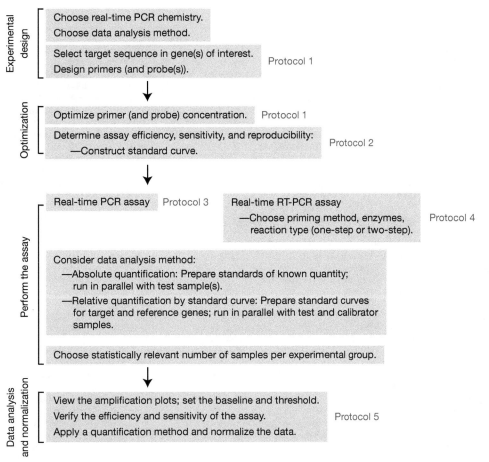

FIGURE 6. Experimental flowchart outlining what parameters need to be considered for a real-time PCR or RT-PCR experiment.

follows the steps as they are organized in Figure 6: First optimize the concentrations of primers and probes for use in real-time PCR (Protocol 1); then construct a standard curve (Protocol 2); then perform a real-time PCR (Protocol 3) or real-time RT-PCR (Protocol 4) assay; and finally analyze the raw data from the real-time PCR or RT-PCR experiment (Protocol 5).

INTERNAL PASSIVE REFERENCE DYES

For some real-time PCR instruments, a passive reference dye (so-named because it does not participate in the PCR) is included in all samples to normalize for possible sample-to-sample variations in non-PCR-related fluorescence signals caused by differences in concentration or volume, which may occur because of pipetting errors, sample evaporation, or instrument limitations. A common passive reference dye is 5-carboxy-X-rhodamine (ROX), which is used in real-time PCR instruments from Applied Biosystems and Stratagene.

When a passive reference dye is used, the instrument will plot fluorescence as a ΔR_n value, which represents the magnitude of the fluorescence signal generated by a given set of PCR conditions:

$$\Delta R_n = (R_n^+) - (R_n^-),$$

where R_n is the fluorescence emission intensity of the reporter dye divided by the fluorescence emission intensity of the passive reference dye. R_n^+ refers to the R_n value of a reaction containing all components, including the template. R_n^- refers to the R_n value of the unreacted sample, which can be obtained either from a sample that does not contain any template or, more commonly, from the early cycles of a real-time PCR run before there is a detectable increase in fluorescence.

REFERENCES

Ahmad AI, Ghasemi JB. 2007. New unsymmetrical cyanine dyes for real-time thermal cycling. *Anal Bioanal Chem* 389: 983–988.

Andersen CL, Jensen JL, Orntoft TF. 2004. Normalization of real-time quantitative reverse transcription-PCR data: A model-based variance estimation approach to identify genes suited for normalization, applied to bladder and colon cancer data sets. *Cancer Res* 64: 5245–5250.

Benes V, Castoldi M. 2010. Expression profiling of microRNA using real-time quantitative PCR, how to use it and what is available. *Methods* 50: 244–249.

Bengtsson M, Karlsson HJ, Westman G, Kubista M. 2003. A new minor groove binding asymmetric cyanine reporter dye for real-time PCR. *Nucleic Acids Res* 31: e45. doi: 10.1093/nar/gng045.

Blomberg J, Andersson M, Fäldt R. 1987. Differential pattern of oncogene and β-actin expression in leukaemic cells from AML patients. *Br J Haematol* 65: 83–86.

Bustin SA. 2000. Absolute quantification of mRNA using real-time reverse transcription polymerase chain reaction assays. *J Mol Endocrinol* 25: 169–193.

Bustin SA, Nolan T. 2004. Pitfalls of quantitative real-time reverse-transcription polymerase chain reaction. *J Biomol Tech* 15: 155–166.

Bustin SA, Benes V, Nolan T, Pfaffl MW. 2005. Quantitative real-time RT-PCR—A perspective. *J Mol Endocrinol* 34: 597–601.

Bustin SA, Benes V, Garson JA, Hellemans J, Huggett J, Kubista M, Mueller R, Nolan T, Pfaffl MW, Shipley GL, et al. 2009. The MIQE guidelines: Minimum information for publication of quantitative real-time PCR experiments. *Clin Chem* 55: 611–622.

Cradic KW, Wells JE, Allen L, Kruckeberg KE, Singh RJ, Grebe SK. 2004. Substitution of 3′-phosphate cap with a carbon-based blocker reduces the possibility of fluorescence resonance energy transfer probe failure in real-time PCR assays. *Clin Chem* 50: 1080–1082.

Crockett AO, Wittwer CT. 2001. Fluorescein-labeled oligonucleotides for real-time PCR: Using the inherent quenching of deoxyguanosine nucleotides. *Anal Biochem* 290: 89–97.

Cusi MG, Valassina M, Valensin PE. 1994. Comparison of M-MLV reverse transcriptase and Tth polymerase activity in RT-PCR of samples with low virus burden. *BioTechniques* 17: 1034–1036.

D'Haene B, Vandesompele J, Hellemans J. 2010. Accurate and objective copy number profiling using real-time quantitative PCR. *Methods* 50: 262–270.

Easton LA, Vilcek S, Nettleton PF. 1994. Evaluation of a 'one tube' reverse transcription-polymerase chain reaction for the detection of ruminant pestiviruses. *J Virol Methods* 50: 343–348.

Eriksson M, Westerlund F, Mehmedovic M, Lincoln P, Westman G, Larsson A, Akerman B. 2006. Comparing mono- and divalent DNA groove binding cyanine dyes—Binding geometries, dissociation rates, and fluorescence properties. *Biophys Chem* 122: 195–205.

Fleige S, Pfaffl MW. 2006. RNA integrity and the effect on the real-time qRT-PCR performance. *Mol Aspects Med* 27: 126–139.

Goidin D, Mamessier A, Staquet MJ, Schmitt D, Berthier-Vergnes O. 2001. Ribosomal 18S RNA prevails over glyceraldehyde-3-phosphate dehydrogenase and β-actin genes as internal standard for quantitative comparison of mRNA levels in invasive and noninvasive human melanoma cell subpopulations. *Anal Biochem* 295: 17–21.

Gupta V, Cobb RR, Brown L, Fleming L, Mukherjee N. 2008. A quantitative polymerase chain reaction assay for detecting and identifying fungal contamination in human allograft tissue. *Cell Tissue Bank* 9: 75–82.

Guy RA, Payment P, Krull UJ, Horgen PA. 2003. Real-time PCR for quantification of *Giardia* and *Cryptosporidium* in environmental water samples and sewage. *Appl Environ Microbiol* 69: 5178–5185.

Heid CA, Stevens J, Livak KJ, Williams PM. 1996. Real time quantitative PCR. *Genome Res* 6: 986–994.

Huggett J, Dheda K, Bustin S, Zumla A. 2005. Real-time RT-PCR normalisation; strategies and considerations. *Genes Immun* 6: 279–284.

Jiang J, Alderisio KA, Singh A, Xiao L. 2005. Development of procedures for direct extraction of *Cryptosporidium* DNA from water concentrates and for relief of PCR inhibitors. *Appl Environ Microbiol* 71: 1135–1141.

Johnson MP, Haupt LM, Griffiths LR. 2004. Locked nucleic acid (LNA) single nucleotide polymorphism (SNP) genotype analysis and validation using real-time PCR. *Nucleic Acids Res* 32: e55. doi: 10.1093/nar/gnh046.

Juhasz A, Ravi S, O'Connell CD. 1996. Sensitivity of tyrosinase mRNA detection by RT-PCR: rTth DNA polymerase vs. MMLV-RT and AmpliTaq polymerase. *BioTechniques* 20: 592–600.

Karlsson HJ, Eriksson M, Perzon E, Akerman B, Lincoln P, Westman G. 2003a. Groove-binding unsymmetrical cyanine dyes for staining of DNA: Syntheses and characterization of the DNA-binding. *Nucleic Acids Res* 31: 6227–6234.

Karlsson HJ, Lincoln P, Westman G. 2003b. Synthesis and DNA binding studies of a new asymmetric cyanine dye binding in the minor groove of [poly(dA-dT)]₂. *Bioorg Med Chem* 11: 1035–1040.

Karrer EK, Lincoln JE, Hogenhout S, Bennett AB, Bostock RM, Martineau B, Lucas WJ, Gilchrist DG, Alexander D. 1995. In situ isolation of mRNA from individual plant cells: Creation of cell-specific cDNA libraries. *Proc Natl Acad Sci* 92: 3814–3818.

Klein D. 2002. Quantification using real-time PCR technology: Applications and limitations. *Trends Mol Med* 8: 257–260.

Kutyavin IV, Afonina IA, Mills A, Gorn VV, Lukhtanov EA, Belousov ES, Singer MJ, Walburger DK, Lokhov SG, Gall AA, et al. 2000. 3′-Minor groove binder-DNA probes increase sequence specificity at PCR extension temperatures. *Nucleic Acids Res* 28: 655–661.

Lefever S, Vandesompele J, Speleman F, Pattyn F. 2009. RTPrimerDB: The portal for real-time PCR primers and probes. *Nucleic Acids Res* 37: D942–D945.

Lefevre J, Hankins C, Pourreaux K, Voyer H, Coutlée F, Canadian Women's HIV Study Group. 2004. Prevalence of selective inhibition of HPV-16 DNA amplification in cervicovaginal lavages. *J Med Virol* 72: 132–137.

Le Guillou-Guillemette H, Lunel-Fabiani F. 2009. Detection and quantification of serum or plasma HCV RNA: Mini review of commercially available assays. *Methods Mol Biol* 510: 3–14.

Lekanne Deprez RH, Fijnvandraat AC, Ruijter JM, Moorman AF. 2002. Sensitivity and accuracy of quantitative real-time polymerase chain reaction using SYBR green I depends on cDNA synthesis conditions. *Anal Biochem* 307: 63–69.

Letertre C, Perelle S, Dilasser F, Arar K, Fach P. 2003. Evaluation of the performance of LNA and MGB probes in 5′-nuclease PCR assays. *Mol Cell Probes* 17: 307–311.

Mackay J, Landt O. 2007. Real-time PCR fluorescent chemistries. *Methods Mol Biol* 353: 237–261.

Mader RM, Schmidt WM, Sedivy R, Rizovski B, Braun J, Kalipciyan M, Exner M, Steger GG, Mueller MW. 2001. Reverse transcriptase template switching during reverse transcriptase-polymerase chain reaction: Artificial generation of deletions in ribonucleotide reductase mRNA. *J Lab Clin Med* 137: 422–428.

Marras SA, Kramer FR, Tyagi S. 2002. Efficiencies of fluorescence resonance energy transfer and contact-mediated quenching in oligonucleotide probes. *Nucleic Acids Res* 30: e122. doi: 10.1093/nar/gnf121.

Martinelli G, Iacobucci I, Soverini S, Cilloni D, Saglio G, Pane F, Baccarani M. 2006. Monitoring minimal residual disease and controlling drug resistance in chronic myeloid leukaemia patients in treatment with imatinib as a guide to clinical management. *Hematol Oncol* 24: 196–204.

Monis PT, Giglio S, Saint CP. 2005. Comparison of SYTO9 and SYBR Green I for real-time polymerase chain reaction and investigation of the effect of dye concentration on amplification and DNA melting curve analysis. *Anal Biochem* 340: 24–34.

Newby DT, Hadfield TL, Roberto FF. 2003. Real-time PCR detection of *Brucella abortus*: A comparative study of SYBR green I, 5′-exonuclease, and hybridization probe assays. *Appl Environ Microbiol* 69: 4753–4759.

Niesters HG. 2001. Quantitation of viral load using real-time amplification techniques. *Methods* 25: 419–429.

Nolan T, Hands RE, Bustin SA. 2006. Quantification of mRNA using real-time RT-PCR. *Nat Protoc* 1: 1559–1582.

Pattyn F, Speleman F, De Paepe A, Vandesompele J. 2003. RTPrimerDB: The real-time PCR primer and probe database. *Nucleic Acids Res* 31: 122–123.

Pattyn F, Robbrecht P, De Paepe A, Speleman F, Vandesompele J. 2006. RTPrimerDB: The real-time PCR primer and probe database, major update 2006. *Nucleic Acids Res* **34**: D684–D688.

Perch-Nielsen IR, Bang DD, Poulsen CR, El-Ali J, Wolff A. 2003. Removal of PCR inhibitors using dielectrophoresis as a selective filter in a microsystem. *Lab Chip* **3**: 212–216.

Pfaffl MW, Tichopad A, Prgomet C, Neuvians TP. 2004. Determination of stable housekeeping genes, differentially regulated target genes and sample integrity: BestKeeper—Excel-based tool using pair-wise correlations. *Biotechnol Lett* **26**: 509–515.

Piqueur MA, Verstrepen WA, Bruynseels P, Mertens AH. 2009. Improvement of a real-time RT-PCR assay for the detection of enterovirus RNA. *Virol J* **6**: 95. doi: 10.1186/1743-422X-6-95.

Rådström P, Knutsson R, Wolffs P, Lövenklev M, Löfström C. 2004. Pre-PCR processing: Strategies to generate PCR-compatible samples. *Mol Biotechnol* **26**: 133–146.

Sanderson IR, Bustin SA, Dziennis S, Paraszczuk J, Stamm DS. 2004. Age and diet act through distinct isoforms of the class II transactivator gene in mouse intestinal epithelium. *Gastroenterology* **127**: 203–212.

Schmittgen TD, Zakrajsek BA, Mills AG, Gorn V, Singer MJ, Reed MW. 2000. Quantitative reverse transcription-polymerase chain reaction to study mRNA decay: Comparison of endpoint and real-time methods. *Anal Biochem* **285**: 194–204.

Spandidos A, Wang X, Wang H, Dragnev S, Thurber T, Seed B. 2008. A comprehensive collection of experimentally validated primers for polymerase chain reaction quantitation of murine transcript abundance. *BMC Genomics* **9**: p633. doi: 10.1186/1471-2164-9-633.

Ståhlberg A, Bengtsson M. 2010. Single-cell gene expression profiling using reverse transcription quantitative real-time PCR. *Methods* **50**: 282–288.

Ståhlberg A, Håkansson J, Xian X, Semb H, Kubista M. 2004. Properties of the reverse transcription reaction in mRNA quantification. *Clin Chem* **50**: 509–515.

Stout JT, Chen HY, Brennand J, Caskey CT, Brinster RL. 1985. Expression of human HPRT in the central nervous system of transgenic mice. *Nature* **317**: 250–252.

Suñén E, Casas N, Moreno B, Zigorraga C. 2004. Comparison of two methods for the detection of hepatitis A virus in clam samples (*Tapes* spp.) by reverse transcription-nested PCR. *Int J Food Microbiol* **91**: 147–154.

Suslov O, Steindler DA. 2005. PCR inhibition by reverse transcriptase leads to an overestimation of amplification efficiency. *Nucleic Acids Res* **33**: e181. doi: 10.1093/nar/gni176.

Tanaka S, Furukawa T, Plotkin SA. 1975. Human cytomegalovirus stimulates host cell RNA synthesis. *J Virol* **15**: 297–304.

Taneyhill LA, Adams MS. 2008. Investigating regulatory factors and their DNA binding affinities through real time quantitative PCR (RT-QPCR) and chromatin immunoprecipitation (ChIP) assays. *Methods Cell Biol* **87**: 367–389.

Tichopad A, Dilger M, Schwarz G, Pfaffl MW. 2003. Standardized determination of real-time PCR efficiency from a single reaction set-up. *Nucleic Acids Res* **31**: pe122. doi: 10.1093/nar/gng122.

Tyagi S, Kramer FR. 1996. Molecular beacons: Probes that fluoresce upon hybridization. *Nat Biotechnol* **14**: 303–308.

Vandesompele J, De Preter K, Pattyn F, Poppe B, Van Roy N, De Paepe A, Speleman F. 2002. Accurate normalization of real-time quantitative RT-PCR data by geometric averaging of multiple internal control genes. *Genome Biol* **3**: presearch0034–research0034.11.

Wang X, Seed B. 2003. A PCR primer bank for quantitative gene expression analysis. *Nucleic Acids Res* **31**: e154. doi: 10.1093/nar/gng154.

Whitcombe D, Theaker J, Guy SP, Brown T, Little S. 1999. Detection of PCR products using self-probing amplicons and fluorescence. *Nat Biotechnol* **17**: 804–807.

Wittwer CT, Herrmann MG, Moss AA, Rasmussen RP. 1997. Continuous fluorescence monitoring of rapid cycle DNA amplification. *BioTechniques* **22**: 130–131, 134–138.

Wittwer CT, Herrmann MG, Gundry CN, Elenitoba-Johnson KS. 2001. Real-time multiplex PCR assays. *Methods* **25**: 430–442.

Wittwer CT, Reed GH, Gundry CN, Vandersteen JG, Pryor RJ. 2003. High-resolution genotyping by amplicon melting analysis using LCGreen. *Clin Chem* **49**: 853–860.

Zhang J, Byrne CD. 1999. Differential priming of RNA templates during cDNA synthesis markedly affects both accuracy and reproducibility of quantitative competitive reverse-transcriptase PCR. *Biochem J* **337**: 231–241.

Zhao S, Fernald RD. 2005. Comprehensive algorithm for quantitative real-time polymerase chain reaction. *J Comput Biol* **12**: 1047–1064.

WWW RESOURCES

Biosearch Technologies   http://www.biosearchtech.com/products/probe_design.asp

BLASTn searching   http://blast.ncbi.nlm.nih.gov/Blast.cgi

Premier Biosoft International   http://www.premierbiosoft.com

PrimerBank   http://pga.mgh.harvard.edu/primerbank/index.html

Primer3Plus   http://www.bioinformatics.nl/cgi-bin/primer3plus/primer3plus.cgi

Real Time PCR Primer Sets   http://www.realtimeprimers.org/

RTPrimerDB   http://medgen.ugent.be/rtprimerdb/

Optimizing Primer and Probe Concentrations for Use in Real-Time PCR

Once primers and probes have been designed and obtained (see the chapter introduction), it is necessary to optimize their concentrations for each real-time PCR assay. A set of PCRs is assembled in which the concentrations of forward and reverse primers are independently varied. Following amplification of the template DNA, amplification plots are compared. A standard curve is generated (Protocol 2) to determine the efficiency, sensitivity, and reproducibility of the assay. If SYBR Green I is used as the probe, then the melting curves are also analyzed.

MATERIALS

It is essential that you consult the appropriate Material Safety Data Sheets and your institution's Environmental Health and Safety Office for proper handling of equipment and hazardous materials used in this protocol.

Recipes for reagents specific to this protocol, marked <R>, are provided at the end of the protocol. See Appendix 1 for recipes for commonly used stock solutions, buffers, and reagents, marked <A>. Dilute stock solutions to the appropriate concentrations.

▲ To reduce the chance of contamination with exogenous DNA, prepare and use a special set of reagents and solutions for PCR only. Bake all glassware for 6 h at 150°C, and autoclave all plasticware. For more information, see Contamination in PCR in Chapter 7.

Reagents

Forward and reverse primers (10 μM)

Nuclease-free water

Probe (if using TaqMan chemistry) (10 μM)

Real-time PCR master mix for SYBR Green I or TaqMan assay <R>

Preformulated real-time PCR master mixes containing all of the reagents required for PCR (except template and primers) in an optimized buffer are available from several vendors (e.g., Applied Biosystems, QIAGEN, Life Technologies, Bio-Rad). The master mixes are designed to simplify experimental setup, decrease the possibility of contamination, and provide optimal performance. Alternatively, it is possible to prepare a homemade master mix as described in the Recipes at the end of this protocol.

Template DNA

As stated in the chapter introduction, the concentration of template DNA will be determined empirically as that which yields a C_T of 20–30. A good starting point is 10–50 ng of genomic DNA or 0.1–1 ng of plasmid DNA.

Equipment

Barrier tips for automatic micropipettes

Microcentrifuge tubes (0.4–1.5 mL, sterile)

PCR plasticware (PCR tubes, strips or 96/384-well plate)

Plasticware should be chosen based on the instructions of the thermal cycler manufacturer. When a block system is used, the plates must fit well to ensure efficient thermal transfer and uniformity between wells. If the signals are detected through the cap of the system, optical-grade caps have to be used. To avoid sample evaporation it is essential to ensure that the seal of the tubes or plates is complete; some systems—but not all—are compatible with heat-sealed film coverings that work very well.

Real-time PCR thermal cycler

METHOD

Assembling and Running the PCRs

1. Using nuclease-free water, prepare 20 µL each of 1 µM, 2 µM, 3 µM, and 6 µM stock solutions for the forward and reverse primers in sterile microcentrifuge tubes.

2. Add 1 µL of each primer stock solution to each tube/well as shown in the primer optimization matrix below, resulting in a final volume of 2 µL in each well. Prepare each reaction in triplicate. In addition, prepare, in triplicate, an additional set of tubes/wells containing the minimum and maximum primer concentrations (i.e., 1 µM forward + 1 µM reverse, and 6 µM forward + 6 µM reverse); these samples will lack template and therefore serve as negative controls in the experiment (generally referred to as the no-template controls).

Forward primer	1 µM	2 µM	3 µM	6 µM
Reverse primer	Final concentration (nM)			
1 µM	50/50	50/100	50/150	50/300
2 µM	100/50	100/100	100/150	100/300
3 µM	150/50	150/100	150/150	150/300
6 µM	300/50	300/100	300/150	300/300

3. In a sterile microcentrifuge tube, prepare the PCR mixture for the template-containing samples. The following tables provide the volume per reaction; multiply each volume by the number of reactions you need.

    In this example, the total number of reactions is 16 × 3 = 48. When preparing PCR mixtures, it is a good idea to prepare enough mixture for one or two extra reactions to ensure that all samples and controls have been accounted for, as pipetting inaccuracies can occur.

    When a hot-start Taq polymerase, such as AmpliTaq Gold DNA Polymerase (Applied Biosystems), is used, a 9–12-min pre-PCR heat step at 92°C–95°C is required to activate the enzyme. Because Ampli-Taq Gold is inactive at room temperature, there is no need to set up the reaction on ice.

    SYBR Green I

Component	Volume added per reaction (µL)
Water	3
Template	5
SYBR Green I Master Mix	10

    TaqMan

Component	Volume added per reaction (µL)
Water	2.5
Probe (10 µM)	0.5
Template	5
TaqMan Master Mix	10

4. In another sterile microcentrifuge tube, prepare the reaction mixture for the no-template control.

    In this example, there are six no-template control reactions; thus, prepare enough reaction mixture for seven reactions.

    SYBR Green I

Component	Volume added per reaction (µL)
Water	8
SYBR Green I Master Mix	10

TaqMan

Component	Volume added per reaction (μL)
Water	7.5
Probe (10 μM)	0.5
TaqMan Master Mix	10

5. Add 18 μL of the appropriate PCR mixture to each well/tube containing the primer pair mix. Mix gently by repeatedly pipetting up and down, ensuring that no bubbles are produced. Cap the tubes/wells carefully. Briefly centrifuge the reaction to collect the contents at the bottom of the well or tube.

To avoid potential cross-contamination, always add the no-template control after all reagents have been dispensed.

6. Place the plate or tubes in the real-time thermocycler. Program and run the machine using the following thermal cycling parameters.

The following protocol is designed for real-time PCRs using AmpliTaq Gold, a hot-start DNA polymerase.

SYBR Green I

	Temperature	Time
Initial steps		
1. AmpErase UNG activation[a]	50°C	2 min
2. AmpliTaq Gold activation	95°C	10 min
PCR (40 cycles)		
3. Melt	95°C	15 sec
4. Annealing/extension	60°C	1 min
Dissociation curve		
5. Melt	From 55°C to 95°C	

[a]This step is required only in the master mixes that contain AmpErase UNG (uracil-*N*-glycosylase).

TaqMan

	Temperature	Time
Initial steps		
1. AmpErase UNG activation[a]	50°C	2 min
2. AmpliTaq Gold activation	95°C	10 min
PCR (40 cycles)		
3. Melt	95°C	15 sec
4. Annealing/extension	60°C	1 min

[a]This step is required only in the master mixes that contain AmpErase UNG.

The PCR should be run as soon as possible after preparing the reaction mixes. Some commercial master mixes contain glycerol and DMSO and can be stored at −20°C for up to 12 h. However, others are more sensitive to storage. Please read the manufacturer's instructions.

Analyzing the Amplification Plot

7. View the amplification plots (ΔR_n against cycle number) (Fig. 1).

First review the amplification plots (in linear view) for all of the samples, and identify any abnormal plots (for further details, see Protocol 5). Most real-time PCR instruments will set the baseline and threshold automatically; at this point, it is usually sufficient to let the machine set these parameters. Second, verify that no signal is produced in the no-template control; if a signal is detected, see Troubleshooting. Finally, identify the primer concentrations that yield optimal assay results. Primer

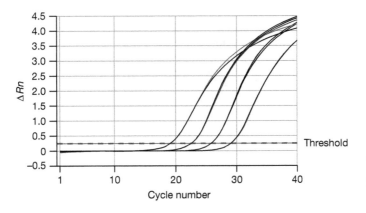

FIGURE 1. **Example of amplification plot (linear view).** Each set of overlapping curves represents a triplicate set of reactions using a different set of primer concentrations. Primer combinations that yield a C_T value between 20 and 30 will yield optimal assay results.

combinations with the lowest C_T (between 20 and 30) and the highest ΔR_n will give the most sensitive and reproducible assays. If the C_Ts are outside of this range, then the concentration of the template must be adjusted. A good rule of thumb is to assume that a 10-fold difference in concentration corresponds to 3.3 cycles and decrease/increase the primer concentration accordingly.

8. Following analysis of the amplification plot, determine the efficiency, sensitivity, and reproducibility of the assay using a standard curve experiment (see Protocol 2).

Analyzing the Melting Curve

9. If using SYBR Green chemistry, also view and evaluate the melting curve (Fig. 2).

   During melting curve analysis, the temperature of the postamplification reaction mixture is gradually increased, causing separation of the double-stranded DNA, which leads to release of the fluorescent dye and a corresponding decrease in fluorescence signal. The temperature at which the DNA strands

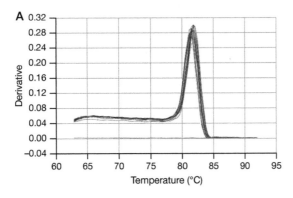

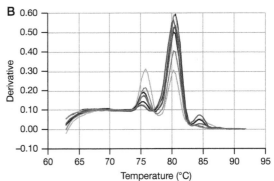

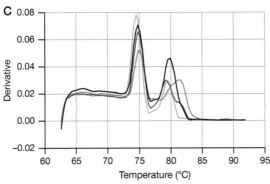

FIGURE 2. **Examples of melt curves.** (*A*) Single peak, resulting from the amplification of a specific amplicon. (*B*) Two peaks, where the lower peak indicates amplification of primer-dimers. (*C*) Two peaks, where the higher peak indicates amplification of genomic DNA.

separate (known as the T_m) and fluorescence decreases depends on amplicon size and sequence; thus, amplified products in the mixture can be distinguished on the basis of their T_ms. The best concentration of primers is one that produces a single, sharp peak in the presence of target nucleic acid and no signal in the corresponding no-template control (see Troubleshooting), indicating that the target sequence has been specifically amplified (Fig. 2A). If primer dimers have formed, the short product generated is characterized by a T_m that is lower than that of the longer target amplicon, thereby generating two different peaks in the melting curve (Fig. 2B). Increasing amounts of primer dimers become apparent at low input RNA concentration.

TROUBLESHOOTING

Problem (Steps 7 and 9): All primer combinations give detectable product in the absence of template.

Solution: If the no-template product melts at a lower temperature than that with template, select the combination that gives the least amount of lower-melting no-template product. The latter is likely primer dimer. Detection can often be avoided, or at least minimized, by adding a 15-sec melting step ~3°C below the melting temperature of the desired PCR product during which fluorescence is measured after the annealing/extension step in each cycle.

RECIPE

It is essential that you consult the appropriate Material Safety Data Sheets and your institution's Environmental Health and Safety Office for proper handling of equipment and hazardous materials used in this protocol.

Real-Time PCR Master Mix for SYBR Green I or TaqMan Assay

This recipe is an alternative to the preformulated real-time PCR master mix. The optimal concentrations of the components are as follows:

200 μM dNTPs, final concentration of each

If dUTP is substituted for dTTP (see UNG below), it should be present at a final concentration of 400 μM.

0.1 μL (0.5 U) *Taq* polymerase

The specificity of the real-time PCR can be enhanced by using a hot-start *Taq* polymerase, such as JumpStart Taq (Sigma-Aldrich), HotStarTaq (QIAGEN), or AmpliTaq Gold (Applied Biosystems). Typically, 0.1 μL (0.5 U) of *Taq* DNA polymerase (5.0 U/μL) is added per 20-μL reaction. If necessary, the reaction can be optimized by increasing the amount by 0.1-U increments. To perform a TaqMan assay, it is necessary to use a *Taq* enzyme with 5'–3' nuclease activity, such as AmpliTaq or AmpliTaq Gold (Applied Biosystems).

4–7 mM $MgCl_2$

UNG

Products from previous PCR runs are a potential source of contamination in real-time PCR assays. If contamination from carryover PCR products is suspected, addition of uracil-*N*-glycosylase (UNG; e.g., AmpErase UNG from Applied Biosystems) can enzymatically destroy contaminants and prevent the reamplification of carryover PCR products. In master mixes containing UNG, dTTP is partially or completely substituted by dUTP. UNG acts by removing uracil incorporated into any contaminating molecules, and contaminating molecules are destroyed by cleavage at the apyrimidinic sites generated by the uracil removal. During subsequent cycling, only target nucleic acid and not contaminating nucleic acid from previous reactions will be amplified.

Passive reference dye

The fluorescent dye ROX serves as an internal reference when carrying out real-time PCR using Applied Biosystems instruments. However, its presence does not interfere with reactions performed using the LightCycler or iCycler. ROX should be at a final concentration of 0.45 nM.

Constructing a Standard Curve

It is essential to prepare a standard curve for every real-time PCR experiment (see the chapter introduction). This protocol is used to construct a standard curve in which the template concentration is unknown. Such a standard curve is suitable for optimization experiments and for performing relative quantification by the standard curve method. To construct a standard curve for absolute quantification, the same principles apply as those presented here, but the concentration of the standards must be determined by an independent method, typically A_{260} absorbance or one of the dye-based methods for DNA, see the section Introduction to Quantifying DNA and the information panel Spectrophotometry, both in Chapter 1; for RNA, see Chapter 6, Protocol 6. Plasmid DNA and in vitro–transcribed RNA are commonly used to prepare absolute standards. For A_{260} absorbance, it is important that the DNA or RNA standards be a single, pure species. For example, plasmid DNA prepared from *Escherichia coli* is often contaminated with RNA, which increases the A_{260} measurement and would consequently inflate the copy number determined for the plasmid. Plasmid DNA or in vitro–transcribed RNA must be concentrated in order to measure an accurate A_{260} value. This concentrated DNA or RNA must then be diluted 100-fold to 1000-fold to be at a concentration similar to the target in biological samples. Finally, it is generally not possible to use DNA as a standard for absolute quantitation of RNA because there is no control for the efficiency of the reverse transcription step.

MATERIALS

It is essential that you consult the appropriate Material Safety Data Sheets and your institution's Environmental Health and Safety Office for proper handling of equipment and hazardous materials used in this protocol.

Reagents

Forward and reverse primers (optimized concentrations; see Protocol 1)

Nuclease-free water

Probe (if using TaqMan chemistry) (optimized concentration; see Protocol 1)

Real-time PCR master mix for SYBR Green I or TaqMan assay

Preformulated real-time PCR master mixes containing all of the reagents required for PCR (except template and primers) in an optimized buffer are available from several vendors (e.g., Applied Biosystems, QIAGEN, Life Technologies, Bio-Rad). The master mixes are designed to simplify experimental setup, decrease the possibility of contamination, and provide optimal performance. Alternatively, it is possible to prepare a homemade master mix (see Recipes at the end of Protocol 1).

Template DNA

Equipment

Barrier tips for automatic micropipettes

Microcentrifuge tubes (0.4–1.5 mL, sterile)

PCR plasticware (PCR tubes, strips or 96/384-well plate)

Plasticware should be chosen based on the instructions of the thermal cycler manufacturer. When a block system is used, the plates must fit well to ensure efficient thermal transfer and uniformity between wells. If the signals are detected through the cap of the system, optical-grade caps have to be used. To avoid sample evaporation, it is essential to ensure that the seal of the tubes or plates is complete; some systems—but not all—are compatible with heat-sealed film coverings that work very well.

Real-time PCR thermal cycler

METHOD

1. Using nuclease-free water, prepare a minimum of five 10-fold serial dilutions of the template in sterile microcentrifuge tubes.

    When preparing the serial dilutions, it is very important to pipette accurately.

2. Add 5 μL of each dilution to reaction tubes or wells, in triplicate. For the no-template control, add 5 μL of water to the tubes or wells, in triplicate.

3. In a sterile microcentrifuge tube, prepare a PCR mixture according to the table below. The total volume is 15 μL per reaction; multiply each volume by the number of reactions you need.

    When preparing PCR mixtures, it is a good idea to prepare enough mixture for one or two extra reactions to ensure that all samples and controls have been accounted for because pipetting inaccuracies can occur.

    When a hot-start Taq polymerase, such as AmpliTaq Gold DNA Polymerase (Applied Biosystems), is used, a 9–12-min pre-PCR heat step at 92°C–95°C is required to activate the enzyme. Because AmpliTaq Gold is inactive at room temperature, there is no need to set up the reaction on ice.

    SYBR Green I

Component	Volume per reaction (μL)
H_2O	3
Forward primer (optimized concentration)	1
Reverse primer (optimized concentration)	1
SYBR Green I Master Mix	10

    TaqMan

Component	Volume per reaction (μL)
H_2O	2.5
Probe (optimized concentration)	0.5
Forward primer (optimized concentration)	1
Reverse primer (optimized concentration)	1
TaqMan Master Mix	10

4. Add 15 μL of the PCR mixture to each sample in the tubes/wells. Mix gently by repeatedly pipetting up and down, ensuring that no bubbles are produced. Cap the tubes/wells carefully. Briefly centrifuge the reaction to collect the contents at the bottom of the well or tube.

5. Place the plate or tubes in the real-time thermocycler. Program and run the machine using the following thermal cycling parameters.

    SYBR Green I

	Temperature	Time
Initial steps		
1. AmpErase UNG activation[a]	50°C	2 min
2. AmpliTaq Gold activation	95°C	10 min
PCR (40 cycles)		
3. Melt	95°C	15 sec
4. Annealing/extension	60°C	1 min
Dissociation curve		
5. Melt	From 55°C to 95°C	

    [a]This step is required only in the master mixes that contain AmpErase UNG.

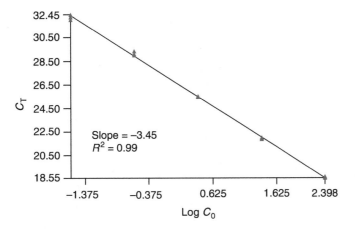

FIGURE 1. **Example of a standard curve.** Each cluster of data points represents a triplicate set of reactions using a different DNA concentration. In this example, the slope of the line is −3.45 (giving a percent efficiency of 94.9%) and the R^2 value is 0.99, indicating that the data are of sufficient quality to yield reliable results.

TaqMan

	Temperature	Time
Initial steps		
1. AmpErase UNG activation[a]	50°C	2 min
2. AmpliTaq Gold activation	95°C	10 min
PCR (40 cycles)		
3. Melt	95°C	15 sec
4. Annealing/extension	60°C	1 min

[a]This step is required only in the master mixes that contain AmpErase UNG.

The PCR should be run as soon as possible after preparing the reaction mixes. Some commercial master mixes contain glycerol and DMSO and can be stored at −20°C for up to 12 h. However, others are more sensitive to storage. Please read the manufacturer's instructions.

6. Using the instrument software instructions, plot C_T against the log of the dilution factor as shown in Figure 1, and determine the slope (to calculate efficiency) and R^2 value.

   As stated above, a perfect slope would be –3.32, but a slope between –3.10 and –3.74 (85%–110% efficiency) is generally acceptable (if the results are not satisfactory, see Troubleshooting). A perfect R^2 value would be 1, but should be >0.98 (if the results are not satisfactory, see Troubleshooting).

7. If you are performing a SYBR Green I assay, analyze the melting curve to ensure that only specific products are amplified in the range of the template concentrations analyzed (see Step 9 of Protocol 1).

Standard Curves for the Comparative C_T Method

To use the comparative C_T method, the amplification efficiencies of the target and endogenous reference genes must be at or near 100% and approximately equal (within 5% of each other). Both of these parameters can be verified using a so-called validation experiment. In this method, serial dilutions of a sample, which expresses both the target and reference genes, are prepared, and real-time PCR is run in separate tubes for the two genes. The C_T values of the target and the reference gene for each dilution are obtained, and their difference is calculated:

$$\Delta C_T = C_T(\text{target}) - C_T(\text{reference}).$$

These ΔC_T values are plotted versus log input amount to create a semi-log regression line. If the amplification efficiencies of the target and reference are equal across the range of initial template amounts, the slope of the line will be 0. If the plot shows unequal efficiency (slope < -0.1 or >0.1), the assays should be reoptimized.

TROUBLESHOOTING

Problem (Step 6): The efficiency of PCR amplification is too low.
Solution: Low efficiency can indicate poor primer design, poor reaction conditions, or pipetting error. To improve efficiency, optimize primer concentrations, design alternative primers, or pipette with more care.

Problem (Step 6): The efficiency of PCR amplification is too high.
Solution: Efficiency clearly above 100% could be due to inhibitors in the sample that cause a delayed C_T in the samples with the highest concentration. PCR inhibitors can originate from the starting material (e.g., proteins or polysaccharides) or be introduced during nucleic acid extraction (e.g., SDS, phenol, ethanol, proteinase K, chaotropic agents, or sodium acetate). Try repurifying the sample or extracting the DNA using a different kit or protocol appropriate for the sample type.

Problem (Step 6): The R^2 value is too low (<0.9).
Solution: Low R^2 values could indicate poor primer design. Redesign the primers.

Problem (Step 6): Reproducibility is low.
Solution: Many factors can cause loss of reproducibility including technical issues such as imprecise pipetting, incomplete mixing of components, air bubbles in the reaction wells, drops on the sides of the wells, and writing on the top of the caps/plate covers. To improve reproducibility, pipette with more care, ensure that reaction components are mixed well, centrifuge samples to remove bubbles, make certain all contents are in the bottom of the wells/tubes, and do not write on the tops of tubes or plates.

Problem (Step 6): One or more points at the lowest concentration of input nucleic acid are shifted away from the linear region of the plot.
Solution: It is likely that the level exceeds assay sensitivity. To improve sensitivity, optimize primer concentrations or design different primers.

Problem (Step 6): One or more points at the highest levels of input nucleic acid are shifted away from the linear region of the plot.
Solution: It is likely that the reaction is saturated and that the level of target exceeds the useful assay range. To address this situation, add less nucleic acid or dilute the sample of nucleic acid.

Problem (Step 6): Several random points are above or below the line.
Solution: Pipetting accuracy may be a problem. Verify that the pipette tips fit the pipettor properly and that the volume dispensed is reproducible.

Quantification of DNA by Real-Time PCR

Before running a real-time PCR experiment, it is important to optimize the concentration of the primers (and probe, if using TaqMan chemistry) and determine the efficiency, sensitivity, and reproducibility of the assay (see Protocols 1 and 2). When designing a real-time PCR experiment, it is important to consider the type of quantitation (data analysis) method that will be used (see the section Extracting Data from a Real-Time PCR Experiment in the chapter introduction). If absolute quantification will be used, a standard curve must be run in parallel with the test samples (see Protocol 2). The standard curve may use plasmid DNA or other forms of DNA in which the absolute concentration of each standard is known. One must be sure, however, that the efficiency of PCR is the same for the standards as for the unknown samples. If either of the relative methods of quantification will be used, an endogenous reference gene must also be analyzed in parallel with the test samples (see the chapter introduction).

MATERIALS

It is essential that you consult the appropriate Material Safety Data Sheets and your institution's Environmental Health and Safety Office for proper handling of equipment and hazardous materials used in this protocol.

Reagents

Forward and reverse primers (optimized concentrations; see Protocol 1)

Nuclease-free water

Probe (if using TaqMan chemistry) (optimized concentration; see Protocol 1)

Real-time PCR master mix for SYBR Green I or TaqMan assay

Preformulated real-time PCR master mixes containing all of the reagents required for PCR (except template and primers) in an optimized buffer are available from several vendors (e.g., Applied Biosystems, QIAGEN, Life Technologies, Bio-Rad). The master mixes are designed to simplify experimental setup, decrease the possibility of contamination, and provide optimal performance. Alternatively, it is possible to prepare a homemade master mix (see Recipes at the end of Protocol 1).

Template DNA

Equipment

Barrier tips for automatic micropipettes

Microcentrifuge tubes (0.4–1.5 mL, sterile)

PCR plasticware (PCR tubes, strips or 96/384-well plate)

Plasticware should be chosen based on the instructions of the thermal cycler manufacturer. When a block system is used, the plates must fit well to ensure efficient thermal transfer and uniformity between wells. If the signals are detected through the cap of the system, optical-grade caps have to be used. To avoid sample evaporation, it is essential to ensure that the seal of the tubes or plates is complete; some systems—but not all—are compatible with heat-sealed film coverings that work very well.

Real-time PCR thermal cycler

METHOD

1. Add 5 μL of the DNA sample(s) to each reaction tube in triplicate. Also include a no-template control, in which 5 μL of H_2O has been added, in triplicate.

2. In a sterile microcentrifuge tube, prepare a PCR mixture according to the table below. The total volume is 15 μL per reaction; multiply each volume by the number of reactions you need.

   When preparing PCR mixtures, it is a good idea to prepare enough mixture for one or two extra reactions to ensure that all samples and controls have been accounted for, because pipetting inaccuracies can occur.

   When a hot-start Taq polymerase, such as AmpliTaq Gold DNA Polymerase (Applied Biosystems), is used, a 9–12-min pre-PCR heat step at 92°C–95°C is required to activate the enzyme. Because AmpliTaq Gold is inactive at room temperature, there is no need to set up the reaction on ice.

   SYBR Green I

Component	Volume per reaction (μL)
H_2O	3
Forward primer (optimized concentration)	1
Reverse primer (optimized concentration)	1
SYBR Green I Master Mix	10

   TaqMan

Component	Volume per reaction (μL)
H_2O	2.5
Probe (optimized concentration)	0.5
Forward primer (optimized concentration)	1
Reverse primer (optimized concentration)	1
TaqMan Master Mix	10

3. Add 15 μL of the PCR mixture to each sample in the tubes/wells. Mix gently by repeatedly pipetting up and down, ensuring that no bubbles are produced. Cap the tubes/wells carefully. Briefly centrifuge the reaction to collect the contents at the bottom of the well or tube.

4. Place the plate or tubes in the real-time thermocycler. Program and run the machine using the following thermal cycling parameters.

   SYBR Green I

	Temperature	Time
Initial steps		
1. AmpErase UNG activation[a]	50°C	2 min
2. AmpliTaq Gold activation	95°C	10 min
PCR (40 cycles)		
3. Melt	95°C	15 sec
4. Annealing/extension	60°C	1 min
Dissociation curve		
5. Melt	From 55°C to 95°C	

   [a]This step is required only in the master mixes that contain AmpErase UNG.

TaqMan

	Temperature	Time
Initial steps		
1. AmpErase UNG activation[a]	50°C	2 min
2. AmpliTaq Gold activation	95°C	10 min
PCR (40 cycles)		
3. Melt	95°C	15 sec
4. Annealing/extension	60°C	1 min

[a]This step is required only in the master mixes that contain AmpErase UNG.

The PCR should be run as soon as possible after preparing the reaction mixes. Some commercial master mixes contain glycerol and DMSO and can be stored at −20°C for up to 12 h. However, others are more sensitive to storage. Please read the manufacturer's instructions.

Following the PCR run, analyze the raw data (Protocol 5).

Quantification of RNA by Real-Time RT-PCR

The following protocol describes a real-time RT-PCR assay, using the two-enzyme, two-tube approach (see the section selecting the Enzyme(s) in the chapter introduction), carried out using either SYBR Green I or TaqMan chemistries. The protocol uses a PCR, volume of 20 μL (although most manufacturers recommend 50-μL reactions). However, if the PCR target is not very abundant (i.e., present at one to 10 copies per sample), a larger volume may yield better reproducibility between samples. For additional information on preparing high-quality RNA, choosing a priming method, and selecting an endogenous reference gene, see the section Performing Real-Time PCR in the chapter introduction.

MATERIALS

It is essential that you consult the appropriate Material Safety Data Sheets and your institution's Environmental Health and Safety Office for proper handling of equipment and hazardous materials used in this protocol.

▲ To reduce the chance of contamination with exogenous DNAs, prepare and use a special set of reagents and solutions for PCR only. Bake all glassware for 6 h at 150°C, and autoclave all plasticware. For more information, please see Contamination in PCR in Chapter 7.

Reagents

DNase I
> There are several options available. First, it is possible to perform the digestion in solution with Ambion's DNase Treatment and Removal Reagent. A less expensive method, but one that is more likely to result in loss of RNA and potential phenolic contamination, is to treat the RNA sample with DNase I and then remove the enzyme using phenol extraction followed by ethanol precipitation (see Chapter 6, Protocol 8). QIAGEN DNase can be used directly on RNA extraction columns, thereby decreasing the opportunity for new contamination and carryovers of PCR inhibitors.

dNTP mix (10 mM)
DTT (0.1 M)
Forward and reverse primers (optimized concentrations; see Protocol 1)
$MgCl_2$ (25 mM)
Nuclease-free water
Polymerases
> For the two-tube procedure, use a thermostable reverse transcriptase (e.g., Superscript III from Life Technologies) and a hot-start Taq polymerase (e.g., AmpliTaq or AmpliTaq Gold from Applied Biosystems).

Primer for reverse transcription [50 μM oligo(dT)$_{20}$, 2 μM gene-specific primer, or 50 ng/μL random hexamers]
Probe (if using TaqMan chemistry) (optimized concentration; see Protocol 1)
Real-time PCR master mix for SYBR Green I or TaqMan assay
> Preformulated real-time PCR master mixes containing all of the reagents required for PCR (except template and primers) in an optimized buffer are available from several vendors (e.g., Applied Biosystems, QIAGEN, Life Technologies, Bio-Rad). The master mixes are designed to simplify experimental setup, decrease the possibility of contamination, and provide optimal performance. Alternatively, it is possible to prepare a homemade master mix (see Recipes at the end of Protocol 1).

Reference template
> This will be used as control material or for the standard curve (e.g., Universal RNA; Stratagene)

Reverse transcriptase (such as SuperScript III from Life Technologies) (200 U/μL)

Reverse transcriptases that are M-MLV RT versions engineered to reduce RNase H activity and provide increased thermal stability are now commercially available (Life Technologies, Stratagene). They are optimized to synthesize first-strand cDNA from purified poly(A)$^+$ or total RNA and to synthesize cDNA at a temperature range of 42°C–55°C, providing increased specificity, higher yields of cDNA, and more full-length product than other reverse transcriptases. The following protocol uses SuperScript III (Life Technologies).

RNase H (such as that available from Life Technologies) (2 U/μL)

RNase inhibitor (such as RNaseOUT from Life Technologies) (40 U/μL)

Please see the information panel Inhibitors of RNases in Chapter 6.

RT buffer (10×)

10× RT buffer for SuperScript III consists of 200 mM Tris-HCl (pH 8.4) and 500 mM KCl.

Total RNA sample, DNA-free, of known concentration and known integrity

RNA extracted using monophasic lysis reagents is the template of choice for real-time RT-PCR (protocols for isolating total RNA using monophasic lysis reagents are given in Chapter 6, Protocols 1–4). Following isolation, the integrity of the RNA sample should be checked (see Checking the Quality of Preparations of RNA in Chapter 6, Protocol 10) and quantified. Quantifying RNA using A_{260} readings is not sufficiently accurate for use with real-time RT-PCR technology. Instead, use a NanoDrop Spectrophotometer or Ribo-Green (see Chapter 6, Protocol 6). It is important that the same method be used to quantify all samples in a single experiment and that data should not be compared from different procedures, because different results can be generated depending on the method used.

Equipment

Barrier tips for automatic micropipettes

Microcentrifuge tubes (0.4–1.5 mL, sterile)

PCR plasticware (PCR tubes, strips or 96/384-well plate)

Plasticware should be chosen based on the instructions of the thermal cycler manufacturer. When a block system is used, the plates must fit well to ensure efficient thermal transfer and uniformity between wells. If the signals are detected through the cap of the system, optical-grade caps have to be used. To avoid sample evaporation, it is essential to ensure that the seal of the tubes or plates is complete; some systems—but not all—are compatible with heat-sealed film coverings that work very well.

Real-time thermal PCR cycler

Water baths or heat blocks, preset to 25°C (as needed), 50°C, 65°C, and 85°C

METHOD

Reverse Transcription

1. Prepare the following reaction mixture in a sterile microcentrifuge tube, in duplicate (one will serve as the no-RT control):

Total RNA	up to 5 μg
Primer	1 μL
dNTP mix (10 mM)	1 μL
H$_2$O	to 10 μL

   If an oligo(dT) primer or random hexamers are used, one reaction will prepare enough material to subsequently analyze multiple different mRNAs. If a gene-specific primer is used, prepare one volume of reaction mixture for each target mRNA to be analyzed.

2. Denature the RNA by incubating the reaction mixture for 5 min at 65°C, followed by rapid chilling for 1 min on ice.

3. To one of the tubes, add each component in the order listed below.

RT buffer (10×)	2 μL
MgCl$_2$ (25 mM)	4 μL
DTT (0.1 M)	2 μL
RNaseOUT (40 U/μL)	1 μL
SuperScript III RT (200 U/μL)	1 μL

4. To the no-RT control tube, add each component in the order listed below.

RT buffer (10×)	2 μL
MgCl$_2$ (25 mM)	4 μL
DTT (0.1 M)	2 μL
RNaseOUT (40 U/μL)	1 μL
H$_2$O	1 μL

Regardless of whether primers span or flank introns, the specificity of the real-time RT-PCR assay should be tested in reactions without reverse transcriptase (no-RT control) to evaluate the specificity of DNA amplification. As mentioned above, DNA sequences with short introns (≤1 kb) may be amplified in RT-PCR. Many genes have additional copies, or pseudogenes, that lack one or more introns. As a result, qRT-PCR assays should be tested for potential DNA-only amplicons by performing reactions that contain RT and the same RNA, but no RT enzyme.

5. Mix the contents of the tubes by gently flicking the tubes, and collect the contents by brief centrifugation. If the reaction was primed using oligo(dT) or gene-specific primers, incubate the reaction for 50 min at 50°C. If the reaction mixture was primed using random hexamers, incubate the reaction mixture for 10 min at 25°C, followed by 50 min at 50°C.

6. Terminate the reactions by incubating the tubes for 5 min at 85°C. Chill the tubes on ice for at least 1 min. Collect the contents of the tubes by brief centrifugation.

7. Remove the RNA template from the cDNA:RNA hybrid molecule by adding 1 μL of RNase H to each tube and incubating for 20 min at 37°C.

The cDNA synthesis reaction can be used immediately for PCR or stored at –20°C for up to 6 mo.

Real-Time PCR

8. Add 5 μL of cDNA sample(s) to each reaction tube in triplicate. Also include a no-template control, in which 5 μL of H$_2$O has been added, in triplicate.

If a standard curve is being performed, prepare a minimum of five 10-fold serial dilutions of the template in sterile microcentrifuge tubes using nuclease-free water. Add 5 μL of each dilution to reaction tubes or wells, in triplicate.

9. In a sterile microcentrifuge tube, prepare a PCR mixture according to the table below. The total volume is 15 μL per reaction; multiply each volume by the number of reactions you need.

When preparing PCR mixtures, it is a good idea to prepare enough mixture for one or two extra reactions to ensure that all samples and controls have been accounted for, because pipetting inaccuracies can occur.

When a hot-start Taq polymerase, such as AmpliTaq Gold DNA Polymerase (Applied Biosystems), is used, a 9–12-min pre-PCR heat step at 92°C–95°C is required to activate the enzyme. Because AmpliTaq Gold is inactive at room temperature, there is no need to set up the reaction on ice.

SYBR Green I

Component	Volume per reaction (μL)
Water	3
Forward primer (optimized concentration)	1
Reverse primer (optimized concentration)	1
SYBR Green I Master Mix	10

TaqMan

Component	Volume per reaction (μL)
H$_2$O	2.5
Probe (optimized concentration)	0.5
Forward primer (optimized concentration)	1
Reverse primer (optimized concentration)	1
TaqMan Master Mix	10

10. Add 15 µL of the PCR mixture to each sample in the tubes/wells. Mix gently by repeatedly pipetting up and down, ensuring that no bubbles are produced. Cap the tubes/wells carefully. Briefly centrifuge the reaction to collect the contents at the bottom of the well or tube.

11. Place the plate or tubes in the real-time thermocycler. Program and run the machine using the following thermal cycling parameters.

SYBR Green I

	Temperature	Time
Initial steps		
1. AmpErase UNG activation[a]	50°C	2 min
2. AmpliTaq Gold activation	95°C	10 min
PCR (40 cycles)		
3. Melt	95°C	15 sec
4. Annealing/extension	60°C	1 min
Dissociation curve		
5. Melt	From 55°C to 95°C	

[a]This step is required only in the master mixes that contain AmpErase UNG.

TaqMan

	Temperature	Time
Initial steps		
1. AmpErase UNG activation[a]	50°C	2 min
2. AmplyTaq Gold activation	95°C	10 min
PCR (40 cycles)		
3. Melt	95°C	15 sec
4. Annealing/extension	60°C	1 min

[a]This step is required only in the master mixes that contain AmpErase UNG.

The PCR should be run as soon as possible after preparing the reaction mixes. Some commercial master mixes contain glycerol and DMSO and can be stored at −20°C for up to 12 h. However, others are more sensitive to storage. Please read the manufacturer's instructions.

Following the PCR run, analyze the raw data (see Protocol 5).

See Troubleshooting.

TROUBLESHOOTING

Problem (Step 11): A signal is present in the no-RT control.

Solution: DNA amplification is not a problem if the C_T values for no-RT reactions are at least five cycles greater (32-fold less) than those for reactions with RT. However, if there are fewer than five cycles between C_T values for reactions with and without RT, DNA amplification may skew attempts at mRNA quantitation. In cases in which DNA amplicons contribute significantly, one should digest the RNA with RNase-free or amplification-grade DNase I before qRT-PCR to allow for reliable mRNA quantification.

Analysis and Normalization of Real-Time PCR Experimental Data

A real-time PCR experiment generates a large amount of raw numerical data, which are generally analyzed by basic software tools provided with the real-time PCR instrument. Because data analysis varies depending on the assay and instrument, it is necessary to refer to the manufacturer's instructions in order to analyze the data in an appropriate manner. However, even with data analysis software, it is good practice to examine the raw fluorescence data, and thereby evaluate the quality and reliability of the data, in order to generate reportable results. This involves three basic steps.

- View the raw data (amplification plots), altering the baseline and threshold if necessary. C_T values are a function of baseline and threshold values. Software default options provide some level of subjectivity, but these settings are not always appropriate and may need to be changed.
- Verify the efficiency and sensitivity of the assay.
- Apply a quantification method and normalize the data.

Once the data have been analyzed, the researcher is faced with an array of options for further processing the data. For example, a wide range of methods has been described for normalizing ChIP-qPCR data (see below). Moreover, the absolute and relative quantification methods are based on the assumption that the amplification efficiencies of the target and reference genes are approximately equal and constant throughout the PCR—which may not always be the case. To circumvent the amplification efficiency problem, several mathematical models have been developed for processing the PCR data (Cikos and Koppel 2009). Such data processing can dramatically affect the interpretation of real-time PCR results and notably influence the final results, making comparisons of published data sets difficult. In the absence of commonly accepted reference procedures (but see the section MIQE Guidelines in the introduction to this chapter), the choice of data processing method is currently at the researcher's discretion.

MATERIALS

It is essential that you consult the appropriate Material Safety Data Sheets and your institution's Environmental Health and Safety Office for proper handling of equipment and hazardous materials used in this protocol.

This protocol uses data generated from either Protocol 3 or 4.

Equipment

Real-time PCR thermal cycler and accompanying software for data management and analysis

METHOD

Viewing the Raw Data (Amplification Plots)

Setting the Baseline

One goal of data analysis is to determine when target amplification is sufficiently above the background signal, facilitating more accurate measurement of fluorescence. The baseline range is usually automatically fixed by the instrument and for most machines is set between cycles 3 to 15. Although this setting is suitable for most experimental conditions, it is extremely important to verify if the baseline is indeed correct or needs to be adjusted. To do this, review the amplification plots for all of the samples (including those used to generate standard curves, if applicable), and identify any abnormal plots. In most cases, unexpected amplification curve distortions are due to incorrect baseline settings and require manual adjustments (which may need to be performed separately for each well or tube).

The correct setting of the baseline is determined empirically. As a good rule of thumb, the minimum value of the baseline (sometimes called the "start cycle") should be after the initial tailing of background noise, and the maximum value of the baseline (called the "end cycle") should be before the signal shift due to amplification. The range does not have to be too large; around five cycles is sufficient, although more is better. The correct baseline setting is best determined in log view, as it amplifies the background activity changes.

1. View the amplification plot in log view.

    If the sample that emerges first (i.e., the most abundant) has an amplification plot that begins just after the maximum value of the baseline, then no adjustments are necessary (Fig. 1A). In general, if the earliest sample has $C_T > 15$ (the superior limit of the baseline), then adjustments are not necessary.

    If the amplification curve of the earliest sample begins before the maximum value of the baseline, then the baseline is too high and adjustments are necessary. If the baseline is too high, the amplification plots will show a characteristic "break" near the middle of the plot (Fig. 1B). In this case, decrease the baseline stop value to one to two cycles before the earliest amplification, and update the analysis.

    If the amplification curve of the earliest sample begins too far to the right of the maximum value of the baseline, then the baseline is too low. In this case, the amplification plots will show a "break" at low cycle numbers (Fig. 1C). To remedy the situation, increase the baseline stop value to one to two cycles after the earliest amplification, and update the analysis.

    If baseline adjustment does not correct the irregularities in the amplification curves, see Troubleshooting.

Setting the Threshold

The threshold is the numerical value assigned for each run, which is calculated as the average standard deviation of R_n for the early PCR cycles, multiplied by an adjustable factor. The threshold is the statistically significant point above the calculated baseline, and it should be set in the region associated with an exponential growth of PCR product. Most instruments automatically calculate the threshold level of fluorescence signal by determining the baseline (background) average signal and setting a threshold 10-fold higher than this average. Once the baseline has been set correctly, it is also possible to adjust the threshold manually. Like baseline setting, adjusting the threshold to the correct value is determined empirically. Note that you should not change the threshold settings when you are taking readings for data that are to be analyzed relative to each other. It is only correct to adjust the threshold to take readings for a different target in the same or different samples and with the same standard curve samples.

2. Viewing the amplification plot in log view, set the threshold in a region that represents exponential amplification across all of the amplification plots—not in the initial linear phase of amplification, and not in the region of background fluorescence.

3. Once the threshold has been changed, check if the slope and R^2 values of the standard curve have improved. If a standard curve is not available, then check the standard deviation of the triplicate; a good threshold setting minimizes the scattering of the C_T values.

Additional Analyses

4. Examine replicates. All replicates should be within 0.5 C_T of each other. Above cycle 35, the variability will be greater and quantification may be unreliable.

5. If a standard curve has been used, ensure that all data points are within the dynamic range defined by the standard curve.

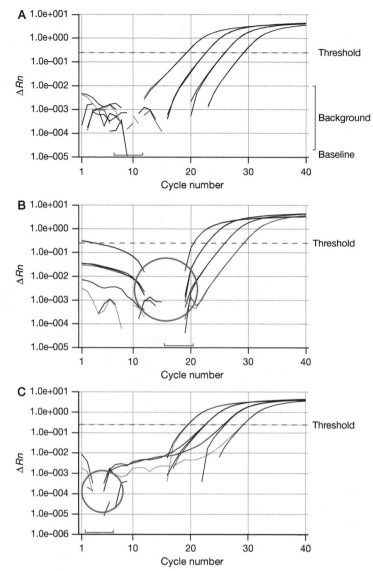

FIGURE 1. Examples of amplification plots (log view). Each set of overlapping curves represents a triplicate set of reactions of different DNA samples. (*A*) A normal plot, in which the amplification plot of the earliest sample begins after the maximum value of the baseline. (*B*) An abnormal plot in which the baseline is too high. This plot shows the characteristic break near the middle of the plot (circled). (*C*) An abnormal plot in which the baseline is too low. This plot shows a characteristic break at low cycle numbers (circled).

6. If a standard curve has been used, verify the efficiency and sensitivity of the PCR by checking the slope of the standard curve and the R^2 value. The slope must be between -3.2 and -3.5 and R^2 must be >0.98.

7. If using SYBR Green I chemistry, check melting curve profiles. Ideally, the melting curve should contain a single peak, as shown in Figure 2A of Protocol 1.

Applying a Quantification Method and Normalizing the Data

Absolute Quantification

As a reminder, to perform absolute quantification, two criteria must be met. First, the test sample can only be quantified if its C_T value falls within the range of C_T values represented by the standard curve. Second, to make accurate comparisons between the test sample and the standard curve, the amplification efficiencies of the test sample(s) and standards must be identical. Ideally, the test sample(s) and standards need to be optimized so that they come close to 100% efficiency.

Making an accurate calculation of the reaction efficiency and the number of starting template molecules depends on being able to ascertain which part of the raw fluorescence intensity data falls in the log-linear phase. A small variation in the slope will cause a significant change in the calculated value; thus, point selection is critical. Several methods to do this have been published, including picking points that appear to fall on the best straight line by eye, calculating the second derivative of the data (Luu-The et al. 2005), and fairly sophisticated statistical analyses of the regression line through the linear phase of the reaction (Ramakers et al. 2003).

The mathematics required for absolute quantification is conceptually simple. A standard curve is constructed, in which the logarithm of the copy number of the standards (x-axis) is plotted versus the C_T value (y-axis). The familiar equation for the linear regression line is

$$y = mx + b,$$

where m is the slope of the line and b is the y-intercept, or

$$C_T = m(\log \text{quantity}) + b.$$

Based on this equation, the following equation can be derived to determine the quantity of the unknown sample:

$$\text{Quantity} = 10^{([C_T - b]/m)}.$$

The quantity of target nucleic acid in each of the sample replicates is calculated using this equation and then normalized to a unit amount of sample, such as number of cells, volume, or total amount of nucleic acid. The values are typically reported as the mean $\pm$ S.D.

Relative Quantification

For both the standard curve and comparative C_T methods of relative quantification, it is important to choose a suitable normalization method. For analysis of target gene expression, samples are most often normalized to the expression of an endogenous reference gene (i.e., housekeeping gene). When quantifying DNA from a chromatin immunoprecipitation experiment (often abbreviated ChIP-qPCR), the data must be normalized for differences in the amount of input chromatin, precipitation efficiency, and variation in the recovery of DNA after ChIP. The normalization methods used currently for ChIP analysis are background subtraction (Mutskov and Felsenfeld 2004), percent of input (%IP) (Nagaki et al. 2003), fold enrichment (Tariq et al. 2003), normalization relative to a control sequence (Mathieu et al. 2005), and normalization relative to nucleosome density (Kristjuhan and Svejstrup 2004). For a discussion of these various methods, along with their most important advantages and disadvantages, see Haring et al. (2007).

Standard Curve Method. In this method, the PCR efficiencies of the target and endogenous control do not have to be equivalent. For each experimental sample, the amount of target and endogenous reference is determined from the appropriate standard curve, similar to the manner described above for absolute quantification. The target amount is then divided by the endogenous reference amount to obtain a normalized target value. One of the experimental samples—such as an untreated control or a sample at time 0 in a time-course study—is selected as the calibrator. Each of the normalized target values is divided by the calibrator-normalized target value to generate the relative levels of expression; in this manner, the level of the calibrator is set at a value of 1. Thus, the normalized amount of target is a unitless number, and all quantities are expressed as an n-fold difference relative to the calibrator.

Comparative C$_T$ Method. As was discussed in detail above, in order to use the comparative C$_T$ method, the amplification efficiencies of the target and endogenous reference must be at or near 100% and approximately equal (within 5% of each other). In addition, the dynamic range of both the target and reference should be similar. Both of these parameters can be verified using a validation experiment (see Protocol 2).

When examining the validation experiment results, it is important to ensure that the data are rigorously analyzed and that real-time PCR has been performed correctly. For instance, PCR inhibitors, inaccurate baseline and threshold settings, or the presence of an outlier in a replicate sample can alter the efficiency calculation. Efficiency variation may be tolerated when high fold-change levels (i.e., 100-fold to 1000-fold) are expected, because a slightly less accurate calculated fold change may not affect the final interpretation of the results. However, tolerance for efficiency variation may be less if relatively small fold differences (i.e., twofold to fourfold) are observed. In this case, it may be prudent to redesign assay(s) to achieve a passing validation experiment, or use the relative standard curve method. The error introduced by the differing PCR efficiencies can be evaluated by comparing the results obtained with the comparative C$_T$ method with results from a relative standard curve method. Because the accuracy of the relative standard curve method is not dependent on the relative efficiencies of the target and reference assays, this comparison can be used to estimate the amount of error in the comparative C$_T$ method experiment. Then it is necessary to evaluate if the error can be tolerated in the specific assay.

As long as the target and reference gene have similar amplification efficiencies and dynamic ranges, the comparative C$_T$ method is the most practical method. Calculations for the quantification begin with normalization of the C$_T$ of the target gene to the C$_T$ of the reference gene, for both the test sample and the calibrator:

$$\Delta C_{T(\text{test})} = C_{T(\text{target,test})} - C_{T(\text{reference,test})},$$

$$\Delta C_{T(\text{calibrator})} = C_{T(\text{target, calibrator})} - C_{T(\text{reference, calibrator})}.$$

The ΔC_T of the test sample is then normalized to that of the calibrator:

$$\Delta\Delta C_T = \Delta C_{T(\text{test})} - \Delta C_{T(\text{calibrator})}.$$

Finally, the normalized expression ratio is calculated using the equation $2^{-\Delta\Delta C_T}$, where 2 indicates 100% amplification efficiency (see the section Constructing a Standard Curve in the chapter introduction). If the target and the reference genes have identical amplification efficiencies but the efficiency is not equal to 2, a modified version of the formula can be used in which the true efficiency replaces 2 (e.g., if the amplification efficiency of both the target and the reference gene is 1.95, the formula $1.95^{-\Delta\Delta C_T}$ should be used).

TROUBLESHOOTING

Problem (Step 1): Adjustment of the baseline does not correct the irregularities in the amplification curves.

Solution: The lamp of the instrument may need replacing (a requirement after more than 2000 h of use for most instruments).

REFERENCES

Cikos S, Koppel J. 2009. Transformation of real-time PCR fluorescence data to target gene quantity. *Anal Biochem* **384:** 1–10.

Haring M, Offermann S, Danker T, Horst I, Peterhansel C, Stam M. 2007. Chromatin immunoprecipitation: Optimization, quantitative analysis and data normalization. *Plant Methods* **3:** 11. doi: 10.1186/1746-4811-3-11.

Kristjuhan A, Svejstrup JQ. 2004. Evidence for distinct mechanisms facilitating transcript elongation through chromatin in vivo. *EMBO J* **23:** 4243–4252.

Luu-The V, Paquet N, Calvo E, Cumps J. 2005. Improved real-time RT-PCR method for high-throughput measurements using second derivative calculation and double correction. *BioTechniques* **38:** 287–293.

Mathieu O, Probst AV, Paszkowski J. 2005. Distinct regulation of histone H3 methylation at lysines 27 and 9 by CpG methylation in *Arabidopsis*. *EMBO J* 24: 2783–2791.

Mutskov V, Felsenfeld G. 2004. Silencing of transgene transcription precedes methylation of promoter DNA and histone H3 lysine 9. *EMBO J* 23: 138–149.

Nagaki K, Talbert PB, Zhong CX, Dawe RK, Henikoff S, Jiang J. 2003. Chromatin immunoprecipitation reveals that the 180-bp satellite repeat is the key functional DNA element of *Arabidopsis thaliana* centromeres. *Genetics* 163: 1221–1225.

Ramakers C, Ruijter JM, Deprez RH, Moorman AF. 2003. Assumption-free analysis of quantitative real-time polymerase chain reaction (PCR) data. *Neurosci Lett* 339: 62–66.

Tariq M, Saze H, Probst AV, Lichota J, Habu Y, Paszkowski J. 2003. Erasure of CpG methylation in *Arabidopsis* alters patterns of histone H3 methylation in heterochromatin. *Proc Natl Acad Sci* 100: 8823–8827.

MULTIPLEX PCR

Multiplex reactions are those in which more than one target is amplified. The availability of a variety of fluorescent dyes spanning the visible spectrum allows multiplex real-time PCR assays to be performed. TaqMan probes, molecular beacons, Scorpion, and LUX fluorogenic primers can be used for multiplex assays by designing each probe with a spectrally unique reporter–quencher pair. It is important to avoid detection of nonspecific signal, and thus the farther apart the spectral ranges used, the better. For example, a duplex reaction is best carried out using FAM quenched by the dark quencher BHQ-1, and HEX quenched by BHQ-1 or BHQ-2. The number of targets that can be detected simultaneously in a single PCR has increased recently through the use of DABCYL, a universal quencher that can replace TAMRA at the 3′ end of probes.

For multiplex reactions, it is critical to design all assays together (e.g., Beacon Designer software [Premier Biosoft International] will design up to four assays to be run in multiplex), because amplicon lengths must be similar (within 5 bp) and annealing temperatures must be the same for all primer sets and probes. In addition, potential oligonucleotide cross-hybridization between reactions must be avoided.

To perform a multiplex experiment, choose a thermal cycler using lasers, halogen, or LED light sources because this permits more precise matching of emission spectra and fluorophores. Moreover, instruments capable of detecting different fluorophores must be used for multiplex experiments.

SNP GENOTYPING

SNP analysis uses two different probes, each matching a specific allele and each labeled with a different reporter–quencher pair. Probes that form a mismatched hybrid with the target DNA melt at a lower temperature than the corresponding perfectly matched hybrid. The resulting differences in melting temperature are sufficient to distinguish between target DNAs carrying wild-type or mutant sequences.

For allelic discrimination assays, it is important to position the polymorphism in the center of the probe and, as in multiplexing, ensure that the two probes have the same T_m. Applied Biosystems recommends using TaqMan MGB probes, especially when conventional TaqMan probes exceed 30 nucleotides. The TaqMan MGB probes contain a nonfluorescent quencher at the $3'$ end (allowing more precise measurement of the reporter dye contribution because the quencher does not fluoresce) and a minor groove binder at the $3'$ end (which increases the T_m of the probe, allowing the use of shorter probes). Consequently, TaqMan MGB probes show greater differences in T_m values between matched and mismatched probes, which provide more accurate allelic discrimination.

Molecular beacons are also a suitable choice for SNP genotyping. The stem probe structure of a molecular beacon enables it to discriminate single base-pair mismatches better than linear probes because the hairpin makes mismatched hybrids thermally less stable than hybrids between the corresponding linear probes and their mismatched target. Before amplification, the complementary bases of the stem hybridize forming the basic "stem-loop" structure of the molecular beacon. The stem raises the probe T_m, and consequently molecular beacons can routinely discriminate between target sequences differing by a single nucleotide.

CHAPTER 10

Nucleic Acid Platform Technologies

Oliver Rando

Department of Biochemistry and Molecular Pharmacology, University of Massachusetts Medical School, Worcester, Massachusetts 01605

EVERY MICROARRAY EXPERIMENT IS BASED ON A COMMON format (Fig. 1). First, a large number of nucleotide "spots" are arrayed onto a substrate, typically a glass slide, a silicon chip, or microbeads. Second, a complex population of nucleic acids (isolated from cells, selected from in vitro–synthesized libraries, or obtained from another source) is labeled, typically with fluorescent dyes. Third, the labeled nucleic acids are allowed to hybridize to their complementary spot(s) on the microarray. Fourth, the hybridized microarray is washed, allowing the amount of hybridized label to then be quantified. Analysis of the raw data generates a readout of the levels of each species of RNA in the original complex population.

Microarrays can be run in "one-color" or "two-color" formats. In a one-color microarray, a single sample (e.g., RNA from liver cells) is labeled, and the abundance of a species of RNA is inferred from the intensity of the spot(s) complementary to the relevant gene. Because, for each spot, hybridization is affected by a panoply of factors, interpretation of single-color experiments can be complicated. Practically, however, many of these complicating factors are resolvable by using the proper bioinformatics tools during data analysis. Two-color microarrays (Fig. 1) use a competitive hybridization in which one nucleic acid sample is labeled with one color (green), and a related sample is

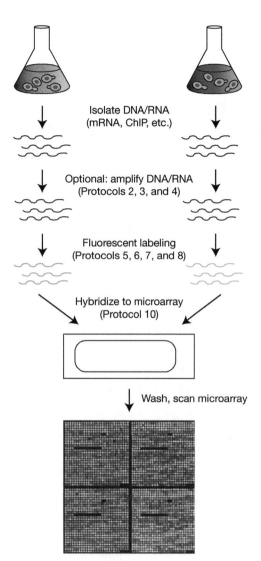

FIGURE 1. **Basic steps in setting up and running a two-color microarray experiment.** The details of each step are explained in the text.

labeled with a second color (red). Following hybridization and removal of unbound nucleic acid, the microarray is scanned with lasers to detect where the red- and green-labeled molecules have bound. The intensity of each spot is determined, and the red/green ratio is measured for each spot. During data analysis, this ratio can be used to measure the ratio of the amount of related nucleic acid molecules in the two samples. For example, if RNA from normal liver is tagged green and RNA from a liver tumor is labeled red, then red spots would represent RNAs that are up-regulated in the tumor, and green spots would represent down-regulated RNAs.

Tiling microarrays consist of regularly spaced oligonucleotides that provide dense coverage of a genome or portion of a genome. For example, yeast chromosome III was tiled using oligonucleotides 50 nucleotides long spaced every 20 bases; that is, spot 1 comprised bases 1–50, spot 2 bases 21–70, and so forth. Tiling microarrays therefore interrogate continuously along a chromosomal path, enabling applications, such as transcript structure and protein localization, to be performed with almost complete genomic coverage. One disadvantage to tiling microarrays is that not all genomic sequence is equally suited to microarray applications. For example, not all 50-mers in a genome are unique, leading to gaps in the tiling path where hybridization-based approaches cannot determine whether the hybridizing material comes from copy 1 or copy 2 (or copy N) in the genome. Furthermore, hybridization-related properties such as the percentage of A and T residues and oligonucleotide melting temperature vary from probe to probe, although many issues like these can be resolved by normalizing two-color arrays. Normalization, which compensates for systematic technical differences between spots and/or arrays, clarifies the systematic biological differences between samples.

MICROARRAY APPLICATIONS

Although the majority of early microarray studies focused on mRNA expression profiling, microarrays can be used for any purpose that involves comparison of two nucleotide populations—for example, comparison of RNAs extracted from normal and tumor cells. In this section, we list several examples (by no means exhaustive) of published microarray analyses.

mRNA Expression Levels

Early gene expression microarrays consisted of arrayed spots each containing a cDNA corresponding to a single gene. For example, one of the first yeast gene expression microarrays consisted of about 6000 spots composed of cDNA corresponding to nearly every gene in the *Saccharomyces cerevisiae* genome (DeRisi et al. 1997). cDNA microarrays have fallen into disuse because of the advent of oligonucleotide microarrays. Advances in oligonucleotide synthesis make it possible to generate gene expression microarrays composed of long oligonucleotides that have been designed to be complementary to a given gene. Oligonucleotides can be synthesized and sold in soluble form, which are then arrayed by a spotting robot, as with cDNA arrays. Alternatively, several commercial microarray suppliers use "in situ" oligonucleotide synthesis, in which oligonucleotides are printed directly onto the microarray substrate. This process allows for much more flexible generation of oligonucleotide microarrays and, thus, is used more often than soluble oligonucleotide arrays because a laboratory need not purchase an excess of oligonucleotides in order to run microarray experiments.

Monitoring Changes in the Transcriptome

A typical mRNA expression experiment begins with the isolation of poly(A)$^+$ RNA from two related populations of cells, such as yeast growing at 30°C and yeast that have been shifted from 30°C to 37°C. Fluorophore-labeled cDNA is prepared from each mRNA population. Typically, the cyanine dyes used are Cy5 (pseudo-colored red in microarray images) and Cy3 (pseudo-colored green). The labeled cDNAs are mixed and hybridized to a microarray. After hybridization and washing, the microarray is scanned with lasers that excite the fluorophores, and the image is processed, generating a Cy5/Cy3 ratio for every spot on the array. The microarray data are normalized, and a list of the Cy5/Cy3 ratios per gene is generated, which can be analyzed for genome-wide changes in the transcriptome.

The microarray data are normalized so that the average $\log_2$(Cy5/Cy3) is 0; this is a necessary step because the absolute intensity of either channel (Cy5 or Cy3) is subject to a wide range of modifying influences, such as input RNA level and labeling efficiency, among others. This normalization should always be kept in mind in interpreting microarray studies. For example, microarray results for a comparison between liver RNA and sperm RNA populations might indicate that a number of RNAs are relatively enriched in sperm, but the absolute numbers of those RNAs per nucleus are still probably lower in sperm than in liver given the disparity in RNA abundance in the two types of cells.

mRNA Abundance

The experiment described above is capable of providing information about the relative abundances of every mRNA expressed in the two cell populations. What about absolute expression levels? These can be assayed in two ways. First, single-color experiments using the Affymetrix platform provide a reasonable measure of mRNA abundance—the intensity of a given spot is related to the absolute abundance of its complementary mRNA in the original labeled population. Second, for two-color experiments, a straightforward way to estimate transcript abundance is to compare labeled mRNA with labeled genomic DNA, which, in principle, should be present in a single copy per gene (unless the gene in question comes from a multicopy family). Thus, gDNA is labeled with Cy3, mRNA is labeled with Cy5, and Cy5/Cy3 ratios provide a relative measure of mRNA abundance. These ratios can be converted to absolute mRNA abundance if the absolute abundance of any RNA is already known. Normalization can be used to set the absolute level of these RNAs, and all remaining RNA abundances can be inferred from their Cy5/Cy3 ratios relative to the Cy5/Cy3 ratios for the known standards.

Comparative Genomic Hybridization

Comparative genomic hybridization (CGH) is used to detect changes in DNA copy number across a genome (Pollack et al. 1999). In a CGH experiment, the two populations of genomic DNA are fluorescently labeled using Klenow DNA polymerase. Labeled DNA is compared using the same experimental design as for analysis of mRNA expression. Changes in copy number are reflected in the Cy5/Cy3 ratios. This method has been used most commonly in the study of tumors, where there is a well-documented role both for gene loss and for amplification in the process of oncogenesis. By analyzing changes in gene copy number, regional amplification or even whole chromosome copy-number changes have been characterized. Variants of this approach have also been used to characterize replication timing, as DNA isolated during S phase shows variations in copy number, based on whether or not the segment of genomic DNA has replicated.

Determination of Transcript Structure

Although changes in mRNA levels can be analyzed easily using microarrays constructed with one probe (spot) per gene, different methods are required to detect variations in transcript structure. These variations include changes in splicing patterns and in the start and stop sites for transcription. These and other changes in transcript structure can be detected and mapped using tiling microarrays. For example, transcriptional start sites can be mapped to ~10–20-nucleotide resolution by a variant of the absolute abundance hybridization method described above. Briefly, mRNA is labeled with Cy5 and hybridized against Cy3-labeled genomic DNA. In principle, an expressed RNA should appear as a long "square wave" of high Cy5/Cy3 ratios: Cy5/Cy3 ratios should be low in spots corresponding to genomic sequences that flank those complementary to the RNA. High Cy5/Cy3 ratios should start at spots containing sequences corresponding to the 5′ end of the RNA and should continue to the end of the RNA (Fig. 2). Indeed, this technique has been used to determine transcriptional start sites to ~20-nucleotide resolution in yeast (Yuan et al. 2005) and to detect and map novel transcripts in human cells (Shoemaker et al. 2001).

Variations in splicing patterns can also be detected and mapped using microarrays. With tiling arrays, expressed exons should appear as peaks of Cy5/Cy3. Another way to determine exon inclusion is to use an exon microarray, in which each spot on the array corresponds to a single exon. When probing an exon microarray, for a given tissue all of a gene's expressed exons will show high Cy5/Cy3 (Shoemaker et al. 2001).

Tiling microarrays have also been used to discover noncoding RNAs and novel genes. The reason for this is obvious—a classical, expression microarray only covers what is known, whereas a tiling microarray can be designed to contain an entire genome. For example, tiling microarrays of chromosomes 21 and 22 revealed the nearly pervasive low-level transcription of the human genome (Kapranov et al. 2002).

Identifying RNA–Protein Interactions

RNA molecules associated with specific RNA-binding proteins have been identified using immunoprecipitation assays to isolate a protein of interest. The associated RNA molecules (or regions of RNA) are then compared with nonspecific RNAs that have been isolated in control immunoprecipitations lacking antibody (Hieronymus and Silver 2003; Gerber et al. 2004).

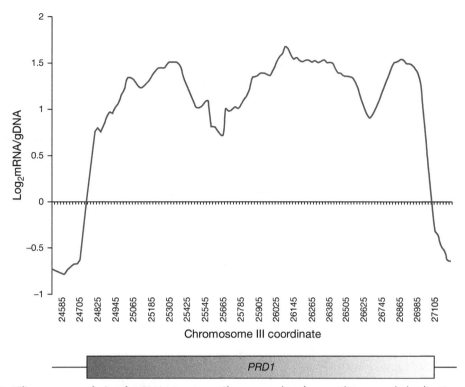

FIGURE 2. **Tiling array analysis of mRNA structure.** Shown are data from a tiling array hybridization of labeled mRNA versus labeled genomic DNA, with the y-axis showing $\log_2$ of the ratio mRNA/gDNA, and the x-axis showing the chromosomal coordinates of microarray probes.

Subcellular Localization of RNA Populations

The functional properties of RNAs (e.g., translation) can be assayed by microarray using more sophisticated fractionation techniques. In an early study, polysomal RNAs were analyzed by microarray techniques (Arava et al. 2003). Subcellular localization of RNAs can also be studied when appropriate purification techniques exist; for example, isolation of dendrites from neurons has been used to identify the relevant RNA populations (Eberwine et al. 2002).

Protein Localization Studies

At present, one of the most popular applications of tiling microarrays is known as "chromatin immunoprecipitation on chip" (or ChIP-on-chip; see Chapter 20). By this method, the location of a specific protein, such as a transcription factor, is assayed through the use of chromatin immunoprecipitation. Briefly, proteins are covalently cross-linked to the genome using formaldehyde, and chromatin is sheared by sonication to small (~500 bp) fragments. Using an antibody specific to a given protein (with either an antibody to an epitope tag, a protein-specific antibody, or even an antibody against a specific modification state), DNA associated with the protein in question is isolated by immunoprecipitation. After washing, the cross-links are dismantled and genomic DNA is isolated. The DNA is typically amplified, and the amplified material is labeled and hybridized competitively against labeled amplification reactions from no-antibody controls or from pre-IP material. Analysis of ChIP-on-chip microarray results provides a snapshot of a transcription factor's location on the genome and has proven to be a tremendously powerful method for studying transcriptional regulation by transcription factors and chromatin (Ren et al. 2000; Iyer et al. 2001).

Nuclease Accessibility as a Structural Probe

Tiling microarrays can also be hybridized with genomic DNA that has been treated diagnostically with a nuclease before labeling the DNA. This method takes advantage of the fact that chromatin packaging of the genome dramatically affects nuclease accessibility. Two nucleases have been used in genome-wide studies. First, DNase I–hypersensitive sites have long been known to occur at genomic regulatory elements such as promoters, enhancers, and insulators. Here, chromatin is lightly digested with DNase I, sites of cleavage are isolated, and DNA surrounding cleavage sites is interrogated by tiling microarray analysis (Sabo et al. 2006). Second, micrococcal nuclease (MNase) is a nonprocessive nuclease with preference for the linker DNA between nucleosomes. Analysis of the DNA protected from MNase cleavage including comparison to whole genomic DNA has proven to be an invaluable tool in genome-wide nucleosome mapping studies (Yuan et al. 2005).

Splicing Microarrays

Although exon microarrays (see above) can monitor expression of each exon represented on the array, the structure of the transcript cannot be discerned. Specifically, exon connectivity is not detected using exon microarrays. For example, expression of exons 1, 2, 3, and 5 could indicate a single species containing all four exons, or two or more distinct species (1-2-5 and 1-3-5, etc.). To determine connectivity of exons requires the use of splicing microarrays, which are comprised of oligonucleotides complementary to specific splice junctions.

Resequencing and SNP Detection

The sequence specificity of DNA hybridization has also been leveraged for genome analyses such as "resequencing" and polymorphism detection. In these applications, each genomic location is represented on the microarray by several oligonucleotides that differ by a single nucleotide. In other words, if the genomic region of interest has the sequence AATGCCA, oligonucleotides containing AATTCCA, AATCCCA, and AATACCA will also be printed. Hybridization of genomic DNA to these oligonucleotides can be used to determine mutations or sequence polymorphisms at the site interrogated by the set of oligonucleotides.

One way to use microarrays for resequencing is the so-called sequence capture protocol (Hodges et al. 2007). A microarray is printed with oligonucleotides corresponding to genomic regions of interest. DNA or RNA is hybridized to the microarray, and all complementary sequences are retained on the microarray through the washing protocols. After retention of desired genomic regions via hybridization, this hybridized material is eluted and sequenced using high-depth sequencing methods (Chapter 11).

PERFORMING MICROARRAY EXPERIMENTS

Although the specific bound oligonucleotides and labeled probes and details of the analysis of microarray hybridizations differ depending on the experimental questions being asked, most microarray experiments involve six steps.

1. Design a microarray.

2. Print or purchase a microarray.

3. Isolate and amplify the DNA or RNA probe material.

4. Label the DNA or RNA with fluorescent groups.

5. Hybridize the labeled probes to the microarray.

6. Analyze the microarray hybridization results.

For the remainder of the chapter, we discuss each of these steps in turn. Protocols are provided for printing a microarray in-house (Protocol 1) and for amplifying DNA and RNA following isolation (Protocols 2–4). Several techniques for adding fluorescent moieties to the nucleic acids are provided (Protocols 5–8). Protocol 9 explains how to block the positive charges of the polylysines bound to homemade microarrays. Finally, there is a detailed generalized protocol for hybridization of labeled probes to a microarray, as well as scanning, formatting, and storing the microarray hybridization data (Protocol 10). A brief guide to microarray analysis is contained in Chapter 8.

Designing a Microarray

A great variety of microarrays are commercially available, and for most researchers, an off-the-shelf product will suffice. If not, customized microarrays can be ordered from several manufacturers or designed (and printed) in-house. The appropriate design of a microarray will depend on the application for which it is to be used. There are, however, some basic design principles that are common to most applications. We consider two types of basic microarray: the gene expression microarray and the tiling microarray. Specialized formats, such as splicing arrays and resequencing arrays, have been designed for different organisms, and we point interested readers to the relevant literature (Hacia 1999; Mockler et al. 2005; Blencowe 2006; Hughes et al. 2006; Calarco et al. 2007; Cowell and Hawthorn 2007; Gresham et al. 2008).

Gene Expression Microarrays

Designing oligonucleotides for microarrays requires expertise in bioinformatics, but some basic design properties are easily understood. First, choose an appropriate oligonucleotide length. The majority of oligonucleotide microarrays today are printed with oligonucleotides 50–70 nucleotides long. Designing oligonucleotides of optimal lengths requires consideration of several factors, including signal strength, specificity, cost, and efficiency of synthesis. Shorter oligonucleotides are more likely to cross-hybridize to many different regions of a given genome and will often have very low melting temperatures, making hybridization technically problematic. A detailed analysis of oligonucleotide length versus sensitivity and specificity can be found in Hughes et al. (2001). A reasonable rule of thumb is that for most applications, 60-nucleotide oligonucleotides provide the best balance between these competing constraints.

The second consideration in oligonucleotide design is the degree of specificity or complementarity between the oligonucleotides on the microarray and the RNA species in the organism of interest. In general, the oligonucleotides should be designed so that they are complementary to only one RNA species. Commonly, BLAST searches (see Chapter 8) are used to screen the sequence of each oligonucleotide against the relevant genomic sequence to identify the potential for multiple hits in the genome of interest. Ideally, rather than focusing on a cut-off BLAST value, the user should select for each gene the oligonucleotide having the lowest-scoring second match in the genome.

The third design consideration is optimization of the hybridization properties. Several features can be optimized, but the most important is the melting temperature (T_m) of the oligonucleotide. Specifically, the T_ms of all of the oligonucleotides on the array should be within as narrow a window as is feasible. Other characteristics, such as entropy (i.e., the complexity of the sequence), GC%, and self-complementarity should be optimized as well. Software tools are available to help with oligonucleotide design, ranging from publicly available tools like ArrayOligoSelector (see below) to commercial services provided by microarray companies (for further details of oligonucleotide design, see Chapters 7 and 8).

As an example, consider a case in which an investigator wishes to design gene expression microarrays for a non-model organism with a recently sequenced genome. The first step in designing gene expression microarrays is, of course, to identify the genes. As this is routinely performed in any genome sequencing effort, we will assume that genes have already been predicted using standard tools (Zhang 2002; Ashurst and Collins 2003; Brent 2005; Solovyev et al. 2006).

ArrayOligoSelector

Once coding regions are identified, oligonucleotides need to be designed for each gene. Many programs exist for oligonucleotide selection, including ArrayOligoSelector (Bozdech et al. 2003a,b), a commonly used program that is freely available at http://arrayoligosel.sourceforge.net/ (for further details, see Chapters 7 and 8).

ArrayOligoSelector is designed to analyze a complete genome and prepare oligonucleotides of a user-defined length. For every oligonucleotide, ArrayOligoSelector calculates scores for uniqueness, sequence complexity, self-annealing, and GC content. Uniqueness is a measure of the theoretical difference in binding energy between a given oligonucleotide and either its perfect match or the next most homologous genomic sequence. Sequence complexity allows the user to filter oligonucleotides with homopolymeric tracts, which otherwise may cause hybridization problems. The self-annealing score is a measure of the secondary structure generated by the self-annealing of an oligonucleotide. Self-annealing is another potential source of hybridization problems. Finally, it is important to minimize variation in GC percentage among the oligonucleotide sequences, both to minimize T_m variation and to minimize variability in the fluorescence intensity among the spots.

When running ArrayOligoSelector, the following features can be specified: oligonucleotide length, GC%, number of oligonucleotides per gene, sequences to mask, and uniqueness cutoff. Common oligonucleotide lengths are 60-mers or 70-mers. GC% will vary depending on the GC% of the genome in question and will typically be chosen as the genomic coding region average. The number of oligonucleotides per gene will depend on whether the microarray is to be purchased or printed in-house. If oligonucleotides are purchased for in-house printing, then cost and the printable spot density will preclude using more than one or two oligonucleotides per gene. Alternatively, if commercial arrays are to be used, then the number of spots available will determine the number of oligonucleotides to be chosen per gene. Masking sequences are not commonly specified, but if a problematic short repeat element is present in the genome, then it is sometimes valuable to mask it out of the microarray oligonucleotides. The uniqueness cutoff is typically left blank, which will result in the default value being used (for additional information, see the ArrayOligoSelector manual).

The output of ArrayOligoSelector can be filtered by the experimenter. For example, it is often desirable to use oligonucleotides located toward the 3′ ends of genes because reverse-transcriptase-based labeling is more efficient in this region.

Tiling Microarrays

Oligonucleotide design for tiling microarrays is more straightforward than for gene expression microarrays because tiling presumes that essentially all of the oligonucleotides within a region of interest will be included on the microarray. The simplest tiling design involves choosing nucleotides 1−50 (say) of some region to be tiled as spot 1, nucleotides 21−70 as spot 2, and so forth. Once all of the oligonucleotides have been designed, BLAST can be used to find oligonucleotides with multiple identical matches in the genome of interest, and these oligonucleotides can be removed. More subtle tiling designs incorporate a small amount of "wiggle" in the oligonucleotide location; thus spot 2 might run from nucleotides 16−65, spot 3 might run from 45−94, and so forth. By doing this, the process of matching hybridization properties, such as T_m and GC%, is better than with a simple tiling microarray.

Printing or Purchasing a Microarray

Once oligonucleotides have been designed, microarrays can be printed commercially. Alternatively, oligonucleotides can be synthesized commercially or by an in-house core facility, and then printed in-house using a spotting robot (see Protocol 1).

Isolating and Amplifying Nucleic Acid Samples for Hybridization to Microarrays

Most of the sample preparation procedures for microarrays follow standard protocols, many of which can be found elsewhere in this manual. For example, comparative genomic hybridization (CGH) analyses use genomic DNA isolated from samples of interest, as described in Chapter 1. Gene expression or splicing studies use either total RNA or mRNA, purified as described in Chapter 6.

Protein localization analysis starts with material isolated via chromatin immunoprecipitation (ChIP), as described in Chapter 20. Typically, however, the intended assays for many of these protocols are based on blotting techniques or quantitative PCR readouts, which often require only nanograms of material. Microarray labeling, on the other hand, typically requires several micrograms of nucleic acid; therefore, an amplification step is necessary before labeling.

General Notes on Amplification

Because amplification of the nucleic acid that will be used to generate labeled microarray probes occurs before the hybridization step, it must not bias representation of any particular sequences in the genome. Thus, unbiased (or minimally biased) whole-genome amplification protocols are a key component of many microarray applications.

Three amplification protocols are included in this chapter: two for DNA (Protocols 2 and 3) and one for RNA (Protocol 4). There are several practical items to keep in mind while performing an amplification protocol. First, when trying an amplification method for the first time, start with a large and easily obtainable pool of material (e.g., liver RNA) and amplify an aliquot of the original pool. Microarray comparisons between the amplified material and the original "bulk" pool then provide a valuable readout of amplification biases. A perfect amplification would result in a "yellow" array with no spots showing any differences between the bulk and amplified material. Second, avoid contaminating your sample with anything that might contain DNA or RNA. Even tiny amounts of foreign nucleic acids will be amplified, contaminating your sample and corrupting your microarray experiments. Always wear gloves and use filter pipette tips when performing the isolation and amplification protocols. Finally, always include a control amplification with water only (no DNA or RNA) to ensure that the reagents are not contaminated with amplifiable material.

RNA Amplification for Expression Profiling

Less common than DNA amplification, RNA amplification is nonetheless required in experiments that use small populations of cells, such as may occur in neurobiological studies using laser-captured cell populations. RNA amplification kits are available from several vendors (e.g., Ambion), or amplification can be performed as described in Protocol 4.

Generating Fluorescently Labeled Nucleic Acid Probes

Labeling nucleic acids for use in microarrays is similar for most microarray platforms, except for Affymetrix microarrays. For most platforms (homemade or commercial), fluorescent molecules are attached to DNA by the Klenow fragment of DNA polymerase I. For labeling RNA, reverse transcriptase is used to prepare labeled cDNA. Labeling methods can include a fluorescent dNTP in the labeling reaction (Protocols 5 and 7). Because this approach is rapid but expensive, a cheaper but lengthier alternative first incorporates aminoallyl nucleotides into the nucleic acid molecules and then couples the aminoallyl group to the fluorophore (Protocols 6 and 8). In all of the labeling protocols, the source nucleic acids can be either unamplified or amplified before labeling.

Hybridizing to a Microarray

Methods for hybridization of labeled probe materials to a microarray differ significantly when using home-printed microarrays versus commercial microarrays. For home-printed microarrays, slides must first be blocked because polylysines remaining on the slide will cause significant background binding to labeled material unless they are neutralized by reaction with succinic anhydride. Protocols 9 and 10 provide methods, respectively, for blocking and hybridizing to homemade arrays. When using commercial microarrays, hybridization protocols are typically provided. Following hybridization, microarrays are scanned in a bench-top scanner. The data are collected, formatted, and stored digitally for subsequent analysis.

Analyzing Microarray Data

The tools used for analysis of microarray data will depend on the experimental question being asked: Localization studies require a different set of tools than do gene expression studies. Some of the available resources are described in Chapter 8. Here, we outline some of the basic steps in data analysis.

The first step in microarray data analysis is to remove bad data (i.e., data from spots that were "flagged" because they were obscured by fluorescent precipitate, etc.), and to normalize the remaining data. Working with a .gpr file, which is the standard output of GenePix software, we typically eliminate all flagged probes and then work exclusively with the Log_2 ratio data. More advanced users may consider using specific features, such as Foreground and Background intensities.

Most two-color microarray studies are normalized to an average Log_2 ratio value of 0. The assumption implicit in this normalization is that there was no overall change in whatever is being measured. Thus, it is important to remember when looking at such normalized data that relative values are being measured, not absolute values. Practically, normalization can be performed by averaging the unflagged log ratio value, then subtracting this value from every one of the entries in the column. This can be done by hand in spreadsheet programs or using common commands in languages such as MATLAB, Perl, or R.

Once normalization is completed, the list can be sorted by log ratio value to identify genes that are dramatically up-regulated or down-regulated (if gene expression) or identify loci with high levels of enrichment of the protein in question. At this point, data analysis paths will diverge depending on the questions being asked. However, because many microarray studies involve multiple microarrays, it is often useful to cluster data to identify genes or loci that share similar behavior.

For clustering and visualization, numerous programs are available online. We use the classic Cluster and TreeView programs (Eisen et al. 1998), available online at SourceForge (http://sourceforge.net/) or via the Eisen laboratory website (http://www.eisenlab.org/). Because sample files that guide the formatting of files for clustering are also available, formatting will not be described here. Briefly, however, data will be loaded into Cluster, various thresholds will be set (fraction of data missing for a given gene, number of genes changing over some threshold, etc.), and one of several clustering algorithms will be used. The output of the clustering can be visualized with TreeView, allowing users to generate the classic "heatmap" view of their microarray data.

ACKNOWLEDGMENTS

The protocols in this chapter were modified from protocols written and generously provided by Ash Alizadeh, Chih Long Liu, Audrey Gasch, Jason Lieb, Bing Ren, and L. Ryan Baugh.

REFERENCES

Arava Y, Wang Y, Storey JD, Liu CL, Brown PO, Herschlag D. 2003. Genome-wide analysis of mRNA translation profiles in *Saccharomyces cerevisiae*. *Proc Natl Acad Sci* **100**: 3889–3894.

Ashurst JL, Collins JE. 2003. Gene annotation: Prediction and testing. *Annu Rev Genomics Hum Genet* **4**: 69–88.

Blencowe BJ. 2006. Alternative splicing: New insights from global analyses. *Cell* **126**: 37–47.

Bozdech Z, Llinas M, Pulliam BL, Wong ED, Zhu J, DeRisi JL. 2003a. The transcriptome of the intraerythrocytic developmental cycle of *Plasmodium falciparum*. *PLoS Biol* **1**: e5. doi: 10.1371/journal.pbio.0000005.

Bozdech Z, Zhu J, Joachimiak MP, Cohen FE, Pulliam B, DeRisi JL. 2003b. Expression profiling of the schizont and trophozoite stages of *Plasmodium falciparum* with a long-oligonucleotide microarray. *Genome Biol* **4**: R9. doi: 10.1186/gb-2003-4-2-r9.

Brent MR. 2005. Genome annotation past, present, and future: How to define an ORF at each locus. *Genome Res* **15**: 1777–1786.

Calarco JA, Saltzman AL, Ip JY, Blencowe BJ. 2007. Technologies for the global discovery and analysis of alternative splicing. *Adv Exp Med Biol* **623**: 64–84.

Cowell JK, Hawthorn L. 2007. The application of microarray technology to the analysis of the cancer genome. *Curr Mol Med* **7**: 103–120.

DeRisi JL, Iyer VR, Brown PO. 1997. Exploring the metabolic and genetic control of gene expression on a genomic scale. *Science* **278**: 680–686.

Eberwine J, Belt B, Kacharmina JE, Miyashiro K. 2002. Analysis of subcellularly localized mRNAs using in situ hybridization, mRNA amplification, and expression profiling. *Neurochem Res* **27**: 1065–1077.

Eisen MB, Spellman PT, Brown PO, Botstein D. 1998. Cluster analysis and display of genome-wide expression patterns. *Proc Natl Acad Sci* **95**: 14863–14868.

Gerber AP, Herschlag D, Brown PO. 2004. Extensive association of functionally and cytotopically related mRNAs with Puf family RNA-binding proteins in yeast. *PLoS Biol* **2:** e79. doi: 10.1371/journal.pbio.0020079.

Gresham D, Dunham MJ, Botstein D. 2008. Comparing whole genomes using DNA microarrays. *Nat Rev Genet* **9:** 291–302.

Hacia JG. 1999. Resequencing and mutational analysis using oligonucleotide microarrays. *Nat Genet* **21:** 42–47.

Hieronymus H, Silver PA. 2003. Genome-wide analysis of RNA–protein interactions illustrates specificity of the mRNA export machinery. *Nat Genet* **33:** 155–161.

Hodges E, Xuan Z, Balija V, Kramer M, Molla MN, Smith SW, Middle CM, Rodesch MJ, Albert TJ, Hannon GJ, McCombie WR. 2007. Genome-wide in situ exon capture for selective resequencing. *Nat Genet* **39:** 1522–1527.

Hughes TR, Mao M, Jones AR, Burchard J, Marton MJ, Shannon KW, Lefkowitz SM, Ziman M, Schelter JM, Meyer MR, et al. 2001. Expression profiling using microarrays fabricated by an ink-jet oligonucleotide synthesizer. *Nat Biotechnol* **19:** 342–347.

Hughes TR, Hiley SL, Saltzman AL, Babak T, Blencowe BJ. 2006. Microarray analysis of RNA processing and modification. *Methods Enzymol* **410:** 300–316.

Iyer VR, Horak CE, Scafe CS, Botstein D, Snyder M, Brown PO. 2001. Genomic binding sites of the yeast cell-cycle transcription factors SBF and MBF. *Nature* **409:** 533–538.

Kapranov P, Cawley SE, Drenkow J, Bekiranov S, Strausberg RL, Fodor SP, Gingeras TR. 2002. Large-scale transcriptional activity in chromosomes 21 and 22. *Science* **296:** 916–919.

Mockler TC, Chan S, Sundaresan A, Chen H, Jacobsen SE, Ecker JR. 2005. Applications of DNA tiling arrays for whole-genome analysis. *Genomics* **85:** 1–15.

Pollack JR, Perou CM, Alizadeh AA, Eisen MB, Pergamenschikov A, Williams CF, Jeffrey SS, Botstein D, Brown PO. 1999. Genome-wide analysis of DNA copy-number changes using cDNA microarrays. *Nat Genet* **23:** 41–46.

Ren B, Robert F, Wyrick JJ, Aparicio O, Jennings EG, Simon I, Zeitlinger J, Schreiber J, Hannett N, Kanin E, et al. 2000. Genome-wide location and function of DNA binding proteins. *Science* **290:** 2306–2309.

Sabo PJ, Kuehn MS, Thurman R, Johnson BE, Johnson EM, Cao H, Yu M, Rosenzweig E, Goldy J, Haydock A, et al. 2006. Genome-scale mapping of DNase I sensitivity in vivo using tiling DNA microarrays. *Nat Methods* **3:** 511–518.

Shoemaker DD, Schadt EE, Armour CD, He YD, Garrett-Engele P, McDonagh PD, Loerch PM, Leonardson A, Lum PY, Cavet G, et al. 2001. Experimental annotation of the human genome using microarray technology. *Nature* **409:** 922–927.

Solovyev V, Kosarev P, Seledsov I, Vorobyev D. 2006. Automatic annotation of eukaryotic genes, pseudogenes and promoters. *Genome Biol* **7:** S10.1–S10.12.

Yuan GC, Liu YJ, Dion MF, Slack MD, Wu LF, Altschuler SJ, Rando OJ. 2005. Genome-scale identification of nucleosome positions in *S. cerevisiae*. *Science* **309:** 626–630.

Zhang MQ. 2002. Computational prediction of eukaryotic protein-coding genes. *Nat Rev Genet* **3:** 698–709.

WWW RESOURCES

Source for Array Oligo Selector http://arrayoligosel.sourceforge.net/.

Source for Cluster and Treeview programs http://sourceforge.net/ and http://www.eisenlab.org/

Printing Microarrays

Most laboratories will choose to order microarrays from a commercial vendor, such as NimbleGen, Agilent, Affymetrix, or Illumina. These vendors sell a number of products encompassing genomes from the three domains of life. In addition, custom microarrays can be made to order by most of these companies. Microarray printing, however, is sufficiently straightforward that even small laboratories with extensive microarray needs may find it cost-effective and worth the effort to produce their own microarrays. Printing microarrays requires a spotting robot. Several commercial spotting robots are available, such as the OmniGrid series (Digilab Genomic Solutions). These instruments are expensive and will most frequently be housed in institutional core facilities or in large laboratories. As an alternative, designs for building a spotting robot have been available from the Brown laboratory at Stanford for many years. Protocols for building a spotting robot are beyond the scope of this chapter, but interested parties can find a protocol at http://cmgm.stanford.edu/pbrown/mguide/.

For typical microarrays, oligonucleotides are prepared at 20–40 μM in aqueous 3× SSC and are distributed into 384-well plates. Oligonucleotides are printed to polylysine-coated microarray slides. It is best to print when the humidity is at least 50%. When the humidity is ~20%–30%, there can be problems with pins drying. To increase or maintain proper humidity, consider building a humidity-controlled enclosure around the robot. For robots lacking an enclosure, it may be adequate to run a humidifier or two during the printing procedure.

The details of a printing protocol vary depending on the robot being used; thus, follow the manufacturer's instructions. However, a typical workflow using a robot equipped with contact steel quill pins is provided in this protocol.

MATERIALS

It is essential that you consult the appropriate Material Safety Data Sheets and your institution's Environmental Health and Safety Office for proper handling of equipment and hazardous materials used in this protocol.

Recipes for reagents specific to this protocol, marked <R>, are provided at the end of the protocol. See Appendix 1 for recipes for commonly used stock solutions, buffers, and reagents, marked <A>. Dilute stock solutions to the appropriate concentrations.

Reagents

Oligonucleotides of desired sequences
   Oligonucleotides are typically prepared at 20–40 μm in aqueous 3× SSC.

Salmon sperm DNA <A>, sheared (150 ng/μL resuspended in 3× SSC)
SSC (3×) <A>

Equipment

Desiccator
Nitrogen gas, compressed
Plates (384 wells)

Polylysine-coated slides

Although polylysine coating can be done in-house, in our experience, the failure rate is significant and purchasing commercial polylysine-coated slides (e.g., Erie Scientific) has proven to be the most cost-effective solution.

Slide box, plastic

Spotting robot

METHOD

Preparing to Print

Day 0

1. Array the oligonucleotides into 384-well plates with identical volumes per well (typically 10–20 μL per well). If the oligonucleotide solutions were frozen, thaw the master plates overnight at 4°C. If the oligonucleotides were dried, resuspend them in 3× SSC, and let them incubate overnight at 4°C.

2. Test that the spotting pins are all printing and that the arrayer is properly calibrated by performing test prints (Fig. 1). To closely approximate the viscosity of the actual print plates, use a test print plate consisting of 150 ng/μL sheared salmon sperm DNA resuspended in 3× SSC. All

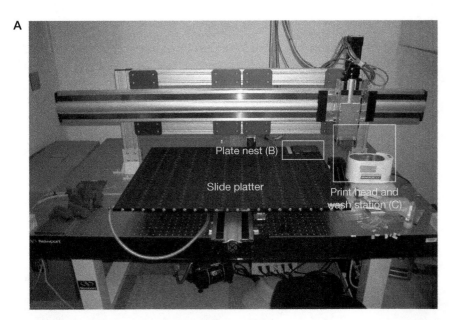

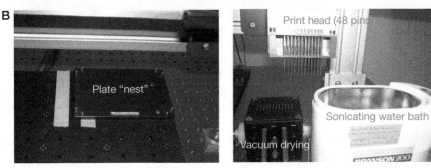

FIGURE 1. **Spotting robot.** Shows the location of the slide platter (*A*), plate nest (*B*), and print head and cleaning/drying station (*C*).

pins should be able to print several hundred consecutive spots. In addition, test print in several locations on the slide platter to determine whether the platter has warped significantly since its last use.

If test prints indicate that the spotting robot is not performing as it should, see Troubleshooting.

Day 1

3. Perform one more test print to make sure that the pins are still printing properly.

4. Place slides gently onto the arrayer platter. Make sure that all slides are sitting flat and are firmly attached (by vacuum, clips, or tape, depending on the arrayer).

   ▲ *Handle the slides carefully to avoid producing minute glass chips, which might settle onto the surface of the slide.*

5. Blow dust off the slides with compressed nitrogen.

6. If the arrayer has a sonicator water bath for cleaning the pins, fill it with fresh water.

7. Transfer four 384-well oligonucleotide-containing print plates from 4°C, and let them come to room temperature for at least 1 h.

8. Centrifuge the plates at ~1000 rpm for 2 min to remove condensation from the plate covers. Carefully remove a plate lid and the adhesive plate cover from one of the print plates.

   ▲ *Be careful not to jolt the plate because cross-contamination between wells will cause serious problems in any downstream data analysis.*

9. Place a plate into the spotting robot's plate holder.

Printing the Microarrays

10. Start the print run. Keep careful notes about plate order and orientation. Knowing the plate order is necessary for creating the ".gal" file that maps spot position to gene name when microarray data are analyzed. Any mistakes during printing such as printing plates out of order (e.g., plate 3 before plate 2) can be corrected later on as long as they are noted.

11. When a plate is finished printing and the print head has come to a complete stop, either let the plate evaporate in a hood if the plates are stored dry, or cover the plate with foil and its lid and store the plate at −80°C.

12. Insert the next print plate into the plate holder.

13. When the print run is done, let the slides dry overnight, unless you are performing back-to-back print runs.

Day 2

14. Transfer all slides from the arrayer into a plastic slide box. Store them in a desiccator.

15. Power down the arrayer.

TROUBLESHOOTING

Problem (Step 2): Printing pins are not printing properly.

Solution: If any pins are not printing, four steps may be taken to improve performance. First, if dirt or other material has clogged the quill tip, wash the pins extensively in a sonicator bath. Second, if examination of a pin under a microscope indicates that the pin is clogged, try clearing the quill tip carefully with a razor blade, which may dislodge the contaminant. Third, subtle variation in pin length may prevent some pins from printing well; swapping pins within the print head may improve printing. Finally, if a pin continues to fail after all of the above steps have been taken, replace it with a fresh pin.

Problem (Step 2): Test plate printing is not uniform everywhere on the arrayer.

Solution: If the arrayer fails to print at particular locations, adjust the print height for that section of the platter. These adjustments can be made using the software that accompanies the robot.

DISCUSSION

A successful print run requires that print plates be handled carefully to prevent cross-contamination, that pins be checked regularly (i.e., after every 384-well plate or two) to confirm that they are still printing, and that the level of water in the bath (or sonicator) for washing pins be always full enough to cover the pins. We typically add 5–10 mL of water to the sonicator every ~6 h during a print run, but this will depend on the ambient humidity.

To check that pins are still printing, shine a flashlight onto the slide at an angle. The salt deposited during printing is white, and when a pin stops printing the pattern of one sector deviates from the others. For example, if five rows of 20 spots have been printed, a perfect 5×20 rectangle can be seen, but if a pin has dropped out, then the rectangle will be incomplete.

If a pin stops printing, it may be possible to remove the pin carefully and clean it under a microscope if any dust is observed in or on the pin. If no dust can be seen, sometimes a pin can be restored by extensive sonication. Finally, replacing the faulty pin with a spare pin, or moving pins around in the print head, can be used as a last resort. After a print run has been paused, replace the first slide with a clean slide and perform another test print to check on the new arrangement. When restarting the spotting robot, be sure that printing recommences at exactly the location where the print run was paused.

WWW RESOURCE

Protocol for building a spotting robot   http://cmgm.stanford.edu/pbrown/mguide/

Round A/Round B Amplification of DNA

The goal of this procedure is to randomly amplify a sample of DNA to achieve the best possible sequence representation. This protocol has been used successfully to amplify genomic representations starting with <10 ng of DNA. The protocol consists of three sets of enzymatic reactions. In Round A, Sequenase is used to extend randomly annealed primers to generate templates for subsequent PCR. During Round B, a specific primer is used to amplify the previously generated templates. Finally, amplified material can be labeled as in Protocol 7 or 8. Alternatively, Round C in this protocol can be used to incorporate either aminoallyl-dUTP or Cy-dye-coupled nucleotides during additional PCR cycles. This protocol may be unsuitable for amplifying material smaller than 250 bp because such material will not be amplified uniformly. In those cases, Protocol 3 is recommended. This protocol was adapted from Bohlander et al. (1992).

MATERIALS

It is essential that you consult the appropriate Material Safety Data Sheets and your institution's Environmental Health and Safety Office for proper handling of equipment and hazardous materials used in this protocol.

Recipes for reagents specific to this protocol, marked <R>, are provided at the end of the protocol. See Appendix 1 for recipes for commonly used stock solutions, buffers, and reagents, marked <A>. Dilute stock solutions to the appropriate concentrations.

Reagents

aa-dNTP/Cy-dNTP mixture ($100\times$) <R>
BSA (500 μg/mL)
DNA, isolated from the sample under study (10–100 ng)
> For example, for CGH analysis, genomic DNA is isolated as described in Chapter 1; for protein localization studies, DNA is isolated by chromatin immunoprecipitation (ChIP) as in Chapter 20.

dNTP mixture (3 mM)
dNTPs ($100\times$; 20 mM each nucleotide)
DTT (0.1 M)
$MgCl_2$ (25 mM)
PCR buffer ($10\times$) (500 mM KCl, 100 mM Tris at pH 8.3)
Primer A: GTTTCCCAGTCACGATCNNNNNNNNN (40 pmol/μL)
Primer B: GTTTCCCAGTCACGATC (100 pmol/μL)
Sequenase (13 units/μL) (US Biochemical, catalog no. 70775)
Sequenase buffer ($5\times$)
Sequenase dilution buffer
Taq polymerase (5 units/μL)

Equipment

Agarose gel (1%)
Microcon 30 spin column (Millipore)
Thermal cycler

METHOD

Round A: Extending Randomly Annealed Primers

1. Prepare Round A reactions as follows:

DNA	7 μL
Sequenase buffer (5×)	2 μL
Primer A (40 pmol/μL)	1 μL

    As little as 10 ng of DNA can be effectively amplified by this protocol. As a control, set up a reaction in which the DNA is replaced with H₂O.

2. Denature the template DNA and anneal the primer by heating for 2 min at 94°C and then rapidly cooling to 10°C. Keep the reaction for 5 min at 10°C.

3. Assemble the reaction mixture:

Sequenase buffer (5×)	1 μL
dNTP (3 mM)	1.5 μL
DTT (0.1 M)	0.75 μL
BSA (500 μg/μL)	1.5 μL
Sequenase (13 U/μL)	0.3 μL

4. Combine the reaction mixture and the template–primer mix. Place the tube(s) into a thermal cycler, and extend the primers as follows.

    i. Ramp from 10°C to 37°C over 8 min.

    ii. Hold at 37°C for 8 min.

    iii. Rapidly ramp to 94°C and hold for 2 min.

    iv. Rapidly ramp to 10°C, add 1.2 μL of diluted Sequenase (1:4 dilution), and hold for 5 min at 10°C.

    v. Ramp from 10°C to 37°C over 8 min.

    vi. Hold at 37°C for 8 min.

5. Dilute samples with water to a final volume of 60 μL.

Round B: PCR Amplification

6. Prepare Round B reactions as follows:

Round A template	15 μL
MgCl₂	8 μL
PCR buffer (10×)	10 μL
dNTP (100×)	1 μL
Primer B (100 pmol/μL)	1 μL
Taq polymerase	1 μL
Water	64 μL

7. Place the tube(s) into a thermal cycler, and amplify the templates as follows.

Cycle number	Denaturation	Annealing	Polymerization
15–35 cycles	30 sec at 94°C	30 sec at 40°C, then 30 sec at 50°C	2 min at 72°C

    Hold at 4°C.

    Run 15–35 cycles, depending on the amount of starting material.

    To optimize the number of cycles, it may be necessary to remove an aliquot every two cycles to monitor the progress of the amplification. It is best to use the minimal number of cycles that generates a visible smear (see Step 8).

    ▲ *Make sure that there is no DNA in the negative control lane!*

8. Run 5 μL of the reaction on a 1% agarose gel. A "smear" of DNA should be present between 500 and 1000 bp.

9. Label the DNA using the Round C procedure, Protocol 7, or Protocol 8.

Round C: Cyanine Labeling or Aminoallyl Activation of DNA

10. Prepare the Round C reaction:

Round B template	10–15 μL
MgCl$_2$	8 μL
PCR buffer (10×)	10 μL
aa-dNTP/Cy-dNTP (100×)	1 μL
Primer B (100 pmol/μL)	1 μL
Taq polymerase	1 μL
Water	63–68 μL

11. Place the tube(s) into a thermal cycler, and label the templates as follows:

Cycle number	Denaturation	Annealing	Polymerization
15–25 cycles	30 sec at 94°C	30 sec at 40°C, then 30 sec at 50°C	2 min at 72°C

Hold at 4°C.

Run 10–25 cycles.

The number of cycles should be determined empirically. The objective is to minimize the number of cycles required to yield ~2–3 μg of material for hybridization.

12. If aa-dNTPs were used in Round C, desalt the sample to remove Tris buffer, which interferes with the dye coupling. Add 400 μL of water to the sample in a Microcon 30, and centrifuge for ~8 min at 12,000 rpm. Repeat again with 500 μL of water.

13. Proceed to Cy-dye coupling as described in Protocol 8, Step 11.

DISCUSSION

The output of this amplification protocol will be double-stranded DNA. Quantify the amount of DNA by absorbance (see Chapter 2), and calculate the fold amplification based on the amount of input DNA used. First-time users of this protocol may want to run some amplified DNA on an agarose gel to observe the size distribution.

RECIPE

It is essential that you consult the appropriate Material Safety Data Sheets and your institution's Environmental Health and Safety Office for proper handling of equipment and hazardous materials used in this protocol.

aa-dNTP/Cy-dNTP Mixture (100×)

Reagent	Final concentration
dATP	25 mM
dCTP	25 mM
dGTP	25 mM
dTTP	10 mM
Aminoallyl-dUTP or CyX-dUTP	15 mM

The ratio of aa-dUTP to dTTP can be altered or optimized.

REFERENCE

Bohlander SK, Espinosa R III, Le Beau MM, Rowley JD, Diaz MO. 1992. A method for the rapid sequence-independent amplification of micro-dissected chromosomal material. *Genomics* **13:** 1322–1324.

T7 Linear Amplification of DNA (TLAD) for Nucleosomal and Other DNA < 500 bp

Protocol 2 has been used extensively in genomic localization analysis and appears to work quite well for typical applications in which DNA is sheared by sonication to ~500 bp. However, when DNA is sheared to a population whose modal size is <500 bp, bias in the PCR step skews representation of some genomic loci (Liu et al. 2003). In addition, a subset of applications requires amplifying DNA populations that are smaller than 500 bp; a notable example is ChIP on mononucleosomal DNA, which is ~150 bp long (Liu et al. 2005). In these circumstances, T7 linear amplification of DNA (TLAD) is preferred because it more accurately maintains uniform representation of short DNA fragments during amplification than does the amplification method described in Protocol 2.

Amplification of double-stranded DNA by TLAD begins with the addition of a 3′ tail of poly-thymidine to DNA by TdT. Second, the Klenow fragment of *E. coli* DNA polymerase is used, along with a T7-poly(A) primer, to generate a complementary strand that carries a 5′ T7 primer. Finally, extension of the original T-tailed DNA strand yields a template suitable for T7-based transcription, which generates amplified RNA (aRNA). This technique avoids the "jackpotting" issues observed with PCR, in which an early amplification event leads to disproportionate representation of a particular sequence, because PCR follows exponential kinetics, whereas transcription is a linear amplification method.

This protocol takes some time to complete. Table 1 provides estimates of the time required for each procedure.

MATERIALS

It is essential that you consult the appropriate Material Safety Data Sheets and your institution's Environmental Health and Safety Office for proper handling of equipment and hazardous materials used in this protocol.

Recipes for reagents specific to this protocol, marked <R>, are provided at the end of the protocol. See Appendix 1 for recipes for commonly used stock solutions, buffers, and reagents, marked <A>. Dilute stock solutions to the appropriate concentrations.

Reagents

β-Mercaptoethanol
Calf intestinal phosphatase (CIP) (New England Biolabs, catalog nos. M0290S or M0290L)
$CoCl_2$ (5 mM)
Dideoxynucleotide tailing solution (8%) (92 mM dTTP, 8 mM ddCTP) (Life Technologies)
dNTP mixture (5 mM) <A>
EDTA (0.5 M, pH 8.0) <A>
Ethanol (95%–100%)
Klenow fragment of DNA polymerase I (New England Biolabs)
Mineral oil
NEB buffer 2 (10×) (New England Biolabs)
NEB buffer 3 (10×) (New England Biolabs)
NTP mixture (75 mM)
Reaction buffer, provided with kit

TABLE 1. T7 linear amplification of DNA (TLAD) of small DNA molecules

Procedure	Time
TdT tailing and cleanup	30–40 min
Second-strand synthesis and cleanup	2–2.5 h
IVT	5.5–20.5 h
IVT cleanup	15–30 min
Assessment of RNA quality and quantity	40–60 min
Total	9.5–25 h

TdT, terminal transferase; IVT, in vitro transcription.

RNase-free H_2O

RNase inhibitor

T7-A_{18}B primer

The primer sequence is 5'-GCATTAGCGGCCGCGAAATTAATACGACTCACTATAGGGAG(A)$_{18}$[B], where B refers to C, G, or T. The primer should be HPLC, PAGE, or equivalent purification grade.

T7 RNA polymerase

Template DNA (maximum 500 ng per 10 μL)

Terminal transferase (TdT) (New England Biolabs, catalog nos. M0315S or M0315L)

Terminal transferase buffer (5×) (Roche, catalog no. 11243276103)

Equipment

MinElute kit (QIAGEN, catalog no. 28204)

RNase/DNase-free tubes (1.5 mL)

RNase-free PCR tubes (0.2 mL)

RNeasy Mini Kit (QIAGEN)

Rotary evaporator (e.g., SpeedVac)

Thermal cycler

Vacuum manifold (optional; see Step 17)

Water bath or heat block set to 37°C

METHOD

CIP Treatment of Samples with Terminal 3'-Phosphate Groups

Treatment of DNA with calf intestinal phosphatase (CIP) is only necessary if the source DNA has been sheared or treated with MNase. Treatment of the DNA with CIP removes 3'-phosphate groups, leaving free hydroxyl groups on the 3' ends, which are necessary for efficient tailing by TdT. Failure to perform this step will likely reduce the yield of amplification products by 50%.

1. Set up the following reaction:

CIP (2.5 U)	0.25 μL
NEB buffer 3 (10×)	1 μL
Template DNA (maximum 500 ng per 10 μL)	8.75 μL

Incubate the reaction for 1 h at 37°C.

Each reaction can be scaled up to 100 μL per tube.

2. Clean up the DNA with a MinElute column. Follow the protocol supplied by the manufacturer. Elute the DNA in 20 μL.

When working with <100 ng of DNA, the 10-μL elution volume in the manufacturer's protocol may yield less than the 80% claimed by QIAGEN. Increase the elution volume to 15–20 μL, and reduce the volume of the DNA by drying, if necessary.

Tailing Reaction with Terminal Transferase

3. Set up the tailing reaction:

TdT buffer (5×)	2 μL
8% ddCTP in 100 μM dTTP mix	0.5 μL
CoCl$_2$ (5 mM)	1.5 μL
Template DNA (maximum 75 ng)	5 μL
TdT enzyme (20 U)[a]	1 μL

[a]Add this last.

Do not use the NEB buffer 4 supplied with the NEB terminal transferase because DTT in the buffer will precipitate the CoCl$_2$ and inhibit the reaction. Use the cacodylate buffer (1 M potassium caco-dylate, 125 mM Tris-HCl, and 1.25 mg/mL BSA at pH 6.6), either supplied with the Roche enzyme or purchased separately. Take precautions in handling this arsenic-containing buffer and use waste-disposal practices appropriate for your institution.

Do not freeze and thaw the dNTP mixes more than three times. Additional freeze–thaw cycles will degrade the dNTPs and will reduce the efficiency of the reaction.

Aim for ~1 pmol of template molecules. The tested range is 2.5–75 ng of DNA per 10 μL of reaction volume. Scale up the reaction volume accordingly for greater starting amounts. For ChIP samples, use a sensitive UV-Vis spectrophotometer or a fluorometer to quantify the amount of sample precisely. If the amount of DNA is unknown, scale up to a 20-μL volume to ensure that there is enough TdT enzyme present for an efficient tailing reaction. Note that if insufficient enzyme is used, the efficiency of subsequent steps in the protocol will be significantly affected and result in significantly reduced yields (as little as 5%–10% of normal expected yields).

We strongly suggest that the NEB terminal transferase be used for this protocol; TdT enzyme from other sources may not perform optimally. If using the Roche recombinant TdT, double the volume of enzyme.

4. Add 1–2 drops of mineral oil to the top of the mixture to prevent evaporation of reactants during incubation. Incubate the reaction for 20 min at 37°C.

5. Stop the reaction by adding 2 μL (per 10 μL of reaction volume) of EDTA (0.5 M, pH 8.0).

6. Clean up the DNA with a MinElute column. Follow the protocol supplied by the manufacturer. Elute the DNA in 20 μL.

   If the starting volume is 10 μL, then add 10 μL of water to bring the volume to 20 μL before adding the reaction to the spin column. When working with <100 ng of DNA, the 10-μL elution volume in the manufacturer's protocol may yield less than the 80% claimed by QIAGEN. Increase the elution volume to 15–20 μL, and dry the volume down if necessary.

Second-Strand Synthesis with Klenow Fragment Polymerase

7. Set up the second-strand synthesis reaction:

T7-A$_{18}$B primer (25 μM)	0.3 μL
NEB buffer 2 (10×)	2.5 μL
dNTP mix (5.0 mM)	1 μL
Water	0.2 μL
T-tailed DNA	20 μL

If production of template-independent product is a significant problem, scale down the reaction volume while keeping the reagent concentrations (except for the T-tailed DNA) constant. See the end of the protocol for an example.

Do not freeze and thaw the dNTP mixes more than three times. Additional freeze–thaw cycles will degrade the dNTPs and will reduce the efficiency of the reaction.

NEB (early 2004) switched the supplied buffer for Klenow enzyme from EcoPol buffer to NEB buffer 2. This buffer should provide at least comparable yields to the old buffer and may actually increase yields up to ~14%.

   ▲ *Do not use mineral oil. Trace amounts of mineral oil appear to interfere with cleanup and in vitro transcription.*

8. Use the following program in a thermal cycler:

   i. 2 min at 94°C.

ii. Ramp from 94°C to 35°C at 1°C/sec, then hold for 2 min to anneal.

iii. Ramp from 35°C to 25°C at 0.5°C/sec.

iv. Hold for 45 sec at 25°C (or up to 6 min).

> *During this time, add 1 µL (5 U) of Klenow DNA polymerase. If necessary, centrifuge the tube to remove condensation from the top and sides of the tube.*

v. 37°C, 90 min.

vi. (Optional) 4°C to temporarily halt enzyme activity until user returns to take reaction tubes out of cycler.

9. Stop the reaction by adding 2.5 µL of EDTA (0.5 M, pH 8.0) (the final concentration will be 45 mM).

10. Clean up the DNA with a MinElute column. Follow the protocol supplied by the manufacturer. Elute the DNA in 20 µL.

> *When working with <100 ng of DNA, the 10-µL elution volume in the manufacturer's protocol may yield less than the 80% claimed by QIAGEN. Increase the elution volume to 15–20 µL, and dry the volume down if necessary. An elution volume of 20 µL at this step increased yields by 30%–40% for a 50-ng sample.*

In Vitro Transcription (IVT)

11. In vitro transcription (IVT) requires that the double-stranded DNA (dsDNA) be in an 8-µL volume. Dry down the eluate from 20 to 8 µL in a rotary evaporator at medium heat for 10–12 min (the drying rate is ~1 µL/min).

12. Set up the in vitro transcription reaction in 0.2-mL RNase-free PCR tubes as follows:

75 mM NTP mix	8 µL
Reaction buffer (*warm to room temperature first!*)	2 µL
RNase inhibitor and T7 RNA polymerase	2 µL
Template dsDNA	8 µL

> *If the IVT kit is new, combine the NTPs in one tube, then realiquot into four tubes. In the first three freeze–thaw cycles, yields drop ~10%–15% after each cycle. If the NTPs go through more than three freeze–thaw cycles, each subsequent freeze–thaw cycle may drop the yield by as much as 50%.*

> *The buffer should be at room temperature. Adding cold buffer and dsDNA may cause the DNA to precipitate. If there is a precipitate, warm the buffer to 37°C until the precipitate dissolves.*

13. Incubate the reaction overnight at 37°C in a thermal cycler with a heated lid or in an air incubator.

> *The incubation can range from 5 to 20 h; typical is overnight, roughly 16 h.*

Amplified RNA (aRNA) Purification Using RNeasy Columns

14. Prepare the buffer (433.5 µL per IVT reaction):

β-Mercaptoethanol (14.2 M stock solution)	3.5 µL
RNase-free H$_2$O	80 µL
Buffer RLT (from RNeasy Mini Kit)	350 µL

15. Aliquot the mix into 1.5-mL RNase/DNase-free tubes.

16. Transfer the contents of the IVT mix (from Step 13) to the RNase/DNase-free tube, and vortex gently and briefly.

17. Add 250 µL of 95%–100% ethanol, and mix well by pipetting. (Do not centrifuge!) Purify the aRNA through an RNeasy column by either the centrifuge method or with a vacuum manifold.

aRNA Purification Using a Centrifuge

i. Apply the entire sample to an RNeasy Mini spin column mounted on a collection tube. Centrifuge the column for 15 sec at $\geq 8000g$. Discard the flowthrough.

ii. Transfer the RNeasy column to a new 2-mL collection tube. Add 500 μL of Buffer RPE (which must have ethanol added before use) and centrifuge for 15 sec at $\geq 8000g$. Discard the flowthrough, but reuse the collection tube.

iii. Add 500 μL of Buffer RPE onto the RNeasy column and centrifuge for 2 min at maximum speed.

iv. Remove the flowthrough, and pipette another 500 μL of Buffer RPE onto the column. Centrifuge for 2 min at maximum speed.

> This additional wash, which is not in the QIAGEN protocol, is necessary because of guanidinium isothyocyanate contamination in the eluted RNA.

aRNA Purification Using a Vacuum Manifold

i. Apply the sample (700 μL) to an RNeasy Mini spin column attached to a vacuum manifold. Apply vacuum.

ii. Shut off the vacuum, and pipette 500 μL of Buffer RPE onto the RNeasy column. Apply vacuum.

iii. Repeat Step ii. Transfer the columns to 2-mL collection tubes. Centrifuge for 1 min at full speed.

iv. Return the column to the vacuum manifold, and add 500 μL of Buffer RPE. Apply vacuum.

v. Transfer the column back to a 2-mL tube. Centrifuge for 1 min at full speed to completely dry the column.

18. Transfer the RNeasy column into a new 1.5-mL collection tube, and add 30 μL of RNase-free water directly onto the membrane. Centrifuge for 1 min at $\geq 8000g$ to elute the RNA. Repeat if expected; the yield is >30 μg.

19. Check the RNA concentration and purity by measuring the A_{260} and A_{260}/A_{280}.

    See Troubleshooting.

20. Proceed to Protocol 5 or 6 to add fluorescent label to the RNA.

TROUBLESHOOTING

In addition to the items below, it may be valuable to consult the troubleshooting section in the manufacturer's manual that accompanies the IVT kit.

Problem (Step 19): Amplified RNA appears to have been damaged by RNase.

Solution: Perform an IVT control, using 250 ng of the pTRI-Xef-linearized plasmid provided with the Ambion IVT kit. If not using the kit, use an appropriate amount of a dsDNA template that contains the pT7 promoter. Be sure that the chosen template has been used successfully as a template for T7 RNA polymerase. Yields should typically range from 100 to 140 μg, limited by the $\sim$100-μg binding capacity of the QIAGEN RNeasy column. If the yield from the IVT control template is poor, contamination with RNases can be assessed by running a 2% nondenaturing agarose gel in Tris-acetate–EDTA (TAE) and ethidium bromide. An RNase-contaminated IVT sample will yield a smear of low-molecular-weight material. If RNase contamination is the cause, ensure that aerosol-barrier, RNase-free pipette tips are used, and that working surfaces are treated with RNase-decontaminating agents (e.g., RNaseZap; Ambion catalog no. 9780). This is particularly important when working with ChIP samples.

Problem (Step 19): Yields of aRNA are poor, but RNase contamination is not the cause.

Solution: If there is no RNase contamination detected, it is likely there may be problems with the IVT reaction conditions. Consider the following.

- The NTP mix has gone through too many freeze–thaw cycles. NTPs are very sensitive to freeze–thaw cycles, and each one decreases the yield. Use a fresh IVT kit, and aliquot the NTP mix before use.

- There has been excessive evaporation of the reaction volume during the incubation. The described IVT conditions (Steps 11–13) were designed to limit evaporation and vapor volume during the long incubation period. Using mineral oil is not recommended because it may interfere with either the IVT reaction, the aRNA cleanup, or both.

DISCUSSION

The output of this protocol is amplified RNA (aRNA) suitable for labeling for microarray studies as described in Protocols 5 and 6. It is important to determine the amount of aRNA produced and calculate the mass amplification obtained. Typical amplifications result in at least a 200-fold increased mass yield. For example, 20 μg of aRNA is synthesized from an input of 75 ng of DNA. Use a 1%–2% agarose gel to assess the composition and quality of the amplified RNA. Unless knowing the size distribution is crucial, it is usually not necessary to run a denaturing gel. Within the resolution limits of an agarose gel, the amplified product may migrate 20–40 bp more slowly on the gel. This shift is to be expected because of the addition of poly(A) tails, the tight size distribution of the poly(A) tail, and the sequence added by the T7 promoter. The size distribution of the poly(A) tail becomes particularly evident in amplification products produced from a single-size template, such as PCR products or a restriction-digested plasmid.

Second-Strand Synthesis with Limiting Primer Amounts

Occasionally a low-molecular-weight band may also appear near the bottom of the gel, at ~100 bp. It has been observed under certain amplification conditions, usually when the concentration of starting material is significantly less than that of the primer during second-strand synthesis. In these cases, a substantial amount of small-molecular-weight material may be generated, which likely represents amplification product produced from IVT-valid template synthesized through the formation of primer dimers during second-strand synthesis. These products represent nonproductive material for downstream analysis, and if substantial amounts of this material are observed, then it is necessary to limit the amount of primer.

Limiting the amount of primer is important when amplifying from very small amounts of starting material. Not only will it decrease the amount of primer-dimer product, it can also increase the

TABLE 2. Second-strand synthesis with limiting primer amounts

DNA (ng)	T7 primer (μL)	NEB 2 buffer (μL)	5 mM dNTPs (μL)	Water (μL)	Tailed DNA (μL)	Klenow (μL)	Total volume (μL)
>75	0.60 (25 μM)	5.0	2.0	20.4	20.0	2.0	50
50–75	0.30 (25 μM)	2.5	1.0	0.20	20.0	1.0	25
25	0.15 (25 μM)	2.5	1.0	0.35	20.0	1.0	25
10[a]	1.50 (1 μM)	1.0	0.4	0.20	6.5	0.4	10
5[a]	0.75 (1 μM)	1.0	0.4	0.95	6.5	0.4	10
2.5[a]	0.38 (1 μM)	1.0	0.4	1.32	6.5	0.4	10

Concentrate the tailed DNA in a vacuum centrifuge to the indicated volume for reaction volumes that total 10 μL.

[a]If using a thermal cycler without a heated lid, then centrifuge the tubes every 30 min during the 37°C incubation step.

yield of the desired amplification product. Table 2 describes the single reaction volumes to use for a suggested mass range of starting material.

REFERENCES

Liu CL, Schreiber SL, Bernstein BE. 2003. Development and validation of a T7 based linear amplification for genomic DNA. *BMC Genomics* **4**: 19. doi: 10.1186/1471-2164-4-19.

Liu CL, Kaplan T, Kim M, Buratowski S, Schreiber SL, Friedman N, Rando OJ. 2005. Single-nucleosome mapping of histone modifications in *S. cerevisiae. PLoS Biol* **3**: e328. doi: 10.1371/journal.pbio.0030328.

Protocol 4

Amplification of RNA

Gene expression profiling typically requires microgram quantities of mRNA, which can be difficult to obtain. In such cases, RNA must be amplified in order to have enough material for microarray labeling and hybridization. Currently, the most popular choice for amplifying RNA is to use a commercial kit, such as MessageAmp II (Ambion), a product with which we have had good success. These kits are expensive, however, and thus this protocol provides an alternative RNA amplification procedure adapted from Baugh et al. (2001).

This protocol generates amplified antisense RNA (aRNA) from limited quantities of total RNA (see Fig. 1). It is designed around maximizing yield and product length while minimizing template-independent side reactions. Template-independent reactions compete with the desired template-dependent reaction, an undesirable situation that becomes more severe as less RNA template is used. Amplification products dominated by template-independent product result in greatly reduced

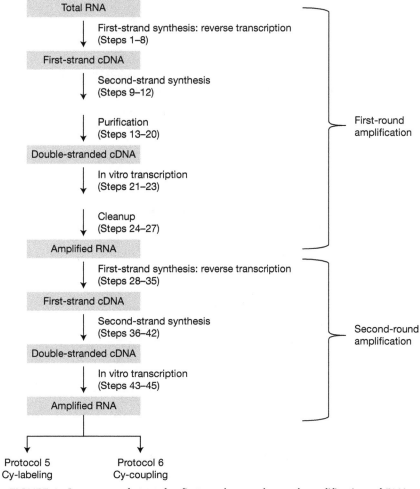

FIGURE 1. Sequence of steps for first- and second-round amplification of RNA.

709

sensitivity and compression of differences in microarray hybridization experiments. Most notably, the oligo(dT) primer used in reverse transcription (RT) yields a high-molecular-weight product in the in vitro transcription (IVT) reaction independent of any cDNA template (Baugh et al. 2001). This reaction occurs under all conditions tested; the protocol is therefore designed to limit the amount of primer used to start with. In addition, high-molecular-weight, template-independent product is generated in the presence of biotinylated NTPs and the absence of any polymer when excessive amounts of T7 RNA polymerase activity are used. Template-dependent product of questionable molecular weight and limited functionality in downstream reactions can also be produced with excessive T7 RNA polymerase activity. Essentially, more yield is not always better. The protocol limits the amount of primer used by employing small cDNA synthesis volumes.

The key consideration in any amplification protocol, as noted in Protocols 2 and 3 for DNA labeling, is preventing representation bias in the amplified material. As with DNA amplification, a valuable initial experiment for investigators who are new to RNA amplification is to compare unamplified RNA and amplified RNA by microarray hybridization. The readout should be designed to reveal which sequences are over- and underrepresented after amplification, and by how much.

At the end of the protocol, quantify the mass yield of amplified material. A single round of amplification typically results in a fivefold to 20-fold mass conversion of starting material. If the first-round aRNA is used as a template for a second round of amplification, 200- to 400-fold amplification is typical.

MATERIALS

It is essential that you consult the appropriate Material Safety Data Sheets and your institution's Environmental Health and Safety Office for proper handling of equipment and hazardous materials used in this protocol.

Recipes for reagents specific to this protocol, marked <R>, are provided at the end of the protocol. See Appendix 1 for recipes for commonly used stock solutions, buffers, and reagents, marked <A>. Dilute stock solutions to the appropriate concentrations.

Reagents

Carrier

Add either 5 μg of linear polyacrylamide (LPA) or 20 μg of glycogen. If a second round of amplification is to be used (or other downstream reverse transcription reactions), then LPA is recommended over glycogen. LPA will slow the Microcon washes (from ~12–14 min to 28–32 min for Microcon 100s at 500g and room temperature), but it does not inhibit reverse transcriptase.

DEPC ddH$_2$O

DEPC-treated TE (pH 8.0)

DNA ligase, from *Escherichia coli*

DNA polymerase I

dNTP (10 mM)

DTT (100 mM)

(dT)-T7 primer

The primer sequence is 5'-GCATTAGCGGCCGCGAAATTAATACGACTCACTATAGGGAGA(T)21V-3' (where V stands for A, C, or G).

Ethanol (70% and 95%)

First-strand buffer (5×), provided with kit

High-yield T7 in vitro transcription kit (e.g., AmpliScribe from Epicentre; similar kits from Ambion, Promega, and others are available)

IVT buffer (10×), provided with IVT kit

NaCl (5 M)

NTP mixture (100 mM)

The NTP mixture contains 100 mM each of ATP, CTP, GTP, and UTP.

Phenol:chloroform
Random primers
Reverse transcriptase (SuperScript II; Life Technologies)
RNase H, from *E. coli*
RNase inhibitor
Second-strand synthesis buffer (5×) <R> (Lifetech or homemade)
T4gp32 (single-strand binding protein; 8 mg/mL)
T4 DNA polymerase
T7 RNA polymerase, high concentration (80 U/μL)
Total RNA (100 ng dissolved in water or TE; the concentration does not matter as the RNA will be dried down in Step 1)

Equipment

Bio-Gel P-6 Micro-Spin Column (Bio-Rad)
Heat block set at 65°C and 70°C
Incubator set at 14°C–16°C
Microcon 100 spin column (Millipore)
Phase Lock Gel Heavy tubes (0.5 mL) (Eppendorf)
Thermal cycler (with a heated lid) or air incubator set at 42°C and 70°C
Tubes (0.6 mL)
Rotary evaporator (e.g., SpeedVac)

METHOD

Reverse Transcription

1. Combine 100 ng of the (dT)-T7 primer with 100 ng of total RNA in a 0.6-mL tube. Reduce the volume under vacuum to 5.0 μL.

   ▲ *Do not allow the RNA to dry out completely.*

2. Prepare the RT premix and place it on ice:

First-strand buffer (5×)	2.0 μL
DTT (100 mM)	1.0 μL
dNTP (10 mM)	0.5 μL
T4gp32 (8.0 mg/mL)	0.5 μL
RNase inhibitor (~20 U)	0.5 μL
SuperScript II (100 U)	0.5 μL

3. Denature the RNA/primer mix for 4 min at 70°C in a thermal cycler with a heated lid.

4. Snap-cool the mix on ice and keep it on ice.

   The volume may drop following denaturation.

5. Add 5.0 μL of ice-cold RT premix to the RNA/primer tube, and mix by pipetting. The final volume should be 10 μL.

   If there was evaporative loss during denaturation (Step 3), then add dH$_2$O to adjust the volume to 10.0 μL. Before committing your RNA, run these initial steps using a control nucleic acid to determine the volume loss.

6. Incubate the RT reaction for 1 h at 42°C in either a thermal cycler with a heated lid or an air incubator, but not in a water bath to reduce the chance of contamination.

7. Heat-inactivate the reaction for 15 min at 65°C.

8. Chill the tube on ice.

Second-Strand Synthesis

9. Prepare second-strand synthesis (SSS) premix:

Second-strand buffer (5×)	15 μL
dNTP (10 mM)	1.5 μL
DNA polymerase I	20 U
RNase H, *E. coli*	1 U
DNA ligase, *E. coli*	5 U
dH₂O	to 65 μL final volume

Chill on ice.

10. Add 65 μL of ice-cold SSS premix to the RT reaction tube, and mix by pipetting. Incubate for 2 h at 14°C–16°C.

11. Add 2 U (10 U) of T4 DNA polymerase, and mix by flicking and gentle vortexing. Incubate for an additional 15 min at 14°C–16°C.

12. Heat-inactivate the reaction for 10 min at 70°C. Go immediately from 15°C to 70°C without letting the tube sit at room temperature to avoid undesirable enzyme activities.

13. Add 75 μL of phenol:chloroform (1:1), and mix by pipetting vigorously. Transfer the mixture to prespun Phase Lock Gel Heavy 0.5-mL tubes, and centrifuge for 5 min at 13,000 rpm.

14. Prepare a Bio-Gel P-6 Micro-Spin Column per the manufacturer's instructions.

15. Transfer the aqueous phase from Step 13 to the prepared P-6 column and centrifuge at 1000*g* for 4 min, recovering the flowthrough (~80 μL) in a clean 1.5-mL tube.

    The flowthrough can be kept in the 1.5-mL tube or transferred to a tube of a different size, depending on the details of the IVT reaction steps. For example, it may be desirable to run the reaction in a thermal cycler; thus, the products can be held at 4°C rather than overincubate them.

16. Add the appropriate carrier and 3.5 μL of 5 M NaCl for precipitating the DNA, and mix by vortexing. Add 2.5 volumes of 95% ethanol (~220 μL) and mix well. Precipitate for at least 2 h at −20°C.

17. Centrifuge to pellet the DNA at 13,000 rpm for 20 min.

18. Carefully remove the supernatant. Wash the pellet with 500 μL of 70% ethanol, and centrifuge for 5 min at 13,000 rpm.

19. Carefully remove the supernatant. Pulse-centrifuge (up to full speed) the tube to collect all residual ethanol at the bottom.

20. Remove the remaining supernatant with a pipette. Allow the pellet to air-dry for 2–3 min.

In Vitro Transcription

21. Prepare the IVT premix at room temperature to avoid forming a precipitate:

DEPC ddH₂O	16.5 μL
IVT buffer (10×)	4.0 μL
ATP (100 mM)	3.0 μL
CTP (100 mM)	3.0 μL
GTP (100 mM)	3.0 μL
UTP (100 mM)	3.0 μL
DTT (100 mM)	4.0 μL
RNase inhibitor (~60 U)	1.5 μL
High-concentration T7 RNA polymerase (80 U/μL)	2.0 μL

22. Add 40 μL of IVT premix to the DNA pellet (from Step 20). Resuspend the pellet in the premix by gently flicking and vortexing the tube. Incubate the reaction for 9 h at 42°C.

    If the starting amount of total DNA was in the microgram range, then IVT yields might improve by using a 60- or 80-μL reaction volume.

23. Proceed with the cleanup and additional amplification, or freeze the IVT reaction at −80°C.

Cleanup

24. Add 480 μL of DEPC-treated TE to the IVT reaction tube.

25. Transfer the 500 μL to a Microcon 100, and centrifuge at 500*g* until the volume is <20 μL (11–15 min at room temperature without LPA, 28–32 min with LPA).

    Processing a sample through a Microcon spin column is slower with LPA in the sample, although the results are fine. Alternatively, RNeasy columns can be used for cleanup.

26. Add another 500 μL of DEPC-treated TE, and centrifuge as before. Repeat this step one more time (three washes total).

    If you intend to proceed with a second amplification, the final wash should be with dH₂O, and with the filtrate should be reduced to a small volume.

27. Measure and, if necessary, adjust the volume for downstream applications.

    It may be comforting or worthwhile to quantify the yield and analyze the products by electrophoresis. Alternatively, proceed to the second round of amplification.

Second Round of Amplification

28. Add 0.5 μg of random primers to the aRNA (from Step 26). Reduce the volume under vacuum to 5.0 μL.

29. Denature the RNA for 5 min at 70°C in a thermal cycler with a heated lid.

30. Snap-cool the mix on ice. Let the tube rest for 5 min at room temperature.

31. Prepare the RT premix and place it on ice:

First-strand buffer (5×)	2.0 μL
DTT (100 mM)	1.0 μL
dNTP (10 mM)	0.5 μL
T4gp32 (8.0 mg/mL)	0.5 μL
RNase inhibitor (~20 U)	0.5 μL
SuperScript II (100 U)	0.5 μL

32. Add 5 μL of room-temperature RT premix to the RNA, and mix by pipetting.

    The final volume should be 10.0 μL. If there was evaporative loss during denaturation (Step 28), then add dH₂O to adjust the volume to 10.0 μL.

33. Incubate the reaction tube in a thermal cycler with a heated lid, programmed as follows:

    i. 20 min at 37°C.

    ii. 20 min at 42°C.

    iii. 10 min at 50°C.

    iv. 10 min at 55°C.

    v. 15 min at 65°C.

    vi. Hold at 37°C.

34. Add 1 U of RNase H, and mix by vortexing gently. Incubate for 30 min at 37°C and then heat for 2 min at 95°C.

35. Chill the tube on ice, and then centrifuge it briefly to collect the condensation. Place the tube on ice.

36. Add 1 μL of 100 ng/μL (dT)-T7 primer while the tube is on ice. Incubate the tube for 10 min at 42°C to anneal the primer.

37. Prepare the SSS premix (minus ligase):

Second-strand buffer (5×)	15 μL
dNTP (10 mM)	1.5 μL
DNA polymerase I	20 U
RNase H, *E. coli*	1 U
dH₂O	to 65 μL final volume

38. Snap-cool the sample (from Step 36) on ice.

39. Add 65 μL of ice-cold SSS premix to the chilled reaction tube. Incubate for 2 h at 14°C–16°C.

40. Add 10 U of T4 DNA polymerase, and mix by gentle flicking and vortexing. Incubate for an additional 15 min at 14°C–16°C.

41. Heat-inactivate the reaction for 10 min at 70°C.

42. Perform phenol:chloroform extraction, Bio-Gel P-6 chromatography, and nucleic acid precipitation as per Steps 13–20.

43. Prepare the IVT premix as in Step 21.

44. Add 40 μL of IVT premix to the DNA pellet (from Step 42). Resuspend the pellet in the premix by gently flicking and vortexing the tube. Incubate the reaction for 9 h at 42°C.

45. Proceed with the cleanup as in Steps 24–27, or freeze the IVT reaction at −80°C.

DISCUSSION

The output of this protocol will be amplified RNA suitable for direct or indirect labeling for microarray hybridization (Protocols 5 or 6). As with DNA amplification, always quantify the mass yield of amplified material, and include control amplifications without input RNA to ensure that there is no contamination of any of your reagents.

First time users of this protocol will find it helpful to analyze amplified product on denaturing or native agarose gels (see Fig. 1 in Baugh et al. 2001). Identification of high-molecular-weight product in the No Template control likely indicates excess primer, and this can be minimized by decreasing primer concentration in the initial in vitro transcription reactions.

Another useful first-time control is to amplify RNA from an abundant RNA source and to competitively hybridize amplified RNA against the original RNA pool, ideally using a microarray containing probes at the $5'$ and $3'$ ends of genes. If the protocol is working well, aRNA and original RNA samples should be highly correlated, and $5'/3'$ ratios should be close to 1. Lower $5'/3'$ ratios indicate poorly processive in vitro transcription, which can be corrected by increasing input RNA mass to the protocol. Alternatively, lower $5'/3'$ ratios may indicate failure to include the single-strand DNA binding protein T4gp32 in the in vitro transcription reaction.

RECIPE

It is essential that you consult the appropriate Material Safety Data Sheets and your institution's Environmental Health and Safety Office for proper handling of equipment and hazardous materials used in this protocol.

Second-Strand Synthesis Buffer (5×)

Reagent	Final concentration
Tris-HCl (pH 6.9)	100 mM
$MgCl_2$	23 mM
KCl	450 mM
β-NAD	0.75 mM
$(NH_4)_2SO_4$	50 mM

REFERENCE

Baugh LR, Hill AA, Brown EL, Hunter CP. 2001. Quantitative analysis of mRNA amplification by in vitro transcription. *Nucleic Acids Res* 29: e29. doi: 10.1093/nar/29.5.e29.

Direct Cyanine-dUTP Labeling of RNA

This is the simplest method to label RNA for use in expression analysis. RNA is reverse-transcribed using both oligo(dT) and random hexamers as primers. The random hexamers improve overall efficiency of labeling, especially at the 5′ end of the RNA. Fluorescently labeled dUTP is incorporated into the cDNA. After reverse transcription, the RNA is degraded, and the labeled cDNA is purified from unincorporated Cy dyes. Finally, samples labeled with Cy3 and Cy5 dyes are mixed and combined with blocking nucleotides and used for hybridization, as described in Protocol 10.

MATERIALS

It is essential that you consult the appropriate Material Safety Data Sheets and your institution's Environmental Health and Safety Office for proper handling of equipment and hazardous materials used in this protocol.

Recipes for reagents specific to this protocol, marked <R>, are provided at the end of the protocol. See Appendix 1 for recipes for commonly used stock solutions, buffers, and reagents, marked <A>. Dilute stock solutions to the appropriate concentrations.

Reagents

Cy3-dUTP or Cy5-dUTP (GE Healthcare Life Sciences)
DEPC-treated H_2O
HCl (0.1 N)
NaOH (0.1 N)
Oligo(dT) (2 μg/μL)
Random hexamers (4 μg/μL)

> N6 random hexamer can be ordered from any oligonucleotide company and made up at 5 mg/mL in RNase-free TE or water.

Reverse transcriptase, 200 U/μL (SuperScript II; Life Technologies)

> First strand buffer <R> and DTT, both required for preparation of the master reagent mix in Step 3, are provided with SuperScript II.

Total RNA (from Protocol 3 or 4)
Unlabeled dNTPs (low dTTP) stock <R>

Equipment

Heat block set at 65°C
MinElute kit (QIAGEN, catalog no. 28004)
Thermal cycler

METHOD

RT Reaction

1. Prepare the following RNA/oligo reaction mixtures in separate microcentrifuge tubes.

	Cy3	Cy5
Total RNA (30 μg recommended)	20–50 μg	20–50 μg
Oligo(dT) (2 μg/μL)	2 μL	2 μL
Random hexamer (4 μg/μL)	1 μL	1 μL
ddH$_2$O (DEPC)	to 15.4 μL	to 15.4 μL

2. Heat the tubes for 10 min to 65°C, and cool on ice to anneal the primers to the RNA.

3. While the RNA/oligo reaction mixtures are incubating, prepare two master reaction mixtures, one containing Cy3-dUTP and one with Cy5-dUTP, in a volume sufficient to transfer 14.6 μL of each to an RNA/oligo reaction mixture.

	Cy3-dUTP	Cy5-dUTP
First-strand buffer (5×)	6.0 μL	6.0 μL
DTT (0.1 M)	3.0 μL	3.0 μL
Unlabeled dNTPs (low dTTP)	0.6 μL	0.6 μL
Cy3-dUTP (1 mM)	3.0 μL	
Cy5-dUTP (1 mM)		3.0 μL
SuperScript II (200 U/μL)	2.0 μL	2.0 μL
Total volume	14.6 μL	14.6 μL

4. To the Cy3 RNA/oligo reaction mixture, add 14.6 μL of the Cy3 master reaction mixture (Step 3) to each tube. To the Cy5 RNA/oligo reaction mixture, add 14.6 μL of the Cy5 master reaction mixture (Step 3) to each tube. Incubate for 1 h at 42°C.

5. Add 1 μL of SuperScript II enzyme (200 U/μL) to each reaction and thoroughly mix the reaction components with a pipette. Incubate for an additional 1 h.

6. Degrade the RNA by adding 15 μL of 0.1 N NaOH to each reaction and incubating for 10 min at 65°C–70°C.

7. Neutralize each reaction by adding 15 μL of 0.1 N HCl.

Cleanup

If you want to determine the amount of fluorophore incorporated, then clean up the Cy5- and Cy3-labeled cDNA samples on separate MinElute columns. However, if thin coverslips and a small probe volume will be used during the hybridization (Protocol 10), it may be necessary to purify the Cy5- and Cy3-labeled cDNA samples together or use vacuum centrifugation to reduce the volume after elution.

8. Add 600 μL of Buffer PB (binding buffer) to each sample.

9. Assemble a MinElute column on a 2-mL collection tube.

10. Add the entire 660 μL or 720 μL (volume depends on whether each color sample is being cleaned up separately or together) to a MinElute column. Centrifuge the column for 1 min at 10,000g. Discard the flowthrough, and reuse the 2-mL tube.

11. Add 750 μL of Wash buffer PE to the column. Centrifuge for 1 min at 10,000g. Discard the flowthrough, and reuse the 2-mL tube.

12. Centrifuge again at maximum speed for 1 min to remove residual ethanol.

13. Place the column in a fresh 1.5-mL tube. Add 10 μL of H$_2$O to elute. Allow the Elution buffer to stand for at least 2 min.

14. Centrifuge at maximum speed for 1 min. Add 10 μL of H$_2$O to elute. Allow the Elution buffer to stand for at least 2 min before spinning.

15. Centrifuge at maximum speed for 1 min.

16. Measure the volume of the eluate for each sample. The volume should be ~18 μL for each column.

DISCUSSION

The output of this protocol will be Cy5-labeled cDNA and Cy3-labeled cDNA. The DNA can be used immediately to hybridize to a microarray, or it can be stored in foil, to prevent bleaching, at 4°C for up to a week. A useful spot check for labeling success is the color of the labeled material after unincorporated nucleotides have been removed (see Step 16). A good labeling will result in blue Cy5-DNA and red Cy3-DNA.

RECIPES

It is essential that you consult the appropriate Material Safety Data Sheets and your institution's Environmental Health and Safety Office for proper handling of equipment and hazardous materials used in this protocol.

First-Strand Buffer

Reagent	Final concentration
Tris-HCl (pH 8.3)	250 mM
KCl	375 mM
MgCl$_2$	15 mM

Unlabeled dNTPs (Low dTTP) Stock

Unlabeled dNTPs	Volume	Final concentration
dATP (100 mM)	25 μL	25 mM
dCTP (100 mM)	25 μL	25 mM
dGTP (100 mM)	25 μL	25 mM
dTTP (100 mM)	15 μL	15 mM
ddH$_2$O	10 μL	
Total volume	100 μL	

Indirect Aminoallyl-dUTP Labeling of RNA

This protocol is slightly longer than the simpler direct-labeling protocol, but it is significantly cheaper because of the high cost of Cy-dNTPs used in Protocol 5. This labeling procedure is called indirect because the fluorescent moiety is not incorporated during the reverse transcription reaction. Instead, a reactive nucleotide analog (aminoallyl-dUTP) is incorporated during reverse transcription, the cDNA is isolated, and then the cyanine dyes are incorporated by binding with the aminoallyl group to produce the desired fluorescent cDNA.

MATERIALS

It is essential that you consult the appropriate Material Safety Data Sheets and your institution's Environmental Health and Safety Office for proper handling of equipment and hazardous materials used in this protocol.

Recipes for reagents specific to this protocol, marked <R>, are provided at the end of the protocol. See Appendix 1 for recipes for commonly used stock solutions, buffers, and reagents, marked <A>. Dilute stock solutions to the appropriate concentrations.

Reagents

aa-dUTP mixture ($50\times$) <R>

Anchored oligo(dT) (Life Technologies)
> Anchored oligo(dT) primer is a mix of 12 primers, each having a sequence of 20 dT residues followed by two nucleotides VN, where V is dA, dC, or dG and N is dA, dC, dG, or dT. Thus, the VN anchor restricts annealing of the primer to the 5' end of the poly(A) tail of mRNA.

Cyanine dyes (typically Cy5 and Cy3) (GE Healthcare Life Sciences)
> Dye comes dried and should be resuspended in 11 μL of DMSO. This amount of material is sufficient for three labeling reactions. If the entire amount will not be used, prepare 3-μL aliquots, dry down in a rotary vacuum, and store them at −20°C.

DTT (0.1 M)

EDTA (0.5 M)

HEPES (1 M, pH 7.5)

$NaHCO_3$ (50 mM, pH 9.0)

NaOH (1 N)

Reverse transcriptase (SuperScript II; Life Technologies)

Reverse transcriptase buffer ($5\times$) (Life Technologies)

RNasin (ribonuclease inhibitor)

Total RNA (from Protocol 3 or 4)

Equipment

Heat block set at 67°C, 70°C, and 95°C

Lightproof box

MinElute kit (QIAGEN, catalog no. 28204)

Thermal cycler (42°C)

Zymo column (Zymo Research, catalog no. D3024)

METHOD

Reverse Transcription to Make aa-dUTP-Labeled cDNAs

1. Combine 30 μg of total RNA and water to a final volume of 14.5 μL. Add 1 μL of 5 μg/μL anchored oligo(dT). Mix by pipetting. Heat for 10 min at 70°C.

2. Cool the tube on ice for 10 min.

3. Pulse-centrifuge to bring the condensate to the bottom of the tube.

4. Prepare the Master mix just before use (be sure to add the enzymes last):

SuperScript II reverse transcriptase buffer (5×)	6.0 μL
DTT (0.1 M)	3.0 μL
aa-dUTP mix (50×)	0.6 μL
dH₂O	2.0 μL
SuperScript II reverse transcriptase	1.9 μL
RNasin	1.0 μL

5. Add 14.5 μL of Master mix to the tube of RNA. Pipette to mix. Incubate the reaction for 2 h at 42°C.

6. Incubate the tube for 5 min at 95°C. Transfer immediately to ice.

7. Add 13 μL of 1 N NaOH and 1 μL of 0.5 M EDTA to hydrolyze the RNA. Mix the reagents and pulse-centrifuge. Incubate the tube for 15 min at 67°C.

8. Neutralize the reaction with 50 μL of 1 M HEPES (pH 7.5). Vortex the tube and pulse-centrifuge.

9. Purify the reaction over a Zymo column to remove unincorporated nucleotides.

   i. Add 1 mL of Binding buffer to the reaction. Mix by pipetting. Load half of the material onto the column. Centrifuge for 10 sec at maximum speed. Discard the flowthrough.

   ii. Add the remainder of the reaction. Centrifuge for 10 sec at maximum speed. Discard the flowthrough.

   iii. Add 200 μL of Wash buffer to the column. Centrifuge for 30 sec at maximum speed.

   > *Typically, multiple individual samples will be labeled simultaneously, with at least two reactions for a given two-color microarray.*

   iv. Add another 200 μL of Wash buffer to the column. Centrifuge for 1 min at maximum speed. Discard the flowthrough, and centrifuge for 1 min at maximum speed.

   v. Add 10 μL of 50 mM NaHCO₃ (pH 9.0) to the filter. Incubate for 5 min at room temperature. Centrifuge for 30 sec at maximum speed to elute the cDNA.

10. Couple cyanine dyes to the aa-dUTP incorporated cDNAs by adding the eluted material directly to 3 μL of Cy5 or Cy3 dye resuspended in DMSO. Incubate for 1 h to overnight at room temperature in a lightproof box or drawer.

Purification of Labeled cDNA

11. Purify the cDNA using a MinElute kit, following the manufacturer's instructions.

DISCUSSION

The products of this protocol will be Cy5-labeled and Cy3-labeled cDNA, which are ready to hybridize to microarrays. This material can be stored in foil, to prevent bleaching, at 4°C for up to 1 wk. A useful check for labeling success is the color of the labeled cDNA after unincorporated nucleotides have been removed: Good labeling will result in visible blue (Cy5) or red (Cy3) color in the cleaned up material.

RECIPE

It is essential that you consult the appropriate Material Safety Data Sheets and your institution's Environmental Health and Safety Office for proper handling of equipment and hazardous materials used in this protocol.

aa-dUTP Mixture (50×)

Dissolve 1 mg of aminoallyl-dUTP (Sigma-Aldrich) with:

Reagent	Quantity	Final concentration
dATP (100 mM)	32.0 µL	12.5 mM
dGTP (100 mM)	32.0 µL	12.5 mM
dCTP (100 mM)	32.0 µL	12.5 mM
dTTP (100 mM)	12.7 µL	5 mM
dH$_2$O	19.3 µL	

Cyanine-dCTP Labeling of DNA Using Klenow

Similar to the direct RNA labeling protocol (Protocol 5), direct DNA labeling with Cy-dCTP is the simplest and fastest method for labeling DNA. This is a standard Klenow labeling protocol in which Cy-dCTP is incorporated during the labeling reaction. After stopping the reaction, labeled nucleotide is separated from unreacted Cy-dCTP, and Cy3- and Cy5-labeled materials are combined for hybridization (Protocol 10). This protocol is suitable for many applications, including detection of copy number variation, nucleosome mapping, and other location analysis (e.g., ChIP-chip).

MATERIALS

It is essential that you consult the appropriate Material Safety Data Sheets and your institution's Environmental Health and Safety Office for proper handling of equipment and hazardous materials used in this protocol.

Recipes for reagents specific to this protocol, marked <R>, are provided at the end of the protocol. See Appendix 1 for recipes for commonly used stock solutions, buffers, and reagents, marked <A>. Dilute stock solutions to the appropriate concentrations.

Reagents

Cy3-dCTP and Cy5-dCTP (GE Life Sciences)
dNTP mixture (10×) <R>
EDTA (0.5 M, pH 8.0)
Genomic DNA (or from Protocol 2)
Human $C_{o}t$-1 DNA (1 μg/μL) (GIBCO)
Klenow fragment of *E. coli* DNA polymerase I (40–50 units/μL)
Poly(dA-dT) (Sigma-Aldrich)
  Make a 5 μg/μL stock in nuclease-free water. Store the stock at −20°C.
Random primer/buffer solution <R>
  This can be obtained from Life Technologies "BioPrime" labeling kit or can be made in the laboratory.
TE (pH 7.4)
Yeast tRNA (GIBCO)
  Make a 5 μg/μL stock in nuclease-free water.

Equipment

Heat block set at 37°C and 95°C–100°C
Microcon 30 spin column (Millipore)

METHOD

1. Add 2–3 μg of DNA to a microcentrifuge tube.

   For high-complexity DNA, such as mammalian genomic DNA, reducing the fragment size by sonication or restriction digestion improves labeling efficiency. Shearing by sonication to ~1000-bp average fragment size is sufficient for this application.

2. Add H$_2$O to the DNA to a final volume of 21 μL. Add 20 μL of 2.5× random primer/buffer mixture. Boil the tube for 5 min in a heat block set to 95°C–100°C.

3. Place the tubes on ice for 5 min.

4. Add reagents in the following order:

dNTP mixture (10×)	5 μL
Cy5-dCTP or Cy3-dCTP	3 μL
Klenow, high concentration (40–50 units/μL)	1 μL

   Cy-UTP also works well, but if using UTP, adjust the 10× dNTP mix accordingly.

5. Incubate the reaction for 1–2 h at 37°C.

   For improved labeling efficiency, add 1 μL of Klenow after the reaction has been going for 1 h. Incubate the reaction for another hour.

6. Stop the reaction by adding 5 μL of 0.5 M EDTA (pH 8.0).

Cleanup of Labeled DNA

7. Combine the Cy3- and Cy5-labeled samples, and add 450 μL of TE (pH 7.4).

8. Add the combined samples to a Microcon 30 spin column. Centrifuge the spin column for 10–11 min at 10,000g in a microcentrifuge.

9. Discard the flowthrough.

10. If hybridizing to homemade microarrays, add blocking nucleotides at this step:

C$_0$t-1 DNA (1 μg/μL)	30–50 μg
Yeast tRNA (5 μg/μL)	100 μg
Poly(dA-dT) (5 μg/μL)	20 μg

   If using a commercial microarray, refer to the manufacturer's protocols for specific hybridization buffers and blocking nucleotides.

11. Add another 450 μL of TE, and centrifuge as in Step 8.

12. Measure the volume of the liquid remaining in the column. If necessary, centrifuge for 1 min. The desired probe volume is ~20 μL. The exact volume will depend on the size of the coverslip that contains the microarray. For small microarrays this volume may be as low as 12 μL (see Protocol 10, Step 3).

13. Discard the flowthrough. Recover the labeled DNA from the filter by inverting the column onto a fresh collection tube. Centrifuge for 1 min at 10,000g.

14. Proceed to hybridization with the microarray (Protocols 9 and 10).

DISCUSSION

The products of this protocol will be Cy5-labeled and Cy3-labeled cDNA, which are ready to hybridize to microarrays. This material can be stored in foil, to prevent bleaching, at 4°C for up to 1 wk. A useful check for labeling success is the color of the labeled cDNA after unincorporated nucleotides have been removed: Good labeling will result in visible blue (Cy5) or red (Cy3) color in the cleaned up material.

RECIPES

It is essential that you consult the appropriate Material Safety Data Sheets and your institution's Environmental Health and Safety Office for proper handling of equipment and hazardous materials used in this protocol.

dNTP Mixture (10×)

Reagent	Final concentration
dATP	1.2 mM
dGTP	1.2 mM
dTTP	1.2 mM
dCTP	0.6 mM
Tris (pH 8.0)	10 mM
EDTA	1 mM

Random Primer/Buffer Solution

Reagent	Final concentration
Tris (pH 6.8)	125 mM
$MgCl_2$	12.5 mM
β-Mercaptoethanol	25 mM
Random octamers	750 μg/mL

Protocol 8

Indirect Labeling of DNA

As with RNA labeling protocols, the main difference between direct and indirect DNA labeling protocols is a trade-off of cost and time. Indirect labeling of DNA takes ~2 h longer than direct labeling but is hundreds of dollars cheaper.

MATERIALS

It is essential that you consult the appropriate Material Safety Data Sheets and your institution's Environmental Health and Safety Office for proper handling of equipment and hazardous materials used in this protocol.

Recipes for reagents specific to this protocol, marked <R>, are provided at the end of the protocol. See Appendix 1 for recipes for commonly used stock solutions, buffers, and reagents, marked <A>. Dilute stock solutions to the appropriate concentrations.

Reagents

aa-dUTP/dNTP mixture (3 mM)
Combine 6 μL of 50× stock <R> used for cDNA synthesis and 44 μL of dH$_2$O.

Cyanine dyes (typically Cy5 and Cy3) (GE Healthcare Life Sciences)
Dyes come dried and should be resuspended in 11 μL of DMSO. This amount of material is sufficient for three labeling reactions. If the entire amount will not be used, prepare 3-μL aliquots, dry down in a rotary vacuum, and store them at −20°C.

EDTA (0.5 M, pH 8.0)

Genomic DNA
Prepare genomic DNA according to your favorite protocol. Because the DNA should be fairly pure, avoid using "quick and dirty" methods. Commercial gDNA isolation kits are suitable, as in Chapter 1, Protocols 12 and 13. As a rule of thumb, the OD$_{260/280}$ ratio should be at least 1.8.

Klenow buffer

Klenow fragment of DNA polymerase I

NaHCO$_3$ (50 mM, pH 9.0)

Random hexamer
N6 random hexamer can be ordered from any oligonucleotide company, made up at 5 μg/μL in RNase-free TE/H$_2$O.

Equipment

Heat block set at 37°C and 100°C
Lightproof box (see Step 11)
MinElute kit (QIAGEN)
Zymo columns (Zymo Research, catalog no. D3024)

METHOD

Genomic DNA Labeling

1. For each array, combine 4 μg of genomic DNA, 10 μg of random hexamer (N6), and dH$_2$O to a final volume of 42 μL. Incubate for 5 min at 100°C.

724

2. Cool the tube(s) quickly on ice for 10 min.

3. Pulse-centrifuge to bring the condensate to the bottom of the tube.

4. On ice, add:

Klenow buffer (10×)	5 μL
aa-dUTP/dNTPs (3 mM)	2 μL
Klenow (40 U/μL)	0.8 μL

Pipette to mix. Incubate for 2 h at 37°C.

5. Add 5 μL of 0.5 M EDTA (pH 8.0) to stop the reaction.

Purification of Random Prime-Labeled DNA

6. Add 1 mL of Binding buffer to the reaction. Mix well, and load half (527 μL) of the reaction mixture onto each of two Zymo columns. Centrifuge the columns for 10 sec at maximum speed.

   The large volume of Binding buffer is necessary to precipitate single-stranded DNA.

7. Discard the flowthrough. Wash the column with 200 μL of Wash buffer. Centrifuge for 1 min at maximum speed.

8. Repeat Step 7.

9. Add 10 μL of 50 mM $NaHCO_3$ (pH 9.0) to each column filter. Incubate for 5 min at room temperature.

10. Elute the DNA from the column by centrifuging for 1 min at maximum speed.

11. Couple cyanine dyes to the aa-dUTP-incorporated cDNAs by adding the eluted material directly to 3 μL of Cy5 dye or Cy3 dye resuspended in DMSO. Incubate for 1 h to overnight at room temperature in a lightproof box or a drawer.

Purification of Labeled Genomic DNA

12. Purify the cDNA using a MinElute kit, following the manufacturer's instructions.

RECIPE

It is essential that you consult the appropriate Material Safety Data Sheets and your institution's Environmental Health and Safety Office for proper handling of equipment and hazardous materials used in this protocol.

aa-dUTP/dNTP Mixture (50×)

Dissolve 1 mg of aminoallyl-dUTP (Sigma-Aldrich) with:

Reagent	Quantity	Final concentration
dATP (100 mM)	32.0 μL	12.5 mM
dGTP (100 mM)	32.0 μL	12.5 mM
dCTP (100 mM)	32.0 μL	12.5 mM
dTTP (100 mM)	12.7 μL	5 mM
dH_2O	19.3 μL	

Blocking Polylysines on Homemade Microarrays

Homemade microarrays are printed on polylysine-coated slides. The lysines form a positively charged surface that can bind nonspecifically to the acidic nucleic acids during hybridization, resulting in significant background fluorescence. Thus, a key step in microarray processing is blocking all of the surface lysines not associated with the oligonucleotides in the microarray spots. The ε-amino group of lysine is succinylated by reacting with succinic anhydride (Fig. 1). Because anhydrides readily hydrolyze in water, use only fresh reagents that have not had the opportunity to absorb much water.

The procedure is straightforward. Microarrays are, if necessary, rehydrated; excess liquid is removed by drying at a moderate temperature; and the succinylation reaction is performed. After the reaction is complete, the slides are washed and dried with ethanol, at which point they are ready for hybridization or they can be stored in a desiccator. Rehydration is necessary if the microarrays were desiccated after printing (see Protocol 1). When microarrays are stored in a desiccator, the spots typically dry out to form "rings." Rehydration is important to restore the full spots.

MATERIALS

It is essential that you consult the appropriate Material Safety Data Sheets and your institution's Environmental Health and Safety Office for proper handling of equipment and hazardous materials used in this protocol.

Recipes for reagents specific to this protocol, marked <R>, are provided at the end of the protocol. See Appendix 1 for recipes for commonly used stock solutions, buffers, and reagents, marked <A>. Dilute stock solutions to the appropriate concentrations.

Reagents

Ethanol (95%) (see Step 11)
1-Methyl-2-pyrrolidinone
    Use only HPLC grade. If it has become slightly yellowed, then it is no longer usable, having absorbed excess water.

Microarrays printed onto glass slides (either homemade [Protocol 1] or purchased)

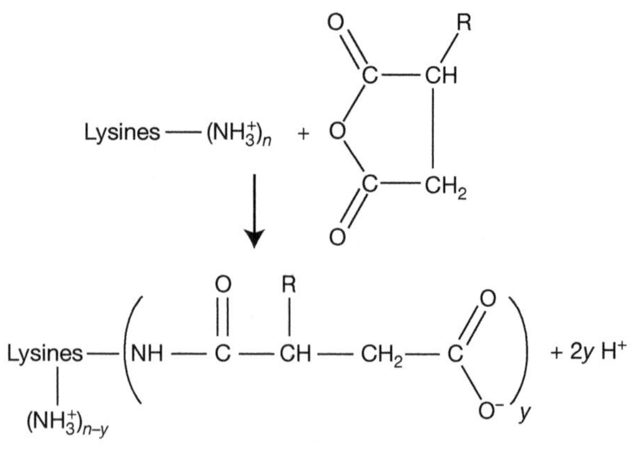

FIGURE 1. **Succinylation of lysines.** (Reproduced from Simpson 2003, with permission from Cold Spring Harbor Laboratory Press.)

Sodium borate (1 M, pH 8.0)

SSC (0.5×) <A>

Succinic anhydride

> *The stock bottle of solid succinic anhydride should be stored under desiccation and vacuum (or under nitrogen).*

▲ *Do not use if exposed to moisture!*

Equipment

Beakers (500 mL and 4 L)

Centrifuge, fitted with microtiter plate carrier

Glass-etching pen

Glass slide racks and wash chambers (e.g., Thermo Scientific Fisher, catalog no. NC9516192)

Gloves, powder-free

Heat block set at 80°C (see Step 5)

Heating plate set at 80°C

Humid chamber, for standard-size glass slides (e.g., Sigma-Aldrich, catalog no. H6644)

Microarray slide box, plastic

Orbital shaker

Stir bar that fits a 500-mL beaker

Water bath set at 80°C

METHOD

1. Select 15 slides for postprocessing. Handle all slides with powder-free gloves. Determine the correct orientation of each slide. With a glass-etching pen, lightly mark the boundaries of the array on the backside of the slide.

   > *Marking the array boundary is important because after processing, the arrays will not be visible.*

2. Add enough distilled water to a large 4-L beaker so that a slide rack will be completely submerged when placed inside. Place the beaker on a heating plate and heat to 80°C.

Rehydrating the Microarrays

3. Fill the bottom of a humidifying chamber with 0.5× SSC. Suspend the arrays face-up over the 0.5× SSC, and cover the chamber with a lid.

4. Rehydrate until all of the microarray spots glisten (usually 15 min at room temperature).

   > *Allow the spots to swell slightly, but do not let them run into each other. Insufficient hydration results in less DNA bound within a spot, and too much hydration will cause spots to run together.*

5. Dry each array by placing each slide, with the DNA side facing up, for 3 sec on an inverted heat block set at 80°C.

6. Place the arrays in a slide rack. Place the slide rack into an empty slide chamber that is sitting on an orbital shaker.

Blocking Free Lysines on the Microarray Slides

7. Prepare the blocking solution by measuring 335 mL of 1-methyl-2-pyrrolidinone into a clean, dry 500-mL beaker. Dissolve 5.5 g of succinic anhydride using a stir bar.

8. *Immediately* after the succinic anhydride dissolves, mix in 15 mL of 1 M sodium borate (pH 8.0). Quickly pour the buffered blocking solution into a clean, dry glass slide dish.

9. Plunge the slides rapidly into the blocking solution, and vigorously shake the slide rack manually, keeping the slides submerged at all times. After 30 sec, place a lid on the glass box, and shake gently on the orbital shaker for 15 min.

10. Drain excess blocking solution from the slides for ~5 sec. Submerge the slide rack into an 80°C water bath. Gently "swish" the rack back and forth under the water for a few seconds. Incubate for 60 sec.

11. Quickly transfer the rack to a glass dish of 95% ethanol, and plunge to mix.

    Make sure that the ethanol is crystal clear. Do not use if it contains particulates or is cloudy.

12. Take the entire glass dish with slide rack still submerged to a bench-top centrifuge. (Be sure to have an equivalently loaded slide rack ready to serve as a balance in the centrifuge.) Drain excess ethanol off the slides for ~5 sec. Quickly, place the slide rack onto a microtiter plate carrier, and centrifuge in a bench-top centrifuge for 3 min at 550 rpm.

13. After centrifugation, the slides should be clean and dry. Remove the slides from the rack and store them in a plastic microscope slide box. Arrays may be used immediately, or can be stored for at least 3–4 mo in a desiccator at room temperature.

REFERENCE

Simpson RJ. 2003. Peptide mapping and sequence analysis of gel-resolved proteins. In *Proteins and proteomics*, pp. 343–424. Cold Spring Harbor Laboratory Press, Cold Spring Harbor, NY.

Hybridization to Homemade Microarrays

Competitive hybridization of labeled probes to a microarray is conceptually similar to other hybridization methods, such as Southern blotting. For massively multiplexed microarrays, the adoption of two-color hybridization schemes has been a significant advance. The use of two colors—typically Cy3- and Cy5-labeled nucleic acids—makes it possible to control for factors that affect hybridization intensity, including the number of labeled nucleotides and the T_m of each oligonucleotide. Thus, the difference in intensity among spots on a microarray can be quantified and analyzed to assess biological phenomena, like changes in gene expression or details of transcript structure.

Hybridization is conceptually straightforward—"cold" (nonfluorescent) blocking nucleotide is added to the mixed probe material, Hybridization buffer is added, and the mixture is applied to the microarray surface. Hybridization occurs overnight, after which the microarray is washed and scanned. The only technically challenging aspects of this protocol are application of the probe solution to the microarray surface and placement of the coverslip (Steps 8 and 9 below). Practice these steps, applying Hybridization buffer containing salmon sperm DNA, before using your fluorescently labeled DNA in an actual experiment. Develop the ability to work rapidly, and avoid introducing bubbles or scratching the array surface.

Many hybridization chambers are available commercially; these typically consist of a two-part metal chamber housing an internal cavity large enough to hold a microarray slide. Often there are small depressions in the internal cavity, which can hold extra buffer that will humidify the chamber. There is also a rubber gasket that forms a watertight seal around the chamber. Other hybridization systems include the "Maui" mixer, which moves hybridization solution back and forth over the array. For the user of homemade microarrays, however, these more expensive hybridization systems are an unnecessary expense.

MATERIALS

It is essential that you consult the appropriate Material Safety Data Sheets and your institution's Environmental Health and Safety Office for proper handling of equipment and hazardous materials used in this protocol.

Recipes for reagents specific to this protocol, marked <R>, are provided at the end of the protocol. See Appendix 1 for recipes for commonly used stock solutions, buffers, and reagents, marked <A>. Dilute stock solutions to the appropriate concentrations.

Reagents

$C_o t$-1 DNA (GIBCO)
> $C_o t$-1 DNA is available commercially at 1 µg/µL. Before hybridization, concentrate $C_o t$-1 DNA from 1 µg/µL to 10 µg/µL in a rotary vacuum device.

Cy5- and Cy3-labeled nucleic acids (from Protocol 5, 6, 7, or 8)

HEPES (1 M, pH 7.0)

Microarrays printed onto glass slides (either homemade [Protocol 1] or purchased)
> If homemade, slides must be blocked as per Protocol 9 before hybridization.

Poly(A) RNA (10 µg/µL) (Sigma-Aldrich, catalog no. P9403)

SDS (10%)

SSC (20×) <A>

Yeast tRNA (10 µg/µL) (GIBCO, catalog no. 15401-011)

Equipment

Coverslips
  Use either 22×60 regular thin coverslips or 22×60 Erie M-series lifter slips.
Dishes for washing microarrays (see Step 13)
Gloves, powder-free
Heat block set at 95°C–100°C
Hybridization chambers
Lightproof box
Microarray scanner
  When using homemade microarrays, the standard scanner is the GenePix 4000B (Molecular Devices).
Microarray slide box
Paper towels
Water bath set at 65°C

METHOD

Preparation of the Hybridization Solution

If you have already combined the Cy dyes and added blocking material and buffer, skip to Step 4.

1. Combine the following blocking nucleic acids:

Cot1 human DNA (10 μg/μL)	2 μL
Poly(A) RNA (10 μg/μL)	2 μL
tRNA (10 μg/μL)	2 μL

2. Combine Cy5- and Cy3-labeled nucleic acids.

3. Prepare the complete hybridization solution as shown in the table below. To avoid introducing bubbles into the solution, do not vortex after adding SDS.

	Coverslip size	
	22×60 regular thin coverslips	22×60 Erie M-series Lifter Slips
Cy5- and Cy3-labeled nucleic acids	21.16 μL	36.68 μL
Cot, poly(A), tRNA mixture	6 μL	6 μL
SSC (20×)	5.95 μL	9.35 μL
SDS (10%)	1.05 μL	1.65 μL
HEPES (1 M, pH 7.0)	0.84 μL	1.32 μL

   For other coverslip sizes, adjust the volumes to maintain the same SSC, SDS, and HEPES concentrations. However, the blocking mix [Cot, poly(A), and tRNA] should be 6 μL in all cases.
   HEPES is recommended for all probes.

4. Denature the probe by heating the hybridization solution in a water-filled heat block for 2 min at 95°C–100°C.

5. Let the probe sit for 10 min in the dark at room temperature.

6. While the probe is sitting at room temperature, set up the necessary number of hybridization chambers. Open each chamber on a clean, flat surface, and place a microarray slide into a chamber with the array side facing up.

7. Centrifuge the solution at 14,000 rpm for 5 min at room temperature.

Application of the Hybridization Solution to the Microarrays

8. Carefully pipette the probe solution as a single drop onto one end of the microarray surface. Avoid creating any bubbles during pipetting, and be careful not to touch the microarray surface with the pipette tip (Fig. 1).

    Leave ~2 μL of probe behind in the tube. If the probe precipitates upon application to the slide, see Troubleshooting.

9. Carefully apply a coverslip by placing one edge of the coverslip on the slide near the probe and slowly lowering the other edge, using another coverslip as a lever and wedge to lower it (Fig. 1).

10. Close the chamber, and immediately submerge it in a 65°C water bath.

    Be careful not to tilt the chamber. Metal tongs may be used to place the hybridization chamber into the water bath.

11. If hybridizing multiple arrays, repeat Steps 8–10.

    Be quick and efficient because the probes should not sit at room temperature for widely varying times. Try to have all the probes onto slides within ~10–15 min.

12. Incubate the chambers for 16–20 h at 65°C.

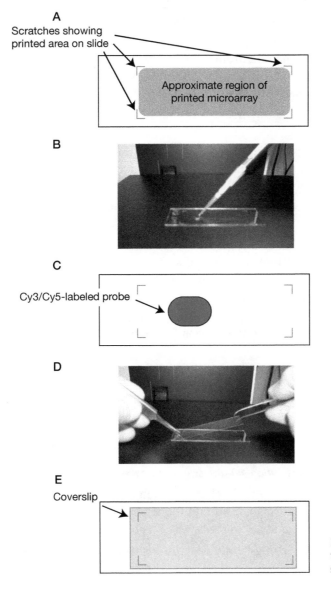

FIGURE 1. Technique for proper placement of the probe solution and coverslip on a microarray slide.

Washing the Microarrays and Setting Up the Scanner

13. Prepare dishes with the following solutions:

    Wash 1A and Wash 1B: 1× SSC containing 0.03% SDS
    Wash 2: 0.2× SSC
    Wash 3: 0.05× SSC

	Wash 1A	Wash 1B	Wash 2	Wash 3
SSC (20×)	20 mL	20 mL	4 mL	1 mL
SDS (10%)	1.2 mL	1.2 mL	—	—
dH₂O	379 mL	379 mL	396 mL	399 mL
Temperature	Room temperature	Room temperature	37°C	Room temperature

14. If there is one, turn on the ozone scrubber connected to the scanner at least 15 min before scanning the arrays.

    This is particularly useful in summer months when ozone levels are high. Bleaching of Cy5 dye becomes a significant problem in urban environments in the summer. If you choose not to use an ozone scrubber, consider performing hybridizations on cool or rainy days.

15. Turn on the scanner, letting it warm up for at least 15 min to allow the laser outputs to stabilize.

16. Launch the GenePix Pro software, and let it connect to the scanner and the network hardware key.

17. Remove a hybridization chamber from the water bath. Working quickly, dry the chamber with a paper towel, open the chamber, remove the slide, and using your gloved hand, tap the slide against the bottom of the Wash 1A dish until the coverslip gently slides off the slide.

18. Transfer the slide to a presubmerged rack in Wash 1B. Keep the slide in Wash 1B until all arrays have been transferred.

19. Quickly transfer the rack from Wash 1B to Wash 2 (tilting the rack back and forth a couple of times to remove excess wash solution). Plunge the rack up and down several times in Wash 2. Incubate for 5 min with occasional plunging up and down.

20. Quickly transfer the rack from Wash 2 to Wash 3 (tilting the rack back and forth a couple of times to remove excess wash solution). Plunge the rack up and down several times in Wash 3. Incubate for 5 min with occasional plunging up and down.

21. As quickly as possible, transfer the rack of slides from Wash 3 to a tabletop centrifuge, with paper towels under the rack, and centrifuge at 500–600 rpm for 5 min.

22. Place the microarrays in a lightproof box, and begin scanning immediately.

    Typically, four to five arrays are washed at a time. If more than five were hybridized, then the next set of slides should be washed as the last array from the first batch is being scanned.

Scanning the Microarrays

23. Open the scanner door, and insert a slide with the array features *facing down* and the barcode/label closest to the edge of the scanner.

24. Run a preview scan using the double-arrow button on top of the right-hand toolbar to visualize the array.

    This performs a low-resolution (40-μm) scan so that the location of array features can be visualized.

25. After the preview scan is completed, select the "Scan Area" button in the left-hand Tools group. Use the mouse to draw a rectangle around the region containing features.

    This limits the data scan to just this area, which reduces the time needed for each scan.

26. Click on the "Hardware Settings" button on the right-hand side of the window to bring up the Settings box.

27. Use the "Auto Scale Brightness/Contrast" button on the left-hand side to adjust the brightness and contrast of the image (both colors will be affected). Switch between the red and green channels by clicking on the respective radio buttons on the top left-hand side of the image tab.

28. Set the Pixel Resolution to 10 μm, then start the data scan. Zoom into the top half of the scanned image, and by eyeballing the image, adjust the PMT gain settings in both the red and green channels so that the average signal intensities balance out.

> *This takes some practice, although Step 29 describes how to do this with the aid of the intensity histograms. By eye, the goal is to balance the array so that the color yellow dominates, with roughly equal numbers of green and red spots. An array with mostly green spots will need the red channel PMT increased, and vice versa.*
>
> *Normally, the PMT gain for the green channel (532 nm) is around 100 less than the red channel (635 nm). Adjust the settings so that any "landing lights" (bright spots printed near the "top" of the array corresponding to highly expressed genes, etc.) and some positive control spots on the array are saturated, but not many of the other features. Saturated pixels are drawn as white on the image.*
>
> *Overall, the goal is to minimize the length of time spent scanning the array, and because some color imbalance can be normalized out later, it is more important to be fast than to be absolutely precise with regard to color balance.*

29. Alternatively (to the previous step), view the "Histogram" tab to check on the red/green ratio. Set the Min and Max Intensity fields in the "Image Balance" area on the left-hand side to 500 and 65,530, respectively. The goal is for the Count Ratio field to be ~1.0, although this is not as important as having the two histogram curves as close to overlapping as possible.

30. Once the PMT Gains for both channels are balanced, stop the scan by hitting the red "Stop Scan" button on the right-hand side of the window. Change the pixel resolution in the "Hardware Settings" window to 5 μm, and ensure that the Lines to Average field is at 1. (More will not be detrimental to the scanned image, but the scanning will take much longer.)

31. Start the scan again with the new settings. Record the PMT gains for each array in an Excel spreadsheet, along with any special observations/notes for the slide (e.g., the red signal is very dim, etc.). Save the spreadsheet in the same folder as the image files.

32. When the entire array is scanned, click the "File" button on the right-hand side of the window. Select "Save Images," and from the Save As Type field, select "Single-Image TIFF Files." Make sure that both wavelengths are selected. Name the arrays with the identifier printed on the label, then hit the "Save" button.

> *There should be two TIFF (*.tif) files, one with the data from the green channel (532 nm) and one with the data from the red channel (635 nm).*

33. Scan the next array, repeating Steps 23–32. Balancing the two PMTs must be done for every microarray, as different RNA or DNA samples label with different efficiency.

34. After all of the slides have been scanned, quit the GenePix Pro program, and turn off the scanner.

Gridding the Microarrays

35. Open the GenePix software, and open the image file in question.

36. Open the ".gal" file, which describes what DNA is found in each spot. This file will either be provided by the microarray manufacturer or it can be generated using the ArrayList generator found in GenePix.

37. Align each pattern block (corresponding to one print pin) over the image so that image spots are roughly aligned with circles in the .gal file.

38. Find features for each block by pressing [F5]. GenePix will move .gal file circles to cover the closest spots and will also flag spots where the intensity is too low.

39. For each block, scan the entire sector by eye, and manually flag spots that are scratched or appear to be fluorescent dust rather than DNA.

40. After all blocks have been flagged, hit the "Analyze" button to extract the .gpr file.

41. At this point, either save the .gpr output file or choose "Flag features" to automatically impose thresholds.

    We typically use this to flag all spots with a signal-to-noise ratio in each channel (532, 635) of <3.

42. Save the .gpr file.

TROUBLESHOOTING

Problem (Step 8): When adding the hybridization solution to the microarray slide, the probe precipitates.

Solution: With some home-printed microarrays, the labeled probe can precipitate when added to the slide at room temperature. This leads to a "pin prick" morphology in which the precipitated probe fills the depressions created during the printing process. To mitigate this phenomenon, place the hybridization chamber onto a 55°C–60°C heat block before the addition of the hybridization solution, and add hybridization solution while the chamber sits on the heat block.

CHAPTER 11

DNA Sequencing

Elaine Mardis[1] and W. Richard McCombie[2]

[1]The Genome Institute, Washington University School of Medicine, St. Louis, Missouri 63108
[2]Cold Spring Harbor Laboratory, Cold Spring Harbor, New York 11724

DNA SEQUENCING HAS BECOME AN ESSENTIAL ELEMENT IN THE toolkit of biologists. Since the last version of this manual, the field has been revolutionized by "next-generation sequencing" (NGS). This chapter includes a section on fluorescent sequencing by capillary instruments, still commonly used, but the primary focus is on next-generation sequencing. Our assumptions are that readers typically will want to know how to prepare samples to be sent out to core labs for sequencing—or that they will have instrument-specific training. We have therefore chosen to minimize descriptions of instrument operation and concentrate instead on sequencing strategies, and to describe the various types of samples and their preparation.

HISTORY OF SANGER/DIDEOXY DNA SEQUENCING

DNA sequencing was first performed using laborious and low-yielding chemical cleavage reactions. Among the first targets for this type of sequencing were the cohesive ends of the λ phage genome (Wu 1970; Wu et al. 1970; Wu and Taylor 1971). This method required weeks or months to produce the sequence of a few nucleotides. In the roughly 40 years since that work, the technology has advanced to the point where 3-billion-base human genomes can be sequenced in a few weeks, and the largest centers measure their data output in terms of trillions of bases per month. This rapid change has transformed areas in biology as disparate as microbiology and human genetics.

The first DNA-sequencing techniques were the enzymatic method of Sanger et al. (1977a) and the chemical degradation method of Maxam and Gilbert (1977). Although very different in principle, both of these methods generate populations of oligonucleotides that differ in length by 1 nucleotide. The oligonucleotides in the population all begin from the same fixed point but terminate at each nucleotide in the fragment of interest. In the simplest case, four populations are created that terminate at each A, G, C, or T residue, respectively. The termination points are nucleotide specific but occur randomly along the length of the target DNA. Each of the four populations therefore consists of a nested set of fragments whose lengths are determined by the distribution of a particular base along the length of the original DNA. These populations of fragments are resolved by electrophoresis under conditions that discriminate between individual DNAs differing in length by as little as 1 nucleotide (Sanger and Coulson 1978). When the four populations are loaded into adjacent lanes of a sequencing gel, the order of nucleotides along the DNA can be read directly from an autoradiographic image of the gel.

Dideoxy Method of DNA Sequencing

The Sanger technique uses controlled synthesis of DNA to generate fragments that terminate at specific points along the target sequence. Sanger's breakthrough came with the introduction into the synthesis reaction of dideoxynucleoside triphosphates (ddNTPs)—nucleotide analogs that (1) could be substituted for deoxynucleotides and incorporated at random positions during template-directed copying of a DNA strand by a DNA polymerase, and (2) upon incorporation, could efficiently terminate DNA synthesis in a base-specific manner (Sanger et al. 1977a). By using ddNTPs corresponding to the conventional nucleotide bases in separate synthesis reactions, four populations of oligonucleotides could be generated that terminate at every position in the template strand. A given sequencing scheme therefore involves the setup of four separate sequencing reactions. Each reaction includes all the materials required to synthesize a new DNA strand (primer, template, polymerase, and deoxynucleotide triphosphates), and each reaction includes one dideoxynucleotide. Thus, for example, the A reaction contains a mixture of deoxy and dideoxy adenosine, as well as the deoxy versions of the other three bases; the C reaction has both deoxy and dideoxy C, but deoxy versions only of the other three bases, and so on (see Fig. 1). The reactants are then incubated to allow new DNA strands to be synthesized on the template DNA. If we follow the A reaction, for example, it is clear that, for some fraction of the template molecules, for any point at which an A should be inserted, a dideoxy A might instead be added, thereby terminating synthesis of that particular chain. Upon completion, the reaction will contain, at least in theory, a population of molecules nearly all of which were terminated by the inclusion of a dideoxy nucleotide. The DNA bands in the four reactions can be separated into "ladders" by gel electrophoresis, with the smallest band corresponding to

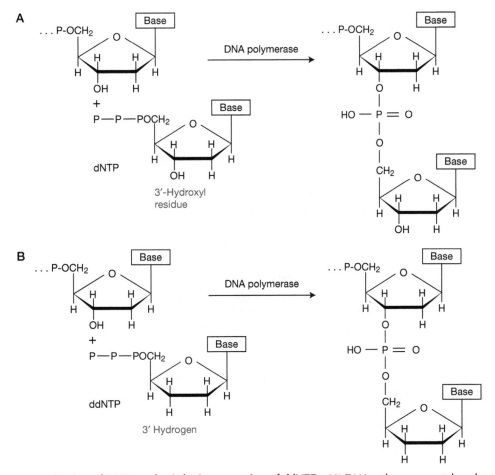

FIGURE 1. **Termination of DNA synthesis by incorporation of ddNTPs.** (*A*) DNA polymerase-catalyzed esterification of the normal dNTP with the 3′-terminal nucleotide of DNA. The reaction, which extends the length of the primer by a single nucleotide, can be recapitulated until all of the bases in the template strand (not shown) are paired with the newly synthesized strand. (*B*) Corresponding reaction with a ddNTP, which lacks a 3′-hydroxyl residue. The ddNTP can be incorporated into the growing chain by DNA polymerase, where it acts as a terminator because it lacks the 3′-hydroxyl group required for formation of further 5′→3′ phosphodiester bonds.

the first base incorporated (blocked in molecules in which the first base incorporated is a dideoxy-nucleotide) and each larger band representing a fragment that is 1 nucleotide longer. The entire DNA sequence can be deduced from the order of bands in the four lanes.

The Sanger DNA sequencing method had several inherent limitations, the most significant being the difficulty of generating the single-stranded DNAs used as templates in the reaction. Reading the gels is also problematic because they are run under voltages that make it difficult to avoid heat-related distortion. This distortion may cause shifting within the fragment separation pattern. However, over the years, ways were found to manage these difficulties, and by the beginning of the 1980s, the Sanger method had become the method of choice for DNA sequencing in general, and in particular for the human genome (Lander et al. 2001; Venter et al. 2001).

Chemical Method of DNA Sequencing

Whereas the dideoxy-mediated chain-termination method grew logically from Sanger's earlier work on protein and RNA sequencing, the chemical method of DNA sequencing was discovered accidentally in Walter Gilbert's laboratory (see Gilbert 1981). In the late 1960s, and early 1970s, Gilbert's work focused on the interaction between the *lac* repressor and the *lac* operator in vitro. He isolated the fragment of DNA protected by the repressor and worked out its sequence by copying the DNA into RNA and then applying the techniques of RNA sequencing developed in Sanger's laboratory (Gilbert and Maxam 1973). This approach was transformed when Andrei Mirzabekov, a Russian scientist, visited Harvard in 1975. Working at the Institute of Molecular Biology, Academy of Science of the U.S.S.R. in Moscow, Mirzabekov had shown that binding of histones and antibiotics to DNA could inhibit methylation of purines by dimethylsulfate. He urged Gilbert to discover whether binding of the repressor by blocking methylation of the operator DNA could identify which bases in the sequence were protected. During discussion of this problem with Mirzabekov, Allan Maxam, and Jay Gralla over a lunch, Gilbert had the idea that ultimately became the basis of the chemical sequencing technique. If a DNA fragment could be labeled at one end with a radioactive phosphate group and then subjected to partial degradation with base-specific chemical reactions, the position of each base in the chain could be determined by measuring the distance between the labeled end and the point of breakage. By 1976, it was clear that the method could be used to locate adenine and guanine residues in short fragments (~40 nucleotides) of DNA. Knowing the locations of the two purine bases on each of the two complementary strands of DNA, one could, in principle, deduce a complete DNA sequence. The next year was spent in developing differential methods to cleave cytosine and thymine residues and thus to position two pyrimidine bases in a DNA sequence. By 1977, Maxam and Gilbert were able to publish a description of the complete sequencing technique (Maxam and Gilbert 1977).

For the next few years, the chemical method of sequencing was continuously refined. The number of base-specific cleavage reactions increased, and their efficiency and discrimination had gradually improved to the point where 200–400 bases could be routinely read from the point of labeling. However, none of these improvements was sufficient to save chemical sequencing from decline. Toxic chemicals, large amounts of radioactivity, sequencing gels that were sometimes ambiguous and all too often ugly, and the lack of automated methods to prepare end-labeled templates have all contributed significantly to the eclipse of the method by the Sanger chain-termination technique.

Evolution of Dideoxy Chain Termination Sequencing: Critical Points

Generation of Single-Stranded Templates for DNA Sequencing

One of the most difficult problems in the original dideoxy sequencing method was the generation of single-stranded DNA templates as a starting point in the procedure. In fact, as a result of this drawback, chemical cleavage sequencing was often used in the very early days of modern sequencing. However, in the late 1970s, Joachim Messing and colleagues at the Max Planck Institute in Munich saw that various biological features of the single-stranded bacteriophage M13 could be exploited for cloning and sequencing of foreign DNA. First, it seemed likely that the phage genome could accommodate additional sequences without compromising growth of the phage. Second, during its life cycle, the phage genome also existed in a double-stranded form—upon infection of *Escherichia coli*, the single-stranded phage undergoes replication in a double-stranded form and then generates single-stranded replicons that bud off the infected cell in high numbers. Thus, Messing reasoned, the double-stranded form of the phage could be isolated and used for cloning foreign genes and the resulting single-stranded form of the recombinant could be used as template in dideoxy-sequencing reactions. A phage vector was therefore engineered to include part of the *E. coli lac* operon to allow for screening of inserted DNA (see the information panels α-Complementation and X-Gal in Chapter 1).

Shotgun Sequencing

The next major advance, accomplished by a group at the MRC laboratory in Cambridge, United Kingdom, was the creation of tools and protocols to generate and use random clones. Initially, the Sanger sequencing method was limited to runs of a few hundred bases. Although a vast improvement over previous methods, this limitation required that most DNA of interest be broken into smaller fragments for sequencing. Initially this was done in a directed fashion—by determining the restriction maps of the targets using multiple restriction fragments, then cloning and sequencing the individual fragments. Bart Barrell and colleagues at Cambridge realized that they could skip the laborious mapping and directed cloning steps by cleaving the target DNA into randomly overlapping fragments by any of various methods, and cloning these into M13 vectors. After sequencing the random fragments, they used a computer program developed by Rodger Staden, also in Cambridge, to determine the overlaps of the fragments from the sequence information itself, without the need for any other mapping information. Thus, the development of "shotgun sequencing" marked the entry of bioinformatics into DNA sequencing.

λ: The First "Genome" Sequence

The initial Sanger sequencing publication demonstrated the method by sequencing the genome of a phage (φX174). But not much was made of this at the time (Sanger et al. 1977b), perhaps because little was known of φX174 biology. In subsequent years, sequencing was focused on defined targets such as small genes, cDNAs, or plasmids such as pBR322 (Sutcliffe 1978, 1979). However, in 1982, making use of the advances of shotgun sequencing and of M13 vectors, the Sanger group produced a technical and biological tour de force—the first genome sequence of an entity important in biological research—bacteriophage λ (Sanger et al. 1982). The significance was twofold. First, the DNA template was about 10 times larger than any previous sequencing project (the λ genome is ∼50 kb); more importantly, the achievement linked the rich biology of λ to the newly produced genome sequence (Sanger et al. 1982). The power of knowing a genome sequence in understanding the biology of an organism—the central principle of genomics—was clearly shown.

Fluorescent Sequencing

The heat generated by passing high voltages through gels to separate DNA fragments at nucleotide resolution had the unfortunate side effect of generating distortions in fragment migration (owing chiefly to uneven heat dissipation). These distortions made reading the gels very difficult to automate because expert human judgment was required to realign the bands on the gel, by compensating for lane-to-lane differences in migration. This problem and several others germane to the use of radioactivity were eliminated when Lee Hood's laboratory at the California Institute of Technology (Caltech) pioneered a novel, fluorescent-based version of Sanger sequencing (Smith et al. 1986).

Their method used a set of four fluorescently labeled primers. Each of the four fluorescent labels emitted light at a different wavelength when exposed to the laser in the sequencing device. Although the emission spectra of the dyes overlapped somewhat, basically each dye could be differentiated

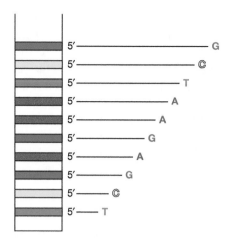

FIGURE 2. **Fluorescent DNA sequencing.**
See text for details.

from the others according to the wavelength of its fluorescent output. With this chemistry, four separate reactions (A, C, T, G) were performed, each with a specific dye-labeled primer and one of the dideoxy nucleotides (Smith et al. 1986). Here, because the four reactions could be differentiated by the fluorescent primer on each extension product, they could be mixed after the reactions were complete and loaded into a single lane for resolution by gel electrophoresis (Smith et al. 1986). Automated software generated the sequence by reading the ladder of colors in a single lane (Fig. 2). Therefore, assuming that a red band in the gel represents all molecules in which dideoxythymidine terminated the reaction, if the first (or shortest) fragment in the lane is red, that band represents a T. Similarly, the next fragment (1 base longer) seen as a yellow band indicates a dideoxycytosine termination, hence a sequence of TC (etc.). In addition to eliminating the need for radioactive materials, the major advance of the fluorescent sequencing was the ability to automate the gel-reading/ base-calling process and to couple it with the separation step. Running all four reactions for a given template in the same lane eliminated the impact of fragment migration distortion in a given sample, and software allowed detection of the order of the alternating four color emissions as they passed the detection system. These advances in automation transformed the field and made possible the sequencing of the human genome.

Dye Terminator Sequencing

Although fluorescent sequencing had great potential, it also had limitations. One of the most serious problems was the need to use a set of primers that would initiate sequencing only from within the cloning vectors; because fluorescent dyes were included in the primer only a few primers were available. With radioactive sequencing, the approach had been to perform a shotgun sequencing phase of the project, assemble the raw data, and then move to a "directed" or finishing phase. In this last phase, sequencing was initiated from the ends of the contiguous fragments (contigs), assembled in the previous phase, by using custom oligonucleotide primers to extend into the gap and finish the complete sequence. This approach was first shown on a large scale by Tom Caskey and colleagues using radioisotope-based sequencing of the human *HGPRT* locus (Edwards et al. 1990). This strategy would eventually become the backbone of the human genome project, but, at that time, finishing was not yet possible with fluorescent-based sequencing. Thus, early users of the new automated systems reverted back to sequencing either restriction fragments (Gocayne et al. 1987) or sets of fragments generated by controlled exonuclease deletion of a larger fragment (Henikoff 1990) to avoid the unfinished regions or gaps left by shotgun sequencing. However, the introduction of fluorescently labeled dideoxynucleotides by Applied Biosystems (ABI), enabled fluorescent sequencing to be performed with any primer. Furthermore, because the fragments were now labeled by a specific dye on a specific dideoxy molecule, all four sequencing reactions could be performed in a single tube. Reducing the reactions from four tubes to one would become very important as the number of templates to be sequenced for the Human Genome Project approached the tens of millions.

Cycle Sequencing

Since the earliest days of radioisotope-based Sanger sequencing, there had been a bit of alchemy about the polymerase, or polymerases, used. Because each polymerase had its own characteristics, they were sometimes mixed or a second polymerase was added as a "chase" to improve the reaction. However, all of these mixes were alike in basic activity—they all catalyzed the extension of a primer bound to a template. The polymerase chain reaction (PCR) came into general use by about 1989 (Saiki et al. 1985; Mullis and Faloona 1987), and multiple heat-resistant polymerases were developed to perform PCR. One of these enzymes was applied to sequencing by using multiple cycles of extension, heating (to denature the nascent strand from the template), renaturation, extension, and so forth. This approach was originally performed using radioactive sequencing (Carothers et al. 1989; Murray 1989; Smith et al. 1990). The idea, of course, was that multiple rounds of extension would boost the signal of the sequencing reaction—and this is exactly what happened. The strategy was rapidly picked up and adapted to fluorescent dye terminator sequencing (McCombie et al. 1992; Craxton 1993). Cycle sequencing benefitted radioisotope-based sequencing, but the benefit to fluorescent sequencing was even greater because fluorescent sequencing had a much lower signal to begin with. Cycle sequencing allowed for a significant reduction in the requirements for template quantity and quality and enabled robust sequencing of double-stranded templates—all features that have become part of the current cycle-sequencing chemistry.

Capillary Sequencing

From its invention in the mid 1970s up until the mid 1990s, Sanger DNA sequencing was performed on polyacrylamide slab gels. Fluorescent sequencing had largely eliminated the worst problems of slab gels, such as those arising from heat distortion and the need for radioactivity/autoradiography, but these gels still had their limitations. One issue was actually in making the gels. Long gels allowed longer read lengths but were harder to pour and required modified instrumentation and software to run them. In addition, fluorescent sequencing required that the area where the gel was scanned be optically clear—not a minor chore because sequencing gels contain a high amount of urea. In addition, to pack more samples per gel, the lanes became smaller and thus were more difficult to load. Lastly, the size of the gels required rather long run times, typically ~8 h for long ABI gels.

The development of capillary sequencing solved these problems and introduced additional reductions in time and error. ABI released two major versions of their capillary sequencer in the mid 1990s. The first of these, the ABI3700, was the instrument used to generate most of the human (Lander et al. 2001; Venter et al. 2001) and mouse genome reference sequences (Waterston et al. 2002). The second, the ABI3730XL, was introduced into laboratories a year or so later and remains in active use today. The Sanger sequencing protocols provided in this chapter are optimized for this instrument. Capillaries provide several benefits over slab gels for sequencing. Each capillary is a separate physical unit and thus can be loaded independently, making automated loading straightforward. The ABI3730 is loaded with 96-well plates containing completed sequencing reactions, and an internal magazine presents plates to a robotic stage that lowers the capillary tips into the wells simultaneously and effects electrokinetic injection of the samples onto the capillary matrix. The heat dissipation properties of capillaries allow them to achieve single-nucleotide separation over a larger number of nucleotides (and hence achieve longer reads) than is possible with a slab gel. Coupled with the automated loading and automated replacement of the capillary matrix between runs, this allows multiple 96-well plates to be loaded into the instrument magazine and run throughout a 24-h period with no human intervention. These characteristics make the instrument ideal for departmental or institutional core facilities.

Improved Assembly Tools

A full description of the development of bioinformatics tools used for sequence analysis is outside the scope of this chapter; however, it is important at least to note the salient developments and highlight their importance. As we described above, the most effective early strategy for large sequencing projects was a shotgun sequencing phase followed by a directed sequencing, or finishing, phase to fill any gaps in the sequence (Edwards et al. 1990). To be performed efficiently at scale, this method required software to align the shotgun sequences and to allow manual editing of the resulting contigs to resolve discrepancies between reads derived from individual templates. Programs to align the sequences and to assemble contigs had been available since Staden's first assembler (Staden 1979, 1996), but they were not very effective on larger-sized clones, such as bacterial artificial chromosomes (BACs), and much less so on complete genomes. Moreover, their editing tools were not optimized for viewing fluorescent sequencing traces to resolve discrepancies. Good editing tools were needed to allow those working on the human genome sequence to skip over to tracts having discrepancies among the random reads. The tools would then allow the finisher to view the raw data (sequence traces) and to resolve the discrepancy. Two sets of software developments in this field were noteworthy. One was the development of PHRED, PHRAP, and CONSED by Phil Green, David Gordon, and colleagues (Ewing and Green 1998; Ewing et al. 1998; Gordon et al. 1998). This suite of bioinformatics tools was used to assemble the clones in the public Human Genome Project. The other was the development of the Celera assembler and assorted tools by Gene Myers and colleagues (Myers 1995). Various commercial as well as free tools are now available to align and view fluorescent DNA sequencing data (see Chapter 8, Protocols 1 and 2).

The Human Genome Reference

Entire books have been written about the Human Genome Project (Davies 2001; Shreeve 2004; McElheny 2010). Here we briefly describe the state of the human genome reference sequence, explaining its utility and its limitations. It is crucial to remember that the human genome reference is continually being updated. It is, thus, a time-based experimental description of the composite of a few human genomes. The reference has evolved over time and, as of this writing, has now reached HG19. In the nomenclature for sequential releases of the reference genome in the database of the National Center for Biotechnology Information, HG19 represents version 19 of the reference sequence. As the sequence is refined, new releases of the genome are becoming less common, and there are on average fewer changes to the sequence with each new release. But changes to the sequence do occur with each release: New clones are identified and sequenced, errors are reduced, gaps are closed, regions that were poorly sequenced and/or assembled initially are refined, and the experimental descriptions of the genome become more complete. That is, as new clones are identified and sequenced, we are continually closing gaps in the reference genome sequence as well as refining regions that initially were poorly sequenced and/or assembled. Another important point is that the reference genome sequence does not represent the genome of any single individual. About 74% of the reference is derived from one person, with the remainder being a composite (Lander et al. 2001). As various large- and small-scale projects continue to define human sequence diversity, the reference will evolve to a statistical representation of the most common alleles at each position in the genome, together with the possible variants and the frequencies at which they are present in any given population.

NEXT-GENERATION SEQUENCING

Sanger dideoxy chain-termination sequencing, used to construct the human genome reference, was essentially the only sequencing method used until 2005. At that point, a series of technical developments yielded new instrumentation and approaches to sequencing that can be collectively called "next-generation sequencing." These changes revolutionized DNA sequencing, and, at the time of this writing, the field continues to undergo dramatic changes. The term "next-generation sequencing," in fact, has blurred in meaning because of rapid and sequential changes in sequencing methods.

Here we deal first with the history, general features, and applications of next-generation sequencing. In the following section, we discuss and compare the various platforms now in wide use as well as newly emerging systems.

History of Next-Generation Sequencing

The origins of next-generation sequencing technology are many and varied. Most instruments reflect an incredible amalgam of micro- or nanotechnology, organic chemistry, optical engineering, and protein engineering. Some credit for funding these instrumentation developments must go to the National Human Genome Research Institute for its "$1,000/$100,000 genome" technology development initiatives; indeed, several ultimately successful technologies were funded in their initial stages of development by this initiative. However, the interest and investment of venture capital and other entrepreneurial groups undoubtedly provided most of the tens of millions of dollars required to enable any one of these technologies to become a viable commercial offering. A further contribution came from large genome centers and other influential genomics laboratories used as initial testing sites for pre-commercial versions of next-generation sequencing technologies. These "guinea pigs" invested in each new technology, evaluated its performance (read lengths, error rates, G+C bias, etc.), and reported their findings in the peer-reviewed literature or at genomics meetings or by personal communication. Their feedback has largely determined the successes and failures in the commercial sphere.

Current Status of Next-Generation Sequencing

By 2005, the genomes of many organisms, including humans, had been mapped and sequenced to a high degree of accuracy by conventional, cloning-based methods such as those described above in this introduction. With these reference genomes in-hand, biologists were poised to study the variation between and within species, to characterize the genomes of individuals with various genetic disease phenotypes, and so on. The emergence of next-generation sequencing instruments that first became commercially available during 2005 has made these types of experiments facile and increasingly inexpensive.

Next-generation sequencing technologies generally share several attributes that distinguish them from clone-based, capillary sequencing. One attribute is the library construction step, which simply requires that DNA to be sequenced be present in relatively short, double-stranded (ds) pieces (100–800 bp) and have blunt ends. Ligation of platform-specific linker/adaptors (synthetic dsDNA adaptors of known sequence) to the blunt-ended fragments effectively generates a library suitable for sequencing (and there is no longer need for laborious cloning of DNA fragments into vectors and amplifying them in bacterial hosts). A second shared attribute is that of surface-based amplification of the library fragments. This fragment amplification is enzymatic and occurs on a solid support (a bead or glass slide, depending on the instrument) that has covalently linked adaptor molecules attached to its surface. These adaptors facilitate the amplification of a single fragment into a population of fragments that provides sufficient signal from the sequencing reactions for the detection of the incorporated nucleotides. Because each group of amplified molecules derives from a single molecule, next-generation sequencing methods provide digital data that are indicative, for a given sequence, of its abundance in the population. Each group therefore represents a single original molecule in the library sample, and the number of times it is read in a sequencing experiment reflects its relative frequency in the original library. This feature can be mined in a variety of different ways during data analysis. Finally, the combination of instrumentation, library construction, and amplification enables the use of "massively parallel" DNA sequencing in next-generation sequencing, as described below. A comparison of features of the various platforms—output per run, read length, type of reaction used, costs, and so on—is given in Table 1.

These next-generation sequencing approaches stand in stark contrast to conventional capillary-based methods, in which the sequencing process takes place in individual microtiter plate wells, and the instrumentation is used to separate and detect the ladder of completed reaction products. In

TABLE 1. Comparison of various next-generation sequencing platforms

Instrument platform	Run time	Read length (bp)	Yield per run (Gbp)	Sequencing method	Error type	Error rate (%)	Reagent cost per run	Purchase cost (×1000)
454 FLX Titanium	10 h	400 (avg)	0.5	Pyrosequencing	Indel	1	$6,000	$500
454 GS Jr. Titanium	10 h	400 (avg)	0.5	Pyrosequencing	Indel	1	$3,000	$108
HiSeq 2000 v3	10–11 d/2 Flow cells	100×100	>600	Sequencing by synthesis	Substitution	<0.7	$23,500	$690
MiSeq	19 h	150×150	>1	Sequencing by synthesis	Substitution	<0.1	$1,000	$125
SOLiD 5500xl	8–14 d/2 FlowChips	75×35 PE 60×60 MP	155	DNA ligase mediated	A-T bias in substitution	<0.01	$13,500	$595
PGM 316 Chip	3 h	100 (avg)	0.1	ΔpH sensing	Indel	1	$750	$49.50
Pacific Biosciences RS	14 h/~8 SMRTCells	1500 (avg)	0.045 per SMRTCell	Fluorescent real time	Insertion	15	$500	$695

PE, paired-end reads; MP, mate-pair reads.

next-generation sequencing, each group of amplified library fragments is sequenced in a repeating series of steps (extension of the primer, detection of the incorporated nucleotide[s], chemical or enzymatic scavenging of reactants or fluorescence sources) that enables all fragment groups to be processed simultaneously. Thus, in "massively parallel" sequencing, hundreds of thousands to hundreds of millions (or more) sequencing reactions are performed and detected simultaneously. Hence, the volume of data from a single sequencer run is quite substantial and requires specialized approaches to data analysis. Another hallmark of next-generation sequencing is the short read lengths that are generated relative to capillary sequencers. Indeed, the challenges presented by the short reads of next-generation sequencing and their increasing use in biomedical research experiments has rejuvenated genome-oriented bioinformatics and algorithm development. The exception here is the Pacific Biosciences platform, which generates very long reads. These reads currently have a higher error rate than other next-generation platforms, invoking their own bioinformatic challenges.

The final attribute of next-generation sequencing is the ability to generate DNA sequence data from both ends of a library fragment (Fig. 3). This dual-end sequencing capability is accomplished by two different approaches, depending on the size of library fragment being used. The resulting paired reads provide nucleotide sequences at a known distance from one another (with some latitude for the inaccuracies of sizing), a feature that is exploited in several ways. In one approach, "paired-end" sequencing, the library consists of relatively short fragments of 300–500 bp, formed by adding distinct sequences to the 5′ and 3′ ends of each strand in the genomic fragments (Fig. 3A). The fragments are sequenced first at one end, leading in from the "forward" common adaptor, and then from the other end using a second primer that corresponds to the "reverse" adaptor. The second type of dual end sequencing results from a "mate-pair" approach, using longer library fragments (1000–10,000 bp) that are circularized by ligation to the ends of a single, common DNA adaptor of known sequence. Once the ends are mated to the adaptor, several clever approaches can be used to obtain a final library of fragments that retain the adaptor and only the ends of the original, circularized piece of DNA (Fig. 3B).

Whole-Genome Sequencing for Variant Discovery

The analysis of individual variations by next-generation read alignment requires a high-quality reference sequence be available for the genome of interest. Standard library protocols for whole-genome sequencing typically consist of DNA fragmentation, adaptor ligation, and amplification before sequencing, as outlined above. The level of genome "coverage" or oversampling to a specific-fold coverage is a function of the read length of the technology (longer reads mean that lower coverage is needed), the average percentage of high-quality reads obtained in each instrument run, and the availability of paired-end or mate-pair data (or both). In general, the coverage required is aimed at obtaining

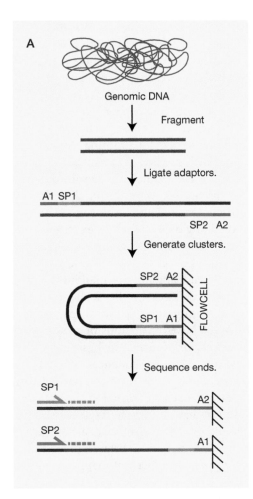

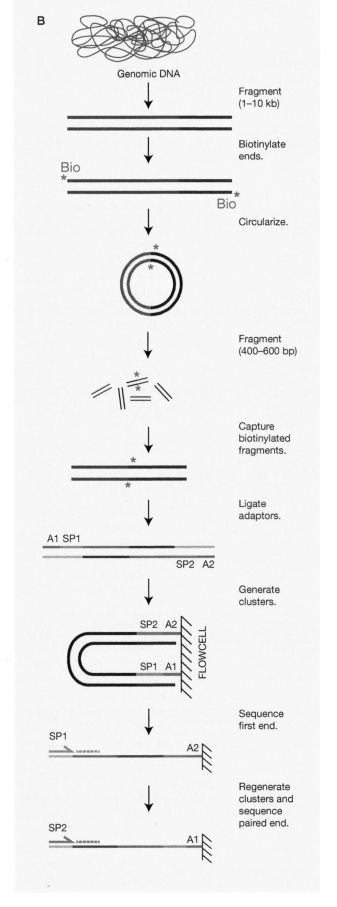

FIGURE 3. **Dual-end sequencing.** Both approaches depicted here employ a two-adaptor strategy. Adaptors ligated to genomic DNA fragments are composed of two sets of sequences. The A1 and A2 adaptor sequences are complementary to oligonucleotides on the flow cell, and thereby allow attachment of the DNA fragments to the flow cell, whereas SP1 and SP2 are complementary to the primers used for the sequencing reactions. (*A*) Paired-end library construction. (*B*) Mate-pair library construction. (Redrawn, with permission, from Illumina.)

(1) sufficient depth of reads to confidently assign variants and (2) sufficient breadth of coverage to ensure that all portions of the genome achieve sufficient read depth for variant discovery. In general, community-accepted standards for whole-genome coverage by short read technologies such as Illumina and SOLiD are around 30-fold, whereas longer read types such as Roche/454 require 12-fold coverage. But "more is better" when it comes to coverage. Thus, the community-accepted standards are sufficient to discover variants with high confidence, but higher levels of confidence can be achieved by additional coverage. Of course, too much coverage can be detrimental to the false discovery rate—for example, at levels of >10,000-fold coverage for Illumina technology, random base-calling errors begin to appear nonrandom and may be identified as variants.

In variant discovery from whole-genome data, different algorithmic analyses are required to detect variants of different types. As we discuss below in detail, all variant discovery begins with alignment of reads to the reference genome. Once aligned, the various algorithms may be applied singly or in combinations, according to the individual laboratory experience and practice. For example, single nucleotide variants (SNVs) are base substitutions found when a sequence is compared with the reference genome. Insertion or deletion of one to several bases is often referred to as "in/del" variants and requires specialized read alignments that introduce a gapped or broken read to accommodate the inserted or deleted bases in the sequence reads. Structural variants (SVs) are the most mathematically challenging mutation type for discovery, because there are several types such as deletions or insertions of >100 bases, inversions of sequence within a chromosome, and translocations either within or between chromosomes. The basic algorithmic approach to structural variant detection, which was worked out in cloned sequences by Eichler and colleagues (Tuzun et al. 2005) and in next-generation sequencing by Snyder and colleagues (Korbel et al. 2007), relies on the detection of altered read placement distance or orientation (or both) in paired-end or mate-pair data. For example, if a paired-end library has an average fragment size of 300 bp, the expected mapping distance of paired-end reads should be at this distance apart on the reference genome, with some leeway for sizing inaccuracies. If the reads map at a distance greater or lesser than the anticipated size range and multiple read pairs support this aberrant mapping, then a deletion or insertion event (respectively) may be suspected.

A further characterization of an SV may be accomplished by using a localized assembly algorithm (of which there are many). In this approach, reads identifying the particular SV become the input data for the localized assembly algorithm, which typically can assemble the region if it is (1) real (as opposed to a false positive) and (2) relatively uncomplex. In this way, SVs may be characterized to nucleotide resolution, allowing, for example, downstream prediction of putative gene fusions that result from a translocation, or identifying the genes affected by a deletion. The detection of SVs remains a very dynamic field and will likely change by the time of the publication of this manual. Those carrying out this type of research will need to avail themselves of the most current methods in the literature.

Although whole-genome sequencing remains the more expensive option from the standpoint of data generation and analysis, it is the most comprehensive approach to understanding all variation. This can be important in diseases such as cancer, which has long been understood to carry extensive structural variations of all types, often leading to differential diagnosis, prognosis, and treatment. (Her2-positive breast cancer and chronic myeloid leukemia are two well-known examples.) However, cancer genomes, in particular, pose challenges because they often are available in limited quantities once the tumor has been used for conventional pathology. In addition, certain tumor types (such as prostate and pancreatic) have diffuse presentations and may require enrichment of the tumor cell population by laser capture microdissection, for example. Studying the earliest neoplastic lesions also is desirable because of the knowledge gained of the earliest genomic variations present in early disease (often known as "drivers" of oncogenesis). For these and other reasons, we have developed low-input methods for tumor whole-genome sequencing that can produce a sufficiently complex library set for adequate coverage (>30×) and without requiring an enzymatic amplification of the genomic DNA, which might introduce errors. These methods essentially condense the steps at which DNA loss tends to occur and obviate the need for ethanol precipitation in order to exchange buffers or enzymes. In addition, sizing of the libraries at the final steps has now become more automated by using free-zone electrophoresis devices such as the Caliper XT that do not require one to visualize the library bands on a gel because the sizing mechanism on the device simply measures library fraction size in comparison to a standard size ladder. These protocols are outlined later in the chapter and permit whole-genome library construction from 100 ng of input DNA.

Targeted Sequencing: An Overview

Historically, the technique used to select regions of a genome for sequencing was the polymerase chain reaction, wherein one or more pairs of primers selective for that region of interest were designed and used as priming sites for discrete DNA polymerase-based copying. The resulting products were treated to remove excess primers and nucleotides and then sequenced directly or subcloned into a TA vector and sequenced with a universal primer. With the advent of next-generation sequencing, however, the scale of the sequencing capacity per instrument quickly exceeded the output of highly automated PCR pipelines. This situation, coupled with a growing need to selectively sequence the known genes of an organism but not the entire genome, brought about the development of several approaches collectively known as "hybrid capture." Hybrid capture was originally envisioned and performed on solid surfaces of oligonucleotide-based microarrays (Albert et al. 2007; Hodges et al. 2007; Okou et al. 2007). Sequences complementary to the sequences being targeted were placed on the microarray at high density using standard microarray printing techniques. Genomic DNA was then hybridized with these arrays and the nonbinding DNA was washed off. The captured DNA was then eluted, made into libraries, and sequenced. Several reports describing this approach were published collectively at about the same time. In addition, a highly multiplexed amplification-based strategy that could target many, but not all, exons was described (Porreca et al. 2007). One of these hybrid capture approaches demonstrated that nearly all of the annotated exons in the genome, the exome, could be captured and sequenced from an individual (Hodges et al. 2007). The protocol, however, required considerable technical skill, and the complicated procedures for washing and eluting DNA from arrays meant that it was difficult to scale to large numbers of samples. Fortunately, rapid development of the concept and the ability to perform the steps in solution solved these problems.

The basic premise of solution-based hybrid capture is that single-stranded synthetic DNA or RNA probes (see Chapter 13) (Gnirke et al. 2009; Bainbridge et al. 2010), when mixed with a molar excess of denatured whole-genome next-generation sequencing library fragments, will hybridize as chimeric probe:library complexes. A second step selectively isolates the library:probe hybrids, allowing the remainder of the genomic fragments and any single-stranded probes to be removed. Here, isolation occurs because the probe sequences carry biotin moieties that react upon mixing with paramagnetic beads carrying streptavidin, and application of a magnet pulls the hybrids from solution (see the information panels Biotin and Magnetic Beads). The resulting captured fragment population can be released from the probes and amplified by PCR in advance of next-generation DNA sequencing. Figure 4 shows the basic scheme for hybrid-capture technology followed by next-generation sequencing. There are presently available reagents for the human and mouse exomes that capture most, but not all, genes annotated in those genomes. Because these reagents differ somewhat in the genes they are designed to capture, each commercial product should be carefully examined to ensure that genes of interest are, indeed, targeted for capture with corresponding probe sets (Clark et al. 2011; Parla et al. 2011). One reason the reagents differ is because probe synthesis may require a fixed-length probe or may accommodate for probes of variable length. With the latter, the T_m is adjustable for all probes, and this may result in an increased number of genes for which probes can be designed successfully. Another reason that a reagent may not efficiently target a gene is because that gene belongs to a gene family, and it may not be possible to design a probe specific for the particular gene of interest. It is also important to recognize that, just because a probe targets a specific gene, it may not capture the gene as efficiently as other probes in the population. In consequence, the target gene sequences may be poorly represented in the resulting library fragment population. Most variant detection algorithms developed for targeted capture will require a minimum level of coverage on the target in order to call variants, and poorer probe performance may not always deliver the required coverage level. For complex mammalian genomes, typically ~85%–90% of

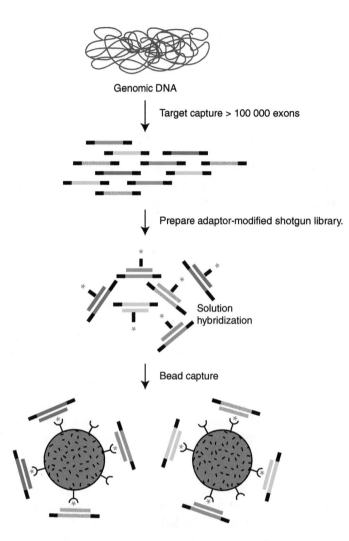

Genomic DNA

Target capture > 100 000 exons

Prepare adaptor-modified shotgun library.

Solution
hybridization

Bead capture

FIGURE 4. **Hybrid capture of exomes or targeted regions.** In hybrid capture, fragments from a whole-genome library, each containing platform-specific adapters on each end, are selected with a set of probes that correspond to targeted genes. These probes are biotinylated, allowing capture of the probe:genomic library fragment hybrids using magnetic beads. The captured hybrids are isolated from solution by the application of a magnet. Denaturing conditions are used to elute the captured genomic library fragment population from the hybrids, and these fragments are prepared for sequencing.

known genes have probes designed for most or all of their exons. Of those probes, ~80%–85% will yield sufficient coverage for variant detection (Clark et al. 2011; Parla et al. 2011).

Of course, one may not wish to target the entire exome, and custom reagents also can be designed for subexomic targets, including unique sequences that lie outside or between exons. The caveat in subexomic targeting is that the lower the total size (in base pairs) of the targeted portion of the genome, the less efficient the procedure becomes, owing to "off-target effects." In essence, off-target effects are randomly captured fragments, acquired through random or partial hybridization events in the capture process that do not yield the targeted sequences in the probe set. These effects are caused by the huge excess of untargeted genome fragments over the targeted sequences and can be detected by aligning sequenced fragments. Many of the sequenced fragments will be shown not to lie in the targeted regions of the genome. Hence, although targeted capture is at its essence a less-expensive sequencing approach than whole-genome sequencing, it can quickly become more expensive than anticipated, especially in very low target-space designs.

Considerations for Sequence Coverage

"Sequence coverage," a relatively simple but often misunderstood concept, refers to the number of sequence reads that are aligned for a given interval (such as per base) of a genome. It is linked to the requirement of sequence coverage necessary to produce data of the quality required for a given project. Coverage is almost always expressed as either an average value over the length of the region being sequenced or, more recently, as a threshold value. An average value, for example, a $10\times$ or 10-fold coverage, means that on average each base in the region (or genome) is covered by 10 independent sequence reads. When used as a threshold, the coverage value is usually expressed as a percent of target covered, for example, as 80% of the target regions with at least $20\times$ coverage. This means that out of 1000 bases in the target regions, 800 were covered by at least 20 independent sequence reads.

Several issues determine the coverage needed for a given project. For example, the coverage required for a haploid genome such as a bacterium is typically much less than for a diploid genome, because in order to discover all of the alleles in the diploid genome, both chromosomes must be covered adequately. Another factor that can lead to a need for higher coverage is sample purity, as in the case either of a metagenomic sample or of a mixed sample of tumor and normal tissue.

It is also important to understand the implications of average coverage. In the case of a human genome that has been sequenced at $20\times$ coverage, we infer that, on average, each base is covered 20 times by independent reads. However, keep in mind that this value is a distribution; therefore, some of the bases will be covered significantly more or significantly less than the 20-fold average. As a result for these experiments, although $20\times$ coverage of a given base may be adequate to accurately call variants in a diploid genome, having $20\times$ coverage actually means that a portion of the genome will be covered with a lower number of reads, and hence variant detection in that part of the genome will not be adequate. As a result researchers doing whole-genome sequencing on pure DNA samples from diploid organisms typically set $30\times$ coverage as their minimum average coverage of short reads for a project. Some factors, such as the typical purity of tumor samples, may drive that average coverage number up to $50\times$ or higher. At the opposite end of the spectrum, when using next-generation sequencing to discover large-scale structural variants, much lower coverage can be used, but the higher the coverage the better the resolution and precision of the resulting analyses.

Whereas whole-genome sequencing provides a relatively predictable range of coverage given a particular average coverage, there are some biases that lead to regions being over- or under-represented in coverage relative to a purely statistical model. In experiments that require much more processing of the samples, as is the case for targeted capture and in exome sequencing, there are potentially very strong biases. These result in some regions of the exome being vastly over- or under-covered (or even completely uncovered) relative to the average. As a result, it is probably best to express exome coverage as a threshold as described above. For example, we may say that 80% of the target (this is a typical value, by the way) was covered at a minimum of either $20\times$ or $30\times$. For further considerations of coverage, see the box Coverage and Multiplexing: The Dilemmas Caused by the Rapid Growth of Sequencing Throughput.

It is also important to recognize the value of quality control in next-generation sequencing, as described below.

COVERAGE AND MULTIPLEXING: THE DILEMMAS CAUSED BY THE RAPID GROWTH OF SEQUENCING THROUGHPUT

When the first commercial Solexa (later to become Illumina) sequencers shipped in early 2007, they produced $\sim$1 billion bases of sequence that was distributed over eight lanes of the microfluidic "flow cell" (roughly 125 million bases per lane). This effort resulted in an impressive increase in sequence compared with previous instruments, but multiple lanes or—in the case of whole human genomes—a large number of entire runs per project were still required. However, as sequencing output has grown from early 2007 to mid 2011, the output from an Illumina HiSeq 2000 is now $\sim$600 billion bases spread over 16 lanes, or $\sim$37.5 billion bases per lane. This very rapid growth of capacity has had the oddly converse effect of making it very difficult to perform all but the largest-sized projects in a cost-efficient manner. Therefore, whereas whole genomes can be sequenced in a small number of lanes on current instruments (compared with the dozens of complete runs required on the first-generation instruments), even projects as large as human exome sequencing cannot efficiently use the capacity in a single lane of the current next-generation sequencers.

(Continued)

One way to address this disparity between actual coverage and need is to include a DNA bar code on the adaptor sequence used in making each sequencing library. The resulting libraries can be combined in equal ratios before sequencing; then, during sequencing, the DNA bar-code sequences are "read" as part of the sequencing data production. The individual fragments from the combined libraries then can be binned together after the sequencing data are analyzed, based on the DNA bar-code identifier. To some extent, this is a tractable solution; however, the rapid growth in sequencing output has not been matched with readily available bar codes to allow high-order multiplexing. As an example, a current HiSeq 2000 lane would produce ~750-fold coverage of a 50-Mb human exome library, which requires only ~100-fold coverage to reliably identify SNPs and indels. Although such idealized calculations do not take into account such factors as the broad dynamic ranges seen in captured targets, this level of coverage is still vastly higher than is needed in virtually any application of exome sequencing.

The challenge therefore becomes to bar-code samples so sufficiently that many can be pooled within a single lane on the Illumina sequencer in order to fully use the sequencing capacity available. This problem becomes worse as the capacity of the sequencers increases or as the region that needs to be sequenced decreases. If, for instance, the goal is to sequence only 100 genes from the human genome with an average size of 50 kb, a single Illumina HiSeq2000 lane would produce ~7500-fold coverage of the targeted genes. In theory, the same lane could generate 100-fold coverage from 75 individuals. Again, it is important to realize that such theoretical calculations are never completely met in real-world practice. Problems such as pipetting inaccuracies in pooling samples that cause one sample to be over- or under-represented means that one must oversequence to a certain degree in order to ensure the minimum coverage of all the samples. However, it should be readily apparent how combining even 50 such sets of genes per lane rather than 50 genes from one sample would drastically reduce the cost of sequencing the genes from each sample. By the time this book is published (mid 2012), it is likely that the major instrument manufacturers will have multiple solutions to this dilemma. These solutions will range from lower-capacity sequencers (see the box Desktop Genome Sequencers: The Future of Next-Generation Sequencing?) to multiplexing reagents that will allow enough bar-coded samples to be pooled in a sequencer lane to take full advantage of the instruments' throughput capacity.

Sequence Validations and Quality Control

Any sequence data, whether based on Sanger sequencing or next-generation sequencing, are experimental data and hence have an associated error. The very large size of the data sets that can be generated with next-generation sequencing require both statistical and specific validation of variants that are found. Next-generation sequencing instruments, when used correctly with adequate coverage of the target, can generate consensus sequences of extraordinary accuracy. However, such data will nevertheless contain several errors. As an example, the whole-genome sequencing of a human with the very low error rate of one in 10 million bases will still generate between 200 and 300 errors. At perhaps a more realistic error rate of one error per 3 million bases, the same data would contain about 1000 errors. In addition, there may be areas that are covered inadequately (or not at all), leading to false-negative data.

There are a number of ways to approach quality control in next-generation sequencing, and it is likely that, as the technology evolves, so will the means for assessing quality. In whole-genome sequencing, one important way to assess quality is to compare the sequencing results with results from whole-genome SNP arrays (Ley et al. 2008). This is done by using the same genomic DNA used in the sequencing process for microarray analysis to detect variants. By comparing the genome sequence following SNP discovery to the results from the microarray experiment at the specific loci on the array, one can determine the percentage of both homozygous and heterozygous variants detected by the microarray that are also detected by sequencing. This approach provides the added value of assessing quality (in terms of the false-negative rate), as well as coverage of the genome, but it does not measure the false-positive rate. It also provides confirmation of the sample identity.

In genome sequencing or in sequencing much smaller regions such as exomes or individual genes, it is advisable to perform some other type of validation of both specific critical variants of interest and of a random selection of other detected variants. The latter gives some idea of the overall error rate inherent in the sequencing and variant discovery processes. Validation can be performed on a small scale by selecting some number of variants, amplifying the regions from the genome, and

OUTSOURCING SEQUENCING

The rapidly decreasing cost of data generation is making human whole-genome sequencing an increasingly viable choice for use as a tool in human genetics. Although it is technically feasible to perform whole-genome sequencing in any competent laboratory, building the required infrastructure can sometimes be daunting—particularly the computational biology analysis. As a result, many human geneticists are likely either to seek outside collaborators who have the infrastructure in place, including the required bioinformatics, or to outsource the project to the increasing number of contract sequencing companies.

Although there are disadvantages to outsourcing, including losing control over one phase of the project and depending on someone else's timelines, outsourcing does provide an answer that may be particularly attractive when only a limited number of whole genomes need to be sequenced in a relatively short period of time. At present, several contract services are available for whole human genome sequencing, for example, Complete Genomics and Illumina, as well as others with less public exposure.

The process starts with the genomic DNA sample being sent to the contract sequencing company; the service provides the whole-genome sequence of individuals, as well as some degree of bioinformatic analysis. There are also companies that will analyze sequence data produced from another source. These options are particularly useful for those who do not need to perform whole-genome sequencing over a long period of time or on a regular basis but, rather, have a need for a finite number of samples to be sequenced and/or analyzed.

Contract services are also available for applications other than whole-genome sequencing (e.g., exome sequencing). However, because exome sequencing requires more extensive up-front manipulations for sample preparation and handling, it is, in terms of cost per base, vastly more expensive than whole-genome sequencing. Because the cost differentials are now quite heavily biased toward whole-genome sequencing, multiple companies are competing very aggressively to capture the market for inexpensive whole-genome sequencing. At the time of writing (2011), it is probably two to three times more expensive to sequence a whole genome than an exome, using a commercial vendor. However, in a laboratory with next-generation sequencing infrastructure already in place, the cost of a whole-genome sequence is probably 10 times more than the cost of an exome. This difference between the contract costs and in-house costs should be taken into consideration, particularly for exome or other targeted sequencing projects.

sequencing these amplicons directly with a capillary sequencer—a rapid and inexpensive way to generate orthogonal data. Validation of larger numbers of variants has used custom capture arrays, such as genome-wide validation for cancer genome studies (E Mardis, pers. comm.), and is likely to evolve to a more rapid turnaround, particularly with the recent development of desktop next-generation sequencers (see the box Desktop Genome Sequencers: The Future of Next-Generation Sequencing? in the following section). The critical issue is the need for some sort of external, preferably orthogonal, quality-control effort that monitors how well next-generation sequencing and all associated steps (which may include capture, long-range PCR, or other methods of isolating a subregion of the genome), as well as the variant detection algorithms discussed elsewhere, are performing in terms of both error and coverage.

Effect of Next-Generation Sequencing on Research

The impact of next-generation sequencing (NGS) technology on biological research has been, and is likely to continue to be, profound. The scale of data generation made possible solely by NGS is unprecedented, exceeding Moore's law predictions by severalfold. Academic laboratories can now sequence the equivalent of a whole human genome in several days for around $10,000 per genome (at the time of this writing), measured in total direct costs to produce and store the data. NGS studies of RNA expression can characterize the repertoire of expressed genes in a sample for roughly the same cost as a defined-content microarray, but also provide much more quantitative data on expression levels and detect RNA-level phenomena such as alternative splicing. In addition, next-generation sequencing has radically increased the breadth and precision of cataloging genome-wide histone- and transcription factor–binding sites (see Chapter 20).

As a result of the expansion in sequencing scope and the reduction in its cost, next-generation sequencing also has enabled the development of new areas of research. One example is added depth to the catalog of noncoding RNAs and the discovery of new classes of these transcriptional products. Similarly, "metagenomics" has flourished as a result of NGS technology. Here, DNA isolated collectively from all species in a particular environmental sample/site is first characterized by sequencing and then identified by database query to the sequenced "universe." Although originally performed with capillary methods using shotgun libraries of DNA isolated from a given community, the low cost and extraordinary magnitude of NGS reads greatly accelerated this field, because of the depth of sequencing data that could be acquired in a relatively short time. And it is not merely nucleic acid sequencing that has benefitted. Profiling methylation sites on genomic DNA now can be accomplished genome-wide with "methyl-seq" approaches such as bisulfite conversion or immunoprecipitation (IP), followed by sequencing of the resulting fragments (see Chapter 12).

Beyond single types of inquiry, moving many formerly array-based experiments to next-generation sequencing platforms has introduced consistency between multiple experiments. This consistency is lacking from hybridization-based methods, which are more prone to experimental vagaries. In addition, using NGS for multiple types of experiments from a given sample or set of samples can greatly enhance integration across data sets, because all assays are sequencing-based and generate data that are digital in nature.

Analysis of the data remains the most daunting, expensive, and time-consuming component of the process. Difficulties in aligning and/or assembling the data are mainly caused by the significantly shorter read lengths and highly different error models of next-generation sequencing reads (compared with capillary sequencers). On the positive side, the renaissance in bioinformatics and computational biology has brought the innovation and interdisciplinary vigor necessary to address these thorny problems. On the negative side, data analysis now represents a significant bottleneck to overall productivity. Chapter 8 deals extensively with the approaches to short read analysis.

OVERVIEW OF NEXT-GENERATION SEQUENCING INSTRUMENTS

There are, at the time of writing (late 2011), three major platforms in common use—the Roche/454 Pyrosequencer, the Illumina/Solexa sequencer, and the Life Technologies SOLiD/5550 sequencer. These platforms are described here, together with a series of newly evolving "personal genome machines" and desktop genome sequencers (for a comparison of features of the various platforms, see Table 1). In addition, we include a brief description of the first of what will likely be a wave of single-molecule sequencers.

Roche/454 Pyrosequencer

The first commercial next-generation sequencing instrument was developed by the 454 Corporation, which later was purchased by Roche Inc. (Margulies et al. 2005). The instrument combines a novel technology, emulsion PCR (emPCR)—to amplify library fragments on the surface of a polystyrene bead that has covalently linked adaptors on its surface—with massively parallel pyrosequencing. Other massively parallel features of the instrument include a PicoTiter Plate, a dual-purpose fused silica consumable used to (1) contain the amplified beads, and auxiliary smaller beads whose surfaces carry enzymes to generate signals and (2) serve as a fluidic conduit for sequencing reagents as they are introduced to the plate wells by diffusion. One side of the PTP is optically clear and provides an interface between the beads used in sequencing and the CCD camera used in image acquisition. Because the sequencing reagents used in 454 pyrosequencing are native nucleotides without fluorescent labels, they must be presented to the PTP in a stepwise fashion, with an imaging step and a wash step to remove unused reactants after each incorporation.

The basic pyrosequencing approach is shown in Figure 5. As each nucleotide flow is presented to the primed library fragments, incorporation by DNA polymerase may occur if the nucleotide presented is complementary to the nucleotide of the template. When incorporation occurs, the

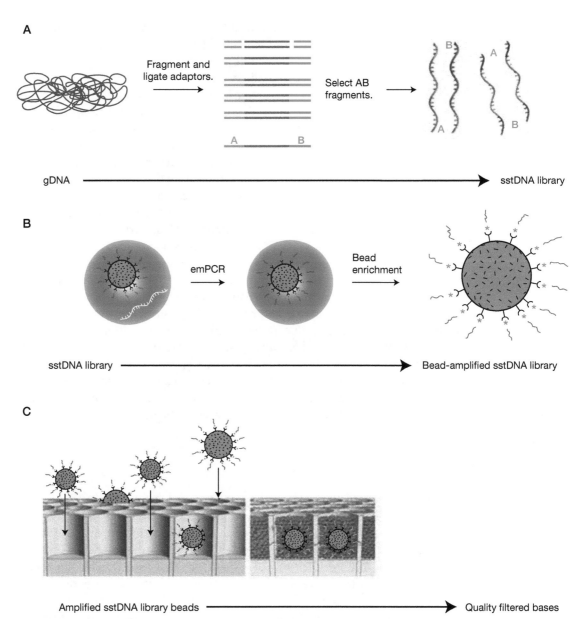

FIGURE 5. Pyrosequencing. The method used by the Roche/454 sequencer to amplify single-stranded DNA copies from a fragment library on agarose beads. (*A*) To prepare a library, genomic DNA is fragmented and adaptors are ligated to the ends of the fragments. (*B*) A mixture of DNA fragments with agarose beads containing complementary oligonucleotides to the adaptors at the fragment ends are mixed in an ~1:1 ratio. The mixture is encapsulated by vigorous vortexing into aqueous micelles that contain PCR reactants surrounded by oil and pipetted into a 96-well microtiter plate for PCR amplification. (*C*) The resulting beads are decorated with approximately 1 million copies of the original single-stranded fragment, which provides sufficient signal strength during the pyrosequencing reaction that follows to detect and record nucleotide incorporation events. sstDNA, single-stranded template DNA. (Reproduced from Mardis 2008.)

resulting pyrophosphate (PPi) triggers a series of downstream reactions catalyzed by enzyme-linked beads containing sulfurylase and luciferase, resulting in the production of light. The emission of light is detected by a charge-coupled device (CCD) camera; light collection occurs following the incorporation of each nucleotide and takes place for the time period required to collect any signals emitted from several hundred thousand beads in the wells of PicoTiter Plates (PTPs). The amount of light produced is proportional to the number of added nucleotides (up to the limit of detector saturation). This proportionality is correlated after the run, using an initial "key" sequence on the adaptor molecule that provides a single T, G, C, and A in a specific order at the beginning of each sequence to quantify the peak heights representing a single base incorporation. The postrun analytical process then translates the signals from each nucleotide flow for each position on the PTP that registered during the initial four flows (the "key" sequence). The resulting "flowgram" consists of base calls and associated quality values.

Because of the sequential introduction of individual nucleotide flows, the substitution error rate of 454 data is quite low. In contrast, insertion or deletion errors occur at a higher rate, especially at homopolymeric tracts. The higher error rate arises because the signal from the insertion of multiple bases of the same type (e.g., a run of A residues) creates a large peak. As a result, the instrument's base-calling software is unable to perfectly integrate the area under the peak and to determine the exact number of bases added. This problem may also occur earlier in the procedure when the detecting CCD reaches a maximum in its linearity of response to light released as a result of incorporation of nucleotides.

Prior to the use of pyrosequencing in the massively parallel instrument offered by Roche/454, a microtiter plate-based system was commercialized. This instrument series, called PyroMark, is offered by QIAGEN Corporation and is suitable for use in mutation detection and/or validation, or in methylation detection applications (see Chapter 12). A major advantage is to provide quantitative data in a single run in sequencing mixed populations of DNA templates (e.g., for determining the allele frequencies of polymorphisms in a population of patient samples or the frequencies of modified sites in DNA methylation analysis). In principle, the instrument operates in a fashion similar to the Roche/454 instruments, producing light with the incorporation of sequentially added nucleotides for each target sequence studied, where each sequence is contained in a separate well of a microtiter plate. The analysis is handled on a well-by-well basis, however, and because of the reduced density of microtiter plate wells compared to the PicoTiterPlate wells used in the 454 Pyrosequencing System, PyroMark does not have significant time-to-analysis nor does it require extensive computational analysis prior to returning a sequence-based answer to the user.

Illumina/Solexa Sequencer

Introduced in 2006, the Solexa sequencer uses an approach that is different from other next-generation sequencing systems, both in the nature of the sequencing reaction and in the number of sequencing reads that can be produced by a single instrument run. Library construction conforms to the general description given above (see Current Status of Next-Generation Sequencing), but from this point onward, the process differs from that used in 454 sequencing. Rather than using emPCR on beads, the DNA is added to a "flow cell" that consists of eight microfluidic channels machined on a silica glass surface and then assembled into a "sandwich" to produce a sealed microfluidic structure. The flow cell lanes are the conduits through which reagents are introduced and removed. The surfaces of the flow cell lanes are precoated by the manufacturer with oligonucleotides complementary to sequences within the common adaptors used for library construction. As the denatured library is passed through a lane or lanes on the flow cell, DNAs in the library hybridize to the oligonucleotides coated onto the surface of the flow cell. The library fragments are then amplified in situ by a process called "bridge amplification": Library fragments hybridized to complementary linkers covalently attached on the flow cell surface are copied multiple times by a DNA polymerase during an isothermal reaction (with denaturation by NaOH between cycles) to produce "clusters" of each amplified fragment. Once amplified, a specific reagent applied to the flow cell lanes releases the forward reaction-specific adaptor from the flow cell surface, while the other adaptor remains attached. This chemical reaction produces linear DNA templates that are first denatured to provide single-stranded templates and are then primed for the DNA sequencing process by a specific oligonucleotide that corresponds to a portion of the released adaptor (Fig. 6). The primer provides a free 3'-OH for DNA polymerase to copy each strand in each cluster. The incoming nucleotide is covalently attached to two chemical side groups. A blocking group attached to the 3' end of the incoming nucleotide restricts the synthetic reaction to addition of a single nucleotide. In addition to the blocking group, a fluorescent group provides the identity of the incorporated nucleotide upon scanning of the flow cell with laser light.

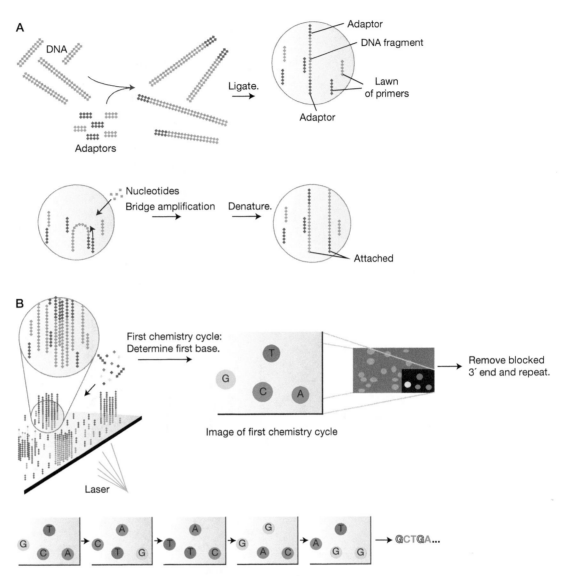

FIGURE 6. The sequencing-by-synthesis approach. (*A*) Cluster strands created by bridge amplification are primed and all four fluorescently labeled, 3′-OH blocked nucleotides are added to the flow cell with DNA polymerase. (*B*) The cluster strands are extended by 1 nucleotide. Following the incorporation step, the unused nucleotides and DNA polymerase molecules are washed away, a scan buffer is added to the flow cell, and the optics system scans each lane of the flow cell by imaging units called tiles. Once imaging is completed, chemicals that effect cleavage of the fluorescent labels and the 3′-OH blocking groups are added to the flow cell, which prepares the cluster strands for another round of fluorescent nucleotide incorporation. (Reproduced from Mardis 2008.)

After both upper and lower flow cell surfaces are scanned by the instrument across all lanes (in the HiSeq 2000 system, the earlier Solexa/Illumina instruments only scanned one surface), the blocking group and the fluorescent label are removed chemically from each incorporated base, allowing the next cycle of sequencing by synthesis to begin. This series of steps proceeds for a user-defined read length, and a second read is initiated at the end of this sequence (if desired). In preparation for the second read, the sequenced strands are denatured from their templates, a brief round of limited cluster regrowth occurs by bridge amplification, and a separate chemical reaction releases the opposite adaptor to generate linear molecules ready for priming. A second oligonucleotide primer is introduced and anneals, and the process begins as previously described for read 1. At the end of the second read, the primary data are analyzed, using a data processing pipeline provided by Illumina to interpret the wavelength signal from each cluster at each step, and to reconstruct each cluster's sequence for reads 1 and 2, including quality values for each base. A series of filters is then

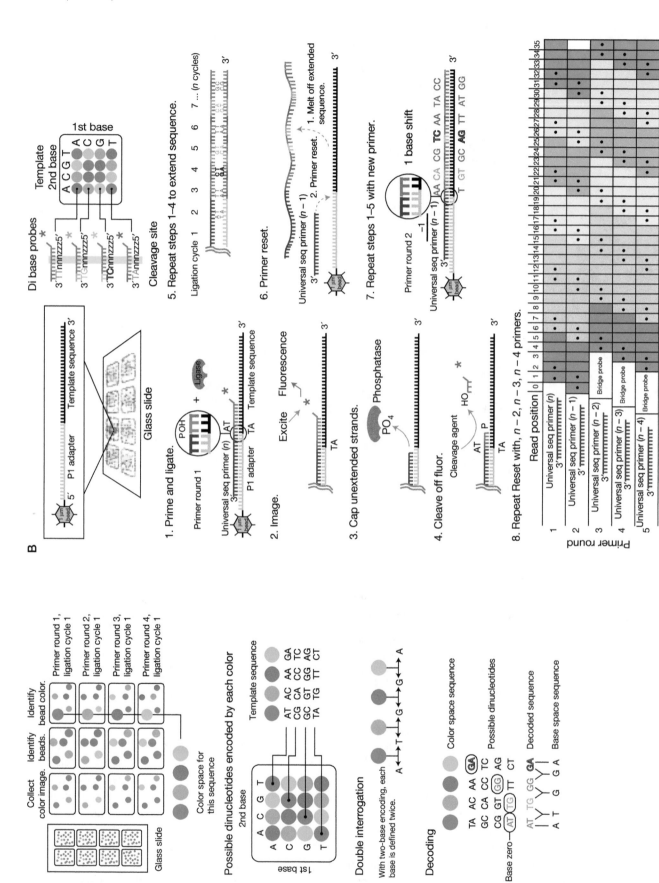

FIGURE 7. See legend on facing page.

applied to remove low-quality sequences, and the high-quality reads are aligned to the specified genome of choice. Based on these alignments and other parameters of the run, a series of run-specific metrics (number of passed filter clusters, number of bases per flow cell, average error rate, etc.) is generated and parsed to a report that allows the user to evaluate the quality of each run.

Life Technologies SOLiD/5500 Sequencer

Rather than rely on DNA polymerase and base-by-base sequencing, the SOLiD technology uses sequencing by ligation. This form of massively parallel sequencing is called "two-base encoding" and uses DNA ligase to identify a specially designed, properly hybridized oligonucleotide and ligate it to the annealed primer's end. SOLiD sequencing libraries, generated either as small-insert or mate-pair libraries, are amplified on ~1-µm-diameter magnetic beads using emPCR and purified to remove beads without DNA on their surfaces. Because of the large numbers of beads that can be sequenced on a single FlowChip and the accompanying large volume emPCR required, three modular preparatory stations can be used to (1) generate the emulsion, (2) emPCR-amplify in bulk solution, and (3) break the emulsion and reclaim the beads. The beads then are loaded onto the FlowChip, and sequencing takes place by first priming the amplified library fragments and then introducing DNA ligase and a set of semidegenerate oligonucleotides. Each oligonucleotide is designed so that the bases at positions 1 and 2, nearest the 3′ end, are "fixed" as one of 16 possible dinucleotide combinations, and the remainder of the oligo is degenerate. At its 5′ end, each oligo also carries one of four fluorescent labels, each of which encodes or corresponds to a set of fixed dinucleotides. In this sequencing approach, hybridization of five bases of the correct oligo to a given library fragment represented on the surface of the bead is followed by its subsequent ligation to the primer.

Fluorescent scanning allows the "decoding" of the nucleotide closest to the 5′ end of the primer (in position 1 of the oligo). Because the first two bases are fixed, the second nucleotide in from the 5′ end of the primer also is identified—hence the term "two-base encoding" (Fig. 7A). After the scanning step, the strands are prepared for the next round of ligation-mediated sequencing, using exonuclease digestion to (1) remove the degenerate portions of the oligonucleotide and the fluorescent group by inclusion and (2) prepare the new 5′ end of the primer for ligation to the incoming oligonucleotide. Subsequent rounds of hybridization, oligonucleotide ligation, and scanning allow bases to be called in a similar fashion, with each round of synthesis initiated at a fixed distance from the 5′ end of the primer (e.g., the fifth base, 10th, 15th, etc.). After the primer has been elongated by ligation to decode every fifth base for the desired number of steps (read length), the synthesized second strand is removed by denaturation and a wash. The next synthetic series relies on a primer that anneals to the template with its 5′ end at the $(n - 1)$ position. The desired number of ligation sequencing steps is then performed. The process is repeated for primers placed at $(n - 2), \ldots,$ allowing a contiguous sequence to be read (Fig. 7B). Typical SOLiD read lengths are 50–75 nucleotides, whereupon the entire ligated second-strand copy can be denatured and a second read primer annealed, followed by additional rounds of ligation-based sequencing. Postrun software is provided with the instrument to deconvolute the data, to filter out poor-quality reads, and to obtain nucleotide sequence and associated quality values for each base.

FIGURE 7. **SOLiD sequencing.** (*A*) Principles of two-base encoding. Because each fluorescent group on a ligated 8-mer identifies a two-base combination, the resulting sequence reads can be screened for base-calling errors versus true polymorphisms versus single base deletions by aligning the individual reads to a known high-quality reference sequence. (*B*) The ligase-mediated sequencing approach of the Applied Biosystems SOLiD sequencer. In a manner similar to Roche/454 emulsion PCR amplification, DNA fragments for SOLiD sequencing are amplified on the surfaces of 1-µm magnetic beads to provide sufficient signal during the sequencing reactions and are then deposited onto a flow cell slide. Ligase-mediated sequencing begins by annealing a primer to the shared adaptor sequences on each amplified fragment, and then DNA ligase is provided along with specific fluorescent-labeled 8-mers, whose fourth and fifth bases are encoded by the attached fluorescent group. Each ligation step is followed by fluorescence detection, after which a regeneration step removes bases from the ligated 8-mer (including the fluorescent group) and concomitantly prepares the extended primer for another round of ligation. (Reproduced from Mardis 2008.)

Pacific Biosciences RS

The DNA sequencing instrument produced by Pacific Biosciences was introduced to beta-test laboratories in mid 2010 and was released commercially in May 2011. This sequencing instrument is based on a nanofabricated structure called a "zero mode waveguide" (ZMW) that is used to provide a focused "look" at the active site of a single DNA polymerase as it synthesizes on a DNA template. The ZMW is basically a tiny pinhole structure that allows the instrument's laser light and its detection optics to focus specifically on the active site of the polymerase molecule (see Fig. 8A). Effectively, tens of thousands of ZMWs can be machined in an ordered array on the silicon-dioxide chip surface to provide a large number of detection chambers. Library construction for Pacific Biosciences sequencing is very similar to that used for other next-generation platforms—the ligation of specific adaptors to the fragment population. In this case, the adaptors are novel, having a "lollipop" shape with one portion double-stranded and the other portion an open loop (Fig. 8B). Ligation of the fragments and adaptors followed by denaturation generates a circular DNA molecule. Each DNA circle is then complexed with DNA polymerases that are specifically modified to attach covalently to the ZMW surface. These complexes are introduced to a SMRTCell device containing 150,000 ZMWs, a time interval is allowed for the deposit of the DNA/polymerase complexes into individual ZMWs, and the sequencing reagents are introduced. In collecting data, the instrument takes real-time "movies" of the polymerase active sites as they incorporate nucleotides in up to 75,000 ZMWs per instrument run. Each movie records the fluorescent-labeled nucleotides that enter the active site and dwell for a sufficient time to be excited by the laser wavelength, and hence are detected. The optics then reorient to the second set of 75,000 ZMWs on the same SMRTCell, and another set of movies is recorded in real time. Unlike the instruments described above, the Pacific Biosciences instrument requires only a single addition of reactants per SMRTCell, and the run duration per SMRTCell is 105 min in length. At this time, average read lengths are ~1500 bp, and reads up to 10,000 bp or longer have been observed. The movies are immediately converted on-instrument (i.e., during the run) to "pulse" files of significantly smaller size, and then to base calls with quality values.

Construction of libraries with very short inserts (average 250 bp), standard length inserts (1–3 kb), or very long inserts (10 kb) can be used to generate different types of reads on the Pacific Biosciences RS, for different applications. In sequencing very short insert libraries, the strands of each fragment can be

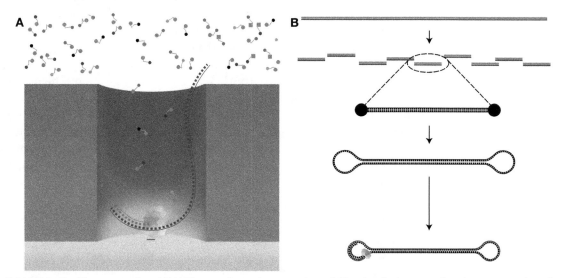

FIGURE 8. **PacBio sequencing.** (*A*) With an active polymerase immobilized at the bottom of each ZMW, nucleotides diffuse into the ZMW chamber. To detect incorporation events and identify the base, each of the four nucleotides A, C, G, and T is labeled with a different fluorescent dye having a distinct emission spectrum. Because the excitation illumination is directed to the bottom of the ZMW, nucleotides held by the polymerase prior to incorporation emit an extended signal that identifies the base being incorporated. (*B*) Shown here is the series of steps for building a SMRTbell sequencing template, in which the DNA fragment is ligated to universal hairpin adaptors. The *bottom* panel depicts DNA polymerase binding to the common polymerase initiation site within one of the adaptor sequences. (Redrawn, with permission, from Pacific Biosciences of California, Inc.)

read multiple times during the run (by repeated sequencing of the circular template) to generate a high-quality consensus sequence. This is called "circular consensus sequencing" (CCS) and is typically used to sequence PCR products for detection or validation of variants. The use of very long insert libraries and extended movie collection times (~75 min) permits the DNA polymerase to generate a very long read from each insert. These sequences can be combined with higher depth coverage from short reads, such as Illumina paired ends, to generate high contiguity, de novo assemblies of genomic sequence. In the "standard mode," fragments 1–3 kb in length can be sequenced in a single pass, using a movie collection time of ~45 min. The initial set of reads can be used to assemble, for example, a bacterial or viral genome. For larger genomes such as simple eukaryotic genomes, a combination of standard length and very long insert libraries could be used to close gaps between standard read length contigs and hence to provide a contiguous assembly of the genome. This system is also able to detect the presence of many DNA-modifying groups such as methyl-cytosine or 5-hydroxymethyl-cytosine, which can be simultaneously detected as the polymerase synthesizes along a template. In particular, this detection relies on the observed longer duration in the time between incorporation signals (so-called "interpulse distance," or IPD) when the polymerase encounters a modified base, relative to the standard IPD for incorporating an unmodified base (Flusberg et al. 2010).

Ion Torrent Personal Genome Machine (PGM)

Recently introduced (in late 2011), the Ion Torrent instrument uses concepts and processes very similar to the Roche/454 instrument (Fig. 9). Instead of detecting luciferase-catalyzed emission of light after nucleotide incorporation, the Ion Torrent sequencer detects the pH changes resulting

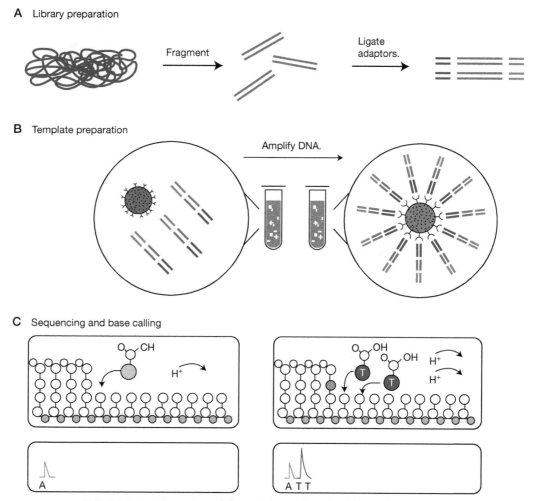

A Library preparation

Fragment

Ligate adaptors.

B Template preparation

Amplify DNA.

C Sequencing and base calling

FIGURE 9. **Ion Torrent workflow.** (Adapted from Ion Torrent Systems, Inc.)

from the release of hydrogen ions generated by the incorporation of nucleotide(s). Many processes shared with the 454 instrument include (1) emPCR amplification of adaptor-ligated library fragments on beads, (2) breaking the emulsion in preparation for sequencing, (3) loading of enriched beads onto the PGM Ion Chip following the emPCR step, and (4) stepwise sequencing by sequential addition of native nucleotides. Current read lengths on the Ion Torrent average 100 nucleotides, and a run requires ~2 h to complete. For further details on Ion Torrent and other "desktop" sequencers, see the box Desktop Genome Sequencers: The Future of Next-Generation Sequencing?

DESKTOP GENOME SEQUENCERS: THE FUTURE OF NEXT-GENERATION SEQUENCING?

Today's very-high-throughput next-generation sequencing instruments have transformed experimentation in fields as diverse as molecular biology, plant genomics, and human genetics; however, they have their limits. One of these is discussed in the box Coverage and Multiplexing: The Dilemmas Caused by the Rapid Growth of Sequencing Throughput. In addition, although the capacity of today's instruments is very high, so is their cost of operation. Likewise, the cost of purchasing the equipment and of the bioinformatics support to use efficiently the data they generate is significant.

As a result, three of the major equipment manufacturers have brought smaller, lower-throughput, lower-cost "desktop sequencers" into the market. These include the 454 GS Junior from Roche, the Ion Torrent PGM from Life Technologies, and the MiSeq from Illumina. As of the time of this writing, the GS Junior has been available to customers for more than a year, the Ion Torrent was released in mid 2011, and the MiSeq began shipping to early-access customers in late summer 2011. As a result of this timing, we are unable to provide a full description of these instruments, because only the GS Junior has been on the market for an extended period of time (see Table 1). However, desktop sequencers are likely to play a very important role in the future of sequencing and hence merit a discussion here. In addition, a section in the Introduction provides information about the Ion Torrent instrument.

All of these "personal" sequencing instruments are designed to produce greater amounts of data than a capillary sequencer. But their capacity is much lower than the corresponding high-end instruments. As a result, these instruments are more likely to be placed in smaller laboratories rather than large sequencing centers, although they will very likely find a role in technology development/optimization and in quality control of libraries in large centers.

Desktop sequencers have several potential uses including small- to medium-size sequencing projects, validation of variants detected using orthogonal chemistry (such as using the Ion Torrent to validate variants found by sequencing an entire genome on an Illumina machine), and testing the quality of libraries and exploring new methods. The Ion Torrent and MiSeq are both designed to run considerably faster than the high-end instruments from their respective manufacturers. As an example, the Illumina MiSeq requires ~1 d from library prep through sequence alignment to a reference genome; by comparison, the sequence run on Illumina HiSeq2000 requires in excess of 10 d for the highest-throughput data format. However, it must be noted that the HiSeq2000 will produce ~600 billion bases in that time, whereas the initial specification announced for the MiSeq is ~1.5 billion bases per run (it is striking that the MiSeq statistic is higher than the output announced for the original Solexa instrument in 2007). In addition to the speed and very useful throughput of these instruments, they will be substantially lower in both initial purchase price and in operating expenses than the high-end instruments. The new instruments have a per-run cost of well under $1,000 (at the time of this writing), compared with tens of thousands of dollars for the higher-end instruments. In addition, the instruments themselves will cost 10%–20% of the price of the high-end instruments from their respective manufacturers. It will be interesting to view the progress as these instruments develop through sequential generations of chemistry and hardware improvements. For many biologists, the majority of their sequencing needs may be addressed by these exciting new instruments.

SANGER SEQUENCING VERSUS NEXT-GENERATION SEQUENCING: WHEN TO DO WHAT?

Sanger-based (capillary) sequencing and next-generation sequencing differ sufficiently that it is generally obvious which technology to use and when. We give here a brief overview of current thinking on this topic and discuss areas in which the choice is more ambiguous. Sanger sequencing excels at generating very high accuracy and very long read length data from a small number of target molecules. The various next-generation sequencing platforms are best at generating shorter (shorter, at least, than Sanger capillary sequence) and less accurate data from a very large number of molecules. With these differences in mind, decisions concerning most pairings of application to instrument become clear.

At one extreme are projects in which the sequencing is used essentially to count the numbers of molecules of different sequence, for example, measuring relative RNA expression levels or mapping binding sites for proteins using ChIP-Seq (see Chapter 20, Protocol 6), and large-scale resequencing projects. These projects clearly require next-generation sequencing. At the other extreme are projects that require the sequencing of a specific clone or sequencing with a custom (i.e., nonuniversal) primer, or undertaking the sequencing of an isolated gene to determine if it contains a mutation relative to a reference sequence or to wild type, or to validate the sequence of a construct. For these projects, dideoxy/capillary sequencing is needed. It should be noted, however, that for sequencing a single gene in large numbers of patient samples, NGS is probably superior. To put this concept in perspective, consider that the target accuracy for the human genome reference, performed with capillary sequencing, was one error per 10,000 bases. The actual accuracy is probably closer to one error in several hundred thousand bases. Current NGS chemistry, coupled with the very high coverage that is readily obtainable with these instruments, results in accuracy better than one error in several million bases. Hence, the higher accuracy of next-generation sequencing in these types of assays may exceed that of the wild-type reference to which they are being aligned.

Perhaps the most complex sequencing application is de novo sequencing of a genome (i.e., sequencing a genome never previously sequenced, and hence for which there is no reference guide). A wealth of experience indicates that the most complete assemblies of genomes come from capillary sequencing. However, these efforts cost vastly more than next-generation sequencing assemblies. Current attempts at sequencing these genomes tend to use either the longest read format NGS instrument (454) and a short read format (Illumina or SOLiD) or, in some cases, 454 alone. The long reads of the Pacific Biosciences instrument may play a role here as well. In other situations, where appropriate, capillary reads are added into the assembly. In general, each investigator must blend the formats to find the best balance for the goals of the project in terms of completeness and the constraints of the project such as cost.

INTRODUCTION TO PROTOCOLS

We have provided here a collection of protocols divided into two parts. The first part covers various approaches for sample preparation—for capillary sequencing as well as for the representative platforms, Illumina and 454. The second part deals with evaluation and analysis of the sequencing results.

Capillary sequencing of DNA remains a mainstay in the toolkit of molecular biologists. Whether the technique is used to verify that a clone is constructed properly or to validate an interesting variant found by next-generation whole-genome sequencing, capillary sequencing is an extremely versatile tool. Capillary sequencing is typically performed on either plasmid DNA isolated from a bacterial clone (Protocol 1) or directly on a PCR product (Protocol 2; see also Chapter 7). In the latter case, primers are designed to amplify the region of interest from the source DNA. The amplified product is sequenced with a custom primer. Primers to amplify the region of interest are typically selected with software developed for this purpose. Primer 3, described in Chapter 8, Protocol 3, is probably the most commonly used software for this purpose. Protocol 3 describes capillary sequencing of plasmid DNA or of PCR products.

At the time of this writing, Illumina has become the most widely used next-generation platform, largely because of its impressive yield in sequencing output and its reduced costs of data production (see Table 1). We therefore include here methods for the preparation of various samples for use with the Illumina/Solexa sequencer. These protocols include preparation of libraries from whole genomes

(Protocols 4–8), from RNA (Protocol 9), and from exomes and targeted sources (Protocol 10). Various approaches to library quantitation are provided in Protocols 12–14.

The 454 platform has not experienced the tremendous decreases in data production costs as the Illumina and SOLiD platforms; however, it is still experiencing usage in two primary areas of genomics. First, the length of reads is advantageous in de novo assembly projects, wherein the genome of an organism is being sequenced and assembled for the first time. Overall, the combination of read lengths aids the primary assembly of the genome, and the inclusion of long distance mate-pair reads such as the 3-kb and 8-kb inserts provided in our protocols (Protocols 7 and 8) aids long-range assembly and contiguity. Second, the platform has a unique advantage in providing "validation" data due to the low incidence of substitution errors that results from the sequencing process. In particular, as bioinformatics-based approaches to variant detection have evolved, it has proven helpful to employ a technology such as the 454 sequencing-by-synthesis approach to provide confirmatory or validation data to these efforts. Because large numbers of PCR products or hybrid capture products can be sequenced as a pool, and because the substitution error rate is quite low, this platform provides a rapid and accurate means of sorting out correct variant calls from false positives. Protocols 15–17 describe, respectively, library preparation, emPCR, and how to carry out a sequencing run using the Roche/454 Pyrosequencer.

The protocols in the second section deal with postsequencing processing and analysis—the validation (Protocol 18) and quality control (Protocol 19) of sequencing data as well as bioinformatics-based analyses (Protocol 20).

ACKNOWLEDGMENTS

We have been ably assisted in compiling the protocols for this chapter by several people in our laboratories: in the Mardis laboratory, Lisa Cook and Henry Bauer; and, in the McCombie laboratory, protocols and valuable comments on the manuscript were provided by Eric Antoniou, Elena Ghiban, Melissa Kramer, Shane McCarthy, Stephanie Muller, Jennifer Parla, and Rebecca Solomon, with valuable assistance by Maureen Bell and Carolann Gundersen. The development and production of this chapter were promoted by the enthusiastic and professional help of those at Cold Spring Harbor Laboratory Press—our Editors, Kaaren Janssen and Alex Gann; Project Manager, Maryliz Dickerson; and Production Editor, Kathleen Bubbeo.

Finally, we would like to thank Jim Watson for his leadership in the Human Genome Project; for his invaluable support of genome sequence analysis at CSHL over nearly two decades, including our course on DNA sequencing; and for his lifelong stewardship of all things genomic.

REFERENCES

Albert TJ, Molla MN, Muzny DM, Nazareth L, Wheeler D, Song X, Richmond TA, Middle CM, Rodesch MJ, Packard CJ, et al. 2007. Direct selection of human genomic loci by microarray hybridization. *Nat Methods* 4: 903–905.

Bainbridge MN, Wang M, Burgess DL, Kovar C, Rodesch MJ, D'Ascenzo M, Kitzman J, Wu YQ, Newsham I, Richmond TA, et al. 2010. Whole exome capture in solution with 3 Gbp of data. *Genome Biol* 11: R62.

Carothers AM, Urlaub G, Mucha J, Grunberger D, Chasin LA. 1989. Point mutation analysis in a mammalian gene: Rapid preparation of total RNA, PCR amplification of cDNA, and *Taq* sequencing by a novel method. *BioTechniques* 7: 494-6–498-9.

Clark MJ, Chen R, Lam HY, Karczewski KJ, Chen R, Euskirchen G, Butte AJ, Snyder M. 2011. Performance comparison of exome DNA sequencing technologies. *Nat Biotechnol* 29: 908–914.

Craxton M. 1993. Cosmid sequencing. *Methods Mol Biol* 23: 149–167.

Davies K. 2001. *Cracking the genome: Inside the race to unlock human DNA.* The Free Press, New York.

Edwards A, Voss H, Rice P, Civitello A, Stegemann J, Schwager C, Zimmermann J, Erfle H, Caskey CT, Ansorge W. 1990. Automated DNA sequencing of the human HPRT locus. *Genomics* 6: 593–608.

Ewing B, Green P. 1998. Base-calling of automated sequencer traces using phred. II. Error probabilities. *Genome Res* 8: 186–194.

Ewing B, Hillier L, Wendl MC, Green P. 1998. Base-calling of automated sequencer traces using phred. I. Accuracy assessment. *Genome Res* 8: 175–185.

Flusberg BA, Webster DR, Lee JH, Travers KJ, Olivares EC, Clark TA, Korlach J, Turner SW. 2010. Direct detection of DNA methylation during single-molecule, real-time sequencing. *Nat Methods* 7: 461–465.

Gilbert W. 1981. DNA sequencing and gene structure (Nobel Lecture, December 8, 1980). *Biosci Rep* 1: 353–375.

Gilbert W, Maxam A. 1973. The nucleotide sequence of the *lac* operator. *Proc Natl Acad Sci* 70: 3581–3584.

Gnirke A, Melnikov A, Maguire J, Rogov P, LeProust EM, Brockman W, Fennell T, Giannoukos G, Fisher S, Russ C, et al. 2009. Solution hybrid selection with ultra-long oligonucleotides for massively parallel targeted sequencing. *Nat Biotechnol* 27: 182–189.

Gocayne J, Robinson DA, FitzGerald MG, Chung FZ, Kerlavage AR, Lentes KU, Lai J, Wang CD, Fraser CM, Venter JC. 1987. Primary structure of rat cardiac β-adrenergic and muscarinic cholinergic receptors obtained by automated DNA sequence analysis: Further evidence for a multigene family. *Proc Natl Acad Sci* 84: 8296–8300.

Gordon D, Abajian C, Green P. 1998. Consed: A graphical tool for sequence finishing. *Genome Res* 8: 195–202.

Henikoff S. 1990. Ordered deletions for DNA sequencing and in vitro mutagenesis by polymerase extension and exonuclease III gapping of circular templates. *Nucleic Acids Res* 18: 2961–2966.

Hodges E, Xuan Z, Balija V, Kramer M, Molla MN, Smith SW, Middle CM, Rodesch MJ, Albert TJ, Hannon GJ, et al. 2007. Genome-wide in situ exon capture for selective resequencing. *Nat Genet* 39: 1522–1527.

Korbel JO, Urban AE, Affourtit JP, Godwin B, Grubert F, Simons JF, Kim PM, Palejev D, Carriero NJ, Du L, et al. 2007. Paired-end mapping reveals extensive structural variation in the human genome. *Science* 318: 420–426.

Lander ES, Linton LM, Birren B, Nusbaum C, Zody MC, Baldwin J, Devon K, Dewar K, Doyle M, FitzHugh W, et al. 2001. Initial sequencing and analysis of the human genome. *Nature* 409: 860–921.

Ley TJ, Mardis ER, Ding L, Fulton B, McLellan MD, Chen K, Dooling D, Dunford-Shore BH, McGrath S, Hickenbotham M, et al. 2008. DNA sequencing of a cytogenetically normal acute myeloid leukaemia genome. *Nature* 456: 66–72.

Mardis ER. 2008. Next-generation DNA sequencing methods. *Annu Rev Genomics Hum Genet* 9: 387–402.

Margulies M, Egholm M, Altman WE, Attiya S, Bader JS, Bemben LA, Berka J, Braverman MS, Chen YJ, Chen Z, et al. 2005. Genome sequencing in microfabricated high-density picolitre reactors. *Nature* 437: 376–380.

Maxam AM, Gilbert W. 1977. A new method for sequencing DNA. *Proc Natl Acad Sci* 74: 560–564.

McCombie WR, Heiner C, Kelley JM, Fitzgerald MG, Gocayne JD. 1992. Rapid and reliable fluorescent cycle sequencing of double-stranded templates. *DNA Seq* 2: 289–296.

McElheney V. 2010. *Drawing the map of life: Inside the Human Genome Project.* Basic Books, New York.

Mullis KB, Faloona FA. 1987. Specific synthesis of DNA in vitro via a polymerase-catalyzed chain reaction. *Methods Enzymol* 155: 335–350.

Murray V. 1989. Improved double-stranded DNA sequencing using the linear polymerase chain reaction. *Nucleic Acids Res* 17: 8889.

Myers EW. 1995. Toward simplifying and accurately formulating fragment assembly. *J Comput Biol* 2: 275–290.

Okou DT, Steinberg KM, Middle C, Cutler DJ, Albert TJ, Zwick ME. 2007. Microarray-based genomic selection for high-throughput resequencing. *Nat Methods* 4: 907–909.

Parla JS, Iossifov I, Grabill I, Spector MS, Kramer M, McCombie WR. 2011. A comparative analysis of exome capture. *Genome Biol* 12:

R97. doi: 10.1186/gb-2011-12-9-r97.

Porreca GJ, Zhang K, Li JB, Xie B, Austin D, Vassallo SL, LeProust EM, Peck BJ, Emig CJ, Dahl F, et al. 2007. Multiplex amplification of large sets of human exons. *Nat Methods* 4: 931–936.

Saiki RK, Scharf S, Faloona F, Mullis KB, Horn GT, Erlich HA, Arnheim N. 1985. Enzymatic amplification of β-globin genomic sequences and restriction site analysis for diagnosis of sickle cell anemia. *Science* 230: 1350–1354.

Sanger F, Coulson AR. 1978. The use of thin acrylamide gels for DNA sequencing. *FEBS Lett* 87: 107–110.

Sanger F, Nicklen S, Coulson AR. 1977a. DNA sequencing with chain-terminating inhibitors. *Proc Natl Acad Sci* 74: 5463–5467.

Sanger F, Air GM, Barrell BG, Brown NL, Coulson AR, Fiddes CA, Hutchison CA, Slocombe PM, Smith M. 1977b. Nucleotide sequence of bacteriophage φX174 DNA. *Nature* 265: 687–695.

Sanger F, Coulson AR, Hong GF, Hill DF, Petersen GB. 1982. Nucleotide sequence of bacteriophage λ DNA. *J Mol Biol* 162: 729–773.

Shreeve J. 2004. *The genome war: How Craig Venter tried to capture the code of life and save the world.* Alfred A. Knopf, New York.

Smith LM, Sanders JZ, Kaiser RJ, Hughes P, Dodd C, Connell CR, Heiner C, Kent SB, Hood LE. 1986. Fluorescence detection in automated DNA sequence analysis. *Nature* 321: 674–679.

Smith DP, Johnstone EM, Little SP, Hsiung HM. 1990. Direct DNA sequencing of cDNA inserts from plaques using the linear polymerase chain reaction. *BioTechniques* 9: 48, 50, 52; passim.

Staden R. 1979. A strategy of DNA sequencing employing computer programs. *Nucleic Acids Res* 6: 2601–2610.

Staden R. 1996. The Staden Sequence Analysis Package. *Mol Biotechnol* 5: 233–241.

Sutcliffe JG. 1978. Nucleotide sequence of the ampicillin resistance gene of *Escherichia coli* plasmid pBR322. *Proc Natl Acad Sci* 75: 3737–3741.

Sutcliffe JG. 1979. Complete nucleotide sequence of the *Escherichia coli* plasmid pBR322. *Cold Spring Harb Symp Quant Biol* 43: 77–90.

Tuzun E, Sharp AJ, Bailey JA, Kaul R, Morrison VA, Pertz LM, Haugen E, Hayden H, Albertson D, Pinkel D, et al. 2005. Fine-scale structural variation of the human genome. *Nat Genet* 37: 727–732.

Venter JC, Adams MD, Myers EW, Li PW, Mural RJ, Sutton GG, Smith HO, Yandell M, Evans CA, Holt RA, et al. 2001. The sequence of the human genome. *Science* 291: 1304–1351.

Waterston RH, Lander ES. 2002. Initial sequencing and comparative analysis of the mouse genome. *Nature* 420: 520–562.

Wu R. 1970. Nucleotide sequence analysis of DNA. I. Partial sequence of the cohesive ends of bacteriophage and 186 DNA. *J Mol Biol* 51: 501.

Wu R, Donelson J, Padmanabhan R. 1970. Nucelotide sequence analysis of DNA. *In 8th International Congress on Biochemistry*, p. 166. Staples Printers, Kent, UK.

Wu R, Taylor E. 1971. Nucleotide sequence analysis of DNA. II. Complete nucleotide sequence of the cohesive ends of bacteriophage DNA. *J Mol Biol* 57: 491.

Preparing Plasmid Subclones for Capillary Sequencing

This protocol describes the preparation of plasmid DNA from bacterial cultures or from archived cultures, based on the standard miniprep method (see Chapter 1, Protocol 1). The resulting DNA is suitable in quantity and quality for use as template in capillary DNA sequencing as described in Protocol 3. An alternative method for preparing template is the amplification of a targeted region of DNA, described in Protocol 2. The protocol relies on the use of the Biomek automated liquid handling system.

MATERIALS

It is essential that you consult the appropriate Material Safety Data Sheets and your institution's Environmental Health and Safety Office for proper handling of equipment and hazardous materials used in this protocol.

Recipes for reagents specific to this protocol, marked <R>, are provided at the end of the protocol. See Appendix 1 for recipes for commonly used stock solutions, buffers, and reagents, marked <A>. Dilute stock solutions to the appropriate concentrations.

Note that recipes are provided for the reagents to be used for sequencing protocols. Buffers P1, P2, and P3 may be prepared in the laboratory according to the recipes or purchased from QIAGEN.

Reagents

Agarose gel (0.8% agarose) cast in $1 \times$ TAE <A>, containing 0.1 μg/mL ethidium bromide, and other reagents and equipment as required for agarose gel electrophoresis, described in Chapter 2, Protocol 1

E. coli cultures carrying plasmid clones to be analyzed

Ethanol (70%)

Glycerol (40%) <R>

Isopropanol (100%)

NaOH (10 M) <R>

P1 buffer <R>

P2 buffer <R>

P3 buffer <R>

RNase H (10 mg/mL) (DNase free) <R>

SDS (10%) <R>

Sodium acetate (10 mM) <R>

Tris Cl (10 mM, pH 8.0) <R>

Equipment

Biomek Fx workstation (Beckman Coulter)

Boxes (1 mL, 96 well), round-well (Beckman Coulter)

Costar assay plates (96 well)

Foil tape

Growth boxes (2 mL), deep well (Whatman)

Jouan centrifuge

Lysate clarification plate (Whatman)

Plate sealer (plastic)
Plate vortexer
Quantification equipment (spectrophotometer, fluorometer, etc.)
Sealing mat (Silicone)
Shaker–incubator
Skan Wash 2000 microplate washer
Timer

METHOD

Preparation of Clones for Archiving and Plasmid Isolation

1. Grow the isolates of *E. coli* cultures carrying clones in Whatman 2-mL deep-well growth boxes for 16–24 h.

   Be sure to incubate bacterial cultures at the correct temperature for as close to 16–24 h as possible. It is very important that there be a 1:1 ratio between the amount of liquid to air (1 mL of culture in a 2-mL box) and that the box be sealed with a breathable membrane. Shaking at the correct speed in the incubator shaker is also crucial to optimal bacterial growth. The temperature and speed will depend on the strain. A "typical" growth setting would be 37°C at 275 rpm.

2. Remove the Whatman 2-mL deep-well growth boxes from the shaker–incubator.

3. Using the Biomek Fx, transfer 50 μL of each overnight culture with 50 μL of sterile 40% glycerol into a well of the Costar 96-well assay plate.

4. Seal the Costar plate with foil tape and vortex for 5 min. Centrifuge the plate briefly and store the archives at −80°C.

5. Place Whatman 2-mL deep-well boxes with remaining cultures (from Step 2) in a Jouan centrifuge. Centrifuge for 5 min at 2700 rpm.

6. Remove the boxes from the centrifuge, and carefully pour off the supernatants into a sink (do not disturb the pellets). Place the boxes upside down on paper towels to allow excess media to drain off (~1 min).

   Before continuing, check the sizes of the pellets:

   Acceptable size   O O O

   Unacceptable size   O or smaller

   Do not proceed if >30% of the pellets are of an unacceptable size. The bacterial pellet and the well must be free of all residual liquid in order for lysis to occur efficiently.

   ▲ *Proceed directly to bacterial lysis (Step 7) to seal the boxes with foil tape and store right-side up in a −20° freezer.*

Bacterial Lysis

7. Using the Biomek FX, add 200 μL of P1 buffer to each well, and seal the box with plastic sealer.

8. Vortex the boxes on a plate vortexer for 10 min at the highest setting.

   It may be necessary to vortex for longer periods to ensure complete resuspension of the bacterial pellets. Complete resuspension of the bacterial pellets is critical for maximum lysis.

9. Begin the lysis step.

   i. Remove the plate sealer from the boxes, and set the timer for 5 min.

   ii. Using the Biomek FX, add 200 μL of P2 buffer to each well, and immediately seal with a plastic plate sealer.

   iii. Start a timer and mark this first box with a "1."

   iv. Gently tilt the box to a 45° angle six times.

   v. Repeat Steps 9.ii–9.iv for all of the boxes, numbering as appropriate.

Bacteria should be exposed to P2 buffer for no more than 5 min. P2 buffer must be made fresh every 2 d. Check the date on the bottle before using.

10. Perform neutralization.

    i. After 5 min, add 200 μL of cold P3 buffer to each well of the box marked "1."

    ii. Seal the box tightly with plate sealer; gently tilt the box six times and place it on ice.

    iii. Repeat Steps 10.i–10.iii for the remaining boxes.

    iv. After the last box is put on ice, start the timer for 10 min.

      Keep the P3 buffer stored at 4°C.

Filtration Setup

11. Add 240 μL of 100% isopropanol to each well of a 96-well, 1-mL round-well box. Fill the appropriate number of boxes to process all lysates.

12. To create the filter assembly, stack a lysate clarification plate on top of the round-well box.

13. After the boxes have been submerged in ice for 10 min, use the Biomek FX to transfer 400 μL of each bacterial lysate into the filter assembly. Transfer the contents of all of the boxes.

    Make sure that a white precipitate is visible inside the wells before transfer.

14. Centrifuge the filter assembly in a Jouan centrifuge at 1000 rpm for 3 min.

15. Discard the top filter plate and seal the bottom round-well box with a silicone sealing mat. Invert the box six times to mix the contents—quickly to avoid contamination between the wells of the plate.

    Make sure that the filter plate does not contain any liquid before discarding.

16. Centrifuge the box at 3800 rpm for 15 min.

17. Immediately pour off the supernatant into an alcohol waste container, and blot the box briefly on paper towels.

18. Using the Skan Wash 2000, add 300 μL of 70% ethanol to each well and reseal the plate using the silicone sealing mat. Invert several times to wash the wells completely.

    The same mat can be used (as in Step 15) or a new one, if preferred, to reduce contamination of wells.

19. Centrifuge at 3800 rpm for 5 min.

20. Pour the supernatant into an alcohol waste container.

21. Place the boxes upside down on paper towels for ~5 min to allow residual liquid to drain off.

22. Repeat Steps 18–21.

    ▲ *Boxes may be stored at room temperature overnight to allow the DNA pellets to dry (DNA resuspension is performed the next day). Or boxes may be spun briefly and incubated for 1 h at 37°C to dry the pellets.*

    In either case, ensure that the wells are completely dry before continuing.

    See Troubleshooting.

DNA Resuspension and Random Quantitation

23. Using the Biomek FX, add ~70 μL of 10 mM Tris Cl (pH 8.0) to each well. (The elution volume varies, depending on your downstream application.)

24. Seal the boxes with a sealing mat and vortex for 10 min at the highest setting.

25. Tap the boxes down gently to collect liquid at the bottom of the wells. Using the Biomek FX, transfer the entire contents of the well into a 96 well Costar assay plate.

26. Seal the plates using foil tape and store at −20°C.

27. To determine if the plasmid quantity and quality are satisfactory, pick three wells from each 96-well Costar assay plate and prepare samples for electrophoresis. Run 3 μL of each sample with 7 μL of 1× loading dye on a 0.8% agarose gel. Include a lane with a 1-kb marker ladder.
    See Troubleshooting.

28. Determine the quantification of template, if desired, by using any form of spectrophotometer, fluorometer, or the like.
    See Troubleshooting.

TROUBLESHOOTING

Problem (Step 27): An additional band (in addition to the plasmid DNA) is seen in the agarose gel.
Solution: If RNAse H is *not* added to the P1 buffer, RNA may appear as an additional band on an agarose gel. This problem can usually be solved by treating the template DNA with RNAse when the entire procedure is completed.

Problem (Steps 22 and 28): The A_{260}/A_{280} ratio (spectrophotometer) of template DNA is higher than expected ($>1.8–2.0$).
Solution: Possibly, ethanol is present in the sample. It is critical that *all* traces of 70% ethanol are removed from the deep-well boxes (in Step 22) before resuspension and elution of template DNA. This process may take up to an hour and a half (in a 37°C incubator) or overnight (at room temperature).

DISCUSSION

Depending on your cloning efficiency (and whether your clones are derived from low- or high-copy-number plasmids), a typical yield from this preparation is somewhere in the range of 100–500 ng/μL. As noted above, the elution volume will depend on your downstream application.

RECIPES

It is essential that you consult the Material Safety Data Sheets and your institution's Environmental Health and Safety Office for proper handling of equipment and hazardous materials used in this protocol.

Glycerol (40%) (1 L)

1. In a clean 1-L beaker add:

H₂O, autoclaved, distilled	600 mL
Glycerol (100%)	400 mL

2. Stir the solution for 5 min. Pour it into a 1-L bottle and autoclave.

NaOH (10 M)

▲ *Use caution when preparing this solution because it may cause severe burns.*

1. In a clean 500-mL beaker, add:

NaOH pellets (wear gloves)	100 g
H₂O, autoclaved, distilled	100 mL up to 250 mL

2. Add a sterile stir bar and stir until the pellets are completely dissolved.

   ▲ *CAUTION: This is a highly exothermic reaction.*

3. Carefully add autoclaved, distilled H_2O until the volume of the solution is 250 mL.

4. Store in glass bottle at room temperature. This is stable for 1 yr at room temperature.

P1 Buffer (1 L)

1. In a clean 2-L beaker, add:

H_2O, autoclaved, distilled	920 mL
Tris-Cl (1 M, pH 8.0)	50 mL
EDTA (0.5 M, pH 8.0)	20 mL
RNase H (DNase free) (10 mg/mL)	10 mL

2. Add a sterile stir bar and stir for 5 min.

3. Filter and store at 4°C. Label the bottle with the date (prepare fresh for each day of use).

P2 Buffer (1 L)

This solution should be made every 2 d.

1. In a clean 1-L beaker, add:

H_2O, autoclaved, distilled	880 mL
NaOH (10 M)	20 mL
SDS (10%)	100 mL

2. Add a sterile stir bar and stir for 5 min. To avoid SDS preacipitation, add the NaOH solution to the water and stir for few minutes before adding the SDS solution. Label the bottle with the date.

3. P2 buffer should be stored at room temperature to avoid SDS precipitation. If precipitation does occur, warm the bottle to 37°C and stir for a few minutes.

P3 Buffer

1. In a clean 1-L beaker, add:

Potassium acetate	294.5 g
H_2O, autoclaved, distilled	500 mL

2. Stir until crystals are dissolved completely, and then add 110 mL of glacial acetic acid.

3. Stir continuously while monitoring pH. Add autoclaved, distilled H_2O to 900 mL. The pH should be between 4.8 and 5.5; if not, titrate with glacial acetic acid. Adjust the final volume of the solution to 1 L.

4. Filter and store at 4°C.

RNase H (10 mg/mL) (DNase free)

1. In a clean 15-mL Falcon tube, add:

Bovine pancreatic RNase H	100 mg
Sodium acetate (10 mM, pH 4.8)	9 mL

2. Dissolve RNase crystals completely by vortexing.

3. Place the tube in a boiling water bath for 15 min. Remove the tube and allow it to cool to room temperature (~5 min).

4. Add 1 mL of 1 M Tris-Cl and vortex for 30 sec.

5. Five-milliliter aliquots of this solution may be stored at −20°C. RNase H is stable for 6 mo at −20°C.

SDS (10%)

▲ *Use caution when preparing this solution. SDS is a potential neurotoxin. Prepare it under a fume hood, wearing a mask.*

1. In a clean 500-mL beaker, add:

SDS	50 g[a]
H2O, autoclaved, distilled[b]	500 mL

[a]Wear gloves.
[b]Add carefully.

2. Add a sterile stir bar and stir until SDS is completely dissolved.

3. Filter and store at room temperature. This is stable for 6 mo at room temperature.

Sodium Acetate (10 mM)

1. In a clean 50-mL Falcon tube, add:

Sodium acetate (3 M, pH 4.8)	167 μL
H_2O, autoclaved, distilled	40 mL

2. Vortex for 30 sec and add autoclaved, distilled H_2O to 50 mL.

3. Store at room temperature. This is stable for 6 mo at room temperature.

Tris Cl (10 mM, pH 8.0)

1. In a clean 1-L beaker, add:

H_2O, autoclaved, distilled	900 mL
Tris-Cl (1 M, pH 8.0)	10 mL

2. Stir the solution for 5 min. Pour into a 1-L bottle and autoclave.

Preparing PCR Products for Capillary Sequencing

This protocol describes the preparation of amplified products for use in Sanger-based capillary DNA sequencing (see Protocol 3), for example, to verify a clone or construct. The approach is based on the use of the LongAmp protocol (see Chapter 7, Protocol 5) for DNA amplification and reagents for BigDye sequencing (Protocol 3).

Amplified samples, subsequently treated with a mixture of exonuclease and shrimp alkaline phosphatase to remove unincorporated primers and dNTPs left from PCR, may be used directly for sequencing. This protocol relies on the use of the Biomek FX Workstation or a multichannel pipettor.

MATERIALS

It is essential that you consult the appropriate Material Safety Data Sheets and your institution's Environmental Health and Safety Office for proper handling of equipment and hazardous materials used in this protocol.

Reagents

Agarose gel (1.0 % agarose) cast in $1\times$ TAE <A>, containing 0.1 μg/mL ethidium bromide, and other reagents and equipment as required for agarose gel electrophoresis, described in Chapter 2, Protocol 1

DNA sample (e.g., clone or construct to be verified)

dNTPs solution, containing all four dNTPs, each at a concentration of 10 mM (NEB)

Ethanol (100 %)

Exonuclease I (USB)

LongAmp *Taq* DNA polymerase (NEB)

LongAmp *Taq* reaction buffer ($5\times$, with Mg^{2+}) (NEB)

NaOAc (3 M, pH 4.8) (Sigma-Aldrich)

Primers (diluted to a concentration of 10 μM; preferably in Tris-HCl)
Left and right (forward and reverse) primers are specific for the segment of target DNA to be amplified.

Shrimp alkaline phosphatase (USB)

Water, PCR grade

Equipment

Biomek FX workstation or multichannel pipettes (electronic or manual)

Centrifuge

Conical tube (50 mL)

Ice bucket or tray with a lid

PCR plate (96 well) (skirted or unskirted)

Plate and/or tube vortexer

Sealing mat (use a silicon capmat or thermal sealing tape)

Thermal cycler

METHOD

Amplification of Template

1. Label a 96-well PCR plate.

2. Deliver 4 µL of DNA sample (at 25 ng/µL) into the appropriate wells.

   Here, a total concentration of 100 ng is used in each reaction. If limited material is available, adjust accordingly, so that the total volume of the reaction is 50 µL.

3. Prepare a "lower mix" as follows. The volumes given below are those required for a single reaction. Make enough lower mix to accommodate all of the samples and controls in the experiment.

LongAmp *Taq* reaction buffer (5×, with Mg²⁺)	5 µL
dNTP solution of four dNTPs, each at a concentration of 10 mM	1.5 µL
Water, PCR grade	10.5 µL (total volume of 17 µL)

4. Add 17 µL of "lower mix" to the appropriate wells containing the DNA sample(s), using a multichannel pipette.

5. Deliver the primers: add 2 µL of each of two 10 µM primers (e.g., 2 µL of 10 µM left primer and 2 µL of 10 µM right primer) to the appropriate wells, being careful to change tips to avoid contamination between wells.

6. Seal the 96-well plate and centrifuge briefly to bring all of the liquid to the bottom of the wells. Preheat a thermal cycler before proceeding to the next step (see the parameters in Step 9).

   The entire program is ~1 h and 30 min of cycling.

7. Prepare an "upper mix" as follows. The volumes given below are those required for a single reaction. Make enough upper mix to accommodate all of the samples and controls in the experiment.

LongAmp *Taq* reaction buffer (5× with Mg²⁺)	5 µL
LongAmp *Taq* DNA polymerase	2 µL
Water, PCR grade	18 µL (total volume of 25 µL)

8. Add 25 µL of "upper mix " to the appropriate wells, using a multichannel pipette. (It is important to "layer" the "upper mix" on top of the "lower mix/template/primer" layer.) Seal the plate with either a silicon capmat or thermal sealing tape; proceed immediately to the thermal cycler.

9. Place the tubes in a thermal cycler, programmed as suggested in the following table (typical cycling conditions are shown here), and perform amplification.

Cycle number	Denaturation	Annealing and polymerization
1	1 min at 94°C	
35 cycles	15 sec at 94°C	1 min+20 sec at 62°C
Last cycle		5 min at 62°C

   Most thermal cyclers have an end routine in which the amplified samples are incubated at 4°C until they are removed from the machine.

   The annealing temperature should be 5°C below the minimum average melting temperature. (Primer information should be provided by your vendor.) Note that when using primers with annealing temperatures above 60°C, a two-step protocol is possible, as shown here.

   The polymerization (extension) time is typically 50 sec/kb.

10. Perform a quality control (QC) check by electrophoresis of a few selected samples through a 1% agarose gel.

    See Troubleshooting.

Treatment with Exonuclease/Shrimp Alkaline Phosphatase

An exonuclease/shrimp alkaline phosphatase cleanup is performed on PCR amplicons before capillary sequencing to eliminate unincorporated primers and dNTPs.

11. Prepare the "Exo-SAP master mix" by combining the following reagents. The volumes given below are those required for a single reaction. Make enough master mix to accommodate all of the samples to be treated.

Exonuclease I	0.3 μL
Shrimp alkaline phosphatase	0.15 μL
Water, PCR grade	1.55 μL (total volume of 2 μL)

12. Add 2 μL of Exo-SAP master mix to the bottom of each well of a PCR plate.

13. Add 8 μL of each PCR product to the appropriate wells for a total reaction volume of 10 μL.

    This volume of PCR product can be scaled up or down, according to the amount of product required for the sequencing reaction, which depends on the quality and quantity of the amplicon.

14. Seal the plate with either a silicon capmat or thermal sealing tape, then centrifuge briefly and place in a thermo cycler. Perform the following program:

    1 h at 37°C

    then 30 min at 72°C

    The plate can be stored at –20°C or can be used directly in sequencing reactions.

TROUBLESHOOTING

Problem (Step 10): The QC assessment suggests that the amplification failed.

Solution: Before amplifying "sample" DNA, perform a test PCR with control DNA to determine the quality of the primers. Purify the amplification products through an agarose gel, and view the results. If the gel image is acceptable (i.e., a single band of the appropriate size is seen), proceed with "sample" DNA. If the gel image is unacceptable (multiple bands or primer dimers), alter (raise or lower) the annealing temperature (for further details, see the introduction to Chapter 7).

Cycle-Sequencing Reactions

Cycle sequencing is used for direct sequencing of clones to identify a mutation or structure of interest. Here the template is typically prepared from plasmid minipreps (described in Protocol 1). For sequencing reactions, amplicons, prepared as described in Protocol 2, are sequenced individually with each of the two primers used in the LongAmp amplification reactions. Both template preparations use primers designed specifically for "targets"—either for direct sequencing or for validation.

The method, which relies on the use of the ABI3730xl capillary sequencer, can also be used to validate genotypic data and to independently test the results obtained from advanced DNA sequencing technologies (e.g., Illumina or Pacific Biosciences; see the discussions of these platforms in the Introduction).

MATERIALS

It is essential that you consult the appropriate Material Safety Data Sheets and your institution's Environmental Health and Safety Office for proper handling of equipment and hazardous materials used in this protocol.

Reagents

ABI BigDye v1.1 (Life Technologies)

ABI sequencing buffer (5×) (Life Technologies)

Agarose gel (1.0% agarose) cast in 1× TAE <A>, containing 0.1 µg/mL ethidium bromide, and other reagents and equipment as required for agarose gel electrophoresis, described in Chapter 2, Protocol 1

dNTPs solution, containing all four dNTPs, each at a concentration of 10 mM (NEB)

Ethanol (100% and 70%)

Primers (diluted to a concentration of 10 µM, preferably in Tris-HCl)

Left and right (forward and reverse) primers are specific for the segment of target DNA to be sequenced.

Template DNA

The template here can be either the plasmid prep from Protocol 1 or the amplified PCR product (from the Exo-SAP cleanup) from Protocol 2. (See Discussion section.)

Equipment

ABI3730xl capillary sequencer

Biomek FX workstation or multichannel pipettes (electronic or manual)

Centrifuge

Conical tube (50 mL)

Ice bucket or tray with a lid

PCR plate (96 well) (skirted or unskirted)

Plate and/or tube vortexer

Sealing mat (use a silicon capmat or thermal sealing tape)

Thermal cycler

METHOD

Cycle Sequencing

1. Prepare a cycle-sequencing reaction mix. The volumes given below are those required for a single reaction. Make enough mix to accommodate all of the samples to be sequenced.

ABI BigDye version 1.1	1 μL
ABI sequencing buffer (5×)	1.5 μL
Water, PCR grade	1.5 μL (total volume of 4 μL)

2. Deliver 4 μL of cycle-sequencing reaction mix into each well of a 96-well PCR plate.

3. Add 1 μL of 10 μM the appropriate primer to each well.

   The PCR products from Protocol 2 are sequenced with the same primers that were used to amplify them.

4. Add the appropriate volume of template DNA. Because the total volume of the reaction is 10 μL, the volume of template DNA product should not exceed 5 μL. If the volume is <5 μL, add the appropriate amount of PCR-grade water to make up the difference in volume.

   Preheat the thermal cycler before proceeding to the next step (see the settings in Step 7).

5. Seal the plate with either a silicon capmat or thermal sealing tape.

6. Centrifuge the plate to collect reagents, and proceed to the thermal cycler.

7. Place the plate in a thermal cycler, programmed as suggested in the following table, and perform the sequencing run:

Cycle number	Denaturation	Annealing	Polymerization
1	2 min at 96°C		
35 cycles	10 sec at 96°C	5 sec at 50°C	4 min at 60°C

   Hold at 4°C.

Precipitation of BigDye Cycle-Sequencing Reactions

8. Immediately after cycling has completed, centrifuge the plate to collect reagents and products at the bottoms of the wells.

9. To prepare a "master precipitation mix," combine 1 mL of 3 M NaOAc with 23 mL of 100% ethanol in a 50-mL conical tube.

10. Deliver 25 μL of the "master precipitation mix" into the appropriate wells of the cycle-sequencing plate.

11. Quickly vortex the plate to mix the contents and centrifuge to collect all of the liquid to the bottoms of the wells.

12. Place the plate on ice in a covered bucket or tray (ABI BigDye reagent is light sensitive) for 30 min.

    Be sure that all wells are submerged in ice.

13. Centrifuge the plate for 30 min at 4000 rpm.

14. Decant the liquid from the plate, and gently tap out any remaining liquid onto a paper towel, being careful not to disturb the pellet.

15. Add 200 μL of 70% ethanol to the appropriate wells, seal the plate with a silicon capmat, and invert it three times to clean the sides of the wells.

16. Centrifuge the plate for 15 min at 4000 rpm.

17. Decant the liquid from the plate, and gently tap the plate on a paper towel.

18. Repeat Steps 15–17, for a total of two 70% ethanol washes.

19. After the second wash, centrifuge the plate again, this time inverted on a folded paper towel, at 200 rpm to remove any excess ethanol.

20. Dry the plate, by placing it in a thermal cycler either for 10 min at 37°C or for ~1 h at room temperature.

   ▲ *It is critical that* all *ethanol be removed from the plate before the pellets are resuspended.*

21. After it is determined that all wells are dry, add 30 μL of PCR-grade water to the appropriate wells of the plate and seal with a capmat.

22. Vortex on a plate vortexer for 10 min at high speed (2500 rpm).

23. To prepare the samples for analysis on the ABI3730xl capillary sequencer, dilute them with formamide in a 384 plate as follows.

   i. Into the appropriate number of wells in a 384 plate, deliver 4 μL of formamide per well.

   ii. To each well, add 3 μL of resuspended precipitated sequencing product, to make up a total volume of 7 μL to be loaded on the sequencer.

      384 well plates are used here so as not to overload the sequencer. Overloading can result in the signal intensity being too high, resulting in unclear or incorrect base-calling.

DISCUSSION

The volume of amplified product (purified by treatment with exonuclease/shrimp alkaline phosphatase) required for these reactions is estimated from the intensity of the band in the gel(s) used to check the quality of amplicons (visual normalization). As a rough guide, if a band appears "weak," 5 μL of Exo-SAP cleaned product would be used in the cycle-sequencing reaction. For a band that appears "bright," 1–2 μL of Exo-SAP cleaned product would be used. Alternatively, for sequencing/validation of results obtained on the ABI 3730×1 capillary sequencer, a 1/8 volume of the BigDye v.1.1 reaction would be used.

Whole Genome: Manual Library Preparation

This protocol describes a manual approach for the preparation of genomic DNA libraries suitable for Illumina sequencing. Genomic DNA fragments produced by shearing by sonication are ligated to adaptors and amplified by PCR. The amplified DNA, separated by size and gel-purified, is suitable for use as template in whole-genome sequencing. Alternative methods using an automated approach are described in Protocols 5 and 6.

MATERIALS

It is essential that you consult the appropriate Material Safety Data Sheets and your institution's Environmental Health and Safety Office for proper handling of equipment and hazardous materials used in this protocol.

Recipes for reagents specific to this protocol, marked <R>, are provided at the end of the protocol. See Appendix 1 for recipes for commonly used stock solutions, buffers, and reagents, marked <A>. Dilute stock solutions to the appropriate concentrations.

Reagents

Acrylamide gels, precast (Life Technologies)
dATP (1 mM)
DNA ladder (100 bp) (NEB)
DNA ladder (50 bp) (TrackIt; Life Technologies)
DNA 1000 reagents (Agilent)
DNA Terminator End Repair Kit (Lucigen)
Ethanol (70% and 100%)
FlashGel DNA Cassette (2.2%, 12+1 well) (Lonza)
FlashGel DNA marker (50 bp–1.5 kb) (Lonza)
FlashGel loading dye (Lonza)
Gel-loading buffer with bromophenol blue (10×) <A>
Genomic DNA
GlycoBlue (15 mg/mL) (Ambion)
Klenow fragment exonuclease (New England Biolabs)
MinElute PCR Purification Kit (QIAGEN)
NaCl (400 mM)
PE forked adaptor duplex (4 μM) (Illumina, Sigma-Aldrich)
PE PCR Primer 1.0 (8 μM) (HPLC purification, 100-nm scale; IDT):
    A*ATGATACGGCGACCACCGAGATCTACACTCTTTCCCTACACGACGCTCTTCCGATCT
PE PCR Primer 2.0 (8 μM) (HPLC purification, 100-nm scale; IDT):
    C*AAGCAGAAGACGGCATACGAGATCGGTCTCGGCATTCCTGCTGAACCGCTCTTCCGATCT
Phusion High-Fidelity PCR Master Mix with HF Buffer (2×) (Finnzymes)
Quick Ligation Kit (NEB)
SYBR Green I nucleic acid gel stain (Life Technologies)
TBE buffer <A>
TBE gel (Novex 4%–12%, 1.0 mm × 12 well) (Life Technologies)

Equipment

DNA 1000 chips and analyzer (Agilent)
Fluorometer (e.g., Qubit; Life Technologies)
HV DuraPore filter columns (0.4 μm) (Millipore)
Microcentrifuge tubes (0.2 mL and 1.7 mL)
microTube (6×16 mm), with AFA fiber and Snap-Cap (Covaris)
Quant-iT dsDNA High Sensitivity Assay Kit (Life Technologies)
Sonicator (Covaris)
THQmicro microTube holder (Covaris)
XCell SureLock Mini-Cell (Life Technologies)

METHOD

Library Sonication and End Repair

1. Prepare a sample of 100 ng, 500 ng, 1 μg, or 3 μg of genomic DNA.

2. Dilute the DNA in a 1.7-mL tube, as follows:

End repair buffer (5×)	10 μL
DNA sample	x μL
Water	$(40-x)$ μL
Total volume	50 μL

   Vortex the tube for 1–2 sec. Centrifuge the tube for 5 sec.

3. Transfer the diluted DNA to a Covaris sample vessel. Centrifuge the vessel to ensure that there are no bubbles in solution.

4. Sonicate the sample according to the manufacturer's instructions.

   Provide your desired fragment size in order to determine the proper program to use.

5. After shearing the DNA, remove the sample vessel from the water bath. Centrifuge to collect the sample at the bottom of the vessel.

6. Add 2 μL of End repair enzyme mix to the sample vessel. Vortex lightly to mix, and centrifuge briefly. Incubate the reaction for 30 min at room temperature.

7. Combine the 50-μL reaction and 250 μL of Buffer PB or PBI in a 1.7-mL tube. Vortex the tube.

8. Purify the DNA using a MinElute column, as follows.

   i. Add the reaction/buffer mix to a MinElute column. Centrifuge the column at 14,000g for 1 min or place on a vacuum manifold and draw the liquid through the column.

   ii. Dispose of waste if the tube was centrifuged.

   iii. Add 750 μL of Buffer PE. Centrifuge the column at 14,000g for 1 min or place on a vacuum manifold and draw the liquid through the column.

   iv. Dispose of waste if the tube was centrifuged.

   v. Add 750 μL of Buffer PE. Centrifuge the column at 14,000g for 1 min or place on a vacuum manifold and draw the liquid through the column.

   vi. Dispose of waste if the tube was centrifuged. If using a vacuum manifold, place the column in a waste collecting tube, and centrifuge the column at 14,000g for 1 min.

   vii. Rotate the tube 180°. Centrifuge the column at 14,000g for 1 min.

   viii. Dispose of the waste tube.

   ix. Transfer the column to a 1.7-mL tube.

   x. Add 16 μL of Buffer EB to the column, and incubate it for 1 min at room temperature. Centrifuge the column at 14,000g for 1 min.

xi. If necessary, add another 16 μL of Buffer EB to the column, and incubate it for 1 min at room temperature. Centrifuge the column at 14,000g for 1 min.

> Two elutions are necessary only if a large amount of DNA was sonicated (e.g., for 3 μg of input DNA).

xii. Discard the column, and transfer the sample to a 0.2-mL PCR tube.

3′ Adenylation (A-Tailing)

9. Set up the A-tailing reaction using the purified DNA from Step 8.xii.

Purified end repair DNA	32 μL
Klenow buffer (10×)	5 μL
dATP (1.0 mM)	10 μL
Klenow exonuclease (5 U/μL)	3 μL
Total volume	50 μL

Pipette up and down to mix, centrifuge briefly, and incubate the reaction for 30 min at 37°C.

10. Combine the 50-μL reaction and 250 μL of Buffer PB or PBI in a 1.7-mL tube. Vortex the tube.

11. Purify the DNA using a MinElute column by following Steps 8.i–8.xii, except alter the Buffer EB elution, using the values in the following table:

Input			
100 ng	500 ng	1 μg	3 μg
10 μL twice	10 μL twice	18 μL twice	18 μL twice

Adaptor Ligation

12. Using the purified product from the A-tailing reaction (purified in Step 11), prepare the adaptor ligation reaction as follows:

	Input			
	100 ng	500 ng	1 μg	3 μg
Purified A-tailed DNA	20 μL	20 μL	20 μL	20 μL
PE forked adaptor duplex	1.0 μL	2.5 μL	5.0 μL	7.0 μL
NEB Quick Ligation Buffer (2×)	25 μL	25 μL	25 μL	25 μL
Water	1.5 μL	—	—	—
NEB Quick Ligase (2000 U/μL)	2.5 μL	2.5 μL	2.5 μL	2.5 μL
Total volume	50 μL	50 μL	52.5 μL	54.5 μL

i. Combine the DNA and the adaptor in a tube. Pipette to mix. Centrifuge briefly.

ii. Add the 2× buffer and water. Pipette to mix. Centrifuge briefly.

iii. Add the ligase. Pipette to mix. Centrifuge briefly.

iv. Incubate the reaction for 15 min at room temperature.

13. Purify the DNA using a MinElute column by following Steps 8.i–8.xii, except use 10 μL of Buffer EB elution in Steps 8.x and 8.xi.

14. Determine the concentration of the DNA by assaying 1 μL of the DNA with the Quant-iT dsDNA HS Assay Kit (for 100-ng input) or the BR Assay Kit (for the others) and reading the results on a fluorometer (Protocol 11 or 12 may be useful).

> The yield of recovered ligated DNA should be ≥20%, relative to the input DNA. For example, a 100-ng input library should generate ≥20 ng of ligated DNA.

If the concentration of the recovered DNA is >10 ng/μL, dilute the sample with Buffer EB as follows:

Input DNA	Dilute to
100 ng	1 ng/μL
500 ng	10 ng/μL
1 μg	10 ng/μL
3 μg	10 ng/μL

A sample calculation follows.

> The elution volume of sample after quantitation is 10.5 μL. The DNA concentration is 21.6 ng/μL. The initial library input was 500 ng; therefore, the DNA needs to be diluted to 10 ng/μL. The following formula will provide the amount of Buffer EB to add to the sample to obtain a 10 ng/μL sample:
>
> Final volume = (10.5 μL × 21.6 ng/μL)/(10 ng/μL) = 22.68 μL.
>
> 22.68 μL − 10.5 μL = 12.18 μL ≈ 12.2 μL ← Add this amount of Buffer EB to sample to dilute it to 10 ng/μL.

PCR Cycle Optimization

15. Set up one amplification reaction for each library construction reaction generated in 0.2-mL PCR tubes. Irrespective of the number of samples used, perform only one negative control (Buffer EB instead of DNA). If desired, a master mix can be made without DNA and aliquoted into 0.2-mL PCR tubes. Add 1 μL of sample or Buffer EB (for negative control), then mix and centrifuge briefly.

Purified DNA sample from ligation (Step 13)	1 μL
Phusion PCR master mix (2×)	25 μL
PCR PE Primer 1.0 (8 μM)	1 μL
PCR PE Primer 2.0 (8 μM)	1 μL
Water	22 μL
Total volume	50 μL
Buffer EB	1 μL
Phusion PCR master mix (2×)	25 μL
PCR PE Primer 1.0 (8 μM)	1 μL
PCR PE Primer 2.0 (8 μM)	1 μL
Water	22 μL
Total volume	50 μL

16. Create two PCR programs.

Program 1

Cycle number	Denaturation	Annealing	Polymerization
1	30 sec at 98°C		
4 or 6 cycles[a]	10 sec at 98°C	30 sec at 65°C	30 sec at 72°C

[a]Run 4 cycles for 10-ng PCR or 6 cycles for 1-ng PCR.

Hold at 4°C.

Program 2

Cycle number	Denaturation	Annealing	Polymerization
2 cycles	10 sec at 98°C	30 sec at 65°C	30 sec at 72°C

Hold at 4°C.

17. Label a 0.2-mL tube for the PCR negative control.

18. Label six 0.2-mL tubes.

for 1-ng PCR	PCR-6, PCR-8, PCR-10, PCR-12, PCR-14, and PCR-16
for 10-ng PCR	PCR-4, PCR-6, PCR-8, PCR-10, PCR-12, and PCR-14

19. Using the 50-μL reaction(s) from Step 15, run PCR Program 1, and collect 5 μL of the reaction in the 0.2-mL tube labeled either PCR-6 (for 1-ng input) or PCR-4 (for 10-ng input).

20. Continue the PCR optimization by running PCR Program 2, and collect 5 μL of the reaction in the second 0.2-mL tube in the series.

21. Repeat Step 20 for the remaining four cycle-collection points.

22. Collect 5 μL of the negative control at the final cycle-collection point.

23. Add 1 μL of loading dye to each 5-μL aliquot. Vortex quickly and centrifuge the tubes.

24. Prepare a 2.2% FlashGel with 4 μL of water in each of the 13 wells and 4 μL of Flash 50–1500-bp gel marker in the seventh lane. Load (+) DNA reactions in Lanes 1–6 and 8–12 and (−) DNA reaction in Lane 13. A duplicate will not be needed for the final cycle run.

25. Load the 6-μL aliquots of DNA amplifications in Lanes 1–6 and 8–12. Load the negative control aliquot in Lane 13. Run the FlashGel for 7 min at 275 V.

26. Photograph the gel. Identify cycles of overamplification. For the full-scale PCR (next section), use the greatest number of cycles that do not show overamplification.

    For small insert libraries, 1.2% FlashGels do not provide sufficient resolution to separate unused primer/primer dimers, PE adaptor artifact, and PCR product smear. The use of 1.2% DNA FlashGels should only be used when 2.2% DNA FlashGels are unavailable.

    For an example of overamplification, see Figure 1 in Protocol 8, Step 83.

PCR Amplification of Ligated Fragments

27. Set up eight (for 500-ng, 1-μg, or 3-μg input libraries) or 16 (for 100-ng input libraries) PCRs using purified sample from the adaptor ligation step (Step 13).

Adaptor ligated DNA (10 ng/μL)	1 μL
Phusion PCR Master Mix (2×)	25 μL
PE Primer 1.0 (8 μM)	1 μL
PE Primer 2.0 (8 μM)	1 μL
Water	22 μL
Total	50 μL

28. Program the thermal cycler based on the optimization results from the previous section.

Cycle number	Denaturation	Annealing	Polymerization
1	30 sec at 98°C		
N cycles[a]	10 sec at 98°C	30 sec at 65°C	30 sec at 72°C
Last cycle			5 min at 72°C

    Hold at 10°C.

    [a]The optimum number of PCR cycles that results in product amplification should be used (*N*).

29. Pool all of the PCRs. Combine the pooled DNA with 5 volumes of Buffer PBI in a 15-mL conical tube.

    It might be necessary to add 5 μL of 3 M sodium acetate (pH 5.0) if the reaction/buffer mix has a violet tinge.

30. Purify the amplified DNA using a MinElute column. Use one column for every eight PCRs that were run.

    i. Add the reaction/buffer mix to the MinElute column(s). If loading two columns, split the mix as evenly as possible between the two columns.

    ii. Centrifuge the column at 14,000g for 1 min or place it on a vacuum manifold and draw the liquid through the column.

iii. Dispose of waste if the tube was centrifuged.

iv. Repeat Steps 30.i–30.iii until all of the reaction has been passed through the MinElute column(s).

v. Wash the column(s) twice with 750 μL of Buffer PE.

vi. If using a vacuum system, place the column in a waste collecting tube, and centrifuge the column at 14,000g for 1 min.

vii. Rotate the tube 180°. Centrifuge the column at 14,000g for 1 min.

viii. Discard the waste tube.

ix. Transfer the column to a 1.7-mL tube.

x. Add 13 μL of Buffer EB. Incubate for 1 min at room temperature.

xi. Centrifuge the column at 14,000g for 1 min.

xii. Run 1 μL of sample on a 2.2% FlashGel, and determine the DNA concentration using 1 μL in a Qubit Quant-iT dsDNA HS DNA assay.

> *Amplification should not require more than 12 cycles. Higher cycle numbers may result in a reduction in the number of unique sequence fragments. If too little ligation DNA was recovered or if too many PCR cycles are required for enrichment, then it may be necessary to prepare a fresh sample aliquot. Determine the sample availability or tolerance for lower percentage of unique fragments within the library before library construction.*

Size Selection

31. Prepare an acrylamide gel made with TBE buffer.

32. Add 3 μL of 10× Gel-loading buffer to the entire purified DNA sample (~10 μL).

33. Load the gel as follows:

Lane	Reagent	Amount
Lane 1	Gel-loading buffer (10×)	
Lane 2	Lonza 50–1500-bp ladder	5 μL
Lane 3	TrackIt 50-bp ladder	5 μL
Lane 4	Empty	
Lane 5	Sample	500 ng (all or half—1 μg vs. 100 ng)
Lane 6	Sample	500 ng (half [100 ng] of sample [15 μL])
Lane 7	Empty	
Lane 8	TrackIt 50-bp ladder	5 μL
Lane 9	Lonza 50–1500-bp ladder	5 μL
Lane 10	Gel-loading buffer (10×)	

> *Care should be taken not to overload the acrylamide gel lanes with excess PCR product. It is possible for high-abundance DNA fragments to migrate at incorrect sizes. For example, if too much 120-bp PE ligation artifact is loaded into a single lane, then it may be present in gel cuts >120 bp.*

> *Acrylamide gel loading is tricky. Diluting the PCR product 1:5 (5× dye:DNA) is likely to be insufficient for the sample to remain at the bottom of the well. Overloading the dye (1:1) seems to allow the sample to rest at the bottom of the well without negatively affecting the run. The larger the amount of DNA, the less dye is required for successful gel loading.*

> *A small amount of DNA+dye (1–2 μL) should be loaded to test if the sample will rest at the bottom of the well. The end of the loading tip should be as low as possible in the well, and the sample should load around the tip. The tip should be slowly removed from the well without introducing an air bubble or dislodging the DNA+dye.*

34. Run the gel at 180 V for ~50 min.

Gel Purification

35. Stain the gel in SYBR Green (diluted 1:10,000) for 10 min.

36. Destain the gel in TBE buffer or water for 3 min.

37. Prepare one or two 0.7-mL tubes for gel shredding by poking a needle through the bottom of the tube. Place each 0.7-mL tube inside a 1.7-mL tube.

38. Excise gel fragments at the desired size range, and place each gel slice in a 0.7-mL tube. Cap and label the tubes.

39. Photograph the gel.

40. Centrifuge the 0.7-mL tubes (within the 1.7-mL tubes) at 14,000 rpm for 5 min.

41. Remove the 0.7-mL tube. Verify that all of the gel has moved through the bottom of the tube. If not, centrifuge it again.

42. Add 400 μL of 400 mM NaCl to the sheared gel. Place the tube(s) on a shaker for at least 2 h (overnight is OK) to elute the DNA from the gel.

43. Using a wide-bore 200-μL pipette tip, transfer all of the gel and solution to a 0.45-μm HV DuraPore filter column.

44. Centrifuge the solution through the filter at 14,000 rpm for 5 min.

45. Dispose of the column. Transfer the solution to a new 1.7-mL tube.

46. Add 1 μL of GlycoBlue reagent. Add 1 mL of ethanol. Incubate for 60 min at −20°C.

47. Centrifuge at 14,000*g* for 30 min.

48. Note the blue precipitate at the bottom of the tube. Remove the ethanol.

49. Add 1 mL of 70% ethanol. Centrifuge at 14,000*g* for 15 min.

50. Note the blue precipitate at the bottom of the tube. Remove the ethanol.

51. Repeat Steps 49 and 50.

52. Using a small pipette tip, carefully remove as much ethanol as possible. Centrifuge the tube briefly if necessary.

53. Air-dry the sample in a vacuum concentrator for 1-min intervals until the tube appears dry.

54. Resuspend the sample in 20 μL of Buffer EB. Vortex well. Repeat vortexing and centrifuging down several times to ensure that all of the precipitate has dissolved.

55. Run two replicates of 1 μL of sample on an Agilent DNA 1000 chip. Record the size and concentration data for dilution.

56. Determine the DNA concentration using a Qubit/Quant-It dsDNA HS kit and measuring on a fluorometer.

57. If the DNA library appears acceptable, dilute it to 10 nM using a spreadsheet calculator and qPCR.

    ▲ *It is more important to have a clean library with a tight size distribution than excess DNA. The newer sequencers require fewer lanes to achieve complete whole-genome sequencing.*

Protocol 5

Whole Genome: Automated, Nonindexed Library Preparation

This protocol describes an automated procedure for constructing a nonindexed Illumina DNA library and relies on the use of a CyBio-SELMA automated pipetting machine, the Covaris E210 shearing instrument, and the epMotion 5075. With this method, genomic DNA fragments are produced by sonication, using high-frequency acoustic energy to shear DNA. Here, double-stranded DNA is fragmented when exposed to the energy of Adaptive Focused Acoustic Shearing (AFA; see the information panel Fragmenting of DNA). The resulting DNA fragments are ligated to adaptors, amplified by PCR, and subjected to size selection using magnetic beads. The product is suitable for use as template in whole-genome sequencing.

Several automated pipetting systems are in wide use; the protocols provided for automation (Protocols 5 and 6) have been developed using the epMotion series of robotic pipetting instruments from Eppendorf. These simple robots provide a reasonable capacity that is scalable, an associated thermal incubation device, and a magnetic separation device that are implemented in many of the preparatory protocols we have outlined in this chapter. In the absence of this robotic platform, the user who wants to automate should refer to the protocol for manual sample preparation (Protocol 4). The Additional Protocol Automated Library Construction presents a general automated procedure for constructing indexed and nonindexed Illumina DNA libraries.

MATERIALS

It is essential that you consult the appropriate Material Safety Data Sheets and your institution's Environmental Health and Safety Office for proper handling of equipment and hazardous materials used in this protocol.

Reagents

AMPure beads

dATP (1 mM)

Elution buffer (EB) (QIAGEN)

> EB, a generic buffer first introduced by QIAGEN, is composed of 10 mM Tris (pH 8.0). Water can also be used as a DNA buffer but can lead to degradation if the DNA concentration is high or if the DNA solution is frozen and thawed many times. The type of buffer to be used as EB is determined by what is appropriate for the procedure.

End repair buffer (NEB)

End repair enzyme (NEB)

Ethanol (70%)

Genomic DNA

Indexes 1–12 (8 μM) (IDT)

Indexes 1–12 (8 μM) (Illumina)

Klenow buffer (10×) (NEB Buffer 2)

Klenow exonuclease (5 U/μL)

PE forked adapter duplex (4 μM) (Illumina)

PE PCR Primer 1.01 (8 μM) (IDT)

PE PCR Primer 2.01 (8 μM) (IDT)

Phusion PCR Master Mix (2×)

Quick Ligase (2000 U/μL) (NEB)

Quick Ligation Buffer (2×) (NEB)

Equipment

DNA shearing instrument (Covaris, catalog no. E210)
Liquid waste reservoir for the epMotion
Magnet plate, Agencourt SPRIPlate 96R (Beckman Coulter)
96 microTUBE plate (Covaris)
PCR tray (96 well)
Pipette tips for the epMotion (50 μL and 300 μL) (epT.I.P.S. Motion; Eppendorf)
Pipetting machine (CyBi-SELMA)
Plate (96 well), round bottom (Costar)
Plate sealer, clear
Reservoir cold block
Reservoirs with lids (30 mL for the epMotion)
Thermal cycler
Thermoplates (Eppendorf)
Tip waste container for the epMotion
TipTrays for CyBi-SELMA

METHOD

Library Sonication

▲ Before running any program on the epMotion, label all plates and reservoirs as they appear on the worktable and be sure to check all PCR plate locations (reagent and sample) for thermo plate requirements.

1. Prepare a sample of 3 μg, 1 μg, or 500 ng of genomic DNA in a 2D tube tray. The tray will contain 38 μL of diluted DNA. Keep on ice until ready for use.

2. On the computer desktop, click the "epBlue Client" icon to open the software. Log in.
   i. Choose **File**.
   ii. Run *Applications*.
   iii. Choose **96 Non-Indexed**.

3. Choose the *End Repair Buffer to 2D* application. This program will add 10 μL of End repair buffer to the 2D tube tray of diluted DNA.

4. Once open, click the "Work Table" tab for labware placement.

5. Double-check "Deck Layout and Labware Orientation" before proceeding.

6. In the software, click the green Start button to begin.
   i. The *Available Devices* prompt will appear.
   ii. Click Run.
   iii. Choose OK.

7. When the *Minimum Volumes* prompt appears, input the same exact volumes that are shown in the Minimum Volume column. The process will start.

8. When the application is finished, seal the Covaris plate with a clear plate seal, and vortex the 2D tubes of DNA for 1–2 sec. Pulse in a centrifuge.

9. Using the CyBi-SELMA (96-well pipette), transfer the 48 μL from the 2D tubes into a Covaris 96 microTUBE plate.

10. Seal the microTUBE plate with foil, vortex, and centrifuge briefly to ensure that there are no bubbles in the wells of the plate.

11. Place the plate into the Covaris E210 instrument. Choose and run the *96 well_plate_prog01_ 200-300bp_20DC_5I_500CB_120s* program.

12. When the program is finished, remove the microTube plate from the shearing instrument.

13. Blot the bottom of the plate dry and centrifuge it briefly.

14. Remove the top layer of foil from the plate. Pierce the remaining foil with a 96-well PCR tray. Using the 96-well pipette, transfer the 48 μL from the Covaris plate into a 96-well round-bottom tray. Centrifuge that plate briefly.

End Repair

This procedure converts the overhangs resulting from fragmentation into blunt ends using the end repair enzyme. End repair running time for 96 samples is ∼2 h and 7 min.

15. For 96 samples, dispense 30 μL of end repair enzyme to each well of Column 1 (Wells A–H) of an Eppendorf PCR plate. For 48 samples, dispense 19 μL in the same manner. Centrifuge briefly and store on ice.

16. Choose the *End Repair* application on the epMotion.

17. Double-check the "Deck Layout and Labware Orientation."

18. Run the program. The epMotion will perform the following steps.

    i. Add 2 μL of end repair enzyme to each sample and mix.

    ii. Incubate the samples for 30 min at room temperature.

    iii. Run an AMPure bead cleanup using a 1.5× ratio of beads to sample (75 μL).

    iv. Wash the samples twice with 165 μL each of 70% ethanol.

    > *After the 70% ethanol wash, you will be prompted to take the bead plate off the magnet to dry the samples. Drying time takes 10–15 min depending on the vacuum concentrator being used. When completely dry, the beads will appear to be cracked. Replace the plate on the magnet. Follow the prompt.*

    v. Elute the dried bead plate with 32 μL of EB buffer, and transfer it to a Costar 96.

3′ Adenylation (A-Tailing)

A single A nucleotide is added to the 3′ ends of the blunt fragments to prevent ligation between fragments. A corresponding single T nucleotide on the 3′ end of the adaptor provides a complementary overhang for an efficient ligation between the adaptor and the fragment. The A-tailing running time for 96 samples is ∼2 h and 6 min.

19. Prepare a master mix of the A-tailing reagents.

	One sample (1×)	48 samples (56×)	96 samples (112×)
Klenow buffer (10×)	5 μL	280 μL	560 μL
dATP (1.0 mM)	10 μL	560 μL	1120 μL
Klenow exonuclease (5 U/μL)	3 μL	168 μL	336 μL
Total volume	18 μL	1008 μL	2016 μL

20. Vortex the master mix and centrifuge it briefly.

21. For 96 samples, dispense 117 μL of the A-tailing master mix into each well of Columns 2 and 3 (Wells A–H) of an Eppendorf PCR plate. For 48 samples, dispense 117 μL of the A-tailing master mix into Column 2 only. Store it on ice.

22. Resuspend the bead solution by mixing several times using a 1-mL pipette.

23. Choose the *A-Tail* application on the epMotion.

24. Double-check the "Deck Layout and Plate Orientation."

25. Run the program. The epMotion will perform the following steps.

   i. Add 18 μL of Klenow buffer, dATP, and Klenow exonuclease cocktail to the end-repaired sample. Mix the samples.

   ii. Incubate the samples for 30 min at 37°C.

   iii. Run an AMPure bead cleanup using a 1.5× ratio of beads to sample (75 μL).

   iv. Wash the samples twice with 165 μL each of 70% ethanol.

   > The beads will bind only DNA fragments above ~100 bp, and the dNTPs from end repair will be washed away.

   > After the 70% ethanol wash, you will be prompted to take the bead plate off the magnet to dry the samples. Drying time takes 10–15 min depending on the vacuum concentrator being used. When completely dry, the beads will appear to be cracked. Replace the plate on the magnet. Follow the prompt.

   v. Elute the dried beads with 20 μL of Buffer EB, and transfer it to a new Costar 96.

Adaptor Ligation

This process ligates adaptors to the ends of the DNA fragments. The reaction adds distinct sequences to the 5′ and 3′ ends of each strand in the genomic fragments. Additional sequences are added later in the protocol, by tailed primers during PCR. These additional sequences are necessary for library amplification on the flow cell during cluster formation. The adaptor ligation running time for 96 samples is ~1 h and 40 min.

26. Prepare a master mix of the adaptor ligation reagents using one of the following tables:

1-μg and 3-μg input	One sample (1×)	48 samples (56×)	96 samples (112×)
NEB Quick Ligation Buffer (2×)	25 μL	1400 μL	2800 μL
III PE forked adaptor duplex (4 μM)	5 μL	280 μL	560 μL
NEB Quick Ligase (2000 U/μL)	2.5 μL	140 μL	280 μL
Total volume	32.5 μL	1820 μL	3640 μL

500-ng input	One sample (1×)	48 samples (56×)	96 samples (112×)
NEB Quick Ligation Buffer (2×)	25 μL	1400 μL	2800 μL
III PE forked adaptor duplex (4 μM)	2.5 μL	140 μL	280 μL
NEB Quick Ligase (2000 U/μL)	2.5 μL	140 μL	280 μL
Total volume	30 μL	1680 μL	3108 μL

27. Vortex the master mix and centrifuge it briefly. Dispense the master mix into the 30-mL reservoir labeled **Adaptor Ligation Mix**. Store on ice.

   > Use the reservoir thermo block to keep the adaptor ligation mix chilled. This will help to prevent dripping.

28. Resuspend the bead solution by mixing several times using a 1-mL pipette.

29. On the epMotion, choose the appropriate *Ligation* application based on input amount (250 ng and 500 ng, or 1 μg and 3 μg).

30. Double-check the "Deck Layout and Plate Orientation."

31. Run the program. The epMotion will perform the following steps.

    i. Add the ligation cocktail to the samples and mix them.

    ii. Incubate the samples for 15 min at room temperature.

    iii. Run an AMPure bead cleanup using a 1.5× ratio of beads to sample (75 μL).

    iv. Wash the samples twice with 165 μL each of 70% ethanol.

    > *After the 70% ethanol wash, you will be prompted to take the bead plate off the magnet to dry the samples. Drying time takes 10–15 min depending on the vacuum concentrator being used. When completely dry, the beads will appear to be cracked. Replace the plate on the magnet. Follow the prompt.*

    v. Elute the dried beads with 30 μL of Buffer EB, and transfer it to a PCR plate.

PCR Amplification of Ligated Fragments

PCR is used to selectively enrich those DNA fragments that have adaptor molecules on both ends and to amplify the amount of DNA in the library for accurate quantification. The PCR is performed with two primers that anneal to the ends of the adaptors. The PCR amplification of ligated fragments running time for 96 samples is ∼38 min.

32. Prepare a master mix of the PCR reagents (each sample will have four reactions):

	One sample (1×)	56 samples (224×)	112 samples (448×)
Phusion PCR Master Mix (2×)	25 μL	5600 μL	11,200 μL
PE Primer 1.0 (8 μM)	1 μL	224 μL	448 μL
PE Primer 2.0 (8 μM)	1 μL	224 μL	448 μL
Water	18 μL	4032 μL	8064 μL
Total volume	45 μL	10,080 μL	20,160 μL

33. Vortex the master mix and centrifuge it briefly. Dispense the master mix to the 30-mL reservoir labeled **PCR Mix**. Store on ice.

34. On the epMotion, choose the appropriate *PCR Amplification* application, based on the number of samples. Programs are available for eight and 96 samples.

35. Double-check the "Deck Layout and Plate Orientation."

36. Run the program. The epMotion will perform the following steps.

    i. Set up four 5-μL PCRs using purified samples from the adaptor ligation step.

    ii. Add 5 μL of sample to 45 μL of Phusion PCR Master Mix/primer cocktail.

    iii. Run the PCR amplification cycles as follows:

Cycle number	Denaturation	Annealing	Polymerization
1	30 sec at 98°C		
8 cycles	10 sec at 98°C	30 sec at 65°C	30 sec at 72°C
Last cycle			5 min at 72°C

Hold at 10°C.

37. When the application finishes, cover and quickly centrifuge the four PCR plates. Store the plates on ice.

PCR Pooling

The PCR pooling running time for 96 samples is ∼30 min.

38. On the epMotion, choose the *PCR Pooling* application.

39. Double-check the "Deck Layout and Plate Orientation."

40. Run the program. The epMotion will pool pairs of like PCRs (pooled volume = 100 μL) in a Costar 96 plate.

41. When the application finishes, quickly centrifuge both pooling plates.

Size Selection by Dual SPRI Beads

Pooled libraries will be size-selected for 300–500-bp fragments. The size selection by dual SPRI beads running time for 96 samples is ~1 h and 50 min.

42. Resuspend the bead solution by mixing several times using a 1-mL pipette.

43. Choose the *Dual SPRI EB* application on the epMotion.

44. Double-check the "Deck Layout and Labware Orientation."

45. Run the program. The epMotion will perform the following steps.

    i. Select larger fragments by the first bead addition of 60 μL (0.6×). Fragments ≥500 bp will be bound to the beads.

    ii. Transfer 60 μL of supernatant, containing fragments <500 bp, into a Costar plate containing 20 μL (0.8×) of concentrated beads. Fragments ≤300 bp will bind to the beads.

    iii. Elute samples of 300–500-bp fragments.

46. When the application is finished, quickly centrifuge the Eluted Samples plate. Store it on ice. The final volume per sample will be ~50 μL.

Determine the Concentration, Size Distribution, and Quality of the Sample

47. Measure the final concentration with a Varioskan Flash Multimode Reader, Agilent, or Caliper GX instrument. See the appropriate instrument protocol for the operating procedure.

48. Prepare a 5 nM dilution of each sample. Clearly indicate on the tube that the library has been diluted to 5 nM.

    For assistance in preparing the dilution, see http://www.promega.com/biomath/Calculators&AdditionalConversions/Dilution.

49. The samples are now ready for qPCR and sequencing on the Illumina Genome Analyzer.

Automated Library Preparation

This protocol describes a generalized automated procedure for constructing indexed and nonindexed Illumina DNA libraries.

ADDITIONAL MATERIALS

It is essential that you consult the appropriate Material Safety Data Sheets and your institution's Environmental Health and Safety Office for proper handling of equipment and hazardous materials used in this protocol.

Reagents

AMPure beads
Buffer EB
dATP (1 mM)
End repair buffer
End repair enzyme
Ethanol (70%)
Indexes 1–12 (8 μM) (IDT)
Indexes 1–12 (8 μM) (Illumina)
Klenow buffer (10×) (NEB Buffer 2)
Klenow exonuclease (5 U/μL)
PE forked adapter duplex (4 μM) (Illumina)
PE PCR Primer 1.01 (8 μM) (IDT)
PE PCR Primer 2.01 (8 μM) (IDT)
Phusion PCR Master Mix (2×)
Quick Ligase (2000 U/μL) (NEB)
Quick Ligation Buffer (2×) (NEB)

Indexed Reagents

IDT PE Indexed Adapter Oligo (4 μM)
Illumina Indexed PE Adapter Oligo Mix (4 μM)
Indexes 1–12 (8 μM) (IDT)
Indexes 1–12 (8 μM) (Illumina)

Equipment

Automated pipettor
DNA shearing instrument (E210; Covaris)
Magnet plate, Agencourt SPRIPlate 96R (Beckman Coulter)
96-microTUBE plate (Covaris)
PCR tray (96 well)
Pipette tips for automated pipettor
Plate (96 well), round bottom (Costar)
Plate sealer, clear

METHOD

Library Sonication

1. Shear the DNA to a consistent size using the Covaris E210 (this can be performed in advance with little additional concern for sample stability). The mean fragment size should be ~250 bp.

End Repair

This procedure converts the overhangs resulting from fragmentation into blunt ends using the end repair enzyme.

2. The automated pipettor will perform the following steps.

    i. Add 2 μL of end repair enzyme to each sample and mix.

    ii. Incubate the samples for 30 min at room temperature.

    iii. Run an AMPure bead cleanup using a 1.5× ratio of beads to sample (75 μL).

    iv. Wash the samples twice with 165 μL each of 70% ethanol.

    v. Elute the dried bead plate with 32 μL of EB buffer and transfer it to a Costar 96.

3′ Adenylation (A-Tailing)

A single A nucleotide is added to the 3′ ends of the blunt fragments to prevent ligation between fragments. A corresponding single T nucleotide on the 3′ end of the adaptor provides a complementary overhang for an efficient ligation between the adaptor and the fragment.

3. Prepare a master mix of the A-tailing reagents.

4. The automated pipettor will perform the following steps.

    i. Add 18 μL of Klenow buffer, dATP, and Klenow exonuclease cocktail to the end-repaired sample. Mix the samples.

    ii. Incubate the samples for 30 min at 37°C.

    iii. Run an AMPure bead cleanup using a 1.5× ratio of beads to sample (75 μL).

    iv. Wash the samples twice with 165 μL each of 70% ethanol.

    > *The beads will bind only DNA fragments above ~100 bp and the dNTPs from end repair will be washed away.*

    v. Elute the dried beads with 20 μL of Buffer EB, and transfer it to a new Costar 96.

Adaptor Ligation

This process ligates adaptors to the ends of the DNA fragments. The reaction adds distinct sequences to the 5′ and 3′ ends of each strand in the genomic fragments. Additional sequences are added later in the protocol, using tailed primers during PCR. These additional sequences are necessary for library amplification on the flow cell during cluster formation.

5. Prepare a master mix of the adaptor ligation reagents.

6. The automated pipettor will perform the following steps.

    i. Add the ligation cocktail to the samples and mix them.

    ii. Incubate the samples for 15 min at room temperature.

    iii. Run an AMPure bead cleanup using a 1.5× ratio of beads to sample (75 μL).

    iv. Wash the samples twice with 165 μL each of 70% ethanol.

    v. Elute the dried beads with 30 μL of Buffer EB, and transfer it to a PCR plate.

PCR Amplification of Ligated Fragments

PCR is used to selectively enrich those DNA fragments that have adaptor molecules on both ends and to amplify the amount of DNA in the library for accurate quantification. The PCR is performed with two primers that anneal to the ends of the adaptors.

7. Prepare a master mix of the PCRs. (Each sample will have four reactions.)

8. The automated pipettor will perform the following steps.

    i. Set up four 5-μL PCRs using purified samples from the adaptor ligation step.

    ii. Add 5 μL of sample to 45 μL of Phusion PCR Master Mix/primer cocktail.

9. Amplify the DNA.

PCR Index Addition

10. Prepare index plates containing either IDT or Illumina indexes in vertical rows.

11. The automated pipettor will add 1 μL of each index (IDT or Illumina Index Primer, 8 μM) to each sample. The Index sequence is added in order to permit sample discrimination in a multiplexed sequencing experiment.

12. Amplify the DNA.

PCR Pooling

13. The automated pipettor will pool pairs of like PCRs (pooled volume = 100 μL) in a Costar 96 plate.

14. When the application finishes, quickly centrifuge the pooling plates.

Size Selection by Dual SPRI Beads

Pooled libraries will be size-selected for 300–500-bp fragments. The size selection by dual SPRI beads running time for 96 samples is ~1 h and 50 min.

15. Resuspend the bead solution by mixing several times using a 1-mL pipette.

16. The automated pipettor will perform the following steps.

    i. Select larger fragments by the first bead addition of 60 μL (0.6×). Fragments ≥500 bp will be bound to the beads.

    ii. Transfer 60 μL of the supernatant, containing fragments <500 bp, into a Costar plate containing 20 μL (0.8×) of concentrated beads. Fragments ≤300 bp will bind to the beads.

    iii. Elute samples of 300–500-bp fragments in 28 μL of EB buffer or water (because PCR pooling samples are in duplicates and will be combined as the final product).

17. When the application is finished, vortex and quickly centrifuge the Eluted Samples plate.

Determine the Concentration, Size Distribution, and Quality of the Sample

18. Continue with Step 46 of the main protocol.

Whole Genome: Automated, Indexed Library Preparation

This protocol describes an automated procedure for constructing an indexed Illumina DNA library. With this method, genomic DNA fragments are produced by sonication, using high-frequency acoustic energy to shear DNA. Double-stranded DNA (dsDNA) will fragment when exposed to the energy of Adaptive Focused Acoustic Shearing (AFA; see the information panel Fragmenting of DNA). The resulting DNA fragments are ligated to adaptors, amplified by PCR, and subjected to size selection using magnetic beads. The product is suitable for use as template in whole-genome sequencing.

In the absence of this robotic platform, the user who wants to automate should refer to the protocol for manual sample preparation (Protocol 4). The Additional Protocol Automated Library Construction in Protocol 5 presents a general automated procedure for constructing indexed and nonindexed Illumina DNA libraries.

MATERIALS

It is essential that you consult the appropriate Material Safety Data Sheets and your institution's Environmental Health and Safety Office for proper handling of equipment and hazardous materials used in this protocol.

Reagents

AMPure beads

dATP (1 mM)

Elution buffer (EB) (QIAGEN)

> EB, a generic buffer first introduced by QIAGEN, is composed of 10 mM Tris (pH 8.0). Water can also be used as a DNA buffer but can lead to degradation if the DNA concentration is high or if the DNA solution is frozen and thawed many times. The type of buffer to be used as EB is determined by what is appropriate for the procedure.

End repair buffer (NEB)

End repair enzyme

Ethanol (70%)

Genomic DNA

IDT PE Indexed Adapter Oligo (4 μM)

Illumina Indexed PE Adapter Oligo Mix (4 μM)

Indexes 1–12 (8 μM) (IDT)

Indexes 1–12 (8 μM) (Illumina)

Klenow buffer (10×) (NEB Buffer 2; New England Biolabs)

Klenow exonuclease (5 U/μL)

PE PCR Primer 1.01 (8 μM) (IDT or Illumina)

PE PCR Primer 2.01 (8 μM) (IDT or Illumina)

Phusion PCR Master Mix (2×)

Quick Ligase (2000 U/μL) (New England Biolabs)

Quick Ligation Buffer (2×) (New England Biolabs)

Equipment

DNA shearing instrument (Covaris, catalog no. E210)
Liquid waste reservoir for the epMotion
Magnet plate, Agencourt SPRIPlate 96R (Beckman Coulter)
96-microTUBE plate (Covaris)
PCR tray (96 well)
Pipette tips for the epMotion (50 μL and 300 μL) (epT.I.P.S Motion; Eppendorf)
Pipetting machine (CyBi-SELMA)
Plate (96 well), round bottom (Costar)
Plate sealer, clear
Reservoir cold block
Reservoirs with lids (30 mL) for the epMotion
Thermal cycler
Thermoplates (Eppendorf)
TipTrays for CyBi-SELMA
Tip waste container for the epMotion

METHOD

Library Sonication

▲ Before running any program on the epMotion, label all plates and reservoirs as they appear on the worktable, and be sure to check all PCR plate locations (reagent and sample) for thermo plate requirements.

1. Prepare a sample of 500 ng, 1 μg, 3 μg, or 5 μg of genomic DNA in a 2D tube tray. The tray will contain 38 μL of diluted DNA. Keep on ice until ready for use.

2. On the computer desktop, click the "epBlue Client" icon to open the software. Log in. Choose File. Run Applications.

3. Choose the "End Repair Buffer to 2D" application. This program will add 10 μL of End repair buffer to the 2D tube tray of diluted DNA.

4. Double-check the "Deck Layout and Labware Orientation" before proceeding.

5. In the software, click the green Start button to begin.

   i. The *Available Devices* prompt will appear.

   ii. Click Run.

   iii. Choose OK.

6. When the *Minimum Volumes* prompt appears, input the same exact volumes that are shown in the Minimum Volume column. The process will start.

7. When the application is finished, seal the Covaris plate with a clear plate seal, and vortex the 2D tubes of DNA for 1–2 sec. Pulse in a centrifuge.

8. Using the CyBi-SELMA (96-well pipette), transfer the 48 μL from the 2D tubes into a Covaris 96 microTUBE plate.

9. Seal the microTUBE plate with foil, vortex, and centrifuge briefly to ensure that there are no bubbles in the wells of the plate.

10. Place the plate into the Covaris E210 instrument. Choose the *96 well_plate_prog01_200-300bp_20DC_5I_500CB_120s* program.

11. When the program is finished, remove the microTube plate from the shearing instrument.

12. Blot the bottom of the plate dry, and centrifuge it briefly.

13. Remove the top layer of foil from the plate. Pierce the remaining foil with a 96-well PCR tray. Using the 96-well pipette, transfer the 48 μL from the Covaris plate into a 96-well round-bottom tray. Centrifuge that plate briefly.

End Repair

This procedure converts the single-stranded DNA overhangs resulting from fragmentation into blunt ends using the end repair enzyme. The end repair running time for 96 samples is ~2 h and 7 min.

14. For 96 samples, dispense 30 μL of end repair enzyme to each well of Column 1 (Wells A–H) of an Eppendorf PCR plate. For 48 samples, dispense 19 μL in the same manner. Centrifuge briefly and store on ice.

15. Choose the *End Repair* application on the epMotion.

16. Double-check the "Deck Layout and Labware Orientation."

17. Run the program. The epMotion will perform the following steps.

 i. Add 2 μL of end repair enzyme to each sample and mix.

 ii. Incubate the samples for 30 min at room temperature.

 iii. Run an AMPure bead cleanup using a 1.5× ratio of beads to sample (75 μL).

 iv. Wash the samples twice with 165 μL each of 70% ethanol.

  After the 70% ethanol wash, you will be prompted to take the bead plate off the magnet to dry the samples. Drying takes 10–15 min depending on the vacuum concentrator being used. When completely dry, the beads will appear to be cracked. Replace the plate on the magnet. Follow the prompt.

 v. Elute the dried bead plate with 32 μL of EB buffer, and transfer it to a Costar 96.

3′ Adenylation (A-Tailing)

A single A nucleotide is added to the 3′ ends of the blunt fragments to prevent ligation between fragments. A corresponding single T nucleotide on the 3′ end of the adaptor provides a complementary overhang for an efficient ligation between the adaptor and the fragment. The A-tailing running time for 96 samples is ~2 h and 6 min.

18. Prepare a master mix of the A-tailing reagents:

	One sample (1×)	48 samples (56×)	96 samples (112×)
Klenow buffer (10×)	5 μL	280 μL	560 μL
dATP (1.0 mM)	10 μL	560 μL	1120 μL
Klenow Exo (5 U/μL)	3 μL	168 μL	336 μL
Total volume	18 μL	1008 μL	2016 μL

19. Vortex the master mix, and centrifuge it briefly.

20. For 96 samples, dispense 117 μL of the A-tailing master mix into each well of Columns 2 and 3 (Wells A–H) of an Eppendorf PCR plate. For 48 samples, dispense 117 μL of the A-tailing master mix into Column 2 only. Store it on ice.

21. Resuspend the bead solution by mixing several times using a 1-mL pipette.

22. Choose the *A-Tail* application on the epMotion.

23. Double-check the "Deck Layout and Plate Orientation."

24. Run the program. The epMotion will perform the following steps.

i. Add 18 μL of Klenow buffer, dATP, and Klenow exonuclease cocktail to the end-repaired sample. Mix the samples.

ii. Incubate the samples for 30 min at 37°C.

iii. Run an AMPure bead cleanup using a 1.5× ratio of beads to sample (75 μL).

iv. Wash the samples twice with 165 μL each of 70% ethanol.

> *The beads will bind only DNA fragments above ~100 bp, and the dNTPs from end repair will be washed away.*
>
> *After the 70% ethanol wash, you will be prompted to take the bead plate off the magnet to dry the samples. Drying takes 10–15 min depending on the vacuum concentrator being used. When completely dry, the beads will appear to be cracked. Replace the plate on the magnet. Follow the prompt.*

v. Elute the dried beads with 20 μL of Buffer EB, and transfer it to a new Costar 96.

Adaptor Ligation

This process ligates adaptors to the ends of the DNA fragments. The reaction adds distinct sequences to the 5′ and 3′ ends of each strand of the genomic fragments. Additional sequences are added later in the protocol, by tailed primers during PCR. These sequences are necessary for library amplification on the flow cell during cluster formation. The adaptor ligation running time for 96 samples is ~1 h and 40 min.

25. Prepare a master mix of the adaptor ligation reagents using one of the following tables:

5-μg input	One sample (1×)	48 samples (56×)	96 samples (112×)
NEB Quick Ligation Buffer (2×)	25 μL	1400 μL	2800 μL
IDT PE Indexed Adapter Oligo (4 μM) *or* Illumina Indexed PE Adapter Oligo Mix (4 μM)	10 μL	560 μL	1120 μL
NEB Quick Ligase (2000 U/μL)	5 μL	280 μL	560 μL
Total volume	40 μL	2240 μL	4480 μL

1-μg and 3-μg input	One sample (1×)	48 samples (56×)	96 samples (112×)
NEB Quick Ligation Buffer (2×)	25 μL	1400 μL	2800 μL
IDT PE Indexed Adapter Oligo (4 μM) *or* Illumina Indexed PE Adapter Oligo Mix (4 μM)	5 μL	280 μL	560 μL
NEB Quick Ligase (2000 U/μL)	2.5 μL	140 μL	280 μL
Total volume	32.5 μL	1820 μL	3640 μL

500-ng input	One sample (1×)	48 samples (56×)	96 samples (112×)
NEB Quick Ligation Buffer (2×)	25 μL	1400 μL	2800 μL
IDT PE Indexed Adapter Oligo (4 μM) *or* Illumina Indexed PE Adapter Oligo Mix (4 μM)	2.5 μL	140 μL	280 μL
NEB Quick Ligase (2000 U/μL)	2.5 μL	140 μL	280 μL
Total volume	30 μL	1680 μL	3108 μL

26. Vortex the master mix and centrifuge it briefly. Dispense the master mix into the 30-mL reservoir labeled **Adapter Ligation Mix**. Store on ice.

> *Use the reservoir thermo block to keep the adapter ligation mix chilled. This will help to prevent dripping.*

27. Resuspend the bead solution by mixing several times using a 1-mL pipette.
28. On the epMotion, choose the appropriate *Ligation* application based on input amount (250 ng and 500 ng, or 1 μg and 3 μg).
29. Double-check the "Deck Layout and Plate Orientation."
30. Run the program. The epMotion will perform the following steps.

   i. Add the ligation cocktail to the samples and mix them.
   ii. Incubate the samples for 15 min at room temperature.
   iii. Run an AMPure bead cleanup using a 1.5× ratio of beads to sample (75 μL).
   iv. Wash the samples twice with 165 μL each of 70% ethanol.

   > After the 70% ethanol wash, you will be prompted to take the bead plate off the magnet to dry the samples. Drying takes 10–15 min depending on the vacuum concentrator being used. When completely dry, the beads will appear to be cracked. Replace the plate on the magnet. Follow the prompt.

   v. Elute the dried beads with 30 μL of Buffer EB, and transfer it to a PCR plate.

PCR Amplification of Ligated Fragments

PCR is used to selectively enrich those DNA fragments that have adaptor molecules on both ends and to amplify the amount of DNA in the library for accurate quantification. The PCR is performed with two primers that anneal to the ends of the adaptors. The PCR amplification of ligated fragments running time for 96 samples is ~38 min.

31. Prepare the appropriate master mix of PCR reagents (see the tables and options below). Each sample will have four reactions.
32. Vortex the master mix and centrifuge it briefly. Dispense the master mix to the 30-mL reservoir labeled **PCR Mix**. Store on ice.

There are three options for amplification (with two different epMotion programs and PCR master mixes).

Option 1: PCR Amplification with Single-Use Plates

PCR amplification is performed using four single-use plates. The epMotion adds DNA and the Phusion master mix to a single-use plate that already contains 8 μL total of the indexes (IDT or Illumina), Primers 1.01 and 2.01, and water. The robot can deliver either 5 μL of DNA and 37 μL of master mix to the four plates or 1 μL of DNA and 41 μL of master mix.

Single-Use Plates

Each well contains the following:

PE Primer 1.01 (8 μM)	1 μL
PE Primer 2.01 (0.5 μM)	1 μL
IDT or Illumina Index Primer (8 μM)	1 μL
Water	5 μL
Total volume	8 μL

Master Mix for Options 1 and 2

	One sample (1×)	48 samples (56 × 4 = 224×)	96 samples (112 × 4 = 448×)
Phusion PCR Master Mix (2×)	25 μL	5600 μL	11,200 μL
Water	12 μL	2688 μL	5376 μL
Total volume	37 μL	8288 μL	16,576 μL

Option 2: PCR Amplification with Index Master Mix

This program contains a water/Phusion only master mix similar to the single-use plate program. It also contains a master mix tray of Primers 1.01, 2.01, and indexes (fully skirted PCR tray). The program will add—in order—the water/Phusion master mix, 5 μL of DNA, and 3 μL of the primers-Index master mix (four reactions for each sample).

Option 3: PCR Amplification with Index Addition

PCR amplification is performed with four empty PCR plates. The epMotion adds the DNA and the master mix containing water, Phusion mix, and Primers 1.01 and 2.01. The robot can deliver either 5 μL of DNA and 44 μL of master mix to the four empty plates or 1 μL of DNA and 48 μL of master mix. The indexes are added by a separate program (*PCR Index*).

Master Mix for Option 3

	One sample (1×)	48 samples (56×4 = 224×)	96 samples (112×4 = 448×)
Phusion PCR Master Mix (2×)	25 μL	5600 μL	11,200 μL
PE Primer 1.01 (8 μM)	1 μL	224 μL	448 μL
PE Primer 2.01 (0.5 μM)	1 μL	224 μL	448 μL
Water	17 μL	3808 μL	7616 μL
Total volume	44 μL	9856 μL	19,712 μL

33. Choose the appropriate *PCR Amplification* application, based on the number of samples. Programs are available for eight and 96 samples from the Eppendorf epMotion website.

34. Double-check the "Deck Layout and Plate Orientation."

35. Run the program.

36. If Option 3 is used, then proceed to the next section, PCR Index Addition. Otherwise, skip to Step 40.

PCR Index Addition (If Not Using Single-Use Plates)

37. Prepare Index plates containing either IDT or Illumina Indexes in vertical rows. For example, IDT Row 1 contains Indexes 1–8 in A1–H1, Row 2 contains Indexes 9–16 in A2–H2, and so on. Illumina Row 1 contains Index 1, A1–H1; Row 2 contains Index 2, A2–H2; and so on.

38. Choose a program based on sample set size, *PCR Index*.

39. Run the program. The epMotion adds 1 μL of each Index (IDT or Illumina Index Primer, 8 μM) to each sample.

> *The Index sequence is added in order to permit sample discrimination in a multiplexed sequencing experiment.*

40. When the application finishes, cover the plates, vortex them, and quickly centrifuge all four PCR plates. The final volume will be ~50 μL.

41. Amplify the DNA using the following PCR program:

Cycle number	Denaturation	Annealing	Polymerization
1	30 sec at 98°C		
18 cycles	10 sec at 98°C	30 sec at 65°C	30 sec at 72°C
Last cycle			5 min at 72°C

Hold at 10°C.

42. When the amplification program is finished, quickly centrifuge the four PCR plates. Store them on ice.

PCR Pooling

The PCR pooling running time for 96 samples is ~30 min.

43. On the epMotion, choose the *PCR Pooling* application.

44. Double-check the "Deck Layout and Plate Orientation."

45. Run the program. The epMotion will pool pairs of like PCRs (pooled volume = 100 μL) in a Costar 96 plate.

46. When the application finishes, quickly centrifuge both pooling plates.

Size Selection by Dual SPRI Beads

Pooled libraries will be size-selected for 300–500-bp fragments. The size selection by dual SPRI beads running time for 96 samples is ~1 h and 50 min.

47. Resuspend the bead solution by mixing several times using a 1-mL pipette.

48. Choose the *Dual SPRI EB* application on the epMotion.

49. Double-check the "Deck Layout And Labware Orientation."

50. Run the program. The epMotion will perform the following steps.

   i. Select larger fragments by the first bead addition of 60 μL (0.6×). Fragments ≥500 bp will be bound to the beads.

   ii. Transfer 60 μL of supernatant, containing fragments <500 bp, into a Costar plate containing 20 μL (0.8×) of concentrated beads. Fragments ≤300 bp will bind to the beads.

   iii. Elute samples of 300–500-bp fragments.

51. When the application is finished, quickly centrifuge the **Eluted Samples** plate. Store it on ice. The final volume per sample will be ~50 μL.

Determine the Concentration, Size Distribution, and Quality of the Sample

52. Measure the final concentration with a Varioskan Flash Multimode Reader, Agilent, or Caliper GX instrument. See the appropriate instrument protocol for the operating procedure.

53. Prepare a 5 nM dilution of each sample. Clearly indicate on the tube that the library has been diluted to 5 nM.

   For assistance in preparing the dilution, see http://www.promega.com/biomath/Calculators&AdditionalConversions/Dilution.

54. The samples are now ready for qPCR and sequencing on the Illumina sequencer.

Preparation of a 3-kb Mate-Pair Library for Illumina Sequencing

This protocol describes how to prepare a 3-kb mate-paired-end library. The resulting amplified DNA is suitable for sequencing on the Illumina sequencer (see the Introduction for a discussion of mate-paired libraries).

MATERIALS

It is essential that you consult the appropriate Material Safety Data Sheets and your institution's Environmental Health and Safety Office for proper handling of equipment and hazardous materials used in this protocol.

Recipes for reagents specific to this protocol, marked <R>, are provided at the end of the protocol. See Appendix 1 for recipes for commonly used stock solutions, buffers, and reagents, marked <A>. Dilute stock solutions to the appropriate concentrations.

Reagents

AMPure XP beads (Beckman Coulter)
Agarose gel (0.8% LE)
Buffer EB (QIAGEN, catalog no. 19086)
Buffer ERC (QIAGEN, catalog no. 1018144)
Buffer PE (QIAGEN, catalog no. 19065)
Buffer QG (QIAGEN, catalog no. 19063)
DNA ladder (1 kb)
DNA polymerase (NEB, catalog no. M0209) and Buffer 2 ($10 \times$)
DNA Terminator End Repair Kit (Lucigen, catalog no. 40035-2)
dNTPs (10 mM)
Ethanol (70%)
Ethidium bromide
FlashGel (2.2%) (Lonza)
Gel-loading buffer IV <A>
Genomic DNA
Internal adaptors (2 μM) (AB)
Internal adaptor duplexes (2 μM = 2 pmol/μL):
    internal_adaptor-A: 5′-Phos/CGTACA(Bio-dT)CCGCCTTGGCCGT-3′
    internal_adaptor-B: 5′-Phos/GGCCAAGGCGGATGTACGGT-3′
Klenow fragment
Library binding buffer ($2 \times$) (Applied Biosystems)
LMP CAP adaptors (50 μM) (Applied Biosystems)
CAP adaptor duplexes (50 μM = 50 pmol/μL)
    LMP Cap_adaptor-A 5′-Phos/CTGCTGTAC-3′
    LMP Cap_adaptor-B 5′-ACAGCAG-3′
NaCl (3 M)
Paired End Sample Prep Kit (Illumina, catalog no. 1001809)
PCR Primers 1.0 and 2.0 (Illumina)
PE forked adaptor duplex (Illumina)

Phusion mix (2×) (Thermo Scientific)
Plasmid-Safe ATP-Dependent DNase (Epicenter, catalog no. E3110K)
QIAquick PCR purification columns (QIAGEN, catalog no. 28183)
Quick Ligation Kit (New England Biolabs, catalog no. M2200L)
S1 Nuclease (Life Technologies, catalog no. 18001-016)
Streptavidin-linked magnetic beads (Dynabeads, catalog no. M-270; Life Technologies, catalog no. 653-05)
Water, molecular grade

Equipment

DNA Shearing instrument (e.g., the S2 system; Covaris)
Fluorometer (e.g., Qubit; Life Technologies, catalog no. Q32857)
Ice-water bath
Magnetic Particle Collector (MPC)
miniTUBE–Blue AFA plastic tube (Covaris, catalog no. 520065)
Quant-iT dsDNA BR Assay Kit (Life Technologies, catalog no. Q32850)
Quant-iT dsDNA HS Assay Kit (Life Technologies)
Thermal cycler
Transilluminator
Tube rotator (e.g., Labquake; Thermo Scientific)
Tubes with caps (0.2 mL, 1.7 mL, and 15 mL), conical
Vacuum concentrator (e.g., SpeedVac)

METHOD

DNA Fragmentation

1. Measure the DNA (10 μg) volume and add Buffer EB to a final volume of 200 μL. Transfer the entire 200-μL volume into a miniTUBE (Blue bottom).

2. Shear the DNA using the Covaris S2 system:

    i. Open the Covaris software.

    ii. Click the **Open** button to choose operating conditions.

    iii. Select the **3Kb_shearing** file and click **Open**.

    iv. On the main software screen, double-check that parameters are as follows:

    Duty Cycle: 20%

    Intensity: 0.1

    Cycles per Burst: 1000

    Total Treatment Time: 600 sec

    v. Insert the DNA sample (from Step 1) into the miniTUBE holder, and place the holder into the water bath.

    vi. Start the shearing program by clicking the **Start** button.

    vii. After shearing is completed, remove the miniTUBE from the holder.

    viii. Transfer the DNA sample to a 1.7-mL microcentrifuge tube.

3. Run a 1.2% gel (FlashGel) to ensure that all of the sample has been completely fragmented to the 3-kb range.

End Repair

4. Add the following reagents—in the order indicated—to the DNA sample (200 μL) in the 1.7-mL tube:

End repair buffer (5×) (Lucigen)	60 μL
Water, molecular grade	38 μL
End Repair Enzyme Mix (Lucigen)	2 μL
Final volume	300 μL

Mix by vortexing, and incubate the end repair reaction for 30 min at room temperature.

5. Purify the end repair fragments using a QIAquick PCR purification column. The capacity of a column is 10 μg of DNA.

   i. Add the end repair reaction to 900 μL of Buffer ERC and vortex.

   ii. Add the reaction/buffer mix to a QIAquick column. Centrifuge the column at 14,000g for 1 min or place on a vacuum apparatus until all of the ERC mix has passed through.

   iii. Dispose of the waste if the tube was centrifuged.

   iv. Add 750 μL of Buffer PE. Centrifuge the column at 14,000g for 1 min or place on vacuum apparatus until all of the PE has passed through.

   v. Dispose of the waste if the tube was centrifuged.

   vi. Rotate the tube 180°. Centrifuge the column at 14,000g for 1 min.

   vii. Transfer the column to a 1.7-mL tube. Dispose of the waste tube.

   viii. Add 30 μL of Buffer EB. Incubate for 1 min at room temperature. Centrifuge the column at 14,000g for 1 min.

   ix. Add an additional 30 μL of Buffer EB. Incubate for 1 min at room temperature. Centrifuge the column at 14,000g for 1 min.

   x. Dispose of the column, and keep the 1.7-mL collection tube containing the eluted DNA.

6. Determine the concentration of the purified DNA by assaying 1 μL of the DNA with the Quant-It dsDNA BR Assay Kit and reading the results on a fluorometer (Protocol 11 or 12 may be useful).

Cap Adaptor Ligation

7. Calculate how many picomoles of LMP CAP adaptors (double-stranded) are needed. First calculate the picomoles of the insert DNA based on its size as follows:

(micrograms of DNA used; from Step 6) × 0.51 pmol of DNA = picomoles of DNA in sample,

(picomoles of DNA in sample) × 100 = picomoles of LM P CAP adaptors needed (100-fold excess),

$$\frac{\text{(picomoles of LMP CAP adaptors needed)}}{50 \text{ pmol/μL of LMP CAP adaptors}} = \text{microliters of LMP CAP adaptors to be used.}$$

For example, for 12.6 μg of DNA with a 3000-bp insert:

$$\frac{12.6 \, \mu g \, \text{DNA used} \left(\dfrac{1 \, \mu g \, \text{DNA} \times (10^6 \, \text{pg/}\mu g) \times \frac{1 \, \text{pmol}}{660 \, \text{pg}}}{3000 \, \text{bp}} \right) \times 100}{50 \, \text{pmol/}\mu l \, \text{LMP CAP adapters}}$$

= 12.73 μl of LMP CAP adapters need to be used for this example.

8. Combine and mix the following components in a 1.7-mL microcentrifuge tube:

DNA from Step 6	60 μL
CAP adaptors from Step 7 calculation	x μL
Quick Ligase Buffer (2 ×)	150 μL
Water, molecular grade	y μL
Ligase	7.5 μL
Total volume	300 μL

Mix by vortexing briefly, and incubate the ligation reaction for 15 min at room temperature.

9. Add the ligation reaction to 900 μL of Buffer ERC and vortex. Purify the ligated fragments using a QIAquick PCR purification column by following Steps 5.ii–5.viii. Retain the eluted DNA in the 1.7-mL collection tube.

Size Selection

10. Prepare a 0.8% LE agarose gel. Add 20 μL of ethidium bromide to the gel buffer before starting the electrophoresis.

11. Load the 1-kb DNA ladder (15 μL of ladder, 1.5 μL of Gel-loading buffer) into the appropriate wells.

12. Add 3 μL of 10× Gel-loading buffer to the DNA sample purified in Step 9. Load the entire sample on the gel. Run the gel at 120 V for 60 min.

13. Cut out the DNA in the 2–4-kb size range from the gel.

    ▲ *Be sure to wash the gel rig thoroughly after the run has completed to remove any residual ethidium bromide.*

14. Weigh the gel slices, and transfer them to a 15-mL conical tube. Add 6 mL of Buffer QG per 2 g of gel.

15. Shake the tube vigorously on a bench-top shaker until the gel is completely dissolved.

16. Purify the DNA using one or more QIAquick Gel Extraction columns. The maximum amount of gel that can be applied to a column is 400 mg. Therefore, the number of columns should be determined based on the weight of the excised gel piece as determined above (e.g., the 2 g of gel above dissolved in 6 mL of QG would need to be divided among five columns).

    i. Add up to 400 mg of gel dissolved in Buffer QG to a QIAquick column. Repeat with as many columns as necessary.

    ii. Centrifuge the column at 14,000*g* for 1 min or place on a vacuum apparatus until all of the QG mix has passed through.

    iii. Dispose of the waste if the tubes were centrifuged.

        Add more QG-gel mix to the column(s) and repeat the centrifugation until all of the QG-gel mix has been loaded.

    iv. Purify the ligated fragments by following Steps 5.iv–5.viii.

    v. Dispose of the columns and retain the 1.7-mL collection tubes containing the eluted DNA. Pool the purified DNA.

17. Determine the concentration of the purified DNA by assaying 1 μL of the DNA with the Quant-iT dsDNA BR Assay Kit and reading the results on a fluorometer (Protocol 11 or 12 may be useful).

    If <1 μg of DNA is recovered, it is recommended that you restart the protocol from the beginning.

DNA Circularization

18. Combine and mix the following components in a 1.7-mL tube. It may be necessary to set up multiple reactions.

Circularization Reaction

DNA from Step 17	1 μg of 2–4 kb
Quick Ligase Buffer (2×)	280 μL
Internal adaptors (2 μM)	0.65 μL
Ligase	14 μL
Water, nuclease free	x μL
Final volume	560 μL

Determine the quantity of nuclease-free water needed (x) to bring the reaction to the respective final volume. These volumes and DNA concentrations are critical for a successful circularization reaction.

19. Mix by vortexing, and incubate the circularization reaction for 10 min at room temperature.

20. Add each circularization reaction to 1680 μL of Buffer ERC (i.e., 3 volumes) in a 15-mL conical tube, and vortex.

21. Purify the circularized fragments using a QIAquick PCR Purification column by following Steps 5.ii–5.x.

Plasmid-Safe

22. Combine the following components in a 1.7-mL microcentrifuge tube:

DNA from Step 21	60 μL
ATP (25 mM)	5 μL
Plasmid Safe Buffer (10×)	10 μL
Water, molecular grade	24 μL
Plasmid-Safe DNase (10 U/μL)	1 μL
Final volume	100 μL

Mix by vortexing, and incubate the reaction for 40 min at 37°C.

23. Pool the reactions if multiple circularization reactions were performed. Add the reaction to 3 volumes of Buffer ERC and vortex.

24. Purify the DNA from the reaction using a QIAquick PCR Purification column by following Steps 5.ii–5.x.

25. Determine the concentration of the purified DNA by assaying 1 μL of the DNA with the Quant-It dsDNA HS Assay Kit and reading the results on a fluorometer (Protocol 11 or 12 may be useful).

Nick Translation

26. For every 200 ng of circular DNA recovered in Step 24, combine and mix the following components on ice. Mix all components and chill on ice.

DNA from Step 24 (200 ng)	x μL
NEB Buffer 2 (10×)	10 μL
dNTPs (10 mM)	10 μL
Water, nuclease free	y μL
Final volume	98 μL

27. Add 2 μL of 10 U/μL DNA polymerase I to each reaction tube, vortex to mix, and immediately incubate the reaction for 20 min in an ice-water bath (0°C).

28. Stop the reaction by adding 300 μL of Buffer ERC (i.e., 3 volumes).

29. Purify the DNA from the reaction using a QIAquick PCR Purification column by following Steps 5.ii–5.x.

Note that for cuts ~300–500 bp only, if the nick translation step produces fragments outside of the 300–500-bp range, then this step may not leave enough library sample for sequencing.

S1 Nuclease Digest

If multiple nick translation reactions were set up, each should be digested separately.

30. Dilute the S1 nuclease to 200 U/μL using the S1 nuclease dilution buffer.

31. Combine and mix the following components in a 1.7-mL microcentrifuge tube:

DNA from Step 29	60 μL
S1 buffer (10×)	10 μL
NaCl (3 M)	10 μL
Water, molecular grade	18 μL
S1 nuclease (200 U/μL)	2 μL
Final volume	100 μL

Mix by vortexing, and incubate the reaction for 15 min at 37°C.

32. Pool the reactions, add 3 volumes of Buffer ERC, and vortex.

33. Purify the DNA using a QIAquick PCR Purification column by following Steps 5.ii–5.viii. Retain the eluted DNA in the 1.7-mL collection tube.

End Repair

34. Combine and mix the following components in a 1.7-mL tube:

DNA from Step 33	30 μL
End repair buffer (5×)	10 μL
Water, molecular grade	8 μL
End repair enzyme mix	2 μL
Final volume	50 μL

Mix by vortexing, and incubate the end repair reaction for 30 min at room temperature.

35. Add the end repair reaction to 3 volumes of Buffer ERC and vortex.

36. Purify the DNA using a QIAquick PCR Purification column as follows.

   i. Perform Steps 5.ii–5.vii.

   ii. Add 25 μL of Buffer EB. Incubate for 1 min at room temperature. Centrifuge the column at 14,000g for 1 min.

   iii. Add an additional 25 μL of Buffer EB. Incubate for 1 min at room temperature. Centrifuge the column at 14,000g for 1 min.

   iv. Dispose of the column, and keep the 1.7-mL collection tube containing the 50 μL of eluted DNA.

Streptavidin Bead Binding

37. Transfer 25 μL of streptavidin beads to a new microcentrifuge tube. Using the Magnetic Particle Collector (MPC), pellet the beads and remove the buffer.

38. Wash the beads twice with 50 μL of 2× Library binding buffer, using the MPC. Vortex well between each wash.

39. Resuspend the beads in 50 μL of 2× Library binding buffer.

40. Add the 50 μL of washed streptavidin beads to the 50 μL of DNA from Step 36.iv. Vortex to mix well, and place the tube on a tube rotator for 15 min at room temperature.

41. Pellet the beads on the MPC and remove the supernatant.

42. Using the MPC, wash the immobilized library three times with 500 μL of Buffer EB. Vortex well between each wash.

43. Remove all of the remaining Buffer EB, and resuspend the pellet in 32 μL of Buffer EB.

A-Tailing Reaction

44. Combine and mix the following components in a 1.7-mL tube:

DNA/beads from Step 43	32 μL
NEB Buffer 2 (10×)	5 μL
dATP (1 mM)	10 μL
Klenow fragment (5 U/μL)	3 μL
Final volume	50 μL

Mix by vortexing, and incubate the reaction for 30 min at 37°C.

45. Using the MPC, wash the immobilized library three times with 500 μL of Buffer EB. Vortex well between each wash.

46. Remove the remaining Buffer EB, and resuspend the pellet in 15 μL of Buffer EB.

Illumina Paired-End Adaptor Ligation

47. Combine and mix the following components in a 1.7-mL tube:

DNA/beads from Step 46	15 μL
Quick Ligase Buffer (2×)	25 μL
PE forked adaptor duplex	5 μL
Ligase	5 μL
Final volume	50 μL

Mix by vortexing, and incubate the reaction for 15 min at room temperature.

48. Using the MPC, wash the immobilized library three times with 500 μL each of Buffer EB. Vortex well between each wash.

49. Remove the remaining Buffer EB, and resuspend the DNA/beads pellet in 30 μL of Buffer EB.

PCR Optimization

50. Combine the following components for one reaction in a 0.2-mL tube. Prepare one reaction per mate-pair library construction.

The cycle optimization is performed to determine the optimal number of PCR cycles. Then, the remainder of the library is amplified in a new PCR set to that predetermined number of cycles.

DNA/beads	2 μL
Phusion mix (2×)	25 μL
PE PCR Primer 1.0	0.5 μL
PE PCR Primer 2.0	0.5 μL
Water, molecular grade	22 μL
Final volume	50 μL

51. Create two PCR programs.

Program 1

Cycle number	Denaturation	Annealing	Polymerization
1	30 sec at 98°C		
10 cycles	10 sec at 98°C	30 sec at 65°C	30 sec at 72°C

Hold at 4°C.

Program 2

Cycle number	Denaturation	Annealing	Polymerization
2 cycles	10 sec at 98°C	30 sec at 65°C	30 sec at 72°C

Hold at 4°C.

52. Label six 0.2-mL tubes: PCR-10, PCR-12, PCR-14, PCR-16, PCR-18, and PCR-20.

53. Using the 50-μL reaction from Step 50, run PCR Program 1, and collect 5 μL of the reaction in the 0.2-mL tube labeled **PCR-10**.

54. Continue the PCR optimization by running PCR Program 2, and collect 5 μL of the reaction in the 0.2-mL tube labeled **PCR-12**.

55. Repeat Step 54 for the remaining four cycle-collection points (i.e., tubes 14, 16, 18, and 20).

56. Add 1 μL of loading dye to each 5-μL aliquot. Vortex quickly and centrifuge the tubes.

57. Prepare a 2.2% FlashGel with 4 μL of water in each of the 13 wells and 2 μL of Flash 50–1500-bp gel marker in the first and eighth lanes.

58. Load the 6-μL PCR aliquots in Lanes 2–7. Run the FlashGel for 5–7 min at 275 V.

59. Photograph the gel. Identify cycles that show overamplification. For the full-scale library PCR step below, use the greatest number of cycles that do not show overamplification. For example, if PCR-18 and PCR-20 show overamplification but none of the other tubes do, then amplify the library for 16 of the indicated cycles.

Full-Scale PCR

60. Prepare a PCR master mix multiplying the quantities per reaction below by the number of reactions. We recommend 12 reactions.

Phusion mix (2×)	25 μL
PE PCR Primer 1.0	0.5 μL
PE PCR Primer 2.0	0.5 μL
Water, molecular grade	22 μL
Final volume per reaction	48 μL

61. Program the thermal cycler based on the results of optimization in the previous section.

Cycle number	Denaturation	Annealing	Polymerization
1	30 sec at 98°C		
N cycles[a]	10 sec at 98°C	30 sec at 65°C	30 sec at 72°C

Hold at 4°C.

[a]The minimal number of PCR cycles that results in product amplification, but not overamplification, should be used (*N*).

62. In each reaction tube, combine 48 μL of PCR master mix and 2 μL of DNA/beads from Step 49. Amplify the DNA.

63. After cycling is complete, pool all of the samples into a 15-mL conical tube, and adjust the volume of the sample to 600 μL with Buffer EB.

64. Add the PCR to 3 volumes of Buffer ERC, and vortex.

65. Purify the DNA using a QIAquick PCR Purification column as follows.

   i. Perform Steps 5.ii–5.vii.

   ii. Add 50 μL of Buffer EB. Incubate for 1 min at room temperature. Centrifuge the column at 14,000*g* for 1 min.

   iii. Add an additional 50 μL of Buffer EB. Incubate for 1 min at room temperature. Centrifuge the column at 14,000*g* for 1 min.

   iv. Dispose of the column, and keep the 1.7-mL collection tube containing the 100 μL of eluted DNA.

Dual AMPure Bead Cleanup

66. Vortex the AMPure XP beads.

67. Add 60 μL of AMPure XP beads to the 100 μL of sample collected in Step 65.iv. Place on a tube rotator for 5 min. This is Tube 1.

68. Aliquot 60 μL of AMPure XP beads into a new 1.5-mL tube. This is Tube 2.

69. Place Tube 2 on the MPC and wait 3 min until the beads have fully separated from the supernatant.

70. Discard the supernatant. Add 20 μL of AMPure XP to the "dry" beads on the MPC. Vortex to resuspend.

71. Transfer Tube 1 from the rotator to the MPC. Let it sit there for 2 min until the beads have fully separated from the supernatant.

72. Transfer the Tube 1 supernatant into Tube 2. Vortex Tube 2 to resuspend, and place it on the tube rotator for 5 min. Discard Tube 1.

73. Place Tube 2 on the MPC for 2 min, until the beads have fully separated from supernatant.

74. Discard the supernatant.

75. Wash the beads with 500 μL of 70% ethanol. Invert the tube several times, and let it incubate for 1 min on the MPC. Remove the supernatant.

    ▲ *Leave on the MPC during entire wash.*

76. Repeat Step 75.

77. Remove as much residual ethanol from the cap and tube as possible.

78. Centrifuge the tube briefly to collect the beads on the bottom of the tube.

79. Place the tube of beads in a vacuum concentrator for 1 min. Check for residual liquid in the bottom of the tube or on the beads (beads should look "cracked" and not shiny).

80. Repeat the drying step in 1-min intervals, if needed.

81. Elute the DNA from the beads in 20 μL of Buffer EB.

    Be diligent about getting all of the beads into solution. Try to remove any beads from the sides of the tube. Tip the tube, and so on.

82. Place the tube on the MPC. Carefully transfer the supernatant to a microcentrifuge tube.

    ▲ *Save the supernatant. This is your final sample.*

Final Library Quantification

83. Run 1 μL of final sample from Step 82 on a 2.2% gel to verify the size of the DNA.

84. Determine the concentration of the purified DNA by assaying 1 μL of the final sample with the Quant-It dsDNA HS Assay Kit and reading the results on a fluorometer (Protocol 11 or 12 may be useful).

85. Enter these values into the Illumina Library Dilution Calculator to prepare 5 nM dilutions of the final paired-end library sample.

Preparation of an 8-kb Mate-Pair Library for Illumina Sequencing

This protocol provides instructions for preparing an 8-kb paired-end library suitable for sequencing on the Illumina instrument. These are large-insert libraries, and the initial shearing will use various parameters and different instruments to achieve the desired high-size fraction (see the information panel Fragmenting of DNA). For further details, see the discussion on paired-end libraries in the introductory section of this chapter.

MATERIALS

It is essential that you consult the appropriate Material Safety Data Sheets and your institution's Environmental Health and Safety Office for proper handling of equipment and hazardous materials used in this protocol.

Recipes for reagents specific to this protocol, marked <R>, are provided at the end of the protocol. See Appendix 1 for recipes for commonly used stock solutions, buffers, and reagents, marked <A>. Dilute stock solutions to the appropriate concentrations.

Reagents

Amplification primers (100 μM)
AMPure XP beads (Agencourt, catalog no. A63880)
ATP, lithium salt (100 mM) (Roche, catalog no. 14470220)
Bst DNA polymerase, large fragment (8000 U/mL) (NEB, catalog no. M0275L)
Buffer EB (10 mM Tris at pH 7.5–8.5)
Buffer ERC (QIAGEN, catalog no. 1018144)
Buffer PBI (QIAGEN, catalog no. 19066)
Buffer PE (QIAGEN, catalog no. 19065)
Buffer QG (QIAGEN, catalog no. 19063)
Carrier DNA
Circularization adaptors (20 μM)
Cre recombinase (NEB, catalog no. M0298L)
dATP (1 mM)
DNA ladder (1 kb) (Life Technologies, catalog no. 15615-024) (material core dilute to 100 ng/μL)
DNA Terminator End Repair Kit (Lucigen, catalog no. 40035-2)
dNTPs (10 mM) (Roche, catalog no. 11581295001)
DTT (1 M)
Ethanol (70%)
Exonuclease I (20,000 U/mL) (NEB, catalog no. M0293L)
Gel-loading buffer with bromophenol blue (10×) <A>
Genomic DNA
GS FLX Titanium Paired End Adapter Kit (Roche, catalog no. 05 463 343 001)
Index PCR primer (8 μM) (IDT)
Index PE forked adaptor duplex (4 μM) (Illumina, Sigma-Aldrich)
Klenow buffer (10×) (NEB Buffer 2) (New England Biolabs)
Klenow exonuclease (5 U/μL) (NEB, catalog no. M0212L)
Library adaptors (20 μM)
Library binding buffer (2×) <R>

Primers

See Table 1 for the primer/oligo sequences.

Multiplexing PCR Primer 1.0 (8 μM) (IDT)
Multiplexing PCR Primer 2.0 (0.5 μM) (IDT)
PCR Primer PE 1.0 (8 μM) (IDT)
PCR Primer PE 2.0 (8 μM) (IDT)

PE forked adaptor duplex (4 μM) (Illumina, Sigma-Aldrich)
Phusion HF PCR master mix (2×) (NEB, catalog no. F-531L)
Plasmid-Safe ATP dependent DNase (Epicenter, catalog no. E3110K)
Quick Ligation Kit (NEB, catalog no. M2200L)
SDS (20%)
SeaKem LE agarose (Lonza, catalog no. 50004)
Sodium acetate (3 M, pH 5.5)
Streptavidin beads (Dynabeads, catalog no. M-270; Life Technologies, catalog no. 653-05)
SYBR Green I nucleic acid gel stain (Life Technologies, catalog no. S7585)
TBE buffer <A>
ThermoPol Reaction Buffer (10×) (NEB, catalog no. B9004S)
Tris-HCl (10 mM, pH 8.0) or Buffer EB (QIAGEN, catalog no. 19086)

Equipment

Bioanalyzer (Agilent 2100, catalog no. 5067-1504)
DNA 1000 kit (Agilent 2100, catalog no. 5067-1504)
DNA shearing device (Hydroshear; Gene Machines, catalog no. JHSH000000-1)
DNA shearing device (Hydroshear Large Assembly; Gene Machines, catalog no. JHSH204007)
DNA shearing instrument (Covaris S2 system)
Electrophoresis apparatus (Chef Mapper)
Magnetic Particle Collector (MPC)
Microcentrifuge tubes with caps (0.2 mL and 1.7 mL)
MinElute Columns (QIAGEN, catalog no. 28006)
miniTUBE−Blue AFA plastic tube (Covaris, catalog no. 520065)
QIAquick columns (QIAGEN, catalog no. 28183)
Quant-iT Assay Kit
Qubit assay tubes
Qubit Fluorometer and Assay Kit
Thermo cycler
Tube rotator (e.g., Labquake)
Vacuum concentrator (e.g., SpeedVac)

TABLE 1. Sequences of adaptors and primers

Adaptors and primers	Sequences
Ill-PE forked adaptor duplex (4 μM) (Sigma-Aldrich)	5′-P-GATCGGAAGAGCGGTTCAGCAGGAATGCCGAG
	5′-ACACTCTTTCCCTACACGACGCTCTTCCGATCT
PE PCR Primer 1.0	5′-AATGATACGGCGACCACCGAGATCTACACTCTTTCCCTACACGACGCTCTTCCGATCT
PE PCR Primer 2.0	5′-CAAGCAGAAGACGGCATACGAGATCGGTCTCGGCATTCCTGCTGAACCGCTCTTCCGATCT
ILL_IndexAdapter_Duplex_IDT	5′-P-GATCGGAAGAGCACACGTCT (5′-phosphorylation)
	5′-ACACTCTTTCCCTACACGACGCTCTTCCGATCT (dephosphorylated)
Multiplexing PCR Primer 1.0	5′-AATGATACGGCGACCACCGAGATCTACACTCTTTCCCTACACGACGCTCTTCCGATCT
Multiplexing PCR Primer 2.0	5′-GTGACTGGAGTTCAGACGTGTGCTCTTCCGATCT

(Continued)

TABLE 1. (*Continued*)

Adaptors and primers	Sequences
PCR Primer, Index 1	CAAGCAGAAGACGGCATACGAGATCGTGATGTGACTGGAGTTC
PCR Primer, Index 2	CAAGCAGAAGACGGCATACGAGATACATCGGTGACTGGAGTTC
PCR Primer, Index 3	CAAGCAGAAGACGGCATACGAGATGCCTAAGTGACTGGAGTTC
PCR Primer, Index 4	CAAGCAGAAGACGGCATACGAGATTGGTCAGTGACTGGAGTTC
PCR Primer, Index 5	CAAGCAGAAGACGGCATACGAGATCACTGTGTGACTGGAGTTC
PCR Primer, Index 6	CAAGCAGAAGACGGCATACGAGATATTGGCGTGACTGGAGTTC
PCR Primer, Index 7	CAAGCAGAAGACGGCATACGAGATGATCTGGTGACTGGAGTTC
PCR Primer, Index 8	CAAGCAGAAGACGGCATACGAGATTCAAGTGTGACTGGAGTTC
PCR Primer, Index 9	CAAGCAGAAGACGGCATACGAGATCTGATCGTGACTGGAGTTC
PCR Primer, Index 10	CAAGCAGAAGACGGCATACGAGATAAGCTAGTGACTGGAGTTC
PCR Primer, Index 11	CAAGCAGAAGACGGCATACGAGATGTAGCCGTGACTGGAGTTC
PCR Primer, Index 12	CAAGCAGAAGACGGCATACGAGATTACAAGGTGACTGGAGTTC
PCR Primer, Index 13	CAAGCAGAAGACGGCATACGAGATAAACCTGTGACTGGAGTTC
PCR Primer, Index 14	CAAGCAGAAGACGGCATACGAGATTTGACTGTGACTGGAGTTC
PCR Primer, Index 15	CAAGCAGAAGACGGCATACGAGATGGAACTGTGACTGGAGTTC
PCR Primer, Index 16	CAAGCAGAAGACGGCATACGAGATTTGACATGTGACTGGAGTTC
PCR Primer, Index 17	CAAGCAGAAGACGGCATACGAGATCTATCTGTGACTGGAGTTC
PCR Primer, Index 18	CAAGCAGAAGACGGCATACGAGATGATGCTGTGACTGGAGTTC
PCR Primer, Index 19	CAAGCAGAAGACGGCATACGAGATAGCGCTGTGACTGGAGTTC
PCR Primer, Index 20	CAAGCAGAAGACGGCATACGAGATAGATGTGTGACTGGAGTTC
PCR Primer, Index 21	CAAGCAGAAGACGGCATACGAGATCTGGGTGTGACTGGAGTTC
PCR Primer, Index 22	CAAGCAGAAGACGGCATACGAGATGGACGGGTGACTGGAGTTC
PCR Primer, Index 23	CAAGCAGAAGACGGCATACGAGATTTCTCGGTGACTGGAGTTC
PCR Primer, Index 24	CAAGCAGAAGACGGCATACGAGATATTCCGGTGACTGGAGTTC
PCR Primer, Index 25	CAAGCAGAAGACGGCATACGAGATAGCTAGGTGACTGGAGTTC
PCR Primer, Index 26	CAAGCAGAAGACGGCATACGAGATGTATAGGTGACTGGAGTTC
PCR Primer, Index 27	CAAGCAGAAGACGGCATACGAGATTCTGAGGTGACTGGAGTTC
PCR Primer, Index 28	CAAGCAGAAGACGGCATACGAGATCAGCAGGTGACTGGAGTTC
PCR Primer, Index 29	CAAGCAGAAGACGGCATACGAGATCTAAGGGTGACTGGAGTTC
PCR Primer, Index 30	CAAGCAGAAGACGGCATACGAGATCCGGTGGTGACTGGAGTTC
PCR Primer, Index 31	CAAGCAGAAGACGGCATACGAGATCGATTAGTGACTGGAGTTC
PCR Primer, Index 32	CAAGCAGAAGACGGCATACGAGATTCCGTCGTGACTGGAGTTC
PCR Primer, Index 33	CAAGCAGAAGACGGCATACGAGATTATATCGTGACTGGAGTTC
PCR Primer, Index 34	CAAGCAGAAGACGGCATACGAGATAGCATCGTGACTGGAGTTC
PCR Primer, Index 35	CAAGCAGAAGACGGCATACGAGATCTCTACGTGACTGGAGTTC
PCR Primer, Index 36	CAAGCAGAAGACGGCATACGAGATCGCGGCGTGACTGGAGTTC
PCR Primer, Index 37	CAAGCAGAAGACGGCATACGAGATTGGAGCGTGACTGGAGTTC
PCR Primer, Index 38	CAAGCAGAAGACGGCATACGAGATTGTGCCGTGACTGGAGTTC
PCR Primer, Index 39	CAAGCAGAAGACGGCATACGAGATCAGGCCGTGACTGGAGTTC
PCR Primer, Index 40	CAAGCAGAAGACGGCATACGAGATGCGGACGTGACTGGAGTTC
PCR Primer, Index 41	CAAGCAGAAGACGGCATACGAGATGCTGTAGTGACTGGAGTTC
PCR Primer, Index 42	CAAGCAGAAGACGGCATACGAGATATTATAGTGACTGGAGTTC
PCR Primer, Index 43	CAAGCAGAAGACGGCATACGAGATGAATGAGTGACTGGAGTTC
PCR Primer, Index 44	CAAGCAGAAGACGGCATACGAGATTGCCGAGTGACTGGAGTTC
PCR Primer, Index 45	CAAGCAGAAGACGGCATACGAGATGGTAGAGTGACTGGAGTTC
PCR Primer, Index 46	CAAGCAGAAGACGGCATACGAGATCATTCAGTGACTGGAGTTC
PCR Primer, Index 47	CAAGCAGAAGACGGCATACGAGATGGAGAAGTGACTGGAGTTC
PCR Primer, Index 48	CAAGCAGAAGACGGCATACGAGATATGGCAGTGACTGGAGTTC

METHOD

DNA Fragmentation

This procedure uses 14 cycles at a speed setting of 9 in the large shearing assembly of the Hydroshear and produces an average fragment size of ~8 kb. However, each assembly should be tested individually for fragmentation size range, as recommened by the manufacturer.

1. Transfer at least 15 μg of sample genomic DNA to a 1.7-mL microcentrifuge tube. Add Buffer EB to a volume of 150 μL.

2. Add 40 μL of 50× Lucigen End Repair Buffer. Mix by inverting the tube.

3. Open the Hydroshear software on the desktop.

4. Switch the Hydroshear to Output. Click on OK.

5. Under "Shearing Parameters," change the settings to the following:

   Volume: 200 μL
   Number of cycles: 14
   Speed Code: 9

6. Click on **Edit Wash Scheme**. Verify that there are four washes for det1, det2, and water each. Click OK.

7. Follow the computer prompts through the washes and drawing of air at the end.

8. Load the sample.

9. Follow the prompts, except for when the sample has just been loaded. Ignore the prompt to turn to Output. Instead, turn the dial vertically. Click OK.

10. After the bubble has been pushed out in the syringe and the arm stops, then turn the dial to Output.

11. Proceed to the shearing step.

12. Once the sample has been sheared and retrieved, proceed through the machine cleaning Steps 6 and 7.

13. Once the machine has been cleaned, click on **Manual Operation**.
    Change values to the following:

    Volume: 500 μL
    Speed Code: 10
    Valve: 0

14. Empty the syringe twice by clicking **Reinitialize**.

15. Draw up 500 μL of water for storage.

End Repair

16. Check the volume of the sheared sample, and adjust the volume to 198 μL with water. Add 2 μL of Lucigen End Repair enzyme mix. Mix by inverting or flicking the tube, centrifuge briefly, and incubate the end repair reaction for 30 min at room temperature.

17. Purify the end repair fragments using a QIAquick PCR Purification column. Use one column.

    i. Add the end repair reaction to 600 μL of Buffer ERC and vortex.

    ii. Add the reaction/buffer mix to a QIAquick column. Centrifuge the column at 14,000g for 1 min or place on a vacuum apparatus until all of the ERC mix has passed through.

    iii. Dispose of the waste if the tube was centrifuged.

    iv. Add 750 μL of Buffer PE. Centrifuge the column at 14,000g for 1 min or place on a vacuum apparatus until all of the PE has passed through.

    v. Dispose of the waste if the tube was centrifuged.

    vi. Rotate the tube 180°. Centrifuge the column at 14,000*g* for 1 min.

    vii. Transfer the column to a 1.7-mL tube. Dispose of the waste tube.

    viii. Add 30 μL of Buffer EB. Incubate for 1 min at room temperature. Centrifuge the column at 14,000*g* for 1 min.

    ix. Add additional 30 μL of Buffer EB. Incubate for 1 min at room temperature. Centrifuge the column at 14,000*g* for 1 min.

    x. Dispose of the column, and keep the 1.7-mL collection tube containing the eluted DNA.

18. Determine the concentration of the purified DNA by assaying 1 μL of the DNA with the Quant-It dsDNA BR Assay Kit and reading the results on a fluorometer (Protocol 11 or 12 may be useful).

Adaptor Ligation for Circularization

19. Combine the following in a 1.7-mL microcentrifuge tube:

DNA	59 μL
Water, molecular grade	21 μL
Circularization adaptors (20 μM)	10 μL
Quick Ligase Buffer (2×)	100 μL
Final volume	190 μL

Flick or invert the tube to mix.

20. Add 10 μL of 2000 units/μL ligase to the reaction. Flick or invert the tube, centrifuge it briefly, and incubate the ligation reaction for 15 min at room temperature.

21. Stop the reaction by adding 20 μL of 10× Gel-loading buffer and 2 μL of 20% SDS.

22. Heat the reaction for 10 min at 65°C, and cool the tube on ice before loading on a gel.

Size Selection Using FIGE

23. Prepare 200 mL of 1% LE agarose gel by combining 200 mL of 0.5× TBE and 2 g of LE agarose.

24. Fill a field inversion gel electrophoresis (FIGE) gel rig with 2.5 L of 0.5× TBE. Turn on the FIGE chiller, and set the temperature to 14°C.

25. Prepare a 1-kb DNA ladder by adding 3 μL of 10× Gel-loading buffer to 30 μL of 100 ng/μL 1-kb DNA ladder.

26. Load 16.5 μL of the ladder into well Position 1 and the remaining DNA ladder into well Position 5.

27. Load the entire sample into a single 3-cm-wide well at Position 3 on the gel.

    Separating the sample from the ladder minimizes the chance of the ladder contaminating the sample.

28. Run the FIGE using the Auto Algorithm selection. Enter **3K** in as the low limit and **15K** in as the high limit. Press the Enter button to cycle through the rest of the program options. Start the run. It should take ∼17.5 h to complete the electrophoresis.

29. Stain the gel with SYBR Green for 45 min.

30. Destain with water for 15 min.

31. Excise the 6.5–9.5-kb size fraction from the gel using a blue-light source (300 nm) to visualize the DNA in the gel.

32. Weigh the gel slices and mix with 3 volumes of Buffer QG in a 15-mL conical tube. For example, combine 3 g of gel with 9 mL of Buffer QG.

33. Gently rock the tube back and forth until the gel is completely dissolved. The gel should be completely dissolved in ~5–10 min.

    ▲ *Do not vortex!*

34. Purify the DNA using one or more QIAquick gel purification columns. The maximum amount of gel that can be applied to each column is 400 mg. Therefore, the number of columns should be determined based on the weight of the excised gel piece as determined above.

    i. Add up to 1 g of gel dissolved in Buffer QG to a QIAquick column. Repeat with as many columns as necessary.

    ii. Centrifuge the column at 14,000g for 1 min or place on a vacuum apparatus until all of the QG mix has passed through.

    iii. Dispose of the waste if the tubes were centrifuged. Add more QG-gel mix to the column(s), and repeat the centrifugation until all of the QG-gel mix has been loaded.

    iv. Purify the ligated fragments by following Steps 17.iv–17.vii.

    v. Add 20 μL of Buffer EB to the first column and 10 μL to each additional column.

    vi. Incubate the columns for 1 min at room temperature with the tube uncapped. Cap the tubes, and centrifuge the first column at 14,000g for 1 min.

    vii. Dispose of the first column, and transfer 20 μL of eluted DNA from the first collection tube into the second column. Centrifuge the second column at 14,000g for 1 min.

    viii. Dispose of the second column, and transfer 30 μL of eluted DNA from the second collection tube into the third column. Centrifuge the third column at 14,000g for 1 min.

    ix. Dispose of the third column, and retain this 1.7-mL collection tube containing ~40 μL of eluted DNA.

Fill-In Reaction

35. Combine and mix the following components in a 1.7-mL tube:

LoxP-adapter ligated DNA	38 μL
ThermoPol Buffer (10×)	5 μL
PCR nucleotide mix (10 mM)	4 μL
Bst DNA polymerase, large fragment (8 U/μL)	3 μL
Final volume	50 μL

    Mix by gently inverting the tube. Incubate the fill-in reaction for 15 min at 50°C.

36. Quantify the DNA using the Qubit dsDNA HS Assay Kit, following the manufacturer's instructions (also see Protocol 11).

    ▲ *It is strongly recommended that fluorescence-based quantification methods be used rather than UV absorbance methods.*

    At this step, the average DNA yield is >400 ng. A minimum of 300 ng of DNA is required to proceed with the preparation. If >300 ng is obtained, use the remaining DNA in simultaneous circularization reactions.

    If the DNA yield is <300 ng, start the procedure again from the beginning.

37. (Optional) The DNA sample can be stored overnight at 4°C.

DNA Circularization

38. Prepare a 300-ng aliquot of the filled-in DNA in a total volume of 80 μL (adjust the volume with molecular biology–grade water).

39. In a 0.2-mL tube, add the following reagents, in the order indicated:

Filled-in *LoxP* DNA (300 ng) (from Step 35)	80 µL
Cre buffer (10×)	10 µL
Cre recombinase (1 U/µL)	10 µL
Total	100 µL

Mix by gently inverting the tube.

40. Incubate the reaction in a thermal cycler using the following program:

    45 min at 37°C
    10 min at 70°C
    4°C forever

41. Prepare fresh 100 mM DTT from a 1 M DTT stock. Vortex and store on ice.

    Discard the 100 mM DTT dilution at the end of the day.

Water, molecular biology grade	18 µL
DTT (1 M)	2 µL
Total	20 µL

42. Upon completion of the incubation period, add 1.1 µL of 100 mM DTT to the Cre-treated DNA reaction. Mix the reaction by inverting the tube, and centrifuge briefly to collect the solution at the bottom of the tube.

43. Add the following reagents to the sample:

ATP (100 mM)	1.1 µL
Plasmid-Safe ATP-Dependent DNase (10 U/µL)	5.0 µL
Exonuclease I (20 U/µL)	3.0 µL

    Mix gently, but completely, by inverting the tube.

44. Incubate the reaction for 30 min at 37°C.

45. Heat-inactivate the reaction for 30 min at 70°C in a thermal cycler.

46. (Optional) The DNA sample can be stored overnight at 4°C.

Fragment *LoxP* Circularized DNA

47. Transfer the 110 µL of sample DNA to a miniTUBE–Blue AFA. Centrifuge the miniTUBE to ensure that there are no bubbles in the solution.

48. Place the sample vessel in the Covaris water bath.
    Run the Covaris program *8kb_illumina*.

    Duty Cycle: 20%
    Intensity: 5
    Cycles/burst: 500
    Time: 30 sec

49. Remove the sample tube from the water bath, and transfer the contents to a 1.7-mL tube.

50. Pool the sheared DNA. If there is only one circularization event, there will be one sheared tube.

51. Add 3 volumes of Buffer ERC to the sheared DNA and vortex.

52. Purify the end-repaired fragments using a QIAquick PCR purification column.

    i. Add the reaction/buffer mix to a column.

    ii. Centrifuge the column at 14,000*g* for 1 min or place on vacuum apparatus until all of the ERC mix has passed through.

    iii. Dispose of waste if the tube was centrifuged.

    iv. Add 750 µL of Buffer PE.

    v. Centrifuge the column at 14,000*g* for 1 min or place on vacuum apparatus until all of the PE has passed through.

vi. Dispose of waste if the tube was centrifuged.

vii. Rotate the tube 180°. Centrifuge the column at 14,000*g* for 1 min.

viii. Transfer the column to a 1.7-mL tube. Discard the waste tube.

ix. Add 20 μL of Buffer EB. Incubate for 1 min at room temperature. Centrifuge the column at 14,000*g* for 1 min.

x. Add an additional 20 μL of Buffer EB. Incubate for 1 min at room temperature. Centrifuge the column at 14,000*g* for 1 min.

xi. Dispose of the column, and retain the 1.7-mL collection tube containing the eluted DNA.

Fragment End Repair

53. Combine and mix the following components in a 1.7-mL tube:

DNA	38 μL
End repair buffer (5×)	10 μL
End repair enzyme mix	2 μL
Final volume	50 μL

Mix by vortexing, and incubate the end repair reaction for 30 min at room temperature.

54. Add the reaction to 3 volumes of Buffer ERC and vortex.

55. Purify the end-repaired fragments using a QIAquick PCR purification column by following Steps 52.i–52.viii.

56. Elute the DNA from the column with 33 μL of Buffer EB. Incubate for 1 min at room temperature.

57. Determine the concentration of the DNA by assaying 1 μL of the DNA with the Quant-It dsDNA HS Assay Kit and reading the results on a fluorometer (Protocol 11 or 12 may be useful). Record the concentration for the adaptor calculations.

Mate-Pair A-Tailing

58. Using the purified product from Step 56, set up the A-tailing reaction:

Purified end repair DNA	32 μL
Klenow buffer (10×) (i.e., NEB Buffer 2)	5 μL
dATP (1.0 mM)	10 μL
Klenow Exo (5 U/μL)	3 μL
Total volume	50 μL

Incubate the reaction in a thermal cycler for 30 min at 37°C.

59. Add the reaction to 3 volumes of Buffer ERC and vortex.

60. Purify the DNA using a QIAquick PCR purification column by following Steps 52.i–52.viii.

61. Elute the DNA from the column with 16 μL of Buffer EB. Incubate for 1 min at room temperature.

Mate-Pair Adaptor Ligation

62. Calculate how many picomoles of adaptors (double-stranded) are needed.

i. First calculate the picomoles of insert DNA based on its size.

$$\text{picomoles of insert DNA} = \text{micrograms of DNA} \times (1 \text{ pmol}/660 \text{ pg}) \times 10^{6} \text{pg}/\mu\text{g}$$
$$\times 1/N,$$

where N is the average insert size;

$$\text{Adaptor needed} = (\text{picomoles of insert DNA}) \times 50.$$

Example: For 30 ng of total DNA with average size of 850 bp:

picomoles of insert DNA

$$= 0.03 \mu g \times (1 \text{ pmol})/(660 \text{ pg}) \times (10^6 \text{ pg}/\mu g) \times 1/850 = 0.053 \text{ pmol};$$

Adaptor needed

$$= (0.053 \text{ pmol}) \times 50 = 2.7 \text{ pmol}.$$

63. Dilute the PE Illumina adaptor oligo mix from 4 μM to 1 μM. Use 2.7 μL of 1 μM adaptor stock.

> Or use the calculator at http://www.promega.com/biomath/Default.htm. Choose dsDNA: Micrograms to Picomoles. Plug in the numbers to get the picomoles of the sample.
> *Example:* For the example from Step 62, after choosing "Micrograms to Picomoles" answer the following questions.
>
> - How long is your DNA (in base pairs)? Enter **850**
> - How many micrograms of DNA do you have? Enter **0.03**
>
>   Then press Calculate, and it will give the answer: *0.053 pmol of DNA.*
>   Multiply the picomoles of the sample with 50 (50-fold excess).
>   For the Index library, use "ILL Index Adapter Duplex" instead of "ILL_IndexAdapter_Duplex_IDT."

64. Combine the following components in a 1.7-mL tube:

A-tailed DNA	15 μL
Quick Ligase Buffer (2×)	25 μL
ILL PE forked adapter duplex *or* ILL Index Adapter Duplex	x μL
Water	y μL
Final volume	45 μL

65. Flick or invert the tube to mix. Add 5 μL of 2000 U/μL ligase. Incubate for 15 min at room temperature.

66. Purify the DNA using the AMPure XP bead cleanup system.

   i. Add 75 μL of calibrated AMPure beads to the adaptor-ligated DNA.

   ii. Incubate on a tube rotator for 5 min at room temperature.

   iii. Incubate the tube on an MPC for 2 min to separate the supernatant from the beads.

   iv. Remove the supernatant from the beads.

   v. Wash the beads twice with 500 μL of 70% ethanol.

   vi. Dry in a vacuum concentrator for 2 min.

   vii. Add 50 μL of Buffer EB. Vortex well. Incubate for 1 min at room temperature.

   viii. Incubate the tube on the MPC for 1 min.

   ix. Transfer the supernatant to a 1.7-mL tube.

Mate-Pair Library Immobilization

67. Transfer 25 μL of streptavidin beads to a fresh microcentrifuge tube. Using the MPC, pellet the beads and remove the buffer.

68. Wash the beads twice with 50 μL of 2× Library binding buffer, using the MPC.

Vortex well between each wash.

69. Resuspend the beads in 50 μL of 2× Library binding buffer. Add the 50 μL of washed streptavidin beads to the 50 μL of DNA from Step 65. Vortex to mix well and place on a tube rotator for 15 min at room temperature.

70. Pellet the beads on the MPC and remove the supernatant.

71. Using the MPC, wash the immobilized library three times with 500 μL of TE buffer. Vortex well between each wash.

72. Remove all of the TE buffer, and resuspend the pellet in 30 μL of TE buffer.

Mate-Pair PCR Optimization

Perform either Step 73 or Step 74.

For a Nonindex Library

73. Combine the following components for one reaction in a 0.2-mL microcentrifuge tube:

DNA/beads	2 μL
Phusion mix (2×)	25 μL
PE PCR Primer 1.0 (8 μM)	0.5 μL
PE PCR Primer 2.0 (8 μM)	0.5 μL
Water, molecular grade	22 μL
Final volume	50 μL

For an Index Library

74. Pick an index PCR primer from Index 1–48. For a different library to run on the same flow cell, pick a different index PCR primer for each library.

DNA/beads	2 μL
Phusion mix (2×)	25 μL
Multiplexing Primer 1.0 (8 μM)	1 μL
Multiplexing Primer 2.0 (0.5 μM)	1 μL
Index PCR Primer (8 μM)	1 μL
Water, molecular grade	20 μL
Final volume	50 μL

75. Create two PCR programs.

Program 1

Cycle number	Denaturation	Annealing	Polymerization
1	30 sec at 98°C		
14 cycles	10 sec at 98°C	30 sec at 65°C	30 sec at 72°C

Hold at 4°C.

Program 2

Cycle number	Denaturation	Annealing	Polymerization
2 cycles	10 sec at 98°C	30 sec at 65°C	30 sec at 72°C

Hold at 4°C.

76. Label six 0.2-mL tubes: PCR-14, PCR-16, PCR-18, PCR-20, PCR-22, and PCR-24.

77. Run the PCRs with PCR Program 1, and collect 5 μL of the reaction in the 0.2-mL tube labeled **PCR-14**.

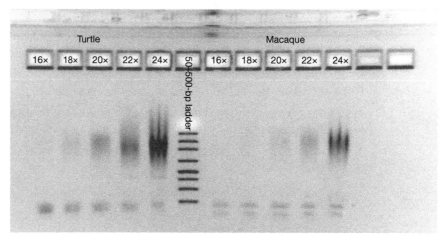

FIGURE 1. **Gel cycle amplification.**

78. Continue the PCR optimization by running PCR Program 2, and collect 5 μL of the reaction in the 0.2-mL tube labeled **PCR-16.**

79. Repeat Step 78 for the remaining four cycle-collection points (i.e., tubes 18, 20, 22, and 24).

80. Add 1 μL of loading dye to each 5-μL aliquot. Vortex quickly and centrifuge the tubes.

81. Prepare a 2.2% FlashGel with 4 μL of water in each of the 13 wells and 2 μL of Flash 50–1500-bp gel marker in the first and eighth lanes.

82. Load the 6-μL PCR aliquots in Lanes 2–7. Run the FlashGel for 5–7 min at 275 V.

83. Photograph the gel. Identify cycles of overamplification. For full-scale PCR, use the greatest number of cycles that do not show overamplification. For example, if PCR-22 and PCR-24 show overamplification but none of the other tubes do, then use the conditions for PCR-20. A sample set of PCR results is shown in Figure 1 (overexpression is shown, for example, in Turtle lane 24×).

Full-Scale PCR

84. Prepare amplification reactions by multiplying the quantities per reaction below by the number of reactions. We recommend 15 reactions.
    PCR components (for nonindex library)

DNA/beads	2 μL
Phusion mix (2×)	25 μL
PE PCR Primer 1.0 (8 μM)	0.5 μL
PE PCR Primer 2.0 (8 μM)	0.5 μL
Water, molecular grade	22 μL
Final volume	50 μL

    PCR components (for index library)

DNA/beads	2 μL
Phusion mix (2×)	25 μL
Multiplexing Primer 1.0 (8 μM)	1 μL
Multiplexing Primer 2.0 (0.5 μM)	1 μL
Index PCR primer (8 μM)	1 μL
Water, molecular grade	20 μL
Final volume	50 μL

    See Table 1 for the primer oligonucleotide sequences.

85. Program the thermal cycler based on the results of optimization in the previous section.

Cycle number	Denaturation	Annealing	Polymerization
1	30 sec at 98°C		
N cycles[a]	10 sec at 98°C	30 sec at 65°C	30 sec at 72°C
Last cycle			5 min at 72°C

Hold at 4°C.

[a]The minimal number of PCR cycles that results in product amplification should be used (N).

86. Amplify the DNA.

87. After cycling is complete, pool all of the samples into a 15-mL conical tube and adjust the volume of the sample to 750 µL with Buffer EB.

88. Add the PCR to 3 volumes of Buffer ERC and vortex.

89. Purify the DNA using a QIAquick PCR purification column by following Steps 52.i–52.viii.

90. Elute the DNA from the column with 50 µL of Buffer EB. Incubate for 1 min at room temperature.

91. Repeat Step 90. Discard the column, and retain the 1.7-mL tube containing the 100 µL of eluted DNA.

Size Selection Using Dual AMPure Bead Cleanup

92. Vortex AMPure XP beads.

> *It is critical to the success of the procedure to use calibrated AMPure beads. The calibration procedure is described in the Additional Protocol AMPure Bead Calibration.*

93. Add x µL of AMPure beads to the tube according to the equation below. Insert the paired-end cutoff value (PE cutoff value) of AMPure beads to the equation. (Determine the PE cutoff value as described in the Additional Protocol AMPure Bead Calibration.)

$$x \text{ µL} = (\text{PE cutoff value}) \times 100,$$
$$x \text{ µL} = 0.65 \times 100 = 65 \text{µL of AMPure beads added to the sample.}$$

94. Mix by vortexing. Incubate without agitation for 5 min at room temperature.

95. Using an MPC, pellet the beads against the wall of the tube.

96. Transfer the supernatant to a new microcentrifuge tube. Ensure that the supernatant volume is equal to 100 µL + x µL (from Step 93). Adjust with Tris-HCl if necessary (e.g., 100 µL + 65 µL = 165 µL). Add 75 µL of 10 mM Tris-HCL (pH 7.5–8.5) to the transferred supernatant. Add y µL of AMPure beads to the tube:

$$y \text{ µL} = x(\text{ from Step 93}) \text{ µL} + 20 \text{ µL}.$$

For example,

$$y = 65 \text{ µL} + 20 \text{ µL} = 85 \text{ µL}.$$

97. Mix by vortexing. Incubate without agitation for 5 min at room temperature.

98. Using an MPC, pellet the beads against the wall of the tube. Leave the tube in the MPC for all washes.

99. Remove the supernatant, and wash the beads twice with 500 µL of 70% ethanol, incubating for 30 sec each time.

100. Quickly centrifuge and remove all of the supernatant.

101. Allow the AMPure beads to air-dry for 2 min at 37°C. The beads are dry when visible cracks start to form in the pellet. Do not overdry.

102. Add 30 μL of 10 mM Tris-HCl (pH 7.5–8.5) and vortex to resuspend the beads. This elutes the paired-end library from the AMPure beads.

103. Using the MPC, pellet the beads against the wall of the tube, and transfer the supernatant to a new tube. Discard the bead pellet.

Final Library Quantification

104. Run 1 μL of the final sample from Step 104 on an Agilent 2100 Bioanalyzer DNA 1000 chip to verify the size range.

105. Determine the concentration of the purified DNA by assaying 1 μL of the final sample with the Quant-It dsDNA HS Assay Kit and reading the results on a fluorometer (Protocol 11 or 12 may be useful).

106. Enter these values into the Illumina Library Dilution Calculator to prepare 5 nM dilutions of the final paired-end library sample.

AMPure Bead Calibration

In this protocol for preparation of libraries, AMPure beads are used to remove both undersized and oversized DNA fragments from the library. Owing to the high variability of AMPure lots in the size-exclusion characteristics, each lot must be calibrated to determine the size-exclusion characteristics of any given AMPure bead lot. This calibration will determine the ratio of AMPure beads to DNA sample (by volume) to use during the library preparation, in order to create the optimal size library for sequencing using the GS FLX Titanium chemistry.

In this calibration assay, a 100–1500-bp DNA ladder is incubated with AMPure beads in a series of beads-to-DNA ratios ranging from 0.5:1 to 1:1 (v:v). Each bead–DNA ratio will provide different fragment size cutoff parameters. The cutoffs are assessed by running the DNA retained by the beads on an Agilent Bioanalyzer DNA 7500 LabChip, and recording the areas under the peaks (DNA concentration) in the 200–500-bp range. The optimal ratio to use for DNA libraries with the bead lot being tested is determined by comparing the results of this calibration assay with the empirically derived optimal data provided below.

METHOD

1. Label 11 1.7-mL tubes for the 11 bead-to-DNA ratios to be included in the assay, from 0.50:1 to 1.00:1, in increments of 0.05, as shown in Table 2.

2. In a fresh microcentrifuge tube, place 48 μL of the 100–1500-bp DNA ladder, and dilute it with 1152 μL of molecular biology–grade water.

3. Aliquot precisely 100 μL of diluted DNA ladder into each labeled microcentrifuge tube.

4. Vortex the tube of AMPure beads vigorously, and aliquot 900 μL into a fresh microcentrifuge tube. This aliquot will be used for the rest of the procedure.

5. Using the same pipette as used to measure the 100-μL aliquots of diluted DNA ladder, add the appropriate amount of beads to each sample. Make sure to do the following.

   • Vortex the bead aliquot between each sample.

   • Change tips between each sample.

TABLE 2. Volumes of diluted DNA ladder and of AMPure beads for each beads:DNA ratio

Beads:DNA ratio (by volume)	Diluted DNA ladder (μL)	AMPure beads (μL)
0.50:1	100	50
0.55:1	100	55
0.60:1	100	60
0.65:1	100	65
0.70:1	100	70
0.75:1	100	75
0.80:1	100	80
0.85:1	100	85
0.90:1	100	90
0.95:1	100	95
1:1	100	100

- Pipette the beads slowly, making sure that no air is aspirated and that there are no beads on the outside of the tip.

- Dispense slowly so that *all* of the beads are delivered to the sample.

6. Vortex all of the tubes, and incubate them for 5 min at room temperature.

7. Using the MPC, pellet the beads against the wall of the tube. This may take several minutes because of the high viscosity of the suspension.

8. Remove the supernatant, and wash the beads twice with 500 μL of 70% ethanol, incubating for 30 sec each time.

   Larger DNA fragments will bind to the AMPure beads, with a decreasing size cutoff as the beads:DNA ratio increases. The DNA species from the DNA ladder that are below the cutoff in each of the incubation conditions will be washed away in the next step.

9. Remove all of the supernatant from each tube, and allow the AMPure beads to air-dry completely. To reduce drying time, place the tubes in a heating block at 37°C. Visible cracks in the pellet are an indication that the beads are dry.

10. Remove the tubes from the MPC. Add 10 μL of Tris-HCl to each tube, and vortex to resuspend the beads.

11. Using the MPC, pellet the beads against the wall of the tube once more, and transfer the supernatants containing the size-selected DNA ladder to a set of fresh, appropriately labeled microcentrifuge tubes.

12. Dilute 4 μL of fresh DNA ladder with 6 μL of molecular biology–grade water. This aliquot of unprocessed, diluted DNA ladder will serve as a control.

13. Run 1 μL of each size-selected DNA ladder and the control ladder on a single Bioanalyzer DNA 7500 LabChip.

Analysis of Bead Calibration

Results will show the gradual removal of small fragments from the DNA ladder samples, as the beads:DNA ratio decreases (i.e., only large fragments bind to the beads at a low bead ratio). To assess this, the DNA concentration of the 200–500-bp peaks is monitored in the 12 LabChip traces (including the nonselected control ladder). The peak at 900 bp should be fully retained in the whole range of beads:DNA ratios tested, and is used to normalize the traces.

- For each of the 12 traces, divide the DNA concentration (in nanograms per microliter) of each of the following four peaks, by the DNA concentration of the 900-bp peak for that trace: 200 bp, 300 bp, 400 bp, and 500 bp.

- Compare the sets of four values for each of the 11 size-selected DNA ladders, with the values for Columns 2 and 3 of Table 3.

TABLE 3. Optimum ratios of the DNA concentration in the low-molecular-weight peaks to the 900-bp peak

Peak ratio (DNA concentrations)	Optimal values for PE cutoff	Values for the control DNA ladder
200/900	N/A	0.7
300/900	0.25	1.1
400/900	0.4	1.4
500/900	1.5	3.4

- For the PE cutoff value used in Step 93, use the ratio of AMPure beads:DNA (by volume) that generated the set of peak ratio values most similar to the values given in Column 2.

- Make sure to also verify that the peak ratios in the control trace match the values given in Column 3.

RECIPE

It is essential that you consult the appropriate Material Safety Data Sheets and your institution's Environmental Health and Safety Office for proper handling of equipment and hazardous materials used in this protocol.

Library Binding Buffer (2×) (2M Sodium Chloride /2× Tris-EDTA Buffer)
For 100 mL:

Sigma water	58 mL
Sodium chloride (5 M)	40 mL
TE buffer (100×, 1:0.1 M)	2 mL

1. Combine reagents in the order above in a 250-mL beaker.

2. Stir for ~30 min to mix. Take conductivity, pH, and density readings.

3. Aliquot 10 mL per 15-mL conical tube. Label (with green label for distinguishing purposes) and add to the IMP database.

   This solution has a 3-mo shelf life. Indicate the expiration date on the label. Store at −20°C. This recipe makes 10 aliquots.

RNA-Seq: RNA Conversion to cDNA and Amplification

The Ovation RNA-Seq Kit provides a fast and simple method for preparing amplified cDNA from total RNA. Amplification is initiated both at the 3′ ends of the transcripts and across the entire range of transcribed sequences; thus, this approach is ideal for next-generation sequencing, because the reads are distributed across the transcript types. The amplified cDNA produced in this protocol is used to create libraries optimized for use in the Illumina Genome Analyzer II platform.

An additional protocol describes the steps for removing small degraded RNAs before the sample is processed into cDNA.

MATERIALS

It is essential that you consult the appropriate Material Safety Data Sheets and your institution's Environmental Health and Safety Office for proper handling of equipment and hazardous materials used in this protocol.

Recipes for reagents specific to this protocol, marked <R>, are provided at the end of the protocol. See Appendix 1 for recipes for commonly used stock solutions, buffers, and reagents, marked <A>. Dilute stock solutions to the appropriate concentrations.

Reagents

DNA 7500 reagents (Agilent, catalog no. 5067-1506)
Ethanol (100% and 70%)
MinElute PCR Purification Kit (QIAGEN, catalog no. 28004)
Ovation RNA-Seq Kit (NuGEN, catalog no. 7100-08)
Quant-iT dsDNA BR Assay Kit (Life Technologies, catalog no. Q32853)
Quant-iT RNA Assay Kit (Life Technologies, catalog no. P11496)
RNA Nano Assay (Agilent)
RNA 6000 Pico Kit (Agilent, catalog no. 5067-1513)
RNA samples to be amplified
TE buffer (1×) <A>

Equipment

2100 Bioanalyzer (Agilent, catalog no. G2939AA)
Fluorometer (e.g., Qubit; Life Technologies)
Qubit assay tubes or similar brand (Life Technologies, catalog no. Q32856)
SPRIPlate 96R ring magnet plate (Beckman Coulter, catalog no. 000219)
   This was formerly the 96-well plate.
Thermal cycler with 0.2-mL tube heat block, heated lid, and 100-μL reaction capacity
Tubes (1.7 mL and 0.7 mL)
Tubes (0.2 mL), thin wall for PCR

METHOD

Evaluate the RNA

1. Prepare to quantify the amount of RNA in the sample, diluting if necessary (modify Protocols 12, 13, or 14 for RNA). If required, process the RNA samples to remove small degraded RNA by following the procedure described in the Additional Protocol RNAClean XP Bead Cleanup (before RNA-Seq).

2. Dilute an aliquot of the RNA sample to within range of the assay: 25–500 ng/μL for the RNA Nano Assay (Agilent) or 50–5000 pg/μL for the RNA 6000 Pico Assay (Agilent). Evaluate the quality of the RNA sample using either the RNA Nano or RNA 6000 Pico Assay (for an example, see Fig. 1).

First-Strand cDNA Synthesis

3. Flick the tube containing the first-strand enzyme mix (blue cap: A3 from the NuGEN Ovation RNA-Seq Kit), pulse in a centrifuge, and place the tube on ice.

4. Thaw the tubes containing the First-Strand Primer Mix (blue cap: A1), First-Strand Buffer Mix (blue cap: A2), and H_2O (green cap: D1) at room temperature. Briefly vortex the tubes to mix, pulse in a centrifuge, and place the tubes on ice.

5. For each cDNA library to be made, combine the following in a 0.2-mL PCR tube:

First-strand primer mix	2 μL
RNA (100 ng)	x μL
H_2O	$(5 - x)$ μL

    The total reaction volume is 7 μL, containing no more than 100 ng of RNA.
    Pulse the reaction tubes in a centrifuge and place them on ice.

    > *RNA inputs above 100 ng per reaction may inhibit amplification. If necessary, the reaction volume can be increased to 8 μL.*

6. Place the tubes in a thermal cycler prewarmed to 65°C. Incubate for 5 min. Cool to 4°C. Remove the tubes from the cycler and place them on ice.

7. For each sample, combine 2.5 μL of First-Strand Buffer Mix and 0.5 μL of First-Strand Enzyme Mix in a 0.7-mL tube. Mix by pipetting, pulse in a centrifuge, and place the tube on ice.

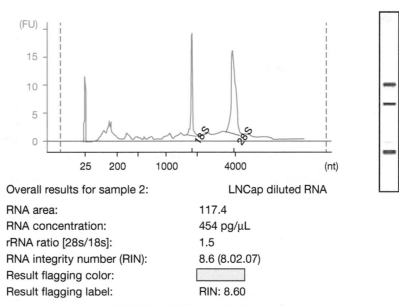

Overall results for sample 2:	LNCap diluted RNA
RNA area:	117.4
RNA concentration:	454 pg/μL
rRNA ratio [28s/18s]:	1.5
RNA integrity number (RIN):	8.6 (8.02.07)
Result flagging color:	
Result flagging label:	RIN: 8.60

FIGURE 1. **RNA quantitation results.**

8. Add 3 μL of diluted enzyme mix to each tube from Step 5 (total reaction volume = 10 μL). Mix by pipetting up and down six to eight times, and pulse in a centrifuge.

9. Place the tube in a thermal cycler precooled to 4°C. Prepare first-strand cDNA using the following conditions:

    i. 1 min at 4°C
    ii. 10 min at 25°C
    iii. 10 min at 42°C
    iv. 15 min at 70°C
    v. Cool to 4°C.

10. Remove the tubes from the cycler and place them on ice. Continue immediately with second-strand cDNA synthesis.

Second-Strand cDNA Synthesis

11. Place the AMPure RNA Clean beads on the bench top, and warm them to room temperature.

12. Flick the tube containing the second-strand enzyme mix (yellow cap: B2), pulse it in a centrifuge, and place the tube on ice.

13. Thaw the tube containing the Second-Strand Buffer Mix (yellow cap: B1) at room temperature, briefly vortex, pulse in a centrifuge, and place the tube on ice.

14. For each sample, combine 9.7 μL of Second-Strand Buffer Mix and 0.3 μL of Second-Strand Enzyme Mix in a 0.7-mL tube. Mix by pipetting up and down, pulse the tube in a centrifuge, and store the mix on ice.

15. Add 10 μL of diluted enzyme mix to each first-strand reaction tube (from Step 10) (total reaction volume = 20 μL). Mix by pipetting up and down six to eight times. Keep the tubes on ice.

16. Place the tube in a thermal cycler precooled to 4°C. Prepare second-strand cDNA using the following conditions:

    i. 1 min at 4°C
    ii. 10 min at 25°C
    iii. 30 min at 50°C
    iv. 20 min at 80°C
    v. Cool to 4°C.

17. Remove the tubes from the cycler and place them on ice. Continue immediately with the purification of unamplified cDNA.

Purification of cDNA

18. Ensure that the RNA Clean beads (from Step 11) have completely reached room temperature before proceeding. Resuspend the beads by inverting the tube.

    ▲ *Do not vortex or centrifuge down the tube of beads.*

19. At room temperature, add 32 μL of beads to each tube. Mix by pipetting up and down 10 times. Incubate the tubes for 10 min at room temperature.

20. Place the tubes on the SPRI 96R magnet plate, and pellet the beads.

21. Remove and discard 42 μL of the Binding buffer from each tube.

22. Wash the beads three times each with 200 μL of freshly prepared 70% ethanol. Remove as much ethanol as possible from the tube.

23. Allow the beads to air-dry on the magnet for at least 15–20 min.

24. Proceed immediately to the SPIA amplification with the cDNA still bound to the dry beads.

SPIA

25. Thaw the SPIA Primer Mix (red cap: C1) and the SPIA Buffer Mix (red cap: C2) at room temperature, briefly vortex to mix, pulse in a centrifuge, and store on ice.

26. Thaw the SPIA enzyme mix (red cap: C3) on ice. Mix by gently inverting the tube five times, pulse in a centrifuge, and store on ice.

27. Prepare an SPIA master mix in a 0.7-mL tube using the table below as a guide.

    ▲ *Make sure to add the enzyme at the last minute. Mix by pipetting, pulse in a centrifuge, place the tube on ice, and use it immediately.*

Reagent	Per reaction
SPIA Primer Mix	10 μL
SPIA Buffer Mix	20 μL
SPIA Enzyme Mix	10 μL

28. Add 40 μL of SPIA master mix to each tube containing the dried beads (from Step 23). Use a pipette set to 30 μL to mix up and down eight to 10 times.

29. Place the tube in a thermal cycler precooled to 4°C. Incubate using the following program:

    i. 1 min at 4°C

    ii. 60 min at 47°C

    iii. 5 min at 95°C

    iv. Cool to 4°C.

30. Remove the tubes from the cycler and place them on ice.

31. Place a tube on the SPRI 96R magnet plate, and allow the beads to pellet for 5 min.

Post–SPIA Modification I

32. Transfer 35 μL of the cleared supernatant containing the amplified cDNA to a new 0.2-mL tube. Discard the beads.

33. Thaw post-SPIA primer mix (violet cap: E1) at room temperature, briefly vortex to mix, pulse in a centrifuge, and store on ice.

34. Add 5 μL of primer mix to the 35-μL sample. Pipette up and down six to eight times to mix, pulse in a centrifuge, and store on ice.

35. Place the tubes in a thermal cycler prewarmed to 98°C, and incubate for 3 min. Cool to 4°C.

36. Remove the tubes from the thermal cycler and place them on ice. Proceed immediately with post–SPIA modification II.

Post–SPIA Modification II

37. Thaw post–SPIA Buffer Mix (violet cap: E2) at room temperature, briefly vortex to mix, pulse in a centrifuge, and store on ice.

38. Pulse the post-SPIA enzyme mix (violet cap: E3) in a centrifuge, and store it on ice.

39. Prepare post-SPIA modification mix in a 0.7-mL tube according to the table below. Mix by pipetting and pulse in a centrifuge. Store on ice.

Reagent	Per reaction
Buffer mix	5 μL
Enzyme mix	5 μL

40. Add 10 μL of the post–SPIA modification mix to each 40-μL reaction (from Step 36).

41. Using a pipette set to 40 μL, mix each reaction up and down 6-eight times, pulse in a centrifuge, and place on ice.

42. When all of the tubes are ready, place them in a thermal cycler precooled to 4°C. Incubate using the following conditions:

    i. 1 min at 4°C

    ii. 10 min at 30°C

    iii. 15 min at 42°C

    iv. 10 min at 75°C

    v. Cool to 4°C.

43. Remove the tubes from the cycler, pulse in a centrifuge, and place them on ice. Store the amplified cDNA at −20°C.

Purification of Amplified cDNA

44. Clean up each library by passing it through a MinElute column, as follows.

    i. Add 5× sample volume of Buffer PB (250 μL).

    ii. Wash twice with freshly prepared 80% ethanol.

    iii. Centrifuge briefly to dry the material.

    iv. Elute the samples with 30 μL of 1× TE buffer (2×15 μL).

    v. Incubate them for 2 min.

    vi. Centrifuge the samples to recover the cDNA.

Quantitate the Yield of Double-Stranded cDNA

45. Make a 1:10 dilution of each double-stranded cDNA sample. Use the diluted sample to perform a Qubit assay (dsHS) as described in Protocol 14.

46. Dilute an aliquot of each sample to ∼50 ng/μL. Assay the diluted samples on an Agilent DNA 7500 instrument. The cDNA will range from <100 bp to 1.5 kb with a strong bias toward products <1 kb (see, e.g., Fig. 2).

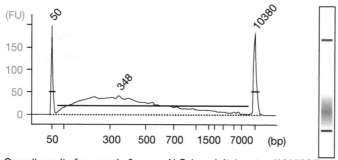

Overall results for sample 3:   ALS 4 occipital cortex (1015268)
Number of peaks found: 1
Peak table for sample 3:   ALS 4 occipital cortex (1015268)

Peak	Size (bp)	Conc. (ng/μL)	Molarity (nmol/L)	Observations
1	◄ 50	8.30	251.5	Lower marker
2	348	48.11	209.5	
3	► 10.380	4.20	0.6	Upper marker

FIGURE 2. **cDNA quantitation results.**

47. Proceed to Illumina library construction (Protocols 4–8). For PE library construction, see the box Illumina PE Library Construction Modification (NuGEN cDNA Only).

> *If the starting RNA is degraded, the cDNA fragments will have a smaller size distribution with few cDNAs larger than 700 bp. In this case, you may skip shearing the cDNA before end repair.*
>
> *Illumina libraries are typically constructed with 500 ng of cDNA before Covaris fragmentation, but up to 1 μg is acceptable.*

ILLUMINA PE LIBRARY CONSTRUCTION MODIFICATION (NuGEN cDNA ONLY)

1. Aliquot the following into a Covaris microTube:

Sample	Amount of cDNA (1 μg)	5× Lucigen DNA Terminator End Repair Buffer	Amount of Nuclease-free water	Total volume
Sample cDNA	x μL	10 μL	y μL	50 μl

2. Briefly vortex to mix and quickly centrifuge.

3. Fragment using the Covaris program (*prog2_5DC_4I_200CB_90s*).

4. Transfer the sample to a 1.7-mL microcentrifuge tube.

End Repair

5. Add 2 μL of Lucigen DNA Terminator End Repair Enzyme to each 1.7-mL tube.

6. Incubate the tubes for 30 min at room temperature, then heat-inactivate for 15 min at 70°C.

AMPure XP Size Selection

7. Perform a 1.3× volume AMPure XP bead cleanup on the fragmented sample (see the Additional Protocol RNAClean XP Bead Cleanup [before RNA-Seq]).

    i. Add 1.3× the volume (67.6 μL) of the AMPure XP beads to the sample.

    ii. Briefly vortex to mix.

    iii. Incubate for 10 min at room temperature.

    iv. Place in the MPC and allow the beads to pellet.

    v. Discard the supernatant.

    vi. Wash twice with 500 μL of 70% ethanol.

    vii. Air-dry.

    viii. Elute in 33 μL of Buffer EB, vortex to resuspend, and pellet in the MPC.

    ix. Transfer the eluted DNA to a 0.2-mL PCR tube.

8. Continue with Illumina PE library construction at the A-tail step.

For Protocol 4, this is Step 9 (these seven steps replace Steps 1–8 in Protocol 4).

RNAClean XP Bead Cleanup (before RNA-Seq)

This process is sometimes used in an attempt to remove the majority of the smaller degraded RNA fragments from an RNA sample (see examples in Fig. 3). It is not always successful and does not guarantee that the NuGEN Ovation RNA-Seq Kit will produce long fragments of cDNA.

METHOD

1. For each sample, aliquot 1 µg of degraded RNA sample to a fresh 1.7-mL microcentrifuge tube.

2. Perform a 1.6× RNA Clean XP bead cleanup (not AMPure XP) on each sample.

   i. Pipette to mix or gently flick (do not vortex RNA).

   ii. Incubate for 10 min at room temperature.

   iii. Place the tube in the MPC, and allow the beads to pellet (5 min).

   iv. Remove the supernatant.

   v. Wash twice with 500 µL of freshly prepared 70% ETOH, incubating each wash for 30 sec, and discard the supernatant.

   vi. Allow beads to air-dry for 15–20 min at room temperature.

   vii. Elute each sample in 13 µL of nuclease-free water.

3. Remove a 1-µL aliquot from each sample to run a Qubit RNA Assay.

4. Remove an additional 1-µL aliquot from each sample. Dilute this aliquot to ~5 ng/µL.

5. Run 1 µL of the diluted sample in duplicate on an Agilent RNA 6000 Pico Assay.

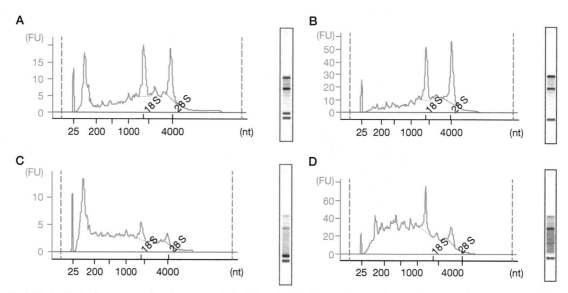

FIGURE 3. RNA clean examples. (*A*) Example 1: Before RNA clean. (*B*) Example 1: After RNA clean. (*C*) Example 2: Before RNA clean. (*D*) Example 2: After RNA clean.

Solution-Phase Exome Capture

This protocol describes the construction of a paired-end library of genomic DNA (gDNA) and subsequent capture of specific regions of a genome using NimbleGen sequence capture probes and Illumina TruSeq oligos. The captured DNA, purified and quantitated, is appropriate for use as template in Illumina sequencing systems. This protocol is the most commonly used approach for exome capture. Solid-phase methods were used at one time but have been superseded by solution-phase capture, in part because of the development of specialized reagents developed by various companies. Further issues and advantages of solution-phase exome capture are discussed in the chapter introduction under Targeted Sequencing: An Overview (see also Fig. 1 in the chapter introduction).

Additional protocols are provided for AMPure Bead Cleanup and for Agarose Gel Size Selection.

MATERIALS

It is essential that you consult the appropriate Material Safety Data Sheets and your institution's Environmental Health and Safety Office for proper handling of equipment and hazardous materials used in this protocol.

Recipes for reagents specific to this protocol, marked <R>, are provided at the end of the protocol. See Appendix 1 for recipes for commonly used stock solutions, buffers, and reagents, marked <A>. Dilute stock solutions to the appropriate concentrations.

Reagents

Agarose (Bio-Rad Low-Melt)
Agilent Bioanalyzer DNA 1000 and DNA 12000 reagents
AMPure XP beads, magnetic beads for concentrating and purifying DNA samples (Beckman Coulter)
Bar codes (Bioo Scientific NEXTflex)
COT human DNA, fluorometric grade (1 mg/mL, 1 mL)
Elution buffer (EB)

EB, a generic buffer first introduced by QIAGEN, is composed of 10 mM Tris (pH 8.0). Water can also be used as a DNA buffer but can lead to degradation if the DNA concentration is high or if the DNA solution is frozen and thawed many times. The type of buffer to be used as EB is determined by what is appropriate for the procedure.

End Repair Buffer Mix (NEXTflex)
End Repair Enzyme Mix (NEXTflex)
Ethanol (100%)
Exome library probes (NimbleGen SeqCap EZ Choice or SeqCap EZ Human Exome Library V2.0 probes)
Fluorometer reagents (Qubit or Quant-It Kit)
Genomic DNA sample (gDNA)
Hybridization and Wash Kit (NimbleGen)
Nuclease-free water
Phusion High-Fidelity Master Mix with HF Buffer (2×) (New England Biolabs)
QIAquick DNA Purification Kit (QIAGEN)
Resuspension Buffer (Bioo Scientific)
Sequencing Kit (Bioo Scientific NEXTflex)

Streptavidin-coupled Dynabeads (Life Technologies)

TAE buffer (1×) <A>

TE (pH 8.0) <A>

TruSeq oligonucleotides (HPLC purified; Illumina)

The following table describes the TruSeq oligonucleotides used in this protocol:

Sequence name	Scale (μmol)	Bases	Sequence
TS-PCR Oligo 1	0.25	22	AATGATACGGCGACCACCGAG*A
TS-PCR Oligo 2	0.25	22	CAAGCAGAAGACGGCATACGA*G
TS-HE Oligo 1	0.25	58	AATGATACGGCGACCACCGAGATCTACACTCTTTC CCTA CACGACGCTCTTCCGATC*T
TS-HE GENERIC Index	0.25	57	GATCGGAAGAGCACACGTCTGAACTCCAGTCAC/ ideoxyl//ideoxyl//ideoxyl/ /ideoxyl//ideoxyl// ideoxyl/ATCTCGTATGCCGTCTTCTGCTT*G

(TS) TruSeq; (HE) hybridization enhancing; (*) 3′-phosphorothioate bond modification. (TS-HE Index Oligos) We use the TS-HE GENERIC Index Oligo as an alternative to the use of different specific TS-HE Index oligos. This generic HE oligo may be substituted for any of the TS-HE Index oligos in the hybridization step to decrease material costs and workflow complexity. However, performance results with Index oligos may be reduced compared with those obtained in experiments using specific TS-HE Index oligos.

Equipment

Bioanalyzer 2100 (Agilent)

The Bioanalyzer provides sizing, quantification, and quality-control analysis for DNA, RNA, and proteins using a single microfluidics-based platform.

Centrifuges for plate and tubes

Covaris tubes for sonication

Dry heat baths (preset to 95°C, 47°C)

Fluorometer (e.g., Qubit, Wallac 1420 multilabel counter)

Gel tanks and trays

Gene catchers or scalpels

Magnetic racks (Ambion or DynaMag-2)

Microcentrifuge tubes

MIDI plates (0.8 mL, 96 well), deep well

PCR plates (96 well)

PCR plate sealers (Bio-Rad)

Pipettes and tips

Sonicator (Covaris)

Spectrophotometer (e.g., NanoDrop)

Thermo cycler

Vortexer, for plate and single tube

Water bath (preset to 47°C)

METHOD

Shearing Genomic DNA for Library Preparation

1. Test the quality of each gDNA sample on a DNA 12000 Bioanalyzer chip.

2. Determine the concentration of the sample in a fluorometer.

3. Add Elution buffer to 1–3 μg of the gDNA sample to a total volume of 80 μL, as follows:

Input DNA (gDNA)	1–3 μg
EB buffer	(80 μL DNA volume)

   Use EB buffer or water in this reagent mix; do *not* use TE because the EDTA in the TE buffer will inhibit the enzymatic reactions downstream.

4. Transfer the dilution to a Covaris tube; take care to avoid creating bubbles.

5. Place the Covaris tube in the tube holder and load into the Covaris sonicator. Run the desired Covaris protocol; to obtain sheared fragments of ~250 bp, use the following settings:

   Duty Cycle: 10%
   Intensity: 5
   Cycle Burst: 200
   Time: 90 sec (Treatments: 30 sec, 30 sec, and 30 sec)

6. (Optional) To check the fragment distribution of the sample, analyze 1 μL of the sheared sample on a DNA 1000 Bioanalyzer chip.

Concentration of Sheared DNA

The DNA is concentrated to 40 μL following Bioo Scientific protocol's requirements.

7. Thaw an aliquot of AMPure XP beads for at least 30 min at room temperature.

8. Vortex the bead sample for 1 min to mix thoroughly.

9. Transfer 144 μL (1.8 × 80 μL) of beads into 1.5-mL tubes or into the wells of a MIDI plate.

10. Transfer the sheared DNA sample into the 1.5-mL tube or MIDI plate well with the AMPure XP beads.

11. Pipette up and down to mix, until the contents are homogenous. Seal the MIDI plate with a plate sealer.

12. Quickly centrifuge the tube or plate (~1 sec) to collect the beads (do not pellet the beads).

13. Allow the DNA to bind to the beads by shaking in a vortex mixer at 2000 rpm for 5 min 30 sec at room temperature.

14. Quickly centrifuge the tube or plate (~1 sec) to collect the beads (do not pellet the beads), and place the tube or rack on a magnetic rack to stand for 2 min.

15. Use a pipette to slowly remove the cleared supernatant, and discard it. Avoid disturbing the bead pellet.

    It is important not to pipette out the beads. It is preferable rather to leave a little amount of supernatant in the tube that will be removed in the ethanol wash steps.

16. Wash the beads with freshly prepared 80% ethanol.

    i. Prepare a fresh solution of 80% ethanol.

       80% ethanol is hygroscopic and should be freshly prepared to achieve optimal results.

    ii. Without removing the tube or plate from the magnetic rack, add 200 μL of 80% ethanol to each sample, and allow it to sit for 30 sec at room temperature.

        Be careful here not to disturb the bead pellet.

    iii. Using a pipette, withdraw and discard the 80% ethanol.

17. Repeat Step 16 once.

18. Remove the residual ethanol by allowing the beads to dry for 5 min at room temperature.

    A 37°C incubator may be used to provide a faster drying time; however, the dry beads tend to flake, thus take extra care in confining the dried beads to their appropriate wells.

19. Add 42 μL of nuclease-free water to the dried beads. Thoroughly resuspend the beads by incubating them for 5 min at room temperature.

20. Centrifuge the sample(s) to pellet beads.

21. Place the tube or plate back on the magnetic rack. Let it stand for 2 min, elute 40 μL of the supernatant, and transfer it to a PCR plate.

    The samples are ready for end repair.

Library Preparation

To make a paired-end (PE) library of your DNA, use the Bioo Scientific NEXTflex DNA Sequencing Kit protocol.

22. Set up the end-repair reaction as follows:

Reagents	Volume
DNA, fragmented, concentrated (Step 21)	40 μL
NEXTflex End Repair Buffer Mix	7 μL
NEXTflex End Repair Enzyme Mix	3 μL
Total	50 μL

Pipette to mix and incubate in a thermocycler for 30 min at 22°C.

23. To purify end-repaired DNA, follow the steps of the Additional Protocol AMPure XP Bead Cleanup. Use $x = 1.8 \times$ (90 μL of AMPure beads) (i.e., $1.8 \times$ of 50 μL) and $y = 17$ μL of Resuspension Buffer (Bioo Scientific).

24. Set up the 3'-adenylation reaction as follows:

Reagents	Volume
Purified end-repaired DNA (Step 23)	17 μL
Adenylation Mix (NEXTflex)	3.5 μL
Total	20.5 μL

Pipette to mix and incubate in a thermocycler for 30 min at 37°C.

Adaptor Ligation

We use Bioo Scientific NEXTflex DNA Barcodes for the ligation step. Currently we use 48 Barcodes (each bar code is 6 bases long). When bar-coded libraries are pooled together, it is important to balance base composition of the bar-code pool as much as possible. For example, position 1 of the bar codes should have all four bases represented in the pool, and so on until position 6 of the bar codes. If this is not possible, each position should at least have an A or C and a G or T.

Table 1 illustrates this principle with an example list for pooling the 48 Barcode set, taking into account what combinations are known.

25. Set up the adaptor reaction as follows:

Reagents	Volume
3'-Adenylated DNA	20.5 μM
NEXTflex Ligation Mix	31.5 μM
NEXTflex DNA Barcode	2.5 μM
Total	54.5 μM

Pipette to mix and incubate in a thermocycler for 15 min at 22°C.

26. Size-select the ligation products, using one of the following methods.

Gel-based selection works well but is slow. Bead size selection is much more amenable to high throughput. Thus, the choice depends on your needs; see the Discussion section for further considerations.

To Perform Size Selection Using Magnetic AMPure XP Beads

i. Perform a first purification by following the steps of the Additional Protocol AMPure XP Bead Cleanup. Use $x = 0.4 \times$ (22 μL of AMPure beads) and $y = 55$ μL of Resuspension Buffer (Bioo Scientific) to elute the DNA.

This step retains DNA fragments of ~400 bp and above.

ii. Perform a second purification by following the steps of the Additional Protocol AMPure XP Bead Cleanup. Use $x = 1.0 \times$ (55 μL of AMPure beads) and $y = 20$ μL of Resuspension Buffer (Bioo Scientific) to elute the DNA.

This step helps to remove any primer dimers that may be present.

iii. Proceed to Step 27.

TABLE 1. NEXTflex Barcode combinations

Barcode number	Sequence	Based on first 3 bases in the Barcode…
1	CGATGT	Pool 1 with any other Barcode.
2	TGACCA	Pool 2 with any other Barcode.
3	ACAGTG	Pool 3 with any other Barcode.
4	GCCAAT	Pool 4 with any other Barcode.
5	CAGATC	Do not pool 5 with 33.
6	CTTGTA	Pool 6 with any other Barcode.
7	ATCACG	Pool 7 with any other Barcode.
8	TTAGGC	Pool 8 with any other Barcode.
9	ACTTGA	Do not pool 9 with 25.
10	GATCAG	Pool 10 with any other Barcode.
11	TAGCTT	Pool 11 with any other Barcode.
12	GGCTAC	Pool 12 with any other Barcode.
13	AGTCAA	Do not pool 13 with 14.
14	AGTTCC	Do not pool 14 with 13.
15	ATGTCA	Do not pool 15 with 26.
16	CCGTCC	Pool 16 with any other Barcode.
17	GTAGAG	Pool 17 with any other Barcode.
18	GTCCGC	Pool 18 with any other Barcode.
19	GTGAAA	Do not pool 19 and 20.
20	GTGGCC	Do not pool 20 and 19.
21	GTTTCG	Pool 21 with any other Barcode.
22	CGTACG	Pool 22 with any other Barcode.
23	GAGTGG	Pool 23 with any other Barcode.
24	GGTAGC	Pool 24 with any other Barcode.
25	ACTGAT	Do not pool 25 with 9.
26	ATGAGC	Do not pool 26 with 15.
27	ATTCCT	Pool 27 with any other Barcode.
28	CAAAAG	Do not pool 28 and 29.
29	CAACTA	Do not pool 29 and 28.
30	CACCGG	Do not pool 30, 31, and 32.
31	CACGAT	Do not pool 31, 30, and 32.
32	CACTCA	Do not pool 32, 30, and 31.
33	CAGGCG	Do not pool with 5.
34	CATGGC	Do not pool 34 and 35.
35	CATTTT	Do not pool 35 and 34.
36	CCAACA	Pool 36 with any other Barcode.
37	CGGAAT	Pool 37 with any other Barcode.
38	CTAGCT	Do not pool 38 and 39.
39	CTATAC	Do not pool 39 with 38.
40	CTCAGA	Pool 40 with any other Barcode.
41	GACGAC	Pool 41 with any other Barcode.
42	TAATCG	Pool 42 with any other Barcode.
43	TACAGC	Pool 43 with anything.
44	TATAAT	Pool 44 with anything.
45	TCATTC	Pool 45 with anything.
46	TCCCGA	Pool 46 with anything.
47	TCGAAG	Do not pool 47 with 48.
48	TCGGCA	Do not pool 48 with 47.

This list was prepared by Stephanie Muller in the McCombie laboratory at Cold Spring Harbor Laboratory.

To Perform Size Selection Using Beads, Followed by Gel Electrophoresis

i. Perform a first purification step by following the steps of the Additional Protocol AMPure XP Bead Cleanup. Use $x = 1.0 \times (55\ \mu L$ of AMPure beads) and $y = 55\ \mu L$ of Resuspension Buffer (Bioo Scientific) to elute the DNA.

ii. Perform a second purification step by following the steps of the Additional Protocol AMPure XP Bead Cleanup. Use $x = 1.0 \times (55\ \mu L$ of AMPure beads) and $y = 20\ \mu L$ of Resuspension Buffer (Bioo Scientific) to elute the DNA.

> *This step helps to remove any primer dimers that may be present.*

iii. In a final purification step, follow the steps of the Additional Protocol Agarose Gel Size Selection.

iv. Proceed to Step 27.

Amplification of the Precapture Sample Library Using Ligation-Mediated PCR

The adaptor-ligated DNA library is amplified using primers complementary to the sequencing adaptors, in preparation for hybridization to SeqCap probe libraries.

27. Combine the following reagents to prepare the ligation-mediated (LM)–PCR master mix:

LM-PCR Master Mix	Amount/reaction	Final concentration
Phusion Master Mix HF Buffer ($2\times$)	50 μL	$1\times$
Water, PCR grade	26 μL	
TS-PCR Oligo 1 (100 μM)	1 μL	1 μM
TS-PCR Oligo 2 (100 μM)	1 μL	1 μM
Total volume	80 μL	

The NG Tech Note recommends using 2 μM TS-PCR oligos with a final concentration of 2 μM. We reduced this amount to lower the primer-dimer peaks.

28. Pipette 80 μL of Master Mix into 0.2-mL strip tubes or into a 96-well PCR plate.

29. To the Master Mix, add 20 μM double-purified sample library from Step 26 (or PCR-grade water for negative control). Pipette to mix.

30. Transfer the strip tubes or PCR plate into a thermo cycler, and perform amplification according to the following program:

Cycle number	Denaturation	Annealing	Polymerization
1	30 sec at 98°C		
8 cycles	10 sec at 98°C	30 sec at 60°C	30 sec at 72°C
Last cycle			5 min at 72°C

Hold at 4°C.

31. Clean up the amplified precapture library sample.

> ▲ *It is critical that the amplified sample library be eluted with PCR-grade water and not Buffer EB or $1\times$ TE.*

To Purify the Library Using Column Purification

i. Load the sample onto a QIAGEN QIAquick DNA column.

ii. Elute the sample in 50 μL of PCR-grade water.

To Purify the Library Using Magnetic Beads

Follow the steps of the Additional Protocol AMPure XP Bead Cleanup. Use $x = 1.8 \times (180\ \mu L$ of AMPure beads) and $y = 50\ \mu L$ of PCR-grade water to elute the DNA.

32. Quantify the libraries using a spectrophotometer (e.g., NanoDrop) or a fluorometer (e.g., Qubit or a multilabel counter such as Wallac 1420 VICTOR[3]).

33. To verify the size and quality of your PCR-enriched fragments, check the template size distribution by running an aliquot of the enriched library on an Agilent Technologies 2100 Bioanalyzer using a DNA 1000 Kit.

    See Troubleshooting.

Preparation of Exome Libraries

34. Hybridize the amplified sample from Step 31 with a human exome library.

    i. Follow the procedure for "Hybridization of Amplified Sample and EZ Probe Libraries," described in the *NimbleGen Technical Note Supplement: Targeted Sequencing with Nimble-Gen SeqCap EZ Libraries and Illumina TruSeq DNA Sample Preparation Kits.*

    ii. As a custom probe source, use either EZ Choice or EZ Human Exome Library v2.0 probes.

        TS-HE Index Oligos: We use TS-HE GENERIC Index oligos as an alternative to the use of different specific TS-HE Index oligos. This generic HE oligo may be used in place of any of the TS-HE Index oligos in the hybridization step to decrease material costs and workflow complexity; however, performance may not be equal to that obtained in experiments using specific TS-HE Index oligos.

        If multiple amplified sample libraries are pooled before hybridization, bring the final concentration of the pooled sample to 1 μg. For example, if you are pooling four samples, then use 250 ng of each sample for pooling to make a total of 1 μg of pooled sample.

35. Recover the template DNA bound to probes using streptavidin Dynabeads, and wash away the unbound DNA. Follow the procedure for "Washing and Recovery of Captured DNA" from Chapter 6 of the *NimbleGen SeqCap EZ Exome Library SR User's Guided Version 2.2.*

Sample Library Postcapture Amplification

Amplify the captured DNA, bound to the streptavidin Dynabeads, using LM-PCR. Two reactions are performed for each sample, and subsequently combined, to reduce PCR bias.

36. To prepare the LM-PCR Master Mix, combine the following reagents:

LM-PCR Master Mix	Amount for two reactions per captured DNA sample	Final concentration
Phusion Master Mix HF Buffer (2×)	100 μL	1×
Water, PCR grade	52 μL	
TS-PCR Oligo 1 (100 μM)	4 μL	2 μM
TS-PCR Oligo 2 (100 μM)	4 μL	2 μM
Total	80 μL	

37. Pipette 80 μL of Master Mix into two 0.2-mL strip tubes or into two wells of a 96-well PCR plate.

38. Add 20 μL of captured DNA sample as template into each of the two tubes containing the 80 μL of Master Mix. Pipette to mix. In addition, prepare a negative control using PCR-grade water instead of DNA.

39. Transfer the strip tubes or PCR plate into a thermo cycler, and perform amplification according to the following program:

Cycle number	Denaturation	Annealing	Polymerization
1	30 sec at 98°C		
18 cycles	10 sec at 98°C	30 sec at 60°C	30 sec at 72°C
Last cycle			5 min at 72°C

Hold at 4°C.

Cleanup and QC of the Postcapture Amplified Sample Library

40. Pool the two reactions for each captured DNA to obtain a pooled postcapture amplified library of 200 μL.

    See note to Step 41.

41. Purify the reaction products.

    For Column Purification

    Use a QIAGEN QIAquick DNA column for sample purification, and elute in 50 μL of EB buffer.

    For AMPure Cleanup

    Follow the steps of the Additional Protocol AMPure XP Bead Cleanup. Use $x = 1.8 \times (360$ μL of AMPure beads) and $y = 50$ μL of EB to elute the DNA.

    The two reactions can be cleaned up separately using AMPure XP beads (i.e., using 180 μL [$1.8 \times$ of 100 μL] for each reaction), and then the cleanup product can be pooled.

Determine the Concentration, Size Distribution, and Quality of the Sample

42. To verify the concentration, size, and quality of your sample library, run an aliquot of the enriched library on an Agilent Technologies 2100 Bioanalyzer using a DNA 1000 Kit.

    See Troubleshooting.

43. Prepare a 10 nM dilution of your library to a total volume of at least 10 μL.

    ▲ *When pooling multiple sample libraries after postcapture, make sure that the final total volume is 10 μL. For example, if pooling four sample libraries, then add 2.5 μL of each library to make a total of 10 μL.*

    The sample is now ready to be sequenced on Illumina sequencers.

TROUBLESHOOTING

Problem (Step 33): Very low or no precapture library is recovered.

Solution 1: Consider increasing the input DNA for library preparation, and repeat the library preparation procedure from the beginning.

Solution 2: Check the reagents, materials, and equipment to ensure that there are no quality issues.

Problem (Steps 33 and 42): Recovered library fragments (precapture and postcapture libraries) do not meet the required size range.

Solution: Check the reagents, materials, and equipment to ensure that there are no quality issues, particularly those items involved in purification steps.

Problem (Step 42): Significant amounts of primer dimers are observed in the postcapture library.

Solution 1: Increase the amount of library used in the hybridization to 1.5–2 μg, and double the input amount of each HE oligo in the hybridization reaction.

Solution 2: Calibrate the AMPure beads using the Additional Protocol AMPure Bead Calibration in Protocol 8, and then modify the beads ratio accordingly.

DISCUSSION

A library produced from 1–3 μg of input genomic DNA should yield 1–4 μg of precapture DNA. The amount of DNA recovered from the capture process will vary depending on the specific regions of the genome targeted by the capture probes. Ultimately, the yield of postcapture DNA must be enough to support adequate sequencing of the targets.

Although fragment size selection of the library is an important aspect of the sample processing workflow, it is also a time-intensive process. There have been many attempts to speed up size selection, including the use of specialized equipment and modified handling procedures. What appears to be the best solution is to exclude the use of gels, even though gel electrophoresis has traditionally been a reliable procedure for producing high-quality libraries. However, newer protocols that involve only liquid handling have the advantage that they can be performed with the aid of robotics (Protocol 11).

AMPure XP Bead Cleanup

The AMPure XP bead cleanup procedure is used throughout Protocol 13 after various enzymatic reaction steps. Follow the general steps of the method provided here. Make sure to use the correct bead ratios (designated here as x and typically determined by the ratio of beads to reagent volume) and elution buffer volume (here indicated by y) that are appropriate for the reaction.

ADDITIONAL MATERIALS

It is essential that you consult the appropriate Material Safety Data Sheets and your institution's Environmental Health and Safety Office for proper handling of equipment and hazardous materials used in this protocol.

Equipment

AMPure XP beads

METHOD

1. Thaw an aliquot of AMPure XP beads for at least 30 min at room temperature.

2. Vortex the bead sample for 1 min to mix thoroughly.

3. Transfer x μL (dependent on the suggested ratio) of vortexed room-temperature AMPure XP beads to a 1.5-mL tube or to the well of a MIDI plate.

4. Transfer the DNA sample from the prior enzymatic reaction step into the tube or well of the MIDI plate (from Step 3).

5. Pipette to mix until the contents are homogeneous, and incubate for 15 min at room temperature to allow beads to bind to the sample. If using a MIDI plate, seal with a plate sealer before incubation.

6. Place the tube or plate on the magnetic rack for 5 min, until the sample appears clear.

7. Gently remove the supernatant, taking care not to disturb beads; some liquid may remain in the wells.

8. Wash the beads with freshly prepared 80% ethanol.

    i. Prepare a fresh solution of 80% ethanol.
       80% ethanol is hygroscopic and should be prepared fresh to achieve optimal results.

    ii. Without removing the tube or plate from the magnetic rack, add 200 μL of 80% ethanol to each sample, and incubate for 30 sec at room temperature.
       Be careful here not to disturb the bead pellet.

    iii. Using a pipette, withdraw and discard the 80% ethanol.

9. Repeat Step 8 once.

10. Remove the tube or plate from the magnetic rack, and let the beads dry for 5 min at room temperature to remove all residual ethanol.

11. Resuspend the dried beads with y μL (dependent on the prior enzymatic reaction step) of recommended elution buffer (EB). Gently pipette to mix thoroughly. Ensure that the beads are no longer attached to the side of the well.

12. Incubate the resuspended beads for 2 min at room temperature.

13. Place tube or plate on a magnetic rack for 5 min until the sample appears clear.

14. Transfer the eluted sample to a fresh tube or PCR plate.

> *If you wish to pause your experiment, the procedure may be safely stopped at this step and samples stored at –20°C. To restart, thaw frozen samples on ice before proceeding.*

Agarose Gel Size Selection

Agarose gel electrophoresis may be used to purify fragmented genomic DNA after ligation of adaptors. After electrophoresis, the region of the gel containing the desired size range of DNA is excised, and the DNA is subsequently extracted from the gel and purified by passage through a spin column.

ADDITIONAL MATERIALS

It is essential that you consult the appropriate Material Safety Data Sheets and your institution's Environmental Health and Safety Office for proper handling of equipment and hazardous materials used in this protocol.

Reagents

Agarose gel (2% agarose; certified low-gelling-temperature agarose) cast in 1× TAE <A>, containing SYBR Gold (15 μL of SYBR Gold to every 150 mL of cooled 1× TAE <A> and agarose gel buffer), and other reagents and equipment as required for agarose gel electrophoresis, described in Chapter 2, Protocol 1

Cleanup spin columns

Column elution buffer

DNA binding buffer

DNA wash buffer (ethanol added; see reagent preparation)

Ethanol (100%), stored at room temperature

Gel-loading dye (6×)

Genomic DNA (gDNA) fragments, adaptor-ligated

MW ladder Ready-to-Load (100 bp)

Reagents provided in the NEXTflex PCR-Free DNA Sequencing Kit (Bioo Scientific)

Equipment

Microcentrifuge tubes (1.5 mL), nuclease free

Razor or scalpel (clean, to excise gel sample)

UV transilluminator or gel documentation instrument

METHOD

1. Add 4 μL of gel loading dye to each gDNA sample, and load the entire sample into one lane of the gel.

   If processing more than one sample, it is recommended that you run separate gels or leave several empty wells between samples to avoid cross-contamination.

2. Load 6 μL of MW ladder into one lane of the gel, skipping at least two lanes between the ladder and the sample.

3. Perform gel electrophoresis at 100–120 V for 60–120 min.

4. Visualize the gel on a UV transilluminator or gel documentation instrument.

5. Use a clean razor or scalpel to cut out a slice of gel from each sample lane corresponding to the 300–500-bp marker.

   DNA in this range includes the gDNA insert size of 200–400 bp (NEXTflex Barcode Adapters, add ~120 bp to each fragment). You may choose other insert sizes when appropriate.

6. Add 400 μL of DNA binding buffer to each gel slice containing the sample gDNA and mix well. Incubate the sample at room temperature, and vortex occasionally until the agarose is completely melted.

7. Add 20 μL of 100% ethanol to each sample and mix well, then transfer the sample to a cleanup spin column.

8. Centrifuge the cleanup spin column in a microcentrifuge at 14,000 rpm for 1 min.

9. Decant the flowthrough, and replace the cleanup spin column in the same collection tube.

10. Add 700 μL of DNA wash buffer to each column, and centrifuge in a microcentrifuge at 14,000 rpm for 1 min.

11. Decant the flowthrough, and replace the cleanup spin column in the same collection tube.

12. Repeat Steps 10 and 11 once.

13. Centrifuge the cleanup spin column in a microcentrifuge at 14,000 rpm for 1 min to remove any residual ethanol.

14. Place the cleanup spin column into a clean 1.5-mL nuclease-free microcentrifuge tube. Add 25 μL of Column elution buffer to the center of the column, and incubate the column for 1 min at room temperature.

15. Centrifuge the cleanup spin column in a microcentrifuge at 14,000 rpm for 1 min to elute the clean DNA.

   If you wish to pause the experiment, the procedure may be safely stopped at this step and samples stored at −20°C. To restart, thaw frozen samples on ice before proceeding.

Automated Size Selection

The Caliper LabChip XT is an automated nucleic acid fractionation instrument that replaces gel-based size-selection methods. By using intersecting microfluidic channels, optical detection, and computer control, the instrument automatically extracts a target band during separation and routes the selected material to a collection well. The resulting sample is accurately sized and is delivered in a sequencing-compatible buffer.

MATERIALS

It is essential that you consult the appropriate Material Safety Data Sheets and your institution's Environmental Health and Safety Office for proper handling of equipment and hazardous materials used in this protocol.

Recipes for reagents specific to this protocol, marked <R>, are provided at the end of the protocol. See Appendix 1 for recipes for commonly used stock solutions, buffers, and reagents, marked <A>. Dilute stock solutions to the appropriate concentrations.

Reagents

Chip Kit (5 chips)
Collection buffer (red vial)
DNA samples
Dye (blue vial)
Ladder (yellow vial)
Reagent kit for LabChip XT (DNA 750 Hi Res Kit)
Sample buffer (6×) (green vial)
Stacking buffer (pink vial)
TE <A>
Water, biology grade

Equipment

LabChip XT (Caliper Life Sciences)
Pipette tips (1 mL) (e.g., MAXYMum Recovery; Axygen Biosciences)
Vacuum tube
This is used to remove excess gel from the sample and collection wells when preparing the DNA chip.

METHOD

Preparation Procedures

1. Equilibrate the reagent kit to room temperature. This will take ~30 min.
2. Prepare the DNA ladder as follows.
   i. Dilute 2 μL of ladder (yellow vial) with 8 μL of water or TE.
      Use the same diluent as in Step 3.

ii. Add 2 μL of 6× Sample buffer (green vial) to the diluted ladder. Vortex the diluted ladder to mix. Centrifuge to collect.

3. Prepare the DNA sample as follows.

   i. Dilute the DNA sample to 10 μL if necessary.

   ii. Mix 10 μL of DNA sample with 2 μL of 6× Sample buffer (green vial). Vortex the sample to mix. Centrifuge to collect.

Create LabChip Run File

▲ *Set up the Run file before completing the chip preparation.*

4. On the LabChip XT Main Window, select **Tools→Run File Editor.** The "Run File Editor" window opens.

5. Select the **XT DNA** method from the *Fractionation Method* dropdown list.

6. Under the Setup tab in the "Run File Editor" window, enter the collection size range and name in each channel.

7. Click the Advanced tab in the "Run File Editor" window. For each channel, select the desired operation for the channel, either *Disabled, Ladder, eXtract and Stop, eXtract and Continue, eXcLuDe Region, Separation, eXtract and Pause,* or *Skip Extraction.* Use the *eXtract and Pause* operation mode for multiple extractions (see Fig. 1).

   • *Disabled.* The channel will not be used in the run. Any channel that has already been used is automatically set to Disabled and cannot be reused.

   • *Ladder.* The channel contains a ladder that is used to calculate the sizes in the samples in the remaining channels during the run.

   • *eXtract and Stop.* The sample in the channel will run until the extraction is complete. After the extraction, no additional sample will move through the channel.

   • *eXtract and Continue.* The sample in the channel will run until the end of the chip range. The size range specified in the Extraction Region will be collected in the collection well.

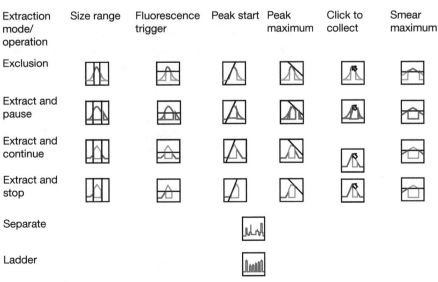

FIGURE 1. **Lab chip display.**

- *eXcLuDe Region.* The sample in the channel will be collected in the waste well on the chip except for the selected region, which moves into the collection well on the chip. The desired output is extracted from the waste well.
- *Separation.* The sample in the channel will be directed into the waste well. No collection will occur.

8. Select the desired extraction mode: (a) Size Range; (b) Fluorescence; (c) Peak Start; (d) Peak Max; (e) Collect On Click; (f) Smear Max (see Fig. 1).

9. Click the "Sizing Table" tab. Use either the default ladder or a custom ladder.

    i. To use the default ladder, select the *Method Ladder* option.

    ii. To use a custom ladder, specify the sizes of the peaks in the ladder, select the *Custom Ladder* option, and type the ladder peak sizes into the size column of the ladder table.

10. Click the Output tab. Confirm the Data Path and input the File Format. Click the Save button to create the run file to the desired directory. Click the Save button to create the run file.

11. For each channel, select the desired extraction mode.

- *Size Range.* The specified size range will be diverted to the chip collection well. The size range can be specified on the Setup tab as a Size $\pm$ a percent of the size, or on the Advanced tab as a specific start and end size in the "Extraction Region" textboxes. The Setup tab and Advanced tab are synchronized to always specify the same settings.
- *Peak Start.* This specifies a slope Threshold, in RFUs (relative fluorescence units) per minute, to use to detect the start of the peak. Collection begins when the slope is greater than the input threshold. Collection ends when the number of base pairs (BP) specified in the "Collection Width" textbox has been collected.
- *Fluorescence.* This specifies an RFU Threshold, which must be exceeded before collection begins. Collection ends when the number of base pairs (BP) specified in the "Collection Width" textbox has been collected.
- *Peak Max.* This specifies a Collection Width, as a percent, that will be collected, centered on the peak maximum. The space on the chip between the detection point and the switch point provides time to divert the sample to the collection well before the peak maximum reaches the switch point. When the software detects the peak maximum in the collection range, the collection start size is calculated by subtracting half the Collection Width from the peak max. If the collection width is very broad and the calculated start size has already passed the switch point, collection will begin immediately but will still collect for a size range equal to the collection width. If no local maximum is found by the end of the specified size range, no extraction will occur. If the signal is decreasing at the beginning of the size range, extraction will begin from the Start size specified. The actual collection range achieved will be reported in the Channel Table.
- *Collect On Click.* This specifies a Collection Width, as a percent, that will be collected. When the user clicks on the **CLICK** text in the LabChip XT Main Window, a confirmation window opens. Collection starts when you click OK in the "Confirmation" window. Collection ends when the "Collection Width (%)" specified in the textbox has been collected.

12. Click on the Output tab and confirm Data Path.

13. Click the Save button and save the created run file to the desired directory.

Prepare the XT DNA Chip

14. Remove a chip from its foil bag, and peel back the top seal from the chip.

15. Gently remove the Sample Well and Collection Well combs.

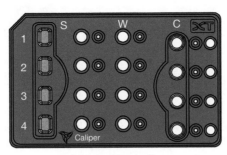

Add stacking buffer

FIGURE 2. **Lab chip collection wells.**

16. Attach a pipette tip to the vacuum line, and aspirate away excess gel from the Sample and Collection wells.

17. Ensure that the top surface of the chip is dry.

18. Remove the chip, and place it on the bench top to complete chip preparation.

19. Add 20 µL of Collection buffer (red vial) to the round collection wells (see Fig. 2).

20. Add 20 µL of Stacking buffer (pink vial) to the rectangular sample wells.

21. Place the chip in the LabChip XT and close the lid. Click on **Instrument** and **Test Chip**. Correct any errors before moving to the next step.

22. Remove the chip from the LabChip XT.

23. Vortex the dye and centrifuge briefly. Pipette 15 µL of XT DNA dye (blue vial) to each of the four waste reservoirs.

24. Tilt the chip sideways and back and forth to ensure homogeneous mixing of the dye and buffer. Before proceeding to the next step, visually check to see that the dye has been evenly distributed throughout the reservoir.

25. Load the sample and prepared ladder into the sample wells.

> ▲ *Position the pipette so that the tip gently touches but is not sealed against the bottom of the well. Pipette the sample very slowly. The goal is to place the sample at the bottom of the well and to avoid mixing the sample with the Stacking buffer.*

Starting the LabChip XT

26. Place the chip in the LabChip XT instrument, lining up the notch on the upper-left corner with the pattern on the instrument.

27. Close the lid and click on **Instrument** and **Start Run**.

28. To import the settings from a previously saved run file, click the Import button at the bottom of the "Start Fractionation" window.

29. Select the desired run file, click the Open button, and click Start.

Collect Fractionated Material

30. After a run is finished, open the instrument lid.

31. Remove the chip and place it on a bench top.

32. Pipette the recovered DNA sample from each collection well and into a clean tube for downstream processing.

Library Quantification Using SYBR Green-qPCR

To quantify complex DNA libraries, Kapa Biosystems has engineered a DNA polymerase specifically for SYBR Green–based qPCR, enabling efficient amplification of targets such as GC-rich DNAs that present a challenge to wild-type DNA polymerases (see the introductory section in Chapter 1 for a discussion of SYBR Green). Kapa Library Quantification Kits contain this engineered polymerase to ensure robust amplification of longer fragments, across a broad range of GC content.

Six DNA standards are provided in Kapa Library Quantification Kits. A set of these DNA standards should be included in triplicate in each qPCR plate. In addition to the DNA standards, each dsDNA library may require at least 12 reactions (triplicate reactions for the 1:1000 library dilution, as well as triplicate reactions for each of the optional 1:2000, 1:4000, and 1:800 dilutions). Figure 1 summarizes the steps necessary to accurately quantify the number of amplifiable molecules in a fragmented library.

MATERIALS

It is essential that you consult the appropriate Material Safety Data Sheets and your institution's Environmental Health and Safety Office for proper handling of equipment and hazardous materials used in this protocol.

Reagents

Elution buffer (EB)
Illumina GA DNA Standards (6×80 µL) (catalog no. KK4804; Kit, catalog no. KK4852)
Illumina GA Primer Premix (10×, 1×1 mL) (catalog no. KK4805; Kit, catalog no. KK4852)
Kapa SYBR FAST qPCR Master Mix (2×)
Library dilution buffer (10 mM Tris-HCl at pH 8.0, 0.05% Tween 20)
Library DNA or DNA standard (5 nM)
Water, PCR grade

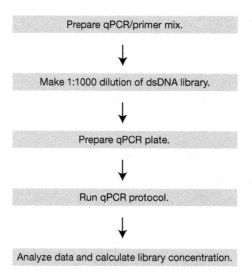

FIGURE 1. Workflow for qPCR.

Equipment

Bench-top minifuge
Kapa SYBR FAST LightCycler 480 qPCR Kit (catalog no. KK4610; Kit, catalog no. KK4852)
Low-bind tips
  Use only low-bind tips throughout this entire protocol.
qPCR instrument (e.g., LightCycler 480)
Vortex mixer

METHOD

For accurate results, ensure that all components are thawed and mixed before use.

Library Sample Preparation

1. Perform three initial 1:1000 dilutions of each 5 nM library stock in Library dilution buffer as follows:

Library dilution buffer	999.0 μL
Library DNA	1.0 μL
Total	1000.0 μL

   Successful library quantification is highly dependent on the accurate dilution of library DNA. Always ensure that proper pipetting techniques are used.

2. Mix the dilutions thoroughly by vortexing for 10 sec.
   Each dilution will be run twice. There will be a total of six readings per library.

qPCR Setup

3. Add 1 mL of 10× Illumina GA Primer Premix to the 2× Kapa SYBR FAST qPCR Master Mix (5 mL).

4. Set up qPCRs in triplicate for the six standards and for each library dilution (from Step 1), and load each sample twice for a total of six points per library. Load each well of the qPCR plate with the reagents shown below, for a total reaction volume of 20 μL.

Kapa SYBR- FAST qPCR Master Mix containing Primer Premix	12.0 μL
Water, PCR grade	4.0 μL
Diluted library DNA or DNA standard (1–6)	4.0 μL

5. Ensure that the qPCR plate is sealed, then mix the reactions gently, and collect all components in the bottom of the wells by brief centrifugation.

6. Perform qPCR.

   i. Program the qPCR instrument with the appropriate parameters (listed below), and perform qPCR on the library DNA samples and on the DNA standards.

   ii. On the desktop, click on the Admin icon, and enter your username and password (found on the machine).
      - Click on **New Experiment From Template**.
      - Under templates, choose **Kapa_Sybr_Illumina_std.curve**.
      - Under subset Template, do not choose anything.
      - Under Sample Editor templates, choose **Matt automation standard a19.**
      - Click the check mark symbol.
      - Place the tray in the machine and check **a1 orientation**.

- Click **Start Run**.
- Choose the experiment folder and name it.

Analysis

7. Annotate the DNA standards as follows before analyzing the data according to the qPCR instrument guidelines:

Sample name	dsDNA concentration (pM)
Std 1	20
Std 2	2
Std 3	0.2
Std 4	0.02
Std 5	0.002
Std 6	0.0002

Each standard must be assayed in triplicate, using 4 μL of the DNA standard supplied in the kit per 20-μL reaction.
Enter the correct concentration of dsDNA (pM) in the standard quantity field.

8. Confirm that the reaction efficiency calculated for the DNA standard dilution series falls within the range of 90%–110%.

9. (Optional) Confirm that the reaction efficiency calculated for the twofold library DNA dilution series (if used) falls in the range of 90%–110%.

10. Calculate the concentration of each library as indicated in the following example.

   i. Obtain the calculated concentration of the 1:1000 dilution of the library (and the calculated concentrations of the optional 1:2000, 1:4000, and 1:8000 dilutions), as determined by qPCR in relation to the concentrations of the correctly annotated DNA Standards 1–6.

   ii. Perform a size adjustment calculation to account for the difference in size between the average fragment length of the library and the DNA standard (452 bp).

   iii. Calculate the concentration of the undiluted library by taking into account the relevant dilution factor (1000) and the volume used per reaction (4 μL).

Library name	Concentration calculated by qPCR instrument (triplicate data points) (pM)	Average concentration (pM)	Size-adjusted concentration (pM)	Concentration of library stock (pM)
Library 1:1000	A1 A2 A3	A	$A \times 452 = W$ average fragment length	$W \times 1000$

11. Use the average of the six data points corresponding to the most concentrated library DNA dilution that falls within the dynamic range of the DNA standard to calculate the concentration of the undiluted library.

12. Discard any point that is not within 0.5 between all six points, average, and move into the nanomolar calculation.

   Repeat qPCR on the 5 nM stock if the value is <3 nM or >11 nM.

   If one of the six replicates appears to be an outlier, it may be omitted from the calculation. If more than one of the three replicates appear to be outliers, the assay should be repeated.

13. If the replicates are within this range (3–11 nM), use the calculated concentration of the undiluted library to prepare a 2 nM dilution of the library.

    Delete nanomolar volumes not being used; that is, if the whole set is 2 nM, delete or hide the 1 nM calculations on the sheet.

14. Run qPCR on all 1 nM and 2 nM stocks to validate what has been created before delivering to sequencing. If the values are ± 0.2 nM, pass into sequencing without further questions; if outside this range, review the library construction process.

15. Proceed with loading of the flow cell and bridge PCR.

    Note that the method described here is likely to yield a higher value for the concentration of the undiluted library than non-qPCR-based methods. The optimal concentration per flow cell calculated with this method is therefore likely to be lower.

Library Quantification Using PicoGreen Fluorometry

This protocol describes the quantification of DNA using PicoGreen and a fluorometer to determine the concentration of DNA sample for downstream processing (see discussion of PicoGreen in the introductory section in Chapter 1). Because dye-based methods will not detect degraded or short DNA fragments, PicoGreen requires DNAs ≥50 bp, but can less reliably detect fragments as small as 20 bp.

MATERIALS

It is essential that you consult the appropriate Material Safety Data Sheets and your institution's Environmental Health and Safety Office for proper handling of equipment and hazardous materials used in this protocol.

Recipes for reagents specific to this protocol, marked <R>, are provided at the end of the protocol. See Appendix 1 for recipes for commonly used stock solutions, buffers, and reagents, marked <A>. Dilute stock solutions to the appropriate concentrations.

Reagents

Control DNA: Human buffy coat (Roche 100 µg Human Genomic DNA) <R>
DNA samples (2 µL)
DNA standard (100 µg/mL) in TE buffer
PicoGreen dsDNA reagent
PicoGreen Reagent Kit (Quant-iT; Life Technologies)
TE buffer (20×) <A>
Water, nuclease-free (e.g., Sigma-Aldrich)

Equipment

Centrifuge with plate adapters
Conical tubes (50 mL)
Fluorometer (Varioskan; Thermo Scientific)
Microplates (96 well) (Microfluor; Thermo Scientific)

METHOD

Before beginning, equilibrate the reagents and DNA samples to room temperature.

Prepare DNA Standards and DNA Controls

1. Prepare a twofold serial dilution series starting with 100 µg/mL DNA standard (supplied with the PicoGreen Assay Kit) and nuclease-free water, using the table below as a guide:

Sample (PicoGreen Std)	Starting concentration (ng/μL)	Sample (μL)	Water (μL)	Final concentration (ng/μL)
100 ng/μL	100	10	10	50
50 ng/μL	50	10	10	25
25 ng/μL	25	10	10	12.5
12.5 ng/μL	12.5	10	10	6.25
6.25 ng/μL	6.25	10	10	3.125
3.13 ng/μL	3.125	10	10	1.5625
1.56 ng/μL	1.5625	10	10	0.78

Prepare PicoGreen Working Solution

One plate of standards and controls requires ~5 mL of diluted PicoGreen. One plate of 96 samples requires ~10 mL of diluted PicoGreen.

2. Prepare 40 mL of 1× TE buffer from the 20× stock.

3. Add 15 mL of 1× TE to a 50-mL conical tube. Add 75 μL of PicoGreen solution. Invert the tube several times to mix.

   ▲ *PicoGreen is light sensitive. Prepare fresh solutions for each batch. Wrap the tube in foil and protect it from light.*

4. Dispense 98 μL of the diluted PicoGreen into Wells A1–H1, A2–H2, A3–C3, and A4–C4 of the standard/control 96-well plate.

5. Dispense 98 μL of the diluted PicoGreen to all 96 wells of a second plate. This will be your Sample plate.

Set Up Standards, Controls, and Samples

6. Add 2 μL of each DNA standard in duplicate and 2 μL of each control in duplicate to the blank microfluor plate, as shown in the table below. Leave two blank wells filled only with PicoGreen.

	1	2	3	4	5	6	7	8	9	10	11	12
A	50 ng/μL standard	50 ng/μL standard	70 ng/μL control	70 ng/μL control								
B	25 ng/μL standard	25 ng/μL standard	30 ng/μL control	30 ng/μL control								
C	12.5 ng/μL standard	12.5 ng/μL standard	BLANK	BLANK								
D	6.25 ng/μL standard	6.25 ng/μL standard										
E	3.125 ng/μL standard	3.125 ng/μL standard										
F	1.5625 ng/μL standard	1.5625 ng/μL standard										
G	0.78 ng/μL standard	0.78 ng/μL standard										
H	0 ng/μL BLANK	0 ng/μL BLANK										

7. Mix the DNA samples well, and gently centrifuge them to collect the samples at the bottoms of the tubes.

8. Add 2 μL of each DNA sample to the wells of the sample microfluor plate.

   When protected from light, the reactions are stable for up to 3 h.

Measure Fluorescence Using the Varioskan Fluorometer

Launch the Software

9. On the desktop, click the "SkanIt RE for Varioskan Flash 2.4.3" icon to open the software. Use the username *admin* (no password is required).

10. Cover the plate with foil tape.

11. Click **Open Session** and choose **Shake**.

   i. Click Open.

   ii. Insert the plate.

   iii. On the bottom-left side of the page, choose **Connect**.

   iv. Allow the plate to shake for 20 sec.

12. Incubate the covered plate for 15 min.

13. Open the session *96 well Varioskan*.

14. Save as **Sample Name**.

15. Go to *Executing Session*.

Protocol Setup (only for a new session)

16. In the *Session Structure* box (upper-left), select **Protocol**.

17. In the *Steps* box, right-click, then select **Fluorometric**. Set the following parameter values:

Emission wavelength	**530 nm**
Measurement time	**900**
Excitation	**485**
Excitation BW	**12**
Dynamic Range	**Auto Range**

Plate Layout Setup (only for a new session)

18. In the *Session* box (upper-left), select **Plate Layout**.

19. At the top of the "Plate Layout" page, name the page **Standards_Controls**.

20. Using the *Fill Wizard* (top-right), fill the first plate with the calibration curve, controls, and blanks.

21. Click the **A1 position** on the 96-well plate diagram. A red dot will appear.

22. Under the *Samples/Type* section, select **Calibrator**. (The standard curve samples are known as "Calibrators.") Point the Fill Order arrow down.

23. Under the *Replicates* section, choose **2**. Point the Fill Order arrow right.

24. In the *Concentrations* section, enter **50**. (Unit=**ng/μL**.)

25. Click **Generate Series**.

26. In the *Series/Operators* section, choose **Divide**. Step by **2**. Click **OK**.

27. Under the *Samples/Type*, choose **Controls**, Number of controls **2**, and Number of replicates **2**. Click **Add/Close**.

28. Under *Samples/Type*, choose **Blank**, Number of replicates **2**. Click **Add/Close**.

29. At the top of the "Plate Layout" page, choose **New**. Name this page **Samples**, or give it a project name.

30. Select the Fill Wizard.

31. Under *Samples/Type*, choose **Unknown**, Number of unknowns **96**. (Samples are known as "Unknowns.") Point the Fill Order arrow down.

32. Under *Dilution/Unit*, enter **ng/μL**. Click **Add/Close**.

33. Once the layout is set, click **Session** and **Save As**. Save the session as the date or plate name.

Results Setup (only for a new session)

Results can be set up or changed after the run, as long as that session has been saved.

34. In the *Session* box (upper-left), select **Results**.

35. In the *Results* box (lower-left), right-click on **Fluorometric** and select **Blank Subtraction**. Right-click on **Blank Subtraction**. Choose **Quantative Curve Fit**.

36. In the *Extrapolation* section, check **Enable Extrapolation** and change the max concentration to **500**.

37. In the *Results* box (lower-left), select **Report**.

38. Select the Parameters tab on the right at the top.

39. Holding the [Ctrl] key, select **Blank Subtraction1**, **Curve Fit1**, and **Fluorometric1**. Click **Add**.

40. From the Added list, select **Curve Fit1**.

   i. Select **Format**.

   ii. Select **List**, **Plate**, **Well**, and **Calculated**.

   iii. Click **OK**.

41. At the bottom, be sure that "Save to File" and "Unique Name" are checked.

42. The file destination should be Z:\varioskan\appropriate year\month folder. The file saves as .xls format only.

Executing the Session

Once your session has been set up and saved, perform the following.

43. Using the "Plate Out" button on the bottom left of the Varioskan software, place your first plate (standard curve) into the plate carrier of the instrument.

44. Click **Start**.

45. When the plate has been scanned, use the "Plate Out" button to retrieve it.

46. Remove the Standard curve plate. Insert the Sample plate.

47. When the run is finished, retrieve the Sample plate from the plate carrier.

48. Click the "Run Plate In" button on the bottom left of the Varioskan software to return the carrier to the home position.

49. View .xls results files in the *My Computer/My Documents* folder of the computer.

RECIPE

It is essential that you consult the appropriate Material Safety Data Sheets and your institution's Environmental Health and Safety Office for proper handling of equipment and hazardous materials used in this protocol.

Control DNA: Human Buffy Coat

Prepare 70 ng/μL and 30 ng/μL human buffy coat controls by diluting stock with nuclease-free water.

Sample	Starting concentration (ng/μL)	Sample (μL)	Water (μL)	Final concentration (ng/μL)
Human buffy coat stock	200	7	13	70
Human buffy coat stock	70	8.6	11.4	30

Library Quantification: Fluorometric Quantitation of Double-Stranded or Single-Stranded DNA Samples Using the Qubit System

Qubit is an accurate and highly sensitive fluorescence-based quantitation system. Several assay kits have been optimized for the Qubit fluorometer, but also function efficiently with other fluorometers. The high-sensitivity dsDNA Qubit Kit has a detection range of 0.2–100 ng. The ssDNA Kit has a detection range of 1–200 ng.

MATERIALS

It is essential that you consult the appropriate Material Safety Data Sheets and your institution's Environmental Health and Safety Office for proper handling of equipment and hazardous materials used in this protocol.

Reagents

Assay kit (Qubit dsDNA or Qubit ssDNA Assay Kit; Life Technologies)
> Each kit contains concentrated dye reagent, Dilution buffer, and DNA standards.

Equipment

Assay tubes (Qubit, Life Technologies)
Fluorometer (Qubit, Life Technologies)
Personal computer running Windows XP
Qubit Data Logger software

METHOD

Reagent and Sample Preparation

1. Label an assay tube on the cap for each sample, as well as, for Standards 1 and 2 and for Controls 1 and 2.
2. Prepare Controls 1 and 2. Control 1 is 1 μL of Standard 2 (high-sensitivity dsDNA kit: 10 ng/μL; ssDNA kit: 20 ng/μL). Control 2 is 1 μL of a 1:10 dilution of Standard 2.
3. Combine 1 μL of concentrated assay dye reagent with 199 μL of Dilution buffer per sample. Prepare enough working solution for each sample, two standards, two controls, and two extra assays.
4. Vortex the working solution for 5 sec.
5. Aliquot 199 μL of working solution into each labeled sample assay tube.
6. Aliquot 190 μL of working solution into each labeled standard assay tube.
7. Add 10 μL of Standard 1 to 190 μL of working solution. Add 10 μL of Standard 2 to 190 μL of working solution.
8. Add 1 μL of Control 1 (actually Standard 2) to 199 μL of working solution. Add 1 μL of Control 2 (diluted Standard 2) to 199 μL of working solution.
9. Add 1 μL of sample to 199 μL of working solution. Repeat for each sample.

10. Vortex each of the standards and samples for 5 sec.

    Handle the samples by holding the tops of the tubes, not the bottoms. This assay is sensitive to temperature changes.

11. Incubate the samples and standards for 2 min at room temperature.

DNA Quantitation

12. Open the Qubit software on the computer. (Make sure that [Caps Lock] is not on.)

13. Click **Start** on the Qubit Data Logger to open the port to the Qubit.

14. Turn on the fluorometer, and follow the instructions on the screen.

15. Choose the assay applications specific to your application. Press **Go**.

16. Choose **Run new calibration**, and press **Go**.

17. Gently agitate standard Tube 1 by rocking it back and forth. Place standard Tube 1 into the Qubit and close the Qubit lid. Press **Go** and wait for reading to complete.

18. Gently agitate standard Tube 2 by rocking it back and forth. Place standard Tube 2 into the Qubit and close the Qubit lid. Press **Go** and wait for reading to complete.

19. In the Qubit recorder, click on the appropriate cell and scan the bar code on the sample tube.

20. Gently agitate a sample tube by rocking it back and forth. Place the sample tube into the Qubit and close the Qubit lid. Press **Go** and wait for reading to complete.

    Do not allow the samples to warm up in your hand; handle them by the cap.

21. Choose **Calculate the concentration**, and press **Go**.

22. Follow the prompt to enter the sample volume used, and press **Go**.

23. Remove the sample from the Qubit.

    See Troubleshooting.

24. Repeat Steps 19–23 for each sample, keeping the samples in order.

25. Save the Qubit recorder file as a .csv file on the desktop.

Clean Up

26. Close the Qubit lid and turn off the Qubit. Press the Home button. Press the arrow up to "Power Off," and then press **Go**.

27. Safely dispose of sample liquids in a dimethyl sulfoxide disposal bottle. Recap and place the tubes in the appropriate hazardous waste container.

TROUBLESHOOTING

Problem (Step 23): Sample reading is beyond the range of the system.
Solution: Prepare a 1:5 dilution of the DNA sample, and repeat Steps 8–22.

Preparation of Small-Fragment Libraries for 454 Sequencing

This protocol describes preparation of libraries of small DNA fragments for use in sequencing on the 454 Sequencing System (FLX or XLR). Fragmented genomic DNA is purified and ligated to adaptor sequences, then immobilized on specialized beads. The library consists of a set of single-stranded DNA fragments representing the complete span of the target DNA. After quality assessment and quantitation, the single-stranded library is ready for amplification by emPCR as described in Protocol 16.

MATERIALS

It is essential that you consult the appropriate Material Safety Data Sheets and your institution's Environmental Health and Safety Office for proper handling of equipment and hazardous materials used in this protocol.

Recipes for reagents specific to this protocol, marked <R>, are provided at the end of the protocol. See Appendix 1 for recipes for commonly used stock solutions, buffers, and reagents, marked <A>. Dilute stock solutions to the appropriate concentrations.

Reagents

Adaptors
AMPure beads (Beckman Coulter)
> *For concentrating and purifying DNA samples.*

ATP
BSA
Buffer EB (room temperature)
Buffer PBI (room temperature)
Buffer PE (room temperature)
DNA sample (in TE buffer <A>)
dNTP mix
Electrode cleaner
Enzymes and Adaptors Kit (−20°C)
Ethanol (70%)
Fill-in polymerase
Fill-in polymerase buffer (10×)
Library binding buffer (2×)
Library immobilization beads (4°C)
Library Prep Buffers Kit (all buffers provided at −20°C) (Roche)
Library wash buffer
Ligase
Ligase buffer (2×)
Melt solution
> *Prepare as 10 N NaOH in microbiology-grade water. See Step 31.*

MinElute PCR Purification Kit
> *MinElute columns are stored at 4°C. Allow the columns to come to room temperature for ~15 min before using.*

Nebulization buffer

Neutralization solution
Prepare as 20% acetic acid in Buffer PB. See Step 33.

Polishing buffer (10×)
RNA Pico Chip Kit (room temperature) (Agilent)
RNA 6000 Pico Chips (Agilent)
RNA 6000 Pico Cond. Solution (Agilent)
RNA 6000 Pico Dye Concentrate (Agilent)
RNA 6000 Pico Gel Matrix (Agilent)
RNA 6000 Pico Kit (4°C) (Agilent)
RNA 6000 Pico Ladder (Agilent) (store at −20°C)
RNA 6000 Pico Marker (Agilent)
Spin columns (4°C)
T4 DNA polymerase
TE buffer
T4 PNK
Tris-HCl (10 mM)
Water, molecular biology grade

Equipment

Aeromist Nebulizer
FlashGel (1.2%) and FlashGel Rig (Lonza)
Incubator
Magnetic Particle Collector (MPC)
Microcentrifuge
Microcentrifuge tube (1.7 mL)
Minicentrifuge (Bench-Top)
Nebulizer condenser tube
Nebulizer holder
Nebulizers Kit: Room Temp. (Roche)
Nebulizer snap cap
Nitrogen tank
Pipettes of appropriate size
Pipette tips of appropriate size, filtered
Safe-Lock Tubes
Spectrophotometer (e.g., NanoDrop)
Spin filters
Syringe kit
Tube rotator
Vacuum manifold
Vented nebulization hood
Vortex Genie

METHOD

Throughout this protocol, room temperature is assumed to be 22°C.

DNA Fragmentation

1. Assess the quality and quantity of the DNA sample on a FlashGel.

2. Using a pipette, transfer 3–5 μg of the sample DNA (in TE) to the bottom of a nebulizer (cup). Add TE buffer to a final volume of 100 μL, then add 500 μL of Nebulization buffer, and mix thoroughly.

3. Assemble the nebulizer and affix the nebulizer tubing to the nebulizer's gas inlet, then transfer the nebulizer to the externally vented nebulization hood.

4. Insert the nebulizer into the nebulizer holder, and connect the loose end of the nebulizer tubing to the nitrogen tank.

5. To fragment the DNA, apply 45 psi of nitrogen for 1 min, then turn off the nitrogen gas flow.

6. Allow the pressure to normalize, and disconnect the tubing.

7. Centrifuge the fragmented sample briefly at 1500 rpm for 30 sec, then carefully unscrew the nebulizer top, and measure the volume of nebulized material.

    Total recovery should be >300 μL.

8. Add 2.5 mL of Buffer PBI to the DNA sample in the nebulizer cup, and swirl to mix.

9. Purify the nebulized DNA using two columns from the MinElute PCR Purification Kit (ensure that the columns have come to room temperature before use). The large sample volume will require that each column be loaded in two aliquots. For each column (used for half of the DNA sample):

    i. Insert the MinElute column into the connector of the vacuum manifold.

    ii. To bind the DNA, pipette 750 μL of the sample into the MinElute column, and apply the vacuum. Add an additional 750 μL of sample to the spin column or until the sample is completely gone.

    iii. Wash the DNA by adding 750 μL of Buffer PE to the MinElute column.

    iv. Turn off the vacuum, and transfer the MinElute column back into the waste tube.

    v. Centrifuge the MinElute column/tube at 13,000 rpm for 1 min.

    vi. To remove any remaining Buffer PE, rotate the MinElute column/tube and centrifuge at 13,000 rpm for an additional 30 sec.

    vii. Transfer the MinElute column to a clean 1.5-mL tube.

    viii. Elute the DNA by adding 25 μL of Buffer EB (room temperature) to the center of the MinElute membrane.

    ix. Allow the column to stand for 1 min at room temperature, then centrifuge at 13,000 rpm for 1 min.

    x. Repeat Steps 9.i–9.viii for the second column and the remaining half sample.

    xi. Pool the eluates of the two columns for a total of ~50 μL.

    xii. Quantify the sample using a spectrophotometer (e.g., NanoDrop).

10. To ensure that the sample was nebulized to the proper size (300–800 bp), analyze 1 μL of sample on a FlashGel.

11. Using a pipette, measure the volume of the eluates and add Buffer EB to a final volume of 50 μL.

12. Add 35 μL of AMPure beads, vortex to mix, and incubate for 5 min at room temperature.

13. Using a Magnetic Particles Collector (MPC), pellet the beads against the wall of the tube.

14. Remove the supernatant, and wash the beads twice with 500 μL of 70% ethanol (incubate for 30 sec each time).

15. Remove all of the ethanol, and allow the AMPure beads to dry completely (you may use a heat block set at 37°C).

16. Remove the tube from the MPC, add 24 μL of EB buffer (pH 8.0), and vortex to resuspend the beads.

    This step elutes the nebulized DNA from the AMPure beads.

17. Using the MPC, pellet the beads against the wall of the tube once more, and transfer the supernatant containing the purified nebulized DNA to a fresh microcentrifuge tube.

18. To assess the quality of the sample, run 1 μL of the pooled, nebulized material on a FlashGel. *The mean size should be between 400 and 800 bp.*

Fragment End Polishing and Adaptor Ligation

19. In a 0.2-mL microcentrifuge tube, combine the following reagents (from the Enzymes and Adaptors Kit), in order:

DNA fragments (Step 15)	~23 μL
Polishing buffer (10×)	5 μL
BSA	5 μL
ATP	5 μL
dNTPs	2 μL
T4 PNK	5 μL
T4 DNA polymerase	5 μL
Total volume	50 μL

Mix well and incubate for 15 min at 12°C; immediately transfer the reaction to 25°C and continue the incubation for an additional 15 min.

20. Purify the polished fragments using one column from the MinElute PCR Purification Kit (ensure that the column has come to room temperature before use).

i. Insert the MinElute column into the connector of the vacuum manifold.

ii. Add 5× the sample volume of Buffer PBI: deliver 100 μL of Buffer PBI into the sample tube, pipette to mix, then transfer the total volume into the MinElute column and apply the vacuum.

iii. Deliver an additional 150 μL of Buffer PBI into the sample tube, pipette to mix, and transfer to the column.

iv. Wash the DNA by adding 750 μL of Buffer PE to the MinElute column.

v. Turn off the vacuum, and transfer the MinElute column back into the 2-mL collection tube.

vi. Centrifuge the MinElute column/tube at 13,000 rpm for 1 min.

vii. To remove any remaining Buffer PE, rotate the MinElute column/tube and centrifuge at 13,000 rpm for an additional 30 sec. Transfer the MinElute column to a clean 1.5-mL tube.

viii. Elute the DNA by adding 16 μL of Buffer EB (room temperature) to the center of the MinElute membrane.

ix. Allow the column to stand for 1 min, then centrifuge at 13,000 rpm for 1 min.

x. Quantify the sample (1 μL) using a spectrophotometer (e.g., NanoDrop).

21. In a 0.2-mL microcentrifuge tube, add the following reagents (from the Enzymes and Adaptors Kit), in order:

▲ *The order of addition of reagents is critically important!*

Polished DNA	~15 μL
Ligase buffer (2×)	20 μL
Adaptors	1 μL
Ligase	4 μL
Total volume	40 μL

Mix well, centrifuge briefly, and incubate the ligation reaction for 15 min at 25°C.

22. Use this time to prepare the immobilization beads.

i. Transfer 50 μL of library immobilization beads to a fresh 1.5-mL tube.

ii. Using an MPC, pellet the beads and remove the buffer.

    iii. Wash the library immobilization beads twice with 100 μL of 2× Library binding buffer, using the MPC.

    iv. Elute the beads with 25 μL of 2× Library binding buffer and set on ice until needed for Step 24 (library immobilization).

23. Purify the ligation reaction using one column from the MinElute PCR Purification Kit. (Ensure that the column has come to room temperature before use.)

    i. Insert the MinElute column into the connector of the vacuum manifold.

    ii. Add 5× sample volume of Buffer PBI. Deliver 100 μL of Buffer PBI into the reaction sample tube (Step 21), pipette to mix, then transfer the total volume into the MinElute column and apply the vacuum.

    iii. Deliver an additional 100 μL of Buffer PBI into the sample tube, pipette to mix, and transfer to the column.

    iv. Wash the DNA by adding 750 μL of Buffer PE to the MinElute column.

    v. Turn off the vacuum, and transfer the MinElute column back into the 2-mL collection tube.

    vi. Centrifuge the MinElute column/tube at 13,000 rpm for 1 min.

    vii. To remove any remaining Buffer PE, rotate the MinElute column/tube and centrifuge at 13,000 rpm for an additional 30 sec.

    viii. Elute the DNA by adding 26 μL of Buffer EB (room temperature) to the center of the MinElute membrane.

    ix. Allow the column to stand for 1 min, and centrifuge it at 13,000 rpm for 1 min.

    x. Quantify the sample using a spectrophotometer (e.g., NanoDrop).

Library Immobilization and Fill-In Reaction

24. Add eluted DNA into washed library immobilization beads.

25. Mix well and place on a tube rotator for 20 min at room temperature.

26. Using the MPC, wash the immobilized Library twice with 100 μL of Library wash buffer.

27. In a 1.5-mL tube, combine the following reagents (from the Enzymes and Adaptors Kit) in order, and mix:

▲ *The order of addition of reagents is critically important!*

Water, molecular biology grade	40 μL
Fill-in polymerase buffer (10×)	5 μL
dNTPs	2 μL
Fill-in polymerase	3 μL
Total volume	50 μL

28. Using an MPC, remove the 100 μL of Library wash buffer from the library-carrying beads from Step 26.

29. To the Library wash buffer, add the 50 μL of fill-in reaction mix prepared in Step 27, mix well, tap out any bubbles, and incubate for 20 min at 37°C.

30. Using the MPC, wash the immobilized library twice with 100 μL of Library wash buffer.

Single-Stranded Template DNA (sstDNA) Library Isolation

31. In a 1.5-mL tube, prepare the Neutralization solution by mixing 500 μL of PBI buffer and 3.8 μL of 20% acetic acid.

32. Using the MPC, remove the 100 μL of Library wash buffer from the library-carrying beads from Step 30.

33. Prepare 1× Melt solution by adding 0.125 mL of 10 N NaOH to 9.875 mL of molecular biology–grade water.

> Melt solution is only good for 7 d after it is prepared. Determine how old the solution is before use and make new if necessary.

34. Add 50 μL of Melt solution to the beads carrying the washed library (Step 30).

35. Vortex well and, using the MPC, pellet the beads away from the 50-μL supernatant.

36. Carefully remove and transfer the supernatant into the freshly prepared neutralization solution.

37. Repeat Steps 34–36 for a total of two 50-μL Melt solution washes of the beads. (Pool these together into the same tube of Neutralization solution.)

38. Purify the neutralization single-stranded template DNA (sstDNA) library using one column from the MinElute PCR Purification Kit (ensure that the column has come to room temperature before use).

   i. Insert the MinElute column into the vacuum manifold's connector.

   ii. To bind the DNA, pipette the sample into the MinElute column and apply the vacuum.

   iii. Wash the DNA by adding 750 μL of Buffer PE to the MinElute column.

   iv. Turn off the vacuum and transfer the MinElute column back into the 2-mL collection tube.

   v. Centrifuge the MinElute column/tube at 13,000 rpm for 1 min.

   vi. To remove any remaining Buffer PE, rotate the MinElute column/tube and centrifuge at 13,000 rpm for an additional 30 sec.

   vii. Transfer the MinElute column to a clean 1.5-mL tube, and elute the DNA by adding 15 μL of TE to the center of the MinElute membrane.

   viii. Allow the column to stand for 1 min, and centrifuge at 13,000 rpm for 1 min.

sstDNA Library Quality Assessment and Quantitation

Transfer Agilent kits to room temperature 30 min before use—the gel matrix must be at room temperature. Note that the RNA marker only should be kept on ice.

39. Run 1 μL of a sample of the library on a RNA Pico 6000 LabChip.

40. Quantitate a sample of the library (1 μL in triplicate) using a spectrophotometer (e.g., NanoDrop or Qubit), and assess the quality of the sstDNA library.

> The average fragment size should be between 400 and 800 bp, with <10% below 300 nt; the total DNA yield should be >10 ng, and there should be no visible dimer peak.

41. Use the calculator software (454 Molecular Calculator) to process the information from the Pico 6000 LabChip assessment in order to calculate the concentration equivalence in molecules per microliter as in the following formula:

$$\frac{\text{molecules}}{\mu L} = \frac{(\text{sample concentration } [ng/\mu L]) \times (6.022 \times 10^{23})}{(328.3 \times 10^9) \times (\text{avg. fragment length } [nt])}.$$

Using 1 μL of the library, make a primary dilution of the library in TE to 1×10^8 molecules/μL. Store the concentrated library (and, if not used immediately, the 1×10^8 diluted stock) at −15°C to −25°C.

42. Thaw the 1×10^8 aliquot of the quantitated sstDNA library (if necessary) to make further dilutions. Vortex to mix all of the dilutions.

   i. To make a 1×10^7 dilution, mix 1 μL of the 1×10^8 stock into 9 μL of TE buffer.

   ii. To make a 1×10^6 dilution, mix 1 μL of the 1×10^7 stock into 9 μL of TE buffer.

   iii. To make a 1×10^5 dilution, mix 1 μL of the 1×10^6 stock into 9 μL of TE buffer.

   These dilutions will be used for the bead:DNA titration in emPCR (Protocol 16).

43. Set up four single-tube emPCRs, titrating the number of molecules of sstDNA per bead as follows:

    - *Tube 1.* 1.5 μL of sstDNA library at $2 \times 10^5/\mu L$ (= 0.5 molecule/bead)
    - *Tube 2.* 6.0 μL of sstDNA library at $2 \times 10^5/\mu L$ (= 2 molecules/bead)
    - *Tube 3.* 1.2 μL of sstDNA library at $2 \times 10^6/\mu L$ (= 4 molecules/bead)
    - *Tube 4.* 4.8 μL of sstDNA library at $2 \times 10^6/\mu L$ (= 16 molecules/bead)

44. Vortex to mix the four tubes for 5 sec.

45. Perform the emPCR procedure as described in Protocol 16.

sstDNA Library Capture and emPCR

This protocol describes how to clonally amplify DNA fragments from a library of single-stranded template DNAs (sstDNA) prepared as described in Protocol 15. Library fragments are first annealed to capture beads, subjected to an emulsification step, and amplified by PCR. Beads carrying amplified single-stranded templates are collected, and the templates are then annealed to sequencing primers. This process results in an immobilized library of amplified DNA fragments suitable for sequencing using the 454 FLX or XLR Titanium Sequencing System.

MATERIALS

It is essential that you consult the appropriate Material Safety Data Sheets and your institution's Environmental Health and Safety Office for proper handling of equipment and hazardous materials used in this protocol.

Reagents

Amplification mix ($5\times$)
Annealing buffer TW ($1\times$)
Bead Recovery Kit
Bleach (10%)
Capture bead wash ($10\times$)
DNA Away
DNA capture beads
emPCR Kit (XLR Titanium) ($-20°C$)
Emulsion oil in specimen cup
Enhancing fluid TW ($1\times$)
Enrichment primer
Enzyme mix
Isopropanol (190 proof)
Melt solution (see Step 42)
Mock amplification mix
NaOH (10 N)
PPIase
Sequencing primer
sstDNA library, quantitated and titrated (prepared in Protocol 15)
Water, molecular biology grade (MBG) (Sigma-Aldrich)

Equipment

Bench-top minifuge
Combitips Plus tip (10.0 mL)
Conical tube caps for SABA unit (50 mL)
Conical tubes (50 mL)
Coulter counter
Custom Blue adaptor

Disposable gloves
Enrichment beads
Eppendorf Repeater Plus
Erlenmeyer flask catch
Filtered pipette tips of appropriate size
Genpro disposable laboratory frock
Heat block (preset to 65°C)
Labquake tube roller
Magnetic Particle Collector (MPC)
Microcentrifuge
Microcentrifuge tubes (1.7 mL)
Pipettes of appropriate size
Plates (96 well), skirted
Semiautomatic braking apparatus (SABA)
Texwipes (alcohol wipes)
Thermal cycler
TissueLyser
TissueLyser tube rack
Transpette
Tubing for SABA
Vacuum port
Versi-Dry pads
Vortex Genie
Waste jar

METHOD

▲ Wear gloves and a Genpro disposable frock for this entire process.

Washing the Capture Beads

1. If necessary, prepare the Capture bead wash by combining the following reagents:

Water (MBG)	9 mL
Capture bead wash buffer	1 mL

2. Pellet the beads in a bench-top minifuge as follows.

   i. Centrifuge the beads for 10 sec.

   ii. Rotate the tube 180° and centrifuge again for 10 sec.

   iii. Remove the supernatant carefully, being careful to avoid disturbing the beads.

3. Dispense 1000 μL of 1× Capture bead wash buffer into the beads, and vortex for 5 sec to resuspend the beads.

4. Pellet the beads in a bench-top minifuge as in Step 2.

5. Repeat Steps 3 and 4 once. Do not resuspend beads after the second wash.

Binding the sstDNA Library Fragments

6. Obtain a sufficient amount of the quantitated and titrated sstDNA library to be amplified.

   Each tube of capture beads contains 35 million beads (2 tubes per kit).

Calculate the amount of library to add as follows:

$$\frac{(\text{Copies per bead} \times 35,000,000 \text{ beads})}{(\text{library molecules}/\mu L)} = \text{microliters of library to add to beads.}$$

7. To each tube of capture beads, add the correct amount of sstDNA from Protocol 15, Step 43.

8. Vortex the tube for 5 sec to mix contents. Place on ice until ready for use.

First Shake of Emulsion Oil

9. Raise the TissueLyser safety shield and remove the tube rack assembly.

10. Place the specimen cup containing *only* emulsion oil on the TissueLyser.

11. Screw the specimen cup firmly in place (be careful not to crack the cup), and lower the safety shields.

12. Set the TissueLyser Shake at 28 cycles/sec for 2 min and press the Start button to begin shaking.

Making the Titanium XLR Live Amplification Mix

13. Allow the frozen kit components to fully thaw on ice, leaving the enzyme mix at −20°C. After thawing, vortex the reagents (except for the enzyme mix) for 5 sec.

14. Prepare the Titanium XLR Live Amplification Mix (table directly below), for up to four Large Volume Emulsions (LVEs) (1 kit = 2 LVEs) and keep on ice until ready to use. Note that if multiple libraries are being amplified, a separate tube of Live Amplification Mix must be made for each library.

Reagent	Volume 1 LVEs	Volume 2 LVEs	Volume 4 LVEs
Water (MBG)	2700	5400	10,800
Amplification mix (5×)	780	1560	3120
Amplification primer	230	460	920
Enzyme mix	200	400	800
PPlase	5	10	20

Volumes are in microliters.

Preparing the Emulsification Cups

15. In a 15-mL conical tube, dilute the 5× Mock amplification mix to 1× by adding 1.0 mL of 5× Mock amplification mix to 4.0 mL of MBG water (for each kit used).

16. Remove the shaken specimen cup(s) with emulsion oil from the TissueLyser.

17. Add 5.0 mL of Mock amplification mix to the specimen cup containing shaken emulsion oil.

18. Place the specimen cup containing emulsion oil and Mock amplification mix back in the TissueLyser. Screw the cup firmly in place, and lower the safety shields. Set the TissueLyser for 28/sec for 5 min, and press the Start button to begin shaking.

Adding the Live Amplification Mix to Beads and Emulsion Oil

19. Add 1.0 mL of Live Amplification Mix to the library/DNA capture bead tube and pipette up and down to mix. Keep the tip to reuse in the next step.

20. Transfer Library/DNA capture bead/Live Amplification Mix to a fresh 15-mL conical tube. Keep the tip to reuse in the next step.

21. Pipette 1.0 mL of Live Amplification Mix into the original Library/Capture bead tube. Quickly vortex and centrifuge briefly. Draw up the remaining bead mix and add to the Library/Capture bead/Live Amplification Mix tube from Step 20.

22. Add another 1.75 mL (for a total of 3.75 mL) of Live Amplification Mix to the 15-mL conical tube from Step 20.

23. Remove the shaken specimen cup(s) with emulsion oil and Mock Amplification Mix from the TissueLyser.

24. Vortex the Library/Capture bead/Live Amplification Mix quickly, and pour it into the specimen cup containing emulsion oil and Mock amplification mix.

25. Place the specimen cup back in the TissueLyser. Screw the cup firmly in place, and lower the safety shields. Set the TissueLyser to 12/sec for 5 min, and press the Start button to begin shaking.

Dispensing the Emulsions for Amplification ("Controlled Room")

26. Place a Versi-Dry pad in the emulsion hood.

27. After emulsification, remove the cup(s) from the TissueLyser, and open the emulsion cup.

28. Use an Eppendorf Repeater Plus to draw and dispense a 200-μL aliquot into a 96-well BioDot PCR tray (~90 wells).

29. Place the plate cover (cap mat or plate seal) on the plate. Remove from the hood.

30. Dispose of the Versi-Dry pads lining the emulsion hood and the bench top. (Pads may be discarded in the wastebasket.) Then cleanse the emulsion hood and bench top with the following products:

    - 10% bleach
    - DNA Away
    - Texwipes (alcohol wipes)

    Dispose of used paper towels and wipes in the wastebasket.

31. Switch on the UV light located on the inside of the emulsion hood. Leave illuminated for 30 min. Ensure that the settings on the emulsion hood are properly shut down after illumination is complete.

Amplification Reaction and Bead Recovery ("Amplicon Room")

32. Place the emulsified amplification reactions in a thermal cycler.

33. Check to ensure that the lid is set to track within 5°C of the block temperature.

34. Set up and launch the 454 Titanium EMBPC thermocycling program.

35. When complete, remove all the plates of amplified material from the thermal cycler.
    Wear gloves and a Genpro disposable laboratory frock for the remainder of the steps.

36. Cleanse the bench top and breaking hood with the following products:

    - 10% bleach
    - DNA Away
    - Texwipes (alcohol wipes)

    Dispose of used paper towels and wipes in the wastebasket.

37. Line the breaking hood and the bench top with Versi-Dry absorbant pads.

38. Switch on the UV light located on the inside of the breaking hood. Leave illuminated for 30 min.

39. Record tracking usage on the breaking hood, and ensure that settings on the hood are properly shut down after illumination is complete.

40. Make 1× Enhancing fluid TW by adding 187.5 mL of MBG water to 4× Enhancing fluid TW.

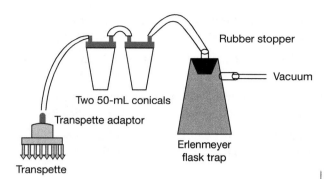

| FIGURE 1. **SABA device.**

41. Make 1× Annealing fluid TW by adding 72 mL of MBG water to 10× Annealing buffer TW.

42. Make 1× Melt solution by adding 500 μL of 10 N NaOH to 40 mL of MBG water. Make fresh after 1 wk.

43. Set up the semiautomatic breaking apparatus (SABA) according to the following specifications (see Fig. 1).

    i. Turn the vacuum pump on.

    ii. Draw the emulsion from each of the wells into the Transpette by placing the Transpette into emulsion slowly in a circular manner. Flip the Transpette upside down after all of the wells have been aspirated to promote better aspiration into the conical tubes.

    iii. Fill the wells again with 150 μL of isopropanol. Pipette up and down to ensure that emulsions and isopropanol have been adequately mixed.

    iv. Again, draw up the isopropanol/emulsion with the Transpette. Be sure to flip the Transpette upside down to facilitate aspiration.

    v. Fill the wells again with 150 μL of isopropanol. Pipette up and down to ensure that emulsions and isopropanol have been adequately mixed.

    vi. Again, draw up the isopropanol/emulsion with the Transpette. Be sure to flip the Transpette upside down to facilitate aspiration.

    vii. Once all of the emulsions have been aspirated, slowly draw up 5 mL of isopropanol into the Transpette from a multichannel reservoir.

    viii. Turn the vacuum pump off.

44. Remove the lids from the 50-mL conicals, and securely replace with the original 50-mL conical lids.

45. Invert the tubes several times to mix beads into an even suspension.

46. Open the tubes and pour the two tubes back and forth until the emulsion oil mix is evenly distributed between the two tubes.

47. Bring the final volume in the tubes to 40 mL with isopropanol.

48. Vortex to resuspend the beads before centrifugation.

Emulsion Breaking and Bead Washing by Centrifugation

49. Centrifuge the 50-mL conical tubes containing emulsion/isopropanol mix in a table top centrifuge for 2000 rpm (930g for Allegra 6) for 5 min.

50. Pour off the supernatant without disturbing pellet, add 35 mL of isopropanol, and vortex until the pellet is completely reconstituted.

51. Centrifuge 50-mL conicals containing emulsion/isopropanol mix in a table top centrifuge for 2000 rpm (930g for Allegra 6) for 5 min.

52. Repeat Steps 50 and 51.

53. Pour off the supernatant without disturbing the pellet.

54. Add 35 mL of 1× Enhancing fluid TW. Vortex until the pellet is completely reconstituted.

55. Centrifuge 50-mL conicals containing emulsion/isopropanol mix in a table top centrifuge for 2000 rpm (930*g* for Allegra 6) for 5 min.

56. Pour off the supernatant without disturbing the pellet, leaving 2 mL of fluid on top of the beads *or* draw up 33 mL of supernatant with a 50-mL steripette.

57. Transfer the beads with a P1000 to a 1.7-mL microcentrifuge tube (two 1.7-mL tubes may be required).

58. Rinse each 50-mL conical tube with 600 μL of enhancing fluid, and pool it with the other beads.
   This step may need to be repeated if some beads are still left. If this is the case, the beads in the 1.7-mL tubes must be pelleted and the supernatant removed to allow for the extra volume.

59. Once all of the beads are transferred, pellet the beads in a minifuge for 10 sec. Rotate the tubes 180° and centrifuge again for 10 sec.

60. Remove the supernatant. Add 1 mL of 1× Enhancing buffer TW and vortex.

61. Pellet the beads in a minifuge for 10 sec. Rotate the tubes 180° and centrifuge again for 10 sec.

62. Remove the supernatant. Add 1 mL of 1× Enhancing buffer TW and vortex.

63. Pellet the beads in a minifuge for 10 sec. Rotate the tubes 180° and centrifuge again for 10 sec.

64. Remove the supernatant. Add 1 mL of 1× Enhancing buffer TW and vortex.

Indirect Enrichment

65. Pellet the beads in a minifuge for 10 sec. Rotate the tubes 180° and centrifuge again for 10 sec.

66. To each 1.7-mL tube, add 1.0 mL of 1× Melt solution and vortex.
   ▲ *Do not leave the beads in the Melt solution for longer than 10 min.*

67. Pellet the beads in a minifuge for 10 sec. Rotate the tubes 180° and centrifuge again for 10 sec.

68. Remove the supernatant. Add another 1.0 mL of 1× Melt solution and vortex.

69. Remove the supernatant. Add 1.0 mL of 1× Annealing fluid TW and vortex.

70. Pellet the beads in a minifuge for 10 sec. Rotate the tubes 180° and centrifuge again for 10 sec.

71. Repeat Steps 69 and 70 for a total of three washes with 1× Annealing fluid TW.

72. After removing the last wash, add 45 μL of 1× Annealing fluid TW and 25 μL of Enrichment Primer to each tube. Vortex to mix completely.

73. Place the tubes on a 65°C heat block for 5 min. Then ice for 2 min.

74. Add 800 μL of 1× Enhancing fluid TW and vortex.

75. Pellet the beads in a minifuge for 10 sec. Rotate the tubes 180° and centrifuge again for 10 sec.

76. Remove the supernatant, and wash twice more with 1.0 mL of 1× Enhancing fluid TW.

77. After removal of the last wash supernatant, add 800 μL of 1× Enhancing fluid TW.

Enrichment

78. Prepare the Enrichment Beads.

   i. Vortex the tube of Enrichment Beads for 1 min to resuspend its contents completely.

   ii. Using a Magnetic Particle Collector (MPC), pellet the paramagnetic Enrichment Beads.

   iii. Remove and discard the supernatant, taking care not to draw off any Enrichment Beads.

   iv. Remove the tube from the MPC, and add 1.0 mL of 1× Enhancing fluid TW to each tube.

v. Vortex for 3 sec to resuspend the beads.

vi. Using an MPC, pellet the paramagnetic Enrichment Beads.

vii. Remove and discard the supernatant, taking care not to draw off any Enrichment Beads.

viii. Repeat Steps 78.iv–78.vii.

ix. Add 160 μL of 1× Enhancing solution TW.

79. Perform enrichment of the DNA-carrying beads.

i. Add 80 μL of washed Enrichment Beads to each of the 1.7-mL tubes of amplified DNA beads. Vortex (vortexing is required for the XLR protocol).

ii. Rotate on a Labquake tube roller at ambient temperature (+15°C to +25°C) for 5 min.

iii. Place the tube in the MPC, wait 2 min, then pellet the paramagnetic Enrichment Beads.
    The supernatant will appear white at this time.

iv. Carefully remove all of the supernatant, taking care not to draw off any pelleted Enrichment Beads.

v. Remove the tube from the MPC and *gently* add 1 mL of 1× Enhancing fluid to the beads. Vortex.

vi. Place the tube in the MPC, wait 2 min, then pellet the paramagnetic Enrichment Beads.
    The supernatant will appear white at this time.

vii. Remove the supernatant.

viii. Repeat Steps 79.v–79.vii until no beads are observed in the supernatant. This may require six or more washes.

    To determine whether beads are still in the supernatant, you can collect the supernatant in a fresh 1.7-mL tube and briefly centrifuge.

80. Collect the enriched DNA beads.

i. Remove the tube from the MPC and resuspend the bead pellet in 700 μL of Melt solution. Do not leave the beads in Melt solution for >10 min.

ii. Vortex for 5 sec, and put the tube back into the MPC to pellet the Enrichment Beads.

iii. Transfer the supernatant, containing enriched DNA beads, to a separate 1.5-mL microcentrifuge tube.

iv. Repeat Steps 80.i–80.iii for better DNA bead recovery, pooling the two melts (total 1400 μL) together.

v. Discard the tube of spent Enrichment Beads.

vi. Pellet the enriched DNA beads by centrifugation as before.

vii. Remove and discard the supernatant, and wash the enriched DNA beads with 1 mL of 1× Annealing buffer. Remove all of the supernatant without disturbing the pellet.

viii. Resuspend once again in 1 mL of 1× Annealing buffer to completely neutralize the Melt solution, centrifuge as above, and remove and discard the supernatant without disturbing the pellet.

ix. Repeat Step 80.viii twice for a total of three washes.

x. Remove the last wash, and add 200 μL of 1× Annealing fluid TW.

Sequencing Primer Annealing Reactions

81. Add 50 μL of Sequencing Primer to each of the tubes of beads. Vortex each tube for 5 sec.

82. Place the tubes on a 65°C heat block for 5 min, then transfer onto ice for 2 min.

83. Add 800 μL of 1× Annealing buffer, pellet the beads as before, and remove the supernatant.

84. Wash the beads once with 1.0 mL of 1× Annealing buffer. Vortex, centrifuge down, and remove the supernatant.

85. Repeat Step 84 for a total of two washes.

86. Resuspend the beads in 950 μL of 1× Annealing fluid TW.

87. Count a 3-μL aliquot of the beads in the Coulter counter, following the manufacturer's instructions.

    The bead count will vary depending on the efficiency of the emPCR and the enrichment.

88. Store the beads, carrying the clonally amplified enriched sstDNA library, at 4°C.

89. Dispose of the Versi-Dry pads lining the bench top. Pads may be thrown in the wastebasket.

90. Cleanse the bench top with the following products:

    - 10% bleach
    - DNA Away
    - Texwipes (alcohol wipes)

    Dispose of used paper towels and wipes in the wastebasket.

Roche/454 Sequencer: Executing a Sequencing Run

This protocol describes (1) the preliminary steps for preparing the clonally amplified enriched sstDNA library (from Protocol 16) for the prewash run and (2) preparation of the PicoTiter Plate devices, in preparation for the sequencing run. Instructions for performing the run are provided, based on the use of the 454 Life Sciences Titanium XLR Sequencing System. Note, however, that the preparatory steps for library construction (Protocol 15) and for emPCR (Protocol 16) also are suitable for use with the 454 FLX sequencer—the differences lie in the numbers of incorporation cycles, hence the read-length. If the FLX system is better suited for the intended purpose, then proceed according to manufacturer's instructions for that instrument.

MATERIALS

It is essential that you consult the appropriate Material Safety Data Sheets and your institution's Environmental Health and Safety Office for proper handling of equipment and hazardous materials used in this protocol.

Reagents

Bleach (10%)
DNA Away
Ethanol (50%)
454 Life Sciences Titanium XLR Sequencing Kit

Box 1:
Apyrase
Bead buffer additive
Control DNA
dATPs
DTT
Enzyme beads
Insert w/assorted buffers
Polymerase
Polymerase cofactors

Box 2:
Bead buffer
Prewash buffer

Box 3:
Sipper tubes

Box 4:
Buffer CB (five bottles chilled at 4°C for 48 h then room temperature after filtering)

Box 5:
Prewash tubes

Box 6:
Packing beads

Box 7:
Bead loading gasket
Cartridge seal
PicoTiter Plate

SparKLEEN laboratory detergent
sstDNA beads from prepared libraries (Protocols 15 and 16)
Texwipes (alcohol wipes)
Versi-Dry pads
Water, distilled, deionized

Equipment

Bead Deposition Device
Bead Deposition Device Base
Bead Deposition Device counterweight
Cartridge seal
Centrifuge
Centrifuge microplate carrier
Centrifuge swinging baskets
Conical tube (50 mL)
Filtered pipette tips of appropriate volume
Gloves
KimWipes
Magnetic Particle Collector (MPC)
Paper towels
PicoTiter Plate (PTP) cartridge
Pipettes of appropriate size, including 50 mL
Prewash insert
Reagents cassette
Zeiss moistened cleaning tissue

METHOD

Laboratory and Reagent Preparation

1. Clean the bench top using the following products:
   - 10% bleach
   - DNA Away
   - Texwipes (alcohol wipes)

   Dispose of used paper towels and wipes in the wastebasket, and line the bench top with Versi-Dry absorbant pads.

2. Bring the Sequencing Reagents Insert of the Titanium XLR Sequencing Kit out of frozen storage.

   Keep the polymerase and polymerase cofactor at −20°C.

3. Open the barrier bag, and allow the concentrated reagents to thaw for 2–3 h at room temperature, or in water in the sink, with the Sequencing Reagents Insert of the kit kept upright and protected from bright light.

4. Place the bottle of Bead buffer on ice.

5. Filter five bottles of Common buffer (Buffer CB) as follows.

   i. Acquire 1 L of 0.22 μm Stericup filter units (one per five bottles).

   ii. Attach the filter units to the vacuum pump.

   iii. Filter Buffer CB.

   iv. As the buffer is filtering, rinse the original bottle with DI water.

   v. Pour the filtered Buffer CB back into the rinse bottle.

   vi. Using a graduated pipette, get a 44-mL aliquot of filtered CB from the fifth bottle and transfer it to a 50-mL Falcon tube. Put the tube in an ice bath.

6. When the contents of the Sequencing Reagents Insert are thawed, transfer the Insert to the 4°C refrigerator to keep the reagents chilled until the run.

   Do not keep the reagents stored for >8 h.

The Prewash Run

7. Bring all components of the GS FLX Sequencing Kit and the GS PicoTiter Plate Kit out of short-term cold storage. Place the −20°C items on ice and the +4°C items at room temperature, except the Wash 2 reagent and Bead buffer 1, which must be kept at +4°C.

8. Close the previous run windows by clicking on the X in the upper-right-hand corner.

9. After the windows have been closed, perform two system stops by double-clicking on the System Stop icon, letting it run, then doing it again.

10. Double-click on the System Start icon, and wait the for the light to turn orange and stop blinking.

11. Open the exterior fluidics door, and raise the sipper manifold completely. Remove the Insert cassette and rotate the manifold.

12. Carry the Reagents cassette to a sink, pour the fluids remaining in the reagent bottles down the drain, and rinse out the Insert cassette.

13. Dry all of the outside surfaces of the Reagents cassette with a paper towel.

14. Wipe down all of the surfaces inside the fluidics door around the manifold 50% ethanol.

15. Replace all of the sipper tubes.

    i. Remove (counterclockwise) all of the old sipper tubes.

    ii. Install (clockwise) all of the new sipper tubes (four long to the left; eight short to the right). Twist them in until you feel the click.

    iii. Rotate the sipper manifold back to its horizontal position.

16. Prepare the Prewash cassette by placing the Prewash cassette insert (tube holder) into the cassette eight small tubes on the right-hand side and the four large tubes on the left-hand side of the cassette.

17. Fill the tubes with Prewash buffer. Slide the Reagents cassette into the fluidics area, lower the sipper carefully, and close the exterior fluidics door.

Launching the Prewash Run

18. If the "Instrument Run" window is not open, double-click on the GS Sequencer icon to launch the application.

19. Make sure that the operator is *adminrig* and click **Sign In**.

20. Click **Start**.

21. After the "Instrument Procedure" screen opens, follow these steps:

Prewash → Next → Start.

PicoTiter Plate Device Preparation

22. Prepare Bead buffer 2 by adding 34 μL of Apyrase solution to the 200-mL bottle of Bead buffer. Label the bottle **Bead Buffer 2** and keep it on ice.

23. Prepare the PicoTiter Plate.

   i. Peel off the lid from the PicoTiter Plate shipping tray.

   ii. Carefully lever the PicoTiter Plate from under the nubs in the shipping tray (do not touch the top of the tray), with a finger or plastic forceps.

   iii. Rest the PicoTiter Plate on top of the nubs in the tray.

     ▲ *Write down the six-digit bar code on the back of the plate for later use.*

24. Pour Bead buffer 2 into the tray until the PicoTiter Plate is completely submerged.

25. Allow the PicoTiter Plate to rest in Bead buffer 2 for at least 10 min at room temperature, until ready to assemble the Bead Deposition Device (Step 44).

26. Remove the bead loading gasket from its packaging, and wash it in SparKLEEN solution. Then rinse the gasket thoroughly with nanopure water, and leave it to air-dry on a paper towel.

Preparation of the Incubation Mix and DNA Beads (Sample)

27. Prepare DNA bead incubation mix by combining the reagents listed below into two 1.7-mL microcentrifuge tubes. Vortex gently; then centrifuge briefly in a microcentrifuge to collect all of the liquid to the bottoms of the tubes.

Loading area size	PicoTiter Plate size	Bead buffer	Polymerase cofactor	DNA polymerase	Total volume (μL)
Large (30×60 mm)	70×75 mm	785 (×2)	75 (×2)	150 (×2)	1010 (x2)

28. Obtain a sufficient quantity of sstDNA beads from the libraries to be sequenced. To determine the volume needed, use the bead count performed at the end of the emPCR procedure.

    For example, if the sstDNA library bead concentration was 2000 beads/μL, use the volumes indicated in the table below, fourth column.

This is only an example; you must determine the correct volume.

Loading area size	PicoTiter Plate size	Number of DNA beads per area	sstDNA beads (example) (μL)	Control DNA beads (μL)
Large (30×60 mm)	70×75 mm	900,000 (×2)	450 (×2)	18 (×2)

29. Vortex the sstDNA library beads to resuspend them, and transfer the appropriate amount of beads into each of two 2-mL tubes.

30. Add the appropriate amount of Control DNA beads to each sstDNA tube (above, fifth column).

31. Centrifuge for 1 min at 10,000 rpm (9300 RCF). Rotate the tube 180° and centrifuge again for 1 min.

32. Calculate the volume of supernatant to remove that will leave 30 μL (the amount of DNA beads mixed with buffer required to proceed).

The table below shows such volumes, continuing the example of an sstDNA library at a concentration of 2000 beads/μL (this is only an example; you must determine the correct volume).

Loading area size	PicoTiter Plate size	Total volume (example) (μL)	Remove (example) (μL)	Remaining (μL)
Large (30×60 mm)	70×75 mm	233 (×2)	203 (×2)	30 (×2)

33. Draw off the volume of the supernatant to be removed and discard it, being careful not to disturb the bead pellet.

34. Transfer the appropriate volume of DNA bead incubation mix, as listed below, into the tube(s) containing the DNA beads, and vortex well. Keep the leftover DNA bead incubation mix for later use.

Loading area size	PicoTiter Plate size	DNA beads (μL)	DNA bead incubation mix (μL)	Total volume (μL)
Large (30×60 mm)	70×75 mm	30 (×2)	870 (×2)	900 (×2)

35. Place the sample(s) on the laboratory rotator, and incubate them for 30 min at room temperature.

Preparation of the Packing Beads

36. Prepare the packing beads.
    i. Vortex the packing beads.
    ii. Add 1 mL of Bead buffer 2 to the tube of packing beads, then vortex until a uniform suspension is achieved.
    iii. Centrifuge at 10,000 rpm (9300 RCF) for 5 min.
    iv. Remove the supernatant carefully without disturbing the pellet (pour out).

37. Repeat Step 36 twice to wash the beads.

38. After the third wash, add the original amount of Bead buffer 2 (550 μL/tube) and vortex to resuspend.

39. With a pipette, aspirate 360 μL of the resuspended packing beads from each tube, and dispense it into separate 2-mL tubes. Add 80 μL of the leftover incubation mix (from Step 34) to each of the tubes.

40. Place the tubes on the rotator for 15–20 min or until ready to use.

Preparation of the Enzyme Beads

41. Vortex enzyme beads and begin by pelleting the beads using a Magnetic Particle Collector (MPC).
    i. Place the tube(s) of enzyme beads in the MPC, and wait 30 sec to pellet them. Invert the MPC several times to wash off any beads that may be lodged inside the cap. Wait 30 sec for the beads to pellet.
    ii. Remove the supernatant, being careful not to bring the pipette tip into contact with the beads.
    iii. Remove the tubes from the MPC.

42. Wash the enzyme beads three times, as follows.

   i. Add 1 mL of Bead buffer 2 to each tube of enzyme beads. Vortex the beads to resuspend them.

   ii. Place the tube(s) in the MPC for 30 sec for the beads to pellet. Invert the MPC several times to wash off beads that may be lodged inside the cap. Wait another 30 sec until all beads are settled.

   iii. Remove the supernatant, being careful not to bring the pipette tip into contact with the beads.

   iv. Remove the tubes from the MPC.

43. After three washes, add the initial amount of Bead buffer 2 (1000 μL/tube), and resuspend the pellet(s) by vortexing. Keep the enzyme bead suspension on ice.

Assembly of the Bead Deposition Device (BDD)

44. Remove the soaked PicoTiter Plate (PTP) from the shipping container, and wipe the back side of the PTP with a KimWipe.

45. Place the PTP onto the Bead Deposition Device (BDD) base. Make sure to align the notched corners of the PTP and the BDD base.

   i. Secure the washed and dried gasket to the BDD base.

   ii. Place the BDD top over the assembled BDD base/PTP/gasket.

   iii. Rotate the two latches from the BDD base into their grooves in the BDD top to firmly secure the assembly.

The Sequencing Run

Wetting of the PicoTiter Plate

46. Fill each loading area with the volume of Bead buffer 2 appropriate for the loading gasket type to be used:

Loading area size	PicoTiter Plate size	Volume to load (μL)
Large (30×60 mm)	70×75 mm	1860 (×2)

47. Place both the assembled BDD and the BDD counterweight into centrifuge swinging baskets, and place the baskets onto the rotor, opposite each other. Check that the microplate carriers are correctly positioned.

48. Centrifuge the PicoTiter Plate in the Bead Deposition Device for 5 min at 2640 rpm (1254 RCF).

   Leave the Bead buffer 2 on the PicoTiter Plate until ready to proceed further.

Deposition of the First Two Bead Layers (DNA and Packing Beads)

49. Return to the BDD, with the wetted PicoTiter Plate. Draw all of Bead buffer 2 back out, and discard it with the pipette tip.

50. Add 960 μL of Buffer 2 to the DNA mix tubes. Vortex the DNA beads mix tubes for 20 sec, and quickly centrifuge down the tubes to draw the liquid down.

51. Pipette up and down three times to resuspend the beads, and load each region of the BDD with 1860 μL of the bead mix.

52. Allow the PicoTiter Plate to sit in the BDD for 10 min to allow the DNA beads to settle into the plate.

53. Retrieve the BDD, and, using a pipette, slowly aspirate out the supernatant from each region and dispense each into separate 2-mL tubes.

54. Centrifuge the tubes at 10,000 rpm (9300 RCF) for 1 min.

55. Using a pipette, aspirate 1460 μL of the supernatant, without disturbing the pellet, out of the tubes and dispense into the 440-μL tubes of packing beads. Vortex to mix.

56. Aspirate 1860 μL out of each tube of the packing beads mix and dispense to each side of the BDD.

57. Centrifuge the BDD for 10 min at 2640 rpm (1430 RCF).

   This is a good time to begin to organize the machine and sequencing reagents (Steps 70–84).

Deposition of the Third Layer (Enzyme Beads)

58. Vortex the washed enzyme beads to achieve a uniform suspension.

59. In two 2-mL tubes, combine enzyme beads and Bead buffer 2 as indicated below. Pipette the beads up and down several times to get a uniform suspension before transferring.

Kit	Enzyme beads (μL)	Bead buffer 2 (μL)	Total volume (μL)
70×75	920 (×2)	980 (×2)	1900 (×2)

60. Vortex the tubes with the mix thoroughly to obtain a uniform suspension.

61. After the centrifugation of the packing bead layer is complete, retrieve the BDD from the centrifuge, and draw out and discard the supernatant from the packing bead layer.

62. Draw the 1860 μL of total diluted enzyme bead suspension (Step 59, last column in table).

63. Pipette up and down three times to resuspend, and load the bead suspension into the BDD and repeat for both regions of the BDD.

64. Once the loading areas are filled, centrifuge the BDD for 10 min at 2640 rpm (1430 RCF), as above.

   Use this 10-min period to finish getting the machine ready (Steps 70–84).

The Sequencing Run: Preparing the Instrument

65. Make sure that the prewash run has completed, then open the exterior fluidics door and raise the sipper manifold completely.

66. Slide out the Reagents cassette, toward the front of the instrument.

67. Remove the prewash tubes and prewash insert from the cassette; empty the tubes and discard them.

68. Wipe all the outside surfaces of the Reagents cassette with a paper towel.

69. Wipe the fluidics area deck with 50% ethanol and a paper towel. Allow the deck to fully air-dry.

Preparing and Loading the Sequencing Reagents Cassette

70. Place the four bottles of CB buffer into the Reagents cassette. Remove the lids, and add 1 mL of DTT to each bottle of CB buffer; replace the caps and gently swirl the bottles to mix.

71. Retrieve the Reagent Insert from 4°C storage, and place it into the Reagent cassette with the Apyrase Buffer tube (yellow sticker top) in the open end of the cassette (this end will go into the back of the fluidics area in the machine).

72. Add 164 μL of Apyrase to the tube of Apyrase buffer and swirl to mix.

73. Add 1.5 mL of dATP to the bottle of dATP buffer (purple sticker top). Swirl the bottle to mix.

74. Remove the caps from all of the tubes, and load the Reagents cassette into the instrument with the large bottles to the right and the tubes to the left. Lower the manifold carefully, and close the exterior fluidics door.

Loading the PicoTiter Plate

75. In the software, select **Unlock Camera Door.** Fluid will be pumped to waste, and then the door will unlock; now open the door.

> *If the PicoTiter Plate cartridge is not completely emptied of liquid after opening the camera cover, close the camera cover, wait 5 sec, and reopen it to remove any remaining liquid (install the Camera Faceplate Guard on the camera face by hanging it from the metal faceplate seating pins). Repeat as necessary until there is no more fluid in the cassette.*

76. Remove the used PicoTiter Plate from the PicoTiter Plate cartridge by pressing the PicoTiter Plate frame spring latch to lift the frame from the cartridge, and then sliding out the used Pico-Titer Plate. Clean the cartridge as follows.

    i. Remove the gasket seal from the cartridge and discard it.

    ii. Squirt 50% ethanol onto a KimWipe and wipe the cartridge. Allow it to air-dry fully.

    iii. Wet a KimWipe with 10% Tween solution, and apply it to the cartridge after it has dried.

    iv. Wipe the camera faceplate with a Zeiss moistened lens tissue, and allow it to air-dry.

Setting Up a Run on the Computer

77. When the Prep Run is completed, a small "Run Complete" window will open. Click OK to continue.

78. Do the following in the open window.

    i. Select **Start.**

    ii. Choose **Custom Sequencing Run.**

    iii. Select **Next.**

    iv. In the PTP ID box, input the six-digit number from the back of the PicoTiter Plate (from Step 23.iii) into the ID box, and select **Next.**

79. From the two available boxes, select **Application 1.**

80. Open a terminal, and type (your Linux login) **ssh@linus170**, then press [Enter].

81. Type your Linux password, then press [Enter].

82. Type **454Loader**, then press [Enter]. An interface will open into which you will enter the appropriate information.

83. Toward the bottom of the 454Loader interface, locate the "DRI Name" box into which you will enter the DRI name from the *Resource Item Order Sheet* provided to you with the order.

    i. Click on the "Add To Filled Region List" button. Verify that the information displayed in the window is correct, and click that button again.

    ii. The information should display in the window again; verify it, and then click on the "Click Here to Generate Run PSE ID To Copy and Paste" button.

    iii. This action generates an ID number; highlight, copy, and then paste this number into the "454 Run Name" box in the "Instrument Procedure" window.

84. To finish setting up the run:

    i. Click Next and select the "LR70" box.

    ii. Click Next and select **2 Region.**

    iii. Click Next and select **100 Cycles.**

    iv. Click Next and select the appropriate analysis state.

    v. Click Next, then Next again.

    The software is now ready.

Installing the PTP into the Machine and Starting the Run

85. After centrifugation of the third bead layer is complete (Step 64), remove the BDD from the centrifuge.

86. Gently draw out and discard the supernatant from the centrifuged second bead layer.

87. Remove the PicoTiter Plate from the Bead Deposition Device as follows.

    i. Rotate down the latches of the BDD.

    ii. Carefully remove the BDD top.

    iii. Gently lift off the gasket.

    iv. Remove the PicoTiter Plate, being careful to handle it only by the edges.

88. Install the PTP into the cartridge.

89. Install the cartridge seal as described below:

    i. Verify that the square ridge on the seal is facing up, and drop the seal into the cartridge groove.

    ii. If necessary, gently tap the seal into place with a gloved hand. *Do not* wipe the seal with anything.

    iii. Lock the PTP down into the cartridge.

    iv. Click on the Finish button at the bottom of the Requirements window to exit it and start the run.

90. Clean up the preparation area.

    i. Dispose of the Versi-Dry pads lining the bench top (they may be discarded in the wastebasket).

    ii. Cleanse the bench top using the following products:
    - 10% bleach
    - DNA Away
    - Texwipes (alcohol wipes)

    iii. Dispose of used paper towels and wipes in the wastebasket.

Validation

Next-generation sequencing technologies have enabled entire new fields of research, but it is important to realize that these technologies are, at their core, approaches that generate experimental data that are subject to error. When performed carefully using best practices for library construction and appropriate coverage levels, the mature NGS platforms (454, Illumina, and SOLiD) all have a very low error rate (approaching or exceeding one error per 10 million bases). However, sequencing errors can be compounded by analytical errors such as read mismapping to a reference genome, base substitutions introduced during amplification, and failure to remove duplicate reads containing the same PCR substitutions. These problems may be magnified during the course of a large-scale experiment and may result in a large number of errors in the final DNA sequence. It is vital that researchers using these technologies (1) appreciate these errors, (2) understand what may cause them, and, most importantly, (3) develop a filtering strategy that removes known sources of error.

It is also critically important to validate the results of next-generation sequencing. Two types of validation that should be undertaken in association with next-generation sequencing are statistical validation and confirmation of the results of particular interest and importance.

Statistical Validation

By "statistical validation," we mean the random selection of variants and their testing either by repetition with similar chemistry or by sequencing with a different chemistry (see capillary sequencing in Protocols 2 and 3). There may come a time when the use of orthogonal chemistry to validate variants is not necessary, but now it seems to be the current community standard. As was the case with capillary sequencing some years ago, researchers preferred to validate their results, and, in consequence, the sequencing community developed a clear understanding of the errors inherent in capillary sequencing.

Validation by capillary sequencing provides a rigorous test of next-generation sequencing data; however, it is often impractical in terms of cost and throughput. Statistical validation of focused data sets (e.g., from targeted capture) can be performed by sequencing a reasonable number of randomly selected variants with several nonvariant regions included to test the false negative rate. These sequencing results can be validated using a combination of targeted PCRs and capillary sequencing. Larger data sets (exomes or whole genomes) may be validated by large-scale PCR or by custom solution-phase capture and read out using Roche/454, Ion Torrent, Pacific Biosciences CCS, or MiSeq platforms.

Another important type of statistical validation is the cross-correlation between the genotypes of polymorphisms identified by sequencing and genotypes derived from arrays. As mentioned above, in whole-genome and exome capture sequencing, it is advisable to run an SNP array on the same samples for validation. The correlation between array and sequencing data (1) gives a good indication of the false negatives as well as the depth of sequence coverage, and (2), as an added benefit, potentially identifies any sample mix-ups. Array results, however, do not identify false positives. Therefore, to estimate the number of false positives, a selection of random variants identified by next-generation sequencing should be confirmed by directed PCR and capillary sequencing, or for larger numbers of variants, by solution-phase capture with a readout on the appropriate scale platform. (The choice is determined by the number of validation loci. See below.)

However, as discussed in the Introduction to this chapter, sequencing technology is changing. Of particular interest is the advent of new methods for targeting small regions of the genome and the development of new personal genome sequencers, such as the Ion Torrent and the MiSeq. These instruments may allow rapid and high-quality assessment of variants and eliminate the need for validation by capillary sequencing. Given the potential value of orthogonal technology, this could mean that one variant set discovered with HiSeq instrumentation would be validated by, for example, Ion Torrent technology.

Another type of validation is confirmation of particular results of interest. The extra effort to validate variants that appear to be biologically interesting are well worth the time and expense in order to avoid the problems of performing extensive experimental follow-up on what turns out to be a false-positive variant.

Biological indicators can also be used to get a better idea of the error rates in a given project. For instance, when sequencing a large number of individuals, the likelihood of a variant that is present in a large percentage of those, say 5% or more, being a false positive is significantly less likely than the chance that a very rare variant is a false positive. That is not to say that rare variants are not real, but a higher percentage of them will be caused by sequencing errors than by common variants in the test population. Similarly, if one is looking at an extended family, the presence of non-Mendelian events may be due to error, or at least a genomic region whose sequence should be validated with a different technology.

In conclusion, next-generation data and new analytical approaches to mine biological discoveries from these data can often combine to produce errors. Investigators must be cognizant of and understand the errors in their particular system, as well as the evolving community standards for validation of variants detected by next-generation sequencing. Because these technologies are very new, it is likely that the community conventions will continue to evolve—at least in the near term, as we better understand the types of errors generated by different analytical approaches.

Quality Assessment of Sequence Data

Next-generation sequencing instruments produce massive amounts of sequence data. It is therefore critical to monitor the quality of each run to ensure that both the instrument and the sequencing chemistry are performing optimally. Quality assessment is quite similar across the major next-generation sequencing platforms. All platforms provide per base quality values that largely evaluate signal-to-noise ratios. In addition, basic read metrics such as read length, total yield, or platform-specific filtering are provided per run. Alignment of reads from standard libraries or control samples to known reference sequences can provide information on error rates and overall success. Real-time metrics are often provided on screen to highlight potential issues with the reagent or instrument performance as the sequencing progresses. The user's guide for each platform typically provides all of the details necessary to interpret metrics, troubleshoot problems, and maintain high quality. Often the true test of the data quality occurs in the post–processing analysis steps that look at overall unique sample representation, adaptor contamination, GC bias, library insert size, read duplication, alignment quality, and base call confidence. We provide here examples of quality assessment specific to the Illumina platform.

Sequencing a control sample can be extremely useful for quality control and troubleshooting. The most common control assay is to use a "spike-in" of the Illumina PhiX Control Library. A small amount (typically 1%–3%) of the PhiX library is added to each lane along with the primary sequencing sample. The Illumina Real Time Analysis (RTA v1.10) software performs on the fly base-calling, quality score assignment, and alignment of the PhiX reads to the reference genome. The summary report of the run can either be viewed on the instrument computer, or metrics can be parsed from the report files as they are transferred to a secondary server. The performance of the control sample can be used to assess instrument function and reagent quality. For a paired-end 100-cycle sequencing run, the PhiX error rate after alignment to the reference should be <1% on each read (shorter read lengths should generate fewer errors). Although a slightly higher error rate is common on read 2, the error rates should not differ greatly if the instrument is performing properly.

The RTA and CASAVA software also provide summary reports, cluster density plots, signal intensity plots, quality score plots, and error rate plots that provide a range of informative quality metrics (see the *CASAVA v1.8 Users Guide*, Chapter 2: "Interpretation of Run Quality"). A variety of graphical reports can be viewed while the run progresses. Reports are available by read, by cycle, by lane, and by tile across the flow cell. Cluster density plots can indicate whether a sample was overloaded or underloaded, or may indicate cluster generation failure. Clusters packed together too densely will cause cluster mapping problems and will fail the Illumina signal-to-noise quality filter, whereas sparse clusters will lead to poor yield. Monitoring signal intensity and focus quality by cycle across the flow cell allows you to determine if there is sufficiently even representation of the four dyes and to detect signal loss that could be the result of blocked reagent flow, camera or laser malfunction, bubbles, or other surface obstructions. Base-call quality score plots are often an early indicator of run performance. A slight drop in base quality toward the 3′ end of the reads is common, but generally quality scores of ~30 or better are desirable over the length of the read. Similarly, if reads are aligned to a reference, tracking error rates by cycle can be useful in determining whether 3′-read trimming will be necessary for postprocessing. The Summary.htm file contains a broad overview of run statistics, per lane or per sample, including cluster density, percentage of clusters passing the Illumina quality filter, percentage of bases at Q30 or better, alignment and error statistics, signal intensity, and phasing/prephasing rates. The percentage of reads passing filter tends to drop with very high cluster density, but generally the pass filter rate for lanes loaded in the recommended density range should be ~80%. Similarly a high percentage (~80%) of Q30 bases should be obtained, a large percentage of the reads (~80%) should map to the reference in the expected orientation, and error rates should be low. (Note that polymorphisms and structural variants will affect the alignment and mismatch rate.) Phasing and prephasing values should be below 1% on both reads. High values may indicate a reagent chemistry issue (flow problems, reagent quality, etc.). Distinct differences between read 1 and read 2 statistics may indicate a paired-end module or resynthesis chemistry malfunction. Retaining "thumbnails" of a subset of images captured during the run is quite valuable. If a drop in data quality is noted in any particular lane, tile, or cycle, manual inspection of the images can often pinpoint or confirm the issue. The *CASAVA v1.8 Users Guide* contains examples of the QC plots and run reports, clarification of the parameters, and hints on generally accepted standards for each of the metrics.

FASTQ files (using Sanger quality score encoding) are generated using CASAVA v1.8 (see the *CASAVA v1.8 Users Guide*, Chapter 3: "BCL Conversion"). FASTQ files are a commonly accepted input file format for many downstream analytical tools for assembly, alignment, and variant calling. Basic Sanger FASTQ is a compact file format consisting of four line records. There is a header line followed by the read sequence, then an additional header line followed by base quality scores (scores are encoded in ASCII plus 33 offset). The .bcl files produced by the Real Time Analysis (RTA v1.10) software on the instrument contain base-call and quality score information in binary format. The .bcl files are transferred to a secondary server for conversion to FASTQ format. If multiple samples are sequenced in each lane using the Illumina 3 read indexing method, samples can be demultiplexed using the CASAVA v1.8 software. Demultiplexing allows the sample files to be sorted and binned by index and provides summary information for each specific index (number of reads per index, quality scores per index, etc.).

For the most up-to-date information on the Illumina analysis pipeline and tools to inspect run quality, refer to the Illumina *CASAVA v1.8 Users Guide*, the *RTA v1.10/ HCS v1.3* guides for HiSeq instruments, and the *RTA v1.9/SCS v2.9* guides for GAIIx instruments.

Protocol 20

Data Analysis

VARIANT DISCOVERY ANALYSIS

The essential first step in discovery of variants is the alignment of next-generation sequencing reads with the reference genome for the organism of interest. There are a multitude of alignment algorithms available either from the instrument manufacturing companies, through SourceForge or other open source sites, or by communicating with one's favorite bioinformaticist. Currently, the field favors the use of Burroughs-Wheeler algorithm aligners like *bwa*. Most alignment algorithms produce a mapping quality metric that should be considered by downstream variant detection algorithms because incorrect mapping to the reference (indicated by a low mapping quality) can be a significant source of false-positive variant detection. As mentioned in the Introduction to this chapter, detection of variants requires different algorithmic approaches, the numbers and types of which are determined by the types of variant one wishes to discover. Several of these approaches are covered in Chapter 8.

DE NOVO ASSEMBLY

Another commonly used application for next-generation sequencing data is assembly of genomes without using a reference (so-called "unguided" assembly). Short-read assemblies can consist of one or many read types/platforms and can combine paired-end and mate-pair reads of different distances to provide the maximum potential for contiguity. Viral and bacterial genomes can be assembled with relative ease and completeness, but large mammalian genomes are too complex and have too many repetitive sequences to assemble fully or even partially, as complete chromosomes.

BIOTIN

Biotin (vitamin H, coenzyme R; FW = 244.31) (for the chemical structure, see Fig. 1) is a water-soluble vitamin that binds with high affinity to avidin, a tetrameric basic glycoprotein, abundant in raw egg white (for review, see Green 1975). Because each subunit of avidin can bind one biotin molecule, 1 mg of avidin can bind $\sim$14.8 μg of biotin. The dissociation constant of the complex is $\sim$1.0$\times$10^{-15} M, which corresponds to a free energy of association of 21 kcal/mol. With such a tight association, the off-rate is extremely slow and the half-life of the complex is 200 d at pH 7.0 (Green and Toms 1973). For all practical purposes, therefore, the interaction between avidin and biotin is essentially irreversible. In addition, the avidin–biotin complex is resistant to chaotropic agents (3 M guanidine hydrochloride) and to extremes of pH and ionic strength (Green and Toms 1972).

Biotin can be attached to a variety of proteins and nucleic acids, often without altering their properties. Similarly, avidin (or streptavidin, its nonglycosylated prokaryotic equivalent) can be joined to reporter enzymes whose activity can be used to locate and/or quantitate avidin–biotin–target complexes. For example, in enzyme immunoassays, a biotinylated antibody bound to an immobilized antigen or primary antibody is often assayed by an enzyme, such as horseradish peroxidase or alkaline phosphatase, that has been coupled to avidin (Young et al. 1985; French et al. 1986). In addition, in nucleic acid hybridization, biotinylated probes can be detected by avidin-conjugated enzymes or fluorochromes. Derivatives of biotin are used to biotinylate proteins, peptides, and other molecules (for review and references, see Wilchek and Bayer 1990). These derivatives include:

- various N-hydroxysuccinimide esters of biotin, which react with free amino groups of proteins or peptides to form amides (N-hydroxysuccinimide is released as a by-product)

- photoreactivatable biotin (photobiotin), which upon activation with a mercury vapor lamp (350 nm) reacts via an aryl nitrene intermediate and binds indiscriminantly to proteins, oligosaccharides, lipids, and nucleic acids

- iodoacetyl biotin, which reacts specifically with thiol groups, generating a stable thioether bond

- biotin hydrazide, which reacts with aldehydes generated by mild oxidation of carbohydrates

- derivatives of biotin equipped with extended spacer arms (e.g., biotinyl-ε-aminocaproyl-N-hydroxysuccinimide ester), which improve the interaction between avidin and biotinylated macromolecules

The strength of the interaction between biotin and avidin provides a bridging system to bring molecules with no natural affinity for one another into close contact.

Biotinylation of Nucleic Acids

Biotinylated nucleotides, which are available commercially, are effective substrates for a variety of polymerases including the DNA polymerases of *E. coli*, *Thermus aquaticus*, bacteriophage T4, and the RNA polymerases of bacteriophages T3 and T7. Biotinylated nucleic acids can therefore be generated in vitro by almost all of the procedures (nick translation, end-filling PCR, random priming, and transcription) that are used to generate radiolabeled probes. In addition, biotin adducts can be introduced into nucleic acids and oligonucleotides by several chemical methods, including attachment to a 5'-amine group. This group is added during the final step of conventional cyanoethyl phosphoroamidite synthesis of

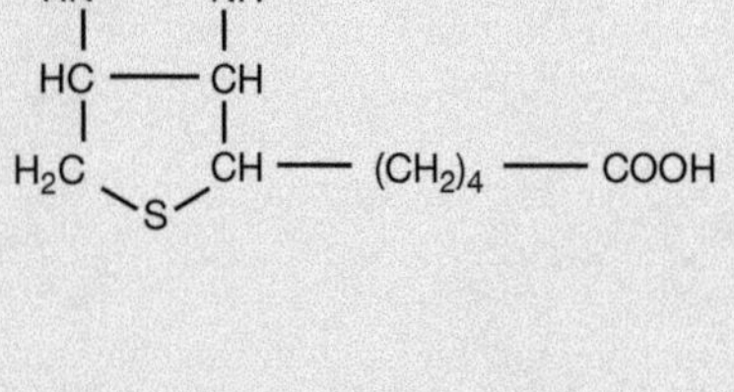

FIGURE 1. The structure of biotin.

oligonucleotides. The amino-oligonucleotide can then be labeled with commercially available biotinyl-*N*-succinimide ester.

As long as their content of biotinylated nucleotide does not exceed a few percent, the resulting probes hybridize to target sequences at approximately the same rate as unsubstituted probes. Incorporation of additional biotin residues is unlikely to increase the sensitivity of detection given that a typical avidin-linked reporter molecule (e.g., avidin–alkaline phosphatase) bound to a single biotin residue will cover between 50 and 100 nucleotides of nucleic acid.

The efficiency of detection of biotinylated probes by avidin-linked reporter systems is greatly improved if the length of the linker between the biotin moiety and the nucleotide is sufficient to overcome steric hindrance and to allow the biotinyl group to penetrate effectively into the binding sites of avidin. Several companies sell biotinylated nucleotides with linkers containing six or more carbon atoms.

Biotin can also be introduced into nucleic acids simply by mixing photoactivatable biotin with double- or single-stranded DNA or RNA and then irradiating the mixture with visible light (350 nm) (Forster et al. 1985). The usual protocol yields a nucleic acid that contains an average of one biotin residue per 200 or so residues. Modification at this modest level certainly does not interfere with the ability of the probe to hybridize to its target and yet is sufficient to allow detection of single-copy sequences in Southern hybridization of mammalian DNA (McInnes and Symons 1989). The following are advantages of biotinylated probes.

- They can be stored for long periods of time without loss of activity.

- They need no special method of disposal.

- Signals from a biotinylated probe can be detected with a variety of avidin-reporter molecules, including those that can be detected by chemiluminescence and fluorescence. This means that a single probe can be used for a variety of different purposes (e.g., Southern blotting and in situ hybridization, selection, and purification).

- The biotinylated nucleic acid can be recovered by affinity purification on avidin columns or on avidin-coated magnetic beads.

Note that the incorporation of bulky biotin groups reduces the electrophoretic mobility of nucleic acids. For example, each addition of biotin-14-dCTP is equivalent to increasing the mass of the nucleic acid by 1.75 cytosine residues (Mertz et al. 1994).

REFERENCES

Forster AC, McInnes JL, Skingle DC, Symons RH. 1985. Non-radioactive hybridization probes prepared by the chemical labelling of DNA and RNA with a novel reagent, photobiotin. *Nucleic Acids Res* **13:** 745–761.

French BT, Maul HM, Maul GG. 1986. Screening cDNA expression libraries with monoclonal and polyclonal antibodies using an amplified biotin-avidin-peroxidase technique. *Anal Biochem* **156:** 417–423.

Green NM. 1975. Avidin. *Adv Protein Chem* **29:** 85–133.

Green NM, Toms EJ. 1972. The dissociation of avidin–biotin complexes by guanidinium chloride. *Biochem J* **130:** 707–711.

Green NM, Toms EJ. 1973. The properties of subunits of avidin coupled to Sepharose. *Biochem J* **133:** 687–700.

McInnes JL, Symons RH. 1989. Preparation and detection of nonradioactive nucleic acid and oligonucleotide probes. In *Nucleic acid probes* (ed. Symons RH), pp. 33–80. CRC Press, Boca Raton, FL.

Mertz LM, Westfall B, Rashtian A. 1994. PCR nonradioactive labeling system for synthesis of biotinylated probes. *Focus* **16:** 49–51.

Wilchek M, Bayer AE. 1990. Biotin-containing reagents. *Methods Enzymol* **184:** 123–138.

Young RA, Bloom BR, Grosskinsky CM, Ivanyi J, Thomas D, Davis RW. 1985. Dissection of *Mycobacterium tuberculosis* antigens using recombinant DNA. *Proc Natl Acad Sci* **82:** 2583–2587.

MAGNETIC BEADS

Since their introduction to molecular cloning in 1989 (Hultman et al. 1989), magnetic beads (also known as microspheres) have been used for a variety of purposes, including purification and sequencing of PCR products, construction of subtractive probes and cDNA libraries, affinity purification of DNA-binding proteins, rescue of shuttle vectors from transfected cells, and hybridization of covalently attached oligonucleotides. Although such tasks can be accomplished by more traditional techniques, these older methods are almost invariably more tedious and less efficient. Among the advantages offered by magnetic

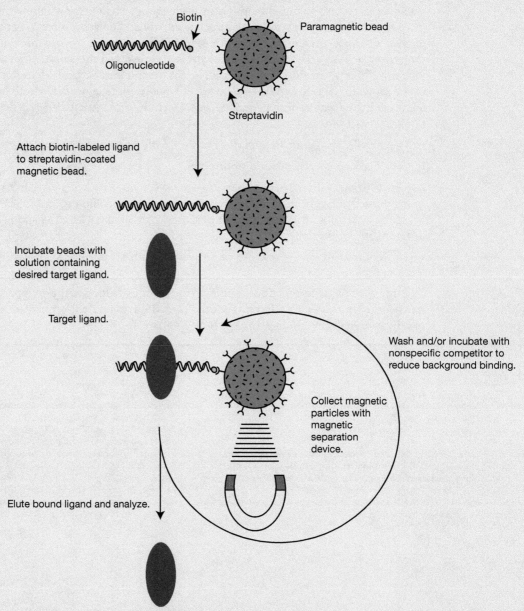

FIGURE 1. **Affinity selection and purification.** A generic approach is illustrated here for the capture of a specific target ligand, using as an example a DNA-binding protein. The biotin-labeled ligand, in this case, an oligonucleotide encoding the putative binding site for the target protein, is attached to streptavidin-coated magnetic beads and incubated in a solution containing the desired target protein. The bound particles are collected with a magnetic separation device and washed repeatedly with a solution containing a nonspecific competitor. Finally, the bound target ligand is eluted from the oligonucleotide and analyzed.

beads are speed of operation and the possibility of working at kinetic rates close to those occurring in free solution. Binding of ligand takes only a few minutes, magnetic separation takes seconds, and washing or elution can be completed in <15 min in most cases.

Magnetic beads are nonporous, monodisperse, superparamagnetic particles of polystyrene and divinyl benzene with a magnetite core ($8 \pm 2 \times 10^{-3}$ cgs units) and a diameter of ~2.8 μm. Different types of beads carry different active groups on their surfaces (OH, NH_2, OH[NH_2], COOH, etc.), which can be used for covalent attachment of protein and nucleic acid ligands (e.g., see Lund et al. 1988). However, most investigators prefer to purchase magnetic beads that are preloaded with covalently bound streptavidin (Lea et al. 1988), which can then be used to tether any biotin-labeled nucleic acid or protein to the surface of the bead (for review, see Haukanes and Kvam 1993). Note that the binding capacity varies with bead size, bead composition, and the size of the binding ligand. Different brands of beads carry different amounts of covalently attached streptavidin so that the binding capacities of beads supplied by different manufacturers are not necessarily equivalent. In addition, some brands of beads can be reused, whereas others must be discarded after a single use. Once attached, the tethered ligand can be used for affinity capture and purification of target molecules from solution. Because the surface area of the particles is very large ($5-8$ m^2/g), streptavidin-coated beads have a very high biotin-binding capacity (>200 pmol/mg). Furthermore, because of the high affinity of streptavidin for biotin ($K_{ass} = 10^{15}$ M^{-1}) (Wilcheck and Bayer 1988), complexes between biotin and streptavidin form very rapidly and once formed are resistant to extremes of pH, organic solvents, and many denaturing agents (Green 1975). Stringent washing does not cause leaching of tethered ligand from the surface of the bead, and enrichment factors of 100,000 can be attained during a single round of affinity capture. Perhaps the major disadvantages of magnetic beads are (1) their high cost and (2) the need for an efficient magnetic particle separator. The best of these devices contain one or more neodymium–iron–boron permanent magnets and are available in a number of formats from several commercial suppliers.

Figure 1 illustrates a generic procedure for affinity capture of a specific DNA-binding protein by a biotin-labeled oligonucleotide ligand tethered to magnetic beads.

REFERENCES

Green NM. 1975. Avidin. *Adv Protein Chem* 29: 85–133.

Haukanes B-I, Kvam C. 1993. Application of magnetic beads in biosassays. *BioTechnology* 11: 60–63.

Hultman T, Stahl S, Hornes E, Uhlén M. 1989. Direct solid phase sequencing of genomic and plasmid DNA using magnetic beads as solid support. *Nucleic Acids Res* 17: 4937–4946.

Lea T, Vartdal F, Nustad K, Funderud S, Berge A, Ellingsen T, Schmid R, Stenstad P, Ugelstad J. 1988. Monosized, magnetic polymer particles: Their use in separation of cells and subcellular components, and in the study of lymphocyte function in vitro. *J Mol Recognit* 1: 9–18.

Lund V, Schmid R, Rickwood D, Hornes E. 1988. Assessment of methods for covalent binding of nucleic acids to magnetic beads, Dynabeads, and the characteristics of the bound nucleic acids in hybridization reactions. *Nucleic Acids Res* 16: 10861–10880.

Wilcheck M, Bayer AE. 1988. The avidin–biotin complex in bioanalytical applications. *Anal Biochem* 171: 1–32.

FRAGMENTING OF DNA

Large-scale whole-genome sequencing projects rely on strategies for random fragmentation of DNA, with the goal of creating a library of overlapping clones that provide sequence redundancy over the entire genome. DNA fragmentation, typically achieved either by enzymatic digestion or by hydrodynamic shearing, can be precisely controlled to generate products of the desired size and that are compatible with the standard sample preparation techniques described in this chapter.

Commercially available enzymatic shearing approaches that produce products suitable for use in standard sample preparation include the Nextera system (Epicentre, now part of Illumina) and NEBNext dsDNA Fragmentase (New England Biolabs). The Nextera system relies on a mutant version of the Tn5 transposase, which fragments DNA and adds sequencing adaptors in a single step. The NEBNext system is composed of two enzymes that work in concert to produce double-strand breaks in DNA. Here, one enzyme generates random nicks in one strand of the DNA sample, and the other enzyme recognizes the nicked site and cuts the DNA strand opposite the nick, to produce a double-strand break.

Hydrodynamic forces are generated in several different ways—typically by sonication or nebulization. In each case, breakage of DNA results from drag forces that first stretch and then break the phosphodiester backbone in the central regions of double-stranded DNA molecules. A series of instruments now commercially available provides automated approaches to fragmenting DNA by sonication under precisely controlled conditions. Covaris, for example, provides an instrument that makes use of high-frequency acoustic energy to shear DNA. This automated sonication device, using Adaptive Focused Acoustic (AFA) technology, is based on shock wave physics and delivers precisely controlled bursts of energy for fragmenting a DNA sample to the desired size range.

For some applications, rather than shearing, it is desirable to cleave DNA with restriction endonucleases or other specific cutters. In other cases, for example, samples derived from formalin-fixed, paraffin-embedded tissue sections or from chromatin immunoprecipitation, and for ancient or degraded samples, shearing is unnecessary because the starting material is already sufficiently short.

CHAPTER 12

Analysis of DNA Methylation in Mammalian Cells

Paul M. Lizardi, Qin Yan, and Narendra Wajapeyee

Yale University School of Medicine, Department of Pathology, New Haven, Connecticut 06520-8005

INTRODUCTION

PROTOCOLS

INFORMATION PANELS

METHYLATION OF DNA, THE MOST EXPERIMENTALLY ACCESSIBLE epigenetic alteration of eukaryotic cells, has generated an extensive literature and an abundance of analytical tools. The term "methylome" (referring to the complete set of cytosine modifications in a genome) is appearing with greater frequency in the literature, reflecting the growing number of researchers in the field. Some readers of this chapter may feel, however, that use of the term "methylome" is perhaps premature, considering that the DNA methylation tools in current use sample only a small fraction of the epigenetic complexity of cell populations. Although there is considerable room for improvement in DNA methylation analysis technology, this chapter contains robust protocols for methods that can be performed routinely for the elucidation of DNA chemical modifications involving methylation of cytosine (Fig. 1). The strengths and limitations of each approach are also discussed. For an

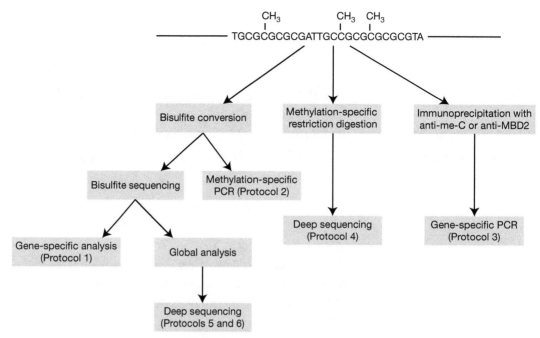

FIGURE 1. Methods for analyzing DNA methylation profiles in genomes.

overview of DNA methylation in mammals, see two concise reviews of the field: Bird (2002) and Klose and Bird (2006). For an overview of the technical issues that arise during DNA methylation analysis, see Laird (2010).

DNA METHYLATION AFFECTS AND REVEALS BIOLOGICAL PHENOMENA

In adult mammals, ~40% of CpG dinucleotides in the genome contain methylation marks at the 5' position of cytosine. A large proportion of methylation marks are erased during germ cell development, and an additional wave of demethylation of the incoming paternal genome occurs in zygotes. The reestablishment of DNA methylation marks begins early in embryonic development and is largely complete at the time of birth. During differentiation, distinct subsets of methylation marks are established and become heritable in different somatic cell lineages, constituting a stable memory of developmental commitment during cell division. Among the most dramatic manifestations of this memory are X-chromosome inactivation and genomic imprinting, both of which are associated with stable, chromatid-specific silencing of gene loci. Somatically heritable DNA-methylation-mediated silencing also occurs in vast regions of the genome containing interspersed repetitive elements. In promoter-associated clusters of methylated cytosines, known as "CpG islands," DNA methylation stabilizes condensed chromatin states initiated by binding of Polycomb group protein complexes. Histone modifications occurring at these sites recruit DNA methyltransferases, and, acting together, these interacting proteins can propagate multiple methylcytosine marks over an entire CpG island. These chemical changes affecting the DNA sequences near each promoter represent a layer of epigenetic control of a specific chromatin state that is somatically heritable, but nonetheless remains susceptible to somatic reprogramming. Loss of subsets of specific methylation marks occurs during normal development, as well as during inflammatory or pathological processes. DNA methylation can be modulated in the absence of cell division. Notable examples are a region in the promoter–enhancer of the interleukin-2 gene, which is demethylated in T lymphocytes following activation (Bruniquel and Schwartz 2003; Murayama et al. 2006), as well as demethylation and transcriptional activation of the neuronal plasticity gene *reelin* during contextual fear conditioning in the hippocampus (Miller and Sweatt 2007). The latter example emphasizes the fact that DNA methylation marks are dynamic and highly responsive to environmental factors.

In summary, DNA methylation can be regarded as a metastable digital record of interactions between the genome and its environment that is stored in the form of a binary string of "zeros" (unmethylated bases) and "ones" (methylated bases). In the context of a rich diversity of responses to biological phenomena, the analysis of DNA methylation in mammalian cells can reveal aspects of developmental lineage history overlaid with changes induced by environmental influences as a cell- and tissue-specific record of an organism's life history. Thus, measurement of methylation marks at the single base level can reveal developmental events, trace normal or abnormal cell lineages, document accidental environmental insults, and generate metrics of cellular aging (Shibata and Tavare 2006). However, genomic methylation is neither perfect nor static. Superimposed on acquired patterns of DNA methylation is noise generated as a result of the imperfect propagation of DNA methylation marks during semiconservative DNA replication. Occasional failures in the replication of the marks occur in dividing cells, and these stochastic variations constitute a biological clock recorded within cell lineages as binary strings in each cell's DNA.

The Chemistry of DNA Methylation and the Bisulfite Reaction

The most common DNA methylation marks involve the addition of a 5′-methyl group to a cytosine that is part of a CpG dinucleotide, denoted as mCG. Less common, but widespread in stem cell progenitors, is 5′-methylation of cytosines in other contexts, denoted as mCHG and mCHH (where H is A, C, or T) (Lister et al. 2009). In addition, 5-hydroxymethylation of cytosine (hmCG) has been reported to occur frequently in neurons (Kriaucionis and Heintz 2009).

The "gold standard" for analysis of cytosine methylation at the sequence level is based on the modification of DNA with sodium bisulfite. This reaction exploits differences in the kinetics of deamination of cytosine and methylcytosine: Under the conditions used, cytosine is deaminated and converted to uracil more rapidly than is methylcytosine. When the deaminated DNA is amplified using the polymerase chain reaction (PCR), the newly created uracil residues in the template DNA then direct incorporation of adenosine in the synthesized strand. At the end of the PCR, the amplified DNA will contain a thymidine residue wherever unmethylated cytosine had been present, whereas cytosine will remain unchanged at those positions where the base was methylated in the original DNA sample. Unfortunately, because 5-hydroxymethylcytosine shows deamination kinetics similar to those of 5-methylcytosine in the presence of bisulfite, the method is unable to discriminate between the two different chemical modifications of the base. Huang et al. (2010) have investigated the kinetics of DNA modification by sodium bisulfite for the three possible states of cytosine. They report that the rate of deamination of 5-methylcytosine is about two orders of magnitude slower than the rate of deamination of cytosine. On the other hand, 5-hydroxymethylcytosine is very rapidly converted to cytosine-5-methylenesulfonate (CMS), and this adduct does not readily undergo deamination. The formation of CMS adducts could be particularly problematic when 5-hydroxymethylcytosine is present in a CC dinucleotide. The persistence of CMS adducts causes reduced PCR amplification yields due to *Taq* polymerase stalling, a result that underscores the limitations of bisulfite chemistry methods, as currently practiced, in the analysis of 5hmC base modifications.

EXPERIMENTAL APPROACHES FOR ANALYSIS OF DNA METHYLATION

In general, analysis based on DNA sequencing will provide superior resolution of modified bases; but in some instances, non-sequencing-based methods may be more cost effective. A general overview of available experimental approaches for analysis of DNA methylation is presented in Table 1.

Bisulfite Sequencing for Single-Base Resolution of DNA Methylation

The method most frequently used for analysis of cytosine methylation at the sequence level is based on modification of DNA with sodium bisulfite (Protocol 1). This approach has the advantage that

TABLE 1. Overview of methods for DNA methylation analysis

Methylation-dependent treatment method	Detection method	Type of analysis	Comments	Cost and time	Protocols
Bisulfite conversion[a]	Sanger sequencing of clones	Locus specific	Single-base resolution	Inexpensive, time consuming and laborious	Protocol 1
	Sanger sequencing of PCR products	Locus specific	Semiquantitative evaluation of average level of DNA methylation; not very sensitive	Inexpensive, quick	
	Methylation-specific PCR	Locus specific	Semiquantitative	Inexpensive, quick	Protocol 2
	Hairpin-bisulfite sequencing	Locus specific	Determination of methylation on both strands	Inexpensive, quick	
	Methylation microarray	Multiple loci	Medium to high multiplexing, single-base resolution, potential hybridization bias	Expensive; analysis is time consuming	
	Direct pyrosequencing	Multiple loci	Medium multiplexing, high-throughput single-base resolution	Expensive, quick	Protocol 5
	Next-generation sequencing after targeted capture	Multiple loci		Expensive; analysis is time consuming	
	Next-generation sequencing of reduced redundancy	Genome-wide	Genome-wide coverage, single-base resolution at high number of locations	Expensive, analysis is time consuming	Protocol 6
	Next-generation sequencing of whole genome	Genome-wide	Whole genome sequencing, single-nucleotide resolution, expensive		Protocol 6
	Mass spectrometry	Multiple loci	High multiplexing	Inexpensive, quick	
Affinity enrichment[b]	Quantitative PCR	Locus specific		Inexpensive, quick	Protocol 3
	Microarray	Genome-wide		Expensive; analysis is time consuming	
	Next-generation sequencing	Genome-wide		Expensive; analysis is time consuming	
Methylation-sensitive enzyme digestion[c]	Gel-based methods	Genome-wide	Low resolution	Inexpensive, quick	
	Microarray	Multiple loci or whole genome	Subject to hybridization artifact	Expensive; analysis is time consuming	
	Next-generation sequencing	Genome-wide		Expensive; analysis is time consuming	Protocol 4

[a]Compatible with ssDNA, accurate and reproducible, provides single-base resolution, could result in significant degradation of sample.
[b]Compatible with ssDNA (MeC IP) or dsDNA (MBD), requires a large amount of DNA, relatively low resolution, cannot distinguish 5mC and 5hmC.
[c]Requires dsDNA of high quality, prone to false positives caused by incomplete digestion.

every candidate cytosine can, in principle, be interrogated for its methylation status. A disadvantage is that bisulfite treatment always results in fragmentation of the DNA sample, limiting the analysis of sequence reads in a continuous strand of DNA to 800 bases, at best.

Targeted bisulfite sequencing typically involves a limited number of loci of interest, which are amplified using PCR. Because these PCR amplicons are usually analyzed by standard DNA sequencing, an interesting and lower-cost alternative is the EpiTYPER platform (Sequenom). EpiTYPER transcribes bisulfite-converted DNA in vitro into RNA molecules, which are subjected to base-specific cleavage followed by analysis of the fragments using mass spectrometry. EpiTYPER is partially automated, and thus this platform is well suited for analysis of medium-sized sets of CpG islands of interest (48–96) in studies involving large numbers of samples. For studies of human DNA that require the analysis of larger numbers of CpG dinucleotides, the Infinium HumanMethylation27 BeadChip (Illumina) is a good compromise between high coverage and cost. This commercial system uses bisulfite-converted DNA to interrogate the methylation status of 27,578 individual CpG sites associated with a total of 14,475 transcription start sites of human genes and 110 miRNA promoters.

For experimental questions that require whole-genome analysis of DNA methylation, DNA treated with bisulfite can be used to generate libraries suitable for sequencing in any of the major second-generation platforms (Roche 454, Illumina, or ABI SOLiD). Any of these sequencing platforms can generate genome-wide coverage, but the high cost of each global analysis experiment can be a limiting factor in deploying the technology. Protocols 5 and 6 have been used by academic laboratories to analyze DNA methylation at the whole-genome level using the Roche 454 and the Illumina Genome Analyzer.

Methylation-Specific PCR for Gene-Specific DNA Methylation Detection

Methylation-specific PCR (MS-PCR) uses bisulfite-converted DNA as the starting material and conditionally generates DNA amplicons based on the use of two different sets of primers, one for methylated DNA and another for unmethylated DNA (Protocol 2). While not providing single-base resolution for the entire DNA amplicon, the method has the advantages of simplicity, flexibility, and low cost.

Immunoprecipitation of Methylated DNA Using Antibodies or Methyl-Binding Protein 2

Proteins or antibodies capable of specifically binding to DNA that contains 5mC are in principle capable of generating DNA fractions enriched in or depleted of methylated cytosines (Protocol 3). These methods work best with single-stranded DNA and typically require relatively large DNA inputs. As pointed out (Laird 2010), the ratio of input DNA to affinity reagent can affect the enrichment efficiencies of genomic regions with varying 5mC density. Because the results obtained from immunoprecipitation methods are affected by the stoichiometry of the reagents, it is important to perform preliminary experiments to optimize and standardize the design of the capture experiments. After immunoprecipitation, a variety of analytical methods are available for generating DNA methylation data. Immunoprecipitation followed by DNA microarray analysis (MeDIP) has been used in a large number of studies, and we expect that the use of immunoprecipitation in combination with high-throughput deep sequencing will become widely used in the future.

Global Analysis of DNA Methylation Using Restriction Endonucleases

Restriction endonucleases are powerful tools for assessing the methylation status of DNA. They have been used as the basis for locus-specific and genome-wide analysis of DNA methylation. Using a single endonuclease limits sequence sampling, but combinations of restriction endonucleases enable genomes to be fractionated into predominantly methylated or unmethylated compartments. An evaluation (Irizarry et al. 2008) of alternative microarray-based methods of DNA methylation analysis reported that approaches based on the use of the methylation-dependent endonuclease McrBC and an optimized bioinformatics approach for data analysis (CHARM) generated results with reasonably high correlation coefficients (0.76) when compared with data generated using the same samples in the Illumina Infinium HumanMethylation 27 BeadChip. Other methods evaluated in this study included MeDIP, HELP (Oda and Greally 2009), and McrBC (without CHARM data analysis); the correlation coefficients relative to the Infinium reference data set were 0.38, 0.48, and 0.63, respectively. No microarray-based methods are included in this chapter because researchers increasingly feel that sequencing-based analysis is the preferred readout for analysis of genomic compartments generated by methylation-sensitive or methylation-dependent restriction endonucleases.

This chapter includes a protocol (Protocol 4) for genome-wide DNA methylation analysis (Edwards et. al. 2010) that is based on enzymatic fractionation of the genome into methylated and unmethylated compartments. Because the method avoids the use of bisulfite modification, DNA fragments of relatively large size are preserved, which permits the generation of paired-end libraries with DNA inserts of known size in the ranges of 0.8–1.5 kb, 1.5–3 kb, and 3–6 kb. In most instances, the paired-end configurations can be uniquely mapped to the genome using software available from Applied Biosystems. This methodological approach delivers a reasonable balance between genome coverage and cost and is uniquely able to analyze the methylation status of repetitive elements in the human genome. Because it samples relatively long sequence domains, this method is also capable, in many instances, of generating valuable strand-specific information from imprinted regions in the mammalian genome.

High-Throughput Deep-Sequencing Technologies to Analyze Genome Partitions or Entire Genomes

Second- and third-generation high-throughput sequencing platforms (see Chapter 11) are especially powerful when used for the analysis of bisulfite-converted DNA, because the availability of multiple reads for each locus in the genome reveals the fine structure of DNA methylation marks within cell populations. The richness of the information provided by deep sequencing is most dramatically apparent when sequencing reads are longer than 150 bases, as is the case with the Roche 454 platform. Protocol 5 describes the use of the Roche 454 platform to analyze bisulfite-converted DNA. Zeschnigk et al. (2009) have reported the DNA methylation status of more than 6000 CpG islands in human blood and human sperm samples. The methylation profiles of CpG islands revealed by this type of analysis often display the different methylation patterns in different reads of the same locus, indicative of imprinting-specific methylation or allele-specific regulatory patterns that remain to be elucidated.

Another advantage of high-throughput deep-sequencing approaches is their greater power for discovery of cytosine modifications that occur outside of the common context of CpG dinucleotides. Protocol 6 presents a method based on the use of the Illumina platform to analyze bisulfite-converted DNA (MethylC-seq, based on the work of Lister et al. 2009 and Popp et al. 2010). This approach revealed that nearly one-quarter of all the methylation identified in embryonic stem cells occurred in a non-CG context and that this non-CG methylation disappeared upon induced differentiation of the embryonic stem cells and was restored in induced pluripotent stem cells. These findings emphasize the remarkable discovery power of this unbiased genome-wide DNA methylation analysis approach.

ADVANTAGES AND LIMITATIONS OF DIFFERENT APPROACHES FOR ANALYZING DNA METHYLATION

The choice of methods and technology platforms for DNA methylation analysis will depend on the specific biological question being addressed, the number and kind of biological samples being analyzed, the breadth and resolution of the genomic information being sought, and the budget constraints of each study.

If the amount of DNA available for analysis is limited, methods that use a DNA amplification step may be necessary. However, if the method involves the use of bisulfite modification, this step must precede the DNA amplification step so that information relevant to pattern of DNA methylation is retained. Some of the methods described in this chapter permit analysis of samples containing as little as 100 ng of genomic DNA, whereas other methods require 2–10 μg of sample. Remarkably, some of the whole-genome sequencing approaches, such as the high-throughput method based on the use of the Illumina platform, have been adapted to work with sample inputs as small as 150 ng (Popp et al. 2010).

If the samples to be analyzed are derived from cancer cells, the presence of deletions, translocations, regions of amplification, and other sequence rearrangements present special challenges. Affinity-based techniques must be used with caution in such cases, because the experimental results can be biased by copy number variation. On the other hand, endonuclease-based methods that measure a ratio of methylated to unmethylated DNA can be used to generate data that are less prone to distortion by locus copy number (Szpakowski et al. 2009).

Desired coverage and resolution are key considerations when choosing a method. If only a few hundred gene promoters need to be analyzed, bisulfite sequencing or MS-PCR is a reasonable choice, the former providing single-nucleotide resolution for regions of up to 800 bases, the latter offering simplicity and lower cost. For genome-wide analysis of DNA methylation, microarray-based approaches may be more economical than sequencing-based approaches. On the other hand, sequencing-based approaches offer improved resolution of methylated bases in DNA. If there is an expectation that the samples to be analyzed contain different cell types or different stages in the development of a cell lineage, sequencing-based approaches that provide information derived from molecular clones will be able to discern different methylation states at the same genomic locus. In other words, clonal sequencing will preserve "digital" information about heterogeneous patterns of methylation, thanks to the molecular sampling advantages of multiple deep-sequencing reads. For example, a rare methylation pattern occurring in one cell (perhaps a progenitor cell of interest?) out of every 200 cells in a sample will be detectable using ultradeep sequencing.

FUTURE PERSPECTIVES

When we wrote this chapter in 2011, single-molecule sequencing technologies were not sufficiently developed and accessible to be included. The ongoing development of these technologies holds promise for accessing DNA methylation data that are generated at the level of single-molecule read-outs occurring in real time and in massively parallel analytical streams. This mode of analysis of DNA methylation may become practical using sequencing platforms that use physically constrained polymerases (e.g., zero mode waveguides [Eid et al. 2009]) or physically constrained DNA strands (e.g., nanopores [Clarke et al. 2009]). Distinguishing between 5mC and 5hmC modifications also seems to be within the realm of technical feasibility. In fact, Flusberg et al. (2010) have described a new method for direct detection of DNA methylation without bisulfite conversion, using single-molecule, real-time (SMRT) sequencing. This methodology was able to detect not only 5-methyl-cytosine but also N^6-methyladenine and 5-hydroxymethylcytosine.

REFERENCES

Bird A. 2002. DNA methylation patterns and epigenetic memory. *Genes Dev* **16**: 6–21.

Bruniquel D, Schwartz RH. 2003. Selective, stable demethylation of the interleukin-2 gene enhances transcription by an active process. *Nat Immunol* **4**: 235–240.

Clarke J, Wu HC, Jayasinghe L, Patel A, Reid S, Bayley H. 2009. Continuous base identification for single-molecule nanopore DNA sequencing. *Nat Nanotechnol* **4**: 265–270.

Edwards JR, O'Donnell AH, Rollins RA, Peckham HE, Lee C, Milekic MH, Chanrion B, Fu Y, Su T, Hibshoosh H, et al. 2010. Chromatin and sequence features that define the fine and gross structure of genomic methylation patterns. *Genome Res* **20**: 972–980.

Eid J, Fehr A, Gray J, Luong K, Lyle J, Otto G, Peluso P, Rank D, Baybayan P, Bettman B et al. 2009. Real-time DNA sequencing from single polymerase molecules. *Science* **323**: 133–138.

Flusberg BA, Webster DR, Lee JH, Travers KJ, Olivares EC, Clark TA, Korlach J, Turner SW. 2010. Direct detection of DNA methylation during single-molecule, real-time sequencing. *Nat Methods* **7**: 461–465.

Huang Y, Pastor WA, Shen Y, Tahiliani M, Liu DR, Rao A. 2010. The behaviour of 5-hydroxymethylcytosine in bisulfite sequencing. *PLoS ONE* **5**: e8888. doi: 10.1371/journal.pone.0008888.

Irizarry RA, Ladd-Acosta C, Carvalho B, Wu H, Brandenburg SA, Jeddeloh JA, Wen B, Feinberg AP. 2008. Comprehensive high-throughput arrays for relative methylation (CHARM). *Genome Res* **18**: 780–790.

Klose RJ, Bird AP. 2006. Genomic DNA methylation: The mark and its mediators. *Trends Biochem Sci* **31**: 89–97.

Kriaucionis S, Heintz N. 2009. The nuclear DNA base 5-hydroxymethylcytosine is present in Purkinje neurons and the brain. *Science* **324**: 929–930.

Laird PW. 2010. Principles and challenges of genome-wide DNA methylation analysis. *Nat Rev Genet* **11**: 191–203.

Lister R, Pelizzola M, Dowen RH, Hawkins RD, Hon G, Tonti-Filippini J, Nery JR, Lee L, Ye Z, Ngo QM, et al. 2009. Human DNA methylomes at base resolution show widespread epigenomic differences. *Nature* **462**: 315–322.

Miller CA, Sweatt JD. 2007. Covalent modification of DNA regulates memory formation. *Neuron* **53**: 857–869.

Murayama A, Sakura K, Nakama M, Yasuzawa-Tanaka K, Fujita E, Tateishi Y, Wang Y, Ushijima T, Baba T, Shibuya K, et al. 2006. A specific CpG site demethylation in the human interleukin 2 gene promoter is an epigenetic memory. *EMBO J* **25**: 1081–1092.

Oda M, Greally JM. 2009. The HELP assay. *Methods Mol Biol* **507**: 77–87.

Popp C, Dean W, Feng S, Cokus SJ, Andrews S, Pellegrini M, Jacobsen SE, Reik W. 2010. Genome-wide erasure of DNA methylation in mouse primordial germ cells is affected by AID deficiency. *Nature* **463**: 1101–1105.

Shibata D, Tavare S. 2006. Counting divisions in a human somatic cell tree: How, what and why? *Cell Cycle* **5**: 610–614.

Szpakowski S, Sun X, Lage JM, Dyer A, Rubinstein J, Kowalski D, Sasaki C, Costa J, Lizardi PM. 2009. Loss of epigenetic silencing in tumors preferentially affects primate-specific retroelements. *Gene* **448**: 151–167.

Zeschnigk M, Martin M, Betzl G, Kalbe A, Sirsch C, Buiting K, Gross S, Fritzilas E, Frey B, Rahmann S, et al. 2009. Massive parallel bisulfite sequencing of CG-rich DNA fragments reveals that methylation of many X-chromosomal CpG islands in female blood DNA is incomplete. *Hum Mol Genet* **18**: 1439–1448.

DNA Bisulfite Sequencing for Single-Nucleotide-Resolution DNA Methylation Detection

DNA methylation plays an important role in multiple biological processes. Therefore, methodologies that can detect changes in DNA methylation are of general importance. Currently, the most popular and reliable method for measuring DNA methylation status is DNA bisulfite sequencing (Clark et al. 1994). Denatured DNA (i.e., single-stranded DNA) is treated with sodium bisulfite under conditions that preferentially convert unmethylated cytosine (C) residues to uracil (U) residues while methylated cytosines remain unmodified (Fig. 1). The converted DNA can then be amplified using a gene-specific primer. U is amplified as thymidine (T) in PCR and detected as T during DNA sequencing. Note that bisulfite conversion does not discriminate between 5-methylcytosine and 5-hydroxymethylcytosine (Fig. 1). 5-Hydroxymethylcytosine is a newly identified cytosine modification observed in embryonic stem cells and Purkinje neurons (Kriaucionis and Heintz 2009; Tahiliani et al. 2009).

DNA bisulfite sequencing uses the same set of primers for the detection of both methylated and unmethylated DNA. Certain criteria need to be followed while designing these primers in order to successfully amplify a gene-specific region. The rules for designing primers for bisulfite sequencing are discussed in the information panel Designing Primers for the Amplification of Bisulfite-Converted Product to Perform Bisulfite Sequencing and MS-PCR. This protocol details the steps required for bisulfite conversion and analysis of either genes or a specific genomic region. A flowchart describing the critical steps of bisulfite conversion is shown in Figure 2.

MATERIALS

It is essential that you consult the appropriate Material Safety Data Sheets and your institution's Environmental Health and Safety Office for proper handling of equipment and hazardous materials used in this protocol.

Recipes for reagents specific to this protocol, marked <R>, are provided at the end of the protocol. See Appendix 1 for recipes for commonly used stock solutions, buffers, and reagents, marked <A>. Dilute stock solutions to the appropriate concentrations.

Reagents

Acetic acid
Agarose gel (please see Chapter 2, Protocol 1)
Ammonium acetate
5-Aza-2′-deoxycytidine (5Aza2dC)
Bacterial cells, ultracompetent
dNTP (10 mM)
EDTA
Ethanol (70% and 100%)
Ethidium bromide solution
Gel buffer (1× TBE)
Genomic DNA

For cells in culture, or peripheral lymphocytes, a QIAGEN Blood and Cell Culture DNA Midi Kit may be used to prepare DNA. Mammalian tissues are pulverized under liquid nitrogen and digested overnight at

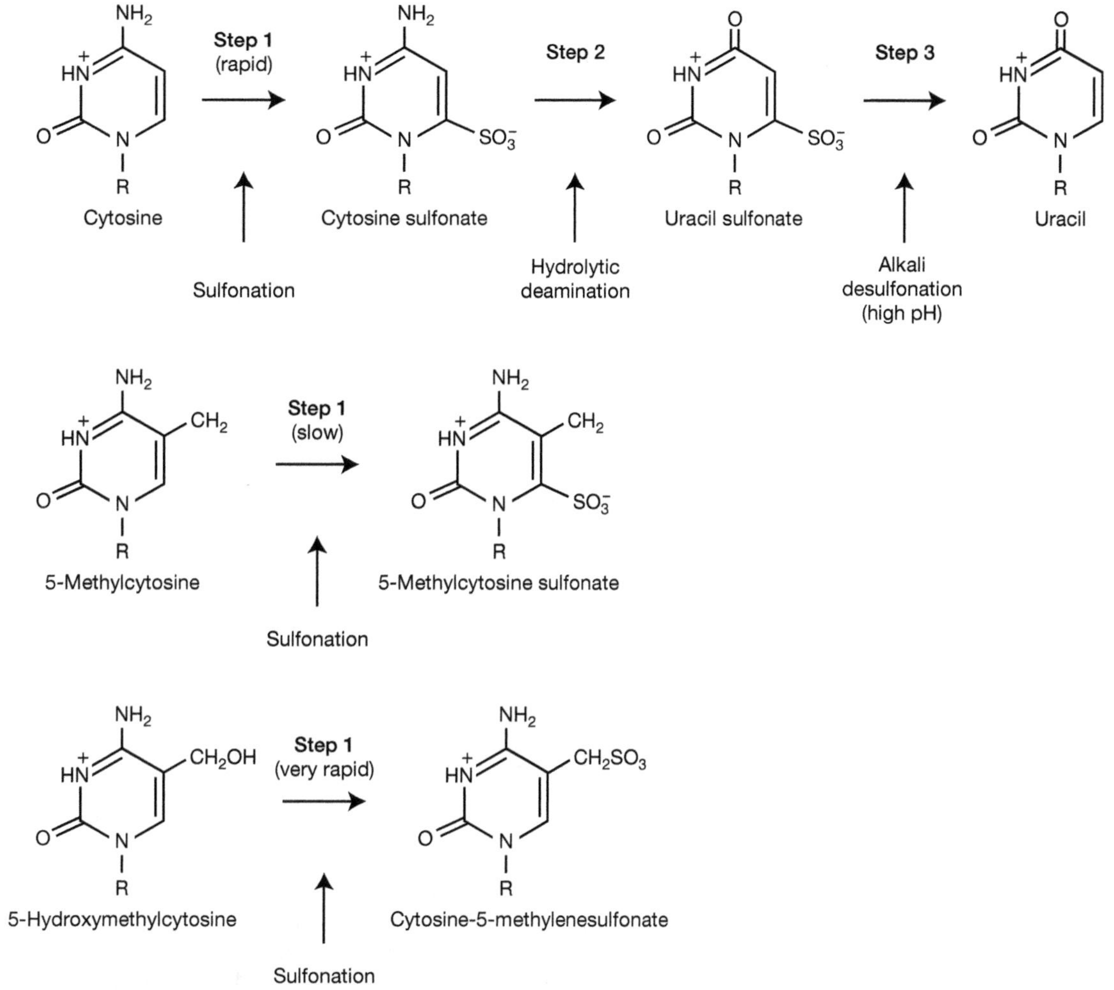

FIGURE 1. Behavior of cytosine, 5-methylcytosine, and 5-hydroxymethylcytosine in bisulfite sequencing. Following treatment with bisulfite, cytosine undergoes rapid conversion of uracil, whereas both 5-methylcytosine and 5-hydroxymethyl cytosine are resistant to this conversion.

55°C (10 mm NaCl, 10 mm Tris at pH 8, 25 mm EDTA at pH 8, 0.5% SDS, 100 ng/mL proteinase K). DNA is purified by phenol:chloroform extraction and RNase A treatment followed by ethanol precipitation. DNA is stored in TE (pH 8) at −20°C. The high-molecular-weight quality of DNA is confirmed by agarose gel electrophoresis, and the DNA is quantified using a NanoDrop 2000 instrument.

Glycogen (1 mg/mL)

Isopropanol

LB ampicillin plates <A> (containing 100 μg/mL ampicillin)

Ligase and ligase buffer

Mineral oil

NaOH (3 M)

PCR buffer

PCR purification column kit (QIAGEN)

Primers for colony PCR

> *If cloning in the pGMT vector, then use:*
>
> *Forward primer, 5′-ATGCGATATAGAGGAGACCGT-3′*
>
> *Reverse primer, 5′-TCAGAAACGTGAGATTGAT-3′*

QIAquick gel extraction kit (QIAGEN)

Quinol (1,4-dihydroxybenzene)

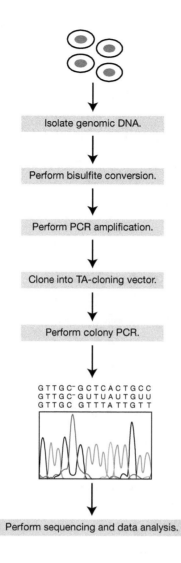

Isolate genomic DNA.

Perform bisulfite conversion.

Perform PCR amplification.

Clone into TA-cloning vector.

Perform colony PCR.

```
GTTGC⁻GCTCACTGCC
GTTGC⁻GUTUAUTGUU
GTTGC GTTTATTGTT
```

Perform sequencing and data analysis.

FIGURE 2. **Flowchart for bisulfite sequencing protocol.** Bisulfite sequencing is considered to be the "gold standard" for DNA methylation analysis.

Sodium bisulfite, saturated <R>
  Prepare saturated sodium bisulfite solution (for Step 3) fresh just before use.
TA Cloning Kit (Promega) or TOPO Cloning Kit (Life Technologies)
Taq polymerase
X-Gal

Equipment

Sequence alignment software
Thermal cycler

EXPERIMENTAL CONTROLS

When performing bisulfite sequencing for the first time, it is important to include controls. Use any human cell line as a source for generating positive control samples. Plate the cells at 40% confluency, and treat them for 3 d with 10 mM 5-aza-2′-deoxycytidine (5Aza2dC). Prepare genomic DNA, follow the steps to treat the DNA with sodium bisulfite, and use the bisulfite-converted DNA as a positive control throughout the remaining steps. 5Aza2dC is a DNA methyltransferase inhibitor that causes cells to lose the methyl tags from DNA in a DNA replication-dependent fashion. Thus, if a cell line has a doubling time of 24 h, a 3-d treatment of cells with 5Aza2dC will erase almost 75% of its genomic DNA methylation.

METHOD

Bisulfite Conversion of DNA

1. Denature the genomic DNA by adding NaOH at a final concentration of 0.3 M (e.g., use 2 μL of 3 M NaOH in 18 μL of diluted DNA).

   Two micrograms of DNA is recommended for a typical bisulfite conversion reaction. However, when the amount of DNA is limiting (e.g., samples derived from patient tissues or paraffin blocks), lower amounts of DNA (<100 ng) can be used.

2. Program a thermal cycler as follows: one cycle for 15 min at 37°C; followed by 2 min at 95°C; and, finally, 5 min at 4°C. Add the DNA from Step 1, and run the program to convert the DNA to single-stranded form.

   Ensure that the lid of the PCR machine is >100°C to avoid evaporation of the diluted DNA.

3. Combine the following:

Denatured DNA	20 μL
Sodium bisulfite (pH 5.0), saturated	208 μL
Quinol (10 mM)	10 μL
Mineral oil	200 μL

   Gently mix the contents of the tube, and centrifuge the reaction to the bottom of the tube. Incubate the reaction in the dark for 4 h to overnight at 55°C.

   Sodium bisulfite solution should be freshly prepared. The reaction should be protected from light to prevent oxidation. It is also important not to perform an overnight bisulfite conversion on DNA that is already degraded (e.g., DNA isolated from paraffin-embedded tissue).

4. Purify the bisulfite-converted genomic DNA using a PCR purification column as per the manufacturer's instruction.

5. Add NaOH to a final concentration of 0.3 M to denature the DNA. Incubate the mix for 20 min at 42°C.

6. Neutralize the solution by adding a 0.25× volume of 5 M ammonium acetate (pH 8.0).

   Ammonium acetate will desulfonate the uracil sulfonate to uracil. This step is crucial because sulfonated uracil derivatives inhibit PCR.

7. Add 2 volumes of ethanol to precipitate the DNA.

   If the amount of DNA used for the bisulfite conversion reaction was low, include 1 μg/mL glycogen to improve the efficiency of precipitation. Using isopropanol rather than ethanol will increase the yield of bisulfite-converted DNA. Use 1 volume of isopropanol.

8. Centrifuge at 10,000 rpm for 30 min at 4°C.

9. Decant the supernatant and wash the pellet with 1 mL of 70% ethanol. Dry the pellet completely by blowing a clean air stream over the tube or by placing the tube in a vacuum desiccator for 20 min. Dissolve the pellet in 40 μL of sterile dH$_2$O. Use the bisulfite-converted DNA for PCR as soon as possible. Otherwise, the DNA can be stored for 1–2 mo at −20°C without affecting its ability to serve as a template in PCR.

   This bisulfite-converted DNA can now be used for gene-specific PCR amplification (this protocol) or global sequencing of bisulfite-converted DNA (Protocols 5 and 6). Bisulfite conversion of DNA can also be performed using any of the commercially available kits described in Table 1. The use of bisulfite conversion kits is convenient but increases the cost of high-throughput experimental work.

Amplification of Bisulfite-Converted DNA

10. Set up a 50-μL PCR, as follows:

PCR buffer (10×) (with 1.5 mM MgCl$_2$)	5 μL
dNTP (10 mM)	1 μL
Forward primer (10 pmol)	1 μL
Reverse primer (10 pmol)	1 μL
Bisulfite-converted DNA	5 μL (10% of the DNA from Step 9)
Taq polymerase (5 units/mL)	1 μL
H$_2$O	36 μL

TABLE 1. Commercial kits for bisulfite conversion of DNA

Vendor	Features
Life Technologies (MethylCode)	All unmethylated cytosines converted to uracil Low template degradation and DNA loss during treatment/cleanup High-yield bisulfate-converted DNA DNA denaturation and bisulfite conversion done in one step No DNA precipitation step required
QIAGEN (EpitTect)	All unmethylated cytosines converted to uracil DNA protection system for sensitive analysis Guaranteed results from ≥1 ng of DNA Spin column or 96-well-based purification and desulfonation Can be fully automated when used with QIAGEN QIAcude
Applied Biosystems (methylSEQr)	Minimal DNA degradation for PCR amplification All unmethylated cytosines converted to uracil Sample DNA can be stored for up to 2.5 yr
Millipore (CpGenome Fast)	All unmethylated cytosines converted to uracil
Millipore (GpGenome Universal)	All unmethylated cytosines converted to uracil
Zymo Research (EZ DNA Methylation)	In-column desulfonation reaction and recovery Low template degradation and DNA loss during treatment/cleanup All unmethylated cytosines converted to uracil
Sigma-Aldrich (Imprint DNA Mod. Kit)	50 pg of DNA or 20 cells are required. Procedure takes <2 h. Greater than 99% conversion rate Extremely low degradation Optional one-step protocol
Human Genetic Signatures (MethylEasy)	>90% reduction in bisulfite DNA loss 100 pg of DNA is required. No DNA pretreatment needed Reduced nonconversion rates
Human Genetic Signatures (MethylEasy Xceed)	50 pg of DNA is required. Column-based protocol Can be incubated in PCR tubes, conventional oven, or incubator Multiple loci can be interrogated after one conversion step Procedure can be completed in 90 min.
ActiveMotif (MethylDetector)	Thermal denaturation and conversion reactions (not NaOH mediated) Works with high G/C sequences and uncut DNA 99% conversion efficiency of unmethylated cytosines Purification columns eliminate precipitation and desulfonation steps
BioChain (DNA Methylation Detection)	Guarantees >99% conversion of cytosine Guarantees >99% CpG protection Sensitive over a range of 2 μg–500 pg

Amplify the DNA.

Amplification conditions are determined empirically. Optimal conditions are affected by several factors, including the size of the region being amplified and the amount of cytosines and guanines within the region being amplified. In general, it is best to use an annealing temperature of 48°C–50°C.

11. Separate the PCR products on a 1.5% agarose gel.

    See Troubleshooting.

12. Purify the amplified PCR product with a QIAquick gel extraction kit.

13. Set up a ligation reaction using 10% of the total DNA or the gel-eluted DNA and a cloning kit. Follow the manufacturer's instructions. Incubate the ligation reaction overnight at 16°C.

14. Transform ultracompetent cells (transformation efficiency $= 10^8$) with half of the ligation reaction.

15. Transfer the cells to two LB-ampicillin plates that have been spread with 10 μL of 1 M IPTG and 50 μL of 50 mg/mL X-Gal. Incubate the plates for 16–20 h at 37°C.

16. Pick individual white bacterial colonies, and disperse each white colony in 15 mL of water.

17. Set up colony PCRs in a total of 25 μL, as follows:

PCR buffer (10×) (with 1.5 mM MgCl$_2$)	2.5 μL
dNTP (10 mM)	0.5 μL
Forward primer (10 pmol)	0.5 μL
Reverse primer (10 pmol)	0.5 μL
Dispersed bacterial colony	2.0 μL
Taq polymerase (5 units/mL)	0.5 μL
H$_2$O	18.5 μL

Place the tubes in a thermal cycler programmed as follows:

Cycle number	Denaturation	Annealing	Polymerization
35 cycles	30 sec at 95°C	45 sec at 56°C	30 sec at 72°C

18. Dilute each colony PCR product to one-tenth in 1× TE (pH 8.0), and use 2 μL of the diluted product for Sanger sequencing with 2 μL of SP6 primers (5 μM).

19. Sequence at least 10 bacterial colonies for each gene locus.

20. Analyze the data by performing pairwise alignment with the in silico converted DNA, wherein all of the cytosines have been replaced by thymidines. In a Linux or an OSX computer, the in silico conversion cab be performed by using the UNIX command<sed 's/C/T/g;s/c/t/g' sequencefile>, where the sequence file is a standard NCBI DNA sequence. Use EBI pairwise alignment (http://www.ebi.ac.uk/Tools/psa/) for performing pairwise alignment. Sequence alignment can be conveniently performed using the program ClustalW or ClustalX (Thompson et al. 1997; Chenna et al. 2003; Larkin et al. 2007).

TROUBLESHOOTING

Problem (Step 11): PCR fails to amplify bisulfite-converted DNA.
Solution: Treatment of the genomic DNA with proteinase K was insufficient. Digest the DNA again with proteinase K and precipitate with ethanol.

Problem (Step 11): PCR fails to amplify bisulfite-converted DNA, and the DNA is degraded.
Solution: Reduce the incubation time for the bisulfite reaction.

Problem (Step 11): PCR fails to amplify bisulfite-converted DNA. This may occur for a number of reasons other than the previous two.
Solution: Try one or more of the following:

• Increase the amount of bisulfite-converted DNA in the PCR.

• Ensure that the bisulfite PCR primers are properly designed by considering that all of the cytosines will be converted to uracils after bisulfite conversion.

• Try a temperature gradient to optimize the annealing temperature.

• Try nested PCR. In our experience, using nested PCR and low annealing temperature (50°C) will often solve this problem.

Problem (Step 11): Nonspecific bands are obtained after PCR.

Solutions: This is likely due to a low annealing temperature. Try different temperature gradients on the thermal cycler. If that does not work, design alternative primer pairs because the primer may have been annealing at a nonspecific genomic locus. In addition, using nested PCR should solve this problem (see Chapter 7, Protocol 7).

RECIPE

It is essential that you consult the appropriate Material Safety Data Sheets and your institution's Environmental Health and Safety Office for proper handling of equipment and hazardous materials used in this protocol.

Sodium Bisulfite, Saturated (pH 5.0)

To prepare 10 mL of saturated sodium bisulfite:

1. Add 4.75 g of sodium metabisulfite (di-sodium disulfite, $Na_2S_2O_5$) to 6.25 mL of sterile, degassed ddH_2O.

2. Add 1.75 mL of 2 M NaOH.

3. Add 1.25 mL of 1 M hydroquinone (0.11 g in 1 mL of H_2O).

4. Heat to 50°C in the dark, inverting the tube frequently.

5. Adjust the pH to 5.0.

Sodium bisulfite solution should be freshly prepared. The reaction should be protected from light to prevent oxidation.

REFERENCES

Chenna R, Sugawara H, Koike T, Lopez R, Gibson TJ, Higgins DG, Thompson JD. 2003. Multiple sequence alignment with the Clustal series of programs. *Nucleic Acids Res* **31:** 3497–3500.

Clark SJ, Harrison J, Paul CL, Frommer M. 1994. High sensitivity mapping of methylated cytosines. *Nucleic Acids Res* **22:** 2990–2997.

Kriaucionis S, Heintz N. 2009. The nuclear DNA base 5-hydroxymethylcytosine is present in Purkinje neurons and the brain. *Science* **324:** 929–930.

Larkin MA, Blackshields G, Brown NP, Chenna R, McGettigan PA, McWilliam H, Valentin F, Wallace IM, Wilm A, Lopez R, et al. 2007. Clustal W and Clustal X version 2.0. *Bioinformatics* **23:** 2947–2948.

Tahiliani M, Koh KP, Shen Y, Pastor WA, Bandukwala H, Brudno Y, Agarwal S, Iyer LM, Liu DR, Aravind L, et al. 2009. Conversion of 5-methylcytosine to 5-hydroxymethylcytosine in mammalian DNA by MLL partner TET1. *Science* **324:** 930–935.

Thompson JD, Gibson TJ, Plewniak F, Jeanmougin F, Higgins DG. 1997. The ClustalX windows interface: Flexible strategies for multiple sequence alignment aided by quality analysis tools. *Nucleic Acids Res* **25:** 4876–4882.

WWW RESOURCE

EBI pairwise alignment http://www.ebi.ac.uk/Tools/psa/

Methylation-Specific PCR for Gene-Specific DNA Methylation Detection

Methylation-specific PCR (MS-PCR) was developed by Herman et al. (1996), who, in their original study, used this method to analyze the methylation status of four tumor-suppressor genes: *p16*, *p15*, *E-cadherin*, and *VHL*. MS-PCR requires two pairs of primer sets, one each for amplification of unmethylated and methylated regions, respectively. MS-PCR primer design is discussed in the information panel Designing Primers for the Amplification of Bisulfite-Converted Product to Perform Bisulfite Sequencing and MS-PCR. Because sequencing is unnecessary, the method provides only a relative difference in the CpG density within a genomic region, rather than the single-nucleotide resolution of CpG methylation that can be achieved using bisulfite sequencing (Protocol 1). MS-PCR is a more rapid way to detect changes in DNA methylation than is bisulfite sequencing. In addition, by incorporating some basic automation, samples can be prepared and analyzed in a 96-well plate format. The method can be used either quantitatively (qRT-PCR-based MethyLight) (Eads et al. 2000) or qualitatively (using agarose gels) to detect changes in DNA methylation; both are described in this protocol. A flowchart outlining the major steps in the MS-PCR protocol is shown in Figure 1. Commercial kits can be used for certain MS-PCR steps; some of these kits are described in Table 1.

MATERIALS

It is essential that you consult the appropriate Material Safety Data Sheets and your institution's Environmental Health and Safety Office for proper handling of equipment and hazardous materials used in this protocol.

Reagents

Agarose gel (3%; containing 1 µg/mL ethidium bromide)

Ammonium acetate

Bisulfite conversion reagents (see Protocol 1)

DNA cleanup kit (QIAGEN)

dNTPs (10 mM)

Ethanol

Ethidium bromide solution

Gel buffer (1× TBE)

Gene-specific TaqMan probes

Genomic DNA

For cells in culture, or peripheral lymphocytes, a QIAGEN Blood and Cell Culture DNA Midi Kit may be used to prepare DNA. Mammalian tissues are pulverized under liquid nitrogen and digested overnight at 55°C (10 mM NaCl, 10 mM Tris at pH 8, 25 mM EDTA at pH 8, 0.5% SDS, 100 ng/mL proteinase K). DNA is purified by phenol:chloroform extraction and RNase A treatment followed by ethanol precipitation. DNA is stored in TE (pH 8) at −20°C. The high-molecular-weight quality of DNA is confirmed by agarose gel electrophoresis, and the DNA is quantified using a NanoDrop 2000 instrument.

Glycogen

Isopropanol

$MgCl_2$

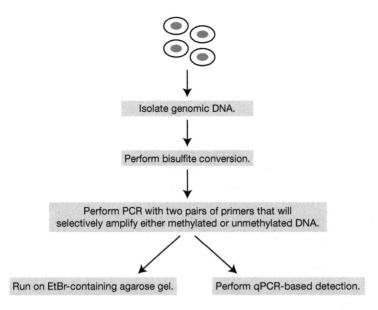

FIGURE 1. **Flowchart for MS-PCR.** Because MS-PCR is more rapid to perform, it can be used to analyze relatively large numbers of samples.

Mineral oil

Oligonucleotide probe with 5′ fluorescent reporter dye (6FAM) and a 3′ quencher dye (TAMRA)

PCR buffer (10×; with 1.5 mM MgCl$_2$)

Primers

The specific primers needed will depend on whether qualitative or quantitative analysis is performed.

Taq polymerase (5 units/μL)

TaqMan Buffer A containing a reference dye and *Taq* polymerase

Equipment

Quantitative thermal cycler

Thermal cycler

EXPERIMENTAL CONTROLS

Similar to bisulfite sequencing, genomic DNA prepared from cells treated with 5Aza2dC can be used as a control for the MS-PCR analysis. See the box Experimental Controls in Protocol 1.

TABLE 1. Commercial kits to aid with MS-PCR

Vendor	Features
QIAGEN (EpiTect MSP)	• Reduced false-positive results
	• Flexibility of primer design
	• Easy assay development
	• "Master mix" solution to simplify reaction setup
	• MSP analysis of even a single CpG site
Millipore (CpGenome and CpG WIZ MSP)	• Detects definitive chances in the CpG methylation of DNA
	• Correlates transcriptional regulation with changes in methylation pattern
	• Universal method for treatment of any gene of interest
	• No restriction digests of Southern blots required for use
	• Can be used with ≥1 ng of DNA

METHOD

Bisulfite Conversion of DNA

1. Follow Steps 1–9 from Protocol 1 to perform bisulfite conversion.

2. For qualitative methylation analysis, complete Method A below. For quantitative analysis of DNA methylation, use Method B.

Method A: Agarose Gel-Based Methylation-Specific PCR Product Detection

i. Set up two PCRs, one for detecting methylated DNA and the other for unmethylated DNA. A typical 25-μL PCR contains:

PCR buffer (10×) (with 1.5 mM MgCl$_2$)	2.5 μL
dNTP (10 mM)	0.5 μL
Forward primer (10 pmol)	0.5 μL
Reverse primer (10 pmol)	0.5 μL
Bisulfite-converted DNA	2.5 μL (5% of the bisulfite-converted DNA)
Taq polymerase (5 units/μL)	0.5 μL (5 units/μL)
H$_2$O	36 μL

ii. Place the reaction tubes in a thermal cycler and amplify the DNA. The optimal amplification conditions will have to be determined empirically.

iii. Resolve the PCR products on a 3% agarose gel containing ethidium bromide. Include a lane containing 100–300-bp molecular size markers. Visualize the results with a UV transilluminator.

Use 5 μL of 10 mg/mL ethidium bromide per 50 mL of gel.

See Troubleshooting.

Method B: Quantitative Analysis of DNA Methylation Using Methylight

Bisulfite-converted DNA can be analyzed using a quantitative methylation assay based on fluorescence-based PCR technology that uses locus-specific PCR primers. This method was initially called MethyLight and is capable of detecting methylated DNA in the presence of a 10,000-fold excess of unmethylated DNA (Eads et al. 2000). In their initial studies, the investigators were able to detect monoallelic versus biallelic methylation of MLH1 (Eads et al. 2000). It is important to note that sequence discrimination can occur at the level of the PCR amplification process and/or at the level of fluorogenic probe hybridization. In both cases, discrimination is based on perfectly matched versus mismatched oligonucleotides. For sequence discrimination to occur during PCR amplification, design primers (or primers and probes) that overlap the potential CpG dinucleotides. Discriminating among sequences by designing the primers at either a methylated or unmethylated region is, in essence, the quantitative PCR-based version of MS-PCR. Perform the following steps for MethyLight analysis to quantitatively discriminate among the differences at the DNA methylation level.

i. Amplify bisulfite-converted genomic DNA using locus-specific PCR primers flanking an oligonucleotide probe with a 5′ fluorescent reporter dye (6FAM) and a 3′ quencher dye (TAMRA). Prepare a 25-μL PCR. A typical reaction contains:

Each primer	600 nM
Probe	200 nM
Each dATP, dCTP, and dGTP	200 μM
dUTP	400 μM
MgCl$_2$	3.5 mM
TaqMan Buffer A containing a reference dye and *Taq* polymerase (1×)	
Bisulfite-converted DNA or unconverted DNA	

Serial dilutions of a control sample such as the one generated by 5Aza2dC treatment should be included to generate a standard curve.

ii. Place the reaction(s) in a real-time thermocycler. Program and run the machine using the following thermal cycling parameters:

2 min at 50°C
10 min at 95°C
40 cycles of 15 sec at 95°C and 1 min at 60°C

iii. Analyze the data using any of the existing qPCR data analysis methods, such as those described in Chapter 9 and Yuan et al. (2006).

TROUBLESHOOTING

Most of the technical problems that are applicable to bisulfite sequencing can also hinder MS-PCR results. Refer to the Troubleshooting section in Protocol 1 to help overcome these problems.

REFERENCES

Eads CA, Danenberg KD, Kawakami K, Saltz LB, Blake C, Shibata D, Danenberg PV, Laird PW. 2000. MethyLight: A high-throughput assay to measure DNA methylation. *Nucleic Acids Res* **28**: e32. doi: 10.1093/nar/28.8.e32.

Herman JG, Graff JR, Myohanen S, Nelkin BD, Baylin SB. 1996. Methylation-specific PCR: A novel PCR assay for methylation status of CpG islands. *Proc Natl Acad Sci* **93**: 9821–9826.

Yuan JS, Reed A, Chen F, Stewart CN Jr. 2006. Statistical analysis of real-time PCR data. *BMC Bioinformatics* **7**: 85. doi: 10.1186/1471-2105-7-85.

Protocol 3

Methyl-Cytosine-Based Immunoprecipitation for DNA Methylation Analysis

In mammalian cells, DNA is methylated at the 5-position of cytosine, usually in the context of CpG dinucleotides, but in some instances, such as human embryonic stem (ES) cells, non-CpG cytosines are methylated (Lister et al. 2009). This modification leads to recruitment of proteins that selectively recognize and bind 5-methylcytosine (5mC). Taking advantage of the structural identity of 5mC, various affinity purification–based protocols have been developed to enrich for either DNA that is modified by 5mC or proteins that recognize 5mC. In this protocol, an antibody against 5mC is used to immunoprecipitate the methylated DNA. Other 5mC immunoprecipitation methods use methyl-binding proteins to analyze DNA methylation patterns (Rauch and Pfeifer 2005, 2009). Interestingly, Huang et al. (2010) have shown that the antibody that recognizes 5mC can distinguish between 5mC and 5hmC (5-hydroxymethylcytosine). Thus, we are optimistic that an antibody will be isolated that uniquely recognizes 5hmC. The method in this protocol can be scaled up to perform genome-wide DNA methylation analysis. A flowchart outlining the major steps in this protocol is shown in Figure 1. Because immunoprecipitation is a straightforward procedure that does not require any prior modification of genomic DNA, several commercial kits are now available to perform the immunoprecipitation-based detection of DNA methylation. Some of these are described in Table 1.

MATERIALS

It is essential that you consult the appropriate Material Safety Data Sheets and your institution's Environmental Health and Safety Office for proper handling of equipment and hazardous materials used in this protocol.

Recipes for reagents specific to this protocol, marked <R>, are provided at the end of the protocol. See Appendix 1 for recipes for commonly used stock solutions, buffers, and reagents, marked <A>. Dilute stock solutions to the appropriate concentrations.

Reagents

Agarose gel (2%)
DNA ladder (1 kb)
Ethanol (100%)
Genomic DNA, proteinase K-treated (prepared from cell lines, tissues, or clinical samples)
IP buffer (10×; 1.4 M NaCl and 0.5% Triton X-100)
M-280 sheep anti-mouse IgG Dynal beads (Life Technologies)
Methyl-C antibody
NaCl (500 mM)
PBS (1×), containing 0.1% BSA <A>
Phenol:chloroform:isoamyl alcohol <A>
Primers for real-time PCR, locus-specific
Proteinase K (stock concentration, 20 mg/mL)
Proteinase K digestion buffer <R>

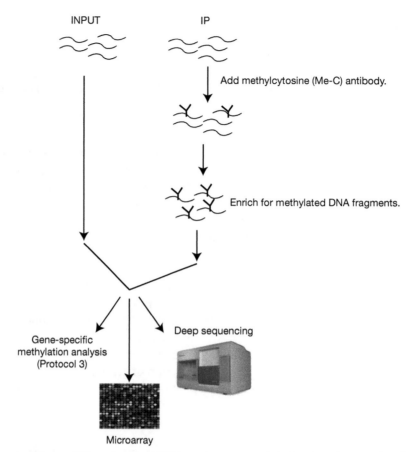

INPUT   IP

Add methylcytosine (Me-C) antibody.

Enrich for methylated DNA fragments.

Gene-specific
methylation analysis
(Protocol 3)

Deep sequencing

Microarray

FIGURE 1. **Flowchart for meDIP protocol.** meDIP is an immunoprecipitation-based protocol and does not require any prior modification of genomic DNA.

Equipment

Magnetic rack for collection of Dynal beads
Microcentrifuge tubes, Phase-Lock (5 PRIME)
Real-time thermal cycler
Sonicator
Water bath, boiling

EXPERIMENTAL CONTROLS

Sonicated genomic DNA from untreated cells and cells treated with 5Aza2dC (see Protocol 1) can be mixed in various ratios (e.g., 1 part methylated DNA to 1/1000 part of unmethylated DNA). These can then be used as positive controls to ensure that the immunoprecipitation using methyl cytosine antibody has worked. When studying a single locus, it is important to ensure that the locus is hypermethylated in untreated cells.

METHOD

Enriching for Hypermethylated DNA

1. Sonicate 100 μg of proteinase K–treated genomic DNA (the total volume of DNA is 200 μL in 1× TE at pH 8.0) to fragments that are ~300–500 bp.
   See the box Experimental Controls.

TABLE 1. Commercial kits for immunoprecipitation-based detection of DNA methylation

Vendor	Features
Epigentek (EpiQuik)	Direct immunoprecipitation of methylated DNA from lysates >95% enrichment of methylated DNA Strip microplate format for manual or high throughput Compatible with all DNA amplification-based methods
Diagenode (MeDIP)	Has available automated kit that will complete procedure in 9 h Completed in 3 d when done manually Contains all reagents for extraction, shearing, IP, and qPCR Allows direct correlation between IP' material and methylation status Procedure is performed in a single tube. Contains methylated and unmethylated BAC clones as internal controls
Zymo Research (Methylated DNA IP Kit)	Includes control DNA and primers for monitoring of the process >100-fold enrichment factor for methylated versus unmethylated DNA Separation of methylated DNA occurs in several hours
Active Motif (MethylCollector Ultra)	Uses magnetic bead-based protocol Procedure completed in <3 h Detects methylated CpGs from 1 ng to 1 μg of fragmented DNA Includes control DNA and PCR primers Uses MBD2b/MBD3L1 complex for a higher affinity for methylated DNA than MBD2b alone
RiboMed (MethylMagnet)	Uses GST-MBD fusion proteins and glutathione magnetic beads to capture methylated DNA Can be used to isolate mCpG DNA from 1 ng to 1 μg Can be used with DNA fragmented by any means or by restriction enzymes
Life Technologies (MethylMiner)	Fourfold greater sensitivity than antibody-based methods dsDNA capture is achieved with MBD2 protein and facilitates ligation of double-strand adaptors Elution is performed with salt; no proteinase K treatment or phenol:chloroform extraction Procedure completed in <4 h

2. Run 200 ng of the sonicated DNA on a 2% agarose gel with a 1-kb DNA ladder to determine the size of the sonicated product.

3. For each sample, dilute 2 μg of sonicated DNA with H_2O to a final volume of 450 μL, and boil it for 10 min.

4. Immediately cool the DNA on ice. Add 50 μL of 10× IP buffer and 10 μg of methyl-C antibody. Reserve a small amount of the sonicated DNA for later analysis.

5. Incubate each sample of DNA and methyl-C antibody overnight at 4°C.

6. For each sample, wash 50 μL of M-280 sheep anti-mouse IgG Dynal beads three times with 1× PBS containing 0.1% BSA. Wash the beads once with 1× IP buffer. Remove the IP buffer. Collect the beads using the magnetic rack rather than by centrifugation.

7. Add the Dynal beads to the tube that has the DNA–antibody complex (from Step 5). Rotate for 2 h at room temperature.

8. Use the magnetic rack to immobilize the Dynal beads. Remove the supernatant as the unbound fraction, which will be depleted of methylated DNA.

9. Wash the beads three times with 500 μL of IP buffer.

10. Resuspend the beads in 500 μL of proteinase K digestion buffer containing 5 μL/mL proteinase K solution (stock concentration, 20 mg/mL). Incubate the reaction for 2 h at 50°C.

11. Phenol:chloroform-extract the DNA in Phase-Lock microcentrifuge tubes.

12. Precipitate the DNA using 500 mM NaCl and 2 volumes of 100% ethanol.

13. Analyze the methylation patterns of the DNA by quantitative PCR using locus-specific PCR primers.

> See Troubleshooting.

TROUBLESHOOTING

Problem (Step 13): The background is high.
Solution: Try one or more of the following.

- Reduce the time of incubation with the antibody from overnight to 2 h at room temperature.

- Reduce the amount of Dynal beads.

- Increase the number of washes after binding the DNA–antibody complex with the secondary antibody–Dynal beads complex.

- Optimize the immunoprecipitation using the genomic DNA prepared from 5Aza2dC-treated cells.

Problem (Step 13): There is no signal in quantitative PCR and/or low yield of immunoprecipitated DNA.
Solution: Try one or more of the following.

- Increase the amount of DNA for immunoprecipitation.

- Increase the number of PCR cycles.

- Ensure that the primers used in the PCR are working by testing the primers in a standard PCR using the genomic DNA. Analyze the amplification products on an agarose gel.

RECIPE

It is essential that you consult the appropriate Material Safety Data Sheets and your institution's Environmental Health and Safety Office for proper handling of equipment and hazardous materials used in this protocol.

Proteinase K Digestion Buffer

Tris (pH 8.0)	50 mM
EDTA	10 mM
SDS	0.5%

REFERENCES

Huang Y, Pastor WA, Shen Y, Tahiliani M, Liu DR, Rao A. 2010. The behaviour of 5-hydroxymethylcytosine in bisulfite sequencing. *PLoS ONE* 5: e8888. doi: 10.1371/journal.pone.0008888.

Lister R, Pelizzola M, Dowen RH, Hawkins RD, Hon G, Tonti-Filippini J, Nery JR, Lee L, Ye Z, Ngo QM, et al. 2009. Human DNA methylomes at base resolution show widespread epigenomic differences. *Nature* 462: 315–322.

Rauch T, Pfeifer GP. 2005. Methylated-CpG island recovery assay: A new technique for the rapid detection of methylated-CpG islands in cancer. *Lab Invest* 85: 1172–1180.

Rauch TA, Pfeifer GP. 2009. The MIRA method for DNA methylation analysis. *Methods Mol Biol* 507: 65–75.

High-Throughput Deep Sequencing for Mapping Mammalian DNA Methylation

A challenge presented by genome-wide analysis of DNA methylation in mammals is striking the right balance between coverage of all relevant CpG dinucleotides and the need to keep the cost of analysis at reasonable levels. An additional challenge is to obtain information about the methylation status of repetitive DNA elements, which comprise a large proportion of the methylated residues in mammalian genomes. The methylation status of repetitive elements often has profound effects on the phenotype and is associated with several diseases, as well as with aging, genome instability, and cellular senescence (for review, see Belancio et al. 2009). The method described in this protocol, methylation mapping analysis by paired-end sequencing (Methyl-MAPS), was developed to address both of these challenges (Edwards et al. 2010). "Paired-end" refers to both how the library is made and how it is sequenced (see Chapter 11). In addition to the sequence information, paired-end sequencing provides information about the physical distance between the two reads in the genome. Methyl-MAPS typically samples ~80% of the CpG dinucleotides in the genome and is also able to report the methylation status of individual genomic loci harboring repetitive elements. This is achieved by enzymatic fractionation of the genome into methylated and unmethylated compartments. Because the method avoids the use of bisulfite modification, DNA fragments of relatively large size are preserved, permitting the generation of paired-end libraries with DNA inserts of known size in the range of 0.8–1.5 kb, 1.5–3 kb, and 3–6 kb. The paired-end configurations can be uniquely mapped to the genome in most instances, because the paired reads will span most repetitive element sequences.

Methyl-MAPS is able to identify sequence domains as predominantly methylated or largely unmethylated and therefore does not provide single-base resolution for DNA methylation patterns. The method has been validated by comparison with the results of bisulfite methylation analysis using the Illumina Infinium HumanMethylation27 BeadChip. The Pearson's correlation coefficient for Methyl-MAPS and the Infinium bisulfite analysis ranges from 0.84 to 0.87, which is superior to the correlation coefficients typically obtained with microarray-base methods (Irizarry et al. 2008). The following procedure, adapted from Edwards et al. (2010), with significant input from Dr. Edwards, to perform Methyl-MAPS analysis is provided. A flowchart describing the major steps of this protocol is shown in Figure 1.

MATERIALS

It is essential that you consult the appropriate Material Safety Data Sheets and your institution's Environmental Health and Safety Office for proper handling of equipment and hazardous materials used in this protocol.

Recipes for reagents specific to this protocol, marked <R>, are provided at the end of the protocol. See Appendix 1 for recipes for commonly used stock solutions, buffers, and reagents, marked <A>. Dilute stock solutions to the appropriate concentrations.

Reagents

Agarose (regular and low-melting)
Agarose loading gel mixture with Ficoll (GenScript, catalog no. D0084)
ATP (10 mM)

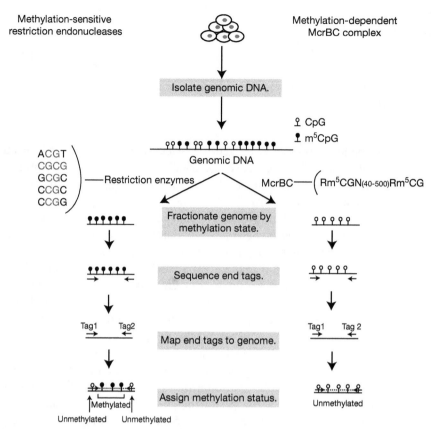

FIGURE 1. Flowchart for Methyl-MAPS. Methyl-MAPS is performed with restriction endonucleases that can be used to generate paired-end libraries for CpG-rich regions. The libraries are then sequenced using an ABI massively parallel sequencing platform. (Adapted, with permission, from Edwards et al. 2010.)

BBW buffer (1×) <R>

Biotinylated internal adaptors T30B and T30 (Applied Biosystems):

    T30B sequence, 5′-phos-CGTACA/iBiodT/CCGCCTTGGCCGT
    T30 sequence, 5′-phos-GGCCAAGGCGGATGTACGGT

Bovine serum albumin (BSA; 1 mg/mL)

BSA buffer for McrBC (100×) (New England Biolabs)

BW buffer (1×) <R>

CAP adaptors containing EcoP15I sites (Applied Biosystems)

CAP sequences 5′-phos-ACAGCAG-3′ and 5′-phos-CTGCTGTAC-3′ (used in equimolar amounts to form double-stranded DNA)

dATP (1 mM)

dATP, dGTP, and dTTP mix (10 mM)

DNA ladder (25 bp) (Life Technologies, catalog no. 10597-011)

DNA polymerase I

DNA retardation PAGE gel (6%)

    Use a precast NOVEX 6% polyacrylamide in 0.5× TBE buffer (Life Technologies).

dNTP mix (2.5 mM dATP, dGTP, dTTP, and dCTP)

Endonuclease EcoP15I

Endonuclease McrBC

Endonucleases HpaII, HpyCH4IV, AciI, HhaI, and BstUI

Ethanol (70% and 100%)

Ethidium bromide

Genomic DNA (isolated as per Protocol 1)

GLASSMILK spin column kit (GENECLEAN)

Glycogen (1 mg/mL)

GTP for McrBC (100×) (100 mM GTP)

Klenow DNA polymerase

LPCR-P1 and LPCR-P2 primers (Applied Biosystems):

    LPCR-P1, 5′-CCACTACGCCTCCGCTTTCCTCTCTATG-3′
    LPCR-P2, 5′-CTGCCCCGGGTTCCTCATTCT-3′

Microcon 100 ultrafiltration cartridge (Amicon)

MinElute Reaction Cleanup Kit (QIAGEN)

NEBuffers 1, 2, and 3 (10×) (New England Biolabs)

Oligonucleotides for SOLiD DNA chemistry

PAGE elution buffer (1:5 mixture of 7.5 M ammonium acetate and 1× TE)

P1ds and P2ds linkers (Applied Biosystems)

Phenol:chloroform:isoamyl alcohol <A>

Plasmid-safe DNase (62.5 μM ATP, 0.0625 U/μL ATP-dependent DNase) (Epicentre)

Platinum PCR Supermix (Life Technologies)

PNK Buffer (10×) (New England Biolabs)

QIAquick gel purification kit (QIAGEN)

Quick Ligase kit (New England Biolabs, catalog no. M2200S)

S-Adenosylmethionine (SAM) (32 mM) (New England Biolabs, catalog no. B9003)

Sephadex G50 column

Sinefungin (adenosyl ornithine) (Enzo Life Sciences, catalog no. ALX-380-070-M001)

Sodium acetate (3 M, pH 5.3)

Streptavidin beads, M280 (Life Technologies)

SYBR Gold or SYBR Green solution (Life Technologies)

T4 DNA polymerase

T4 polynucleotide kinase

TAE buffer (50×) <A>

TBE buffer (5×) <A>

TE (pH 8.0) <A>

TE with low EDTA (10 mM Tris and 0.01 mM EDTA at pH 8.0)

Unmethylated *cl857 Sam7* λ DNA (Promega)

Equipment

Bioanalyzer DNA 1000 LabChip (Agilent)

Sequencing system, high-throughput (e.g., Applied Biosystems SOLiD 3 or SOLiD 4 instrument)

Software for sequence analysis

Spectrophotometer (e.g., NanoDrop 1000 or 2000; Thermo Scientific)

Water baths (or heat blocks) (set at 37°C, 60°C, 65°C, and 70°C)

METHOD

Enzymatic Fractionation of Genomic DNA by Methylation State

The unmethylated DNA compartment is obtained by limit digestions of 10–15 μg of DNA with endonuclease McrBC. The methylated DNA compartment is generated by digestion of a similar amount of DNA with five tetranucleotide methylation-sensitive restriction endonucleases (HpaII, HpyCH4IV, AciI, HhaI, and BstUI; referred to as "RE" below).

Preparation of Unmethylated DNA

1. Prepare the unmethylated DNA compartment in a 300-μL reaction, as follows:

Genomic DNA (0.5 μg/mL)	30 μL (total 15 μg)
NEBuffer 2 (10×)	30 μL
GTP (100×)	3 μL
BSA (100×)	3 μL
Water	219 μL
McrBC enzyme (10,000 units/mL)	15 μL

   Digest the DNA for 4–6 h at 37°C.

2. Extract the DNA with phenol:chloroform:isoamyl alcohol [25:24:1] (pH 8).

3. Add 0.01 volume of 1 mg/mL glycogen, 0.1 volume of 3 M sodium acetate (pH 5.3), and 2 volumes of ethanol to the DNA. Incubate to precipitate the DNA for 15 min at −20°C.

4. Centrifuge to pellet the DNA at 20,800*g* for 15 min at 4°C.

5. Discard the supernatant, and wash the pellet with 600 μL of cold 70% ethanol. Centrifuge the DNA at 20,800*g* for 15 min at 4°C.

6. Discard the wash supernatant. Air-dry the pellet and resuspend the DNA in 1× TE (pH 8.0).

7. Remove excess GTP by purification over a Sephadex G50 column as described in the manufacturer's instructions. Measure DNA concentration using a NanoDrop spectrophotometer, and take an appropriate aliquot containing 10 μg of DNA. Add 0.1 volume of 3 M sodium acetate (pH 5.3) and 2 volumes of ethanol to the DNA. Incubate to precipitate the DNA for 15 min at −20°C. Save the partially dry DNA pellet for use in the paired-end library preparation step.

Preparation of Methylated DNA

8. Digest 15 μg of genomic DNA with 10 U of HpaII and HpyCH4IV restriction enzyme per microgram of starting DNA. Include 1× NEBuffer 1 in the reaction. Incubate the digest for 4–6 h at 37°C.

9. Extract the digest with phenol:chloroform:isoamyl alcohol [25:24:1] (pH 8). Repeat Steps 3–6, but resuspend the DNA in 30 μL of TE with low EDTA.

10. Add 10 U each of AciI and HhaI per microgram of starting DNA. Include 1× NEBuffer 3 and 1× BSA in the reaction. Incubate the digest for 4–6 h at 37°C.

11. Extract the digest with phenol:chloroform:isoamyl alcohol (pH 8). Repeat Steps 3–6, but resuspend the DNA in 30 μL of TE with low EDTA.

12. Add 10 U of BstUI per microgram of starting DNA. Include 1× NEBuffer 2 in the reaction. Incubate the digest for 2–3 h at 60°C.

13. Extract the digest with phenol:chloroform:isoamyl alcohol (pH 8). Repeat Steps 3–6, but resuspend the DNA in 30 μL of TE with low EDTA. Purify the DNA over a Sephadex G50 column as described in the manufacturer's instructions. Measure DNA concentration using a NanoDrop spectrophotometer. Add 0.1 volume of 3 M sodium acetate (pH 5.3) and 2 volumes of ethanol to the DNA. Incubate to precipitate the DNA for 15 min at −20°C. Save the partially dry DNA pellet for use in the paired-end library preparation step.

Paired-End Library Preparation

The following procedure is an adaptation, with some modifications, of the SOLiD System Mate-Paired Library Preparation Guide (Applied Biosystems, Part Number 4407413B). For further information about paired-end sequencing, please see Chapter 11.

Methylation of Endogenous EcoP15I Sites

14. For each DNA compartment, perform separate but identical reactions to repair DNA fragment ends with T4 DNA polymerase and T4 polynucleotide kinase, in a volume of 100 μL with final concentrations as indicated in the first column:

DNA	12 μg
PNK buffer (1×)	10 μL of 10× PNK buffer
BSA (0.1 mg/mL)	10 μL of 1 mg/mL BSA
ATP (0.4 mM)	4 μL of 10 mM ATP
dNTPs (0.4 mM)	4 μL of 10 mM dNTPs
Enzyme	50 U
ddH₂O	to a final volume of 100 μL

Incubate the reaction for 15 min at 12°C followed by an additional 15 min at 25°C.

15. Remove excess unincorporated dNTPs by passing the reaction through a Sephadex G50 column followed by phenol:chloroform:isoamyl alcohol extraction.

16. For each DNA compartment, methylate the endogenous EcoP15I sites using 12 μg of end-repaired DNA (from Step 14) in a volume of 125 μL as follows:

10 U of EcoP15I per measured microgram of DNA	12 μL of enzyme at 10 U/μL
NEBuffer 3 (1×)	12.5 μL of 10× NEBuffer 3
BSA (0.1 mg/mL)	12.5 μL of 1 mg/mL BSA
SAM (360 μM)	1.25 μL of 36 mM SAM
ddH₂O	to 125 μL

Incubate the methylation reaction for 2 h or overnight at 37°C. Include a control reaction using 12 μg of DNA (see Step 18).

17. Purify the DNA with the QIAquick Gel Extraction Kit, following the kit instructions, or as follows:

    i. Add 3 volumes of Buffer QG and 1 volume of isopropanol to the methylated DNA. The correct color of the mixture is yellow. If the color of the mixture is orange or violet, add 10 μL of 3 M sodium acetate (pH 5.5) and mix.

       The pH required for efficient adsorption of the DNA to the membrane is ≤7.5.

    ii. Apply 750 μL of methylated DNA in Buffer QG to the column(s). The maximum amount of DNA that can be applied to a QIAquick column is 10 μg. Use multiple columns if necessary.

    iii. Let the column(s) stand for 2 min at room temperature. Centrifuge the column(s) at ≥10,000*g* (13,000 rpm) for 1 min and discard the flowthrough.

    iv. Repeat Steps 17.ii and 17.iii until the entire sample has been loaded onto the column(s).

    v. Place the QIAquick column(s) back into the same collection tube(s). Add 750 μL of Buffer PE to wash the column(s). Centrifuge the column(s) at ≥10,000*g* (13,000 rpm) for 2 min, then discard the flowthrough. Repeat to remove residual wash buffer.

    vi. Air-dry the column(s) for 2 min to evaporate any residual alcohol.

    vii. Transfer the column(s) to clean 1.5-mL LoBind tube(s). Add 30 μL of Buffer EB to the column(s) to elute the DNA, and let the column(s) stand for 2 min. Centrifuge the column(s) at ≥10,000*g* (13,000 rpm) for 1 min.

    viii. Repeat the EB elution step a second time, and pool the eluted DNA to yield a total sample volume of 60 μL.

    ix. Quantitate the purified DNA by using 2 μL of the sample on the NanoDrop 1000 Spectrophotometer.

18. Confirm whether the EcoP15I methylation is successful by measuring resistance to digestion by EcoP15I of a control EcoP15I-methylated sample performed in parallel with library samples.

Addition of Adaptors and Size Fractionation of the DNA

19. Ligate a 100-molar ratio excess of double-stranded CAP adaptors (containing EcoP15I sites) onto the end-polished, EcoP15I-methylated DNA per reaction. A typical reaction can be performed in a volume of 300 μL as follows:

Eluted DNA (in buffer EB, from the previous QIAquick purification step)	58 μL
EcoP15I adaptors (double-stranded, 50 pg/μL)	20 μL
Quick ligase buffer (2×) (contains 2 mM ATP)	150 μL
Quick ligase	7.5 μL
ddH$_2$O	64.5 μL

Incubate the reaction for 10 min at room temperature (20°C–25°C).

20. Extract the DNA with an equal volume of phenol:chloroform:isoamyl alcohol.

21. Run the samples on a 1% TAE low-melting agarose gel at 45 V for 2.5 h.

22. Size-fractionate the gel at 800 bp to remove digested DNA and unligated CAP adaptors. Collect the DNA fragments into several size fractions (i.e., McrBC: 0.8–2 kb, 2–5 kb, >5 kb; RE: 0.8–1.5 kb, 1.5–3 kb, 3–6 kb, >6 kb).

23. Extract the DNA from each size fraction using a GLASSMILK spin column kit. Quantify the DNA using a 1-μL aliquot in a NanoDrop spectrophotometer.

Preparation of Circular DNA Molecules

24. For each size fraction, circularize the DNA by combining the size-fractionated DNA (typically 2–3 μg) with a 3 M excess of double-stranded T30B biotinylated internal adaptor in a total volume of 730 μL. A typical reaction mixture will contain:

DNA (2 μg)	60 μL
Quick ligase buffer (2×) (contains 2 mM ATP)	365 μL
Internal adaptor (2 μM stock solution of annealed T30B and T30 oligonucleotides, each 2 mM)	3 μL
Quick ligase	18 μL
ddH$_2$O	284 μL

Incubate the ligation reaction for 12 min at 20°C, under conditions that promote 95% circularization efficiency of DNA molecules of each length range.

For more detailed information, refer to page 92 of ABI Document No. 4407413B.

25. Degrade the noncircularized DNA molecules with Plasmid-Safe DNase. For example, for 1 μg of circularized DNA, use a reaction volume of 100 μL:

DNA (1 μg)	60 μL
ATP (10 mM)	6.25 μL
Plasmid-Safe buffer (10×)	10 μL
Plasmid-Safe DNase (10 U/μL)	3 μL
ddH$_2$O	20.8 μL

Incubate the reaction mixture for 40 min at 37°C.

26. Heat-inactivate the DNase for 20 min at 70°C.

27. Clean up the DNA by phenol:chloroform:isoamyl alcohol extraction.

28. Combine the four restriction enzyme fractions into <2-kb and >2-kb fractions. Remove excess nucleotides using a G50 Sephadex column.

Adding Primers for PCR

29. Digest the circularized DNA with EcoP15I as follows. Combine in a volume of 60 μL:

EcoP15I	10 U/100 ng of circularized DNA
NEBuffer 3	1×
BSA	1×
ATP	2 mM
Sinefungin	0.1 mM

Incubate the reaction overnight at 37°C.

30. Add fresh ATP, BSA, and sinefungin the following morning 1 h before termination of the reaction.

    Sinefungin blocks the methylation of bases in DNA and RNA, such as 5-methylcytosine or N^6-methyladenosine.

31. Repair the ends of the digested DNA with 5 U of Klenow DNA polymerase in the presence of 25 μM dNTPs for 30 min at room temperature.

32. Inactivate the enzyme by heating for 20 min at 65°C.

33. Ligate a 60-molar excess of P1ds and P2ds linkers containing primer sequences for PCR onto the ends of the DNA molecules, in a volume of 200 μL:

1 μg of circularized DNA (average size 1.5 kb)	70 μL
ds P1 linker (10 μM)	6 μL
ds P2 linker (10 μM)	6 μL
Quick ligase buffer (2×) (contains 2 mM ATP)	100 μL
Quick ligase	5.0 μL
ddH$_2$O	13 μL

Incubate the reaction for 10 min at room temperature (20°C–25°C).

Binding DNA to Beads

34. Wash M280 Strepavidin beads sequentially with 1× BBW buffer, 1× BSA, and then 1× BW buffer.

35. Combine the DNA (from Step 33) with 15 μL of washed M280 Strepavidin beads for 15 min at room temperature in a final concentration of 1× BW buffer. The DNA will bind to the beads via the biotin tag on the internal adaptor.

36. Wash the beads once with 500 μL of 1× BBW buffer, twice with 1× BW buffer, and once with 1× NEBuffer 2. Each wash is performed by vortexing for 15 sec, followed by a pulse-spin. The tube is then placed in a magnetic rack for 1.5 min to bind the beads. The supernatant is removed and discarded.

37. Resuspend the beads in 100 μL of 1× NEBuffer 2 containing 500 μM dNTP and 15 U of DNA polymerase I. Incubate the reaction for 30 min at 16°C.

Amplifying the DNA

38. Perform a test PCR, using 1 μL of DNA (or 1% of the material in the bead suspension) to confirm the presence of the expected 154–156-bp DNA product. This PCR product will comprise sequences from the P1 (or P2) adaptor and the CAP adaptor, which flank the "mate pair" DNA sequences, each of which encompasses ~100 bases of genomic DNA. The 154–156-bp product results from the sum of 100 bp of genomic DNA and the primer sequences used for PCR. Titrate the PCR cycles up and down as needed to produce sufficient product as assessed by the presence of a strong stained band in a 4% agarose gel.

39. Set up 50-μL PCRs in order to amplify all of the desired sequence(s) of DNA bound to the beads:

Bead template	1 μL
LPCR- P1 primer (50 μM)	1 μL
L-PCR-P2 primer (50 μM)	1 μL
Platinum PCR Supermix	47 μL

    Run the PCR as follows:

Cycle number	Denaturation	Annealing	Polymerization
1	5 min at 95°C		
18–22 cycles	15 sec at 95°C	15 sec at 62°C	1 min at 70°C
Last cycle			5 min at 70°C

40. Combine all of the like reactions, and precipitate the DNA using ethanol. Alternatively, the PCRs can be stored indefinitely at 4°C.

Isolating and Purifying the DNA

41. Prepare a 6% polyacrylamide nondenaturing DNA retardation gel. Prerun the gel for 5 min at 115 V.

42. Add Ficoll loading dye to the samples and a 25-bp ladder. Run them on the 6% gel in $1 \times$ TBE at 115 V for 45 min. For each 50 μL of PCR product, use 6 μL of loading buffer. When using a 1.0-mm thick gel, the five-well comb can accommodate up to 60 μL of load volume.

    ▲ *The sample should* not *be boiled before loading, to avoid DNA denaturation.*

43. To visualize the bands, stain the gel in 1:6000 ethidium bromide (10 mg/mL) in $1 \times$ TBE for 5 min and destain twice in water.

44. Dissect the 156-bp library band from the gel with a spatula, and shred the gel slices by centrifugation at 13,000 rpm for 3 min in a microcentrifuge.

45. Extract the DNA from the shredded gel in 200 μL of PAGE elution buffer for 20 min at room temperature. Centrifuge for 1 min at 13,000 rpm, remove and save the supernatant, and add 200 μL of PAGE elution buffer for a second elution step for 40 min at 37°C. Centrifuge again for 1 min at 13,000 rpm, remove the supernatant, and pool it with the first supernatant to yield 400 μL of eluate.

46. Clean the eluate by centrifugation on a 0.45-μm filter spin column to remove residual gel particulates. Add 0.1 volume of 3 M sodium acetate (pH 5.3) and 2 volumes of ethanol, incubate at −20°C, and centrifuge to recover the DNA. Resuspend the DNA in 150 μL of TE (pH 8.0).

47. Purify the resuspended DNA with a MinElute Reaction Cleanup Kit following the manufacturer's instructions. Elute the DNA in 20 μL of 10 mM Tris (pH 8.0). Elute any remaining DNA with a second round of 20 μL of 10 mM Tris (pH 8.0). Combine the two eluates to bring the final library volume to 40 μL.

48. Quantitate the DNA on a NanoDrop 2000 spectrophotometer, and determine the quality and quantity of the DNA on a Bioanalyzer DNA 1000 LabChip or on a standard nondenaturing 6% polyacrylamide gel in $0.5 \times$ TBE buffer.

DNA Sequencing

49. Amplify individual library fragments on 1-μm beads using emulsion-PCR according to the SOLiD emulsion PCR standard protocol.

50. Perform paired-end sequencing on the samples using a SOLiD 3, SOLiD 4, or SOLiD 5500 DNA sequencer.

    Genomic sequence reads of either 25 bp or 50 bp are obtained from the libraries, depending on the specific instrument and sequencing chemistry used.

Data Analysis: Tag Mapping, Data Filtering, and CpG Analysis

Initial tag mappings can be performed with the SOLiD software analysis package. Paired tags are each individually mapped in color space allowing up to two mismatches in each 25-bp tag to the reference genome sequence obtained from the UCSC Genome Browser (http://genome.ucsc.edu). A mate pair is reported if a single uniquely placed pair can be made from the hits of each tag that met the constraints of order, orientation, and distance (0–15 kb).

A set of custom Perl scripts (developed by Dr. John Edwards, Washington University, St. Louis, MO 63130; jedwards@dom.wustl.edu) are used to parse the output files from the SOLiD system and filter sequences that do not have at least one restriction site (McrBC or RE, respectively) at the fragment ends. CpG island, RepeatMasker, RefSeq gene data, and other genomic annotation information can be downloaded from the UCSC Genome Browser website. Other details relevant to sequence data analysis are included in the original publication by Edwards et al. (2010). The Perl scripts are shown below, and the latest versions can be found at http://epigenomics.wustl.edu. The scripts assume that data were first mapped to a reference genome and start with the F3_R3.mates file produced by the Applied Biosystems mate-pair mapping tool. See http://solidsoftwaretools.com for more information concerning the SOLiD System Color Space Mapping Tool (file "cms_058717.pdf"). The F3_R3.mates file reports the SOLiD paired tags (as a single line) and the category that each pairing falls into:

TAG_ID   F3_SEQUENCE   R3_SEQUENCE   F3_MISMATCHES
R3_MISMATCHES   TOTAL_MISMATCHES   F3_CHROMOSOME
R3_CHROMOSOMEF3_POSITION   R3_POSITION   CATEGORY

Example of .mates file output (a few rows):

1279_39_68	T00223131101333201000000110	G22123030211132321302011	0	0	0	20	20	30495426	30494696	AAA
1279_39_443	T02021113102211230022222231	G00220321220312132203013 20	0	0	0	11	11	−119924797	−119925140	AAA
1279_39_491	T01032211330211303200123 11	G311200100220301023330033 0	2	1	3	24	24	57590725	57579600	AAA
1279_39_524	T212302100002022002020221 00	G21001211302001211110312 20	0	1	1	1	1	−76034016	−76034838	AAA

Other requirements for running the Perl scripts are:

1. installation of the nibFrag tool available as part of the BLAT suite from Jim Kent at UCSC
2. .nib formatted files for each chromosome
3. cmap file used for the AB mapping software
4. CpG position file. Tab-delimited file in which the first column is the chromosome (chr1, etc.) and the second column is the coordinate of each CpG on the chromosome

Although the filtering and CpG overlap steps (Steps 3–5) can be performed with the scripts for the whole genome, it is recommended that each chromosome be processed separately and jobs be split across multiple CPUs or across a compute cluster.

- **Step 1:** Parse mates files (run for both McrBC and RE .mates files).

    USAGE: parseMates. pl <cmapFile><inFile><outDir><libDescriptor[M, R]>

This converts .mates file from AB mate-pair mapping software and creates BED files (one per chromosome).

<cmapFile> = cmap file used for tag mapping
<inFile> = .mates file from tag mapping
<outDir> = location for bedFiles
<libDescriptor> = M for McrBC, R for RE

Here is an example of a few annotation tracks that use a complete BED (Browser Extensible Data) file definition (http://genome.ucsc.edu/goldenPath/help/customTrack.html#BED) generated by Step 1 output:

track name = "chr20 M" description = " paired data M" RGB color = 0, 0, 255,

chr20	30494696	30495451	1279_39_68	1	+	30494696	30495451	0,0,255
chr20	46956060	46957272	1279_39_1695	1	+	46956060	46957272	0,0,255
chr20	47756101	47756710	1279_40_77	1	–	47756101	47756710	0,0,255
chr20	61157032	61157498	1279_40_592	1	–	61157032	61157498	0,0,255
chr20	45670721	45672414	1279_41_341	1	–	45670721	45672414	0,0,255

- **Step 2:** Filter McrBC fragments, selecting those with at least one McrBC recognition site.

USAGE: filterMcrBC.pl <bedFile><outFile>

BED files from McrBC data from Step 1 can be concatenated or each chromosome file processed separately.

<bedFile> = bedFile or concatenated bedFile from Step 1

<outFile> = filtered output file

- **Step 3:** Filter RE fragments.

USAGE: filterRE.pl <bedFile><outFile>

Bed files from RE data from Step 1 can be concatenated, or each chromosome file can be processed separately.

<bedFile> = bedFile or concatenated bedFile from Step 1

<outFile> = filtered output file

- **Step 4:** Normalize McrBC and RE fragments. This step is not fully automated. To find the correct ratio of McrBC and RE fragments, different ratios are selected for several chromosomes, and the scripts from Steps 5 and 6 are used to compute the overall coverage values.

- **Step 5:** Overlap fragments with CpG sites.

USAGE: overlapCpGsites.pl <CpGfile><McrBCfile> <REfile><outFile>

This overlaps McrBC and RE fragments with genomic CpG positions. The output is a text file that contains the number of McrBC and RE fragments overlapping each CpG site.

<CpGfile> = CpG position file. First column is chromosome, second is coordinate

<McrBCfile> = Filtered McrBC fragment bed file from step 2.

<REfile> = Filtered RE fragment bed file from step 3.

<outfile> = output file

output file format:

col1- chromosome

col2- CpG position

col3- Number of McrBC fragments where the CpG is well interior (and thus was protected from digestion and unmethylated)

col4- Number of McrBC fragments overlapping where this CpG was just inside the fragment (probable McrBC recognition site). Distance inside is set as parameter in script. Default is 50 bp.

col5- Similar to col4, but CpGs just outside the end of the McrBC fragment

col6- Number of RE fragments where CpG is interior (and thus was protected from digestion and methylated)

col7- Number of RE fragments where this CpG sits at the end of the fragment and was part of the RE cleavage site (and thus unmethylated)

Here is an output file example after Step 5 (a few rows):

chr20	8179	0	0	0	0	39
chr20	8551	0	0	0	41	0
chr20	8578	0	0	0	41	0
chr20	8807	0	0	0	28	26
chr20	8963	0	0	0	40	1
chr20	9070	0	0	1	40	0
chr20	9182	1	1	0	36	4
chr20	9237	2	0	0	35	0

- **Step 6:** Calculate methylation scores. Use data from Step 5 to calculate a methylation score at every CpG site. A score is computed at every CpG site. Scores at non-McrBC/non-RE sites are effectively estimated scores. It is recommended that scores only be used at McrBC and RE sites and caution be used when using scores at other sites.

USAGE: calculateMethylationScores.pl < cpgFile > < outFile >

< cpgFile > = output file from step 5

< outFile > = output file

outFile format

col1- chromosome

col2- CpG position

col3- Methylation score. 0 = fully unmethylated, 1 = fully methylated

col4- Unmethylation score. 1 = fully unmethylated, 0 = fully methylated

col5- coverage, Number of McrBC and RE fragments covering any given CpG site

Several Open source tools are available at http://solidsoftwaretools.com/gf/; these can be used to facilitate the analysis of raw data generated by the Applied Biosystems SOLiD 4 System. Most relevant among these tools are the following:

- SOLiD Accuracy Enhancement Tool (SAET) (http://solidsoftwaretools.com/gf/project/saet/). This tool uses raw data generated by the SOLiD system to correct miscalls within reads either before mapping or contig assembly.

- SOLiD BaseQV Tool (http://solidsoftwaretools.com/gf/project/sam/). This tool converts SOLiD output files to base sequences data with associated quality values.

- SOLiD System Alignment Browser (SAB) (http://solidsoftwaretools.com/gf/project/sab/). This tool is a Genome Annotation Viewer and Editor, based on the Apollo Genome Annotation Curation Tool. SAB runs on Windows, LINUX, and Mac OS.

- SOLiD System Color Space Mapping Tool (mapreads) (http://solidsoftwaretools.com/gf/project/mapreads/). This tool enables researchers to map SOLiD color space reads to whole human genomes.

- SOLiD System Analysis Pipeline Tool (Corona Lite) (http://solidsoftwaretools.com/gf/project/corona/). This tool can be used to map color space reads to large or small genomes, to place and annotate paired reads, and to call SNPs.

Methyl-Analyzer (Xin et al. 2011) is a Python package that implements a pipeline for the downstream analysis of genome-wide DNA methylation data generated by the Methyl-MAPS paired-end sequencing procedure documented in Protocol 4. It takes as input paired-end reads that have been mapped to a reference genome. It represents an alternative to the Perl scripts, developed by Dr. John Edwards and included in this chapter. Methyl-Analyzer is available for download at https://github.com/epigenomics/methylmaps.

RECIPES

It is essential that you consult the appropriate Material Safety Data Sheets and your institution's Environmental Health and Safety Office for proper handling of equipment and hazardous materials used in this protocol.

BBW Buffer (1×)

Reagent	Final concentration
Tween 20	2%
Triton X-100	2%
EDTA	10 mM

BW Buffer (1×)

Reagent	Final concentration
Tris-HCl, (pH 7.8)	10 mM
NaCl	1 M
EDTA	1 mM

REFERENCES

Belancio VP, Deininger PL, Roy-Engel AM. 2009. LINE dancing in the human genome: Transposable elements and disease. *Genome Med* 1: 97. doi: 10.1186/gm97.

Edwards JR, O'Donnell AH, Rollins RA, Peckham HE, Lee C, Milekic MH, Chanrion B, Fu Y, Su T, Hibshoosh H, et al. 2010. Chromatin and sequence features that define the fine and gross structure of genomic methylation patterns. *Genome Res* 20: 972–980.

Irizarry RA, Ladd-Acosta C, Carvalho B, Wu H, Brandenburg SA, Jeddeloh JA, Wen B, Feinberg AP. 2008. Comprehensive high-throughput arrays for relative methylation (CHARM). *Genome Res* 18: 780–790.

Xin Y, Ge Y, Haghighi FG. 2011. Methyl-Analyzer—Whole genome DNA methylation profiling. *Bioinformatics* 27: 2296–2297.

WWW RESOURCES

Methyl-Analyzer https://github.com/epigenomics/methylmaps

Open source tools for analyzing SOLiD 4 data http://solidsoftwaretools.com/gf/

Perl scripts to parse the output files from the SOLiD system http://epigenomics.wustl.edu

SOLiD-Related Information concerning the SOLiD System Color Space Mapping Tool http://solidsoftwaretools.com

UCSC Genome Browser http://genome.ucsc.edu

Roche 454 Clonal Sequencing of Bisulfite-Converted DNA Libraries

The Roche 454 sequencing technology uses emulsion PCR to amplify clonally single DNA molecules bound to 28-μm beads, which are then deposited into picoliter-scale wells for pyrosequencing (for further discussion, see Chapter 11, introduction and Protocols 15–17). During each sequencing cycle, the addition of nucleotides complementary to the template strand results in a chemiluminescent signal recorded by the CCD camera of the Genome Sequencer FLX Instrument. The advantage of this platform is the read length, which is typically in the range of 300–800 bp. However, this platform has a greater error rate than other platforms for sequencing of homopolymer regions and is more expensive than other competing platforms. DNA sequencing reads generated by the Roche 454 platform are used for interrogating DNA methylation status over the entire length of CpG islands, often revealing long patterns of variation in methylation that may contain allele-specific information. To achieve near whole-genome coverage, bioinformatics analysis of any fully sequenced mammalian genome can be used to generate reduced representation libraries through judicious selection of restriction endonucleases.

The following procedure, adapted from Zeschnigk et al. (2009), is used for genome-wide sequencing of CpG islands using the Roche 454 platform. By evaluating different sets of restriction enzymes, Zeschnigk et al. found a combination of enzymes that, when used to digest the human genome, generates a DNA fragment set with optimally enriched CpG island content. After digestion, size-fractionated DNA is treated with bisulfite, and libraries are generated for the Roche 454 GS FLX platform. The resulting DNA methylation information contains multiple reads, and relatively long read lengths. A flowchart indicating the crucial steps of this protocol is presented in Figure 1.

MATERIALS

It is essential that you consult the appropriate Material Safety Data Sheets and your institution's Environmental Health and Safety Office for proper handling of equipment and hazardous materials used in this protocol.

Recipes for reagents specific to this protocol, marked <R>, are provided at the end of the protocol. See Appendix 1 for recipes for commonly used stock solutions, buffers, and reagents, marked <A>. Dilute stock solutions to the appropriate concentrations.

Reagents

Agarose
Agarose loading gel mixture with Ficoll (GenScript)
ATP (10 mM)
Bisulfite conversion reagents (see Protocol 1)
Bst DNA polymerase (Epicentre Biotechnologies)
dATP, dGTP, and dTTP mix (10 mM)
DNA ladder (100 bp) (Life Technologies, catalog no. 15628-019)
dNTP mix (2.5 mM dATP, dGTP, and dTTP, but no dCTP)
Endonucleases MseI, Tsp509I, NlaIII, HpyCH4V (New England Biolabs)
Ethanol (70% and 100%)

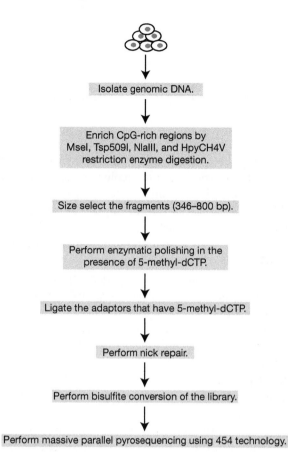

FIGURE 1. Flowchart for Roche 454 clonal sequencing of bisulfite-converted DNA libraries. This method uses restriction endonucleases for enrichment of CpG-rich regions followed by bisulfite conversion of the library. These bisulfite-converted libraries are then sequenced using massively parallel pyrosequencing.

Genomic DNA
GoTaq Green Master Mix (Promega)
Klenow DNA polymerase
5-Methyl-dCTP (Jena Bioscience GmbH)
MethylEasy Xceed kit (Human Genetic Signatures)
MinElute or QIAquick PCR purification kit (QIAGEN)
MinElute or QIAquick gel purification kit (QIAGEN)
Microcon 100 ultrafiltration device (Amicon)
NEBuffer 2 (10×)
Oligonucleotides for Roche 454 DNA chemistry (adaptors A and B, primers GSFLXA and GSFLXB)
  Sequences are from Margulies et al. (2005).

  To preserve the sequence during bisulfite modification, the 44-bp adaptors A and B contain 5-methyl-C instead of C.

  44-bp adaptor A, CCATCTCATCCCTGCGTGTCCCATCTGTTCCCTCCCTGTCTCAG
  44-bp adaptor B, 5BioTEG/CCTATCCCCTGTGTGCCTTGCCTATCCCCTGTTGCGTGTCTCAG
  Primer GSFLXA, 20 bp, complementary to the 5′ end of 44-bp adaptor A
  Primer GSFLXB, 20 bp, complementary to the 5′ end of 44-bp adaptor B
Phenol:chloroform:isoamyl alcohol <A>
Phusion Hot Start High-Fidelity DNA Polymerase (New England Biolabs)
QIAGEN Blood and Cell Culture DNA Midi Kit (QIAGEN; catalog no. 13343)
Quick Ligase kit (New England Biolabs, M2200S)
Sodium acetate (3 M, pH 5.3)
SYBR Green solution (Life Technologies)
TAE buffer (50×) <A>
T4 DNA ligase and 1× ligation buffer
T4 DNA polymerase

TE (pH 8.0) <A>
T4 polynucleotide kinase
Unmethylated *cl857 Sam7* λ DNA (Promega)
Wizard SV gel and a PCR cleanup system (Promega)

Equipment

Bioanalyzer 2100 (Agilent)
454 GS FLX Instrument (Roche)
RNA 6000 Pico LabChip (Agilent Technologies)
Spectrophotometer, NanoDrop 1000 or 2000 (Thermo Scientific)
Thermal cycler
Water bath (or heat block)

METHOD

DNA Extraction

1. Isolate DNA from tissues using proteinase K treatment.

   A commercial kit using proteinase K, such as the QIAamp DNA Mini Kit, can be used for this purpose. For cells in culture, or peripheral lymphocytes, use a QIAGEN Blood and Cell Culture DNA Midi Kit to prepare genomic DNA.

DNA Digestion and Size Fractionation

2. Digest 40 μg of genomic DNA sequentially with the restriction endonucleases MseI and Tsp509I, following the manufacturer's recommendations. For each enzyme, add 40 units and incubate for 2 h; then add an additional 40 units and incubate for another 2 h.

3. After finishing the second incubation with Tsp509I, purify the DNA by extraction with phenol:chloroform:isoamyl alcohol, followed by standard ethanol precipitation.

4. Resuspend the DNA and digest further (as per Step 2) using the enzymes NlaIII and then HpyCH4V.

5. Following sequential digestion with NlaIII and HpyCH4V, purify the DNA by extraction with phenol:chloroform:isoamyl alcohol, followed by ultrafiltration and concentration using a Microcon 100 device, centrifuging at 500*g* for 5 min.

6. Fractionate the digested DNA in a 1.8% agarose gel, loading 1.2 μg per lane in a total of eight lanes. Load a molecular weight marker lane with a 100-bp DNA ladder. Run the gel until the blue dye marker has migrated two-thirds of the way down the gel.

7. Stain the gel using SYBR Green (see Chapter 2, Protocol 2) and excise the region of 350–800 bp using a scalpel.

8. Recover the size-selected DNA from the agarose using a Wizard SV gel and PCR cleanup system.

9. Concentrate the DNA by ultrafiltration using a Microcon 100 device, centrifuging at 500*g* for 5 min.

DNA Library Preparation

10. Prepare DNA libraries as described in the Roche *GS FLX Titanium General Library Preparation Method Manual* (Roche Diagnostics GmbH, April 2009a) and the *GS FLX Titanium General Library Preparation Quick Guide* (Roche Diagnostics GmbH, April 2009b) with the following modifications:

    i. *Enzymatic polishing* (Section 3.4 of Roche Manual). Perform the polishing reaction in the presence of 5-methyl-dCTP instead of dCTP in the dNTP mixture.

ii. *Adaptor ligation* (Section 3.5 of Roche Manual). The A and B adaptors each comprise double-stranded oligonucleotides, which are ligated to the polished ends of the fragments. To preserve the sequence in the following bisulfite modification step, both adaptors A and B contain 5-methylcytosine instead of C. The base modification of the B adaptor with a biotin tag remains unaltered.

iii. *Fill-in reaction* (Section 3.8 of the Roche manual). In this step, nicks are repaired by the strand-displacement activity of a strand-displacing DNA polymerase. To preserve the methylated cytosines in the adaptor sequence, perform the reaction in the presence of 5-methyl-dCTP instead of dCTP in the dNTP mixture.

Bisulfite Modification of the DNA Library

11. Perform bisulfite modification on the single-stranded DNA fragments as described in Protocol 1.

12. Resuspend the bisulfite-treated products in 10 μL of autoclaved ddH$_2$O.

Massively Parallel Prosequencing in the Roche 454 GS FLX System

13. Set up PCRs using 2 μL of the bisulfite-converted library in a total reaction volume of 25 μL using GoTaq Green Master Mix and 200 nM each of the primers GSFLXA and GSFLXB.

    The primers GSFLXA and GSFLXB are each 20 nt in length and bind to the 5′ portion of adaptors A and B, respectively.

14. Perform PCR amplification as follows:

Cycle number	Denaturation	Annealing	Polymerization
1	2 min at 95°C		
20 cycles	30 sec at 95°C	30 sec at 62°C	1 min at 72°C
Last cycle			5 min at 72°C

15. To remove nonspecific reaction products, purify the PCR products in 1.8% agarose gels as described in Steps 6–9. However, recover DNA fragments in the size range 440–900 bp.

16. Quantify the DNA using the Agilent 2100 BioAnalyzer on an RNA 6000 Pico LabChip.

17. Perform bead preparation and sequence analysis according to the shotgun protocol of the GS emPCR Kit User's Manual for the Roche Genome Sequencer FLX.

18. Perform massively parallel pyrosequencing of the amplification products according to the protocol provided by the manufacturer (see also Chapter 11, Protocol 17).

Data Analysis

All 454 sequence reads are searched for overlaps with any of the adaptor sequences using a simple sequence alignment algorithm. If an adaptor is found, it is removed from the read. These functions are performed by the software module "cutadapt" (http://code.google.com/p/cutadapt/), developed by Marcel Martin. The sequence reads can then be mapped back to the genome using VerJInxer software, also developed by Marcel Martin (Zeschnigk et al. 2009). The VerJInxer software is publicly available as source code under the Artistic License/GPL and can be obtained uncompiled at http://verjinxer.googlecode.com. For mapping of reads from CGI-enriched fragments, a CGI reference sequence is generated, by extracting CGI sequences as defined by the UCSC browser from the RepeatMasked human genome reference sequence including an additional 200 nt at both ends of each CGI.

To map bisulfite reads, the VerJInxer software creates a bisulfite q-gram index (where a q-gram is a sequence substring of length q) of the CGI-RefSeq, which indexes not the sequence itself but a simulated bisulfite-treated version of it. For each q-gram found in the target sequence, all possible methylation patterns are simulated. All of the resulting q-grams are recorded in the index as belonging to this position. As an example, for q = 5, let the reference sequence contain CGACA at some position. Because the first cytosine can be either methylated or not, bisulfite treatment results in a read containing either CGATA or TGATA. Both of these q-grams are included in the index. The VerJInxer software also includes the non-bisulfite-treated q-gram in the index in order to detect entirely or partially unmodified reads. The analysis takes into account that after PCR amplification of bisulfite-treated DNA, every G opposite an unmethylated C is replaced by an A (relative to the RefSeq sequence). The respective q-grams with A substituting G in all possible combinatorial instances created by possible DNA methylation patterns are also simulated and included in the index. For each read, its reverse complement is computed and separately mapped to the reference sequence. The number of q-grams resulting from the simulation can be large. The bisulfite q-gram index (q = 10) of CGI-RefSeq contains 4.36 q-grams per sequence position on average. Matches are reported in BED format. An example of BED format output specifying read numbers with genomic coordinates, strand information, and html color track specification (last column) is shown below:

chr12	50687522	50687614	read no. 9	0	−	50687522	50687614	0,255,0
chr19	2311464	2311570	read no. 53	0	+	2311464	2311570	0,255,0
chr19	2311464	2311570	read no. 83	0	+	2311464	2311570	0,255,0

For additional details pertaining to sequence data analysis, refer to Zeschnigk et al. (2009).

REFERENCES

Margulies M, Egholm M, Altman WE, Attiya S, Bader JS, Bemben LA, Berka J, Braverman MS, Chen YJ, Chen Z, et al. 2005. Genome sequencing in microfabricated high-density picolitre reactors. *Nature* **437**: 376–380.

Roche Diagnostics GmbH. 2009a. *GS FLX Titanium General Library Preparation Method Manual: April 2009.* http://dna.uga.edu/docs/GS-FLX-Titanium-General-Library-Preparation-Method-Manual%20 (Roche).pdf. Roche Applied Science, Mannheim, Germany.

Roche Diagnostics GmbH. 2009b. *GS FLX Titanium General Library Preparation Quick Guide (April 2009).* ftp://ftp.genome.ou.edu/pub/for_broe/titanium/GS%20FLX%20Titanium%20General%20Library%20Preparation%20Quick%20Guide%20(April%202009).pdf. Roche Applied Science, Mannheim, Germany.

Zeschnigk M, Martin M, Betzl G, Kalbe A, Sirsch C, Buiting K, Gross S, Fritzilas E, Frey B, Rahmann S, et al. 2009. Massive parallel bisulfite sequencing of CG-rich DNA fragments reveals that methylation of many X-chromosomal CpG islands in female blood DNA is incomplete. *Hum Mol Genet* **18**: 1439–1448.

WWW RESOURCES

cutadapt tool http://code.google.com/p/cutadapt/

VerJInxer software http://verjinxer.googlecode.com

Illumina Sequencing of Bisulfite-Converted DNA Libraries

The Illumina Solexa sequencing technology uses cluster PCR to amplify the clonal sequencing features. Each sequencing cycle involves single-base extension with a modified DNA polymerase and a mixture of four modified nucleotides. Each of these nucleotides is marked with a different fluorescent label and a cleavable terminating moiety. Each cycle of sequencing is imaged in four channels, followed by chemical cleavage of the fluorescent labels and terminating moiety. The read lengths are typically up to 36 bp, but longer reads of up to 100 bp can be generated. In addition, it can be adopted for paired-end reads as with the other platforms (for further discussion, see Chapter 11, introduction and Protocols 4–10).

The relative low cost of this platform has made it an ideal platform for genome-wide methylome sequencing of mammalian cells. The initial attempt was made with reduced representation bisulfite sequencing (Meissner et al. 2008), in which MspI-digested mouse genomic DNA of 40–220 bp was sequenced after bisulfite conversion. Two groups have mapped mammalian DNA methylomes at base resolution using MethylC-seq (Lister et al. 2009) or BS-seq (Popp et al. 2010) methods. The major difference between these two methods is the usage of adaptors. In the MethylC-seq protocol, methylated adaptors are used to build the sequencing library, whereas two different unmethylated adaptors are used in BS-seq protocol.

We describe below a standard MethylC-seq protocol (Lister et al. 2009) using single-read sequencing on an Illumina Genome Analyzer II platform. The protocol involves ligation of methylated sequencing adaptors to sonicated genomic DNA, gel purification, sodium bisulfite conversion, PCR amplification, and sequencing. A flowchart describing the crucial steps in the protocol is presented in Figure 1.

MATERIALS

It is essential that you consult the appropriate Material Safety Data Sheets and your institution's Environmental Health and Safety Office for proper handling of equipment and hazardous materials used in this protocol.

Recipes for reagents specific to this protocol, marked <R>, are provided at the end of the protocol. See Appendix 1 for recipes for commonly used stock solutions, buffers, and reagents, marked <A>. Dilute stock solutions to the appropriate concentrations.

Reagents

Agarose, low-range ultra
Bisulfite conversion reagents (see Protocol 1)
Cytosine-methylated adapter oligo mix (Illumina)
dATP, dGTP, dTTP (100 mM) (Promega)
DNA ladder (50 bp) (New England Biolabs)
DNA loading buffer (6×) (New England Biolabs)
EB Buffer (from QIAGEN MinElute Purification Kit)
End-It DNA End-Repair Kit (Epicentre Technologies)
Genomic DNA isolated from cell lines or tissues
Klenow (3′–5′ Exo⁻) fragment of DNA polymerase
MinElute Gel Purification Kit (QIAGEN)
MinElute PCR Purification Kit (QIAGEN)

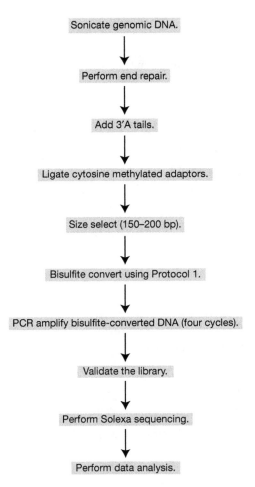

Sonicate genomic DNA.

Perform end repair.

Add 3'A tails.

Ligate cytosine methylated adaptors.

Size select (150–200 bp).

Bisulfite convert using Protocol 1.

PCR amplify bisulfite-converted DNA (four cycles).

Validate the library.

Perform Solexa sequencing.

Perform data analysis.

FIGURE 1. **Flowchart for MethylC-seq.** MethylC-seq uses libraries prepared from bisulfite-converted DNA that is then sequenced using an Illumina platform for deep sequencing.

NEBuffer 2 (10×)
PCR primer PE 1.1 (25 μM) (Illumina)
PCR primer PE 2.1 (25 μM) (Illumina)
PfuTurbo Cx Hotstart DNA Polymerase (Stratagene)
Quick Ligase kit (New England Biolabs, M2200S)
SYBR Gold solution (Life Technologies)
TAE buffer (50×) <A>
Unmethylated *c*I857 *Sam7* λ DNA (Promega)

Equipment

Bioruptor (Diagenode)
Dark Reader (BioExpress)
Genome Analyzer II (Illumina)
Thermal cycler
Water bath (or heat block)

METHOD

Fragmentation of Genomic DNA

1. In a 0.5-mL tube, add 5 μg of genomic DNA and 25 ng of unmethylated *c*I857 *Sam7* λ DNA. Add TE so that the total volume of the DNA sample is 100 μL.

   Genomic DNA can be extracted from cells or tissues using a DNeasy Mini Kit. Unmethylated λ DNA serves as the control for estimation of error rate, which takes into account both nonconversion and sequencing error frequency.

2. Set the multitimer of the Bioruptor on 30 sec "ON" and 30 sec "OFF."

3. Switch the output selector on "High."

4. Fill a water bath with ice-cold water to 0.5 cm below the maximum fill line.

5. Add ~0.5 cm of ice.

6. Put the DNA sample into the Bioruptor.

7. Using 0.5-mL tubes, fill all empty positions in the holder with tubes containing 100 μL of water.

8. Set the timer to 15 min to sonicate the DNA samples.

9. After the first run, put the DNA sample on ice, pump out warm water, and repeat Steps 4–8.

   The DNA samples should be ~50–500 bp after Steps 2–9. Optimize the sonication procedure for specific samples in order to generate fragments of this range.

10. Purify the DNA using a MinElute PCR Purification Kit as per the manufacturer's instructions. Elute the column with 34 μL of EB buffer.

End Repair of the Genomic DNA

11. Prepare the following 50-μL end-repair reaction:

Column purified genomic DNA	34 μL
End-repair buffer (10×) (from End-It DNA Repair Kit)	5 μL
dATP, dGTP, and dTTP mix (2.5 mM)	5 μL
ATP (10 mM)	5 μL
End-It enzyme mix (T4 DNA polymerase and T4 PNK)	1 μL

    Incubate the reaction for 45 min at room temperature.

12. Purify the DNA using a MinElute PCR Purification Kit as per the manufacturer's instructions. Elute the column with 32 μL of EB buffer.

Addition of 3'-A

13. Prepare the following 50-μL tailing reaction:

End-repaired DNA	32 μL
NEBuffer 2 (10×)	5 μL
dATP (1 mM)	10 μL
Klenow fragment (5 U/μL)	3 μL

    Incubate the reaction for 30 min at 37°C.

14. Purify the DNA using a MinElute PCR Purification Kit as per the manufacturer's instructions. Elute the column with 10 μL of EB buffer.

Ligation of Cytosine-Methylated Adaptors

15. Prepare the following 50-μL ligation reaction:

Column purified DNA	10 μL
Quick ligation buffer (2×)	25 μL
Cytosine-methylated adaptor oligo mix	10 μL
Quick T4 DNA ligase (1 U/μL)	5 μL

    Incubate for 15 min at room temperature.

    This procedure uses a 10:1 molar ratio of adaptor to genomic DNA insert, based on a starting quantity of 5 μg of DNA before sonication. The volume of adaptor needs to be adjusted accordingly if a different amount of starting material is used.

16. Purify the DNA using a MinElute PCR Purification Kit as per the manufacturer's instructions. Elute the column with 40 μL of EB buffer.

Size Selection

17. Prepare a gel from 50 mL of 2% low-range ultra agarose in 1× TAE.

18. Add 8 μL of 6× DNA loading dye to the linker-ligated genomic DNA.

19. Load 10 μL of a 50-bp DNA ladder into a well of the gel.

20. Load the entire genomic DNA sample in another lane of the gel, leaving at least one empty lane between the ladder and sample.

21. Run the gel at 120 V until the leading dye has migrated two-thirds of the way down the gel.

22. Stain the gel with 100 mL of 1× TAE buffer containing 10 μL of 10,000× SYBR Gold solution. Cover the staining dish with foil and incubate with gentle rocking for 15–20 min at room temperature.

23. Visualize the gel on a Dark Reader transilluminator. With a clean scalpel, excise the bands in the range 150–200 bp.
    See Troubleshooting.

24. Purify DNA using a MinElute Gel Extraction Kit as per the manufacturer's instructions. Elute the column with 30 μL of EB buffer.

Bisulfite Conversion

25. Perform bisulfite conversion of the DNA following the steps in Protocol 1.

PCR Amplification of Bisulfite-Converted DNA

26. Divide the bisulfite-converted DNA into three equal samples, each in 15 μL of EB buffer. Set up three separate amplification reactions by adding the following reagents to each one-third DNA sample:

PfuTurbo reaction buffer (10×)	5 μL
dNTPs (2.5 mM)	4 μL
PCR primer PE 1.1	1 μL
PCR primer PE 2.1	1 μL
dH₂O	23 μL
PfuTurbo Cx Hotstart DNA polymerase (2.5 U/μL)	1 μL

    Each one-third sample of the bisulfite-converted DNA is used to create an amplified library; that is, three libraries are generated from the DNA converted in Step 25. This step, combined with much lower numbers of amplification cycles, will reduce the incidence of "clonal" reads resulting from PCR amplification.

27. Program the thermal cycler as follows:

Cycle number	Denaturation	Annealing	Polymerization
1	2 min at 95°C		
2	45 sec at 98°C	30 sec at 60°C	4 min at 72°C
3, 4, and 5	15 sec at 98°C	30 sec at 60°C	4 min at 72°C
Last cycle			10 min at 72°C
Hold at 4°C			

28. When the block temperature reaches 95°C, transfer the three reaction tubes (from Step 26) to the thermal cycler, and begin the run.

29. Purify each amplified DNA sample using a MinElute PCR Purification Kit as per the manufacturer's instructions. Elute the column with 40 μL of EB buffer.

30. Gel-purify the PCR products as described in Steps 17–24.
    See Troubleshooting.

Validation of the Library

31. Determine the absorbance of the gel-purified DNA at 260 nm and 280 nm. Ideally, the 260/280 ratio should be ~1.8.

32. Reconfirm the size and concentration of the purified DNA sample by 2% agarose gel electrophoresis.

Solexa Sequencing and Data Analysis

33. Sequence the library on an Illumina Genome Analyzer II as described in the manufacturer's instructions. Single-read sequencing assays can be performed for up to 87 cycles to yield longer sequences for unambiguous mapping to the human genome.

34. Preprocess the sequence reads produced by the standard Illumina pipeline by removing the low-quality base and adaptor base and replacing all of the cytosine reads with thymine.

35. Align all of the reads to two computationally converted reference genomic sequences. In one, the cytosines were replaced with thymines, and in the other, the guanines were replaced with adenines.

36. Add back the cytosines that were originally replaced by thymines to the aligned reads for the identification of methylated cytosines.

TROUBLESHOOTING

Problem (Step 23): The DNA is too large after sonication.
Solution: Sonicate longer.

Problem (Step 30): There is no PCR product after PCR amplification (degradation of DNA due to overheating during sonication).
Solution: Keep the sample cool at all times.

Problem (Step 30): There is no PCR product after PCR amplification (degradation of DNA due to bisulfite treatment).
Solution: Reduce the time for bisulfite conversion.

Problem (Step 30): There is no PCR product after PCR amplification (unsuccessful ligation of adaptors).
Solution: Adjust the ratio of adaptor to genomic DNA to obtain optimal ligation reaction, and increase the time of the ligation reaction.

Problem (Step 30): There is no PCR product after PCR amplification (inefficient end repair and addition of 3'-A).
Solution: Increase the reaction time.

REFERENCES

Lister R, Pelizzola M, Dowen RH, Hawkins RD, Hon G, Tonti-Filippini J, Nery JR, Lee L, Ye Z, Ngo QM, et al. 2009. Human DNA methylomes at base resolution show widespread epigenomic differences. *Nature* 462: 315–322.

Meissner A, Mikkelsen TS, Gu H, Wernig M, Hanna J, Sivachenko A, Zhang X, Bernstein BE, Nusbaum C, Jaffe DB, et al. 2008. Genome-scale DNA methylation maps of pluripotent and differentiated cells. *Nature* 454: 766–770.

Popp C, Dean W, Feng S, Cokus SJ, Andrews S, Pellegrini M, Jacobsen SE, Reik W. 2010. Genome-wide erasure of DNA methylation in mouse primordial germ cells is affected by AID deficiency. *Nature* 463: 1101–1105.

PUBLIC DOMAIN SOFTWARE FOR IDENTIFYING CpG ISLANDS IN PROMOTER AND CODING REGIONS OF MAMMALIAN GENES

Several programs are freely available over the Internet that simplify the task of identifying CpG islands. Some of these programs are listed in Table 1. Here we describe briefly the two computational methods that are used most frequently for the prediction of CpG islands (CGIs).

CpGcluster CpG Island Prediction

The source code is available (Perl) from http://www.biomedcentral.com/content/supplementary/1471-2105-7-446-S7.zip.

CpGcluster is a fast and computationally efficient algorithm that uses only integer arithmetic to predict statistically significant clusters of CpG dinucleotides. All predicted CGIs start and end with a CpG dinucleotide, which should be appropriate for a genomic feature whose functionality is based precisely on CpG dinucleotides. The only search parameter required in CpGcluster is the distance between two consecutive CpGs. None of the statistical properties of CpG islands (e.g., G+C content, CpG fraction, or length threshold) that are often required in other programs are needed as search parameters, which

TABLE 1. Programs for the identification of CpG islands

Organization	URL	Availability	OS compatibility	Browser compatibility
European Bioinformatics Institute EMBOSS CpG Plot/ Report/Isochore	http://www.ebi.ac.uk/Tools/ emboss/cpgplot/index.html	Free (online)		
University of Southern California CpG Island Searcher	http://cpgislands.usc.edu/	Free (online)	Mac or Windows	IE
University of Lyon CpGProD	http://pbil.univ-lyon1.fr/ software/cpgprod_query.html	Free (online)	Mac or Windows	
Iowa State University Finding CpG Islands	http://deepc2.psi.iastate.edu/aat/ mavg/cg.html	Free (online/ download)	Mac or Windows	
Hong Kong University CpG Island Explorer	http://bioinfo.hku.hk/cpgieintro. html	Free (download)	Mac, Windows, Linux, UNIX	
UHN Microarray Centre[a] Human CpG Island Database	http://data.microarrays.ca/cpg/ index.htm	Free (online)		
Bioinformatics Sequence Manipulation Suite	http://www.bioinformatics.org/ sms/	Free (online)	Mac or Windows	Firefox, Safari, Netscape, IE
NCBI QUMA	http://quma.cdb.riken.jp/	Free (online)	Mac or Windows	Firefox, Safari, Opera, IE
Gregor Mendel Institute CyMATE	http://www.gmi.oeaw.ac.at/en/ cymate-index/	Free (online)	Mac or Windows	
CpG PatternFinder[b]			Windows	
CpG Analyzer[b]			Windows	

[a]The website for this software was found, but it did not allow access to the program.
[b]Although articles exist that reference these two programs, neither could be found.

may lead to the high specificity and low overlap with spurious *Alu* elements observed for CpGcluster predictions (Hackenberg et al. 2006, 2010).

HMM Model-Based CpG Island Prediction

See http://rafalab.jhsph.edu/CGI/index.html. The Irizarry laboratory has developed software (make-CGI) that uses a hidden Markov model (HMM)–based approach to CpG island prediction. They have fit the HMM model to genomes from 30 species, and the results are available at the University of California Santa Cruz genome browser (http://genome.ucsc.edu/goldenPath/customTracks/cust-Tracks.html#Multi). Their results support a new view toward the development of DNA methylation in species diversity and evolution. The observed-to-expected ratio (O/E) of CpG residues in islands and nonislands segregated closely along phylogenetic lines and shows substantial loss in both groups in animals of greater complexity, while maintaining a nearly constant difference in CpG O/E ratio between islands and nonisland compartments. Lists of CGIs for some species are available at http://www.rafalab.org (Irizarry et al. 2009; Wu et al. 2010).

REFERENCES

Hackenberg M, Previti C, Luque-Escamilla PL, Carpena P, Martínez-Aroza J, Oliver JL. 2006. CpGcluster: A distance-based algorithm for CpG-island detection. *BMC Bioinformatics* 7: 446. doi: 10.1186/1471-2105-7-446.

Hackenberg M, Barturen G, Carpena P, Luque-Escamilla PL, Previti C, Oliver JL. 2010. Prediction of CpG-island function: CpG clustering vs. sliding-window methods. *BMC Genomics* 11: 327. doi: 10.1186/1471-2164-11-327.

Irizarry RA, Wu H, Feinberg AP. 2009. A species-generalized probabilistic model-based definition of CpG islands. *Mamm Genome* 20: 674–680.

Wu H, Caffo B, Jaffee HA, Irizarry RA, Feinberg AP. 2010. Redefining CpG islands using hidden Markov models. *Biostatistics* 11: 499–514.

WWW RESOURCES

CpGcluster CpG island prediction http://www.biomedcentral.com/content/supplementary/1471-2105-7-446-S7.zip

HMM Model-Based CpG island prediction http://rafalab.jhsph.edu/CGI/index.html:

Rafael Irizarry research page http://www.rafalab.org

DESIGNING PRIMERS FOR THE AMPLIFICATION OF BISULFITE-CONVERTED PRODUCT TO PERFORM BISULFITE SEQUENCING AND MS-PCR

Both bisulfite sequencing (Protocol 1) and MS-PCR (Protocol 2) require primers to amplify the genomic DNA that has been treated with sodium bisulfite. Bisulfite sequencing uses only one pair of primers to amplify both methylated and unmethylated loci. MS-PCR uses two different primer pairs that selectively amplify either unmethylated or methylated regions. We strongly recommend nested PCRs for bisulfite sequencing because this enhances the success and specificity of the genomic region being analyzed. Typically, if only a few loci are being analyzed, the following primer design procedure can be used. When designing primers for multiple genomic loci, use one of the programs described in Table 1.

1. Cut and paste the nucleotide sequence of interest into a Word document.

2. Mark the CpGs in the nucleotide sequence.

3. Replace all of the Cs with Ts.

4. For bisulfite sequencing, design two sets of nested primers of at least 20 nucleotides or more in length. Do not design primers that overlap with CpG dinucleotides in the DNA sequence because these primers will not anneal properly after bisulfite conversion of DNA.

TABLE 1. Programs for the design of primers for multiple genomic loci

Organization	URL	Availability	MSP or BSP
Institute of Enzymology BiSearch	http://bisearch.enzim.hu/	Free (online)	Both
Urogene MethPrimer	http://www.urogene.org/methprimer/index1.html	Free (online)	Both
Sequenom EpiDesigner Beta	http://www.epidesigner.com/	Free (online)	BSP
Ghent University methBLAST	http://medgen.ugent.be/methBLAST/	Free (online)	Both
Ghent University MethPrimerDB	http://medgen.ugent.be/methprimerdb/index.php	Free (online)	Both
Murdoch Children's Research Institute PerlPrimer	http://perlprimer.sourceforge.net/	Free (online)	BSP
Applied Biosytems Methyl Primer Express 1.0	https://products.appliedbiosystems.com/ab/en/US/adirect/ ab?cmd=catNavigate2&catID=602121&tab=Overview	Free (online)	Both
University of Helsinki FastPCR	http://www.biocenter.helsinki.fi/bi/Programs/fastpcr.htm	Free (online)	BSP
Chang Bioscience Primo MSP 3.4	http://www.changbioscience.com/primo/primom.html	Free (online)	MSP

POSTSEQUENCE PROCESSING OF HIGH-THROUGHPUT BISULFITE DEEP-SEQUENCING DATA

High-throughput deep sequencing generates terabytes of sequence data. These deep-sequencing technologies are fast, reliable, and relatively affordable platforms for obtaining genome-wide information. If the resolution is high enough, they can also provide information at single-nucleotide resolution. The technology presents several challenges to obtaining highly reproducible data, and, furthermore, the analyses are expensive to perform. Thus, high-throughput deep sequencing is not the method to use if a project requires only small-scale, low-resolution sequence information. In addition, owing to the lack of universal software for efficient and appropriate data management, the downstream processing and analysis of terabytes of data generated by each run become extremely difficult. Table 1 describes some of the commonly used software programs used in the analysis, alignment, assembly, and visualization of deep-sequencing data.

As of now, there is no standard method for analyzing genome-scale high-throughput data obtained from deep-sequencing methodologies. In the future, developing a unified data analysis and management pipeline system will be very important, because this will allow reproducible analysis of genome-scale DNA methylation data across different laboratories. The National Center for Biotechnology Information (NCBI) has started an initiative in this direction. NCBI has created a portal, called the Sequence Read Archive (SRA), which has been designed to meet the challenges presented by high-throughput deep-sequencing

TABLE 1. Summary of data and statistical analysis methods used by different genome-wide high-throughput deep-sequencing DNA methylation studies

Sequencing platform	Aim of the study	Methods for sequencing data analysis and statistical analysis	Reference
Roche/454	Analysis of global DNA methylation in the tissue and the sera of breast cancer patients	The data and efficacy of bisulfite mutagenesis for this study were analyzed using MethylMapper. The data set was also examined to ensure that each amplicon and each patient had balanced representation. T-tests and discriminant analysis were performed to identify significantly changed amplicons in cancer-free versus cancer samples. All the statistical analyses were carried out using SAS (SAS Institute) and R (R Foundation for Statistical Computing) software.	Bormann et al. 2010
	Analysis of global CpG islands methylation of sperm and female white blood cells	The data was first preprocessed to remove all the sequences that had adaptor sequence. The remaining sequences were then mapped using VerJInxer software available at http://verjinxer.googlecoe.com. For mapping of the CpG island–enriched fragments, a CpG island reference sequence was generated by extracting CpG island sequences and defined by the UCSC browser from the Repeat Masked human genome reference sequence.	Zeschnigk et al. 2009
ABI SOLiD	Evaluating the utility of SOLiD for bisulfite sequencing of large and complex genomes	First, two bisulfite reference genomes were created by replacing all he C's by T's in silico in both the DNA strands of DH10B genome. Sequence reads were then aligned to both bisulfite-converted reference genomes and to the normal DH10B genome using the SOLiD System Analysis Pipeline Tool (http://solidsoftwaretools.com/gf/project/corona/), allowing up to five mismatches per read.	Lister et al. 2008
	Development of a method (Methyl-MAPS) that can globally detect DNA methylation status for both unique and repetitive DNA sequences	Initial tag mapping was performed using the SOLiD System Analysis Pipeline Tool (http://solidsoftwaretools.com/gf/project/corona/). Paired-end tags were each individually mapped in color space, allowing up to two mismatches in each 25-bp tag to the human hg18 sequence obtained from the UCSC Genome Browser. A custom Perl script was used for further identifying the methylated versus unmethylated regions in the sequences. CpG island, RepeatMasker, and RefSeq gene data were all downloaded from the UCSC Genome Browser. Each CpG island was annotated according to its genomic location. Promoter islands were defined as islands that occur within 1 kb of a gene transcription start site.	Edwards et al. 2010

(Continued)

TABLE 1. (*Continued*)

Sequencing platform	Aim of the study	Methods for sequencing data analysis and statistical analysis	Reference
Illumina/SOLEXA	Genome-wide DNA methylation analysis of pluripotent and differentiated cells	Sequence reads from bisulfite-converted libraries were identified using standard Illumina base-calling software and analyzed using a custom computational pipeline.	Meissner et al. 2008
	Single-base resolution DNA methylation map of *Arabidopsis*	Sequence information was extracted from the image files with the Illumina Firecrest and Bustard applications and mapped to the *Arabidopsis* (Col-0) reference genome sequence (TAIR 7) with the Illumina ELAND algorithm. Reads were mapped against computationally bisulfite-converted and nonconverted genome sequences. Reads that aligned to multiple positions in the three genomes were aligned to an unconverted genome using the cross_match algorithm. To identify the presence of a methylated cytosine, a significance threshold was determined at each base position using the binomial distribution, read depth, and precomputed error rate based on the combined bisulfite conversion failure rate and sequencing error. Methylcytosine calls that fell below the minimum required threshold of percent methylation at a site were rejected. This approach ensured that no more than 5% of methylcytosine calls were false positives. The investigators also developed an open source web-based application called Anno-J for visualization of genomic data.	Ruike et al. 2010
	Human DNA methylation analysis at single-base resolution	MethylC-seq sequencing data were processed using the Illumina analysis pipeline, and FastQ format reads were aligned to the human reference genome (hg18) using the Bowtie alignment algorithm46. The base calls per reference position on each strand were used to identify methylated cytosines at a 1% false discovery rate.	Lister et al. 2009

Reproduced, with permission, from Gupta et al. 2010.

technologies. SRA provides a central repository for short read sequencing data and provides links to other resources referring to or using these data. It also allows retrieval based on ancillary information and sequence comparison. Finally, it establishes the basis for user-interactive submission and retrieval.

REFERENCES

Bormann Chung CA, Boyd VL, McKernan KJ, Fu Y, Monighetti C, Peckham HE, Barker M. 2010. Whole methylome analysis by ultra-deep sequencing using two-base encoding. *PLoS ONE* **5:** e9320. doi: 10.1371/journal.pone.0009320.

Edwards JR, O'Donnell AH, Rollins RA, Peckham HE, Lee C, Milekic MH, Chanrion B, Fu Y, Su T, Hibshoosh H, et al. 2010. Chromatin and sequence features that define the fine and gross structure of genomic methylation patterns. *Genome Res* **20:** 972–980.

Gupta R, Nagarajan A, Wajapeyee N. 2010. Advances in genome-wide DNA methylation analysis. *BioTechniques* **49:** iii–xii.

Lister R, O'Malley RC, Tonti-Filippini J, Gregory BD, Berry CC, Millar AH, Ecker JR. 2008. Highly integrated single-base resolution maps of the epigenome in *Arabidopsis*. *Cell* **133:** 523–536.

Lister R, Pelizzola M, Dowen RH, Hawkins RD, Hon G, Tonti-Filippini J, Nery JR, Lee L, Ye Z, Ngo QM, et al. 2009. Human DNA methylomes at base resolution show widespread epigenomic differences. *Nature* **462:** 315–322.

Meissner A, Mikkelsen TS, Gu H, Wernig M, Hanna J, Sivachenko A, Zhang X, Bernstein BE, Nusbaum C, Jaffe DB, et al. 2008. Genome-scale DNA methylation maps of pluripotent and differentiated cells. *Nature* **454:** 766–770.

Ruike Y, Imanaka Y, Sato F, Shimizu K, Tsujimoto G. 2010. Genome-wide analysis of aberrant methylation in human breast cancer cells using methyl-DNA immunoprecipitation combined with high-throughput sequencing. *BMC Genomics* **11:** 137. doi: 10.1186/1471-2164-11-137.

Zeschnigk M, Martin M, Betzl G, Kalbe A, Sirsch C, Buiting K, Gross S, Fritzilas E, Frey B, Rahmann S, et al. 2009. Massive parallel bisulfite sequencing of CG-rich DNA fragments reveals that methylation of many X-chromosomal CpG islands in female blood DNA is incomplete. *Hum Mol Genet* **18:** 1439–1448.

WWW RESOURCES

SOLiD System Analysis Pipeline Tool http://solidsoftwaretools.com/gf/project/corona/

VerJInxer software http://verjinxer.googlecode.com

CHAPTER 13

Preparation of Labeled DNA, RNA, and Oligonucleotide Probes

LABELED NUCLEIC ACIDS AND OLIGONUCLEOTIDES ARE USED IN molecular cloning either as reagents or as probes.

- Labeled fragments of cloned DNA and oligonucleotides of defined size are used as reagents in a variety of methods to analyze RNA, including digestion with S1 nuclease, RNase protection, and primer extension. In these cases, the labeled reagent is converted to products of different sizes that can be detected by a variety of methods, the most common being gel electrophoresis followed by autoradiography or phosphorimaging. Labeled DNA and RNA can also be used as size markers in gel electrophoresis.

- Labeled DNA, RNA, LNA, and oligonucleotide probes are used in hybridization-based techniques to locate and bind DNAs and RNAs of complementary sequence. These techniques include Southern and northern analyses and in situ hybridization.

Success in all these applications depends on the facile introduction of label(s) into the nucleic acid or oligonucleotide, typically by enzymatic methods such as end-labeling, random priming, nick translation, in vitro transcription, and variations of the polymerase chain reaction (PCR). Some of these methods place the label in specific locations within the nucleic acid (e.g., at the 5' or 3' terminus); others generate molecules that are labeled internally at multiple sites. Some methods yield labeled single-stranded products, whereas others generate double-stranded nucleic acids. Finally, some generate probes of defined length, whereas others yield a heterogeneous population of labeled molecules. Table 1 summarizes the options available for labeling nucleic acids and guides investigators to a method of labeling that suits the task at hand.

RADIOACTIVE VERSUS NONRADIOACTIVE LABELING OF NUCLEIC ACIDS

In conventional in vitro radiolabeling, radioactive isotopes—usually ^{32}P—are woven into the natural fabric of the probe in place of their nonradioactive homologs. In 5'-terminal-labeling reactions of DNAs and oligonucleotides, [γ-^{32}P]ATP is used to provide the ^{32}P moiety, which is transferred to the 5'-most deoxyribonucleotide triphosphate (dNTP) using bacterial T4 polynucleotide kinase. This phosphorylation reaction results in the incorporation of one atom of ^{32}P per nucleic acid molecule. In internal labeling methods, radioactive derivatives of [α-^{32}P]dNTP are substituted for their nonradioactive counterparts in reactions that are typically performed in the presence of one ^{32}P-dNTP and three unlabeled dNTPs, resulting in the incorporation of multiple ^{32}P atoms per nucleic acid molecule. Radioactive nucleotides can be incorporated internally using one of several enzymatic methods, such as random priming or primer extension (using the Klenow fragment of *Escherichia coli* DNA polymerase I [Pol I]), nick translation (using a combination of DNase I and

TABLE 1. Methods used to label nucleic acids in vitro

Starting material	Method	Enzyme used	Type of probe generated	Position of label	Size of probe	Application: S1 nuclease mapping	RNase protection	Primer extension	Size markers	Southern and northern blotting	In situ hybridization	Differential screening	Specific activity if radiolabeled	Can be nonradio-labeled?	Protocol number
DNA	Random priming	Klenow	dsDNA	Internal	~400–600 nucleotides					•			5×10^8 to 5×10^9 dpm/μg	Yes	1,2
	Nick translation	DNase I+DNA Pol I	dsDNA	Internal	~400 nucleotides					•	•		5×10^8 to 5×10^9 dpm/μg	Yes	3
	PCR	Thermostable DNA polymerase (e.g., *Taq*)	ss or dsDNA	Internal	Depends on the positions of primers; generates probes of defined length	•				•		•	~1×10^9 to 2×10^9 dpm/μg	Yes	4
	Primer extension	T4 polynucleotide kinase+Klenow	ssDNA	Internal	Generates probes of defined length following restriction enzyme digestion	•				•			~1×10^9 dpm/μg	No	Chapter 6, Protocol 16
	In vitro transcription	DNA-dependent RNA polymerase (e.g., T3, T7, or SP6)	ssRNA	Internal	Generates probes of defined length equivalent to template		•			•	•		~1×10^9 dpm/μg	Yes	5 (see also Chapter 6, Protocol 17)
	Addition of nucleotides to recessed 3' termini	Klenow	dsDNA	3' terminus	Generates probes of defined length approximately equivalent to template		•	•	•				1×10^8 to 2×10^8 dpm/μg	Yes	8

(Continued)

TABLE 1. *(Continued)*

Starting material	Method	Type of probe generated	Position of label	Size of probe	S1 nuclease mapping	RNase protection	Primer extension	Size markers	Southern and northern blotting	In situ hybridization	Differential screening	Specific activity if radiolabeled	Can be nonradiolabeled?	Protocol number
	Phosphorylation of 5' termini	dsDNA	5' terminus	Generates probes of defined length equivalent to template	•	•	•	•				1×10^8 to 2×10^8 dpm/μg	No	9,10,11
RNA	Random priming	ssDNA or DNA–RNA hybrids	Internal	~400–600 nucleotides							•	~1×10^9 dpm/μg	No	6,7
Oligonucleotide or LNA	Phosphorylation of 5' termini	ssDNA	5' terminus	Generates probes of defined length equivalent to oligo					•			1×10^8 to 4×10^8 dpm/μg	No	12
	Addition of nucleotides to 3' termini	ssDNA	3' terminus	Generates probes of defined length approximately equivalent to template					•			3×10^{10} dpm/μg	Yes	13
	Primer extension	dsDNA, whose strands must then be separated	Internal	Generates probes of defined length approximately equivalent to template					•			2×10^{10} dpm/μg	No	14 (see also Chapter 6, Protocol 18)

Enzyme used (by method):
- Phosphorylation of 5' termini: T4 polynucleotide kinase
- Random priming: Reverse transcriptase, Klenow
- Phosphorylation of 5' termini: T4 polynucleotide kinase
- Addition of nucleotides to 3' termini: Terminal deoxynucleotidyl transferase
- Primer extension: Klenow

DNA Pol I), PCR (using a thermostable DNA polymerase such as *Taq*), and in vitro transcription (using a bacteriophage DNA-dependent RNA polymerase). Finally, DNAs can be labeled at their 3′ end by the addition of one or more radiolabeled [α-^{32}P]dNTPs to recessed 3′ termini using the Klenow fragment of *E. coli* DNA polymerase, whereas oligonucleotides are labeled at their 3′ termini using terminal deoxynucleotidyl transferase (TdT), which catalyzes non-template-directed nucleotide incorporation onto the 3′-hydroxyl end of single-stranded DNA (ssDNA).

In contrast to radiolabeling methods, nonradioactive labeling methods involve the incorporation of a chemical group or compound that is not normally found in nucleic acids. The nonradioactive label can be conjugated to the probe via a photochemical or synthetic reaction (for reviews, see Kricka 1992; Mansfield et al. 1995) or, more commonly, a labeled nucleotide can be incorporated into the probe using the same enzymatic reactions (e.g., random priming, nick translation, in vitro transcription) used to incorporate a radioactive label. However, as described in more detail below, unlike radiolabeling, nonradioactive labeling of the 5′ terminus cannot be done enzymatically. After hybridization to the target nucleic acid, the modifying groups on the probe are detected by an appropriate indicator system.

Choosing a Label: Radioactive or Nonradioactive?

The main advantage of radioactive probes is their high sensitivity; the chief disadvantage is the use of hazardous radioactive isotopes. The exposure of personnel to radiation from the energetic β particles released by decaying ^{32}P may be significant. Not surprisingly, the amount of paperwork and equipment required to order and monitor the use and disposal of radiolabeled materials has increased significantly over the years to the point where it has become a real burden. In addition, the physical and political difficulties in the storage and disposal of low-level radioactive waste have become significant and continue to grow. Finally, there are also practical inconveniences; for example, experiments must be scheduled around the delivery of radiolabeled compounds.

The advantages claimed for nonradioactive methods include faster results, more stable probes, and lower cost. However, choosing a nonradioactive detection method is not a trivial task: A plethora of nonradioactive detection kits are available on the commercial market and more are appearing all the time. Claims on the sensitivity and signal-to-noise ratios made on their behalf should be viewed with caution. Many of these claims are based on experiments performed under idealized conditions that cannot be immediately or easily reproduced by inexperienced investigators in the hurly-burly of a working laboratory. Frequently, the signal-to-noise ratio is higher than advertised and the neophyte will likely struggle to match the sensitivity of more traditional methods of radiolabeling. Many of the kits are expensive, and nonradioactive detection methods only become cost-effective after they have been optimized for the particular task at hand.

Perhaps the chief factor to consider when choosing between a radioactive or nonradioactive label is the type of experiment being performed. For some techniques, the use of radioactivity is still the predominant labeling method. For example, differential display, subtractive hybridization, and RNA analyses using S1 nuclease, RNase protection, or primer extension almost exclusively rely on radioactively labeled probes (see Chapter 6). In these methods, the radiolabeled products are first separated by gel electrophoresis and then detected by autoradiography or phosphorimaging. It should be noted, however, that nonradioactive modifications of these methods have been described (see, e.g., Turnbow and Garner 1993; Chan et al. 1996).

For Southern and northern hybridization–based techniques (see Chapters 2 and 6, respectively), researchers have a choice between using radioactive and nonradioactive labels. ^{32}P-labeled probes offer high sensitivity—they can detect minute quantities of immobilized target DNA (<1 pg)— but also suffer from a short half-life (14.3 d) and the inability to be used for high-resolution imaging. Nonradioactive probes offer several advantages: Not only are they less hazardous and less costly than radiochemical techniques, but their half-life and shelf life are longer, and detection of hybrids is much faster than for radioactive probes.

Nowadays, in situ hybridization techniques rely exclusively on the use of nonradiolabeled probes. Traditionally, in situ hybridization for the detection of messenger RNA (mRNA) in tissues was performed on sectioned material using probes that were labeled with ^{3}H or ^{35}S, which required the use of a photographic film emulsion covering the sections for the detection of the signal. In addition to the inherent problems associated with radioactive probes, such as safety and limited shelf life, the scatter associated with radioactive decay greatly limited the spatial resolution of the technique. The development of highly sensitive nonradioactively labeled probes has allowed in situ hybridization to be performed directly in whole embryos, tissues, and, in the case of fluorescence in situ hybridization (FISH), in single cells with incredibly high resolution.

The protocols in this chapter provide detailed methods for generating radiolabeled probes. Where applicable, the materials and steps for synthesizing nonradioactive probes are integrated into the protocol.

TYPES OF NONRADIOACTIVE DETECTION SYSTEMS

Indirect Detection Systems

The first nonradioactive methods of detection developed in the 1980s were based on the labeling of the nucleic acid probe with dinitrophenol (Keller et al. 1988), bromodeoxyuridine (Traincard et al. 1983), or biotin (Langer et al. 1981; Chu and Orgel 1985; Reisfeld et al. 1987). It was this latter compound that provided the most durable and sensitive of these prototypic systems and remains one of the most commonly used nonradioactive labels today. After hybridization, the biotinylated probe is detected via interaction with streptavidin that has been conjugated to a reporter enzyme—usually alkaline phosphatase. The membrane is then exposed to an enzyme capable of hydrolyzing a colorigenic, fluorogenic, or chemiluminescent substrate. The sensitivity of this system derives from the rapid enzymatic conversion of the substrate to a product that is colored or emits visible or UV light at a specific wavelength. The distribution and intensity of the product correspond to the spatial distribution and concentration of target sequences on the membrane.

Because the reporter enzyme is not conjugated directly to the probe but is linked to it through a bridge (in this case, streptavidin-biotin), this type of nonradioactive detection is known as an indirect system. Many of the kits marketed today by commercial companies rely on indirect detection systems that are elaborations of the original biotin:streptavidin:reporter enzyme method. Most of these kits use biotinylated nucleotides (e.g., biotin-11-dUTP, biotin-12-dUTP, or biotin-16-dUTP) (UTP, uridine triphosphate) as labeling agents. The biotin moiety is conjugated to the nucleotide via a spacer that is 11–16 carbon atoms long and that reduces the possibility of steric hindrance during synthesis and detection of biotinylated nucleic acid probes. Under most circumstances, biotinylated probes work very well and are free from problems. However, because biotin (vitamin H) is a ubiquitous constituent of mammalian tissues and because biotinylated probes tend to stick doggedly to certain types of nylon membranes, high levels of background can occur during in situ, northern, and Southern hybridization. These difficulties can be avoided by labeling probes with a compound that (1) has a low affinity for membranes of all types and (2) does not occur naturally in eukaryotic animals. The cardenolide digoxigenin (DIG) (Pataki et al. 1953), which is synthesized exclusively by *Digitalis* plants (Hegnauer 1971), fulfills both of these criteria (see Fig. 1).

DIG, like biotin, can be chemically coupled to linkers and nucleotides and, as described below, DIG-substituted nucleotides can be incorporated into nucleic acid probes by standard enzymatic methods. These probes generally yield significantly lower backgrounds than those labeled with biotin. DIG-labeled hybrids are detected with an anti-DIG Fab (fragment antigen-binding) fragment conjugated to a reporter enzyme (usually, alkaline phosphatase; less frequently, horseradish peroxidase (HRP); and even less frequently, β-galactosidase, xanthine oxidase, or glucose oxidase) (Höltke et al. 1990, 1995; Seibl et al. 1990; Kessler 1991).

A third class of nonradioactive probes, fluorescein-labeled probes (fluorescein-12-UTP or -dUTP), are detected by antifluorescein antibodies conjugated to alkaline phosphatase. Labeling

FIGURE 1. *Digitalis purpurea.* (Redrawn from Culbreth 1927.)

of nucleic acids with fluorescein, biotin, or DIG is equally easy, and the sensitivity of indirect systems based on fluorescein is at least equal to those based on biotin and DIG. All three labeling systems are capable of detecting subpicogram quantities of target nucleic acids on Southern and northern blots. Nevertheless, for reasons that are unclear, indirect systems based on fluorescein are not as commonly used as those based on biotin and DIG.

Methods to Label Probes with Biotin, Digoxigenin, or Fluorescein

Enzymatic Methods

Biotinylated, digoxigeninated, or fluoresceinated nucleotides can be used to internally label DNA by any of the standard enzymatic techniques such as random priming, nick translation, and PCR (see Table 1). Labeled RNA can be synthesized in vitro in transcription reactions catalyzed by SP6, T3, or T7 RNA polymerases. Labeling of the 3′ terminus can be performed by a 3′-end-filling reaction for DNA probes or via TdT for oligonucleotide probes. It is important to note that, unlike radiolabeling, nonradioactive labeling of the 5′ terminus cannot be done enzymatically. Instead, nonradioactive 5′ labeling of oligonucleotides is performed directly during oligonucleotide synthesis by chemical addition of a biotin- or DIG-succinimide ester. Most oligonucleotide synthesis companies provide a service for direct labeling of oligonucleotides with biotin or DIG.

The label in all of these reactions is provided as modified uridine ribotriphosphates or deoxyribotriphosphates that are used in place of UTP or deoxythymidine triphosphate (dTTP) in the labeling reaction. Commonly used modified nucleotides include biotin-11-dUTP, biotin-16-dUTP, DIG-11-dUTP, and fluorescein-12-dUTP. In this nomenclature, the number preceding "dUTP" refers to the number of carbon atoms in the backbone of the linker between dUTP and the label; a linker length of "11" is optimal for the majority of applications.

Biotin-, DIG-, or fluorescein-conjugated nucleotides typically are purchased from a commercial source, either individually or in combination with a labeling kit that provides all reagents required for probe synthesis. Such kits are marketed by many commercial suppliers (e.g., Ambion, Enzo Life Sciences, PerkinElmer, Roche Applied Science, Thermo Scientific). It is worth noting that it is also possible to synthesize conjugated nucleotides "in house"; for example, labeled nucleotides can be synthesized by chemically coupling allylamine-dUTP to succinimidyl-ester derivatives of biotin or DIG (Henegariu et al. 2000).

In contrast to radiolabeling with [^{32}P]NTPs or [^{32}P]dNTPs, where the efficiency of incorporation can be monitored and the specific activity of the final probe easily calculated, there is no simple or rapid method to monitor the progress of the nonisotopic labeling reactions or to calculate their efficiency. In the case of biotin- and DIG-labeled probes, the manufacturers recommend that no more than 30%–35% of the thymidine residues in the probe be replaced by labeled uridine residues (see Gebeyehu et al. 1987; Lanzillo 1990). Higher levels of replacement cause a reduction in the sensitivity with which target sequences can be detected, presumably because of the steric hindrance between closely spaced adducts during the hybridization step. However, because of a lack of an effective monitoring system, synthesis of nonradioactive probes remains an ad hoc affair. It is therefore best to perform a pilot reaction in which samples are withdrawn at intervals during the course of the labeling reaction. The labeled products are tested in dot or slot blots against dilutions of the target DNA. Large-scale labeling reactions can then be designed and performed with some confidence that the labeled product will have the required sensitivity.

Photolabeling

Biotin and DIG can also be attached to nucleic acids in photochemical reactions. In this method, the label is linked to a nitrophenyl azido group that is converted by irradiation with UV or strong visible light to a highly reactive nitrene that can form stable covalent linkages to DNA and RNA (Forster et al. 1985; Habili et al. 1987). Photochemical labeling is far less efficient than enzymatic labeling: At best, only 1 base in 150 becomes modified, and the resulting probe is not potent enough to detect single-copy sequences in Southern analyses of mammalian DNA. However, newer versions of biotinylated probes combine biotin with psoralen, a nucleic acid intercalating moiety (e.g., BrightStar Psoralen-Biotin labeled probes from Applied Biosystems). The psoralen-biotin moiety intercalates within the RNA and is then covalently bound by irradiation with UV light; such probes are reported to be two to four times more sensitive than enzymatically labeled biotinylated nucleotides.

Detection of Nonradioactively Labeled Probes after Hybridization

Southern and Northern Blotting

Two steps are required to detect probes labeled with biotin, fluorescein, or DIG residues after hybridization to Southern and northern blots. The membrane is first exposed to a synthetic conjugate consisting of the reporter enzyme and a ligand that binds tightly and with high specificity to the labeling moiety. In the case of biotin, most of the detection procedures exploit the high-affinity reaction between biotin and enzyme-conjugated streptavidin. Alternative procedures rely on specific monoclonal or polyclonal antibiotin antibodies. Fluoresceinated or digoxigeninated probes are routinely detected by interaction with specific enzyme-conjugated antibodies. The reporter enzyme is then assayed with colorimetric, fluorogenic, or chemiluminescent substrates.

In colorimetric assays (e.g., see Leary et al. 1983; Urdea et al. 1988; Rihn et al. 1995), a combination of two dyes is used to detect alkaline phosphatase that has been captured by labeled hybrids. The enzyme catalyzes the removal of the phosphate group from BCIP (5-bromo-4-chloro-3-indolyl phosphate) (Horwitz et al. 1966), generating a product that oxidizes and dimerizes to dibromodichloro indigo. The reducing equivalents produced during the dimerization reaction reduce NBT (nitroblue tetrazolium) (McGadey 1970) to an insoluble purple dye, diformazan, that becomes visible at sites where the labeled probe has hybridized to its target (Leary et al. 1983; Chan et al. 1985). The results can be analyzed visually and recorded on conventional photographic film. Colorimetric detection works well with nitrocellulose and PVDF membranes, but not as well with nylon and charged nylon. Unfortunately, even when working at their best, colorimetric methods are two or more orders of magnitude less sensitive than other nonradioactive detection methods, for example, chemiluminescence (Bronstein et al. 1989a; Beck and Köster 1990; Kerkhof 1992; Kricka 1992). Detection of single-copy sequences in mammalian genomes is therefore barely within reach of colorimetric assays (Düring 1993). A final disadvantage is that the colored precipitates are difficult to remove from membranes, making reprobing difficult or impossible. In some cases, the procedures used to strip the precipitates (100% formamide at 50°C–60°C) either dissolve the membrane or greatly increase its fragility.

Fluorescent assays for alkaline phosphatase make use of HNPP (2-hydroxy-3-naphthoic acid 2'-phenylanilide phosphate) (Kagiyama et al. 1992). After dephosphorylation, HNPP generates a fluorescent precipitate on membranes that can be excited by irradiation at 290 nm from a transilluminator. Light emitted at 509 nm can be captured by charged couple device (CCD) cameras or on Polaroid film using the same photographic setup (filters, etc.) as that used for conventional ethidium-bromide–stained gels. Fluorescent assays using HNPP are more sensitive than colorimetric methods and are capable of detecting single-copy sequences in Southern blots of mammalian genomic DNA when the membrane is incubated with substrate for 2–10 h (Diamandis et al. 1993; Höltke et al. 1995). The fluorescent precipitate can be easily removed by washing in ethanol and the filters used for several more rounds of hybridization before the signal-to-noise ratio increases to unacceptable levels.

Chemiluminescence is the fastest and most sensitive assay to detect DNA labeled with biotin, fluorescein, or DIG, via conjugates containing HRP or alkaline phosphatase (Beck et al. 1989; Bronstein et al. 1989b,c; Schaap et al. 1989). An HRP-luminol detection system (Thorpe and Kricka 1986) is marketed by several companies (e.g., BioRad, Life Technologies, Sigma-Aldrich). HRP catalyzes the oxidation of luminol in the presence of hydrogen peroxide, generating a highly reactive endoperoxide that emits light at 425 nm during its decomposition to its ground state. A wide range of compounds, including benzothiazoles, phenols, naphthols, and aromatic amines, can significantly enhance the amount of light generated during the reaction (Thorpe and Kricka 1986; Pollard-Knight et al. 1990; Kricka and Ji 1995). This signal amplification, which can be as much as 1000-fold (Whitehead et al. 1983; Hodgson and Jones 1989), increases the sensitivity of the HRP-luminol system to the point where it can detect single-copy genes in Southern blots of mammalian DNA. Light is produced at a rapid rate for the first few minutes of the reaction but thereafter declines and is no longer detectable after 1–2 h. Under ideal conditions, $\sim$1 pg (5×10^{-17} M) of target DNA can be detected on Southern blots after 1 h of exposure on gray-tinted, blue-light-sensitive X-ray film or a cooled CCD camera (Beck and Köster 1990; Durrant et al. 1990; Kessler 1992). The probes can be stripped from filters by heating them to 80°C (Dubitsky et al. 1992).

Many commercial suppliers (e.g., Applied Biosytems, Life Technologies, Roche Applied Science, Sigma-Aldrich) sell alkaline phosphatase detection systems that use dioxetane-based chemiluminescent substrates such as AMPPD and Lumigen-PPD and their more recent derivatives CSPD and CDP-Star. One of the most sensitive chemiluminescent substrates available (Edwards et al. 1994; Höltke et al. 1994), CDP-Star differs from AMPPD in that it carries halogen substituents at the 5 positions in its adamantyl and phenyl rings. This suppresses the tendency of the 1,2-dioxetane to aggregate and reduces background caused by thermal degradation (Bronstein et al. 1991). Cleavage of the phosphate ester of CDP-Star by alkaline phosphatase drastically reduces the thermal stability of the dioxetane ring, which consequently decomposes with emission of light at 465 nm. Nylon membranes (either amphoteric or positively charged) strongly enhance the signal (Tizard et al. 1990) by providing hydrophobic domains into which the dephosphorylated intermediate produced in the chemiluminescent reaction becomes sequestered. This stabilizes the intermediate and reduces its nonluminescent decomposition. In turn, the intensity of the light emitted from the excited dioxetane anion begins as a glow that increases in intensity for several minutes and then persists for several hours (Martin and Bronstein 1994). The hydrophobic interactions between the nylon and the anion also cause a "blue shift" of $\sim$10 nm in the emitted light, that is, from 477 to 466 nm (Beck and Köster 1990; Bronstein 1990).

In most experimental situations, the extended kinetics of chemiluminescence on nylon filters are advantageous because they allow time to capture images at several exposures. However, the slow kinetics may be disadvantageous when alkaline phosphatase–triggered chemiluminescence is used to detect extremely low concentrations of DNA, RNA, or protein, for example, when the target band on a filter is expected to contain only 10^{-18} mol. In such cases, a halogen-substituted derivative of AMPPD, for example, CSPD, may be a better choice. The addition of a chlorine atom to the

5 position of the adamantyl group restricts its interactions with nylon filters. With this compound, the time to reach maximum light emission is markedly reduced so that very small quantities of target molecules can be detected rapidly (e.g., see Martin et al. 1991).

Under ideal conditions, as little as 1 zmol (10^{-21} M) of alkaline phosphatase or 1×10^{-19} mol of DNA can be detected (Schaap et al. 1989; Beck and Köster 1990; Bronstein 1990; Kricka 1992). Such sensitivity means that a single-copy gene in a Southern blot of mammalian genomic DNA can be detected after the nylon membrane has been exposed for ~5 min to gray-tinted, blue-light-sensitive X-ray film or a cooled CCD camera (Höltke et al. 1995). The probes can be stripped by using 50% formamide at 65°C (Dubitsky et al. 1992) and the nylon membranes can then be used for several additional rounds of hybridization.

In Situ Hybridization

Detection methods for in situ hybridization are similar to those used in Southern and northern hybridization and generally rely on colorimetric assays following detection of biotin or DIG using streptavidin or a specific antibody. Following hybridization, samples are incubated with alkaline phosphatase color reagents NBT and BCIP, followed by visual detection using a bright field microscope.

For FISH, detection of the biotin or DIG moiety is performed using a fluorochrome-conjugated detection reagent, such as streptavidin, or an antibiotin or anti-DIG antibody conjugated to fluorescein isothiocyanate (FITC), rhodamine, or Texas Red. The probe is then detected by immunofluorescence microscopy.

Direct Detection Systems

In this type of detection, nucleic acid probes are labeled directly with fluorescent dyes such as fluorescein, Texas Red, rhodamine, Cy3, or Cy5 (Agrawal et al. 1986); lanthanide chelates (europium) (Templeton et al. 1991; Dahlén et al. 1994); acridinium esters (Nelson et al. 1992); or enzymes, for example, alkaline phosphatase (Jablonski et al. 1986) or HRP (Renz and Kurz 1984). The presence of such covalently attached adducts is assayed by any one of a number of techniques, including colorimetry, chemiluminescence, bioluminescence, time-resolved fluorometry, or energy transfer/fluorescence quenching. Of course, direct detection methods can be converted to indirect methods if an enzyme-antibody conjugate specific for the adduct is available.

Because it forms the basis of automated DNA sequencing methods (Smith et al. 1986; Ansorge et al. 1987; Prober et al. 1987), direct labeling of nucleic acids with fluorescent compounds has revolutionized molecular biology (for further discussion, see the introduction to Chapter 11). A single fluorescent adduct per DNA molecule is sufficient for detection by automated DNA sequencers when the DNA is in the gel. Furthermore, the ability of ABI-type automated sequencers to discriminate accurately among different fluors signifies that all four sequencing reactions can be analyzed in a single lane of a gel. The stunning success of direct labeling in automated DNA sequencing has, however, not spread to other areas of molecular cloning, where direct nonradioactive methods of detection are rarely used. This is because of the following.

- Traditional methods used for the direct conjugation of enzymes such as alkaline phosphatase require the extensive use of column chromatography, polyacrylamide gel electrophoresis, and/or high-performance liquid chromatography (HPLC) to obtain a pure product (Jablonski et al. 1986; Urdea et al. 1988; Farmar and Castaneda 1991). Commercial kits are available but are expensive and typically yield an impure product. Newer methods of conjugation (e.g., see Reyes and Cockerell 1993) are simpler but slightly less efficient.

  Whatever method is used, a separate conjugation step is required for each different probe. Once attached, large enzymes such as alkaline phosphatase may reduce the rate of annealing of probes and may be labile under the conditions conventionally used for annealing of nucleic acids. Hybridization and washing must therefore be performed at low temperature (usually <50°C) in aqueous solvents. Under such nonstringent conditions, high backgrounds are an inevitable problem. Because of the presence of bulky, charged adducts, probes directly labeled with enzymes may bind nonspecifically but tightly to certain types of membranes. For example, the background signal from oligonucleotide probes directly conjugated to alkaline phosphatase is considerably greater on charged nylon than on neutral nylon membranes (Benzinger et al. 1995).

- The detection of single-copy sequences in Southern hybridization of mammalian DNA lies close to the limit of sensitivity of many direct methods unless specialized equipment is available (Urdea et al. 1988). Detection of fluorescent adducts, for example, requires irradiating with light of one wavelength and acquiring data at another. The signal must then be enhanced electronically and stored digitally.

  Because of these problems, it is not surprising that direct detection techniques have fallen from favor and that indirect detection has become the dominant nonradioactive method to locate target nucleic acids in Southern and northern hybridization. However, despite these problems, probes directly labeled with alkaline phosphatase are now produced commercially for forensic and pathological analyses. The sensitivity of these probes can approach that of isotopically labeled probes (Dimo-Simonin et al. 1992; Klevan et al. 1993; Benzinger et al. 1995). The use of such standardized nonisotopically labeled probes is expected to increase greatly in laboratories involved, for example, in constructing large numbers of DNA profiles or in diagnosing infectious diseases.

One area in which direct detection techniques have become more popular is in the use of FISH, which allows probe-target hybrids to be visualized under a fluorescence microscope immediately after the hybridization reaction. Direct fluorescent labeling can be performed in one of two ways: fluorochromes can be terminally added to the 5′ or 3′ end of the nucleic acid (Bauman et al. 1980) or fluorochome-conjugated dNTPs can be used in enzyme labeling reactions (Renz and Kurz 1984). Fluorescently labeled probes offer several advantages over biotin- or DIG-labeled probes including increased specificity (because shorter probes can be used) and reduced background signal. However, they also tend to generate a lower signal intensity and, as noted above, are costly to synthesize, which has somewhat hampered their popularity as reagents in FISH.

DESIGNING OLIGONUCLEOTIDES FOR USE AS PROBES

Success with oligonucleotide probes begins with the design of an oligonucleotide (1) that hybridizes specifically to its target and (2) whose physical properties (e.g., length and GC content) do not impose unnecessary constraints on the experimental protocol. In a few cases, the investigator has little latitude in the design of the oligonucleotide. For example, when oligonucleotides are used as allele-specific probes for the presence or absence of defined mutations, the salient properties of the oligonucleotides are defined to a very large extent by the particular sequences in which the mutation is embedded. In most other circumstances, however, the chance of success can be increased by careful adjustment of parameters defined by the investigator, such as the length, sequence, and base composition of the oligonucleotide. The rules outlined below were developed to optimize the design of oligonucleotide hybridization probes.

Melting Temperature and Hybridization Temperature

An ideally designed oligonucleotide probe should form a perfect duplex only with its target sequence. These duplexes should be sufficiently stable to withstand the posthybridization washing steps used to remove probes nonspecifically bound to nontarget sequences. For oligonucleotides of 200 nucleotides or less, the reciprocal of the melting temperature of a perfect hybrid (T_m^{-1}, measured in degrees Kelvin) is approximately proportional to n^{-1}, where n is the number of bases in the oligonucleotide (see Gait 1984).

Several equations are available to calculate the melting temperature of hybrids formed between an oligonucleotide primer and its complementary target sequence. Because none of these is perfect, the choice among them is largely a matter of personal preference (for additional information on calculating melting temperatures, see the information panel Melting Temperatures). The melting temperatures of each member of a primer:target duplex should obviously be calculated using the same equation.

An empirical and convenient equation known as the "Wallace rule" (Suggs et al. 1981b; Thein and Wallace 1986) can be used to calculate the T_m of perfect duplexes, 15–20 nucleotides in length, in solvents of high ionic strength (e.g., 1 M NaCl or 6× SSC):

$$T_m \text{ (in °C)} = 2(A + T) + 4(G + C),$$

where $(A+T)$ is the sum of the A and T residues in the oligonucleotide and $(G+C)$ is the sum of G and C residues in the oligonucleotide.

The equation derived originally by Bolton and McCarthy (1962) and later modified by Baldino et al. (1989) predicts reasonably well the T_m of oligonucleotides, <100 nucleotides long, in cation concentrations of 0.5 M or less and with a $(G+C)$ content between 30% and 70%. It also compensates for the presence of base-pair mismatches between the oligonucleotide and the target sequence:

$$T_m \text{ (in °C)} = 81.5°C + 16.6(\log_{10}[Na^+]) + 0.41(\%[G + C]) - 675/n - 1.0m,$$

where n is the number of bases in the oligonucleotide and m is the percentage of base-pair mismatches. This equation can also be used to calculate the melting temperature of an amplified product whose sequence and size are both known. When PCR amplification is performed under standard conditions, the calculated T_m of the amplified product should not exceed ~85°C, thus ensuring complete separation of its strands during the denaturation step. Note that the denaturation temperature in PCR is more accurately defined as the temperature of irreversible strand separation of a homogeneous population of molecules. The temperature of irreversible strand separation is several degrees higher than the T_m (typically, 92°C for DNA whose content of $G+C$ is 50%) (Wetmur 1991).

Neither of the above equations takes into account the effect of base sequence (as opposed to base composition) on the T_m of oligonucleotides. A more accurate estimate of the T_m can be obtained by incorporating nearest-neighbor thermodynamic data into the equations (Breslauer et al. 1986; Freier et al. 1986; Kierzek et al. 1986; Rychlik et al. 1990; Wetmur 1991; Rychlik 1994). A relatively straightforward equation, described by Wetmur (1991), is

$$T_m \text{ (in °C)} = (T^0\Delta H^0)/(\Delta H^0 - \Delta G^0 + RT^0 \ln[c]) + 16.6 \log_{10}([Na^+]/\{1.0 + 0.7[Na^+]\}) - 269.3,$$

where

$$\Delta H^0 = \sum_{nn}(N_{nn}\Delta H^0_{nn}) + \Delta H^0_p + \Delta H^0_e,$$

$$\Delta G^0 = \sum_{nn}(N_{nn}\Delta G^0_{nn}) + \Delta G^0_i + \Delta G^0_e,$$

N_{nn} is the number of nearest neighbors (e.g., 13 for a 14-mer), $R=1.99$ cal/mol (in degrees Kelvin), $T^0=298.2°K$, c is the total molar strand concentration, and $[Na^+]\leq 1$ M.

Some of the thermodynamic terms can be approximated as follows: average nearest-neighbor enthalpies,

$$\Delta H^0_{nn} = -8.0 \text{ kcal/mol};$$

average nearest-neighbor free energies,

$$\Delta G^0_{nn} = -1.6 \text{ kcal/mol};$$

initiation term,

$$\Delta G^0_i = +2.2 \text{ kcal/mol};$$

average dangling end enthalpy,

$$\Delta H_e^0 = -8.0 \, \text{kcal/mol/end};$$

average dangling end free energy,

$$\Delta G_e^0 = -1 \, \text{kcal/mol/end};$$

and average mismatch/loop enthalpy,

$$\Delta H_p^0 = -8.0 \, \text{kcal/mol/mismatch}.$$

Note that for each mismatch or loop, N_{nn} is reduced by 2.

When the oligonucleotide is used as a probe, hybridization is usually performed at ~5°C–12°C below the calculated T_m, and stringent posthybridization washing is performed at ~5°C below the T_m. Working so close to the calculated T_m has two consequences: (1) It reduces the number of mismatched hybrids and (2) less desirably, it reduces the rate at which perfect hybrids form.

For all practical purposes, hybrid formation between complementary DNA molecules >200 nucleotides in length is an irreversible reaction when annealing is performed at ~10°C below the calculated T_m. The chance that such a long stretch of perfect double helix will unwind under the conditions conventionally used for hybridization (68°C and ~1 M [Na$^+$]) is small. However, hybrids (even perfect hybrids) formed between short oligonucleotides and their target sequences at 5°C–10°C below the T_m are far easier to unwind; thus, hybridization reactions of this type can be considered reversible. This behavior has important practical consequences. The concentration of oligonucleotide in the hybridization reaction should be high (0.1–1.0 pmol/mL) so that the annealing reaction rapidly reaches equilibrium within 3–8 h. Posthybridization washing, however, should be brief (1–2 min) and should be initially performed under conditions of low stringency and then under conditions of stringency that are at least equal to those used for hybridization (Miyada and Wallace 1987).

Length of Oligonucleotide Probes

The longer an oligonucleotide, the higher its specificity for the target sequence. The following equations can be used to calculate the probability that a specific string of nucleotides will occur within a sequence space (Nei and Li 1979):

$$K = \frac{[g/2]^{G+C} \times [(1-g)/2]^{A+T}}{N},$$

where K is the expected frequency of occurrence within the sequence space, g is the relative G+C content of the sequence space, and G, C, A, and T are the number of specific nucleotides in the oligonucleotide. For a double-stranded genome of size N (in nucleotides), the expected number (n) of sites complementary to the oligonucleotide is $n = 2NK$.

These equations predict that an oligonucleotide of 14–15 nucleotides would be represented only once in a mammalian genome where N is ~3.0×10^9. In the case of a 16-mer, there is only one chance in 10 that a typical mammalian complimentary DNA (cDNA) library (complexity of ~10^7 nucleotides) will fortuitously contain a sequence that exactly matches that of the oligonucleotide. However, these estimates are based on the assumption that the distribution of nucleotides in mammalian genomes is random. Unfortunately, this is not the case, due to bias in codon usage (Lathe 1985) and because a significant fraction of the genome is composed of repetitive DNA sequences and gene families. Because of the presence of these elements, no more than 85% of the mammalian genome can be targeted precisely, even by probes and primers that are 20 nucleotides in length (Bains 1994).

To minimize problems of nonspecific hybridization, it advisable to use oligonucleotides longer than the statistically indicated minimum. Any clones that then hybridize to the probe are likely to be derived from the gene of interest. Bear in mind that when a cDNA library is screened with an oligonucleotide probe, no relationship exists between the observed number of positive clones and their frequency predicted by statistics. For example, if by chance the oligonucleotide should match an expressed sequence that is abundantly represented in mRNA, the number of clones that hybridize to the probe will be much larger than theory predicts. Before synthesizing an oligonucleotide probe or primer, it is advisable to scan DNA databases to ensure that the proposed sequence occurs only in the desired gene and not in vectors, undesired genes, or repetitive elements (e.g., see Mitsuhashi et al. 1994).

The effect of hybridization between imperfectly matched sequences cannot be easily quantified because different types of mismatching (mispairing between single bases, loopouts on either strand, or multiple mismatches, whether closely or distantly spaced) have different effects on the stability of double-stranded DNA (dsDNA). For example, a single mismatch in the center of a short oligonucleotide (of ~16 nucleotides in length) will destabilize the hybrid by ~7°C (Wetmur 1991) and, consequently, may attenuate the hybridization signal to a very considerable degree. On the other hand, a mismatch at the 3′ end of a probe will have little effect on hybridization signal (Ikuta et al. 1987) but greatly diminish the ability of an oligonucleotide to prime DNA synthesis.

Locked Nucleic Acids

Locked nucleic acids (LNAs) are a class of nucleic acid analogs in which the ribose ring is "locked" into a particular conformation by virtue of a methylene bridge connecting the 2′-oxygen atom and the 4′-carbon atom (reviewed in Vester and Wengel 2004). This restricted confirmation promotes optimal Watson–Crick binding. When incorporated into a DNA oligonucleotide, LNAs increase the stability of the resulting duplex, thereby increasing the hybridization properties (i.e., melting temperature) of the oligonucleotide. Incorporation of an LNA into an oligonucleotide increases the T_m of the oligonucleotide by 2°C–8°C per LNA. The high binding affinity of LNA-enhanced oligonucleotides allows for the use of shorter probes.

First synthesized by Jesper Wengel and colleagues in 1998 (Koshkin et al. 1998), LNA-enhanced oligonucleotides are now used in hybridization-based assays when extra specificity and sensitivity are required. They have become particularly prevalent in probes used in in situ hybridization assays (Thomsen et al. 2005; Kubota et al. 2006). The only disadvantage of LNAs is their high cost relative to DNA oligonucleotides; for this reason, DNA oligonucleotides are still widely preferred for most applications.

LNAs are commercially available from many sources (e.g., Exiqon, Integrated DNA Technologies, Bio-Synthesis Inc.). The synthesis and incorporation of LNA bases are achieved using standard DNA synthesis chemistry, and LNA-enhanced oligonucleotides can be purified and analyzed using the same standard methods as those used for DNA. Due to the high affinity and thermal stability of the LNA:DNA duplex, it is not advisable to have >15 LNA bases in an oligonucleotide, because it induces strong self-hybridization; for hybridization-based assays, it is generally recommended to substitute 4–6 bases with LNAs. Placement of the LNA analogs within the oligonucleotide is important. To assist the researcher, several software programs are available for LNA design. One of the most popular programs is OligoDesign (Tolstrup et al. 2003), which is freely accessible at http://www.exiqon.com/oligo-tools. OligoDesign software features recognition and filtering of the target sequence by genome-wide BLAST analysis in order to minimize cross-hybridization with nontarget sequences. Furthermore, it includes algorithms for prediction of melting temperature, self-annealing, and secondary structure for LNA-substituted oligonucleotides, as well as secondary structure prediction of the target nucleotide sequence.

Like DNA oligonucleotides, LNA oligonucleotide probes are labeled enzymatically at either the 5′ or 3′ end: 5′-end-labeling reactions are performed using T4 polynucleotide kinase in the presence of [γ-^{32}P]ATP, whereas 3′-end-labeling is performed using terminal transferase and a radioactive, biotin, DIG, or fluorescent nucleotide.

Degenerate Pools of Oligonucleotides

In molecular cloning, the situation frequently arises in which a short sequence of amino acids has been obtained by sequencing a purified protein. Because of the degeneracy of the genetic code, many different oligonucleotides can potentially code for a given tract of amino acids. For example, 64 possible 18-mers can code for the sequence Asn Phe Tyr Ala Trp Lys. However, only one of these oligonucleotides will exactly match the coding sequence used in the gene of interest. Because there is no way to know a priori which of these oligonucleotides has its true counterpart in the gene, pools of oligonucleotides containing all potential coding combinations are synthesized and used as probes. Depending on the length of the amino acid sequence and the amount of degeneracy at each position, pools can contain up to several hundred oligonucleotides. If hybridization conditions can be found under which only perfectly matched sequences form stable duplexes, cloned copies of the gene of interest can be readily isolated from genomic or cDNA libraries (e.g, see Goeddel et al. 1980; Agarwal et al. 1981; Sood et al. 1981; Suggs et al. 1981b; Wallace et al. 1981; Toole et al. 1984; Jacobs et al. 1985; Lin et al. 1985).

Pools of degenerate oligonucleotides are generally used to select a series of candidate clones that can then be analyzed further to identify the true target. To ensure that the target clone is represented in the initial group of candidates, screening conditions are usually chosen that allow detection of perfect hybrids formed between the most A/T-rich oligonucleotide in the pool and its potential target. Other hybrids, both perfect and imperfect, that are stable under such conditions will identify additional candidate clones.

It is easy to calculate accurately the T_m of a perfectly matched hybrid formed between a single oligonucleotide and its target sequence. However, when using pools of oligonucleotides whose members have greatly different contents of G+C, it is not feasible to estimate a consensus T_m. Because it is impossible to know which member of the pool will match the target sequence perfectly, conditions must be used that allow the oligonucleotide with the lowest content of G+C to hybridize efficiently. Usually, conditions are chosen to be 2°C below the calculated T_m of the most A/T-rich member of the pool (Suggs et al. 1981a). However, the use of such "lowest common denominator" conditions can generate false positives: Mismatched hybrids formed by oligonucleotides of higher G+C may be more stable than a perfectly matched hybrid formed by the correct oligonucleotide. In most cases, this problem is not serious, because the number of positive clones obtained by screening cDNA libraries with pools of nucleotides is quite manageable. It is therefore easy to distinguish false positives from true positives by another test (e.g., DNA sequencing or hybridization with a second pool of oligonucleotides corresponding to another segment of amino acid sequence).

Quarternary Alkylammonium Salts

When the number of apparent positives is unacceptably high, it is worthwhile using hybridization solvents that contain quaternary alkylammonium salts such as tetraethylammonium chloride (TEACl) or tetramethylammonium chloride (TMACl) (Melchior and von Hippel 1973; Jacobs et al. 1985, 1988; Wood et al. 1985; Gitschier et al. 1986; DiLella and Woo 1987). By binding to A/T-rich polymers (Shapiro et al. 1969), quaternary alkylammonium salts reduce the preferential melting of A:T versus G:C base pairs (Melchior and von Hippel 1973; Riccelli and Benight 1993). In TEACl or TMACl, the T_m of an oligonucleotide–DNA hybrid becomes less dependent on its base composition and more dependent on its length. By choosing a hybridization temperature appropriate for the lengths of the oligonucleotides in a pool, the effects of potential mismatches can be minimized.

It is important to obtain an accurate estimate of the T_m in TMACl or TEACl before using pools of oligonucleotides to screen cDNA or genomic libraries. Jacobs et al. (1988) measured the T_i (irreversible melting temperature) of the hybrid formed between the probe and its target as a function of chain length for a number of oligonucleotides of different (G+C) content in solvents containing either sodium ions or tetraalkylammonium ions. Hybrids involving oligonucleotides 16 and 19 nucleotides in length melt over a smaller range of temperatures in solvents containing TMACl than in solvents containing sodium salts. For example, 16-mers melt over a 3°C temperature range in solvents containing TMACl, compared with 17°C in SSC.

The optimal temperature for hybridization is usually chosen to be 5°C below the T_i for the given chain length. The recommended temperature for 17 mers in 3 M TMACl is 48°C–50°C; for 19-mers, 55°C–57°C; and for 20-mers, 58°C–66°C. Two points are worth emphasizing. First, the T_i of hybrids is uniformly 15°C–20°C higher in solvents containing TMACl than in solvents containing TEACl. The higher T_i in solvents containing TMACl allows hybridization to be performed at temperatures that suppress nonspecific adsorption of the probe to solid supports (such as nylon membranes). Second, hybridization solvents containing TMACl do not have significant advantages over those containing sodium ions until the length of the oligonucleotide exceeds 16 nucleotides.

Guessmers

A guessmer is a long synthetic oligonucleotide (usually 30–75 nucleotides in length) whose sequence corresponds to an amino acid sequence of 15–25 residues. Because of the degeneracy of the genetic code, many thousands of possible nucleotide sequences could code for such a sizable tract of amino acids. Unless the designer is incredibly lucky and chooses the correct codon in every single position, the guessmer will not form a perfect hybrid with the target gene. However, because most amino acids are specified by codons that differ only in the third position, at least 2 of the 3 nucleotides of each codon are likely to be perfectly matched. In addition, the number of mismatches in the third position can be minimized by selecting codons that are used preferentially by a particular organism, organelle, or cell type. In this way, it is possible to make guessmers of sufficient length so that the detrimental effects of mismatches are outweighed by the stability of extensive tracts of perfectly matched bases. For further details, see the box Designing a Guessmer.

DESIGNING A GUESSMER

- Eliminate codon choices that generate the sequence CpG between codons. This dinucleotide is significantly underrepresented in mammalian DNA (Bird 1980), and CpG (C-phosphate-G) occurs at the junction between codons at about one-half of the expected frequency (Lathe 1985).

- Choose the codon that most commonly codes for a particular amino acid in the species under study. Comprehensive lists of relative codon frequencies in genes have been published by Wada et al. (1992) or may be found at http://www.kazusa.or.jp/codon.

- Remember that specific gene families occasionally display a marked bias for or against certain codons. For example, histone genes show a marked preference for codons enriched in A and T residues, whereas proteins expressed abundantly in yeast are biased toward a set of highly preferred codons (Bennetzen and Hall 1982) and, for example, hardly ever use the codons UCG and UAU for serine and isoleucine (Ogden et al. 1984).

- Different mammalian tissues display different patterns of codon usage (e.g., see Newgard et al. 1986). It would therefore be worthwhile to determine the codon usage in any genes that have already been cloned and are known to be expressed in the same tissue as the gene of interest.

- At positions where the choice is between C and T and there is no strong codon preference, choose T. This rule is still somewhat controversial. There is good evidence indicating that rG:rU base pairs are more stable than rA:rC base pairs (Uhlenbeck et al. 1971). However, the extension to dG:dT base pairs (Wu 1972) has been both challenged (Smith 1983) and upheld (Martin et al. 1985).

- Examine the sequence of the proposed nucleotide for regions of internal complementarity that might reduce hybridization efficiency. Wherever possible, avoid sequences that could form stable duplexes under the conditions used for hybridization.

- Ensure that the 5′-terminal position of the oligonucleotide is not occupied by a cytosine residue. For reasons that are unknown, the efficiency of radiolabeling of oligonucleotides depends on the sequence of the oligonucleotide (van Houten et al. 1998). Oligonucleotides with a cytosine residue at their 5′ termini are labeled fourfold less efficiently than oligonucleotides beginning with A or T and sixfold less efficiently than oligonucleotides beginning with G.

No rules guarantee selection of the correct codon at a position of ambiguity; neither is it possible to predict how many correct choices must be made to achieve success. On the basis of mathematical calculations (Lathe 1985), a probe would be expected to have at least 76% identity with the authentic gene even if all codon choices were made on a random basis. If substitutions are chosen on the statistical basis of known codon usage in the species of interest, the expected homology increases to 82%; it rises still further (to 86%) if regions lacking leucine, arginine, and serine are chosen (each of these three amino acids is specified by six codons). Clearly, as the degree of homology increases, the guessmer becomes a more specific probe for the gene of interest and forms hybrids that are stable under a wider range of hybridization conditions. If the sequences of selected genes, successfully isolated by probing with guessmers, are compared with their corresponding guessmer sequences, the (respective) homologies range from a low of 71% to a high of 97%. Unless the choice of codons is extremely unfortunate, guessmers can be designed whose homology with their target genes lies near the upper end of this range. However, in addition to overall homology, the presence of a perfectly matching sequence of contiguous nucleotides within the guessmer may also be important for success. Almost all guessmers that have been used successfully contain regions that exactly match sequences in their target genes. Although it is not possible to guarantee that a guessmer will contain an exactly matching sequence, the chances increase dramatically as the degree of overall homology rises.

It is sometimes possible to synthesize a small pool containing two to eight guessmers that includes all possible codon choices at certain amino acid positions. This kind of limited substitution is most useful when an amino acid, whose codon is highly degenerate, separates two tracts containing only a few potential mismatches. In this way, it may be possible to generate a continuous sequence within the guessmer that is a perfect match for the target gene. However, when using a mixture of guessmers as probes, the strength of the hybridization signal generated by the "correct" probe is usually reduced. For guessmers labeled by phosphorylation or by filling of recessed ends, an eightfold reduction in signal strength typically does not compromise the identification of clones of interest.

Perhaps the most critical step in the use of guessmers is the choice of conditions for hybridization. The temperature should be high enough to suppress hybridization of the probe to incorrect sequences but must not be so high as to prevent hybridization to the correct sequence, even though it may be mismatched. Before using an oligonucleotide to screen a library, perform a series of trial experiments in which a series of northern or genomic Southern hybridizations are performed under different degrees of stringency (Anderson and Kingston 1983; Wood et al. 1985). Lathe (1985) presents a set of theoretical curves relating the temperature of the washing solution to the length and homology of the probe. Using these curves as a guide, determine the optimal conditions for detection of sequences complementary to the probe by hybridizing the oligonucleotide to a series of nitrocellulose or nylon membranes at different temperatures. The membranes are washed extensively in $6\times$ SSC at room temperature and then briefly (5–10 min in $6\times$ SSC) at the temperature used for hybridization. This method, in which both hybridization and washing are performed under the same conditions of temperature and ionic strength, appears to be more discriminating than the more commonly used procedure of hybridizing under conditions of lower stringency and washing under conditions of higher stringency. If trial experiments are not possible, estimate the melting temperature (T_m) as follows.

1. Calculate the minimum G+C content of the oligonucleotide, assuming that A or T is present at all positions of ambiguity.

2. Use the following formula to calculate the T_m of a dsDNA with the calculated G+C content:

$$T_m = 81.5°C + 16.6(\log_{10}[Na^+]) + 0.41(\%[G + C]) - (500/n),$$

where n is the number of nucleotides. The term $500/n$ is derived from a compilation (Hall et al. 1980) of measurements, from several laboratories, of the effect of duplex length on melting temperature. This formula only works for Na^+ concentrations of 1.0 M or less.

3. Calculate the maximum amount of possible mismatches assuming that all choices of degenerate codons are incorrect. Subtract 1°C from the calculated T_m for each 1% of mismatch. The resulting number should be the T_m of a maximally mismatched hybrid formed between the probe and its target DNA sequence.

Almost certainly, the actual T_m will be higher than that predicted by this worst-case calculation. If the bases used at positions of ambiguity were chosen at random, one out of four should be correct, and approximately half of these would be expected to be G or C. The observed T_m should therefore be significantly higher than that estimated. However, to minimize the risk of missing the clone of interest, it is best to hybridize and wash at 5°C–10°C below the T_m estimated as described above. If, under these conditions, the probe hybridizes indiscriminately, repeat the hybridization at a higher temperature or wash under conditions of higher stringency.

Guessmers reached their height of popularity in the mid 1980s, a time when the fledgling recombinant DNA companies were desperately trying to clone genes of commercial value. cDNA and genomic clones of many useful genes were obtained during this period. In recent years, however, guessmers have been eclipsed by probes generated in PCRs using sets of redundant primers. This is chiefly because PCR-based methods require far shorter tracts of unambiguous amino acid sequence. Nevertheless, when an amino acid sequence of sufficient length is available, guessmers remain the probes of choice. They have a strong track record of success and are free of the artifacts that PCR can visit upon even the best of investigators.

Universal Bases

Universal bases have reduced hydrogen-bonding specificity and can therefore "pair" with natural bases at positions of ambiguity without disrupting the DNA duplex. For several years, the "universal" base of choice has been the purine nucleoside inosine, whose neutral base, hypoxanthine, forms stable base pairs with cytosine, thymine, and adenine. Because of the absence of a 2-amino group, base pairing between inosine and cytosine involves two hydrogen bonds (as in A:T base pairs) instead of the three that occur in C:G base pairs (Corfield et al. 1987; Xuan and Weber 1992). In most other respects, however, inosine behaves like guanosine.

- Inosine occurs naturally in the first (wobble) position of some tranfer RNAs (tRNAs), where it pairs with adenosine in addition to cytidine and uridine, the nucleosides that normally pair with guanosine in that position (Crick 1966).

- Inosine is able to occupy the middle position of the anticodon, where it again pairs with adenosine (Davis et al. 1973). However, in inosine:adenosine pairs in B-DNA duplexes, inosine adopts a *trans* (anti) orientation with respect to the furanose moiety, whereas adenosine is in the syn configuration (Corfield et al. 1987). This is similar to the arrangement that is formed when guanosine mispairs with adenosine (Brown et al. 1986).

- Poly(rI) and poly(dI) form helices with poly(rC) and poly(dC) that are stable enough to serve as templates used by various RNA and DNA polymerases. The enzymes can incorporate cytosine into the polymerization products (for review, see Felsenfeld and Miles 1967; Hall et al. 1985).

Pools of synthetic oligonucleotides and guessmers containing inosine have been used extensively as hybridization probes to screen cDNA or genomic libraries for genes encoding proteins whose amino acid sequence is only partly known. However, although inosine is able to form hydrogen bonds with the three other nucleotides, the resulting base pairs are less stable than the equivalent guanosine-containing base pairs (Martin et al. 1985). The melting temperatures of duplexes containing inosine vary widely depending on the base to which the analog is paired and on the surrounding sequence. In the worst case, the melting temperature of synthetic oligonucleotide duplexes containing inosine opposite any base may be depressed by 15°C, which corresponds to a difference in base-pair stability of 2–3 kcal/mol^{-1} (Martin et al. 1985; Kawase et al. 1986). Although this situation is

TABLE 2. Inosine-containing codons recommended for use in oligonucleotide probes

Amino acid	Codon	Amino acid	Codon
A (Ala)	GCI	M (Met)	ATG
C (Cys)	TGC[a]	N (Asn)	AAC[a]
D (Asp)	GAT	P (Pro)	CCI
E (Glu)	GAI	Q (Gln)	CAI
F (Phe)	TTC[a]	R (Arg)	CGI[c]
G (Gly)	GGI	S (Ser)	TCC[b,c]
H (His)	CAC[a]	T (Thr)	ACI
I (Ile)	ATI	V (Val)	GTI
K (Lys)	AAI	W (Trp)	TGG
L (Leu)	CTI[c]	Y (Tyr)	TAC[a]

[a]If the first nucleotide of the succeeding codon is G, use T in the third position.
[b]If the first nucleotide of the succeeding codon is G, use I in the third position.
[c]Try to avoid amino acids with six codons if at all possible.

less than ideal, it has surprisingly few practical consequences: Inosine-containing oligonucleotides have been used successfully to clone many genes from genomic and cDNA libraries of high complexity (e.g., see Jaye et al. 1983; Ohtsuka et al. 1985; Takahashi et al. 1985; Bray et al. 1986; Nagata et al. 1986).

Table 2 contains a list of codons that are recommended when designing oligonucleotides that are to be used to screen primate or mammalian cDNA libraries and that contain inosine at positions of ambiguity. The recommendations take into account the natural usage of codons in human genes and the fact that the sequence CpG is underrepresented in human DNA.

Hybridization of Oligonucleotides that Contain Inosine at Positions of Degeneracy

Although the conditions for hybridization of probes that contain the neutral base inosine have not been extensively explored, it is possible to make a conservative estimate of the T_m as follows.

- Subtract the number of inosine residues from the total number of nucleotides in the probe to give a value S.

- Calculate the G+C content of S.

- Estimate the T_m of a perfect hybrid involving S using either the Wallace rule, the Baldino algorithm, or the Wetmur equation.

- Use conditions for hybridization that are 15°C–20°C below the estimated T_m.

The T_m of hybrids involving oligonucleotides that contain neutral bases can also be estimated empirically as described in Protocol 8. Oligonucleotides containing inosine can be used with washing buffers containing TMACl or TEACl (e.g., see Andersson et al. 1989).

REFERENCES

Agarwal KL, Brunstedt J, Noyes BE. 1981. A general method for detection and characterization of an mRNA using an oligonucleotide probe. *J Biol Chem* **256**: 1023–1028.

Agrawal S, Christodoulou C, Gait MJ. 1986. Efficient methods for attaching non-radioactive labels to the 5′ ends of synthetic oligodeoxyribonucleotides. *Nucleic Acids Res* **14**: 6227–6245.

Anderson S, Kingston IB. 1983. Isolation of a genomic clone for bovine pancreatic trypsin inhibitor by using a unique-sequence synthetic DNA probe. *Proc Natl Acad Sci* **80**: 6838–6842.

Andersson S, Davis DL, Dahlbäck H, Jörnvall H, Russell DW. 1989. Cloning, structure, and expression of the mitochondrial cytochrome P-450 sterol 26-hydroxylase, a bile acid synthetic enzyme. *J Biol Chem* **264**: 8222–8229.

Ansorge W, Sproat B, Stegemann J, Schwager C, Zenke M. 1987. Automated DNA sequencing: Ultrasensitive detection of fluorescent bands during electrophoresis. *Nucleic Acids Res* **15**: 4593–4602.

Bains W. 1994. Selection of oligonucleotide probes and experimental conditions for multiplex hybridization experiments. *Genet Anal Tech Appl* **11:** 49–62.

Baldino FJ, Chesselet MF, Lewis ME. 1989. High resolution in situ hybridization. *Methods Enzymol* **168:** 761–777.

Bauman JG, Wiegant J, Borst P, van Duijn P. 1980. A new method for fluorescence microscopical localization of specific DNA sequences by in situ hybridization of fluorochromelabelled RNA. *Exp Cell Res* **128:** 485–490.

Beck S, Köster H. 1990. Applications of dioxetane chemiluminescent probes to molecular biology. *Anal Chem* **62:** 2258–2270. (Erratum *Anal Chem* [1991] **63:** 848.)

Beck S, O'Keeffe T, Coull JM, Koster H. 1989. Chemiluminescent detection of DNA: Application for DNA sequencing and hybridization. *Nucleic Acids Res* **17:** 5115–5123.

Bennetzen JL, Hall BD. 1982. The primary structure of the *Saccharomyces cerevisiae* gene for alcohol dehydrogenase. *J Biol Chem* **257:** 3018–3025.

Benzinger EA, Riech AK, Shirley RE, Kucharik KR. 1995. Evaluation of the specificity of alkaline phosphatase-conjugated oligonucleotide probes for forensic DNA analysis. *Appl Theor Electrophor* **4:** 161–165.

Bird AP. 1980. DNA methylation and the frequency of CpG in animal DNA. *Nucleic Acids Res* **8:** 1499–1504.

Bolton ET, McCarthy BJ. 1962. A general method for the isolation of RNA complementary to DNA. *Proc Natl Acad Sci* **48:** 1390.

Bray P, Carter A, Simons C, Guo V, Puckett C, Kamholz J, Spiegel A, Nirenberg M. 1986. Human cDNA clones for four species of Gα$_s$ signal transduction protein. *Proc Natl Acad Sci* **83:** 8893–8897.

Breslauer KJ, Frank R, Blöcker H, Marky LA. 1986. Predicting DNA duplex stability from the base sequence. *Proc Natl Acad Sci* **83:** 8893–8897.

Bronstein I. 1990. Chemiluminescent 1,2-dioxetane-based enzyme substrates and their applications. In *Luminescence immunoassays and molecular applications* (ed Van Dyke K, Van Dyke R), pp. 255–274. CRC, Boca Raton, FL.

Bronstein I, Edwards B, Voyta JC. 1989a. 1,2-dioxetanes: Novel chemiluminescent enzyme substrates. Applications to immunoassays. *J Biolumin Chemilumin* **4:** 99–111.

Bronstein I, Voyta JC, Edwards B. 1989b. A comparison of chemiluminescent and colorimetric substrates in a hepatitis B virus DNA hybridization assay. *Anal Biochem* **180:** 95–98.

Bronstein I, Cate RL, Lazzarri K, Ramachandran KL, Voyta JC. 1989c. Chemiluminescent 1,2-dioxetane based enzyme substrates and their application in the detection of DNA. *Photochem Photobiol* **49:** S9–S10.

Bronstein I, Joa RR, Voyta JC, Edwards B. 1991. Novel chemiluminescent adamantyl 1,2-dioxetane enzyme substrates. In *Proceedings of the 6th international symposium on bioluminescence and chemiluminescence: Current status* (ed Stanley PE, Kricka LJ), pp. 73–82. Wiley, Chichester, UK.

Brown T, Hunter WN, Kneale G, Kennard O. 1986. Molecular structure of the G·A base pair in DNA and its implications for the mechanism of transversion mutations. *Proc Natl Acad Sci* **83:** 2402–2406.

Chan SD, Dill K, Blomdahl J, Wada HG. 1996. Nonisotopic quantitation of mRNA using a novel RNase protection assay: Measurement of erbB-2 mRNA in tumor cell lines. *Anal Biochem* **242:** 214–220.

Chan VT, Fleming KA, McGee JO. 1985. Detection of sub-picogram quantities of specific DNA sequences on blot hybridization with biotinylated probes. *Nucleic Acids Res* **13:** 8083–8091.

Chu BC, Orgel LE. 1985. Detection of specific DNA sequences with short biotin-labeled probes. *DNA* **4:** 327–331.

Corfield PW, Hunter WN, Brown T, Robinson P, Kennard O. 1987. Inosine·adenine base pairs in a B-DNA duplex. *Nucleic Acids Res* **15:** 7935–7949.

Crick FH. 1966. Codon–anticodon pairing: The wobble hypothesis. *J Mol Biol* **19:** 548–555.

Dahlén P, Liukkonen L, Kwiatkowski M, Hurskainen P, Iitia A, Siitari H, Ylikoski J, Mukkala VM, Lovgren T. 1994. Europium-labeled oligonucleotide hybridization probes: Preparation and properties. *Bioconjug Chem* **5:** 268–272.

Davis BD, Anderson P, Sparling PF. 1973. Pairing of inosine with adenosine in codon position two in the translation of polyinosinic acid. *J Mol Biol* **76:** 223–232.

Diamandis EP, Hassapoglidou S, Bean CC. 1993. Evaluation of nonisotopic labeling and detection techniques for nucleic acid hybridization. *J Clin Lab Anal* **7:** 174–179.

DiLella AG, Woo SL. 1987. Hybridization of genomic DNA to oligonucleotide probes in the presence of tetramethylammonium chloride. *Methods Enzymol* **152:** 447–451.

Dimo-Simonin N, Brandt-Casadevall C, Gujer HR. 1992. Chemiluminescent DNA probes; evaluation and usefulness in forensic cases. *Forensic Sci Int* **57:** 119–127.

Dubitsky A, Brown J, Brandwein H. 1992. Chemiluminescent detection of DNA on nylon membranes. *BioTechniques* **13:** 392–400.

Düring K. 1993. Non-radioactive detection methods for nucleic acids separated by electrophoresis. *J Chromatogr* **618:** 105–131.

Durrant I, Benge LC, Sturrock C, Devenish AT, Howe R, Roe S, Moore M, Scozzafava G, Proudfoot LM, Richardson TC, et al. 1990. The application of enhanced chemiluminescence to membrane-based nucleic acid detection. *BioTechniques* **8:** 564–570.

Edwards B, Sparks A, Voyta JC, Bronstein I. 1994. New chemiluminescent dioxetane enzyme substrates. In *Proceedings of the 8th international symposium on bioluminescence and chemiluminescence: Fundamentals and applied aspects* (ed Campbell AK, et al.), pp. 56–59. Wiley, Chichester, UK.

Farmar JG, Castaneda M. 1991. An improved preparation and purification of oligonucleotide-alkaline phosphatase conjugates. *BioTechniques* **11:** 588–589.

Felsenfeld G, Miles HT. 1967. The physical and chemical properties of nucleic acids. *Annu Rev Biochem* **31:** 407–448.

Forster AC, McInnes JL, Skingle DC, Symons RH. 1985. Non-radioactive hybridization probes prepared by the chemical labelling of DNA and RNA with a novel reagent, photobiotin. *Nucleic Acids Res* **13:** 745–761.

Freier SM, Kierzek R, Jaeger JA, Sugimoto N, Caruthers MH, Neilson T, Turner DH. 1986. Improved free-energy parameters for predictions of RNA duplex stability. *Proc Natl Acad Sci* **83:** 9373–9377.

Gait MJ, ed. 1984. An introduction to modern methods of DNA synthesis. In *Oligonucleotide synthesis: A practical approach*, pp. 1–22. IRL, Oxford, UK.

Gebeyehu G, Rao PY, SooChan P, Simms DA, Klevan L. 1987. Novel biotinylated nucleotide: Analogs for labeling and colorimetric detection of DNA. *Nucleic Acids Res* **15:** 4513–4534.

Gitschier J, Wood WI, Shuman MA, Lawn RM. 1986. Identification of a missense mutation in the factor VIII gene of a mild hemophiliac. *Science* **232:** 1415–1416.

Goeddel DV, Yelverton E, Ullrich A, Heyneker HL, Miozzari G, Holmes W, Seeburg PH, Dull T, May L, Stebbing N, et al. 1980. Human leukocyte interferon produced by *E. coli* is biologically active. *Nature* **287:** 411–416.

Habili N, McInnes JL, Symons RH. 1987. Nonradioactive, photobiotin-labelled DNA probes for the routine diagnosis of barley yellow dwarf virus. *J Virol Methods* **16:** 225–237.

Hall K, Cruz P, Chamberlin MJ. 1985. Extensive synthesis of poly[r(G-C)] using *Escherichia coli* RNA polymerase. *Arch Biochem Biophys* **236:** 47–51.

Hall TJ, Grula JW, Davidson EH, Britten RJ. 1980. Evolution of sea urchin non-repetitive DNA. *J Mol Evol* **16:** 95–110.

Hegnauer R. 1971. Pflanzenstoffe und Pflanzensystematik (plant constituents and plant taxonomy). *Naturwissenschaften* **58:** 585–598.

Henegariu O, Bray-Ward P, Ward DC. 2000. Custom fluorescent-nucleotide synthesis as an alternative method for nucleic acid labeling. *Nat Biotechnol* **18:** 345–348.

Hodgson M, Jones P. 1989. Enhanced chemiluminescence in the peroxidase-luminol-H$_2$O$_2$ system: Anomalous reactivity of enhancer phenols with enzyme intermediates. *J Biolumin Chemilumin* **3:** 21–25.

Höltke H-J, Seibl R, Burg J, Mühlegger K, Kessler C. 1990. Non-radioactive labeling and detection of nucleic acids. II. Optimization of the digoxygenin system. *Biol Chem Hoppe-Seyler* **371:** 929–938.

Höltke H-J, Schneider S, Ettl I, Binsack R, Obermaier I, Seller M, Sagner G. 1994. Rapid and highly sensitive detection of dioxigenin-labelled nucleic acids by improved chemiluminescent substrates. In *Proceedings of the 8th symposium on bioluminscence and chemiluminscence: Fundamentals and applied aspects* (ed Campbell AK, et al.), pp. 273–276. Wiley, Chichester, UK.

Höltke H-J, Ankenbauer W, Mühlegger K, Rein R, Sagner G, Seibl R, Walter T. 1995. The digoxigenin (DIG) system for non-radioactive labelling and detection of nucleic acids: An overview. *Cell Mol Biol* 41: 883–905.

Horwitz JP, Chua J, Noel M, Donatti JT, Freisler J. 1966. Substrates for cytochemical demonstration of enzyme activity. II. Some dihalo-3-indolyl phosphates and sulfates. *J Med Chem* 9: 447.

Ikuta S, Takagi K, Wallace RB, Itakura K. 1987. Dissociation kinetics of 19 base paired oligonucleotide-DNA duplexes containing different single mismatched base pairs. *Nucleic Acids Res* 15: 797–811.

Jablonski E, Moomaw EW, Tullis RH, Ruth JL. 1986. Preparation of oligodeoxynucleotide-alkaline phosphatase conjugates and their use as hybridization probes. *Nucleic Acids Res* 14: 6115–6128.

Jacobs K, Shoemaker C, Rudersdorf R, Neill SD, Kaufman RJ, Mufson A, Seehra J, Jones SS, Hewick R, Fritsch EF, et al. 1985. Isolation and characterization of genomic and cDNA clones of human erythropoietin. *Nature* 313: 806–810.

Jacobs KA, Rudersdorf R, Neill SD, Dougherty JP, Brown EL, Fritsch EF. 1988. The thermal stability of oligonucleotide duplexes is sequence independent in tetraalkylammonium salt solutions: Application to identifying recombinant DNA clones. *Nucleic Acids Res* 16: 4637–4650.

Jaye M, de la Salle H, Schamber F, Balland A, Kohli V, Findeli A, Tolstoshev P, Lecocq JP. 1983. Isolation of a human anti-haemophilic factor IX cDNA clone using a unique 52-base synthetic oligonucleotide probe deduced from the amino acid sequence of bovine factor IX. *Nucleic Acids Res* 11: 2325–2335.

Kagiyama N, Fujita S, Momiyama M, Saito H, Shirahama H, Hori SH. 1992. A fluorescent detection method for DNA hybridization using 2-hydroxy-3-naphthoic acid-2-phenylanilide phosphate as a substrate for alkaline phosphatase. *Acta Histochem Cytochem* 25: 467–471.

Kawase Y, Iwai S, Inoue H, Miura K, Ohtsuka E. 1986. Studies on nucleic acid interactions. I. Stabilities of mini-duplexes (dG2A4XA4G2-dC2T4YT4C2) and self-complementary d(GGGAAXYTTCCC) containing deoxyinosine and other mismatched bases. *Nucleic Acids Res* 14: 7727–7736.

Keller GH, Cumming CU, Huang DP, Manak MM, Ting R. 1988. A chemical method for introducing haptens onto DNA probes. *Anal Biochem* 170: 441–450.

Kerkhof L. 1992. A comparison of substrates for quantifying the signal from a nonradiolabeled DNA probe. *Anal Biochem* 205: 359–364.

Kessler C. 1991. The digoxigenin:anti-digoxigenin (DIG) technology: A survey on the concept and realization of a novel bioanalytical indicator system. *Mol Cell Probes* 5: 161–205.

Kessler C. 1992. Nonradioactive labeling for nucleic acids. In *Nonisotopic DNA probe techniques* (ed Kricka LJ), pp. 28–92. Academic, New York.

Kierzek R, Caruthers MH, Longfellow CE, Swinton D, Turner DH, Freier SM. 1986. Polymer-supported RNA synthesis and its application to test the nearest-neighbor model for duplex stability. *Biochemistry* 25: 7840–7846.

Klevan L, Horton E, Carlson DP, Eisenberg AJ. 1993. Chemiluminescent detection of DNA probes in forensic analysis. *The 2nd international symposium on the forensic aspects of DNA analysis*. FBI Academy, Quantico, VA (http://www.fbi.gov).

Koshkin A, Singh SK, Nielsen PS, Rajwanshi VK, Kumar R, Meldgaard M, Olsen CE, Wengel J. 1998. LNA (locked nucleic acids): Synthesis of the adenine, cytosine, guanine, 5-methylcytosine, thymine and uracil bicyclonucleoside monomers, oligomerisation, and unprecedented nucleic acid recognition. *Tetrahedron* 54: 3607–3630.

Kricka LJ ed. 1992. Nucleic acid hybridization test formats: Strategies and applications. In *Nonisotopic DNA probe techniques*, pp. 3–27. Academic, New York.

Kricka LJ, Ji X. 1995. 4-Phenylylboronic acid: A new type of enhancer for the horseradish peroxidase catalysed chemiluminescent oxidation of luminol. *J Biolumin Chemilumin* 10: 49–54.

Kubota K, Ohashi A, Imachi H, Harada H. 2006. Improved in situ hybridization efficiency with locked-nucleic-acid-incorporated DNA probes. *Appl Environ Microbiol* 72: 5311–5317.

Langer PR, Waldrop AA, Ward DC. 1981. Enzymatic synthesis of biotin-labeled polynucleotides: Novel nucleic acid affinity probes. *Proc Natl Acad Sci* 78: 6633–6637.

Lanzillo JJ. 1990. Preparation of digoxigenin-labeled probes by the polymerase chain reaction. *BioTechniques* 8: 620–622.

Lathe R. 1985. Synthetic oligonucleotide probes deduced from amino acid sequence data. Theoretical and practical considerations. *J Mol Biol* 183: 1–12.

Leary JJ, Brigati DJ, Ward DC. 1983. Rapid and sensitive colorimetric method for visualizing biotin-labeled DNA probes hybridized to DNA or RNA immobilized on nitrocellulose: Bio-blots. *Proc Natl Acad Sci* 80: 4045–4049.

Lin FK, Suggs S, Lin CH, Browne JK, Smalling R, Egrie JC, Chen KK, Fox GM, Martin F, Stabinsky Z, et al. 1985. Cloning and expression of the human erythropoietin gene. *Proc Natl Acad Sci* 82: 7580–7584.

Mansfield ES, Worley JM, McKenzie SE, Surrey S, Rappaport E, Fortina P. 1995. Nucleic acid detection using non-radioactive labelling methods. *Mol Cell Probes* 9: 145–156.

Martin CS, Bronstein I. 1994. Imaging of chemiluminescent signals with cooled CCD camera systems. *J Biolumin Chemilumin* 9: 145–153.

Martin FH, Castro MM, Aboul-ela F, Tinoco I Jr. 1985. Base pairing involving deoxyinosine: Implications for probe design. *Nucleic Acids Res* 13: 8927–8938.

Martin CS, Bresnick L, Juo RR, Voyta JC, Bronstein I. 1991. Improved chemiluminescent DNA sequencing. *BioTechniques* 11: 110–113.

McGadey J. 1970. A tetrazolium method for non-specific alkaline phosphatase. *Histochemie* 23: 180–184.

Melchior WB Jr, von Hippel PH. 1973. Alteration of the relative stability of dA-dT and dG-dC base pairs in DNA. *Proc Natl Acad Sci* 70: 298–302.

Mitsuhashi M, Cooper A, Ogura M, Shinagawa T, Yano K, Hosokawa T. 1994. Oligonucleotide probe design: A new approach. *Nature* 367: 759–761.

Miyada CG, Wallace RB. 1987. Oligonucleotide hybridization techniques. *Methods Enzymol* 154: 94–107.

Nagata S, Tsuchiya M, Asano S, Kaziro Y, Yamazaki T, Yamamoto O, Hirata Y, Kubota N, Oheda M, Nomura H, et al. 1986. Molecular cloning and expression of cDNA for human granulocyte colony-stimulating factor. *Nature* 319: 415–418.

Nei M, Li WH. 1979. Mathematical model for studying genetic variation in terms of restriction endonucleases. *Proc Natl Acad Sci* 76: 5269–5273.

Nelson NC, Reynolds MA, Arnold LJ Jr. 1992. Detection of acridinium esters by chemiluminescence. In *Nonisotopic DNA probe techniques* (ed Kricka LJ), pp. 276–311. Academic, New York.

Newgard CB, Nakano K, Hwang PK, Fletterick RJ. 1986. Sequence analysis of the cDNA encoding human liver glycogen phosphorylase reveals tissue-specific codon usage. *Proc Natl Acad Sci* 83: 8132–8136.

Ogden RC, Lee MC, Knapp G. 1984. Transfer RNA splicing in *Saccharomyces cerevisiae*: Defining the substrates. *Nucleic Acids Res* 12: 9367–9382.

Ohtsuka E, Matsuki S, Ikehara M, Takahashi Y, Matsubara K. 1985. An alternative approach to deoxyoligonucleotides as hybridization probes by insertion of deoxyinosine at ambiguous codon positions. *J Biol Chem* 260: 2605–2608.

Pataki S, Meyer K, Reichstein T. 1953. Die Konfiguration des Digoxygenins Glykoside und Algycone 116, Mitteling. *Helv Chim Acta* 36: 1295–1308.

Pollard-Knight D, Read CA, Downes MJ, Howard LA, Leadbetter MR, Pheby SA, McNaughton E, Syms A, Brady MA. 1990. Nonradioactive nucleic acid detection by enhanced chemiluminescence using probes directly labeled with horseradish peroxidase. *Anal Biochem* 185: 84–89.

Prober JM, Trainor GL, Dam RJ, Hobbs FW, Robertson CW, Zagursky RJ, Cocuzza AJ, Jensen MA, Baumeister K. 1987. A system for rapid DNA sequencing with fluorescent chain-terminating dideoxynucleotides. *Science* 238: 336–341.

Reisfeld A, Rothenberg JM, Bayer EA, Wilchek M. 1987. Nonradioactive hybridization probes prepared by the reaction of biotin hydrazide with DNA. *Biochem Biophys Res Commun* 142: 519–526.

Renz M, Kurz C. 1984. A colorimetric method for DNA hybridization. *Nucleic Acids Res* 12: 3435–3444.

Reyes RA, Cockerell GL. 1993. Preparation of pure oligonucleotide-alkaline phosphatase conjugates. *Nucleic Acids Res* 21: 5532–5533.

Riccelli PV, Benight AS. 1993. Tetramethylammonium does not universally neutralize sequence dependent DNA stability. *Nucleic Acids Res* 21: 3785–3788.

Rihn B, Bottin MC, Coulais C, Martinet N. 1995. Evaluation of non-radioactive labelling and detection of deoxyribonucleic acids. II. Colorigenic methods and comparison with chemiluminescent methods. *J Biochem Biophys Methods* 30: 103–112.

Rychlik R. 1994. New algorithm for determining primer efficiency in PCR and sequencing. *J NIH Res* 6: 78.

Rychlik W, Spencer WJ, Rhoads RE. 1990. Optimization of the annealing temperature for DNA amplification in vitro. *Nucleic Acids Res* 18: 6409–6412. (Erratum *Nucleic Acids Res* [1991] 19: 698.)

Schaap AP, Akhavan H, Romano LJ. 1989. Chemiluminescent substrates for alkaline phosphatase: Application to ultrasensitive enzyme-linked immunoassays and DNA probes. *Clin Chem* 35: 1863–1864.

Seibl R, Höltke HJ, Rüger R, Meindl A, Zachau HG, Rasshofer R, Roggendorf M, Wolf H, Arnold N, Wienberg J, et al. 1990. Non-radioactive labeling and detection of nucleic acids. III. Applications of the digoxigenin system. *Biol Chem Hoppe-Seyler* 371: 939–951.

Shapiro JT, Stannard BS, Felsenfeld G. 1969. The binding of small cations to deoxyribonucleic acid. Nucleotide specificity. *Biochemistry* 8: 3233–3241.

Smith LM, Sanders JZ, Kaiser RJ, Hughes P, Dodd C, Connell CR, Heiner C, Kent SB, Hood LE. 1986. Fluorescence detection in automated DNA sequence analysis. *Nature* 321: 674–679.

Smith M. 1983. Synthetic oligodeoxyribonucleotides as probes for nucleic acids and as primers in sequence determination. In *Methods of DNA and RNA sequencing* (ed Weissman SM), pp. 23–68. Praeger, New York.

Sood AK, Pereira D, Weissman SM. 1981. Isolation and partial nucleotide sequence of a cDNA clone for human histocompatibility antigen HLA-B by use of an oligodeoxynucleotide primer. *Proc Natl Acad Sci* 78: 616–620.

Suggs SV, Wallace RB, Hirose T, Kawashima EH, Itakura K. 1981a. Use of synthetic oligonucleotides as hybridization probes: Isolation of cloned cDNA sequences for human β_2-microglobulin. *Proc Natl Acad Sci* 78: 6613–6617.

Suggs SV, Hirose T, Miyake T, Kawashima EH, Johnson MJ, Itakura K, Wallace RB. 1981b. Use of synthetic oligodeoxyribonucleotides for the isolation of specific cloned DNA sequences. In *Developmental biology using purified genes* (ed Brown DB, et al.), pp. 683–693. Academic, New York.

Takahashi Y, Kato K, Hayashizaki Y, Wakabayashi T, Ohtsuka E, Matsuki S, Ikehara M, Matsubara K. 1985. Molecular cloning of the human cholecystokinin gene by use of a synthetic probe containing deoxyinosine. *Proc Natl Acad Sci* 82: 1931–1935.

Templeton EF, Wong HE, Evangelista RA, Granger T, Pollak A. 1991. Time-resolved fluorescence detection of enzyme-amplified lanthanide luminescence for nucleic acid hybridization assays. *Clin Chem* 37: 1506–1512.

Thein SL, Wallace RB. 1986. The use of synthetic oligonucleotides as specific hybridization probes in the diagnosis of genetic disorders. In *Human genetic diseases: A practical approach* (ed Davies KE), pp. 33–50. IRL, Oxford.

Thomsen R, Nielsen PS, Jensen TH. 2005. Dramatically improved RNA in situ hybridization signals using LNA-modified probes. *RNA* 11: 1745–1748.

Thorpe GH, Kricka LJ. 1986. Enhanced chemiluminescent reactions catalyzed by horseradish peroxidase. *Methods Enzymol* 133: 331–353.

Tizard R, Cate RL, Ramachandran KL, Wysk M, Voyta JC, Murphy OJ, Bronstein I. 1990. Imaging of DNA sequences with chemiluminescence. *Proc Natl Acad Sci* 87: 4514–4518.

Tolstrup N, Nielsen PS, Kolberg JG, Frankel AM, Vissing H, Kauppinen S. 2003. OligoDesign: Optimal design of LNA (locked nucleic acid) oligonucleotide capture probes for gene expression profiling. *Nucleic Acids Res* 31: 3758–3762.

Toole JJ, Knopf JL, Wozney JM, Sultzman LA, Buecker JL, Pittman DD, Kaufman RJ, Brown E, Shoemaker C, Orr EC, et al. 1984 Molecular cloning of a cDNA encoding human antihaemophilic factor. *Nature* 312: 342–347.

Traincard F, Ternynck T, Danchin A, Avrameas S. 1983. An immunoenzyme technique for demonstrating the molecular hybridization of nucleic acids (translation). *Ann Immunol* 134D: 399–404.

Turnbow MA, Garner CW. 1993. Ribonuclease protection assay: Use of biotinylated probes for the detection of two messenger RNAs. *Bio-Techniques* 15: 267–270.

Uhlenbeck OC, Martin FH, Doty P. 1971. Self-complementary oligoribonucleotides: Effects of helix defects and guanylic acid-cytidylic acid base pairs. *J Mol Biol* 57: 217–229.

Urdea MS, Warner BD, Running JA, Stempien M, Clyne J, Horn T. 1988. A comparison of non-radioisotopic hybridization assay methods using fluorescent, chemiluminescent and enzyme labeled synthetic oligodeoxyribonucleotide probes. *Nucleic Acids Res* 16: 4937–4956.

van Houten V, Denkers F, van Dijk M, van den Brekel M, Brakenhoff R. 1998. Labeling efficiency of oligonucleotides by T4 polynucleotide kinase depends on 5'-nucleotide. *Anal Biochem* 265: 386–389.

Vester B, Wengel J. 2004. LNA (locked nucleic acid): High-affinity targeting of complementary RNA and DNA. *Biochemistry* 43: 13233–13241.

Wada K, Wada Y, Ishibashi F, Gojobori T, Ikemura T. 1992. Codon usage tabulated from the GenBank genetic sequence data. *Nucleic Acids Res* (suppl) 20: 2111–2118.

Wallace RB, Johnson MJ, Hirose T, Miyake T, Kawashima EH, Itakura K. 1981. The use of synthetic oligonucleotides as hybridization probes. II. Hybridization of oligonucleotides of mixed sequence to rabbit β–globin DNA. *Nucleic Acids Res* 9: 879–894.

Wetmur JG. 1991. DNA probes: Applications of the principles of nucleic acid hybridization. *Crit Rev Biochem Mol Biol* 26: 227–259.

Whitehead TP, Thorpe GHG, Carter TJN, Groucutt C, Kricka LJ. 1983. Enhanced luminescence procedure for sensitive determination of peroxidase-labelled conjugates in immunoassay. *Nature* 305: 158–159.

Wood WI, Gitschier J, Lasky LA, Lawn RM. 1985. Base composition-independent hybridization in tetramethylammonium chloride: A method for oligonucleotide screening of highly complex gene libraries. *Proc Natl Acad Sci* 82: 1585–1588.

Wu R. 1972. Nucleotide sequence analysis of DNA. *Nat New Biol* 236: 198–200.

Xuan JC, Weber IT. 1992. Crystal structure of a B-DNA dodecamer containing inosine, d(CGCIAATTCGCG), at 2.4 Å resolution and its comparison with other B-DNA dodecamers. *Nucleic Acids Res* 20: 5457–5464.

WWW RESOURCES

List of relative codon frequencies   http://www.kazusa.or.jp/codon

Source of OligoDesign software   http://www.exiqon.com/oligo-tools

Random Priming: Labeling of Purified DNA Fragments by Extension of Random Oligonucleotides

Oligonucleotides can serve as primers for initiation of DNA synthesis on single-stranded templates by DNA polymerases (Goulian 1969). If the oligonucleotides are heterogeneous in sequence, they will form hybrids at many positions, so that the complement of every nucleotide of the template (except those at the extreme 5' terminus) will be incorporated at equal frequency into the product. These products of DNA synthesis can be radiolabeled by using one $[\alpha$-^{32}P]dNTP and three unlabeled dNTPs as precursors, generating probes with specific activities of 5×10^8 to 5×10^9 dpm/ μg. This labeling method is also suitable for nonradioactive labeling using biotin-, DIG-, or fluorescein-labeled dUTP.

Taylor et al. (1976) reported the first use of random priming to generate radiolabeled probes for hybridization. However, the method did not find wide acceptance until the mid 1980s, when the ready availability of commercial DNA polymerases and oligonucleotide primers allowed Feinberg and Vogelstein (1983, 1984) to develop a set of standardized and hardy reaction conditions. These conditions have since been incorporated into labeling kits that are marketed by several commercial manufacturers. However, random priming reactions are so robust and simple that kits are an unnecessary luxury. Labeling can be performed with equal efficiency and greater economy using components purchased individually. For a discussion of each element in a random priming reaction, see the box Components of Random Priming Reactions.

Random priming is inherently simpler than nick translation (see Protocol 3) because the requirements for two nuclease activities (DNase I and 5'→3' exonuclease) are eliminated. The labeled products of random priming reactions are therefore more homogeneous in size and behave more reproducibly in hybridization reactions. In nick translation, the average size of the labeled products cannot be controlled with great accuracy. In random priming, however, the average size of the probe DNA is under the control of the investigator because probe length is an inverse function of the primer concentration (Hodgson and Fisk 1987). These advantages have proven to be so decisive that random priming has almost completely replaced nick translation as the standard method of labeling of dsDNA probes.

COMPONENTS OF RANDOM PRIMING REACTIONS

DNA Polymerase

Random priming reactions may be catalyzed by any of several DNA polymerases including, for example, *Taq* (Sayavedra-Soto and Gresshoff 1992). However, by virtue of its efficiency and tolerance of a wide range of conditions, the enzyme of choice is the Klenow fragment of *E. coli* DNA Pol I. As discussed in the information panel *E. coli* DNA Polymerase I and the Klenow Fragment, the Klenow fragment lacks 5'→3' exonuclease activity, so that the radioactive product is synthesized exclusively by primer extension rather than by nick translation and is not subject to exonuclease degradation. The reaction can be performed at pH 6.6, where the 3'→5' exonuclease activity of the standard Klenow enzyme is much reduced (Lehman and Richardson 1964). Alternatively, Sequenase Version 2.0 (Affymetrix) or variants of the Klenow fragment that lack 3'→5' exonuclease activity (Stratagene/Agilent Technologies or New England Biolabs) can be used. Both of these enzymes lack the 3'→5' exonuclease activity of the Klenow fragment.

(Continued)

Radiolabel

Random priming reactions typically contain one radiolabeled [α-^{32}P]dNTP (sp. act. 3000 Ci/mmol) and three unlabeled dNTPs as precursors. Under the reaction conditions described in this protocol, 40%–80% of the [α-^{32}P]dNTP is incorporated into DNA and the specific activity of the radiolabeled product varies from 1×10^9 to 4×10^9 dpm/μg, depending on the amount of template DNA in the reaction. Although higher specific activities can be generated using two [α-^{32}P]dNTPs as precursors, the resulting probes degrade very rapidly due to radiochemical decay (Stent and Fuerst 1960) and must therefore be used without delay. When [α-^{32}P]dNTP of lower specific activity is used as a precursor (e.g., 800 Ci/mmol), the yield of DNA is increased fourfold and its specific activity is reduced to $<10^9$ dpm/μg. Such probes are more than adequate for most purposes and can be stored for several days frozen before radiolytic degradation becomes a problem.

Nonradioactive Label

In random primed DNA labeling, biotin-, DIG- or fluorescein-dUTP replaces dTTP in a reaction that uses a ratio of 35% dUTP:65% dTTP. Under these conditions, the labeled dUTP is incorporated every 25–36 nucleotides in the newly synthesized DNA probe (Kessler et al. 1990).

Attachment of a biotin molecule to nucleic acids by random priming was first described using biotin-7-dATP (Roy et al. 1988, 1989) and was shown to provide equivalent or higher levels of sensitivity than probes using ^{32}P. Further modifications for the biotinylation of DNA probes using the random primer method were reported to allow the detection of subfemtogram quantities of DNA (Eweida et al. 1989). Note that DIG-dUTP is alkali labile, so it is necessary to avoid exposing DIG-labeled probes to strong alkali (e.g., 0.2 mM NaOH). Alkali-stable versions are available, but the use of an alkali-labile formulation is recommended when stripping and reprobing Southern blots.

Primers

Oligonucleotide primers are typically purchased from a commercial source but can also be generated by DNase I digestion of calf thymus DNA or synthesized on an automated DNA synthesizer. The length of the primers is crucial. Oligonucleotides shorter than 6 bases in length are very poor primers, whereas those longer than 7 bases in length have progressively greater tendency to self-anneal and self-prime (Suganuma and Gupta 1995). Ideally, therefore, populations of oligonucleotides used in random priming should be either 6 or 7 nucleotides in length. All possible sequences should be represented in the population at equal frequencies.

The average size of the probe generated during random priming reactions is an inverse function of the concentration of primer (Hodgson and Fisk 1987): The length of radiolabeled product $= k/\sqrt{\ln P_c}$, where P_c is the concentration of primer.

The standard priming reaction described in this protocol contains between 60 and 125 ng of random hexamers or heptamers and generates labeled products that are ~400–600 nucleotides in length, as determined by electrophoresis through either an alkaline agarose gel or a denaturing polyacrylamide gel. Higher concentrations of primer lead to steric hindrance or exhaustion of the [α-^{32}P]dNTP precursor with a concomitant depression of yield; lower concentrations of primer generate populations of labeled products that are so heterogeneous in length (0.4–4 kb) (Hodgson and Fisk 1987) that they would be expected to hybridize with anomalous kinetics.

Template DNA

The reaction conditions given in this protocol, adapted from Feinberg and Vogelstein (1983, 1984), are optimized for labeling of linear dsDNA templates up to 1 kb in length. Shorter DNA templates generate probes of low specific activity that do not hybridize well under stringent conditions. Closed circular dsDNAs are inefficient templates and should be converted to linear molecules by digestion with a restriction enzyme before use in a random priming reaction. Wherever possible, a purified segment of target DNA should be used as a template rather than an entire plasmid. This greatly suppresses the level of background when the labeled DNA is used as a hybridization probe (Feinberg and Vogelstein 1983, 1984).

MATERIALS

It is essential that you consult the appropriate Material Safety Data Sheets and your institution's Environmental Health and Safety Office for proper handling of equipment and hazardous materials used in this protocol.

Recipes for reagents specific to this protocol, marked <R>, are provided at the end of the protocol. See Appendix 1 for recipes for commonly used stock solutions, buffers, and reagents, marked <A>. Dilute stock solutions to the appropriate concentrations.

Reagents

Alkaline agarose gel or denaturing polyacrylamide gel (see Step 4)

Ammonium acetate (10 M) <A> (optional; see Step 5)

Ethanol (optional; see Step 5)

Gel-filtration matrix (Sephadex G-50 or Bio-Gel), equilibrated with 1× TEN (pH 8.0) <A>

Klenow fragment of *E. coli* DNA Pol I

NA stop/storage buffer <R>

Nonradioactive labeling solutions

> dNTP solution containing dATP, dCTP, and dGTP, each at 1 mM and 0.65 mM dTTP
>
> *For advice on making and storing stock solutions of dNTPs, see the information panel Preparation of Stock Solutions of dNTPs.*
>
> Biotin-dUTP, fluorescein-dUTP, or DIG-dUTP (0.35 mM)

or

Radiolabeling solutions

> dNTP solution containing three unlabeled dNTPs, each at 5 mM
>
> *The composition of this solution depends on the [α-^{32}P]dNTP to be used. If radiolabeled dATP is used, the mix should contain dCTP, dTTP, and dGTP, each at a concentration of 5 mM. If two radiolabeled dNTPs are used, the solution should contain the other two dNTPs, each at a concentration of 5 mM. For advice on making and storing stock solutions of dNTPs, see the information panel Preparation of Stock Solutions of dNTPs.*
>
> [α-^{32}P]dNTP (10 mCi/mL; sp. act. >3000 Ci/mmol)
>
> *To minimize problems caused by radiolysis of the precursor, it is best, whenever possible, to prepare radiolabeled probes on the day that the [^{32}P]dNTP arrives in the laboratory.*

Random deoxynucleotide primers 6 or 7 bases in length (125 ng/μL in TE; pH 7.6)

> *Because of their uniform length and lack of sequence bias, synthetic oligonucleotides of random sequence are the primers of choice. Oligonucleotides of optimal length (hexamers and heptamers) (Suganuma and Gupta 1995) can be purchased from a commercial source (e.g., Sigma-Aldrich or Life Technologies) or synthesized locally on an automated DNA synthesizer. Store the solution of primers at –20°C in small aliquots.*

Random priming buffer (5×) <R>

Template DNA (5–25 ng/μL) in TE (pH 7.6)

> *Purify the DNA to be labeled by one of the methods described in Chapter 1.*
>
> *This protocol works best when 25 ng of template DNA is used in a standard 50-μL reaction. Larger amounts of DNA generate probes of lower specific activity; smaller amounts of DNA require longer reaction times. See note to Step 4.*

Equipment

Boiling water bath or heating block (95°C)

Equipment for spun-column chromatography (glass wool and 1-mL disposable syringe)

Sephadex G-50 spun column, equilibrated in TE (pH 7.6) (optional; see Step 5)

> *The Sephadex G-50 gel-filtration matrix can be used to separate DNA (which passes through the matrix) from lower-molecular-weight substances (e.g., radioactive precursors, which are retained on the column). Sephadex G-50 columns for radiolabled DNA purification are available commercially from several sources (e.g., Quick Spin Columns [Roche Applied Science]). However, it is also easy to set one up using a 1-mL disposable syringe plugged with siliconized glass wool (see this protocol, Steps 6–14).*

METHOD

1. In a 0.5-mL microcentrifuge tube, combine template DNA (25 ng) in 30 μL of H$_2$O with 1 μL of random deoxynucleotide primers (~125 ng). Close the top of the tube tightly and place the tube in a boiling water bath for 2 min.

2. Remove the tube and leave it for 1 min on ice. Centrifuge the tube for 10 sec at 4°C in a microcentrifuge to concentrate the mixture of primer and template at the bottom of the tube. Return the tube to the ice bath.

3. Set up a labeling reaction containing one of the following reaction mixtures.

   For Preparing Radioactively Labeled Probe
   To the mixture of primer and template, add

dNTP solution (5 mM)	1 μL
Random priming buffer (5×)	10 μL
[α-^{32}P]dNTP (10 mCi/mL)	5 μL (sp. act. 3000 Ci/mmol)
H$_2$O to 50 μL	

   For Preparing Probe Labeled with DIG, Biotin, or Fluorescein
   To the mixture of primer and template, add

dNTP solution (1 mM)	1 μL (contains 0.65 mM dTTP)
Random priming buffer (5×)	10 μL
Labeled dUTP (0.35 mM)	5 μL
H$_2$O to 50 μL	

4. Add 5 U (~1 μL) of the Klenow fragment. Mix the components by gently tapping the outside of the tube. Centrifuge the tube at maximum speed for 1–2 sec in a microcentrifuge to transfer all of the liquid to the bottom of the tube. Incubate the reaction mixture for 60 min at room temperature.

   To label larger amounts of DNA, assemble reaction mixtures as described in Steps 3 and 4 and then incubate the reaction for 60 min. To label smaller amounts of DNA, incubate the reactions for times that are in inverse proportion to the amount of template added. For example, random priming reactions containing 10 ng of template DNA should be incubated for 2.5 h.

   To monitor the course of the reaction, measure the proportion of radiolabeled dNTPs that is incorporated into material precipitated by trichloroacetic acid (TCA) (see Appendix 2).

   Under these reaction conditions, the length of the radiolabeled product is ~400–600 nucleotides, as determined by electrophoresis through an alkaline agarose gel (Chapter 6, Protocol 11) or a denaturing polyacrylamide gel (Chapter 6, Protocol 11).

5. Add 10 μL of NA stop/storage buffer to the reaction and proceed with one of the following options as appropriate.

   - Store the radiolabeled probe at –20°C until it is needed for hybridization.
     or

   - Separate the radiolabeled probe from unincorporated dNTPs by either spun-column chromatography (using a commercially available column or a homemade column according to Steps 6–14 below) or selective precipitation of the radiolabeled DNA with ammonium acetate and ethanol (see Chapter 1, Protocol 4). This step is generally not required if >50% of the radiolabeled dNTP has been incorporated during the reaction.

     Assuming that 50% of the radioactivity has been incorporated into TCA-precipitable material during the random priming reaction and that 90% of this material has been generated by random priming events on the template DNA (rather than self-priming of oligonucleotides), the probe DNA would contain a total of ~4.5 × 10^7 dpm (enough for two to five Southern hybridizations of mammalian genomic DNA). The specific activity of the probe would be ~1.8 × 10^9 dpm/μg and the weight of DNA synthesized during the reaction would be 9.7 ng, which is enough to detect single-copy sequences in mammalian DNA by Southern analysis.

Spun-Column Chromatography

6. Plug the bottom of a 1-mL disposable syringe with a small amount of sterile glass wool. This is best accomplished by using the barrel of the syringe to tamp the glass wool in place.

7. Fill the syringe with Sephadex G-50 or Bio-Gel P-60, equilibrated in 1× TEN buffer (pH 8.0). Start the buffer flowing by tapping the side of the syringe barrel several times. Add more resin until the syringe is completely full.

   Note that not all resins are suitable for spun-column centrifugation: DEAE Sephacel forms an impermeable lump during centrifugation, and the larger grades of Sephadex (G-100 and up) cannot be used because the beads are crushed by centrifugation. If a coarser-sieving resin is required, use Sepharose CL-4B.

8. Insert the syringe into a 15-mL disposable plastic tube. Centrifuge at 1600g for 4 min at room temperature in a swinging-bucket rotor in a bench-top centrifuge. Do not become alarmed by the appearance of the column. The resin packs down and becomes partially dehydrated during centrifugation. Continue to add more resin and recentrifuge until the volume of the packed column is ~0.9 mL and remains unchanged after centrifugation.

9. Add 0.1 mL of 1× TEN buffer to the columns and recentrifuge as in Step 8.

10. Repeat Step 9 twice more.

    Spun columns may be stored at this stage if desired. Several spun columns can be prepared simultaneously and stored for periods of 1 mo or more at 4°C before being used. Fill the syringes with 1× TEN buffer and wrap Parafilm around them to prevent evaporation. Store the columns upright at 4°C. Spun columns stored in this way should be washed once with sterile 1× TEN buffer, as described in Step 9, just before they are used.

11. Apply the DNA sample to the column in a total volume of 0.1 mL (use 1× TEN buffer to make up the volume). Place the spun column in a fresh disposable tube containing a decapped microcentrifuge.

12. Centrifuge again as in Step 8, collecting the effluent from the bottom of the syringe (~100 μL) into the decapped microcentrifuge tube.

13. Remove the syringe, which will contain unincorporated radiolabeled dNTPs or other small components. Using forceps, carefully recover the decapped microcentrifuge tube containing the eluted DNA and transfer its contents to a capped, labeled microcentrifuge tube.

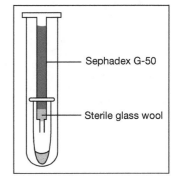

    A rough estimate of the proportion of radioactivity that has been incorporated into nucleic acid may be obtained by holding the syringe and the eluted DNA to a handheld minimonitor.

14. If the syringe is radioactive, carefully discard it in the radioactive waste. Store the eluted DNA at –20°C until needed.

RECIPES

It is essential that you consult the appropriate Material Safety Data Sheets and your institution's Environmental Health and Safety Office for proper handling of equipment and hazardous materials used in this protocol.

NA Stop/Storage Buffer

Reagent	Quantity (for 1 mL)	Final concentration
Tris-Cl (1 M, pH 7.5)	50 μL	50 mM
NaCl (5 M)	10 μL	50 mM
EDTA (0.5 M, pH 8.0)	10 μL	5 mM
SDS (10%, w/v)	50 μL	0.5% (w/v)

Random Priming Buffer (5×)

Reagent	Quantity (for 1 mL)	Final concentration
Tris-Cl (1 M, pH 8.0)	250 μL	250 mM
MgCl$_2$ (1 M)	25 μL	25 mM
NaCl (5 M)	20 μL	100 mM
Dithiothreitol (DTT) (1 M)	10 μL	10 mM
HEPES (2 M, adjusted to pH 6.6 with 4 M NaOH)	500 μL	1 M

Use a fresh dilution in H$_2$O of 1 M DTT stock, stored at −20°C. Discard the diluted DTT after use.

REFERENCES

Eweida M, Sit TL, Sira S, AbouHaidar MG. 1989. Highly sensitive and specific non-radioactive biotinylated probes for dot-blot, Southern and colony hybridizations. *J Virol Methods* **26:** 35–43.

Feinberg AP, Vogelstein B. 1983. A technique for radiolabeling DNA restriction endonuclease fragments to high specific activity. *Anal Biochem* **132:** 6–13.

Feinberg AP, Vogelstein B. 1984. A technique for radiolabeling DNA restriction endonuclease fragments to high specific activity. Addendum. *Anal Biochem* **137:** 266–267.

Goulian M. 1969. Initiation of the replication of single-stranded DNA by *Escherichia coli* DNA polymerase. *Cold Spring Harbor Symp Quant Biol* **33:** 11–20.

Hodgson CP, Fisk RZ. 1987. Hybridization probe size control: Optimized "oligolabelling." *Nucleic Acids Res* **15:** 6295.

Kessler C, Holtke HJ, Seibl R, Burg J, Muhlegger K. 1990. Non-radioactive labeling and detection of nucleic acids. I. A novel DNA labeling and detection system based on digoxigenin: anti-digoxigenin ELISA principle (digoxigenin system). *Biol Chem Hoppe Seyler* **371:** 917–927.

Lehman IR, Richardson CC. 1964. The deoxyribonucleases of *Escherichia coli*. IV. An exonuclease activity present in purified preparations of deoxyribonucleic acid polymerase. *J Biol Chem* **239:** 233–241.

Roy BP, AbouHaidar MG, Sit TL, Alexander A. 1988. Construction and use of cloned cDNA biotin and ^{32}P-labeled probes for the detection of papaya mosaic potexvirus RNA in plants. *Phytopathology* **78:** 1425–1429.

Roy BP, AbouHaidar MG, Alexander A. 1989. Biotinylated RNA probes for the detection of potato spindle tuber viroid (PSTV) in plants. *J Virol Methods* **23:** 149–155.

Sayavedra-Soto LA, Gresshoff PM. 1992. *Taq* DNA polymerase for labeling DNA using random primers. *BioTechniques* **13:** 568, 570, 572.

Stent GS, Fuerst CR. 1960. Genetic and physiological effects of the decay of incorporated radioactive phosphorus in bacterial viruses and bacteria. *Adv Biol Med Phys* **7:** 2–71.

Suganuma A, Gupta KC. 1995. An evaluation of primer length on random-primed DNA synthesis for nucleic acid hybridization: Longer is not better. *Anal Biochem* **224:** 605–608.

Taylor JM, Illmensee R, Summers J. 1976. Efficient transcription of RNA into DNA by avian sarcoma virus polymerase. *Biochim Biophys Acta* **442:** 324–330.

Random Priming: Labeling of DNA by Extension of Random Oligonucleotides in the Presence of Melted Agarose

A variation of the method described in Protocol 1 can be used to radiolabel DNA in slices cut from gels cast with low-melting-temperature agarose (Feinberg and Vogelstein 1983, 1984). Most of the materials required for this protocol are the same as those used in Protocol 1. However, the labeling buffer has been slightly modified to include unlabeled dNTPs and random oligonucleotide primers. According to the investigator's needs, the labeling reaction can be assembled from its individual components, as described in Protocol 1, or from the composite buffer, as described here. For additional details concerning materials required for the method and the specific activity and length of the generated probe, see the introduction to Protocol 1.

MATERIALS

It is essential that you consult the appropriate Material Safety Data Sheets and your institution's Environmental Health and Safety Office for proper handling of equipment and hazardous materials used in this protocol.

Recipes for reagents specific to this protocol, marked <R>, are provided at the end of the protocol. See Appendix 1 for recipes for commonly used stock solutions, buffers, and reagents, marked <A>. Dilute stock solutions to the appropriate concentrations.

Reagents

Alkaline agarose gel or denaturing polyacrylamide gel (see Step 4)

[α-^{32}P]dNTP (10 mCi/mL; sp. act. >3000 Ci/mmol)

> *To minimize problems caused by radiolysis of the precursor, it is best, whenever possible, to prepare radiolabeled probes on the day that the [^{32}P]dNTP arrives in the laboratory.*

Ammonium acetate (10 M) <A> (optional: see Step 5)

Bovine serum albumin (BSA) (10 mg/mL)

Ethanol (optional; see Step 5)

Ethidium bromide (10 mg/mL) or SYBR Green staining solution

Klenow fragment of *E. coli* DNA polymerase

> *The Klenow fragment (5 U) is required in each random priming reaction.*

NA stop/storage buffer <R>

Oligonucleotide labeling buffer (5×) <R>

> *The composition of the 5× oligonucleotide labeling buffer depends on the [α-^{32}P]dNTP to be used. If radiolabeled dATP is to be used, the buffer should contain dCTP, dTTP, and dGTP. If two radiolabeled dNTPs are used, the buffer should contain two unlabeled dNTPs. For advice on making and storing stock solutions of dNTPs, see the information panel Preparation of Stock Solutions of dNTPs.*

Random deoxynucleotide primers 6 or 7 bases in length (125 ng/μL in TE; pH 7.6)

> *Because of their uniform length and lack of sequence bias, synthetic oligonucleotides of random sequence are the primers of choice. Oligonucleotides of optimal length (hexamers or heptamers) (Suganuma and Gupta 1995) can be purchased from a commercial source (e.g., Sigma-Aldrich or Life Technologies) or synthesized locally on an automated DNA synthesizer. Store the solution of primers at −20°C in small aliquots. Primers are incorporated into the 5× oligonucleotide labeling buffer.*

Template DNA

> *The DNA to be labeled is recovered after electrophoresis through a low-melting-temperature (LMT) agarose gel (e.g., FMC SeaPlaque LMT Agarose) (see this protocol, Steps 1–3). The gel should be cast and run*

in 1× tris-acetate electrophoresis buffer. For details on casting and running low-melting-temperature gels, see Chapter 2, Protocol 9.

This protocol works best when 25 ng of template DNA is used in a standard 50-μL reaction. Larger amounts of DNA generate probes of lower specific activity; smaller amounts of DNA require longer reaction times. See the note to Step 4.

Equipment

Boiling water bath
Sephadex G-50 spun column, equilibrated in TE (pH 7.6) (optional; see Step 5)

The Sephadex G-50 gel-filtration matrix can be used to separate DNA (which passes through the matrix) from lower-molecular-weight substances (e.g., radioactive precursors, which are retained on the column). Sephadex G-50 columns for radiolabeled DNA purification are available commercially from several sources (e.g., Quick Spin Columns [Roche Applied Science]). However, it is also easy to set one up using a 1-mL disposable syringe plugged with siliconized glass wool (see Protocol 1, Steps 6–14).

METHOD

1. After electrophoresis, stain the gel with ethidium bromide (final concentration 0.5 μg/mL) or SYBR Green, and excise the desired band, eliminating as much extraneous agarose as possible.

2. Place the band in a preweighed microcentrifuge tube and measure its weight. Add 3 mL of H_2O for every gram of agarose gel.

3. Leave the microcentrifuge tube for 7 min in a boiling water bath to melt the gel and denature the DNA.

   If performing radiolabeling immediately, store the tube at 37°C until the template is required. Otherwise, store the tube at –20°C. After each removal from storage, reheat the DNA/gel slurry for 3–5 min to 100°C and then store at 37°C until the radiolabeling reaction is initiated.

4. To a fresh microcentrifuge tube in a 37°C water bath or heating block, add the following ingredients in the following order:

Oligonucleotide labeling buffer (5×)	10 μL
BSA solution (10 mg/mL)	2 μL
DNA (in a volume no greater than 32 μL)	20–50 ng
[α-^{32}P]dNTP (10 mCi/mL)	5 μL (sp. act. >3000 Ci/mmol)
Klenow fragment (5 U)	1 μL
H_2O to 50 μL	

   Mix the components completely with a micropipettor. Incubate the reaction for 2–3 h at room temperature or for 60 min at 37°C.

   To label larger amounts of DNA, adjust the volume of the reaction mixture proportionately and incubate the reaction for 60 min. To label smaller amounts of DNA, incubate the reactions for times that are in inverse proportion to the amount of template added. For example, random priming reactions containing 10 ng of template DNA should be incubated for 2.5 h.

   To monitor the course of the reaction, measure the proportion of radiolabeled dNTPs that are incorporated into material precipitated by TCA (see Appendix 2).

   Under these reaction conditions, the length of the radiolabeled product is ~400–600 nucleotides, as determined by electrophoresis through an alkaline agarose gel (Chapter 2, Protocol 9 or a denaturing polyacrylamide gel (Chapter 6, Protocol 11).

5. Add 50 μL of NA stop/storage buffer to the reaction and proceed with one of the following options as appropriate:

   • Store the radiolabeled probe at –20°C until it is needed for hybridization.

     or

   • Separate the radiolabeled probe from unincorporated dNTPs by either spun-column chromatography (using a commercially available column or a homemade column according to Protocol 1, Steps 6–14) or selective precipitation of the radiolabeled DNA with ammonium acetate and ethanol (see Chapter 1, Protocol 4). This step is not required if >50% of the radiolabeled dNTP has been incorporated during the reaction.

> *Assuming that 50% of the radioactivity has been incorporated into TCA-precipitable material during the random priming reaction and that 90% of this material has been generated by random priming events on the template DNA (rather than self-priming of oligonucleotides), the probe DNA would contain a total of ~4.5 × 10⁷ dpm (enough for two to five Southern hybridizations of mammalian genomic DNA). The specific activity of the probe would be ~1.8 × 10⁹ dpm/μg and the weight of DNA synthesized during the reaction would be 9.7 ng, which is enough to detect single-copy sequences.*

RECIPES

It is essential that you consult the appropriate Material Safety Data Sheets and your institution's Environmental Health and Safety Office for proper handling of equipment and hazardous materials used in this protocol.

NA Stop/Storage Buffer

Reagent	Quantity (for 1 mL)	Final concentration
Tris-Cl (1 M, pH 7.5)	50 μL	50 mM
NaCl (5 M)	10 μL	50 mM
EDTA (0.5 M, pH 8.0)	10 μL	5 mM
SDS (10%, w/v)	50 μL	0.5% (w/v)

Oligonucleotide Labeling Buffer (5×)

Reagent	Quantity (for 1 mL)	Final concentration
Tris-Cl (1 M, pH 8.0)	250 μL	250 mM
MgCl₂ (1 M)	25 μL	25 mM
DTT (1 M)	20 μL	20 mM
Unlabeled dNTPs (100 mM)	20 μL	2 mM (each)
HEPES (2 M, adjusted to pH 6.6 with 4 N NaOH)	500 mL	1 M
Random deoxynucleotide primers, 6 bases in length (10 mg/mL)	100 μL	1 mg/mL

Store the buffer in small aliquots at −20°C. The buffer may be frozen and thawed several times without harm.

REFERENCES

Feinberg AP, Vogelstein B. 1983. A technique for radiolabeling DNA restriction endonuclease fragments to high specific activity. *Anal Biochem* **132:** 6–13.

Feinberg AP, Vogelstein B. 1984. A technique for radiolabeling DNA restriction endonuclease fragments to high specific activity. Addendum. *Anal Biochem* **137:** 266–267.

Suganuma A, Gupta KC. 1995. An evaluation of primer length on random-primed DNA synthesis for nucleic acid hybridization: Longer is not better. *Anal Biochem* **224:** 605–608.

Labeling of DNA Probes by Nick Translation

Nick translation requires the simultaneous activity of two different enzymes (see Fig. 1). DNase I is used to cleave (nick) phosphodiester bonds at random sites in both strands of a double-stranded target DNA (in the presence of Mg^{2+}, DNase I becomes a single-stranded endonuclease). *E. coli* DNA Pol I is used to add deoxynucleotides to the 3′-hydroxyl termini created by DNase I. In addition to its polymerizing activity, DNA Pol I carries a 5′→3′ exonucleolytic activity that removes nucleotides from the 5′ side of the nick. The simultaneous elimination of nucleotides from the 5′ side and the addition of labeled nucleotides to the 3′ side result in movement of the nick (nick translation) along the DNA, which becomes labeled to high specific activity (Kelly et al. 1970). The reaction produces double-stranded probes that can be used for a variety of purposes including screening genomic and cDNA libraries and Southern, northern, and in situ hybridizations. Both radioactive and nonradioactive probes can be synthesized using this method. This method requires a large quantity of DNA (1 μg). The reaction time is relatively long (2 h), and the temperature (15°C) is important.

After occupying a dominant position for many years as the method of choice for radiolabeling dsDNA, nick translation was largely displaced by random oligonucleotide priming, which requires only one enzyme. One problem with nick translation has always been the difficulty in balancing the activities of DNA Pol I and DNase I used in the procedure. With the advent of nick-translation kits (e.g., NIK-IT, Worthington Biochemical Inc.), however, this difficulty has been solved. Tips for optimizing nick-translation reactions are given in the box Optimizing Nick-Translation Reactions.

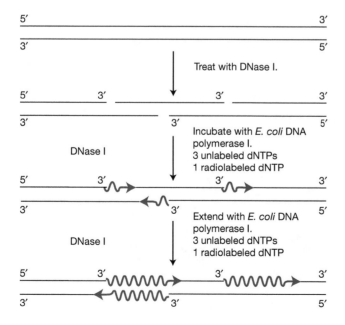

FIGURE 1. **Nick translation using *E. coli* DNA polymerase.** (Adapted from Krieg and Melton 1987, with permission, from Elsevier.)

OPTIMIZING NICK-TRANSLATION REACTIONS

The amount of label incorporated during nick translation depends on the number of 3'-hydroxyl termini created in the template DNA by DNase I. Too much nicking creates an excess of initiation sites that leads to maximal incorporation of label but yields DNA fragments that are too short to be useful as hybridization probes. Too little nicking, on the other hand, restricts the number of sites available for initiation of nick translation, resulting in a product of low specific activity. The activity of DNase I varies among preparations, and the amount of DNase contaminating preparations of *E. coli* DNA Pol I can also differ. It is therefore necessary to titrate each new batch of DNase I to find a concentration that yields probes with the desired specific activity and length. The nicking and polymerization reactions can be performed simultaneously; in this case, a slight lag is observed before incorporation of nucleotides begins (Rigby et al. 1977), reflecting the time required for DNase I to generate nicks in the template. This lag, which is usually of little practical significance, can be eliminated by performing the nicking and polymerization reactions sequentially rather than simultaneously (Koch et al. 1986).

For radiolabeled probes, the aim is to establish empirically a concentration of DNase I that results in incorporation of ~40% of the [α-^{32}P]dNTP. A standard nick-translation reaction performed under these conditions yields a product whose specific activity exceeds 10^8 cpm/μg and whose radiolabeled DNA strands are ~400–750 nucleotides in length. It is thus necessary to determine empirically the optimum amounts of each enzyme required to obtain both high specific activity and appropriate probe length. This optimization is best accomplished by keeping the amount of DNA polymerase added constant (e.g., at 2.5 U) and varying the concentration of input DNase I.

Although *E. coli* DNA Pol I works adequately with concentrations of dNTPs as low as 2 μM, the enzyme catalyzes DNA synthesis much more efficiently when supplied with higher concentrations of substrates. For reasons of cost, nick-translation reactions usually contain minimal concentrations of radiolabeled dNTPs (0.5–5 μM) and much greater concentrations of unlabeled dNTPs (1 mM). For radiolabeled probes, the specific activity of the final product depends in large part on the specific activity of the radiolabeled dNTP used in the reaction. By diluting the radiolabeled dNTP with the homologous unlabeled dNTP, it is possible to prepare DNA labeled to different specific activities. If high specific activities (i.e., >5 × 10^8 dpm/μg) are required (e.g., for screening recombinant DNA libraries or for detecting single-copy sequences in Southern hybridizations of complex mammalian genomes), the nick-translation reaction should contain all four radiolabeled dNTPs (sp. act. >800 Ci/mmol) and no unlabeled dNTPs. For most other purposes, it is adequate to use one dNTP labeled with [α-^{32}P] (800 Ci/mmol) and three unlabeled dNTPs or to dilute each [α-^{32}P]dNTP with an appropriate amount of the unlabeled dNTP. The amount of radiolabeled dNTP in a solution of specific activity 800 Ci/mmol (~12 pmol/μL) is ~3.75-fold higher than that in a solution of specfic activity 3000 Ci/mmol (~3.3 pmol/μL).

MATERIALS

It is essential that you consult the appropriate Material Safety Data Sheets and your institution's Environmental Health and Safety Office for proper handling of equipment and hazardous materials used in this protocol.

Recipes for reagents specific to this protocol, marked <R>, are provided at the end of the protocol. See Appendix 1 for recipes for commonly used stock solutions, buffers, and reagents, marked <A>. Dilute stock solutions to the appropriate concentrations.

Reagents

[α-^{32}P]dATP (800 Ci/mmol), 0.5 mM biotin-11-dUTP, or 0.5 mM DIG-11-dUTP

To minimize problems caused by radiolysis of the precursor, it is best, whenever possible, to prepare radiolabeled probes on the day that the [^{32}P]dNTP arrives in the laboratory.

DNase I (1 mg/mL stock solution, diluted either 1000- or 10,000-fold; see the note below)

DNase I is commonly sold as a solution (as RNase- and/or protease-free forms) but is sometimes sold as a lyophilized powder at a concentration of 2000–3000 U/mg protein. To prepare a stock solution from the lyophilized powder, dissolve 1 mg of DNase 1 in a 1-mL solution containing 20 mM Tris-Cl (pH 7.5) and 1 mM MgCl$_2$. Aliquot into microcentrifuge tubes, quick-freeze on dry ice, and store at −80°C. This solution *can be stored for up to 1 yr without loss of enzyme activity. To use, thaw the solution on ice; unused portions should not be refrozen.*

When preparing radiolabeled probes, dilute the stock 10,000-fold in a solution containing 20 mM Tris-Cl (pH 7.5), 0.5 mg/mL BSA, and 10 mM β-mercaptoethanol. This solution can be stored for up to 1 mo at 4°C. Therefore, when preparing nonradioactive probes, dilute the stock DNase I only 1000-fold.

dNTP solution containing dTTP, dCTP, and dGTP, each at a concentration of 0.5 mM

Therefore, when preparing nonradioactive probes, replace the dNTP solution containing dTTP, dCTP, and dGTP with one that contains dUTP, dCTP, and dGTP (0.5 mM each).

E. coli DNA Pol I (10 U/μL)

EDTA (0.5 M, pH 8.0) <A>

Nick-translation reaction buffer (10×) <R>

Template DNA (50–500 ng/μL) in TE (pH 7.6)

This protocol works best when 50 μg of DNA is used in a standard 50-μL reaction.

Equipment

Container with ice

Sephadex G-50 spun column, equilibrated in TE (pH 7.6) (optional; see Step 4)

The Sephadex G-50 gel-filtration matrix can be used to separate DNA (which passes through the matrix) from lower-molecular-weight substances (e.g., radioactive precursors, which are retained on the column). Sephadex G-50 columns for radiolabeled DNA purification are available commercially from several sources (e.g., Quick Spin Columns [Roche Applied Science]). However, it is also easy to set one up using a 1-mL disposable syringe plugged with siliconized glass wool (see Protocol 1, Steps 6–14).

Water bath (15°C)

METHOD

1. In a 1.5 mL microcentrifuge tube on ice, mix the following components:

Nick-translation reaction buffer (10×)	5 μL
dNTP solution	5 μL
[α-^{32}P]dATP (50 μCi), biotin-11-dUTP (0.5 mM), or DIG-11-dUTP (0.5 mM)	5 μL
Diluted DNase I	2 μL
E. coli DNA Pol I	2 μL
H$_2$O to a final volume of 50 μL after addition of DNA in Step 2	

2. Add either 0.5 μg of DNA for radiolabeling or 1 μg of DNA for biotin or DIG labeling. Cap the tube and mix the components by gently tapping the outside of the tube. Centrifuge the tube at maximum speed for 1–2 sec in a microcentrifuge to transfer all the liquid to the bottom of the tube.

3. Incubate the reaction for 1 h at 15°C when preparing radiolabeled probe. For DIG or biotin labeling, increase the incubation to 2 h at 15°C.

   For nonradioactive probes, the incubation time of the reaction is increased to allow optimal incorporation of the modified deoxynucleotide.

4. Stop the reaction by adding 2 μL of 0.5 M EDTA and proceed with one of the following options as appropriate.

   - Store the radiolabeled probe at −20°C until it is needed for hybridization.

     or

   - Separate the radiolabeled probe from unincorporated dNTPs by either spun-column chromatography (using a commercially available column or a homemade column according to Protocol 1, Steps 6–14) or selective precipitation of the radiolabeled DNA with ammonium

acetate and ethanol (see Chapter 1, Protocol 4). This step is generally not required if >50% of the radiolabeled dNTP has been incorporated during the reaction.
See Troubleshooting.

5. Load an aliquot of the reaction mixture on an agarose minigel along with suitable size markers. Ideally, the digested DNA should be between 100 and 500 bp. If the probe size is between 500 and 1000 bp (or larger), increase the amount of DNase I added to the reaction.
See Troubleshooting.

TROUBLESHOOTING

Problem (Step 4): Insufficient label is incorporated into the probe.
Solution: The reaction time is too short or the amount of DNase I is too low. To optimize the nick-translation reaction, it may be necessary to optimize both the reaction time and the amount of DNase I added to the reaction. To do this, set up a series of parallel reactions, vary the reaction time and amount of enzyme, and measure the incorporation of $[\alpha\text{-}^{32}P]dATP$ into the probe. Ideally, the probe should have 30%–40% incorporation.

Problem (Step 4): The probe is degraded.
Solution: The reaction time or the amount of DNase I is too high. Set up a series of parallel reactions as per above.

Problem (Step 5): Insufficient label is incorporated into the probe.
Solution: The DNA template fragment may be too short (<500 bp). Increase the amount of DNase I in order to ensure that the DNA is nicked at least once.

RECIPE

It is essential that you consult the appropriate Material Safety Data Sheets and your institution's Environmental Health and Safety Office for proper handling of equipment and hazardous materials used in this protocol.

Nick-Translation Reaction Buffer (10×)

Reagent	Quantity (for 1 mL)	Final concentration
Tris-Cl (1 M, pH 7.5)	500 μL	0.5 M
MgCl$_2$ (1 M)	100 μL	0.1 M
DTT (1 M)	10 μL	10 mM
BSA (10 mg/mL)	50 μL	0.5 mg/mL

Prepare aliquots and store at −20°C.

REFERENCES

Kelly RB, Cozzarelli NR, Deutscher MP, Lehman IR, Kornberg A. 1970. Enzymatic synthesis of deoxyribonucleic acid. XXXII. Replication of duplex deoxyribonucleic acid by polymerase at a single strand break. *J Biol Chem* 245: 39–45.

Koch J, Kolvraa S, Bolund L. 1986. An improved method for labelling of DNA probes by nick translation. *Nucleic Acids Res* 14: 7132.

Krieg PA, Melton DA. 1987. In vitro RNA synthesis with SP6 RNA polymerase. *Methods Enzymol* 155: 397–415.

Rigby PW, Dieckmann M, Rhodes C, Berg P. 1977. Labeling deoxyribonucleic acid to high specific activity in vitro by nick translation with DNA polymerase I. *J Mol Biol* 113: 237–251.

Labeling of DNA Probes by Polymerase Chain Reaction

The polymerase chain reaction (see Chapter 7) can be used to produce both nonradiolabeled DNA probes and radiolabeled DNA probes with high specific activity. The advantages of the method include the following.

- Defined segments of the target DNA can be amplified and labeled independently of the location or type of restriction sites.

- There is no need to isolate fragments of DNA or to subclone them into vectors containing bacteriophage promoters.

- Only small amounts of template DNA are required (2–10 ng or ~1 fmol).

- The specific activity of the radiolabeled amplified DNA can exceed 10^9 dpm/μg.

Four different methods have been described for simultaneous amplification and radiolabeling of DNA. The first three methods require some information about the sequence of the target DNA, but the fourth method does not.

- dsDNA probes can be produced in conventional PCRs containing equal concentrations of two primers. Radioactive probes are synthesized in reactions containing three unlabeled dNTPs at concentrations exceeding the K_m (200 μM) and one [α-^{32}P]dNTP at a concentration at or slightly above the K_m (2–3 μM) (Jansen and Ledley 1989; Schowalter and Sommer 1989). Non-radioactive probes are synthesized in reactions that include nonradiolabled deoxynucleotides, such as biotin-11-dUTP or DIG-11-dUTP, at a concentration of one-third of normal levels (66.7 μM) and in which the amount of dTTP is reduced to about two-thirds of normal levels (133 μM).

- Labeled probes biased heavily in favor of one strand of DNA can be produced in PCRs in which the concentration of one primer exceeds the other by a factor of 20–200. During the initial cycles of the PCR, dsDNA is synthesized in a conventional exponential fashion. However, when the concentration of one primer becomes limiting, the reaction generates ssDNA that accumulates at an arithmetic rate. By the end of the reaction, the concentration of one strand of DNA is three to five times greater than the concentration of the other (Scully et al. 1990). (For more information, see the Additional Protocol Asymmetric Probes below.)

- Labeled probes consisting entirely of one strand of DNA can be synthesized in thermal cycling reactions that contain a dsDNA template but only one primer. Double-stranded template DNA (20 ng) generates ~200 ng of single-stranded probe during the course of 40 cycles. The length of the probe can be defined by cleaving the template DNA at a restriction site downstream from the binding site of the primer (e.g., see Stürzl and Roth 1990; Finckh et al. 1991). Note that uniformly labeled ssDNA probes can also be generated by a primer-extension approach using a combination of bacterial T4 polynucleotide kinase and Klenow enzyme (see Chapter 6, Protocol 16).

- The target DNA may be digested with CviJI, a restriction enzyme that uses adenosine triphosphate (ATP) as a cofactor to cleave the recognition sequence GC (except YGCR, where Y is a pyrimidine and R a purine) (Swaminathan et al. 1996). Because the dinucleotide GC occurs frequently in most DNAs, the modal size of the resulting blunt-ended DNA fragments is small—between 20 and 60 nucleotides. These fragments can be used as sequence-specific primers in

PCRs in which an aliquot of the target DNA is used as template. The product is a heterogeneous population of double-stranded molecules whose size ranges from a minimum of ∼60 bp to a maximum that exceeds the size of the target (Swaminathan et al. 1994). These larger molecules are thought to be complex scrambled versions of the target DNA that are generated from chimeric templates and/or primers. Investigators who feel uneasy about using a probe whose sequence is not colinear with the original template DNA may wish to avoid this method; others wanting to use a PCR-based technique in the absence of DNA sequence information may embrace it.

The following protocol, provided by Mala Mahendroo and Galvin Swift (University of Texas Southwestern Medical Center, Dallas), remains close to the original procedure of Schowalter and Sommer (1989) for the generation of double-stranded radiolabeled probes. A modification that describes how to generate asymmetric probes by PCR is presented in the Additional Protocol Asymmetric Probes below.

MATERIALS

It is essential that you consult the appropriate Material Safety Data Sheets and your institution's Environmental Health and Safety Office for proper handling of equipment and hazardous materials used in this protocol.

Recipes for reagents specific to this protocol, marked <R>, are provided at the end of the protocol. See Appendix 1 for recipes for commonly used stock solutions, buffers, and reagents, marked <A>. Dilute stock solutions to the appropriate concentrations.

Reagents

[α-^{32}P]dCTP (10 mCi/mL; sp. act. 3000 Ci/mmol)
   To minimize problems caused by radiolysis of the precursor, it is best, whenever possible, to prepare radiolabeled probes on the day that the [^{32}P]dNTP arrives in the laboratory.

Ammonium acetate (10 M) <A>

Amplification buffer (10×) <R>

Biotin-11-dUTP or DIG-11-dUTP

Carriers used during ethanol precipitation of radiolabeled probe (see Step 5)
   Use either glycogen (stock solution = 50 mg/mL in H$_2$O) or yeast tRNA (stock solution 10 mg/mL in H$_2$O).

Chloroform

dCTP (0.1 mM; for use with radiolabeling)
   Dilute 1 volume of a stock solution of 10 mM dCTP with 99 volumes of 10 mM Tris-Cl (pH 8.0). Store the diluted solution at −20°C in 50-μL aliquots.

dNTP solution

dNTP solution 1, containing dATP, dGTP, and dTTP, each at 10 mM
   or

dNTP solution 2, containing dATP, dGTP, and dCTP, each at 10 mM
   Solution 1 is for use with radioactive dCTP, and solution 2 is for use with a nonradioactive dUTP. For advice on making and storing stock solutions of dNTPs, see the information panel Preparation of Stock Solutions of dNTPs.

dTTP (5 mM; for use with DIG or biotin labeling)

Ethanol

Forward primer (20 μM) in H$_2$O and reverse primer (20 μM) in H$_2$O
   Each primer should be 20–30 nucleotides in length and contain approximately equal numbers of the four bases, with a balanced distribution of G and C residues and a low propensity to form stable secondary structures. Store the stock solutions of primers at −20°C.

Oligonucleotide primers synthesized on an automated DNA synthesizer can generally be used in stand-ard PCR without further purification. However, amplification and radiolabeling of single-copy sequences from mammalian genomic templates is often more efficient if the oligonucleotide primers are purified by chromatography on commercially available resins (e.g., NENSORB, PerkinElmer Life Science Inc. products) or by denaturing polyacrylamide gel electrophoresis, as described in Chapter 2, Protocol 3.

TE (pH 7.6) <A>

Template DNA (2–10 ng)

A variety of templates can be used in this protocol, including crude minipreparations of plasmid DNAs (see Chapter 1, Protocol 1), purified DNA fragments isolated from agarose or polyacrylamide gels (see Chapter 2, Protocol 1 or 3), and genomic DNAs from organisms with low complexity (e.g., bacteria and yeast).

When using recombinant plasmids or other types of recombinants as templates, a small proportion of the radiolabeled probe may be derived from the vector sequences. This contamination arises if the oligonucleotide primers are complementary to flanking vector sequences or because the PCR will generate long "read-through" strands on the starting template DNA. For most hybridization applications, the presence of low levels of vector sequences will not interfere with the experiment. However, when screening libraries made in high-copy-number plasmid vectors (e.g., pUC, pBluescript, and pGEM vectors), the presence of vector sequences in the probe can lead to a substantial background. To manage this problem, (1) use forward and reverse primers that are located within the target region, (2) use a gel-purified DNA fragment as a starting template for the amplification, and (3) cleave the template DNA with restriction enzyme(s) to prevent the synthesis of read-through strands.

Thermostable DNA polymerase (e.g., *Taq* DNA polymerase)

Thermostable DNA polymerase (2.5 U) is required for each amplification/labeling reaction.

Equipment

Barrier tips for automatic micropipettes

Microcentrifuge tubes (0.5-mL thin-walled tubes designed for PCR)

Positive-displacement pipette

Sephadex G-75 spun column, equilibrated in TE (pH 7.6)

The Sephadex G-75 gel-filtration matrix can be used to separate DNA (which passes through the matrix) from lower-molecular-weight substances (e.g., radioactive precursors, which are retained on the column). It is easy to set up a column using a 1-mL disposable syringe plugged with siliconized glass wool (see Protocol 1, Steps 6–14, but substitute Sephadex G-50 with Sephadex G-75, available from companies such as Sigma-Aldrich or GE Healthcare Life Sciences).

Thermal cycler programmed with desired amplification protocol

If the thermal cycler is not equipped with a heated lid, use either mineral oil or a bead of paraffin wax to prevent evaporation of liquid from the reaction mixture during PCR.

METHOD

1. In a 0.5-mL thin-walled microcentrifuge tube, set up an amplification/labeling reaction containing either of the following sets of reactions.

   For Radiolabeling

Amplification buffer (10×)	5.0 μL
dNTP solution 1 (10 mM)	1.0 μL
dCTP (0.1 mM)	1.0 μL
Forward oligonucleotide primer (20 μM)	2.5 μL
Reverse oligonucleotide primer (20 μM)	2.5 μL
Template DNA (2–10 ng or ∼1 fmol)	5–10 μL
[α-^{32}P]dCTP (10 mCi/mL) (sp. act. 3000 Ci/mmol)	5.0 μL
H$_2$O	to 48 μL

For Labeling with DIG or Biotin

Amplification buffer (10×)	5.0 μL
dNTP solution 2 (10 mM)	1.0 μL
dTTP (5 mM)	1.33 μL
Forward oligonucleotide primer (20 μM)	2.5 μL
Reverse oligonucleotide primer (20 μM)	2.5 μL
Template DNA (2–10 ng or ~1 fmol)	5–10 μL
Biotin-11-dUTP or DIG-11-dUTP	to 66.7 μM
H$_2$O	to 48 μL

Add 2.5 U of thermostable DNA polymerase to the reaction mixture. Gently tap the side of the tube to mix the ingredients.

If more than one DNA fragment is to be labeled using a single pair of primers, make up and dispense a master mix consisting of all of the reaction components except the DNA templates to the PCR tubes. Individual DNA templates can then be added to each tube just before addition of enzyme and initiation of the reaction.

2. If the thermal cycler is not fitted with a heated lid, overlay the reaction mixture with 1 drop (50 μL) of light mineral oil or a bead of paraffin wax to prevent evaporation of the samples during repeated cycles of heating and cooling. Place the tubes in a thermal cycler.

3. Amplify the samples using the denaturation, annealing, and polymerization times listed in the table.

Cycle Number	Denaturation	Annealing	Polymerization
30 cycles	30–45 sec at 94°C	30–45 sec at 55–60°C	1–2 min at 72°C
Last cycle	1 min at 94°C	30 sec at 55°C	1 min at 72°C

These times are suitable for 50-μL reactions assembled in thin-walled 0.5-mL tubes. Times and temperatures may need to be adapted to suit other types of equipment and reaction volumes.

Polymerization should be performed for 1 min for every 1000 bp of length of the target DNA.

Most thermal cyclers have an end routine in which the amplified samples are incubated at 4°C until they are removed from the machine. Samples can be left overnight at this temperature but should be stored thereafter at –20°C.

4. Remove the tubes from the thermal cycler. Use a micropipettor to remove as much mineral oil from the top of the reaction mixture as possible. Extract the reaction mix with 50 μL of chloroform to remove the remaining mineral oil. Separate the aqueous and organic layers by centrifugation for 1 min at room temperature in a microcentrifuge.

5. Remove the upper, aqueous layer to a fresh microcentrifuge tube, add carrier tRNA (10–100 μg) or glycogen (5 μg), and precipitate the DNA with an equal volume of 4 M ammonium acetate and 2.5 volumes of ethanol. Store the tube for 1–2 h at –20°C or for 10–20 min at –70°C. Collect the precipitated DNA by centrifugation at maximum speed for 5–10 min at 4°C.

6. Dissolve the DNA in 20 μL of TE (pH 7.6) and remove remaining unincorporated dNTPs and the oligonucleotide primers by spun-column chromatography through Sephadex G-75 as described in Protocol 1, Steps 6–14.

   Approximately 60% of the radioactivity should have been incorporated into DNA that elutes in the void volume during spun-column chromatography.

7. When preparing radiolabeled probes, use a liquid scintillation counter to measure the amount of radioactivity in 1.0 μL of the void volume of the spun column. Store the remainder of the radiolabeled DNA at –20°C until required.

   The yield of radiolabeled DNA ranges from 20 to 50 ng, sp. act. 1×10^9 to 2.5×10^9 dpm/μg.

Asymmetric Probes

By limiting the amount of one primer in the amplification reaction, a preponderance of one strand of the dsDNA template will be synthesized during the amplification reaction (Gyllensten and Erlich 1988; Innis et al. 1988; Shyamala and Ames 1989, 1993; McCabe 1990; Scully et al. 1990). The resulting asymmetric probes can be used in northern hybridization to determine the strand of an unknown DNA that represents the sense strand of a gene and the strand that represents the antisense strand of a gene.

To synthesize an asymmetric probe in the above reaction, substitute a 0.4-μM solution of either forward oligonucleotide primer or reverse primer for one of the standard 20-μM primer solutions (see Protocol 4, Step 1). Perform the remainder of the protocol exactly as described.

Keep in mind that asymmetric amplification initially proceeds at an exponential rate and then slows to an arithmetic rate when the amount of one oligonucleotide primer becomes limiting. The specific activity of the asymmetric probe is the same as that produced in the normal PCR, but the amount of DNA synthesized in the reaction will be much less. Setting up multiple reactions at Step 1 can help to compensate for the decrease in total yield of probe.

In addition, remember that the bias in favor of one radiolabeled strand over the other is usually not more than a factor of 5. If necessary, this ratio can be improved by (1) separating the ssDNA and dsDNA products of the PCR by gel electrophoresis or anion-exchange chromatography or (2) using a more complex, two-stage amplification procedure that enables a single-stranded probe to be produced in >20-fold excess (Finckh et al. 1991).

RECIPE

It is essential that you consult the appropriate Material Safety Data Sheets and your institution's Environmental Health and Safety Office for proper handling of equipment and hazardous materials used in this protocol.

Amplification Buffer (10×)

Reagent	Quantity (for 50 mL)	Final concentration
KCl (1 M)	25 mL	0.5 M
Tris-Cl (1 M, pH 8.3)	5 mL	0.1 M
MgCl$_2$ (1 M)	750 μL	15 mM

Autoclave the 10× buffer for 10 min at 15 psi (1.05 kg/cm^2) on liquid cycle. Divide the sterile buffer into aliquots and store them at −20°C.

REFERENCES

Finckh U, Lingenfelter PA, Myerson D. 1991. Producing single-stranded DNA probes with the *Taq* DNA polymerase: A high yield protocol. *BioTechniques* **10:** 35–39. (Erratum *BioTechniques* [1992] **12:** 382.)

Gyllensten UB, Erlich HA. 1988. Generation of single-stranded DNA by the polymerase chain reaction and its application to direct sequencing of the *HLA-DQA* locus. *Proc Natl Acad Sci* **85:** 7652–7656.

Innis MA, Myambo KB, Gelfand DH, Brow MAD. 1988. DNA sequencing with *Thermus aquaticus* DNA polymerase and direct sequencing of polymerase chain reaction–amplified DNA. *Proc Natl Acad Sci* **85**: 9436–9440.

Jansen R, Ledley FD. 1989. Production of discrete high specific activity DNA probes using the polymerase chain reaction. *Gene Anal Tech* **6**: 79–83.

McCabe PC. 1990. Production of single-stranded DNA by asymmetric PCR. In *PCR protocols: A guide to methods and applications* (ed Innis MA, et al.), pp. 76–83. Academic, New York.

Schowalter DB, Sommer SS. 1989. The generation of radiolabeled DNA and RNA probes with polymerase chain reaction. *Anal Biochem* **177**: 90–94.

Scully SP, Joyce ME, Abidi N, Bolander ME. 1990. The use of polymerase chain reaction generated nucleotide sequences as probes for hybridization. *Mol Cell Probes* **4**: 485–495.

Shyamala V, Ames GF. 1989. Genome walking by single-specific-primer polymerase chain reaction: SSP-PCR. *Gene* **84**: 1–8.

Shymala V, Ames GF. 1993. Genome walking by single specific primer-polymerase chain reaction. *Methods Enzymol* **217**: 436–446.

Stürzl M, Roth WK. 1990. Run-off synthesis and application of defined single-stranded DNA hybridization probes. *Anal Biochem* **185**: 164–169.

Swaminathan N, George D, McMaster K, Szablewski J, Van Etten JL, Mead DA. 1994. Restriction generated oligonucleotides utilizing the two base recognition endonuclease *Cvi*JI*. *Nucleic Acids Res* **22**: 1470–1475.

Swaminathan N, Mead DA, McMaster K, George D, Van Etten JL, Skowron PM. 1996. Molecular cloning of the three base restriction endonuclease R.CviJI from eukaryotic *Chlorella* virus IL-3A. *Nucleic Acids Res* **24**: 2463–2469.

Synthesis of Single-Stranded RNA Probes by In Vitro Transcription

Strand-specific ssRNA probes are not only easier to make than DNA probes but also generally yield stronger signals in hybridization reactions than do DNA probes of equal specific activity. This is probably a result of the innately higher stability of hybrids involving RNA (Casey and Davidson 1977). DNA probes continue to be of general utility in, for example, northern and Southern hybridizations, but labeled RNAs are now the probes of choice when analyzing transcripts of mammalian genes. Instead of digesting DNA–RNA hybrids with the idiosyncratic nuclease S1, RNA–RNA hybrids are digested with RNase A, a durable and obedient enzyme that can be used at a wide range of concentrations without compromising the results of the experiment (Zinn et al. 1983; Melton et al. 1984) (see Chapter 6, Protocol 17). Probes generated by in vitro transcription can be either radiolabeled or nonradiolabeled.

The templates for transcribing RNA probes are generated by linearizing recombinant plasmids that carry a powerful bacteriophage promoter immediately upstream of the DNA fragment of interest (Fig. 1, and see the box Plasmid Vectors Used for In Vitro Transcription) or by using PCR to generate templates whose 5′ ends encode a bacteriophage promoter (see the Additional Protocol Using PCR to Add Promoters for Bacteriophage-Encoded RNA Polymerases to Fragments of

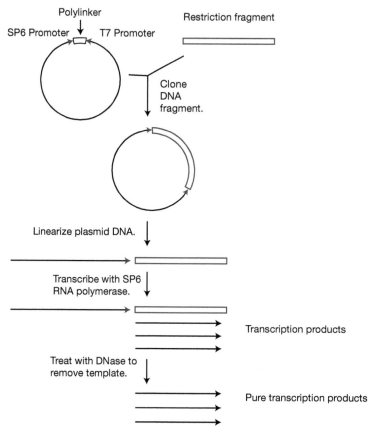

FIGURE 1. **In vitro synthesis of RNA by bacteriophage-encoded RNA polymerases.** (Adapted from Krieg and Melton 1987, with permission from Elsevier.)

DNA at the end of this protocol). Kits for in vitro transcription are sold by many manufacturers (e.g., MAXIscript and MEGAscript [Ambion] and Riboprobe Gemini Systems [Promega]). These kits are convenient for investigators who are using in vitro transcription methods for the first time, and they are certainly a marvelous help if something goes wrong with the technique. However, the reagents and buffers supplied in the kits can be easily assembled in the laboratory by any competent worker, and the enzymes can be purchased inexpensively as separate items.

In this protocol, we describe procedures for synthesizing RNA probes of high specific activity from plasmids containing promoters for bacteriophage-encoded RNA polymerases. An Additional Protocol below deals with the generation and transcription of PCR products. Guidance and background information on both methods are provided in the following information panels.

- Enzymatic properties of RNA polymerases used in the transcription reactions are described in the information panel In Vitro Transcription Systems at the end of this chapter.

- Necessary precautions to reduce problems with contaminating RNases are outlined in the information panel How to Win the Battle with RNase in Chapter 6.

PLASMID VECTORS USED FOR IN VITRO TRANSCRIPTION

Many plasmid and phagemid vectors that contain various combinations of bacteriophage promoters and polycloning sites are available commercially (e.g., the pGEM series [Promega]). Some of these vectors also encode the lacZ α-complementing fragment, which allows selection by color of recombinants on plates containing X-Gal (5-bromo-4-chloro-3-indolyl-β-D-galactopyranoside). The choice among these vectors is largely a matter of personal preference. However, if transcripts of both strands of the template are required, it is better to use a vector carrying two different bacteriophage promoters than to clone the template in two orientations in a vector carrying a single promoter. It is also important to consider the disposition of restriction sites within the template DNA and downstream from it. The 5′ terminus of the transcript is fixed by the bacteriophage promoter, but the 3′ terminus is defined by the downstream site of cleavage by the restriction enzyme. By using different restriction enzymes, RNAs of various lengths can be synthesized from a series of linear templates generated from the same plasmid. However, the presence of plasmid sequences in RNA probes can increase the level of background hybridization to levels that are unacceptable if the probe is to be used to screen a plasmid or cosmid library.

During generation of template DNAs, complete cleavage of superhelical plasmid DNA by a restriction enzyme is essential. Small amounts of circular plasmid DNAs will dramatically reduce yields by producing multimeric transcripts. Restriction enzymes that generate blunt or 5′-protruding termini produce the best linear templates. However, both types of termini yield RNA products that show heterogeneity at their 3′ ends (Melton et al. 1984; Milligan and Uhlenbeck 1989). Transcription of templates with 3′-protruding termini results in the synthesis of significant amounts of RNA molecules that are aberrantly initiated at the termini of the templates and thus in the production of dsRNA molecules (Schenborn and Mierendorf 1985). Restriction enzymes that generate protruding 3′ termini should therefore be avoided.

In addition to plasmids, some bacteriophage and cosmid vectors also contain bacteriophage promoters, usually arranged in opposite orientations on either side of the cloning site for foreign DNA. When a recombinant constructed in a vector of this type is digested with a restriction enzyme that cleaves many times within the foreign DNA, a large number of fragments are generated, one of which contains a particular bacteriophage promoter and the foreign sequences that lie immediately adjacent to it. If the fragments do not carry protruding 3′ termini, only the fragment bearing the bacteriophage promoter serves as a template for in vitro transcription. The resulting radiolabeled RNA, which is complementary to sequences located at one end of the original segment of foreign DNA, can then be used as a probe to isolate overlapping clones from genomic DNA or cDNA libraries. These vectors greatly simplify the task of "walking" from one recombinant clone to another along the chromosome. The following are two highly efficient methods available to generate strand-specific RNA probes.

- The relevant DNA fragment may be cloned or subcloned into specialized plasmids that contain promoters for bacteriophage-encoded DNA-dependent RNA polymerases. The recombinant plasmids are a source of double-stranded templates that can be transcribed in vitro into ssRNAs of defined length and strand specificity (Zinn et al. 1983; Melton et al. 1984).

(Continued)

- The DNA fragment to be transcribed may be amplified in PCRs with primers whose 5' ends encode synthetic promoters for bacteriophage-encoded DNA-dependent RNA polymerases. Following purification, the products of the PCRs are used as double-stranded templates for in vitro transcription reactions (Logel et al. 1992; Bales et al. 1993; Urrutia et al. 1993).

In both cases, the synthesis of RNA is remarkably efficient. When in vitro transcription reactions are saturated with ribonucleoside triphosphates (rNTPs), the templates can be transcribed many times, yielding a mass of RNA that exceeds the weight of the template severalfold. In addition, because bacteriophage-encoded DNA-dependent RNA polymerases function efficiently in vitro in the presence of relatively low concentrations of rNTPs (1–20 μM), full-length probes of high specific activity can be synthesized relatively inexpensively. Finally, RNA probes can be freed from template DNA by treating the reaction products with RNase-free DNase I. Probes usually do not need to be purified by gel electrophoresis. However, when using RNA probes of high specific activity to detect rare mRNA transcripts, background hybridization can be kept to a minimum using probes purified by denaturing gel electrophoresis.

MATERIALS

It is essential that you consult the appropriate Material Safety Data Sheets and your institution's Environmental Health and Safety Office for proper handling of equipment and hazardous materials used in this protocol.

Recipes for reagents specific to this protocol, marked <R>, are provided at the end of the protocol. See Appendix 1 for recipes for commonly used stock solutions, buffers, and reagents, marked <A>. Dilute stock solutions to the appropriate concentrations.

▲ Prepare all reagents used in this protocol with DEPC-treated H_2O (see the information panel How to Win the Battle with RNase in Chapter 6).

Reagents

Agarose gel (0.8%–1.0%) (see Step 1)

$[\alpha\text{-}^{32}P]rGTP$ (10 mCi/mL; sp. act. 400–3000 Ci/mmol)
   To minimize problems caused by radiolysis of the precursor, it is best, whenever possible, to prepare radiolabeled probes on the day that the [^{32}P]dNTP arrives in the laboratory.

Ammonium acetate (10 M) <A> (optional; see Step 8)

Appropriate restriction enzymes (see Step 1)

Bacteriophage T4 DNA polymerase (2.5 U/μL) (optional; see Step 2)

Bacteriophage T4 DNA polymerase buffer (10×) (optional; see Step 2)

Biotin-11-UTP or DIG-11-UTP (3.5 mM)

BSA (2 mg/mL; Fraction V, Sigma-Aldrich)

Dithiothreitol (DTT) (1 M) <A>

DNA-dependent RNA polymerase of bacteriophage T3, T7, or SP6
   These enzymes are available from several companies and are usually supplied at concentrations of 10–20 U/μL. Most manufacturers also supply a 10× transcription buffer that has presumably been optimized for their particular preparation of the DNA-dependent RNA polymerase. A generic 10× transcription buffer (see below) can be used if the manufacturer's buffer is in short supply.

dNTP solution containing dATP, dCTP, dGTP, and dTTP, each at 2 mM (optional; see Step 2)
   For advice on making and storing solutions of dNTPs, see the information panel Preparation of Stock Solutions of dNTPs.

Ethanol

Phenol:chloroform (1:1, v/v)

Placental RNase inhibitor (20 U/μL)

rGTP (0.5 mM) (optional; see Step 5)

RNase-free pancreatic DNase I (1 mg/mL)
   This enzyme is available from several manufacturers (e.g., RQ1 RNase-free DNase I [Promega]).

rNTP solution 1, containing rATP, rCTP, and rUTP, each at 5 mM

or

rNTP solution 2, containing rATP, rCTP, and rGTP, each at 10 mM, and rUTP (6.5 mM)

Sodium acetate (3 M, pH 5.2) <A>

Template DNA

The DNA fragment to be transcribed should be cloned into one of the commercially available plasmids containing bacteriophage RNA polymerase promoters on both sides of the polycloning sequence (e.g., pGEM [Promega] or pBluescript [Stratagene]). Purify the superhelical recombinant plasmid by one or more of the methods described in Chapter 1.

The DNA used as a template in the in vitro transcription reaction need not be highly purified; crude minipreparations work well. The essential requirement is that the template be free of RNase, a criterion that can usually be fulfilled by extracting the preparation of plasmid DNA twice with phenol:chloroform. However, if RNase was added to the plasmid at a late stage in the purification process (i.e., after deproteinization), it should be removed by treatment with proteinase K, as follows.

1. Add to the plasmid DNA preparation 0.1 volume of 10× proteinase K buffer (100 mM Tris-Cl [pH 8.0]/ 50 mM EDTA [pH 8.0]/500 mM NaCl), 0.1 volume of 5% (w/v) SDS, and proteinase K (20 mg/mL to a final concentration of 100 µg/mL.

2. Incubate the reaction for 1 h at 37°C.

3. Extract the DNA with phenol:chloroform and recover the DNA by standard precipitation with ethanol.

4. Resuspend the DNA in RNase-free TE (pH 7.6) at a concentration of ≥100 µg/mL.

Transcription buffer (10×) <R>

Equipment

Microcentrifuge tubes (0.5 mL)

Sephadex G-50 spun column, equilibrated with 10 mM Tris-Cl (pH 7.5) (optional; see Step 8)

The Sephadex G-50 gel-filtration matrix can be used to separate DNA (which passes through the matrix) from lower-molecular-weight substances (e.g., radioactive precursors, which are retained on the column). Sephadex G-50 columns for radiolabled DNA purification are available commercially from several sources (e.g., Quick Spin Columns [Roche Applied Science]). However, it is also easy to set one up using a 1-mL disposable syringe plugged with siliconized glass wool (see Protocol 1, Steps 6–14).

Water bath (40°C) (needed if you use bacteriophage SP6 DNA-dependent RNA polymerase in Step 4)

METHOD

1. Prepare 5 pmol of linear template DNA by digestion of superhelical plasmid DNA with a suitable restriction enzyme. Analyze an aliquot (100 ng) of the digested DNA by agarose gel electrophoresis. If necessary, add more restriction enzyme and continue incubation until there is no trace of the undigested DNA.

    Approximately 2 µg of a plasmid 3 kb in length is ~1 pmol.

    ▲ *It is essential that plasmid DNA templates be cleaved to completion, because trace amounts of closed-circular plasmid DNA result in the generation of extremely long transcripts that include plasmid sequences. Because of their length, these transcripts can account for a substantial proportion of the incorporated labeled rNTP.*

2. If restriction enzymes, such as PstI or SstI that generate protruding 3′ termini, must be used, treat the digested DNA with bacteriophage T4 DNA polymerase in the presence of all four dNTPs to remove the 3′ protrusion.

    Protruding 3′ termini create an opportunity for the DNA-dependent RNA polymerase to transfer to the complementary strand of the template and to generate long U-turn transcripts with extensive secondary structure. T4 DNA polymerase carries a 3′→5′ exonucleolytic activity that is more active against single-stranded substrates than double-stranded substrates. The enzyme rapidly digests 3′-protruding termini and then continues at a slower pace to remove 3′ nucleotides from the doublestranded portion of the DNA substrate. In the presence of high concentrations of dNTPs, however, recessed 3′-hydroxyl termini generated by exonucleolytic activity act as primers for template-

directed addition of mononucleotides by the 5'→3' polymerase. Because the synthetic capacity of T4 DNA polymerase exceeds its exonucleolytic abilities, protruding 3' termini are converted to termini with flush ends (Richardson et al. 1964).

i. Following digestion, purify the DNA by extraction with phenol:chloroform and precipitation with ethanol in the presence of 2.5 M ammonium acetate (see Chapter 2, Protocol 4).

ii. Dissolve the DNA pellet in

Bacteriophage T4 DNA polymerase buffer (10×)	2 μL
Solution of unlabeled dNTPs (2 mM)	1 μL
Bacteriophage T4 DNA polymerase (2.5 U/μL)	1 μL
H$_2$O to 20 μL	

Incubate the reaction for 5 min at 37°C.

iii. Stop the reaction by heating it for 5 min to 70°C.

3. Purify the template DNA by extraction with phenol:chloroform and standard precipitation with ethanol. Dissolve the DNA in H$_2$O at a concentration of 100 nM (i.e., 200 μg/mL for a 3-kb plasmid).

4. Choose one of the two recipes below, warm the first six components listed to room temperature, and in a sterile 0.5-mL microcentrifuge tube mix in the following order at room temperature.

For Radiolabeling

Template DNA	0.2 pmol (400 ng for a 3-kb plasmid)
RNase-free H$_2$O	to 6 μL
rNTP solution 1 (5 mM)	2 μL
DTT (100 mM)	2 μL
Transcription buffer (10×)	2 μL
BSA (2 mg/mL)	1 μL
[α-^{32}P]rGTP (10 mCi/mL)	5 μL (sp. act. 400–3000 Ci/mmol)

For Labeling with DIG or Biotin

Template DNA	0.2 pmol (400 ng for a 3-kb plasmid)
RNase-free H$_2$O	to 9 μL
rNTP solution 2 (10 mM)	2 μL (contains 6.5 mM rUTP)
DTT (100 mM)	2 μL
Transcription buffer (10×)	2 μL
BSA (2 mg/mL)	1 μL
Biotin-11-UTP or DIG-11-UTP (3.5 mM)	2 μL

Mix the components of the mixture by gently tapping the outside of the tube. Then, add

Placental RNase inhibitor (10 U)	1 μL
Bacteriophage DNA-dependent RNA polymerase (~10 U)	1 μL

▲ *Add the components in the order shown and at room temperature to avoid the possibility that the template DNA may be precipitated by the high concentration of spermidine in the transcription buffer.*

Mix the reagents by gently tapping the outside of the tube. Centrifuge the tube for 1–2 sec to transfer all of the liquid to the bottom. Incubate the reaction for 1–2 h at 37°C (bacteriophages T3 and T7 DNA-dependent RNA polymerases) or 40°C (bacteriophage SP6 DNA-dependent RNA polymerase).

The reaction may be scaled from 20 to 50 μL to accommodate more dilute reagents.

When the reaction is performed as described above, 80%–90% of the radiolabel will be incorporated into RNA. The yield of RNA will be ~20 ng (sp. act. 4.7 × 10^9 dpm/μg) when the specific activity of the [α-^{32}P]GTP is 3000 Ci/mmol and ~150 ng (sp. act. 6.2 × 10^8 dpm/μg) when the specific activity of the precursor is 400 Ci/mmol.

When using biotin- or DIG-labeled rNTPs, the modified nucleotide is used at a low concentration in the reaction to ensure that only a few labeled nucleotides are incorporated per molecule of probe (~1 labeled nucleotide for every 20–25 nucleotides polymerized). The posthybridization recognition of these labels by antibodies is most efficient when only a few bases are replaced by their labeled counterparts. Greater incorporation of the modified nucleotide can also decrease hybridization of the probe to target mRNAs or DNAs.

5. (Optional) When preparing radiolabeled probes, if full-length transcripts are desired, add 2 μL of 0.5 mM rGTP and incubate the reaction mixture for an additional 60 min at the temperature appropriate for the polymerase.

6. Terminate the in vitro transcription reaction by adding 1 μL of 1-mg/mL RNase-free pancreatic DNase I to the reaction tube. Mix the reagents by gently tapping the outside of the tube. Incubate the reaction mixture for 15 min at 37°C.

7. Add 100 μL of RNase-free H_2O and purify the RNA by extraction with phenol:chloroform.

    If the probe will be used in experiments where length is important (e.g., RNase protection), purify the labeled RNA by polyacrylamide gel electrophoresis (see Chapter 2, Protocol 3). The aim of this extra step is to eliminate truncated labeled molecules from the preparation.

8. Transfer the aqueous phase to a fresh microcentrifuge tube and separate the labeled RNA from undesired small RNAs and rNTPs by one of the following three methods.

To Purify RNA by Ethanol Precipitation

i. Add 30 μL of 10 M ammonium acetate to the aqueous phase. Mix and then add 250 μL of ice-cold ethanol to the tube. After storage for 30 min on ice, collect the RNA by centrifugation at maximum speed for 10 min at 4°C in a microcentrifuge.

ii. Remove as much of the ethanol as possible by gentle aspiration and leave the open tube on the bench for a few minutes to allow the last visible traces of ethanol to evaporate. Dissolve the RNA in 100 μL of RNase-free H_2O.

iii. Add 2 volumes of ice-cold ethanol to the tube and store the RNA at −70°C until needed.

    To recover the RNA, transfer an aliquot of the ethanolic solution to a fresh microcentrifuge tube. Add 0.25 volume of 10 M ammonium acetate, mix, and then store the tube for at least 15 min at −20°C. Centrifuge the solution at maximum speed for 10 min at 4°C in a microcentrifuge. Remove the ethanol by aspiration and dissolve the RNA in the desired volume of the appropriate RNase-free buffer.

To Purify RNA by Spun-Column Chromatography

i. If not using a commercially available column, prepare a Sephadex G-50 spun column (see Protocol 1) that has been autoclaved in 10 mM Tris-Cl (pH 7.5).

ii. Purify the RNA by spun-column chromatography according to Protocol 1.

iii. Store the eluate in a microcentrifuge tube at −70°C until the RNA is needed.

To Purify RNA by Gel Electrophoresis

i. Prepare a neutral polyacrylamide gel according to Chapter 6, Protocol 11.

ii. Add the appropriate gel-loading buffer to the aqueous phase and purify the RNA by gel electrophoresis.

iii. Locate the RNA by autoradiography according to Chapter 2, Protocol 7.

iv. Purify the RNA from the gel slice using the crush and soak method according to Chapter 2, Protocol 10.

v. Store the RNA at −70°C until needed.

Any of these purification schemes should remove >99.0% of unincorporated rNTPs from the RNA.

See Troubleshooting

TROUBLESHOOTING

Problem (Step 8): No RNA is synthesized.

Solution: The most common cause of an apparent lack of RNA synthesis is contamination of tubes or reagents with RNase. This contamination can be avoided by taking the precautions described in the introduction to Chapter 6.

Less frequently, failure to synthesize RNA is a consequence of precipitation of the DNA template by the spermidine in the 10× transcription buffer. Make sure that the components of the reaction are assembled at room temperature and in the stated order. If necessary, the presence of soluble template can be confirmed by analyzing an aliquot of the reaction by electrophoresis through an agarose gel.

Problem (Step 8): Transcription is from the wrong strand of DNA.

Solution: Usually, >99.8% of transcripts synthesized in vitro by bacteriophage DNA-dependent RNA polymerases are derived from the correct DNA strand (Melton et al. 1984). However, this high degree of specificity is only achieved provided that templates are both linear and lacking protruding 3′ termini. Contamination of the template with superhelical plasmid DNA causes an increase in aberrant initiation of RNA chains on both strands of the DNA. The presence of a protruding 3′ terminus downstream from the bacteriophage promoter leads to the synthesis of transcripts complementary to the full length of the wrong strand of DNA (i.e., U-turn transcripts). Both of these problems can be avoided by careful preparation of the template (see Steps 1 and 2).

Problem (Step 8): Synthesized transcripts are shorter than the desired length.

Solution: Synthesis of abbreviated transcripts can be due to the chance occurrence in the template of sequences that terminate transcription by the particular DNA-dependent RNA polymerase being used. Another contributing factor can be limiting concentrations of precursor (usually the radiolabeled rNTP).

The first of these obstacles may be resolved by constructing a new plasmid in which transcription of the desired sequence is driven by a polymerase from a different bacteriophage. Transcription terminator sequences are not always recognized equally by the various bacteriophage DNA-dependent RNA polymerases.

Although the strengths of transcription termination signals vary greatly, all but the strongest of them can be overcome, at least in part, by increasing the concentration of the limiting rNTP. In most cases, it is impractical to increase the concentration of the rNTP that is radiolabeled in the reaction, because the improvement in yield of full-length product is gained at the expense of reducing the specific activity of the probe. In this case, the following additional steps can be taken:

- Lower the temperature of incubation of the transcription reaction to 30°C (Krieg and Melton 1987).

- Pare the sequences to be transcribed to the minimum. In this way, it may be possible to eliminate the termination sequences from the clone.

- Purify the desired product by electrophoresis through a polyacrylamide or agarose gel as described in Chapter 2, Protocol 1 or 3. The transcription reactions are so efficient that it is often possible to purify sufficient quantities of the desired RNA, even if it is only a relatively minor proportion of the total RNA synthesized in the reaction.

Using PCR to Add Promoters for Bacteriophage-Encoded RNA Polymerases to Fragments of DNA

Templates for bacteriophage-encoded RNA polymerases may be generated by cloning target DNA fragments into plasmids carrying bacteriophage promoters (Protocol 5). Alternatively, templates can be synthesized in PCRs using gene-specific primers whose 5′ ends encode synthetic promoters for bacteriophage-encoded DNA-dependent RNA polymerases. Following purification, the products of the PCRs are used as double-stranded templates for in vitro transcription reactions (Logel et al. 1992; Bales et al. 1993; Urrutia et al. 1993). By using pairs of primers that encode different promoters, DNA fragments are generated that may be transcribed in a strand-specific manner by the appropriate RNA polymerase. Advantages of the PCR method include the following.

- Probes may be generated directly from DNA templates amplified from a heterogeneous population of DNA fragments.

- The need for cloning and preparation of plasmids is obviated.

- Probes contain no plasmid or polylinker sequences.

- Probes of high specificity and of virtually any size can be created.

RNA polymerases encoded by bacteriophages T3 and T7 transcribe PCR-amplified DNA carrying the appropriate promoters with high specificity. However, the yields of RNA are generally three-fold to fourfold lower than can be obtained from linearized plasmid templates. When using amplified DNA templates, between 20% and 30% of the labeled rNTP is converted into acid-insoluble material, compared to >75% in reactions containing linearized plasmid DNAs as templates (Logel et al. 1992). Nevertheless, the yield and specific activity of the RNA generated from the amplified products of PCRs are more than sufficient for most purposes. The RNA polymerase encoded by bacteriophage SP6 is reported to transcribe PCR-amplified DNA much less efficiently than linearized plasmid DNAs (Logel et al. 1992). For this reason, we recommend using primers that encode promoters for bacteriophages T3 and T7.

Primer Design

The primers are usually quite long (>50 nucleotides) and consist of three regions:

5′ Clamp (~10 nucleotides)	Core promoter (~22 nucleotides)	Gene-specific sequence 3′ (20 nucleotides)

The sequences of the core promoters recognized by bacteriophage T3 and T7 RNA polymerases are taken from Jorgensen et al. (1991) (see the information panel In Vitro Transcription Systems):

Bacteriophage T7 core promoter: 5′TAATACGACTCACTATAGGGAGA3′
Bacteriophage T3 core promoter: 5′ATTAACCCTCACTAAAGGGAGA3′

For the T3 promoter, the 3′-most dinucleotide can be GA or AG.

Use a different clamp sequence for each member of a primer pair. Suggested clamp sequences are (Logel et al. 1992) 5′CAGAGATGCA3′ and 5′CCAAGCCTTC3′. Design the gene-specific segment of the primer according to the usual rules (see the discussion on primer design in the introduction to Chapter 7).

Amplification Conditions

The amplification reactions contain 10–20 pg of a single species of template DNA or proportionately more of complex populations of DNAs. The remainder of the reagents in the amplification reactions are used at standard concentrations (see Chapter 7, Protocol 1). The denaturation, annealing, and polymerization times listed below are suitable for 50-μL reactions assembled in thin-walled 0.5-mL tubes and incubated in thermal cyclers such as the Perkin-Elmer 9600 or 9700, Master Cycler (Eppendorf), and PTC 100 (MJ Research). Times and temperatures may need to be adapted to suit other types of equipment and reaction volumes.

Cycle Number	Denaturation	Annealing	Polymerization
1–4	1 min at 94°C	2 min at 54 °C	3 min at 72°C
5–36	1 min at 94°C	1 min at 65 °C	3 min at 72°C

Polymerization should be performed for 1 min for every 1000 bp of length of the target DNA.

Most thermocyclers have an end routine in which the amplified samples are incubated at 4°C until they are removed from the machine. Samples can be left overnight at this temperature but should be stored thereafter at –20°C.

Purification of Amplified DNA

Although amplified DNA may be used without purification as a template in the in vitro transcription reaction (Bales et al. 1993; Urrutia et al. 1993), the efficiency of RNA synthesis is improved if unused primers and the by-products of the amplification reaction are removed by electrophoresis through a gel cast with low-melting-temperature agarose, by spun-colum chromatography through Sephadex G-75, or by absorbtion/elution on a commercial resin such as Wizard PCR Preps Purification System (Promega) or QIAquick (QIAGEN).

In Vitro Transcription of Amplified DNA

Approximately 0.5 μg of purified amplified DNA may be used as template in standard transcription reactions catalyzed by the appropriate bacteriophage-encoded DNA-dependent RNA polymerase (see Protocol 5).

RECIPE

It is essential that you consult the appropriate Material Safety Data Sheets and your institution's Environmental Health and Safety Office for proper handling of equipment and hazardous materials used in this protocol.

Transcription Buffer (10×)

Reagent	Quantity (for 1 mL)	Final concentration
Tris-Cl (1 M, pH 7.5)	400 μL	400 mM
MgCl$_2$ (1 M)	60 μL	60 mM
Spermidine HCl (100 mM)	200 μL	20 mM
NaCl (5 M)	10 μL	50 mM

Sterilize the 10× buffer by filtration and then store it in small aliquots at –20°C. Discard each aliquot after use.

REFERENCES

Bales KR, Hannon K, Smith CK II, Santerre RF. 1993. Single-stranded RNA probes generated from PCR-derived DNA templates. *Mol Cell Probes* 7: 269–275.

Casey J, Davidson N. 1977. Rates of formation and thermal stabilities of RNA:DNA and DNA:DNA duplexes at high concentrations of formamide. *Nucleic Acids Res* 4: 1539–1552.

Jorgensen ED, Durbin RK, Risman SS, McAllister WT. 1991. Specific contacts between the bacteriophage T3, T7, and SP6 RNA polymerases and their promoters. *J Biol Chem* 266: 645–651.

Krieg PA, Melton DA. 1987. In vitro RNA synthesis with SP6 RNA polymerase. *Methods Enzymol* 155: 397–415.

Logel J, Dill D, Leonard S. 1992. Synthesis of cRNA probes from PCR-generated DNA. *BioTechniques* 13: 604–610.

Melton DA, Krieg PA, Rebagliati MR, Maniatis T, Zinn K, Green MR. 1984. Efficient in vitro synthesis of biologically active RNA and RNA hybridization probes from plasmids containing a bacteriophage SP6 promoter. *Nucleic Acids Res* 12: 7035–7056.

Milligan JF, Uhlenbeck OC. 1989. Synthesis of small RNAs using T7 RNA polymerase. *Methods Enzymol* 180: 51–62.

Richardson CC, Schildkraut CL, Aposhian HV, Kornberg A. 1964. Enzymatic synthesis of deoxyribonucleic acid. XIV. Further purification and properties of deoxyribonucleic acid polymerase of *Escherichia coli. J Biol Chem* 239: 222–232.

Schenborn ET, Mierendorf RC Jr. 1985. A novel transcription property of SP6 and T7 RNA polymerases: Dependence on template structure. *Nucleic Acids Res* 13: 6223–6236.

Urrutia R, McNiven MA, Kachar B. 1993. Synthesis of RNA probes by the direct in vitro transcription of PCR-generated DNA templates. *J Biochem Biophys Methods* 26: 113–120.

Zinn K, DiMaio D, Maniatis T. 1983. Identification of two distinct regulatory regions adjacent to the human β-interferon gene. *Cell* 34: 865–879.

Synthesis of cDNA Probes from mRNA Using Random Oligonucleotide Primers

This protocol describes the generation of radiolabeled cDNA probes from poly(A)$^+$ RNA in a random priming reaction. Probes of this type are used for differential screening of cDNA libraries.

MATERIALS

It is essential that you consult the appropriate Material Safety Data Sheets and your institution's Environmental Health and Safety Office for proper handling of equipment and hazardous materials used in this protocol.

Recipes for reagents specific to this protocol, marked <R>, are provided at the end of the protocol. See Appendix 1 for recipes for commonly used stock solutions, buffers, and reagents, marked <A>. Dilute stock solutions to the appropriate concentrations.

▲ Prepare all reagents used in this protocol with DEPC-treated H$_2$O (see the information panel How to Win the Battle with RNase in Chapter 6).

Reagents

[α-^{32}P]dCTP (10 mCi/mL; sp. act.>3000 Ci/mmol)

To minimize problems caused by radiolysis of the precursor and the probe, it is best, whenever possible, to prepare radiolabeled probes on the day that the [^{32}P]dNTP arrives in the laboratory.

Ammonium acetate (10 M) <A>

dCTP (125 μM)

Add 1 μL of a 20 mM stock solution of dCTP to 160 μL of 25 mM Tris-Cl (pH 7.6). Store the diluted solution in small aliquots at –20°C.

dNTP solution containing dATP, dGTP, and dTTP, each at 20 mM

For advice on making and storing solutions of dNTPs, see the information panel Preparation of Stock Solutions of dNTPs.

DTT (1 M)

EDTA (0.5 M, pH 8.0)

Ethanol

HCl (2.5 N)

NaOH (3 N)

Phenol:chloroform (1:1, v/v)

Placental RNase inhibitor (20 U/μL)

These inhibitors are sold by several manufacturers under various trade names (e.g., RNasin [Promega]; RNaseOUT [Life Technologies]). For more details, see the information panel Inhibitors of RNases in Chapter 6.

Random deoxynucleotide primers, 6 or 7 bases in length

Because of their uniform length and lack of sequence bias, synthetic oligonucleotides of random sequence are the primers of choice. Oligonucleotides of optimal length (hexamers or heptamers) (Suganuma and Gupta 1995) can be purchased from a commercial source (e.g., Sigma-Aldrich and Boehringer Mannheim) or synthesized locally on an automated DNA synthesizer. Store the solution of primers at 0.125 μg/μL in TE (pH 7.6) at –20°C in small aliquots.

Reverse transcriptase

Reverse transcriptase derived from the pol gene of the Moloney murine leukemia virus (Mo-MLV) is more efficient in cDNA synthesis than that obtained from the avian myeloblastosis virus (e.g., see Fargnoli et al. 1990). The cloned version of reverse transcriptase encoded by the Mo-MLV enzyme is the enzyme of choice in this protocol. Mutants of the enzyme that lack RNase H activity (e.g., StrataScript [Stratagene] or Superscript II [Life Technologies]) have some advantages over the wild-type enzyme because they (1) produce higher yields of full-length extension product and (2) work equally well at both 47°C and 37°C (Gerard and D'Alessio 1993).

Mo-MLV reverse transcriptase is temperature sensitive and should be stored at −20°C until needed at Step 2.

Reverse transcriptase buffer (10×) <A>

SDS (10%, w/v) <A>

Template mRNA

Prepare poly(A)⁺ RNA as described in Chapter 6, Protocol 9, and dissolve in RNase-free H₂O at a concentration of 250 μg/mL.

Tris-Cl (1 M, pH 7.4) <A>

Equipment

Ice-water bath

Sephadex G-50 spun column, equilibrated in TE (pH 7.6) (optional; see Step 7)

The Sephadex G-50 gel-filtration matrix can be used to separate DNA (which passes through the matrix) from lower-molecular-weight substances (e.g., radioactive precursors, which are retained on the column). Sephadex G-50 columns for radiolabled DNA purification are available commercially from several sources (e.g., Quick Spin Columns [Roche Applied Science]). However, it is also easy to set one up using a 1-mL disposable syringe plugged with siliconized glass wool (see Protocol 1, Steps 6–14).

Water baths or heating blocks (45°C, 68°C, and 70°C)

METHOD

1. Transfer 1 μg of poly(A)⁺ RNA to a sterile microcentrifuge tube. Adjust the volume of the solution to 4 μL with RNase-free H₂O. Heat the closed tube for 5 min at 70°C and then quickly transfer the tube to an ice-water bath.

2. To the chilled solution in the microcentrifuge tube, add

DTT (10 mM)	2.5 μL
Placental RNase inhibitor	20 U
Random deoxyoligonucleotide primers	5 μL
Reverse transcriptase buffer (10×)	2.5 μL
dGTP, dATP, and dTTP (20 mM solution)	1 μL
dCTP (125 μM solution)	1 μL
[α-³²P]dCTP (10 mCi/mL) (sp. act. >3000 Ci/mmol)	10 μL
RNase-free H₂O	to 24 μL
Reverse transcriptase (200 U)	1 μL

▲ *Add the reverse transcriptase last.*

Reverse transcriptase supplied by different manufacturers varies in its activity per unit. When using a new batch of enzyme, set up a series of extension reactions containing equal amounts of poly(A)⁺ RNA and oligonucleotide primer and different amounts of enzyme. If possible, the primer should be specific for an mRNA present at moderate abundance in the preparation of poly(A)⁺ RNA. Assay the products of each reaction by gel electrophoresis, as described in this protocol. Use the minimal amount of enzyme required to produce the maximum yield of extension product. The units used in this protocol work well with most batches of StrataScript and Superscript II.

Mix the components by gently tapping the side of the tube. Remove bubbles by brief centrifugation in a microcentrifuge. Incubate the reaction mixture for 1 h at 45°C.

As an alternative, [α-³²P]dCTP of specific activity 800 Ci/mmol can be substituted in this reaction. If this substitution is made, omit the 125 μM dCTP from the reaction mixture.

3. Stop the reaction by adding

EDTA (0.5 M, pH 8.0)	1 μL
SDS (10%, w/v)	1 μL

Mix the reagents in the tube completely.

The single-stranded radiolabeled cDNA is quite sticky and adheres nonspecifically to glass, filters, and some plastics. For this reason, it is important to maintain a minimum of 0.05% (w/v) SDS in all buffers after Step 3 of the protocol and 0.1%–1.0% SDS in hybridization buffers.

4. Add 3 μL of 3 N NaOH to the reaction tube. Incubate the mixture for 30 min at 68°C to hydrolyze the RNA.

5. Cool the reaction mixture to room temperature. Neutralize the solution by adding 10 μL of 1 M Tris-Cl (pH 7.4), mixing well, and then adding 3 μL of 2.5 N HCl. Check the pH of the solution by spotting a very small amount on pH paper.

6. Purify the cDNA by extraction with phenol:chloroform.

7. Separate the radiolabeled probe from the unincorporated dNTPs by either spun-column chromatography (using a commercially available column or a homemade column according to Protocol 1, Steps 6–14) or selective precipitation by ethanol in the presence of 2.5 M ammonium acetate (see Chapter 1, Protocol 4).

8. Measure the proportion of radiolabeled dNTPs that are incorporated into material precipitated by TCA (see Appendix 2).

   Assuming that 30% of the radioactivity has been incorporated into TCA-precipitable material during the random priming reaction and that 90% of this material has been generated by random priming events on the template mRNA (rather than self-priming of oligonucleotides), the probe DNA would contain a total of ~6 × 10^7 dpm. The specific activity of the probe would be ~5 × 10^9 dpm/μg, and the weight of DNA synthesized during the reaction would be 11.7 ng.

   If a larger amount of radiolabeled cDNA is required, scale up the reaction by increasing the volumes of all components proportionally. It is important to maintain a ratio of 200 U of reverse transcriptase/ microgram of input mRNA to ensure maximum yield.

   The purified radiolabeled cDNA can be used for hybridization without denaturation. Use 5 × 10^7 dpm of radiolabeled cDNA for each 150-mm filter and 5 × 10^6 to 1 × 10^7 dpm for each 90-mm filter.

REFERENCES

Fargnoli J, Holbrook NJ, Fornace AJ Jr. 1990. Low-ratio hybridization subtraction. *Anal Biochem* 187: 364–373.

Gerard GF, D'Alessio JM. 1993. Reverse transcriptase (EC2.7.7.49): The use of cloned Moloney murine leukemia virus reverse transcriptase to synthesize DNA from RNA. In *Enzymes of molecular biology* (ed Burrell MM), pp. 73–94. Humana, Totowa, NJ.

Suganuma A, Gupta KC. 1995. An evaluation of primer length on random-primed DNA synthesis for nucleic acid hybridization: Longer is not better. *Anal Biochem* 224: 605–608.

Radiolabeling of Subtracted cDNA Probes by Random Oligonucleotide Extension

In this procedure, synthesis of cDNA is performed in the presence of saturating concentrations of all four dNTPs and trace amounts of a single radiolabeled dNTP. After subtraction hybridization, the enriched single-stranded cDNA is radiolabeled to high specific activity in a second synthetic reaction by extension of random oligonucleotide primers using the Klenow fragment of *E. coli* DNA Pol I. Because the concentrations of dNTP in the first reaction are nonlimiting, both the amounts and size of cDNA generated are greater than those achieved in standard labeling protocols. The subtractive hybridization step can therefore be performed with higher efficiency. Because the resulting population of cDNA is not vulnerable to radiolytic cleavage, it can be stored indefinitely and radiolabeled to higher specific activity when needed.

The protocol works best when the cDNA synthesized in the initial synthetic reaction is full length or close to it. For this reason, synthesis of cDNAs is primed by oligo(dT) rather than random hexanucleotide primers. In contrast, the subsequent radiolabeling reaction is primed by random oligonucleotides, yielding shorter DNA products whose size is ideal for hybridization.

The cDNA prepared as described in this protocol, Steps 1–10 can be converted into dsDNA and cloned into a bacteriophage or plasmid vector to produce a subtracted cDNA library (e.g., see Sargent and Dawid 1983; Davis 1986; Rhyner et al. 1986; Fargnoli et al. 1990). The cDNA library can then be screened with a subtracted probe. Because subtracted libraries are enriched for cDNA clones corresponding to nonabundant mRNAs, the amount of screening required to find a clone corresponding to a rare mRNA may be reduced by a factor of 10. Subtracted probes radiolabeled as described here can detect cDNAs corresponding to mRNAs expressed at a level as low as five molecules/mammalian cell.

MATERIALS

It is essential that you consult the appropriate Material Safety Data Sheets and your institution's Environmental Health and Safety Office for proper handling of equipment and hazardous materials used in this protocol.

Recipes for reagents specific to this protocol, marked <R>, are provided at the end of the protocol. See Appendix 1 for recipes for commonly used stock solutions, buffers, and reagents, marked <A>. Dilute stock solutions to the appropriate concentrations.

▲ Prepare all reagents used in this protocol with DEPC-treated H_2O (see the information panel How to Win the Battle with RNase in Chapter 6).

Reagents

$[\alpha\text{-}^{32}P]$dATP (10 mCi/mL; sp. act.>3000 Ci/mmol)
$[\alpha\text{-}^{32}P]$dCTP (10 mCi/mL; sp. act. 800 – 3000 Ci/mmol)
   To minimize problems caused by radiolysis of the precursor and the probe, it is best, whenever possible, to prepare radiolabeled probes on the day that the $[^{32}P]$dNTP arrives in the laboratory.

Ammonium acetate (10 M) <A>
dNTP solution (complete) containing four dNTPs, each at 5 mM
dNTP solution containing dCTP, dGTP, and dTTP, each at 5 mM
   ▲ *Omit the dATP from this solution.*
   For advice on making and storing solutions of dNTPs, see the information panel Preparation of Stock Solutions of dNTPs.

Driver mRNA (see Step 8)

DTT (1 M) <A>

EDTA (0.5 M, pH 8.0) <A>

Ethanol

HCl (2.5 N)

Isobutanol

Klenow fragment of *E. coli* DNA Pol I

NaOH (3 N)

Oligo(dT)$_{12-18}$

> *Purchase and dissolve oligo(dT)$_{12-18}$ at 1 mg/mL in TE (pH 7.6). Store at –20°C.*

Phenol:chloroform (1:1, v/v)

Placental RNase inhibitor

> *These inhibitors are sold by several manufacturers under various trade names (e.g., RNasin [Promega]; RNaseOUT [Life Technologies]). For more details, see the information panel Inhibitors of RNases in Chapter 6.*

Random deoxynucleotide primers, 6 or 7 bases in length

> *Because of their uniform length and lack of sequence bias, synthetic oligonucleotides of random sequence are the primers of choice. Oligonucleotides of optimal length (hexamers and heptamers) (Suganuma and Gupta 1995) can be purchased from a commercial source (e.g., Sigma-Aldrich and Life Technologies) or synthesized locally on an automated DNA synthesizer. Store the solution of primers at 0.125 μg/μL in TE (pH 7.6) at –20°C in small aliquots.*

Random priming buffer (5×) <R>

Reverse transcriptase

> *Reverse transcriptase derived from the pol gene of Mo-MLV is more efficient in cDNA synthesis than that obtained from the avian myeloblastosis virus (e.g., see Fargnoli et al. 1990). The cloned version of reverse transcriptase encoded by the Mo-MLV enzyme is the enzyme of choice in this protocol. Mutants of the enzyme that lack RNase H activity (e.g., StrataScript [Stratagene] or Supercript II [Life Technologies]) have some advantages over the wild-type enzyme because they (1) produce higher yields of full-length extension product and (2) work equally well at both 47°C and 37°C (Gerard and D'Alessio 1993).*

Reverse transcriptase buffer (10×) <A>

SDS (20%, w/v) <A>

SDS/EDTA solution

> EDTA (30 mM, pH 8.0)
> SDS (1.2%)

Sodium acetate (3 M, pH 5.2) <A>

Sodium phosphate buffer (2 M, pH 6.8) <A>

SPS buffer

> Sodium phosphate buffer (0.12 M, pH 6.8)
> SDS (0.1%, w/v)

Template RNAs:

> Two-pass poly(A)$^+$-enriched mRNA prepared from cells or tissue that expresses the mRNA(s) of interest
> Two-pass poly(A)$^+$-enriched mRNA prepared from cells or tissue that does not express the mRNA(s) of interest
>
> *For mRNA preparation and isolation by magnetic oligo(dT) beads, see Chapter 6, Protocol 9. Both RNAs should be dissolved in H$_2$O at a concentration of ~1 mg/mL.*

Tris-Cl (1 M, pH 7.4) <A>

Water, RNase-free

Equipment

Sephadex G-50 spun column, equilibrated in TE (pH 7.6)

> *The Sephadex G-50 gel-filtration matrix can be used to separate DNA (which passes through the matrix) from lower-molecular-weight substances (e.g., radioactive precursors, which are retained on the column). Sephadex G-50 columns for radiolabled DNA purification are available commercially from several sources*

Siliconized microcentrifuge tubes (1.5 mL)
Water baths or heat blocks (45°C, 60°C, and 68°C)

METHOD

1. To synthesize first-strand cDNA, mix the following ingredients at 4°C in a sterile microcentrifuge tube:

Template RNA (1 mg/mL)	10 μL
Oligo(dT)$_{12-18}$ (1 mg/mL)	10 μL
dNTP solution (complete) (5 mM)	10 μL
DTT (50 mM)	1 μL
Reverse transcriptase buffer (10×)	5 μL
[α-^{32}P]dCTP (10 mCi/mL) (sp. act. 800 or 3000 Ci/mmol)	5 μL
Placental RNase inhibitor	25 U
RNase-free H$_2$O	to 46 μL
Reverse transcriptase (~800 U)	4 μL

   ▲ *Add the reverse transcriptase last.*

   Reverse transcriptase supplied by different manufacturers varies in its activity per unit. When using a new batch of enzyme, set up a series of extension reactions containing equal amounts of poly(A)$^+$ RNA and oligonucleotide primer and different amounts of enzyme. If possible, the primer should be specific for an mRNA present at moderate abundance in the preparation of poly(A)$^+$ RNA. Assay the products of each reaction by gel electrophoresis as described in this protocol. Use the minimal amount of enzyme required to produce the maximum yield of extension product. The units used in this protocol work well with most batches of StrataScript and Superscript II.

   Mix the components by gently tapping the side of the tube. Collect the reaction mixture in the bottom of the tube by brief centrifugation in a microcentrifuge. Incubate the reaction for 1 h at 45°C.

   [α-^{32}P]dCTP is used as a tracer to measure the synthesis of the first strand of cDNA.

2. Measure the proportion of radiolabeled dNTPs that are incorporated into material precipitated by TCA (see Appendix 2). Calculate the yield of cDNA using the equation below. In a reaction containing 50 nmol of each dNTP,

   $$\frac{\text{cpm incorporated}}{\text{total cpm}} \times 200 \text{ nmol dNTP} \times 330 \text{ ng nmol} = \text{ng of cDNA synthesized.}$$

3. Stop the reaction by adding

EDTA (0.5 M, pH 8.0)	2 μL
SDS (20%, w/v)	2 μL

   Mix the reagents in the tube completely.

   The single-stranded, radiolabeled cDNA is quite sticky and adheres nonspecifically to glass, filters, and some plastics. For this reason, it is important to maintain a minimum of 0.05% (w/v) SDS in Step 3 of the protocol and 0.1%–1.0% SDS in hybridization buffers.

4. Add 5 μL of 3 N NaOH to the reaction tube. Incubate the mixture for 30 min at 68°C to hydrolyze the RNA.

5. Cool the mixture to room temperature. Neutralize the solution by adding 10 μL of 1 M Tris-Cl (pH 7.4), mixing well, and then adding 5 μl of 2.5 N HCl. Check the pH of the solution by spotting <1 μL on pH paper.

6. Purify the cDNA by extraction with phenol:chloroform.

7. Separate the radiolabeled probe from the unincorporated dNTPs by chromatography through a Sephadex G-50 spun column (see Protocol 1, Steps 6–14).

   ▲ *Perform this step and all subsequent steps with siliconized tubes (see Appendix 2).*

8. Perform two rounds of subtractive hybridization as follows.

   i. To the radiolabeled cDNA, add tenfold excess by weight of the driver mRNA that will be used to subtract the cDNA probe, 0.2 volume of 10 M ammonium acetate, and 2.5 volumes of ice-cold ethanol. Incubate the mixture for 10–15 min at 0°C and then recover the nucleic acids by centrifugation at maximum speed for 5 min at 4°C in a microcentrifuge.

   ii. Remove all of the ethanol by aspiration and store the open tube on the bench to allow most of the remaining ethanol to evaporate. Dissolve the nucleic acids in 6 μL of RNase-free H$_2$O.

   iii. To the dissolved nucleic acids, add

Sodium phosphate (2 M, pH 6.8)	2 μL
SDS/EDTA solution	2 μL

   iv. Cover the solution with a drop of light mineral oil and leave the microcentrifuge tube for 5 min in a boiling water bath. Transfer to a water bath set at 68°C and allow the nucleic acids to hybridize to $C_{r_0}t = 1000$ mol-sec/L.

   To calculate the time required to reach this $C_{r_0}t$, solve the following equation for t:

   $$D/D_0 = e^{-kC_{r_0}t},$$

   where D is the remaining single-stranded cDNA at time t, D_0 is the total amount of input cDNA, e is the natural logarithm, k is a rate constant for the formation of RNA–DNA hybrids that is dependent on the complexity of the mRNA population and may be assumed to be $\sim 6.7 \times 10^{-3}$ L/mol-sec, and $C_{r_0}t$ is the initial concentration of the RNA driver (which does not change appreciably during the hybridization reaction). (For a lucid description of this equation, see Sargent 1987 and for additional information, see page 538 in Davidson 1986.)

   v. Remove the microcentrifuge tube from the water bath. Use a drawn-out pipette tip attached to a micropipettor to remove the hybridization solution from the microcentrifuge tube. Transfer the hybridization mixture into a tube containing 1 mL of SPS buffer.

   vi. Separate the single- and double-stranded nucleic acids by chromatography on hydroxyapatite at 60°C.

   Measure the amount of radioactivity in each fraction by liquid scintillation counting. At least 90% of the input [^{32}P]cDNA should have hybridized to the mRNA and be present in the >0.36 M sodium phosphate wash.

   vii. Pool the fractions containing the single-stranded cDNA and concentrate them by repeated extractions with isobutanol extraction; add an equal volume of isobutanol. Mix the two phases by vortexing and centrifuge the mixture at maximum speed for 2 min at room temperature in a microcentrifuge. Discard the upper (organic) phase. Repeat the extraction with isobutanol until the volume of the aqueous phase is <100 μL.

   viii. Remove salts from the cDNA by spun-column chromatography through Sephadex G-50 equilibrated in TE (pH 8.0) containing 0.1% SDS (see Protocol 1, Steps 6–14).

   ▲ *Do not use ethanol precipitation to concentrate the cDNA because the presence of phosphate ions interferes with precipitation. Do not use dialysis to remove phosphate ions because the cDNA will adhere to the dialysis bag.*

   ix. Measure the amount of radioactivity in the sample and calculate the weight of DNA in the subtracted probe.

   x. Repeat Steps i–ix.

   Between 10% and 30% of the cDNA will form hybrids with the driver RNA during the second round of hybridization.

   It is not necessary to concentrate or remove salts from the final preparation of cDNA if it is to be used to probe a cDNA library. The radiolabeled cDNA can be used for hybridization without denaturation. Probes radiolabeled to high specific activity are rapidly damaged by radiochemical decay. The subtractive hybridizations should therefore be performed as rapidly as practicable, and the probe should be used without delay. Use 5×10^7 dpm of radiolabeled cDNA for each 150-mm filter and 5×10^6 to 1×10^7 dpm for each 90-mm filter.

If a genomic DNA library is screened with the radiolabeled subtracted probe, oligo(dA) can be added to the prehybridization and hybridization reactions at 1 μg/mL to prevent nonspecific hybridization between the oligo(dT) tails of the cDNA and oligo(dA) tracts in the genomic DNA.

9. Concentrate the final preparation of cDNA by sequential extractions with isobutanol and remove salts by chromatography on Sephadex G-50 as described in Steps 8.vii and 8.viii above.

10. Recover the cDNA by standard precipitation with ethanol (see Chapter 1, Protocol 4). Dissolve the cDNA in H_2O at a concentration of 15 ng/μL.

    ▲ *Do not attempt to precipitate the cDNA with ethanol before removing the phosphate ions by spun-column chromatography. The presence of phosphate ions interferes with DNA precipitation.*

11. To radiolabel the subtracted cDNA to high specific activity, mix the following in a 0.5-mL microcentrifuge tube:

Subtracted cDNA	5 μL
Random deoxynucleotide primers (125 μg/mL)	5 μL

12. Heat the mixture for 5 min to 60°C and then cool it to 4°C.

13. To the primer:cDNA template mixture, add

Random primer buffer (5×)	10 μL
dNTP solution of dCTP, dGTP, and dTTP (5 mM)	5 μL
[α-^{32}P]dATP (10 mCi/mL) (sp. act. >3000 Ci/mmol)	25 μL
Klenow fragment (12.5 U)	2.5 μL
H_2O	to 50 μL

    Incubate the reaction for 4–6 h at room temperature. In each random priming reaction, 10–15 units of the Klenow fragment are required.

14. Stop the reaction by adding

EDTA (0.5 M, pH 8.0)	1 μL
SDS (20%, w/v)	2.5 μL

15. Separate the radiolabeled cDNA from the unincorporated dNTPs by spun-column chromatography through Sephadex G-50 (see Protocol 1, Steps 6–14).

    The radiolabeled cDNA should be denatured by heating for 5 min to 100°C before it is used for hybridization. Use 5×10^7 dpm of radiolabeled cDNA for each 138-mm filter and 5×10^6 to 1×10^7 dpm for each 82-mm filter. Once radiolabeled, use the probe immediately to avoid damage by radiochemical decay.

RECIPE

It is essential that you consult the appropriate Material Safety Data Sheets and your institution's Environmental Health and Safety Office for proper handling of equipment and hazardous materials used in this protocol.

Random Priming Buffer (5×)

Reagent	Quantity (for 1 mL)	Final concentration
Tris-Cl (1 M, pH 8.0)	250 μL	250 mM
$MgCl_2$ (1 M)	25 μL	25 mM
NaCl (5 M)	20 μL	100 mM
DTT (1 M)	10 μL	10 mM
HEPES (2 M, adjusted to pH 6.6 with 4 M NaOH)	500 μL	1 M

Use a fresh dilution in H_2O of 1 M DTT stock, stored at −20°C. Discard the diluted DTT after use.

REFERENCES

Davidson EH. 1986. *Gene activity in early development*, 3rd ed., pp. 538–540. Academic, New York.

Davis MM. 1986. Subtractive cDNA hybridization and the T-cell receptor genes. In *Handbook of experimental immunology* (ed Weir DM, et al.), vol. 2, pp. 76.1–76.13. Blackwell Scientific, Oxford.

Fargnoli J, Holbrook NJ, Fornace AJ Jr. 1990. Low-ratio hybridization subtraction. *Anal Biochem* **187:** 364–373.

Gerard GF, D'Alessio JM. 1993. Reverse transcriptase (EC2.7.7.49): The use of cloned Moloney murine leukemia virus reverse transcriptase to synthesize DNA from RNA. In *Enzymes of molecular biology* (ed Burrell MM), pp. 73–94. Humana, Totowa, NJ.

Rhyner TA, Biguet NF, Berrard S, Borbély AA, Mallet J. 1986. An efficient approach for the selective isolation of specific transcripts from complex brain mRNA populations. *J Neurosci Res* **16:** 167–181.

Sargent TD. 1987. Isolation of differentially expressed genes. *Methods Enzymol* **152:** 423–432.

Sargent TD, Dawid IB. 1983. Differential gene expression in the gastrula of *Xenopus laevis*. *Science* **222:** 135–139.

Suganuma A, Gupta KC. 1995. An evaluation of primer length on random-primed DNA synthesis for nucleic acid hybridization: Longer is not better. *Anal Biochem* **224:** 605–608.

Labeling 3' Termini of Double-Stranded DNA Using the Klenow Fragment of *E. coli* DNA Polymerase I

The simplest way to label linear dsDNA is to use the Klenow fragment of *E. coli* DNA Pol I to cata-lyze the incorporation of one or more $[\alpha\text{-}^{32}P]$dNTPs into a recessed 3' terminus (Telford et al. 1979; Cobianchi and Wilson 1987). For a summary of other methods that may be used to label the termini of DNA, see Table 1. Fragments suitable as templates for the end-filling reaction are produced by digestion of DNA with an appropriate restriction enzyme. The Klenow enzyme is then used to cata-lyze the attachment of dNTPs to the recessed 3'-hydroxyl groups (see the information panel *E. coli* DNA Polymerase I and the Klenow Fragment). The labeling reaction is versatile and quick and has the advantage that it can be performed in the restriction digest, without any change of buffer. Because the Klenow enzyme has an indolent 3'→5' exonuclease activity, its ability to incorporate nucleotides at blunt or 3'-protruding termini is limited. One or both of the 3' ends of a linear dsDNA molecule can therefore be labeled, depending on the nature of the termini and radiolabeled nucleo-tide included in the labeling reaction (see the box Labeling Recessed and Blunt-Ended 3' Termini). However, under most circumstances, it is best to include unlabeled dNTPs in the labeling reaction. This allows the labeling of the recessed 3' terminus at any position. In addition, unlabeled dNTPs incorporated downstream from the labeled dNTP shield the labeled nucleotide from the action of the indolent 3'→5' exonuclease. Fragments labeled at both 3' termini are used as the following:

- molecular-weight standards in Southern blotting (see the box Preparing Radiolabeled Size Markers for Gel Electrophoresis, following Step 4)
- tracers for small quantities of DNAs on gels

  Fragments labeled at only one terminus are used as the following:
- probes for RNA mapping with nuclease S1 (Chapter 6, Protocol 16)
- primers in primer-extension reactions

MATERIALS

It is essential that you consult the appropriate Material Safety Data Sheets and your institution's Environmental Health and Safety Office for proper handling of equipment and hazardous materials used in this protocol.

Recipes for reagents specific to this protocol, marked <R>, are provided at the end of the protocol. See Appendix 1 for recipes for commonly used stock solutions, buffers, and reagents, marked <A>. Dilute stock so-lutions to the appropriate concentrations.

Reagents

$[\alpha\text{-}^{32}P]$dNTP (10 mCi/mL; sp. act. 800–3000 Ci/mmol)

To minimize problems caused by radiolysis of the precursor and the probe, it is best, whenever possible, to prepare radiolabeled probes on the day that the $[^{32}P]$dNTP arrives in the laboratory.

Ammonium acetate (10 M) <A> (optional; see Step 4)

Appropriate restriction enzyme(s)

Choose enzymes that produce 3'-recessed termini.

TABLE 1. End-labeling of DNA

Template	Desired position of label	Preferred methods
ssDNA or RNA	5' Terminus	*Kinase reaction:* A hydroxyl group is first created by removing the unlabeled phosphate residue from the 5' end of the nucleic acid with an alkaline phosphatase such as CIP or SAP (Protocol 9). Bacteriophage T4 polynucleotide kinase is then used to transfer the label from the γ position of ATP to the newly created 5'-hydroxyl group (Wu et al. 1976) (Protocols 10 and 11). These reactions can be performed sequentially in one tube but require inhibition of phosphatase with inorganic phosphate (Chaconas and van de Sande 1980; Cobianchi and Wilson 1987). An alternative procedure has been described in which digestion with a restriction endonuclease and 5'-end labeling with polynucleotide kinase can be performed in one step without the benefit of alkaline phosphatase (Oommen et al. 1990). The mechanism of this reaction is obscure.

Exchange reaction: The unlabeled phosphate residue is transferred from the 5' terminus of the nucleic acid to ADP and is then replaced by labeled phosphate from the γ position of ATP. Both of these reactions are catalyzed by bacteriophage T4 polynucleotide kinase and proceed simultaneously in the same test tube in the presence of excess ADP and limiting amounts of γ-labeled ATP (Berkner and Folk 1977). When labeling ssDNAs that are >300 nucleotides in length, the overall efficiency of the exchange reaction is greatly improved by including a macromolecular crowding agent such as PEG (4%–10%) in the reaction mixture (Harrison and Zimmerman 1986a,b). Even so, the specific activity of the 5'-labeled DNA is always less than can be achieved with the kinase reaction.

Note: Bacteriophage T4 polynucleotide kinase carries a 3'-phosphatase activity that degrades ATP. Several commercial suppliers prepare the enzyme from cells infected with a strain of bacteriophage T4 (amN81 pseT1) that carries a mutated version of the gene encoding polynucleo-tide kinase. This mutant form of the enzyme lacks 3'-phosphatase activity. |
| ssDNA | 3' Terminus | Template-independent polymerization of [α-³²P]NTP to the 3' terminus of ssDNA is catalyzed by calf thymus terminal deoxy-DNA nucleotidyl transferase and requires Co²⁺ as a cofactor (Protocol 13; Deng and Wu 1983). The radiolabeled DNA is then exposed to alkali and treated with alkaline phosphatase to generate a uniformly labeled DNA fragment of defined length (Roychoudhury et al. 1976, 1979; Wu et al. 1976). Alternatively, incorporation at the 3' terminus can be limited to just one nucleotide by using [α-³²P]ddATP (Yousaf et al. 1984) or [α-³²P]cordycepin triphosphate (Tu and Cohen 1980) as a substrate for terminal transferase. Because neither of these molecules carries a 3'-hydroxyl group, no additional molecules can be incorporated. Under most circumstances, [α-³²P]ddATP is the preferred substrate for labeling 3' termini because it is incorporated more efficiently than [α-³²P]cordycepin triphosphate. However, DNA molecules carrying cordycepin residues at their 3' termini have one advantage: They are resistant to digestion with contaminating exonucleases that may remove 3'-terminal ribonucleotide labels from ssDNA.

Terminal transferase can also be used to add biotinylated or fluoresceinated deoxynucleotides or dideoxynucleotides to the 3' termini of ssDNA fragments (see Protocol 13, Alternative Protocol Synthesizing Nonradiolabeled Probes Using TdT; e.g., see Vincent et al. 1982; Schneider et al. 1994). |
dsDNA with protruding 5' terminus	5' Termini	Labeling is performed with polynucleotide kinase as described above for labeling of the 5' termini of ssDNA (Protocol 10).
dsDNA with blunt-ended or recessed 5' termini	5' Termini	Labeling is performed with polynucleotide kinase as described above for labeling of the 5' termini of ssDNA (Protocol 11). However, dsDNA molecules with blunt ends or recessed 5' termini are labeled less efficiently by polynucleotide kinase than molecules with protrud-ing 5' termini. Even a partial reaction at recessed 5' termini requires large amounts of enzyme and [γ-³²P]ATP or the addition of a crowding agent such as PEG (Lillehaug and Kleppe 1975; Harrison and Zimmerman 1986a).
dsDNA with blunt-ended or recessed 3' termini	3' Termini	The Klenow fragment, which retains the template-dependent deoxynucleotide polymerizing activity and the 3'→5' exonuclease of the holo-enzyme but lacks its powerful 5'→3' exonuclease activity, is used to fill recessed 3' termini of dsDNA (Protocol 8). Usually, only one of the four dNTPs present in the reaction mixture is radiolabeled. The presence of the three unlabeled dNTPs serves two purposes: It blocks exonucleolytic removal of nucleotides from the 3' terminus of the template, and it allows the radiolabeled dNTP to be the second, third, or fourth nucleotide added to the recessed 3' terminus. The presence of only one radiolabeled dNTP in the reaction mixture sometimes permits dsDNA that has been cleaved with two different restriction endonucleases to be labeled selectively at only one end. For example, a DNA that has been cleaved with

G   G

BamHI CCTAG and EcoRI CTTAA

can be selectively labeled at the BamHI terminus by using [α-^{32}P]dGTP as the sole source of radioactivity in the fill-in reaction. Alternatively, DNA can be cleaved with one restriction enzyme, radiolabeled, and then cleaved with a second enzyme to generate two DNA fragments, each of which carries radiolabel at only one end.

Specialized vectors have been constructed that facilitate the asymmetric labeling of DNA fragments. For example, pSP64CS and pSP65CS contain sites that are cleaved by the restriction enzyme Tth111I, which recognizes the redundant sequence

↓
5' GACNNNGTC 3'
3' CTGNNNCAG 5'
↑

and generates a terminus with a single protruding 5' nucleotide. The DNA of interest is cloned between two Tth111I sites that yield DNAs with different protruding nucleotides at their termini. After cleavage, the DNA can be selectively labeled at one or both ends by the Klenow fragment and the appropriate [α-^{32}P]dNTP(s) (Volkaert et al. 1984; Eckert 1987).

Blunt-ended DNA fragments can be labeled by template-independent addition of a single nucleotide (usually (α-^{32}P]dATP) catalyzed by Taq polymerase.

| dsDNA with protruding 3' termini | 3' Termini |

Bacteriophage T4 DNA polymerase carries a 5'→3' polymerase and 3'→5' exonuclease activity that is more active on ssDNA than dsDNA. The exonuclease of bacteriophage T4 DNA polymerase is ~200 times more potent than the equivalent exonuclease of the Klenow fragment. T4 DNA polymerase is used to end-label DNA molecules with protruding 3' termini. This reaction works in two stages. First the powerful 3'→5' exonuclease removes the protruding tails from the DNA and creates recessed 3' termini. Then, in the presence of high concentrations of radiolabeled precursor(s), exonucleolytic degradation is balanced by incorporation of dNTPs at the 3' termini. This reaction, which consists of cycles of removal and replacement of the 3'-terminal nucleotides from recessed or blunt-ended DNA, is sometimes called an exchange or replacement reaction (O'Farrell et al. 1980).

Alternatively, protruding 3' termini can be labeled by template-independent addition of nucleotides catalyzed by calf thymus terminal deoxynucleotidyl transferase, poly(A) polymerase (see 3'-terminal labeling of ssDNA) or certain thermostable DNA polymerases, (e.g., Taq).

| RNA | 3' Termini |

Bacteriophage T4 RNA ligase catalyzes the joining of a terminating radioactive bisphosphate nucleotide ([5'-^{32}P]pNp) to a 3'-hydroxyl terminus of RNA (Uhlenbeck and Gumport 1982). The reaction extends the length of the RNA molecule by one nucleotide and generates a phosphorylated 3' terminus with a [^{32}P]phosphate in the last internucleotide linkage (England et al. 1977; Kikuchi et al. 1978; Tyc and Steitz 1989). The terminal 3'-phosphate group acts as a chain terminator by preventing formation of further phosphodiester bonds. This method of labeling has has been chiefly used with small RNAs. However, perhaps because of steric effects caused by secondary structure, the amount of label incorporated seriously misrepresents the relative abundance of the RNAs used as substrates.

Poly(A) polymerase catalyzes the template-independent addition of A residues to the 3' terminus of RNA. When [α-^{32}P]ATP is used as a substrate, the phophodiester bonds in the poly(A) tract contain ^{32}P atoms (Lingner and Keller 1993). Substitution of cordycepin triphosphate (3'-dATP) for ATP results in the addition of a single 3'-dA residue to the end of the RNA, which is a useful method for radiolabeling RNA at the 3' end. Yeast poly(A) polymerase preferentially labels longer RNA molecules, whereas short RNA molecules are labeled more efficiently by RNA ligase.

Terminal transferase has been used to add a biotin-labeled dideoxynucleotide to the 3' terminus of RNA (Schneider et al. 1994). However, the efficiency of this reaction has not been thoroughly investigated.

LABELING RECESSED AND BLUNT-ENDED 3' TERMINI

This protocol requires that the DNA to be end-labeled contain a sequence recognized and cleaved by a restriction endonuclease that generates 3'-recessed ends. The choice of which $[\alpha\text{-}^{32}P]dNTPs$ to use for the reaction depends on the sequence of the protruding 5' termini at the ends of the DNA and on the objective of the experiment. For example, ends created by cleavage of DNA with EcoRI can be labeled with $[\alpha\text{-}^{32}P]dATP$:

5'-G OH3'   5'P AATTC-3'   **Klenow enzyme**   5'-GAA OH3'   5'P AATTC-3'
3'-CTTAA P5' + 3'HO G-5' ——————→ 3'-CTTAA P5' + 3'HO AAG-5' .
                $[\alpha\text{-}^{32}P]dATP$

The two proximal 5'-protruding nucleotides are both thymidine residues, and the Klenow enzyme can therefore insert two adenine residues. Termini created by cleavage of DNA with BamHI can be labeled with $[\alpha\text{-}^{32}P]dGTP$:

5'-G OH3'   5'P GATCC-3'   **Klenow enzyme**   5'-GG OH3'   3'P GATCC-3'
3'-CCTAG P5' + 3'HO G-5' ——————→ 3'-CCTAG P5' + 3'HO GG-5' .
                $[\alpha\text{-}^{32}P]dGTP$

Note that only one radiolabeled nucleotide can be incorporated per end in a reaction using $[\alpha\text{-}^{32}P]dGTP$ as the radiolabeled substrate. However, if unlabeled dGTP, dATP, and dTTP were included in the polymerization reaction, the DNA fragment could be radiolabeled with $[\alpha\text{-}^{32}P]dCTP$. The addition of unlabeled dNTPs (1) allows the use of any available radiolabeled dNTP whose complement is included in the protruding 5' end, (2) eliminates the possibility of exonucleolytic removal of nucleotides from the 3' terminus of the template by the Klenow enzyme's $3' \rightarrow 5'$ exonuclease activity, and (3) ensures that all radiolabeled DNA products will be the same length.

By choosing the appropriate $[\alpha\text{-}^{32}P]dNTP$, it is possible to label only one end of a dsDNA molecule. For example, DNA cleaved by EcoRI at one end and by BamHI at the other end can be labeled selectively by including either $[\alpha\text{-}^{32}P]dATP$ (to radiolabel the EcoRI site) or $[\alpha\text{-}^{32}P]dGTP$ (to radiolabel the BamHI site). Furthermore, if the DNA sequence of a fragment to be end-labeled is known, the N nucleotide in the recognition sequence of certain restriction enzymes can be used to label one end of a DNA molecule preferentially. For example, the restriction enzyme HinfI recognizes the sequence 5'-GANTC-3' and cleaves between the G and A to yield a 3'-recessed end. If the HinfI recognition sequence at the 5' end of a DNA fragment is 5'-GACTC-3' and at the 3' end the sequence is 5'-GAGTC-3', the DNA fragment can be selectively labeled at the 5' end by including unlabeled dATP and $[\alpha\text{-}^{32}P]dGTP$ in the reaction or at the 3' end by including unlabeled dATP and $[\alpha\text{-}^{32}P]dCTP$ in the reaction. Numerous variations on this strategy can be used with restriction enzymes such as DdeI, Fnu4H, Bsu36I, and EcoO109I.

The terminal nucleotide of a blunt-ended DNA fragment can be replaced by exploiting the weak $3' \rightarrow 5'$ exonucleolytic and strong polymerizing activities of the Klenow fragment. For example, incubation of DNA fragments produced by the restriction enzyme AluI, which recognizes the sequence 5'-AGCT-3' and cleaves between the G and C, with the Klenow enzyme and $[\alpha\text{-}^{32}P]dGTP$ results in exonucleolytic removal of the dGMP and replacement with the radiolabeled guanine nucleotide:

5'-AG OH3' 5'P CT-3' **Klenow enzyme** 5'-A OH3' 5'P CT-3' **Klenow enzyme** 5'-AG OH3' 5'P CT-3'
3'-TC P5' + 3'OH GA-5' ————→ 3'-TC P5' + 3'HO A-5' ————→ 3'-TC P5' + 3'OH GA-5' .
                                        $[\alpha\text{-}^{32}P]dGTP$

Although the specific activity of radiolabeled blunt-ended DNA fragments is not high, it is generally sufficient to allow the fragments to be used as size standards in gel electrophoresis. Blunt-ended molecules can be labeled more efficiently using bacteriophage T4 DNA polymerase, which has a much more powerful $3' \rightarrow 5'$ exonuclease than the Klenow fragment.

Finally, the end-filling reaction catalyzed by the Klenow enzyme is not limited to radiolabeled dNTPs. Deoxynucleotides modified with haptens such as biotin, fluorescein, and DIG can also be used.

dNTP solution containing the appropriate unlabeled dNTPs, each at 1 mM
    For advice on making and storing solutions of dNTPs, see the information panel Preparation of Stock Solutions of dNTPs.

Ethanol
Klenow fragment of *E. coli* DNA Pol I
Template DNA (0.1–5 μg)
    Use linear dsDNA carrying appropriate recessed 3' termini.

Equipment

Sephadex G-50 spun column, equilibrated in TE (pH 7.6) (optional; see Step 4)

The Sephadex G-50 gel-filtration matrix can be used to separate DNA (which passes through the matrix) from lower-molecular-weight substances (e.g., radioactive precursors, which are retained on the column). Sephadex G-50 columns for radiolabeled DNA purification are available commercially from several sources (e.g., Quick Spin Columns [Roche Applied Science]). However, it is also easy to set one up using a 1-mL disposable syringe plugged with siliconized glass wool (see Protocol 1, Steps 6–14).

Water bath (75°C)

METHOD

1. Digest up to 5 μg of template DNA with the desired restriction enzyme in 25–50 μL of the appropriate restriction enzyme buffer.

   The labeling reaction may be performed immediately after digesting the DNA with a restriction enzyme. There is no need to remove or inactivate the restriction enzyme. Because the Klenow fragment works well under a variety of conditions (as long as Mg^{2+} is present in millimolar concentrations), it is unnecessary to change buffers. At the end of the restriction reaction, the Klenow enzyme, unlabeled dNTPs, and the appropriate [α-^{32}P]dNTP are added and the reaction is incubated for an additional 15 min at room temperature, as described below. The procedure works well even on relatively crude DNA preparations (e.g., minipreparations of plasmids).

2. To the completed restriction digest, add

[α-^{32}P]dNTP (10 mCi/mL) (sp. act. 800–3000 Ci/mmol)	2–50 μCi
Unlabeled dNTPs	to a final concentration of 100 μM
Klenow fragment	1–5 U

   To modify this protocol for nonradioactive labeling, eliminate the [α-^{32}P]dNTP, substituting a part of the dTTP with the modified nucleotide (e.g., biotin-, DIG- or fluorescein-dUTP) at a ratio of 35% modified dUTP to 65% dTTP.

   Incubate the reaction for 15 min at room temperature.

   Approximately 0.5 unit of the Klenow enzyme is required for each microgram of template DNA (1 μg of a 1000-bp fragment is equivalent to ~3.1 pmol of termini of dsDNA; see the box Calculating the Amount of 5' Ends in a DNA Sample in Protocol 9). The Klenow enzyme works well in almost all buffers used for digestion by restriction enzymes.

   Reverse transcriptase (1–2 U) can be used in place of the Klenow enzyme in this protocol. However, reverse transcriptase is not as forgiving of buffer conditions as the Klenow enzyme and is therefore used chiefly to label purified DNA fragments in reactions containing conventional reverse transcriptase buffer.

   When the labeled DNA is to be used for mapping mRNA by the nuclease S1 method (see Chapter 6, Protocol 16), the concentration of labeled dNTP in the reaction should be increased to the greatest level that is practicable. After the reaction has been allowed to proceed for 15 min at room temperature, add all four unlabeled dNTPs to a final concentration of 0.2 mM for each dNTP and continue the incubation for an additional 5 min at room temperature. This cold chase ensures that each recessed 3' terminus will be completely filled and that all labeled DNA molecules will be exactly the same length.

3. Stop the reaction by heating it for 10 min at 75°C.

4. Separate the radiolabeled DNA from unincorporated dNTP by spun-column chromatography through Sephadex G-50 (using a commercially available column or a homemade column according to Protocol 1, Steps 6–14) or by two rounds of precipitation with ethanol in the presence of 2.5 M ammonium acetate (see Chapter 1, Protocol 4).

PREPARING RADIOLABELED SIZE MARKERS FOR GEL ELECTROPHORESIS

This protocol generates labeled DNA fragments that can be used as size markers during gel electrophoresis. Because the fragments are labeled in proportion to their molar concentrations and not their sizes, both small and large fragments in a restriction enzyme digest become labeled to an equal extent. Bands of DNA that are too small to be visualized by staining with ethidium bromide or other dyes can therefore be located by autoradiography.

It is desirable, but not necessary, to remove unincorporated [α-^{32}P]dNTP from the radiolabeled DNA before gel electrophoresis. [α-^{32}P]dNTP migrates faster than bromophenol blue on both polyacrylamide and agarose gels and does not interfere with the detection of anything but the smallest of DNA fragments. However, removing the unincorporated radiolabel has some advantages: It avoids the possibility of contaminating the anodic buffer chamber of the electrophoresis tank with radioactivity, allows the amount of radiolabeled DNA applied to the gel to be estimated by a handheld minimonitor, and reduces the possibility of radiochemical damage to the DNA.

REFERENCES

Berkner KL, Folk WR. 1977. Polynucleotide kinase exchange reaction: Quantitative assay for restriction endonuclease-generated 5′-phosphoryl termini in DNA. *J Biol Chem* **252:** 3176–3184.

Chaconas G, van de Sande JH. 1980. 5′-^{32}P labeling of RNA and DNA restriction fragments. *Methods Enzymol* **65:** 75–85.

Cobianchi F, Wilson SH. 1987. Enzymes for modifying and labeling DNA and RNA. *Methods Enzymol* **152:** 94–110.

Deng G, Wu R. 1983. Terminal transferase: Use of the tailing of DNA and for in vitro mutagenesis. *Methods Enzymol* **100:** 96–116.

Eckert RL. 1987. New vectors for rapid sequencing of DNA fragments by chemical degradation. *Gene* **51:** 247–254.

England TE, Gumport RI, Uhlenbeck OC. 1977. Dinucleoside pyrophosphate are substrates for T4-induced ligase. *Proc Natl Acad Sci* **74:** 4839–4842.

Harrison B, Zimmerman SB. 1986a. T4 polynucleotide kinase: Macromolecular crowding increases the efficiency of reaction at DNA termini. *Anal Biochem* **158:** 307–315.

Harrison B, Zimmerman SB. 1986b. Stabilization of T4 polynucleotide kinase by macromolecular crowding. *Nucleic Acids Res* **14:** 1863–1870.

Kikuchi Y, Hishinuma F, Sakaguchi K. 1978. Addition of mononucleotides to oligoribonucleotide acceptors with T4 RNA ligase. *Proc Natl Acad Sci* **75:** 1270–1273.

Lillehaug JR, Kleppe K. 1975. Effect of salts and polyamines polynucleotide kinase. *Biochemistry* **14:** 1225–1229.

Lingner J, Keller W. 1993. 3′-End labeling of RNA with recombinant yeast poly(A) polymerase. *Nucleic Acids Res* **21:** 2917–2920.

O'Farrell PH, Kutter E, Nakanishi M. 1980. A restriction map of the bacteriophage T4 genome. *Mol Gen Genet* **179:** 421–435.

Oommen A, Ferrandis I, Wang MJ. 1990. Single-step labeling of DNA using restriction endonucleases and T4 polynucleotide kinase. *BioTechniques* **8:** 482–486.

Roychoudhury R, Jay E, Wu R. 1976. Terminal labeling and addition of homopolymer tracts to duplex DNA fragments by terminal deoxynucleotidyl transferase. *Nucleic Acids Res* **3:** 101–116.

Roychoudhury R, Tu CP, Wu R. 1979. Influence of nucleotide sequence adjacent to duplex DNA termini on 3′ terminal labeling by terminal transferase. *Nucleic Acids Res* **6:** 1323–1333.

Schneider GS, Martin CS, Bronstein I. 1994. Chemiluminescent detection of RNA lableled with biotinylated dideoxynucleotides and terminal transferase. *J NIH Res* **6:** 90.

Telford JL, Kressmann A, Koski RA, Grosschedl R, Müller F, Clarkson SG, Birnstiel ML. 1979. Delimitation of a promoter for RNA polymerase III by means of a functional test. *Proc Natl Acad Sci* **76:** 2590–2594.

Tu CP, Cohen SN. 1980. 3′-End labeling of DNA with [α-^{32}P]cordycepin-5′-triphosphate. *Gene* **10:** 177–183.

Tyc K, Steitz JA. 1989. U3, U8 and U13 comprise a new class of mammalian snRNPs localized in the cell nucleolus. *EMBO J* **8:** 3113–3119.

Vincent C, Tchen P, Cohen-Solal M, Kourilsky P. 1982. Synthesis of 8-(2-4 dinitrophenyl 2-6 aminohexyl) aminoadenosine 5′ triphosphate: Biological properties and potential uses. *Nucleic Acids Res* **10:** 6787–6796.

Volkaert G, de Vleeschouwer E, Blöcker H, Frank R. 1984. A novel type of cloning vector for ultrarapid chemical degradation sequencing of DNA. *Gene Anal Tech* **1:** 52–59

Wu R, Jay E, Roychoudhury R. 1976. Nucleotide sequence analysis of DNA. *Methods Cancer Res* **12:** 87–176.

Yousaf SI, Carroll AR, Clarke BE. 1984. A new and improved method for 3′-end labelling DNA using [α-^{32}P]ddATP. *Gene* **27:** 309–313.

Dephosphorylation of DNA Fragments with Alkaline Phosphatase

The removal of 5′ phosphates from nucleic acids is used to enhance subsequent labeling with [γ-^{32}P]ATP, reduce the circularization of plasmid vectors in ligation reactions, and render DNA susceptible or resistant to other enzymes that act on nucleic acids (e.g., λ exonuclease). Essentially, any nucleotide phosphatase (e.g., bacterial alkaline phosphatase, calf intestinal alkaline phosphatase [CIP], placental alkaline phosphatase, shrimp alkaline phosphatase [SAP], or several acid phosphatases such as sweet potato and prostate acid phosphatase) will catalyze the removal of 5′ phosphates from nucleic acid templates. In fact, these enzymes prefer small substrates such as p-nitrophenyl phosphate (PNPP) and the exposed 5′ phosphates of nucleic acids to bulky globular protein substrates. For additional details on alkaline phosphatases, see the information panel Alkaline Phosphatase.

Because CIP and SAP are commercially available and readily inactivated, they are the most widely used phosphatases in molecular cloning. Although CIP is less expensive per unit of activity, SAP enzyme has the advantage of being readily inactivated in the absence of chelators. DNA modification reactions (e.g., phosphorylation and ligation) can therefore be performed in serial fashion in the same reaction tube, thereby avoiding extraction with phenol:chloroform and precipitation with ethanol.

MATERIALS

It is essential that you consult the appropriate Material Safety Data Sheets and your institution's Environmental Health and Safety Office for proper handling of equipment and hazardous materials used in this protocol.

Recipes for reagents specific to this protocol, marked <R>, are provided at the end of the protocol. See Appendix 1 for recipes for commonly used stock solutions, buffers, and reagents, marked <A>. Dilute stock solutions to the appropriate concentrations.

Reagents

Chloroform
CIP or SAP (Affymetrix, Roche Applied Science, or Worthington Biochemicals)
Dephosphorylation buffer (for use with CIP or SAP) (10×) <A>
DNA sample (0.1–10 μg [1–100 pmol])
Dephosphorylation reactions are usually performed in a volume of 25–50 μL containing 1–100 pmol of 5′-phosphorylated termini of DNA.

EDTA (0.5 M, pH 8.0) <A> or EGTA (0.5 M, pH 8.0) <A>, if using CIP
Ethanol
Phenol:chloroform (1:1, v/v)
Proteinase K
Restriction enzyme(s)
SDS (10%, w/v) <A>, if using CIP
Sodium acetate (3 M, pH 7.0 [if using CIP] and pH 5.2) <A>
The 3 M sodium acetate solution at pH 7.0 is used because EDTA precipitates from solution at acid pH.

TE (pH 7.6) <A>
Tris-Cl (1 M, pH 8.5) <A>

Equipment

Water baths or heating blocks (56°C, 65°C, or 75°C [CIP] or 70°C [SAP])

CALCULATING THE AMOUNT OF 5′ ENDS IN A DNA SAMPLE

Use the following formulas to calculate the number of picomoles of 5′ ends present in a given DNA sample.

For dsDNA,

Amount of 5′ ends (in picomoles) = $[X/(Y \times 660 \, \text{g/mol/bp})] \times 10^{12}$ pmol/mol $\times$ 2 ends/molecule, where X is the mass of DNA fragment in grams and Y is the length of DNA fragment in base pairs. For example, 1 μg of linearized 3-kb plasmid DNA = 1 pmol 5′ ends, and 1 μg of a 1-kb dsDNA fragment = 3 pmol 5′ ends.

For ssDNA,

$$\text{pmol 5′ ends} = [X/(Y \times 330 \, \text{g/mol/nucleotide})] \times 10^{12} \, \text{pmol/mol},$$

where X is the mass of DNA fragment in grams and Y is the length of DNA fragment in nucleotides. For example, 1 μg (~0.03 OD_{260}) of a 25-mer oligodeoxynucleotide is 120 pmol 5′ ends.

METHOD

1. Use the restriction enzyme of choice to digest to completion 1–10 μg (10–100 pmol) of the DNA to be dephosphorylated.

    Follow the restriction enzyme manufacturer's recommendations for incubation time and temperature. The progress of the digest can be analyzed by agarose gel electrophoresis.

    CIP and SAP will dephosphorylate DNA at a slightly reduced efficiency in restriction buffers that have been adjusted to pH 8.5 with 10× CIP or 10× SAP buffer, as is done in the next step. If this is unacceptable, the restricted DNA may be purified by extraction with phenol:chloroform and standard precipitation with ethanol and then dissolved in a minimal volume of 10 mM Tris-Cl (pH 8.5).

2. Dephosphorylate the 5′ ends of the restricted DNA with either CIP or SAP.

    To Dephosphorylate DNA Using CIP

    i. Add to the DNA

CIP dephosphorylation buffer (10×)	5 μL
H_2O	to 48 μL

    ii. Add the appropriate amount of CIP.

    1 U of CIP will dephosphorylate ~1 pmol of 5′-phosphorylated termini (5′-recessed or blunt-ended DNA) or ~50 pmol of 5′-protruding termini. These amounts may vary slightly from one manufacturer to the next.

    iii. Incubate the reaction for 30 min at 37°C, add a second aliquot of CIP, and continue incubation for an additional 30 min.

    iv. To inactivate CIP at the end of the incubation period, add SDS and EDTA (pH 8.0) to final concentrations of 0.5% and 5 mM, respectively. Mix the reagents well and add proteinase K to a final concentration of 100 μg/mL. Incubate for 30 min at 56°C.

    v. Cool the reaction to room temperature and purify the DNA by extracting it twice with phenol:chloroform and once with chloroform alone.

    Proteinase K and SDS used to inactivate and digest CIP must be completely removed by extraction with phenol:chloroform before subsequent enzymatic treatments (phosphorylation by polynucleotide kinase, ligation, etc.).

    ▲ *A single extraction with chloroform alone is insufficient to reduce the concentration of SDS to the point at which it no longer will inhibit polynucleotide kinase. Kurien et al. (1997) suggest that up to four extractions with chloroform may be required.*

Glycogen or linear polyacrylamide (Gaillard and Strauss 1990) can be added as a carrier before phenol:chloroform extraction if small amounts of DNA (<100 ng) were used in the reaction. Do not add carrier nucleic acid (tRNA, salmon sperm DNA, etc.), as it will compete with the dephosphorylated DNA for the radiolabeled ATP during the kinasing reaction.

Alternatively, CIP can be inactivated by heating for 30 min to 65°C (or for 10 min to 75°C) in the presence of 10 mM EGTA (pH 8.0).

▲ *Use EGTA, not EDTA.*

To Dephosphorylate DNA Using SAP

i. Add to the DNA

SAP dephosphorylation buffer (10×)	5 μL
H$_2$O	to 48 μL

ii. Add the appropriate amount of SAP.

One unit of SAP will dephosphorylate ~1 pmol of 5'-phosphorylated termini (3' recessed or 5' recessed) or ~0.2 pmol of blunt-ended DNA. These amounts may vary slightly from one enzyme manufacturer to the next.

iii. Incubate the reaction for 1 h at 37°C.

iv. To inactivate SAP, transfer the reaction to 70°C, incubate for 20 min, and cool to room temperature.

3. Transfer the aqueous phase to a clean microcentrifuge tube and recover the DNA by standard ethanol precipitation in the presence of 0.1 volume of 3 M sodium acetate (pH 5.2), if SAP was used, or 0.1 volume of 3 M sodium acetate (pH 7.0), if CIP was used.

4. Allow the precipitate to dry at room temperature before dissolving it in TE (pH 7.6) at a DNA concentration of >2 nmol/mL.

To remove 5' phosphates from RNA, use 0.01 U of CIP/pmol of 5' termini. Incubate for 15 min at 37°C, followed by 30 min at 55°C. Alternatively, use 0.01–0.1 U of SAP/pmol of 5' ends to dephosphorylate RNA in a 1-h incubation at 37°C.

If the dephosphorylated DNA is to be used as a substrate for polynucleotide kinase, it must be rigorously purified by spun-column chromatography, gel electrophoresis, density gradient centrifugation, or chromatography on columns of Sepharose CL-4B to free it from low-molecular-weight nucleic acids. Although such contaminants may comprise only a small fraction of the weight of the nucleic acid in the preparation, they contribute a much larger proportion of the 5' termini. Unless steps are taken to remove them, contaminating low-molecular-weight DNA and RNA molecules can be the predominant species of nucleic acids that are labeled in reactions catalyzed by bacteriophage T4 polynucleotide kinase.

Because ammonium ions are strong inhibitors of bacteriophage T4 polynucleotide kinase, dephosphorylated DNAs should not be dissolved in, or precipitated from, buffers containing ammonium salts before treatment with polynucleotide kinase.

REFERENCES

Gaillard C, Strauss F. 1990. Ethanol precipitation of DNA with linear polyacrylamide as carrier. *Nucleic Acids Res* **18:** 378.

Kurien BT, Scofield RH, Broyles RH. 1997. Efficient 5' end labeling of dephosphorylated DNA. *Anal Biochem* **245:** 123–126.

Phosphorylation of DNA Molecules with Protruding 5'-Hydroxyl Termini

The removal of 5' phosphates from nucleic acids with phosphatases and their readdition in radiolabeled form by bacteriophage T4 polynucleotide kinase is a widely used technique for generating ^{32}P-labeled probes (see the box Labeling the 5' Termini of DNA with Bacteriophage T4 Polynucleotide Kinase, below). When the reaction is performed efficiently, 40%–50% of the protruding 5' termini in the reaction becomes radiolabeled (Berkner and Folk 1977). However, the specific activity of the resulting probes is not as high as that obtained by other radiolabeling methods because only one radioactive atom is introduced per DNA molecule. Nevertheless, the availability of [γ-^{32}P]ATP with specific activities in the 3000–7000 Ci/mmol range allows the synthesis of probes suitable for many purposes, including the following:

- as primers in primer-extension experiments to map the 5' ends of mRNAs
- as substrates in the analysis of RNA structure by nuclease S1

This protocol describes a method to label dephosphorylated protruding 5' termini. For a summary of methods used to label the termini of DNA, see Protocol 8, Table 1.

LABELING THE 5' TERMINI OF DNA WITH BACTERIOPHAGE T4 POLYNUCLEOTIDE KINASE

The enzyme of choice to catalyze the labeling of 5' termini of nucleic acids is bacteriophage T4 polynucleotide kinase, which facilitates transfer of γ-phosphate residues from [γ-^{32}P]ATP to the 5'-hydroxyl termini of dephosphorylated ssDNA, and dsDNA, and RNA. In addition, the enzyme will, with much lower efficiency, restore phosphate residues to 5'-hydroxyl groups located at nicks in dsDNA. T4 polynucleotide kinase is a tetrameric protein of M_r ~142,000 composed of four identical subunits encoded by the structural gene *pseT* of bacteriophage T4 (for reviews, see Richardson 1971; Maunders 1993). Transfer of ^{32}P to the 5' terminus of DNA catalyzed by bacteriophage T4 polynucleotide kinase can be performed in two ways (Fig. 1).

- In the forward reaction, ssDNA or dsDNA is first treated with a phosphatase (CIP or SAP) to remove phosphate residues from the 5' termini (Protocol 9). The resulting 5'-hydroxyl termini are then rephosphorylated by transferring the γ-phosphate from [γ-^{32}P]ATP in a reaction catalyzed by polynucleotide kinase (see this protocol and Protocol 11). The efficiency of radiolabeling depends on the purity of the

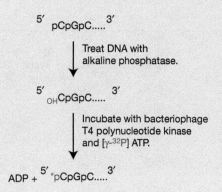

FIGURE 1. Bacteriophage T4 polynucleotide kinase can be used to label the 5' termini of DNA molecules. For details, see text.

(Continued)

DNA (due to kinase inhibitors present in an unpurified DNA solution) and the sequence at the 5′ terminus of the DNA. For unknown reasons, oligonucleotides with a cytosine residue at their 5′ termini are labeled fourfold less efficiently than oligonucleotides beginning with A or T and sixfold less efficiently than oligonucleotides beginning with G (van Houten et al. 1998).

- Some DNA templates, such as synthetic oligonucleotides, do not need to be treated with a phosphatase enzyme before kinasing. Oligonucleotides are almost invariably synthesized with a free 5′-hydroxyl group. Because the phosphorylation reaction is reversible, T4 polynucleotide kinase will catalyze dephosphorylation of 5′ termini in the presence of a molar excess of a nucleotide diphosphate acceptor, such as adenosine diphosphate (ADP) (van de Sande et al. 1973). The second method of labeling (the exchange reaction) uses both the forward and reverse reactions catalyzed by the enzyme.

Because T4 polynucleotide kinase will work, albeit inefficiently, in restriction buffers, the exchange reaction can be performed at the same time as restriction of DNA with enzymes that generate protruding 5′ termini. The resulting DNA can be used as radiolabeled size markers for gel electrophoresis (Oommen et al. 1990).

MATERIALS

It is essential that you consult the appropriate Material Safety Data Sheets and your institution's Environmental Health and Safety Office for proper handling of equipment and hazardous materials used in this protocol.

Recipes for reagents specific to this protocol, marked <R>, are provided at the end of the protocol. See Appendix 1 for recipes for commonly used stock solutions, buffers, and reagents, marked <A>. Dilute stock solutions to the appropriate concentrations.

Reagents

Ammonium acetate (10 M) <A> (optional; see Step 3)

Bacteriophage T4 polynucleotide kinase

Wild-type polynucleotide kinase is a 5′ phosphotransferase and a 3′ phosphatase (Depew and Cozzarelli 1974; Sirotkin et al. 1978). However, mutant forms of the enzyme (Cameron et al. 1978) are commercially available (e.g., Boehringer Mannheim) that lack the phosphatase activity but retain a fully functional phosphotransferase. We recommend that the mutant form of the enzyme be used for 5′ labeling whenever possible; 10–20 U of the enzyme are required to catalyze the phosphorylation of 10–50 pmol of dephosphorylated 5′-protruding termini.

Bacteriophage T4 polynucleotide kinase buffer (10×) <A>

DNA (10–50 pmol)

The DNA should be dephosphorylated as described in Protocol 9 or synthesized with a 5′-hydroxyl moiety. See the box Calculating the Amount of 5′ Ends in a DNA Sample, in Protocol 9.

EDTA (0.5 M, pH 8.0) <A>

Ethanol

$[\gamma\text{-}^{32}\text{P}]$ATP (10 mCi/mL; sp. act. 3000–7000 Ci/mmol)

To minimize problems caused by radiolysis of the precursor and the probe, it is best, whenever possible, to prepare radiolabeled probes on the day that the [³²P]dNTP arrives in the laboratory.

Equipment

Liquid scintillation spectrometer, capable of quantifying ^{32}P by Cerenkov radiation

Sephadex G-50 spun column, equilibrated in TE (pH 7.6)

or

Sephadex G-50 column (1 mL), equilibrated in TE (pH 7.6) (both optional; see Step 3)

The Sephadex G-50 gel-filtration matrix can be used to separate DNA (which passes through the matrix) from lower-molecular-weight substances (e.g., radioactive precursors, which are retained on the column). Sephadex G-50 columns for radiolabeled DNA purification are available commercially from several sources (e.g., Quick Spin Columns [Roche Applied Science]). However, it is also easy to set one up using a 1-mL disposable syringe plugged with siliconized glass wool (see Protocol 1, Steps 6–14).

METHOD

1. In a microcentrifuge tube, mix these reagents in the following order:

Dephosphorylated DNA	10–50 pmol
Bacteriophage T4 polynucleotide kinase buffer (10×)	5 µL
[γ-^{32}P]ATP (10 mCi/mL) (sp. act. 3000–7000 Ci/mmol)	50 pmol
Bacteriophage T4 polynucleotide kinase	10 U
H$_2$O	to 50 µL

   Incubate the reaction for 1 h at 37°C.

   Ideally, ATP should be in a fivefold molar excess over DNA 5′ ends, and the concentration of DNA termini should be ≥0.4 µM. The concentration of ATP in the reaction should therefore be >2 µM, but this is rarely achievable in practice. To increase the specific activity of the radiolabeled DNA product, increase the amount of [γ-^{32}P]ATP used in the phosphorylation reaction. Decrease the volume of H$_2$O to maintain a reaction volume of 50 µL.

2. Terminate the reaction by adding 2 µL of 0.5 M EDTA (pH 8.0). Measure the total radioactivity in the reaction mixture by Cerenkov counting in a liquid scintillation counter.

3. Separate the radiolabeled probe from unincorporated dNTPs using one of the following methods:

   • spun-column chromatography through Sephadex G-50 (using a commercially available column or a homemade column according to Protocol 1, Steps 6–14)

   • conventional size-exclusion chromatography through 1-mL columns of Sephadex G-50 (equilibrated in TE)
   or

   • two rounds of selective precipitation of the radiolabeled DNA with ammonium acetate and ethanol (see Chapter 1, Protocol 4).

4. Measure the amount of radioactivity in the probe preparation by Cerenkov counting. Calculate the efficiency of transfer of the radiolabel to the 5′ termini by dividing the amount of radioactivity in the probe by the total amount in the reaction mixture (Step 2).

REFERENCES

Berkner KL, Folk WR. 1977. Polynucleotide kinase exchange reaction: Quantitative assay for restriction endonuclease-generated 5′-phosphoryl termini in DNA. *J Biol Chem* **252:** 3176–3184.

Cameron V, Soltis D, Uhlenbeck OC. 1978. Polynucleotide kinase from a T4 mutant which lacks the 3′ phosphatase activity. *Nucleic Acids Res* **5:** 825–833.

Depew RE, Cozzarelli NR. 1974. Genetics and physiology of bacteriophage T4 3′-phosphatase: Evidence for involvement of the enzyme in T4 DNA metabolism. *J Virol* **13:** 888–897.

Maunders MJ. 1993. Polynucleotide kinase. *Methods Mol Biol* **16:** 343–356.

Oommen A, Ferrandis I, Wang MJ. 1990. Single-step labeling of DNA using restriction endonucleases and T4 polynucleotide kinase. *BioTechniques* **8:** 482–486.

Richardson CC. 1971. Polynucleotide kinase from *Escherichia coli* infected with bacteriophage T4. In *Procedures in nucleic acid research* (ed Cantoni GL, Davies DR), vol. 2, pp. 815–828. Harper and Row, New York.

Sirotkin K, Cooley W, Runnels J, Snyder LR. 1978. A role in true-late gene expression for the T4 bacteriophage 5′ polynucleotide kinase 3′ phosphatase. *J Mol Biol* **123:** 221–233.

van de Sande JH, Kleppe K, Khorana HG. 1973. Reversal of bacteriophage T4 induced polynucleotide kinase action. *Biochemistry* **12:** 5050–5055.

van Houten V, Denkers F, van Dijk M, van den Brekel M, Brakenhoff R. 1998. Labeling efficiency of oligonucleotides by T4 polynucleotide kinase depends on 5′-nucleotide. *Anal Biochem* **265:** 386–389.

Phosphorylation of DNA Molecules with Dephosphorylated Blunt Ends or Recessed 5′ Termini

DNA substrates with blunt ends, recessed 5′ termini, or internal nicks are labeled less efficiently in the forward reaction catalyzed by T4 polynucleotide kinase than are protruding 5′ termini of dsDNA. For example, the incorporation of phosphate residues at internal nicks in DNA is 30-fold less efficient than transfer to 5′ termini (Lillehaug et al. 1976; Berkner and Folk 1977). However, the difficulty of labeling such substrates can be overcome by (1) increasing the concentration of ATP to astronomical levels (>100 µM) (Lillehaug and Kleppe 1975a) or (2) including polyamines or polyethylene glycol 8000 (PEG 8000) in the reaction (Lillehaug and Kleppe 1975b; Harrison and Zimmerman 1986a). In the presence of PEG and magnesium, DNA collapses into a highly condensed state (Lerman 1971). The increased efficiency of the subsequent phosphorylation reaction is thought to be a consequence of the resulting macromolecular crowding (Harrison and Zimmerman 1986b; for review, see Zimmerman and Minton 1993). The amount of stimulation is crucially dependent on the concentration of PEG. It may therefore be useful to test the efficiency of the reaction in concentrations of PEG between 4% and 10%. The beneficial effects of PEG only become apparent when using DNAs longer than 300 bp. Smaller fragments of DNA are probably too rigid to collapse into a condensed state. For additional information on labeling DNA with T4 polynucleotide kinase, see the box Labeling the 5′ Termini of DNA with Bacteriophage T4 Polynucleotide Kinase, in the introduction to Protocol 10. For a summary of methods used to label the termini of DNA, see Table 1 of Protocol 8.

In this protocol, PEG is included, along with high concentrations of ATP and polynucleotide kinase, to enhance the labeling of DNA fragments with blunt-ended or 5′-recessed termini.

MATERIALS

It is essential that you consult the appropriate Material Safety Data Sheets and your institution's Environmental Health and Safety Office for proper handling of equipment and hazardous materials used in this protocol.

Recipes for reagents specific to this protocol, marked <R>, are provided at the end of the protocol. See Appendix 1 for recipes for commonly used stock solutions, buffers, and reagents, marked <A>. Dilute stock solutions to the appropriate concentrations.

Reagents

Ammonium acetate (10 M) <A> (optional; see Step 5)
Bacteriophage T4 polynucleotide kinase

Wild-type polynucleotide kinase is a 5′ phosphotransferase and a 3′ phosphatase (Depew and Cozzarelli 1974; Sirotkin et al. 1978). However, mutant forms of the enzyme (Cameron et al. 1978) are commercially available (e.g., Boehringer Mannheim) that lack the phosphatase activity but retain a fully functional phosphotransferase. We recommend that the mutant form of the enzyme be used for 5′ labeling whenever possible.

DNA (10–50 pmol in a volume of ≤11 µL)

Dephosphorylate the DNA as described in Protocol 9 or synthesize with a 5′-hydroxyl moiety. See the box Calculating the Amount of 5′ Ends in a DNA Sample in Protocol 9.

EDTA (0.5 M, pH 8.0) <A>

Ethanol

Imidazole buffer (10×) <R>

[γ-^{32}P]ATP (10 mCi/mL; sp. act. 3000 Ci/mmol)

> *To minimize problems caused by radiolysis of the precursor and the probe, it is best, whenever possible, to prepare radiolabeled probes on the day that the [^{32}P]dNTP arrives in the laboratory*

PEG 8000 (24%, w/v) in H$_2$O

Equipment

Liquid scintillation spectrometer, capable of quantifying ^{32}P by Cerenkov radiation

Sephadex G-50 spun column, equilibrated in TE (pH 7.6)

or

Sephadex G-50 column (1 mL), equilibrated in TE (pH 7.6) (both optional; see Step 5)

> *The Sephadex G-50 gel-filtration matrix can be used to separate DNA (which passes through the matrix) from lower-molecular-weight substances (e.g., radioactive precursors, which are retained on the column). Sephadex G-50 columns for radiolabled DNA purification are available commercially from several sources (e.g., Quick Spin Columns [Roche Applied Science]). However, it is also easy to set one up using a 1-mL disposable syringe plugged with siliconized glass wool (see Protocol 1, Steps 6–14).*

METHOD

1. In a microcentrifuge tube, mix in the following order:

Dephosphorylated DNA	10–50 pmol
Imidazole buffer (10×)	4 µL
H$_2$O	to 15 µL
PEG (24%, w/v)	10 µL

2. Add 40 pmol of [γ-^{32}P]ATP (10 mCi/mL; sp. act. 3000 Ci/mmol) to the tube and bring the final volume of the reaction to 40 µL with H$_2$O.

   > *Ideally, ATP should be in a fivefold molar excess over DNA 5′ ends, and the concentration of DNA termini should be ≥0.4 µM. The concentration of ATP in the reaction should therefore be >2 µM, but this is rarely achievable in practice. To increase the specific activity of the radiolabeled DNA product, increase the amount of [γ-^{32}P]ATP used in the phosphorylation reaction. Decrease the volume of H$_2$O to maintain a reaction volume of 40 µL.*

3. Add 40 U of bacteriophage T4 polynucleotide kinase to the reaction. Mix the reagents gently by tapping the side of tube and incubate the reaction for 30 min at 37°C.

4. Terminate the reaction by adding 2 µL of 0.5 M EDTA (pH 8.0). Measure the total radioactivity in the reaction mixture by Cerenkov counting in a liquid scintillation counter.

5. Separate the radiolabeled probe from unincorporated dNTPs using one of the following options as appropriate:

   - spun-column chromatography through Sephadex G-50 (using a commercially available column or a homemade column according to Protocol 1, Steps 6–14),

   - conventional size-exclusion chromatography through 1-mL columns of Sephadex G-50 (equilibrated in TE),
     or

   - two rounds of selective precipitation of the radiolabeled DNA with ammonium acetate and ethanol (see Chapter 1, Protocol 4).

6. Measure the amount of radioactivity in the probe preparation by Cerenkov counting. Calculate the efficiency of transfer of the radiolabel to the 5′ termini by dividing the amount of radioactivity in the probe by the total amount in the reaction mixture (Step 4).

RECIPE

It is essential that you consult the appropriate Material Safety Data Sheets and your institution's Environmental Health and Safety Office for proper handling of equipment and hazardous materials used in this protocol.

Imidazole Buffer (10×)

Reagent	Quantity (for 1 mL)	Final concentration
Imidazole·HCl (1 M, pH 6.4)	500 µL	500 mM
MgCl$_2$ (1 M)	180 µL	180 mM
DTT (1 M)	50 µL	50 mM
Spermidine HCl (100 mM)	10 µL	1 mM
EDTA (0.5 M, pH 8.0)	2 µL	1 mM

REFERENCES

Berkner KL, Folk WR. 1977. Polynucleotide kinase exchange reaction: Quantitative assay for restriction endonuclease-generated 5'-phosphoryl termini in DNA. *J Biol Chem* 252: 3176–3184.

Cameron V, Soltis D, Uhlenbeck OC. 1978. Polynucleotide kinase from a T4 mutant which lacks the 3' phosphatase activity. *Nucleic Acids Res* 5: 825–833.

Depew RE, Cozzarelli NR. 1974. Genetics and physiology of bacteriophage T4 3'-phosphatase: Evidence for involvement of the enzyme in T4 DNA metabolism. *J Virol* 13: 888–897.

Harrison B, Zimmerman SB. 1986a. T4 polynucleotide kinase: Macromolecular crowding increases the efficiency of reaction at DNA termini. *Anal Biochem* 158: 307–315.

Harrison B, Zimmerman SB. 1986b. Stabilization of T4 polynucleotide kinase by macromolecular crowding. *Nucleic Acids Res* 14: 1863–1870.

Lerman LS. 1971. A transition to a compact form of DNA in polymer solutions. *Proc Natl Acad Sci* 68: 1886–1890.

Lillehaug JR, Kleppe K. 1975a. Kinetics and specificity of T4 polynucleotide kinase. *Biochemistry* 14: 1221–1225.

Lillehaug JR, Kleppe K. 1975b. Effect of salts and polyamines polynucleotide kinase. *Biochemistry* 14: 1225–1229.

Lillehaug JR, Kleppe RK, Kleppe K. 1976. Phosphorylation of double-stranded DNAs by T4 polynucleotide kinase. *Biochemistry* 15: 1858–1865.

Sirotkin K, Cooley W, Runnels J, Snyder LR. 1978. A role in true-late gene expression for the T4 bacteriophage 5' polynucleotide kinase 3' phosphatase. *J Mol Biol* 123: 221–233.

Zimmerman SB, Minton AP. 1993. Macromolecular crowding: Biochemical, biophysical, and physiological consequences. *Annu Rev Biophys Biomol Struct* 22: 27–65.

Protocol 12

Phosphorylating the 5′ Termini of Oligonucleotides Using T4 Polynucleotide Kinase

Synthetic oligonucleotides lack phosphate groups at their 5′ termini and are therefore easily radiolabeled by transferring the γ-^{32}P from [γ-^{32}P]ATP in a reaction catalyzed by bacteriophage T4 polynucleotide kinase. Oligonucleotides radiolabeled in this way may be used as hybridization probes and primers for DNA sequencing, 5′-end mapping of mRNAs.

When the reaction is performed under standard conditions, >50% of the oligonucleotide molecules become radiolabeled. However, for reasons that are unknown, the efficiency of radiolabeling of oligonucleotides depends on the sequence of the oligonucleotide (van Houten et al. 1998). Oligonucleotides with a cytosine residue at their 5′ termini are labeled fourfold less efficiently than oligonucleotides beginning with A or T and sixfold less efficiently than oligonucleotides beginning with G. When designing oligonucleotide probes, it clearly pays to ensure that the 5′-terminal position is not occupied by a cytosine residue.

The reaction described here is designed to label 10 pmol of an oligonucleotide to high specific activity (1×10^8 to 4×10^8 dpm/μg). Labeling of different amounts of oligonucleotide can easily be achieved by increasing or decreasing the size of the reaction and keeping the concentrations of all components constant.

MATERIALS

It is essential that you consult the appropriate Material Safety Data Sheets and your institution's Environmental Health and Safety Office for proper handling of equipment and hazardous materials used in this protocol.

Recipes for reagents specific to this protocol, marked <R>, are provided at the end of the protocol. See Appendix 1 for recipes for commonly used stock solutions, buffers, and reagents, marked <A>. Dilute stock solutions to the appropriate concentrations.

Reagents

Bacteriophage T4 polynucleotide kinase

Wild-type polynucleotide kinase is a 5′ phosphotransferase and a 3′ phosphatase (Depew and Cozzarelli 1974; Sirotkin et al. 1978). However, mutant forms of the enzyme (e.g., see Cameron et al. 1978) are commercially available (e.g., Boehringer Mannheim) that lack the phosphatase activity but retain a fully functional phosphotransferase. We recommend that the mutant form of the enzyme be used for 5′ labeling whenever possible; 10–20 U of the enzyme are required to catalyze the phosphorylation of 10–50 pmol of dephosphorylated 5′-protruding termini.

▲ *Be aware that preparations of polynucleotide kinase sold by different manufacturers have been reported to vary widely in their ability to catalyze the phosphorylation of the 5′ termini of single-stranded synthetic oligonucleotides (van Houten et al. 1998).*

Bacteriophage T4 polynucleotide kinase buffer (10×) <A>

[γ-^{32}P]ATP (10 mCi/mL; sp. act. >5000 Ci/mmol) in aqueous solution

10 pmol of [γ-^{32}P]ATP is required to label 10 pmol of dephosphorylated 5′ termini to high specific activity. To minimize problems caused by radiolysis of the precursor and the probe, it is best, whenever possible, to prepare radiolabeled probes on the day that the [^{32}P]ATP arrives in the laboratory.

Oligonucleotide

To achieve maximum efficiency of labeling, the oligonucleotide should be PAGE purified. Crude preparations of oligonucleotides are labeled with lower efficiency in reactions catalyzed by polynucleotide

kinase (van Houten et al. 1998). When using unpurified preparations of oligonucleotide, make sure that the last cycle of synthesis was programmed to be "trityl-off," that is, that the dimethoxytrityl blocking group at the 5′ end of the oligonucleotide primer was removed before release of the DNA from the solid synthesis support. The dimethoxytrityl group efficiently protects the 5′-hydroxyl group of the oligonucleotide from 5′ modification.

Tris-Cl (1 M, pH 8.0) <A>

Equipment

Microcentrifuge tubes (0.5 mL)
Water bath (68°C)

METHOD

1. Set up a reaction mixture in a 0.5-mL microcentrifuge tube containing the following:

Synthetic oligonucleotide (10 pmol/μL)	1 μL
Bacteriophage T4 polynucleotide kinase buffer (10×)	2 μL
[γ-^{32}P]ATP (10 pmol; sp. act.>5000 Ci/mmol)	5 μL
H$_2$O	11.4 μL

   Mix the reagents well by gentle but persistent tapping on the outside of the tube. Place 0.5 μL of the reaction mixture in a tube containing 10 μL of 10 mM Tris-Cl (pH 8.0). Set the tube aside for use in Step 4.

   The reaction contains equal concentrations of [γ-^{32}P]ATP and oligonucleotide. Generally, only 50% of the radiolabel is transferred to the oligonucleotide. The efficiency of transfer can be improved by increasing the concentration of oligonucleotide in the reaction by a factor of 10. This increase results in transfer of ~90% of the radiolabel to the oligonucleotide. However, the specific activity of the radiolabeled DNA is reduced by a factor of ~5. To label an oligonucleotide to the highest specific activity,

   - *increase the concentration of [γ-^{32}P]ATP in the reaction by a factor of 3 (i.e., use 15 μL of radiolabel and decrease the volume of H$_2$O to 1.4 μL)*

   - *decrease the amount of oligonucleotide to 3 pmol*

   Under these circumstances, only ~10% of the radiolabel is transferred, but a high proportion of the oligonucleotide becomes radiolabeled.

   Ideally, ATP should be in a fivefold molar excess over DNA 5′ ends, and the concentration of DNA termini should be ≥0.4 μM. The concentration of ATP in the reaction should therefore be >2 μM, but this is rarely achievable in practice.

2. Add 10 U (~1 μL) of bacteriophage T4 polynucleotide kinase to the remaining reaction mixture. Mix the reagents well and incubate the reaction mixture for 1 h at 37°C.

3. At the end of the incubation period, place 0.5 μL of the reaction in a second tube containing 10 μL of 10 mM Tris-Cl (pH 8.0). Heat the remainder of the reaction for 10 min at 68°C to inactivate the polynucleotide kinase. Store the tube containing the heated reaction mixture on ice.

4. Before proceeding, determine whether the labeling reaction has worked well by measuring the fraction of the radiolabel that has been transferred to the oligonucleotide substrate in a small sample of the reaction mixture. Transfer a sample of the reaction (exactly 0.5 μL) to a fresh tube containing 10 μL of 10 mM Tris-Cl (pH 8.0). Use this sample (along with the two aliquots set aside in Steps 1 and 3) to measure the efficiency of transfer of the α-^{32}P from ATP by one of the following methods:

   - Measure the proportion of radiolabeled dNTPs that are incorporated into material precipitated by TCA (see Appendix 2).

     or

- Measure the efficiency of the labeling reaction by estimating the fraction of label that migrates with the oligonucleotide during size-exclusion chromatography through Sephadex G-15 or Bio-Rad P-60 columns. For details of this method, see Protocol 17. In some ways, this is the easier of the two methods because the relative amounts of incorporated and unincorporated radioactivity can be estimated during chromatography on a handheld minimonitor.

    If the specific activity is too low, see Troubleshooting.

5. If the specific activity of the oligonucleotide is acceptable, purify the radiolabeled oligonucleotide as described in Protocol 15, 16, or 17.

TROUBLESHOOTING

Problem (Step 4): The specific activity of the probe is too low.

Solution: Add an additional 8 U of polynucleotide kinase, continue incubation for an additional 30 min at 37°C (i.e., a total of 90 min), heat the reaction for 10 min at 68°C to inactivate the enzyme, and analyze the products of the reaction again, as described in Step 4.

Problem (Step 4): The specific activity of the probe is too low and an additional round of phosphorylation fails to yield an oligonucleotide of sufficiently high specific activity.

Solution: Check whether the oligonucleotide contains a cytosine residue at its 5′ terminus. If so, consider redesigning the oligonucleotide so that it contains a G, A, or T residue at its 5′ end (see the introduction to this protocol). Alternatively, try purifying the original preparation of oligonucleotide by Sep-Pak chromatography (Protocol 17) and then repeat this protocol.

REFERENCES

Cameron V, Soltis D, Uhlenbeck OC. 1978. Polynucleotide kinase from a T4 mutant which lacks the 3′ phosphatase activity. *Nucleic Acids Res* 5: 825–833.

Depew RE, Cozzarelli NR. 1974. Genetics and physiology of bacteriophage T4 3′-phosphatase: Evidence for involvement of the enzyme in T4 DNA metabolism. *J Virol* 13: 888–897.

Sirotkin K, Cooley W, Runnels J, Snyder LR. 1978. A role in true-late gene expression for the T4 bacteriophage 5′ polynucleotide kinase 3′ phosphatase. *J Mol Biol* 123: 221–233.

van Houten V, Denkers F, van Dijk M, van den Brekel M, Brakenhoff R. 1998. Labeling efficiency of oligonucleotides by T4 polynucleotide kinase depends on 5′-nucleotide. *Anal Biochem* 265: 386–389.

Labeling the 3′ Termini of Oligonucleotides Using Terminal Deoxynucleotidyl Transferase

Terminal deoxynucleotidyl transferase (TdT, also simply called terminal transferase) is a template-independent polymerase that catalyzes the addition of deoxynucleotides and dideoxynucleotides to the 3′-hydroxyl terminus of a DNA molecule. Cobalt (Co^{2+}) is a necessary cofactor for the activity of this enzyme. Terminal transferase exhibits a substrate preference for ssDNA but will also add nucleotides to protruding, recessed, and blunt-ended dsDNA fragments, albeit with a lower efficiency. The reaction can be limited to the addition of a single nucleotide when used in the presence of a chain-terminating analog such as a dideoxynucleotide (ddNTP), typically [α-^{32}P]ddATP (Yousaf et al. 1984). Alternatively, the enzyme is capable of adding several (2–100) nucleotides to 3′ ends in a so-called homopolymeric "tailing" reaction (Deng and Wu 1983) (for more information, see the Additional Protocol Tailing Reaction at the end of this protocol). In addition to adding radioactive nucleotides, terminal transferase can also be used to add nonradioactive labels such as biotin-, DIG-, or fluorescein-labeled nucleotides (Kumar et al. 1988; Igloi and Schiefermayr 1993). Nonradioactive labeling is typically performed under conditions that promote the addition of several nucleotides to increase the sensitivity of the probe. Although the 3′-end-labeling reaction is relatively easy to perform (see the Alternative Protocol Synthesizing Nonradiolabeled Probes Using TdT at the end of this protocol), biotin and DIG 3′-end-labeling kits are available from several manufacturers (e.g., Thermo Scientific, Roche Applied Science).

MATERIALS

It is essential that you consult the appropriate Material Safety Data Sheets and your institution's Environmental Health and Safety Office for proper handling of equipment and hazardous materials used in this protocol.

Recipes for reagents specific to this protocol, marked <R>, are provided at the end of the protocol. See Appendix 1 for recipes for commonly used stock solutions, buffers, and reagents, marked <A>. Dilute stock solutions to the appropriate concentrations.

Reagents

[α-^{32}P]ddATP (10 mCi/mL; 3000 Ci/mmol)
CoCl$_2$ solution (25 mM)
DNA oligonucleotide (~10 pmol 3′ DNA ends, 10–100 ng depending on length)
    See the box Calculating the Amount of 5′ Ends in a DNA Sample in Protocol 9.

EDTA (0.2 M, pH 8.0) <A> (optional; see Step 3)
Recombinant terminal transferase (400 U/μL)
    The recombinant form of terminal transferase has greater activity than the native enzyme and is available from companies such as Roche Applied Science and New England Biolabs.

Terminal transferase reaction buffer (5×) <R>
Water, sterile and nuclease-free

Equipment

Water baths or heat blocks (37°C and 70°C) (optional)

1021

METHOD

1. In a microcentrifuge tube on ice, add the following:

Terminal transferase reaction buffer	10 μL
CoCl$_2$ (25 nM)	5 μL
DNA oligonucleotide (10 pmol 3′ ends)	x μL[a]
[α-^{32}P]ddATP	5 μL
Terminal transferase	1 μL
H$_2$O	to a final volume of 50 μL

   [a]Determine the appropriate volume based on the DNA oligonucleotide concentration.

   Mix the reagents well by gentle but persistent tapping on the outside of the tube. Briefly centrifuge to collect the contents in the bottom of the tube. Place 0.5 μL of the reaction mixture in a tube containing 0.02 M EDTA and set aside the tube for use in Step 4 to determine incorporation of the radiolabel.

2. Incubate the reaction at 37°C for 60 min.

3. Stop the reaction by heating for 10 min to 70°C or by adding 5 μL of 0.2 M EDTA (pH 8.0).

4. Determine whether the labeling reaction has worked well by measuring the fraction of the radiolabel that has been transferred to the oligonucleotide substrate in a small sample of the reaction mixture. Transfer a sample of the reaction (exactly 0.5 μL) to a fresh tube containing 0.02 M EDTA. Use this sample (along with the aliquot set aside in Step 1) to measure the efficiency of incorporation of the ^{32}P-ddATP by one of the following methods.

   • Measure the proportion of radiolabeled dNTPs that are incorporated into material precipitated by TCA (see Appendix 2).

     or

   • Measure the efficiency of the labeling reaction by estimating the fraction of label that migrates with the oligonucleotide during size-exclusion chromatography through Sephadex G-15 or Bio-Rad P-60 columns. For details of this method, see Protocol 17. In some ways, this is the easier of the two methods because the relative amounts of incorporated and unincorporated radioactivity can be estimated during chromatography on a handheld minimonitor. An incorporation rate of at least 30% should be obtained.

5. If the specific activity of the oligonucleotide is acceptable, purify the radiolabeled oligonucleotide as described in Protocol 15, 16, or 17.

Synthesizing Nonradiolabeled Probes Using TdT

METHOD

To modify the above protocol for the use of nonradioactive nucleotides, increase the number of 3′ ends in the reaction to ∼100 pmol and double the amount of $CoCl_2$ in the reaction mixture to 5 mM.

1. Place the following in a microcentrifuge tube on ice:

Terminal transferase reaction buffer	4 μL
$CoCl_2$ (25 mM)	4 μL
DNA oligonucleotide (100 pmol 3′ ends)	x μL[a]
Biotin-, DIG-, or fluoroscein-ddUTP	1 μL
Terminal transferase	1 μL
H_2O	to a final volume of 20 μL

   [a]Determine the appropriate volume based on the DNA oligonucleotide concentration.

   Mix the reagents well by gentle but persistent tapping on the outside of the tube. Briefly centrifuge to collect the contents in the bottom of the tube.

2. Incubate the reaction for 15 min at 37°C.

3. Stop the reaction by heating for 10 min to 70°C or by adding 2 μl of 0.2 M EDTA (pH 8.0).

4. Purify the probe from unincorporated label as described in Protocol 15, 16, or 17.

Tailing Reaction

A tailing reaction is performed in the presence of a mixture of labeled and unlabeled dATP, dTTP, dGTP, or dCTP. The rate of addition of dNTPs, and thus the length of the tail, is a function of the ratio of 3′ DNA ends to dNTP concentration and, in addition, the specific dNTP that is used. The following values are approximate and are given for a 15-min incubation at 37°C.

Ratio of pmol 3′ ends (μM dNTP)	Tail length (for a 15-min incubation at 37°C) (nucleotides)			
	dATP	dTTP	dGTP	dCTP
1:0.1	1–5	1–5	1–3	1–3
1:1	10–20	10–20	5–10	10–20
1:5	100–300	200–300	10–25	50–200
1:10	300–500	250–350	15–25	100–150

The reaction conditions described in this protocol generate tail lengths of 75–125 nucleotides for radiolabeled probes tailed with dATP and dTTP and 15–30 nucleotides for probes tailed with dGTP and dCTP.

ADDITIONAL MATERIALS

Reagents

dNTP labeling mix (see Step 1)

METHOD

1. For radioactive labeling, prepare the following:

   CoCl$_2$ solution (15 mM stock)

   Radioactive labeling mix

   For a dATP or dTTP labeling mix, mix 1 volume of a 2.5 mM solution of dATP or dTTP with 4 volumes of [α-^{32}P]dATP or [α-^{32}P]dTTP (800 Ci/mmol) and 15 volumes of H$_2$O. For a dGTP or dCTP labeling mix, mix 1 volume of a 2 mM solution of dGTP or dTTP with 4 volumes of [α-^{32}P]dATP or [α-^{32}P]dTTP (800 Ci/mmol) and 15 volumes of H$_2$O.

2. In a microcentrifuge tube on ice, add the following:

Terminal transferase reaction buffer	4 μL
CoCl$_2$ (15 mM)	2 μL (for a tailing reaction using dATP or dTTP)
	or
	1 μL (for a tailing reaction using dGTP or dCTP)
DNA oligonucleotide (1 pmol 3′ ends)	x μL[a]
Radioactive dNTP labeling mix	1 μL
Terminal transferase	1 μL
H$_2$O	to a final volume of 20 μL

   [a]Determine the appropriate volume based on the DNA oligonucleotide concentration.

Mix the reagents well by gentle but persistent tapping on the outside of the tube. Briefly centrifuge to collect the contents in the bottom of the tube.

3. Incubate the reaction for 15 min at 37°C.

4. Stop the reaction by heating to 70°C for 10 min or by adding 2 μL of 0.2 M EDTA (pH 8.0).

Modifications for Synthesizing Nonradiolabeled Probes

Tailing reactions for nonradiolabeled probes are performed in the presence of biotin-, DIG-, or fluorescein-dUTP and dATP, dTTP, dGTP, or dCTP. The tail length reflects the number of incorporated labeled dUTPs, which is dependent on the type and concentration of dNTPs and the ratio of labeled dUTP nucleotides to unlabeled nucleotides. The following protocol uses a ratio of 1:10 of labeled dUTP:dNTP and gives the following results.

	dATP	dTTP	dGTP	dCTP
Average tail length	50	10	15	25
Range of tail length	10–100	1–20	5–10	10–40
Amount of labeled dUTP molecules/ tail	5	1	1.5	2.5

METHOD

1. In a microcentrifuge tube on ice, add the following:

Terminal transferase reaction buffer	(4 µL)
$CoCl_2$ (25 mM)	4 µL
DNA oligonucleotide (100 pmol 3' ends)	x µL[a]
Biotin-, DIG-, or fluorescein-dUTP (1 mM)	1 µL
dATP, dTTP, dGTP, or dCTP (10 mM)	1 µL
Terminal transferase	1 µL
H_2O	to a final volume of 20 µL

   [a]Determine the appropriate volume based on the DNA oligonucleotide concentration.

   Mix the reagents well by gentle but persistent tapping on the outside of the tube. Briefly centrifuge to collect the contents in the bottom of the tube.

2. Incubate the reaction for 15 min at 37°C.

3. Stop the reaction by heating for 10 min to 70°C or by adding 2 µL of 0.2 M EDTA (pH 8.0).

RECIPE

It is essential that you consult the appropriate Material Safety Data Sheets and your institution's Environmental Health and Safety Office for proper handling of equipment and hazardous materials used in this protocol.

Terminal Transferase Reaction Buffer (5×)

Reagent	Quantity (for 10 mL)	Final concentration
Potassium cacodylate	1.76 g	1 M
Tris-HCl (1 M, pH 6.6)	1.25	0.125 M
BSA (10 mg/mL)	12.5 mL	1.25 mg/mL

REFERENCES

Deng G, Wu R. 1983. Terminal transferase: Use of the tailing of DNA and for in vitro mutagenesis. *Methods Enzymol* **100**: 96–116.

Igloi GL, Schiefermayr E. 1993. Enzymatic addition of fluorescein- or biotin-riboUTP to oligonucleotides results in primers suitable for DNA sequencing and PCR. *BioTechniques* **15**: 486–488, 490–482, 494–487.

Kumar A, Tchen P, Roullet F, Cohen J. 1988. Nonradioactive labeling of synthetic oligonucleotide probes with terminal deoxynucleotidyl transferase. *Anal Biochem* **169**: 376–382.

Yousaf SI, Carroll AR, Clarke BE. 1984. A new and improved method for 3′-end labelling DNA using [α-^{32}P]ddATP. *Gene* **27**: 309–313.

Labeling of Synthetic Oligonucleotides Using the Klenow Fragment of *E. coli* DNA Polymerase I

As an alternative to 5′- and 3′-end-labeling, probes of high specific activities can be obtained using the Klenow fragment of *E. coli* DNA Pol I to synthesize a strand of DNA complementary to the synthetic oligonucleotide (Studencki and Wallace 1984; Ullrich et al. 1984b; for review, see Wetmur 1991). In this method, a short primer is hybridized to an oligonucleotide template whose sequence is the complement of the desired radiolabeled probe. The primer is then extended using the Klenow fragment to incorporate [α-^{32}P]dNTPs in a template-directed manner. After the reaction, the template and product are separated by denaturation followed by electrophoresis through a polyacrylamide gel under denaturing conditions (see Fig. 1). With this method, it is possible to generate oligonucleotide probes that contain several radioactive atoms per molecule of oligonucleotide and to achieve specific activities as high as 2×10^{10} cpm/μg of probe. Because the end product of the reaction is dsDNA, whose strands must be separated and the labeled product isolated (see below), this method is generally not used to prepare nonradiolabeled oligonucleotides. To obtain the best results, consider the points discussed in the box Optimizing Oligonucleotide Labeling Using the Klenow Fragment.

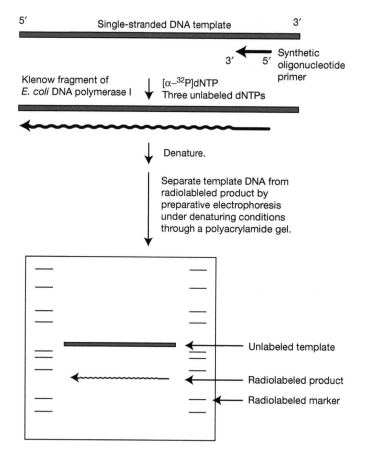

FIGURE 1. Labeling of synthetic oligonucleotide using the Klenow fragment of *E. coli* DNA Pol I.

OPTIMIZING OLIGONUCLEOTIDE LABELING USING THE KLENOW FRAGMENT

When planning a labeling experiment of this type, consider the following points.

- *The specific activity of the [α-^{32}P]dNTPs in the reaction.* The specific activity of a dNTP whose α-phosphate has been completely substituted by ^{32}P is ~9000 Ci/mmol. Preparations with lower specific activities contain a mixture of both ^{32}P-labeled and -unlabeled molecules. Thus, if a single radiolabeled dNTP with a specific activity of 3000 Ci/mmol is included in the synthetic reaction and there are only three positions at which that nucleotide can be incorporated into the final product, an average of only one atom of ^{32}P will be present in each molecule of probe (i.e., the specific activity will be approximately equal to that of a probe labeled by bacteriophage T4 polynucleotide kinase). If the sequence of the desired probe and the specific activity of the available precursors are known, it is possible to predict the probe's specific activity when one, two, three, or all four [α-^{32}P]dNTPs are included in the reaction.

- *The concentration of dNTPs in the reaction.* Polymerization of nucleotides will not proceed efficiently unless the concentration of each dNTP in the reaction remains at 1 μM or greater (1 μM = ~0.66 ng/μL) throughout the course of the reaction. Calculate the total quantity of each dNTP that would be incorporated into the probe, assuming that all of the single-stranded sequences in the reaction are used as templates. The total amount of each dNTP in the reaction should be the sum of the amount that could be incorporated into the probe plus 0.66 ng/μL.

- *The primer must be of sufficient length and specificity to bind to the template and promote synthesis at the appropriate position.* Primers used for this purpose are usually 7–9 nucleotides long and are complementary to sequences at or close to the 3' terminus of the template oligonucleotide. Because the stability of such short hybrids is difficult to predict, the ratio of primer to template that gives the maximum yield of full-length probe should be determined empirically before embarking on large-scale labeling reactions.

- *The final product is a double-stranded fragment of DNA whose length is equal to or slightly less than that of the template (depending on the location of the sequences complementary to the primer).* To be maximally effective as a probe, the unlabeled template strand must be efficiently separated from the complementary radiolabeled product; otherwise, the two complementary strands anneal to each other and the efficiency of hybridization to the desired target sequence is reduced. For oligonucleotides <30 nucleotides in length, electrophoresis through a 20% polyacrylamide gel is the most effective method to separate the complementary strands. The efficiency of the separation is improved if the lengths of the two strands are not exactly the same. Whenever possible, the primer should therefore be designed so that it is complementary to the subterminal nucleotides of the template. If the primer cannot hybridize to the 2–3 nucleotides at the 3' terminus of the template, the radiolabeled product will be shorter than the unlabeled template and separate from it more effectively during electrophoresis. However, even when their lengths are identical, there is a good chance that the template and product strands will separate to some extent during electrophoresis because the rate of migration of single-stranded nucleic acids through gels is dependent not only on their length but also on their base composition and sequence.

  The extent of separation can often be increased by either phosphorylating one of the two strands (template or primer) with nonradioactive ATP before synthesis of the probe (see Protocol 12) or leaving the dimethoxytrityl group attached to the 5' terminus of the primer (Studencki and Wallace 1984). Because it is impossible to predict which method will be most effective for a given oligonucleotide, it is usually necessary to perform a series of trial experiments to determine in each case the efficiency with which an extended product can be separated from the template.

MATERIALS

It is essential that you consult the appropriate Material Safety Data Sheets and your institution's Environmental Health and Safety Office for proper handling of equipment and hazardous materials used in this protocol.

Recipes for reagents specific to this protocol, marked <R>, are provided at the end of the protocol. See Appendix 1 for recipes for commonly used stock solutions, buffers, and reagents, marked <A>. Dilute stock solutions to the appropriate concentrations.

TABLE 1. Percent polyacrylamide required to resolve oligonucleotides

Length of oligonucleotide	Polyacrylamide (%)
12–15 nucleotides	20
25–35 nucleotides	15
35–45 nucleotides	12
45–70 nucleotides	10

Reagents

[α-^{32}P]dNTPs

To keep the substrate concentration high, perform the extension reaction in as small a volume as possible. Thus, it is best to use radiolabeled dNTPs supplied in ethanol/H$_2$O rather than those supplied in buffered aqueous solvents. Appropriate volumes of the ethanolic [α-^{32}P]dNTPs can be mixed and evaporated to dryness in the microcentrifuge tube that will be used to perform the reaction. To minimize problems caused by radiolysis of the precursor and the probe, it is best, whenever possible, to prepare radiolabeled probes on the day that the [^{32}P]dNTP arrives in the laboratory.

Denaturing polyacrylamide gel

The percentage of polyacrylamide in the solution used to form the gel and the conditions under which electrophoresis is performed vary according to the size of the oligonucleotides in the reaction mixture. Table 1 provides useful guidelines.

Polyacrylamide gels are usually cast in 1× TBE (89 mM TBE, 2 mM EDTA), and electrophoresis is performed in the same buffer. For additional details of the methods used to cast and handle polyacrylamide gels, see Chapter 1, Protocol 3.

Formamide-loading buffer <A>
Klenow buffer (10×) <A>
Klenow fragment of *E. coli* DNA Pol I
Oligonucleotide primer

The oligonucleotide primer should be PAGE purified. To ensure efficient radiolabeling, the primer should be in threefold to tenfold molar excess over the template DNA in the reaction mixture.

Template oligonucleotide

The template oligonucleotide should be PAGE purified. The sequence of the oligonucleotide template should be the complement of the desired radiolabeled probe.

Equipment

Phosphorescent adhesive labels (available from commercial sources) or adhesive labels marked with very hot radioactive ink <A>

Water bath or heating block (80°C)

METHOD

1. Transfer to a microcentrifuge tube the calculated amounts of [α-^{32}P]dNTPs necessary to achieve the desired specific activity and sufficient to allow complete synthesis of all template strands (see the introduction to this protocol).

   The concentration of dNTPs should not drop below 1 μM at any stage during the reaction. To keep the substrate concentration high, perform the extension reaction in as small a volume as possible.

2. Add to the tube the appropriate amounts of oligonucleotide primer and template oligonucleotide.

   To ensure efficient radiolabeling, the primer should be in threefold to tenfold molar excess over the template DNA in the reaction mixture.

3. Add 0.1 volume of 10× Klenow buffer to the tube. Mix the reagents well.

4. Add 2–4 U of the Klenow fragment per 5 μL of reaction volume. Mix well. Incubate the reaction for 2–3 h at 14°C.

> *If desired, the progress of the reaction may be monitored by removing small (0.1 μL) aliquots and measuring the proportion of radioactivity that has become precipitable with 10% trichloroacetic acid (TCA) (see Appendix 2).*

5. Dilute the reaction mixture with an equal volume of formamide-loading buffer, heat the mixture for 3 min to 80°C, and load the entire sample on a denaturing polyacrylamide gel.

6. Following electrophoresis, disassemble the electrophoresis apparatus, leaving the polyacrylamide gel attached to one of the glass plates (for details, see Chapter 6, Protocol 11).

> ▲ *Unincorporated [α-³²P]dNTPs may have migrated into the lower buffer reservoir, making it radioactive. Treat the gel, glass plates, buffers, and electrophoresis apparatus as potential sites of radioactivity. Handle them appropriately, behind a Plexiglas shield.*

7. Wrap the gel and its backing plate in plastic wrap. Note the position of the tracking dyes and use a handheld minimonitor to check the amount of radioactivity in the region of the gel that should contain the oligonucleotide. Attach a set of adhesive dot labels, marked with either very hot radioactive ink or phosphorescent spots, around the edge of the sample on the plastic wrap. Cover the radioactive dots with adhesive tape to prevent contaminating the film holder or intensifying screen with the radioactive ink.

8. Expose the gel to autoradiographic film (see Chapter 2, Protocol 5).

> *Usually, the amount of radioactivity incorporated into the probe is so great that the time needed to obtain an image on film is no more than a few seconds.*

9. After developing the film, align the images of the radioactive ink with the radioactive marks on the labels and locate the position of the probe in the gel. Excise the band and recover the radioactive oligonucleotide as described in Chapter 2, Protocol 11.

DISCUSSION

In a modification of this radiolabeling procedure, longer probes can be synthesized from two oligonucleotides that contain complementary sequences at their 3′ termini (Ullrich et al. 1984a). Here, the two oligonucleotides are annealed via their complementary 3′ sequences and then extended with the Klenow fragment. Both strands are extended (and radiolabeled), resulting in a duplex that may be considerably longer than either of the two original oligonucleotides. Although this method requires two oligonucleotides, it has the advantage that both strands of the product are radiolabeled. Strand separation and purification by polyacrylamide gel electrophoresis are therefore no longer mandatory. Instead, the radiolabeled probes can be purified by chromatography through Sephadex G-50 or Bio-Gel P-60, essentially as described in Protocol 1.

REFERENCES

Studencki AB, Wallace RB. 1984. Allele-specific hybridization using oligonucleotide probes of very high specific activity: Discrimination of the human β A- and β S-globin genes. *DNA* 3: 7–15.

Ullrich A, Berman CH, Dull TJ, Gray A, Lee JM. 1984a. Isolation of the human insulin-like growth factor I gene using a single synthetic DNA probe. *EMBO J* 3: 361–364.

Ullrich A, Gray A, Wood WI, Hayflick J, Seeburg PH. 1984b. Isolation of a cDNA clone coding for the α-subunit of mouse nerve growth factor using a high-stringency selection procedure. *DNA* 3: 387–392.

Wetmur JG. 1991. DNA probes: Applications of the principles of nucleic acid hybridization. *Crit Rev Biochem Mol Biol* 26: 227–259.

Purification of Labeled Oligonucleotides by Precipitation with Ethanol

If labeled oligonucleotides are to be used only as probes in hybridization experiments, complete removal of unincorporated label is generally not necessary. However, to reduce background to a minimum, the bulk of the unincorporated label should be separated from the labeled oligonucleotide. Most of the residual unincorporated precursors can be removed from the preparation by differential precipitation with ethanol if the oligonucleotide is >18 nucleotides in length. If complete removal of the unincorporated radiolabel is required (e.g., when the radiolabeled oligonucleotide will be used in primer-extension reactions), chromatographic methods (Protocols 17 and 18) or gel electrophoresis (essentially as described in Chapter 2, Protocol 3) should be used.

MATERIALS

It is essential that you consult the appropriate Material Safety Data Sheets and your institution's Environmental Health and Safety Office for proper handling of equipment and hazardous materials used in this protocol.

Recipes for reagents specific to this protocol, marked <R>, are provided at the end of the protocol. See Appendix 1 for recipes for commonly used stock solutions, buffers, and reagents, marked <A>. Dilute stock solutions to the appropriate concentrations.

Reagents

Ammonium acetate (10 M) <A>
Ethanol
Radiolabeled oligonucleotide

The starting material for purification is the reaction mixture from Protocol 12 (either Step 3 or Step 5) or Protocol 13, after the enzyme has been heat inactivated or chemically inactived.

TE (pH 7.6) <A>

METHOD

1. Add 40 μL of H_2O to the tube containing the labeled oligonucleotide. After mixing, add 240 μL of a 5 M solution of ammonium acetate. Mix the reagents again and then add 750 μL of ice-cold ethanol. Mix the reagents once more and store the ethanolic solution for 30 min at 0°C.

   Ammonium acetate is used in place of sodium acetate to ensure more effective removal of unincorporated ribonucleotides. The solubility constant of ribonucleotides (and deoxyribonucleotides) is higher in ethanolic ammonium acetate solutions than in ethanolic sodium acetate solutions. Unincorporated ribonucleotides remain in the ethanolic phase of the precipitation reaction.

2. Recover the labeled oligonucleotide by centrifugation at maximum speed for 20 min at 4°C in a microcentrifuge.

3. Use a micropipettor equipped with a disposable tip to carefully remove all of the supernatant from the tube.

   ▲ *For radiolabeled probes, the supernatant contains most of the unincorporated radioactive nucleotides. Exercise care and be diligent about disposal of unincorporated radioactivity, pipette tips, and microcentrifuge tubes.*

4. Add 500 μL of 80% ethanol to the tube, tap the side of the tube to rinse the nucleic acid pellet, and centrifuge the tube again at maximum speed for 5 min at 4°C in a microcentrifuge.

5. Use a micropipettor equipped with a disposable tip to carefully remove the supernatant (which, for radiolabled probes, will contain appreciable amounts of radioactivity) from the tube. Stand the open tube on the bench behind a Plexiglas screen (if radiolabeled) until the residual ethanol has evaporated.

6. Dissolve the labeled oligonucleotide in 100 μL of TE (pH 7.6).

   Radiolabeled oligonucleotides may be stored for a few days at –20°C. However, during prolonged storage, decay of ^{32}P causes radiochemical damage that can impair the ability of the oligonucleotide to hybridize to its target sequence. Nonradiolabeled oligonucleotides can be stored longer.

Purification of Labeled Oligonucleotides by Size-Exclusion Chromatography

When labeled oligonucleotides are to be used in enzymatic reactions such as primer extension, virtually all of the unincorporated label must be removed from the oligonucleotide. For this purpose, chromatographic methods (this protocol and Protocol 17) or gel electrophoresis (essentially as described in Chapter 2, Protocol 3) are superior to differential precipitation of the oligonucleotide with ethanol or cetylpyridinium bromide (CPB). This protocol describes a method to separate labeled oligonucleotides from unincorporated label that takes advantage of differences in mobility between oligonucleotides and mononucleotides during size-exclusion chromatography. It should be noted that several companies (e.g., Biosearch Technologies, Dionex, Roche Molecular Biochemicals) offer ready-to-use spin columns that provide a rapid and simple method for the purification of unincorporated nucleotides from labeling reactions.

Although size-exclusion chromatography can, in principle, be used to purify either radiolabeled or nonradiolabeled oligonucleotides, this protocol is geared toward purifying radiolabeled oligonucleotides, whose elution from the column is monitored using a minimonitor and whose separation from unincorporated nucleotides is monitored by liquid scintillation counting.

MATERIALS

It is essential that you consult the appropriate Material Safety Data Sheets and your institution's Environmental Health and Safety Office for proper handling of equipment and hazardous materials used in this protocol.

Recipes for reagents specific to this protocol, marked <R>, are provided at the end of the protocol. See Appendix 1 for recipes for commonly used stock solutions, buffers, and reagents, marked <A>. Dilute stock solutions to the appropriate concentrations.

Reagents

Chloroform (optional; see Step 7)
EDTA (0.5 M, pH 8.0) <A>
Ethanol
Phenol:chloroform (optional; see Step 7)
Radiolabeled oligonucleotide
  The starting material for purification is the reaction mixture from Protocol 12 (either Step 3 or Step 5) or Protocol 13, after the enzyme has been heat or chemically inactived.

Sodium acetate (3 M, pH 5.2) <A> (optional; see Step 7)
TE (pH 7.6) <A>
Tris-Cl (1 M, pH 8.0) <A> (optional; see Step 7)
Tris-SDS chromatography buffer
  Tris-Cl (pH 8.0) 10 mM
  SDS (0.1%, w/v)

Equipment

Gel-filtration resin (e.g., Bio-Gel P-60 [fine grade] or Sephadex G-15)

Bio-Gel P-60 (fine grade) may be purchased from Bio-Rad. Sephadex G-15 is available from most suppliers of laboratory chemicals (e.g., Sigma-Aldrich). Bio-Gel P-60 is supplied as a preswollen gel; Sephadex G-15 must be swollen and equilibrated before use.

Glass wool

Wrap a small amount of glass wool in aluminum foil and autoclave the package at 15 psi (1.05 kg/cm²) for 15 min on wrapped item cycle.

Microcentrifuge tubes (1.5 mL) in a rack or fraction collector

These are used for the collection of radiolabeled oligonucleotides eluting from the column.

Pasteur pipette

METHOD

1. Add 30 μL of 20 mM EDTA (pH 8.0) to the tube containing the radiolabeled oligonucleotide. Store the solution at 0°C while preparing a column of size-exclusion chromatography resin.

   For convenience, Bio-Gel P-60 is used throughout this protocol as an example of a suitable resin. However, the method works equally well with Sephadex G-15.

2. Prepare a Bio-Gel P-60 column in a sterile Pasteur pipette.

   i. Equilibrate the slurry of Bio-Gel P-60 supplied by the manufacturer in 10 volumes of Tris-SDS chromatography buffer.

   If a centrifugal evaporator (Savant SpeedVac or its equivalent) is available, the Bio-Gel P-60 column may be poured and run in a solution of 0.1% ammonium bicarbonate. The pooled fractions containing the radiolabeled oligonucleotide (see Step 6) can then be evaporated to dryness in a centrifugal evaporator, thereby eliminating the need for extraction of the oligonucleotide preparation with organic solvents and precipitation with ethanol.

   ii. Tamp a sterile glass wool plug into the bottom of a sterile Pasteur pipette.

   A glass capillary tube works well as a tamping device.

   iii. With the plug in place, pour a small amount of Tris-SDS chromatography buffer into the column and check that the buffer flows at a reasonable rate (one drop every few seconds).

   iv. Fill the pipette with the Bio-Gel P-60 slurry. The column forms rapidly as the gel matrix settles under gravity and the buffer drips from the pipette. Add additional slurry until the packed column fills the pipette from the plug of glass wool to the constriction near the top of the pipette.

   v. Wash the column with 3 mL of Tris-SDS chromatography buffer.

   ▲ *Do not allow the column to run dry. If necessary, seal the column by wrapping a piece of Parafilm around the bottom of the pipette.*

3. Use a pipette to remove excess buffer from the top of the column and then rapidly load the radiolabeled oligonucleotide (in a volume of 100 μL or less) onto the column.

4. Immediately after the sample has entered the column, add 100 μL of buffer to the top of the column. As soon as the buffer has entered the column, fill the pipette with buffer. Replenish the buffer as necessary so that it continuously drips from the column; do not allow the column to run dry.

5. Use a handheld minimonitor to follow the progress of the radiolabeled oligonucleotide. When the radioactivity first starts to elute from the column, begin collecting two-drop fractions into microcentrifuge tubes.

   ▲ *Because phosphorylation reactions are often performed with >100 μCi of radiolabeled ATP, the amount of radioactivity in these column fractions may be considerable. Exercise care and be diligent about disposal of unincorporated radioactivity, pipette tips, and microcentrifuge tubes.*

6. When nearly all of the radioactivity has eluted from the column, use a liquid scintillation counter to measure the radioactivity in each fraction by Cerenkov counting. If there is a clean separation of the faster-migrating peak (the radiolabeled oligonucleotide) from the slower peak of unincorporated nucleotide, pool the samples containing the radiolabeled oligonucleotide. If the peaks are not well separated, analyze ~0.5 μL of every other fraction by thin-layer chromatography. Pool those fractions containing radiolabeled oligonucleotide that do not contain appreciable amounts of unincorporated nucleotide.

7. If the radiolabeled oligonucleotide is to be used in enzymatic reactions, proceed as follows. Otherwise, proceed to Step 8.

   i. Extract the pooled fractions with an equal volume of phenol:chloroform.

   ii. Back-extract the organic phase with 50 μL of 10 mM Tris-Cl (pH 8.0) and combine the two aqueous phases.

   iii. Extract the combined aqueous phases with an equal volume of chloroform.

   iv. Add 0.1 volume of 3 M sodium acetate (pH 5.2), mix well, and add 3 volumes of ethanol. Incubate the sample for 30 min at 0°C and then centrifuge at maximum speed for 20 min at 4°C in a microcentrifuge. Use a micropipettor equipped with a disposable tip to remove the ethanol (which should contain very little radioactivity) from the tube.

8. Add 500 μL of 80% ethanol to the tube, vortex briefly, and centrifuge the tube again at maximum speed for 5 min in a microcentrifuge.

9. Use a micropipettor equipped with a disposable tip to remove the ethanol from the tube. Stand the open tube behind a Plexiglas screen until the residual ethanol has evaporated.

10. Dissolve the precipitated oligonucleotide in 20 μL of TE (pH 7.6) and store at −20°C.

Purification of Labeled Oligonucleotides by Chromatography on a Sep-Pak C$_{18}$ Column

This protocol, which is a modification of procedures described by Lo et al. (1984), Sanchez-Pescador and Urdea (1984), and Zoller and Smith (1984), describes a method for the separation of radiolabeled oligonucleotides from unincorporated radiolabel that takes advantage of the reversible affinity of oligonucleotides for silica gel. This protocol can be used only to purify oligonucleotides carrying a 5′-phosphate group, radiolabeled or unlabeled. It should be noted that ready-to-use Sep-Pak C$_{18}$ chromatography cartridges are commercially available (e.g., Science Kit, Waters Division of Millipore).

MATERIALS

It is essential that you consult the appropriate Material Safety Data Sheets and your institution's Environmental Health and Safety Office for proper handling of equipment and hazardous materials used in this protocol.

Recipes for reagents specific to this protocol, marked <R>, are provided at the end of the protocol. See Appendix 1 for recipes for commonly used stock solutions, buffers, and reagents, marked <A>. Dilute stock solutions to the appropriate concentrations.

Reagents

Acetonitrile (5%, 30%, and 100%)
Use 10 mL HPLC-grade acetonitrile (100%) for each Sep-Pak column. Prepare the diluted solutions of acetonitrile in H$_2$O just before use.

Ammonium bicarbonate (25 mM, pH 8.0)

Ammonium bicarbonate (25 mM, pH 8.0) containing 5% (v/v) acetonitrile
Mix 5 mL of acetonitrile with 95 mL of 25 mM ammonium bicarbonate.

Radiolabeled oligonucleotide
The starting material for purification is the reaction mixture from Protocol 12 (either Step 3 or Step 5) or Protocol 13, after the enzyme has been heat or chemically inactivated.

TE (pH 7.6) <A>

Equipment

Centrifugal evaporator (Savant SpeedVac or equivalent)

Microcentrifuge tubes (1.5 mL) in a rack or fraction collector
These are used for the collection of radiolabeled oligonucleotides eluting from the column.

Sep-Pak classic columns, short body
Sep-Pak classic columns (available from the Waters Division of Millipore) contain 360 mg/column of a hydrophobic (C$_{18}$) reversed-phase chromatography resin. The separation principle makes use of the fact that the oligonucleotide adsorbs to the column when the polarity of the solvent is high (e.g., aqueous buffers) and elutes from the column when the polarity of the solvent is reduced (e.g., a mixture of methanol and H$_2$O). A separate column is required for each phosphorylation reaction.

Syringe (10-cc polypropylene)

METHOD

1. Prepare a Sep-Pak C$_{18}$ reversed-phase column as follows:

   i. Attach a polypropylene syringe containing 10 mL of acetonitrile to a Sep-Pak C$_{18}$ column.

   ii. Slowly push the acetonitrile through the Sep-Pak column.

   iii. Remove the syringe from the Sep-Pak column and then take the plunger out of the barrel. This prevents air from being pulled back into the column. Reattach the barrel to the column.

   iv. Flush out the organic solvent with two 10-mL aliquots of sterile H$_2$O. Repeat Step 1.iii after each wash.

2. Dilute the labeled oligonucleotide preparation to 1.5 mL with sterile H$_2$O and apply the entire sample to the column through the syringe.

3. Wash the Sep-Pak column with the following four solutions. Repeat Step 1.iii after each wash.

Ammonium bicarbonate (25 mM, pH 8.0)	10 mL
Ammonium bicarbonate/5% acetonitrile (25 mM)	10 mL
Acetonitrile (5%)	10 mL
Acetonitrile (5%)	10 mL

4. Elute the labeled oligonucleotide with three 1-mL aliquots of 30% acetonitrile. Collect each fraction in a separate 1.5-mL microcentrifuge tube. Repeat Step 1.iii after each elution.

   ▲ *For radiolabeled probes, because phosphorylation reactions are often performed with >100 μCi of radiolabeled ATP, the amount of radioactivity in these column fractions may be considerable. Exercise care and be diligent about disposal of unincorporated radioactivity, pipette tips, and microcentrifuge tubes.*

5. Recover the oligonucleotide by evaporating the eluate to dryness in a centrifugal evaporator (Savant SpeedVac or its equivalent).

6. Dissolve the radiolabeled oligonucleotide in a small volume (10 μL) of TE (pH 7.6).

REFERENCES

Lo K-M, Jones SS, Hackett NR, Khorana HG. 1984. Specific amino acid substitutions in bacterioopsin: Replacement of a restriction fragment in the structural gene by synthetic DNA fragments containing altered codons. *Proc Natl Acad Sci* **81:** 2285–2289.

Sanchez-Pescador R, Urdea MS. 1984. Use of unpurified synthetic deoxynucleotide primers for rapid dideoxynucleotide chain termination sequencing. *DNA* **3:** 339–343.

Zoller MJ, Smith M. 1984. Oligonucleotide-directed mutagenesis: A simple method using two oligonucleotide primers and a single-stranded DNA template. *DNA* **3:** 479–488.

Hybridization of Oligonucleotide Probes in Aqueous Solutions: Washing in Buffers Containing Quaternary Ammonium Salts

Jacobs et al. (1988) describe methods and theoretical reasons to use hybridization buffers containing quaternary alkylammonium salts. The following protocol is a simple variation of these methods. Hybridization is first performed in conventional aqueous solvents at a temperature well below the melting temperature, and the hybrids are then washed at higher stringency in buffers containing quaternary alkylammonium salts. TMACl is used with probes that are 14–50 nucleotides in length, whereas TEACl is used with oligonucleotides that are 50–200 nucleotides in length.

The graph in Figure 1 can be used to estimate a washing temperature when using oligonucleotide probes of a given length in TMACl buffers. To estimate a washing temperature when using TEACl buffers, calculate the value from the TMACl curve in the figure and then subtract 33°C.

MATERIALS

It is essential that you consult the appropriate Material Safety Data Sheets and your institution's Environmental Health and Safety Office for proper handling of equipment and hazardous materials used in this protocol.

Recipes for reagents specific to this protocol, marked <R>, are provided at the end of the protocol. See Appendix 1 for recipes for commonly used stock solutions, buffers, and reagents, marked <A>. Dilute stock solutions to the appropriate concentrations.

Reagents

Nitrocellulose or nylon filters or membranes containing the immobilized target nucleic acids of interest (e.g., Southern or northern blots, lysed bacterial colonies filters, or bacteriophage plaques)
Oligonucleotide hybridization solution <R>
Oligonucleotide prehybridization solution <R>
Radiolabeled oligonucleotide probe

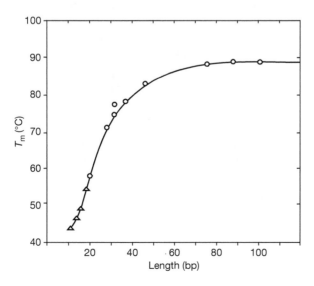

FIGURE 1. Estimating T_m in buffers containing 3.0 M TMACl. (Adapted, with permission, from Wood et al. 1985.)

1039

Prepared as described in Protocol 12 or 14. We recommend that phosphorylated probes be purified by precipitation with CPB as described in Protocol 16 before use in hybridization. This method of purification removes unincorporated [^{32}P]ATP and cuts down on the number of intense radioactive spots (pepper spots) on the autoradiographs that can be mistaken for positive hybridization signals.

SSC or SSPE (6×) <A>

Place these solutions on ice before using in Steps 4 and 5 of the protocol.

TEACl wash solution <R>

Use this wash solution for probes 50–200 nucleotides in length. It is essential to warm an aliquot of the solution to the desired temperature before use (see Steps 6 and 7).

TMACl (5 M) or TEACl (3 M)

Prepare a 5 M solution of TMACl or a 3 M solution of TEACl in H$_2$O. TMACl is used with probes that are 14–50 nucleotides in length, whereas TEACl is used with oligonucleotides that are 50–200 nucleotides in length. Both chemicals are available from Sigma-Aldrich. To the solution of TMACl or TEACl, add activated charcoal to a final concentration of ~10% and stir for 20–30 min. Allow the charcoal to settle and then filter the solution of quaternary alkylammonium salts through Whatman No. 1 paper. Sterilize the solution by passage through a nitrocellulose filter (0.45-μm pore size). Measure the refractive index of the solution and calculate the precise concentration of the quaternary alkylammonium salts from the equation C=(n − 1.331)/0.018, where C is the molar concentration of quaternary alkylammonium salts and n is the refractive index.

Store the filtered solution in dark bottles at room temperature.

TMACl wash solution <R>

Use this wash solution for probes 14–50 nucleotides in length. It is essential to warm an aliquot of the solution to the desired temperature before use (see Steps 6 and 7).

Equipment

Hybridization device (see Step 1)

Shaking incubator, water bath, or hybridization apparatus (initially preset to 37°C [Steps 1 and 3] and later adjusted to a temperature appropriate for washing [Step 7])

METHOD

1. Prehybridize the filters or membranes for 4–16 h at 37°C in oligonucleotide prehybridization solution.

   Prehybridization, hybridization, and washing of circular filters are best performed in Seal-A-Meal bags (Rival) or plastic boxes with tight-fitting lids. For Southern and northern blots, a hybridization device equipped with sealable glass tubes could be used.

2. Discard the prehybridization solution and replace it with oligonucleotide hybridization solution containing a radiolabeled oligonucleotide probe at a concentration of 180 pM.

   When hybridizing with several oligonucleotides simultaneously, each probe should be present at a concentration of 180 pM and the specific activity of the radiolabeled probe should be 5 × 10^5 to 1.5 × 10^6 cpm/pmol.

3. Incubate the filters for 12–16 h at 37°C.

4. Discard the radiolabeled hybridization solution into an appropriate disposable container. Rinse the filters three times at 4°C with ice-cold 6× SSC or 6× SSPE to remove most of the dextran sulfate.

5. Wash the filters twice for 30 min at 4°C in ice-cold 6× SSC or 6× SSPE.

6. Rinse the filters at 37°C in two changes of the TMACl or TEACl wash solution.

   The aim of this step is to replace the SSPE and SSC with the solution of quaternary alkylammonium salts. Unless this step is diligently performed, the full benefits of using TEACl or TMACl will not be realized.

7. Wash the filters twice in TMACl or TEACl wash solution for 20 min each at a temperature that is 2°C–4°C below the T_m indicated in Figure 1.

> *Note that the T$_m$ of a hybrid is 33°C lower in a buffer containing TEACl than in a buffer containing TMACl. Make sure that the buffers are prewarmed to the desired temperature and that fluctuations in temperature are less than ±1°C.*

8. Remove the filters from the washing solution. Blot them dry at room temperature and autoradiograph them as described in Chapter 2, Protocol 7.

RECIPES

It is essential that you consult the appropriate Material Safety Data Sheets and your institution's Environmental Health and Safety Office for proper handling of equipment and hazardous materials used in this protocol.

Oligonucleotide Hybridization Solution

Reagent	Quantity (for 1 L)	Final concentration
SSC or SSPE (20×)	300 mL	6×
Sodium phosphate (0.1 M, pH 6.8)	500 mL	0.05 M
EDTA (0.5 M, pH 8.0)	2 mL	1 mM
Denhardt's solution (50×) <A>	100 mL	5×
Denatured, fragmented salmon sperm DNA (10 mg/mL)	10 mL	100 μg/mL
Dextran sulfate	100 g	100 mg/mL

Denatured, fragmented salmon sperm DNA can be purchased from Pharmacia or prepared as described in Appendix 1.

Oligonucleotide Prehybridization Solution

Reagent	Quantity (for 1 L)	Final concentration
SSC or SSPE (20×)	300 mL	6×
Sodium phosphate (0.1 M, pH 6.8)	500 mL	0.05 M
EDTA (0.5 M, pH 8.0)	2 mL	1 mM
Denhardt's solution (50×) <A>	100 mL	5×
Denatured, fragmented salmon sperm DNA (10 mg/mL)	10 mL	100 μg/mL

TEACl Wash Solution

Reagent	Quantity (for 1 L)	Final concentration
TEACl (5 M)	480 mL	2.4 M
Tris-Cl (1 M, pH 8.0)	50 mL	50 mM
EDTA (0.5 M, pH 7.6)	400 μL	0.2 mM
SDS	1 g	1 mg/mL

TMACl Wash Solution

Reagent	Quantity (for 1 L)	Final concentration
TMACl (6 M)	500 mL	3 M
Tris-Cl (1 M, pH 8.0)	50 mL	50 mM
EDTA (0.5 M, pH 7.6)	400 μL	0.2 mM
SDS	1 g	1 mg/mL

REFERENCES

Jacobs KA, Rudersdorf R, Neill SD, Dougherty JP, Brown EL, Fritsch EF. 1988. The thermal stability of oligonucleotide duplexes is sequence independent in tetraalkylammonium salt solutions: Application to identifying recombinant DNA clones. *Nucleic Acids Res* 16: 4637–4650.

Wood WI, Gitschier J, Lasky LA, Lawn RM. 1985. Base composition-independent hybridization in tetramethylammonium chloride: A method for oligonucleotide screening of highly complex gene libraries. *Proc Natl Acad Sci* 82: 1585–1588.

PREPARATION OF STOCK SOLUTIONS OF dNTPs

pH-adjusted solutions of dNTPs are available from many commercial manufacturers; however, it is also possible to prepare dNTP stocks from powdered sodium salts. Stock solutions usually contain a single dNTP at a concentration of 10 mM or 20 mM, according to the needs of the investigator. These stocks are stored at −20°C in small aliquots, and they may be diluted to generate, for example, solutions that contain all four dNTPs at a concentration of 5 mM.

A microbalance is used to weigh out the required amounts of dNTP into sterile microcentrifuge tubes. Either use a disposable spatula or clean the spatula well with ethanol between each weighing when making up solutions of different dNTPs. The table below shows the amount of solid, anhydrous deoxynucleotide required to make 1 mL of a 20 mM stock solution.

Deoxynucleotide	F.W.	Amount (in milligrams) required to make 1 mL of a 20 mM solution
dATP	491.2	9.82
dCTP	467.2	9.34
dTTP	482.2	9.64
dGTP	507.2	10.14

Dissolve the deoxynucleotides in a small volume of H_2O and then use pH paper and an automatic pipetting device to adjust the pH to ~8.0 by adding small amounts of 2 N NaOH until the pH reaches 8.0. Alternatively, the stock solutions may be generated from monosodium salts of the deoxynucleotide triphosphates by dissolving an appropriate amount of the solid in 1.0 mL of H_2O. Mix the contents of the tube well and store the stock solution in small aliquots at −20°C.

E. COLI DNA POLYMERASE I AND THE KLENOW FRAGMENT

DNA Pol I (Kornberg et al. 1956) consists of a single polypeptide chain (M_r ~103,000; Joyce et al. 1982) encoded by the *E. coli polA* gene (De Lucia and Cairns 1969). In addition to a phosphate-exchange activity, Pol I carries out three enzymatic reactions that are performed by three distinct functional domains (see Table 1).

During DNA replication in *E. coli*, the enzymatic activities of Pol I act in a coordinated fashion to remove the RNA primers from the 5′ termini of nascent DNA and to fill gaps between adjacent tracts of DNA. For detailed descriptions of these processes, see Kornberg and Baker (1992). Pol I can be cleaved by mild treatment with subtilisin into two fragments; the larger fragment (comprising residues 326–928) is known as the Klenow fragment. The DNA polymerase and the 3′→5′ exonuclease activities of Pol I are carried on the Klenow fragment (Brutlag et al. 1969; Klenow and Henningsen 1970; Klenow and Overgaard-Hansen 1970; Klenow et al. 1971), whereas the 5′→3′ exonuclease activity of the holoenzyme is carried on the smaller amino-terminal fragment (residues 1–325), which is nameless (for review, see Joyce and Steitz 1987; also see Table 1).

The Pol I gene has been sequenced (Joyce et al. 1982) and expressed from prokaryotic expression vectors (Murray and Kelley 1979), making it possible to purify large amounts of the protein for commercial purposes (Murray and Kelley 1979). The segment of the Pol I gene encoding the Klenow fragment has also been cloned into various expression vectors (e.g., see Joyce and Grindley 1983; Pandey et al. 1993), thereby providing pure protein in amounts large enough both for commercial purposes and for biophysical and biochemical studies.

Knowledge of the three-dimensional structure of the Klenow fragment, which was obtained in 1985 by X-ray diffraction (Ollis et al. 1985), stimulated a series of elegant structural investigations (Beese and Steitz 1991; Beese et al. 1993a,b), kinetic studies (Kuchta et al. 1987, 1988; Cowart et al. 1989; Catalano et al. 1990; Guest et al. 1991), and mutational analyses (Freemont et al. 1986; Derbyshire et al. 1988, 1991; Polesky et al. 1990, 1992). The results of these various approaches are remarkably consonant, and they confirm that the active sites for polymerization of dNTPs and for 3′→5′ exonucleolytic digestion reside on different structural domains of the Klenow fragment and are separated by 30–35 Å.

The exonuclease activity, which is carried on the smaller (22 kDa) domain (Joyce and Steitz 1987), performs a proofreading function by resecting DNA that contains either mismatched bases or frameshift errors (Bebenek et al. 1990). 3′→5′ digestion of dsDNA is thought to follow a three-step pathway: binding of the substrate to a cleft in the polymerase domain, translocation of the 3′ terminus to the active site of the exonuclease domain, and, finally, chemical catalysis (Catalano et al. 1990). During translocation, which is the rate-limiting step of the reaction, melting of the 4 or 5 terminal base pairs of duplex DNA (Cowart et al. 1989) generates a frayed 3′ terminus that binds to the active site of the exonuclease domain, with K_m values in the nanomolar range (Kuchta et al. 1988). Three carboxylic acid residues in the active site interact directly with one or both of the divalent metal ions that are bound to the exonuclease domain and are crucial for enzymatic cleavage. The metals bound to the exonuclease domain interact with the terminal phosphate residue of the substrate (Derbyshire et al. 1991; Han et al. 1991), thereby

TABLE 1. The domains of *E. coli* DNA polymerase I

Domain	Activity	Biochemical function
Carboxy-terminal domain (residues 543–928; ~46 kDa)	5′→3′ DNA polymerase	Addition of mononucleotide residues generated from dNTPs to the 3′-hydroxyl termini of RNA or DNA primers. These termini can be provided by nicks or gaps in a duplex DNA as well as by short segments of RNA or DNA that are base-paired to an ssDNA molecule.
Central domain (residues 326–542; ~22 kDa)	3′→5′ exonuclease	Cleavage of nucleotide residues from 3′-hydroxyl termini creates recessed 3′ termini.
Amino-terminal domain (residues 1–325)	5′→3′ exonuclease	Cleavage of oligonucleotides from base-paired 5′ termini.

The carboxy-terminal and central regions constitute the Klenow fragment of Pol I.

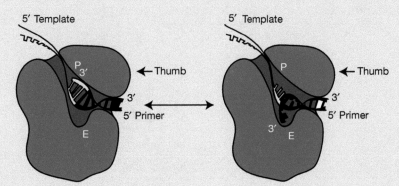

FIGURE 1. A model of DNA bound to the Klenow DNA polymerase. (Adapted from Beese et al. 1993a; reprinted with permission from AAAS.)

allowing nucleophilic attack (probably from a hydroxide ion) on the exposed phosphodiester bond (Freemont et al. 1988). This reaction generates a product, dNMP, that at high concentrations can inhibit the $3' \rightarrow 5'$ exonuclease reaction (Que et al. 1978) by occupying the same site in the enzyme as the frayed terminus of ssDNA (Ollis et al. 1985). The exonucleolytic degradation of double-stranded substrates is very slow, with a k_{cat} of $\sim 10^{-3}$ sec^{-1}. Digestion of single-stranded substrates, however, is ~ 100 times faster, with a k_{cat} of 0.09 sec^{-1} (Kuchta et al. 1988; Derbyshire et al. 1991). Using site-directed mutagenesis to change the amino acids within the active site, mutant forms of the Klenow enzyme have been constructed that retain normal polymerase activity but are essentially devoid of exonuclease activity (Derbyshire et al. 1988, 1991). The average error rate for these enyzmes is ~ 1 base substitution for each 10,000–40,000 bases polymerized. This rate is between 7- and 30-fold times higher than the error rate of the wild-type Klenow enzyme (Bebenek et al. 1990; Eger et al. 1991). Therefore, the high fidelity of the Klenow enzyme results more from highly accurate selection of bases by the polymerase domain than from postreplication editing by the exonuclease domain.

A large cleft in the polymerase domain is the binding site for the primer:template DNA (Beese et al. 1993a). The amino acid residues directly involved in substrate binding and catalysis have been identified by site-directed mutagenesis (Polesky et al. 1990, 1992) and confirmed by X-ray crystallographic analysis of the Klenow fragment bound to duplex DNA (Beese et al. 1993a). The polymerase domains of Pol I and several other type-B DNA polymerases, including those from *Thermus aquaticus* and bacteriophages T5 and T7, share a high degree of homology within the polymerase domain (Delarue et al. 1990; Blanco et al. 1991). This homology includes (1) a cluster of residues directly involved in catalysis and (2) a tyrosine residue involved in binding of dNTPs (Beese et al. 1993b). The current model of the orientation of the primer:template and the two domains of the Klenow fragment is shown in Figure 1.

Uses of the Klenow Fragment of E. coli *Polymerase I in Molecular Cloning*

Filling the Recessed 3' Termini Created by Digestion of DNA with Restriction Enzymes

In many cases, a single buffer can be used both for cleavage of DNA with a restriction enzyme and for the subsequent filling of recessed 3' termini. The end-filling reaction can be controlled by omitting one, two, or three of the four dNTPs from the reaction, thereby generating partially filled termini that contain novel cohesive ends (see Fig. 2 and Protocol 8).

Labeling the 3' Termini of DNA Fragments by Incorporation of Radiolabeled dNTPs

In general, these labeling reactions contain three unlabeled dNTPs each at a concentration in excess of the K_m and one radiolabeled dNTP at a far lower concentration that is usually below the K_m. Under these conditions, the proportion of label that is incorporated into DNA can be very high, even though the rate of the reaction may be far from maximal. The presence of high concentrations of three unlabeled dNTPs lessens the possibility of exonucleolytic removal of nucleotides from the 3' terminus of the template. Which of the four α-labeled dNTPs that is added to the reaction depends on the sequence and nature of the termini of the DNA.

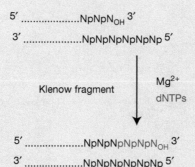

5′ ....NpNpN$_{OH}$ 3′

3′ ....NpNpNpNpNpNp 5′

Klenow fragment   Mg^{2+}

dNTPs

5′ ....NpNpNpNpNpN$_{OH}$ 3′

3′ ....NpNpNpNpNpNp 5′

FIGURE 2. Filling in recessed 3′ termini of DNA fragments using Klenow DNA polymerase.

- Recessed 3′ termini can be labeled with any dNTP whose base is complementary to an unpaired base in the protruding 5′ terminus. Radioactivity can therefore be incorporated at any position within the rebuilt terminus depending on the investigator's choice of radiolabeled dNTP. To ensure that all of the radiolabeled molecules are the same length, it may be necessary to complete the end-filling reaction by performing a "chase" reaction containing high concentrations of all four unlabeled dNTPs.

- Blunt-ended and protruding 3′ termini may be labeled in an enzymatic reaction that uses both domains of the Klenow fragment. First, the 3′→5′ exonuclease activity removes any protruding tails from the DNA and creates a recessed 3′ terminus. Then, in the presence of high concentrations of one radiolabeled precursor, exonucleolytic degradation is balanced by incorporation of radiolabeled dNTP at the 3′ terminus. This reaction, which consists of cycles of removal and replacement of the 3′ terminus from recessed or blunt-ended DNA, is sometimes called an exchange or replacement reaction. The specific activities that can be achieved with this reaction are modest because the 3′→5′ exonuclease of the Klenow enzyme is rather sluggish, especially on double-stranded substrates. T4 DNA polymerase carries a more potent 3′→5′ exonuclease that is ~200-fold more active than the Klenow fragment and is the enzyme of choice for this type of reaction.

If a [^{35}S]dNTP is used instead of the conventional [α-^{32}P]dNTP, the reaction is limited to one cycle of removal and replacement because the 3′→5′ exonuclease of *E. coli* DNA Pol I, unlike the exonuclease of T4 DNA polymerase, cannot attack thioester bonds (Kunkel et al. 1981; Gupta et al. 1984).

Labeling Single-Stranded DNA by Random Priming

For details, see Feinberg and Vogelstein (1983, 1984) and Protocols 1 and 2.

Production of Single-Stranded Probes by Primer Extension

For details, see Meinkoth and Wahl (1984), Studencki and Wallace (1984), and Chapter 6, Protocol 18. For many years, the Klenow fragment of Pol I was the highest-quality DNA polymerase that was commercially available and, in consequence, was the enzyme of first and last resort for in vitro synthesis of DNA. However, as polymerases that are better suited to various synthetic tasks have been discovered or engineered, Klenow has been gradually replaced and is no longer the enzyme of first choice for a variety of procedures in molecular cloning. These procedures include the following.

- *DNA sequencing by the Sanger method.* Klenow has been replaced by bacteriophage or thermostable polymerases that give longer read lengths.

- *Synthesis of dsDNA from single-stranded templates during in vitro mutagenesis.* Although the Klenow fragment is still widely used for in vitro synthesis of circular DNAs using mutagenic primers, it is not always the best enzyme for this purpose. Unless large quantities of ligase are present in the polymerization/extension reaction mixture, the Klenow enzyme can displace the mutagenic oligonucleotide primer from the template strand, thereby reducing the number of mutants obtained. This problem can be solved by using DNA polymerases that are unable to perform strand displacement, including bacteriophage T4 DNA polymerase (Nossal 1974; Lechner et al. 1983; Geisselsoder et al. 1987), bacteriophage T7 DNA polymerase (Bebenek and Kunkel 1989), and Sequenase (Schena 1989).

Bacteriophage T4 gene 32 protein (a ssDNA-binding protein) can be used in primer-extension reactions catalyzed by DNA polymerases (including the Klenow enzyme) to alleviate stalling problems caused by templates rich in secondary structure (Craik et al. 1985; Kunkel et al. 1987).

- *Polymerase chain reactions.* The Klenow fragment was the enzyme used in the first PCRs (Saiki et al. 1985). However, it has now been completely replaced by thermostable DNA polymerases that need not be replenished after each round of synthesis and denaturation.

Facts and Figures about the Klenow Fragment

The standard assay used to measure the polymerase activity of the Klenow fragment is that of Setlow (1974) with poly(d[A-T]) as template. One unit of polymerizing activity is the amount of enzyme that catalyzes the incorporation of 3.3 nmol of dNTP into acid-insoluble material in 10 min at 37°C. A sample of pure Klenow fragment has a specific activity of ~10,000 U/mg protein (Derbyshire et al. 1993). The reaction is usually performed in the presence of Mg^{2+}. Substitution of Mn^{2+} for Mg^{2+} increases the rate of misincorporation and decreases the accuracy of proofreading (Carroll and Benkovic 1990). The K_m of the enzyme for the four dNTPs varies between 4 and 20 μM. For methods to assay the exonuclease activity of the Klenow fragment, see Freemont et al. (1986) and Derbyshire et al. (1988).

Uses of Pol I in Molecular Cloning

- *Labeling of DNA by nick translation* (Maniatis et al. 1975; Rigby et al. 1977). In this reaction, the enzyme binds to a nick or short gap in duplex DNA. The $5' \rightarrow 3'$ exonuclease activity of Pol I then removes nucleotides from one strand of the DNA, creating a template for simultaneous synthesis of the growing strand of DNA. The original nick is therefore translated along the DNA molecule by the combined action of the $5' \rightarrow 3'$ exonuclease and the $5' \rightarrow 3'$ polymerase. In Figure 3, the nick in the upper strand of duplex DNA is translated from left to right and the patch of newly synthesized DNA is represented by the shaded arrow. Nick-translation reactions are usually carried out at 16°C to reduce the synthesis of "snap-back" DNA that is produced when the 3'-hydroxyl terminus of a growing strand of DNA loops back on itself and primes synthesis of hairpin-shaped molecules of DNA (Richardson et al. 1964a).

- *Replacement synthesis of second-strand cDNA* (e.g., see Gubler and Hoffman 1983). In this method, the product of first-strand synthesis—a cDNA–mRNA hybrid—is used as a template for a nick-translation reaction. RNase H is used to generate nicks and gaps in the mRNA strand of the hybrid, creating a series of RNA primers that are used by Pol I to initiate the synthesis of second-strand cDNA.

- *End-labeling of DNA molecules with protruding 3' tails.* This reaction works in two stages: First, the $3' \rightarrow 5'$ exonuclease activity of the holoenzyme removes protruding 3' tails from the cDNA and creates a recessed 3' terminus. Then, in the presence of high concentrations of one radiolabeled precursor,

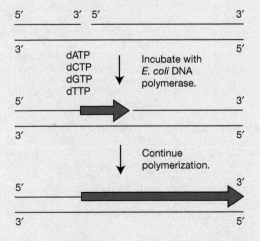

FIGURE 3. **Nick-translation schematic.**

continuing exonucleolytic degradation is balanced by incorporation of dNTPs at the 3′ terminus. Although Pol I catalyzes this exchange or replacement reaction fairly efficiently, bacteriophage T4 DNA polymerase remains the enzyme of choice because of its more potent 3′→5′ exonuclease activity.

Facts and Figures About DNA Polymerase I

- Most of the Pol I distributed by commercial manufacturers is isolated (Kelley and Stump 1979) from a strain of *E. coli* that is lysogenic for a transducing bacteriophage λ carrying a copy of the *polA* gene (e.g., NM 964) (Murray and Kelley 1979).

- One unit of DNA polymerase is the amount of enzyme required to catalyze the conversion of 10 nmol of total dNTPs to an acid-insoluble form in 30 min at 37°C using poly(d[A-T]) as the template primer (Richardson et al. 1964b). The specific activity of the commercial enzyme is usually ~5000 U/mg of protein.

- Like all other DNA polymerases, Pol I requires divalent cations for activity. Mg^{2+} is preferred for accurate replication, whereas Mn^{2+} may be used to deliberately increase the frequency of errors.

REFERENCES

Bebenek K, Kunkel TA. 1989. The use of native T7 DNA polymerase for site-directed mutagenesis. *Nucleic Acids Res* 17: 5408.

Bebenek K, Joyce CM, Fitzgerald MP, Kunkel TA. 1990. The fidelity of DNA synthesis catalyzed by derivatives of *Escherichia coli* DNA polymerase I. *J Biol Chem* 265: 13878–13887.

Beese LS, Steitz TA. 1991. Structural basis for the 3′-5′ exonuclease activity of *Escherichia coli* DNA polymerase I: A two metal ion mechanism. *EMBO J* 10: 25–33.

Beese LS, Derbyshire VC, Steitz TA. 1993a. Structure of DNA polymerase I Klenow fragment bound to duplex DNA. *Science* 260: 352–355.

Beese LS, Friedman JM, Steitz TA. 1993b. Crystal structures of the Klenow fragment of DNA polymerase I complexed with deoxynucleoside triphosphate and pyrophosphate. *Biochemistry* 32: 14095–14101.

Blanco L, Bernad A, Blasco MA, Salas M. 1991. A general structure for DNA-dependent DNA polymerases *Gene* 100: 27–38. (Erratum *Gene* [1991] 108: 165.)

Brutlag D, Atkinson MR, Setlow P, Kornberg A. 1969. An active fragment of DNA polymerase produced by proteolytic cleavage. *Biochem Biophys Res Commun* 37: 982–989.

Carroll SS, Benkovic SJ. 1990. Mechanistic aspects of DNA-polymerases: *Escherichia coli* DNA polymerase I (Klenow fragment) as a paradigm. *Chem Rev* 90: 1291–1307.

Catalano CE, Allen DJ, Benkovic SJ. 1990. Interaction of *Escherichia coli* DNA polymerase I with azido DNA and fluorescent DNA probes: Identification of protein-DNA. *Biochemistry* 29: 3612–3621.

Cowart M, Gibson KJ, Allen DJ, Benkovic SJ. 1989. DNA substrate structural requirements for the exonuclease and polymerase activities of procaryotic and phage DNA polymerases. *Biochemistry* 28: 1975–1983.

Craik CS, Largman C, Fletcher T, Roczniak S, Barr PJ, Fletterick R, Rutter WJ. 1985. Redesigning trypsin: Alteration of substrate specificity. *Science* 228: 291–297.

Delarue M, Poch O, Tordo N, Moras D, Argos P. 1990. An attempt to unify the structure of polymerases. *Protein Eng* 3: 461–467.

De Lucia P, Cairns J. 1969. Isolation of an *E. coli* strain with a mutation affecting DNA polymerase. *Nature* 224: 1164–1166.

Derbyshire V, Freemont PS, Sanderson MR, Beese L, Friedman JM, Joyce CM, Steitz TA. 1988. Genetic and crystallographic studies of the 3′,5′-exonucleolytic site of DNA polymerase I. *Science* 240: 199–201.

Derbyshire V, Grindley NDF, Joyce CM. 1991. The 3′-5′ exonuclease of DNA polymerase I of *Escherichia coli*: Contribution of each amino acid at the active site to the reaction. *EMBO J* 10: 17–24.

Derbyshire V, Astatke M, Joyce CM. 1993. Re-engineering the polymerase domain of Klenow fragment and evaluation of overproduction and purification strategies. *Nucleic Acids Res* 23: 5439–5448.

Eger BT, Kuchta RD, Carroll SS, Benkovic PA, Dahlberg ME, Joyce CM, Benkovic SJ. 1991. Mechanism of DNA replication fidelity for three mutants of DNA polymerase I: Klenow fragment KF(exo⁺), KF(polA5) and KF(exo⁻). *Biochemistry* 30: 1441–1448.

Feinberg AP, Vogelstein B. 1983. A technique for radiolabeling DNA restriction endonuclease fragments to high specific activity. *Anal Biochem* 132: 6–13.

Feinberg AP, Vogelstein B. 1984. A technique for radiolabeling DNA restriction endonuclease fragments to high specific activity. Addendum. *Anal Biochem* 137: 266–267.

Freemont PS, Ollis DL, Steitz TA, Joyce CM. 1986. A domain of the Klenow fragment of *Escherichia coli* DNA polymerase I has polymerase but no exonuclease activity. *Proteins* 1: 66–73.

Freemont PS, Friedman JM, Beese LS, Sanderson MR, Steitz TA. 1988. Cocrystal structure of an editing complex of Klenow fragment with DNA. *Proc Natl Acad Sci* 85: 8924–8928.

Geisselsoder J, Witney F, Yuckenberg P. 1987. Efficient site-directed in vitro mutagenesis. *BioTechniques* 5: 786–791.

Gubler U, Hoffman BJ. 1983. A simple and very efficient method for generating cDNA libraries. *Gene* 25: 263–269.

Guest CR, Hochstrasser RA, Dupuy CG, Allen DJ, Benkovic SJ, Millar DP. 1991. Interaction of DNA with the Klenow fragment of DNA polymerase I studied by time-resolved fluorescence spectroscopy. *Biochemistry* 30: 8759–8770.

Gupta AP, Benkovic PA, Benkovic SJ. 1984. The effect of the 3′,5′ thiophosphoryl linkage on the exonuclease activities of T4 polymerase and the Klenow fragment. *Nucleic Acids Res* 12: 5897–5911.

Han H, Rifkind JM, Mildvan AS. 1991. Role of divalent cations in the 3′,5′-exonuclease reaction of DNA polymerase I. *Biochemistry* 30: 11104–11108.

Joyce CM, Grindley NDF. 1983. Construction of a plasmid that overproduces the large proteolytic fragment (Klenow fragment) of DNA polymerase I of *Escherichia coli*. *Proc Natl Acad Sci* 80: 1830–1834.

Joyce CM, Steitz TA. 1987. DNA polymerase I: From crystal structure to function via genetics. *Trends Biochem Sci* 12: 288–292.

Joyce CM, Kelley WS, Grindley NDF. 1982. Nucleotide sequence of the *Escherichia coli polA* gene and primary structure of DNA polymerase I. *J Biol Chem* 257: 1958–1964.

Kelley WS, Stump KH. 1979. A rapid procedure for isolation of large quantities of *Escherichia coli* DNA polymerase utilizing a λ*polA* transducing phage. *J Biol Chem* 254: 3206–3210.

Klenow H, Henningsen I. 1970. Selective elimination of the exonuclease activity of the deoxyribonucleic acid polymerase from *Escherichia coli* B by limited proteolytic digestion. *Proc Natl Acad Sci* 65: 168–175.

Klenow H, Overgaard-Hansen K. 1970. Proteolytic cleavage of DNA polymerase from *Escherichia coli* B into an exonuclease unit and a polymerase unit. *FEBS Lett* **6**: 25–27.

Klenow H, Overgaard-Hansen K, Patkar SA. 1971. Proteolytic cleavage of native DNA polymerase into two different catalytic fragments. Influence of assay conditions on the change of exonuclease activity and polymerase-activity accompanying cleavage. *Eur J Biochem* **22**: 371–381.

Kornberg A, Baker TA. 1992. *DNA replication*, 2nd ed. W.H. Freeman, New York.

Kornberg A, Lehman IR, Bessman MJ, Simms ES. 1956. Enzymic synthesis of deoxyribonucleic acid. *Biochim Biophys Acta* **21**: 197–198.

Kuchta RD, Benkovic P, Benkovic SJ. 1988. Kinetic mechanism whereby DNA polymerase I (Klenow) replicates DNA with high fidelity. *Biochemistry* **27**: 6716–6725.

Kuchta RD, Mizrahi V, Benkovic PA, Johnson KA, Benkovic SJ. 1987. Kinetic mechanism of DNA polymerase I (Klenow). *Biochemistry* **26**: 8410–8417.

Kunkel TA, Eckstein F, Mildvan AS, Koplitz RM, Loeb LA. 1981. Deoxynucleoside [1-thio]triphosphates prevent proofreading during in vitro DNA synthesis. *Proc Natl Acad Sci* **78**: 6734–6738.

Kunkel TA, Roberts JD, Zakour RA. 1987. Rapid and efficient site-specific mutagenesis without phenotypic selection. *Methods Enzymol* **154**: 367–382.

Lechner RL, Engler MJ, Richardson CC. 1983. Characterization of strand displacement synthesis catalyzed by bacteriophage T7 DNA polymerase. *J Biol Chem* **258**: 11174–11184.

Maniatis T, Jeffrey A, Kleid DG. 1975. Nucleotide sequence of the rightward operator of phage λ. *Proc Natl Acad Sci* **72**: 1184–1188.

Meinkoth J, Wahl G. 1984. Hybridization of nucleic acids immobilized on solid supports. *Anal Biochem* **138**: 267–284.

Murray NE, Kelley WS. 1979. Characterization of λ*polA* transducing phages; effective expression of the *E coli polA* gene. *Mol Gen Genet* **175**: 77–87.

Nossal NG. 1974. DNA synthesis on a double-stranded DNA template by the T4 bacteriophage DNA polymerase and the T4 gene 32 DNA unwinding protein. *J Biol Chem* **249**: 5668–5676.

Ollis DL, Brick P, Hamlin R, Xuong NG, Steitz TA. 1985. Structure of the large fragment of *Escherichia coli* DNA polymerase complexed with dTMP. *Nature* **313**: 762–766.

Pandey VN, Kaushok N, Sanzgiri RP, Patil M, Modak M, Barik S. 1993. Site-directed mutagenesis of DNA polymerase I (Klenow) from *Escherichia coli*. *Eur J Biochem* **214**: 59–65.

Polesky AH, Dahlberg ME, Benkovic SJ, Grindley NDF, Joyce CM. 1992. Side chains involved in catalysis of the polymerase reaction of DNA polymerase I from *Escherichia coli*. *J Biol Chem* **267**: 8417–8428.

Polesky AH, Steitz TA, Grindley NDF, Joyce CM. 1990. Identification of residues critical for the polymerase activity of the Klenow fragment of DNA polymerase I from *Escherichia coli*. *J Biol Chem* **265**: 14579–14591.

Que BG, Downey KM, So A. 1978. Mechanisms of selective inhibition of 3′ to 5′ exonuclease activity of *Escherichia coli* DNA polymerase I by nucleoside 5′-monophosphates. *Biochemistry* **17**: 1603–1606.

Richardson CC, Inman RB, Kornberg A. 1964a. Enzymic synthesis of deoxyribonucleic acid. XVIII. The repair of partially single-stranded DNA templates by DNA polymerase. *J Mol Biol* **9**: 46–69.

Richardson CC, Schildkraut CL, Aposhian HV, Kornberg A. 1964b. Enzymatic synthesis of deoxyribonucleic acid. XIV. Further purification and properties of deoxyribonucleic acid polymerase of *Escherichia coli*. *J Biol Chem* **239**: 222–232.

Rigby PW, Dieckmann M, Rhodes C, Berg P. 1977. Labeling deoxyribonucleic acid to high specific activity in vitro by nick translation with DNA polymerase I. *J Mol Biol* **113**: 237–251.

Saiki RK, Scharf S, Faloona F, Mullis KB, Horn GT, Erlich HA, Arnheim N. 1985. Enzymatic amplification of α-globin genomic sequences and restriction site analysis for diagnosis of sickle cell anemia. *Science* **230**: 1350–1354.

Schena M. 1989. High efficiency oligonucleotide mutagenesis using Sequenase. Comments (U.S. Biochemical Corp.) **15**: 23.

Setlow P. 1974. DNA polymerase I from *Escherichia coli*. *Methods Enzymol* **29**: 3–12.

Studencki AB, Wallace RB. 1984. Allele-specific hybridization using oligonucleotide probes of very high specific activity: Discrimination of the human β A- and β S-globin genes. *DNA* **3**: 7–15.

IN VITRO TRANSCRIPTION SYSTEMS

Virtually all in vitro transcription of DNA into RNA is performed with bacteriophage-encoded DNA-dependent RNA polymerases. The value of using bacteriophage-encoded DNA-dependent RNA polymerases arises because (1) these enzymes initiate efficient and selective transcription from well-defined cognate promoters (Melton et al. 1984) and (2) being composed of a single polypeptide chain ($\sim$900 amino acids), these enzymes need no auxiliary transcription factors, perhaps because they have evolved to transcribe a small number of genes in a specialized genome with high efficiency.

The best-characterized bacteriophage RNA polymerases are those encoded by the *Salmonella typhimurium* bacteriophage SP6 (Butler and Chamberlin 1982; Green et al. 1983) and the *E. coli* phages T3 and T7 (Studier and Rosenberg 1981; Davanloo et al. 1984; Tabor and Richardson 1985). The genes encoding these three RNA polymerases have been isolated (Davanloo et al. 1984; Morris et al. 1986), sequenced (Moffatt et al. 1984; McGraw et al. 1985: Kotani et al. 1987), and expressed (Davanloo et al. 1984; Morris et al. 1986). In the case of the T7 polymerase, the crystal structure of the enzyme has been elucidated (Doublie et al. 1998). The SP6, T3, and T7 DNA-dependent RNA polymerases behave in a similar fashion, and there are no distinct biochemical advantages to using one RNA polymerase over the other. The SP6 enzyme is typically four to five times more expensive than the T7- and T3-encoded enzymes, even though it is the easiest of the three to prepare from bacteriophage-infected cells.

Although all three RNA polymerases have the ability to transcribe ssDNA (Salvo et al. 1973; Milligan et al. 1987), virtually all in vitro transcription is performed with double-stranded linear DNA templates that contain an appropriate promoter (see Table 1).

Because the minimal promoter for bacteriophage RNA polymerases is just 21 bp in length (Jorgensen et al. 1991), portable bacteriophage promoters can easily be manufactured from synthetic oligonucleotides. dsDNA linkers containing a bacteriophage RNA polymerase promoter can also be ligated directly to purified DNA fragments (Loewy et al. 1989) or PCR products. By synthesizing the appropriate linker/adaptor, one fragment out of a mixture can be modified and subsequently transcribed in vitro.

Synthetic promoters can also be added to the 5′ ends of primers used to amplify DNA by PCR (see the Additional Protocol Using PCR to Add Promoters for Bacteriophage-Encoded RNA Polymerases to Fragments of DNA following Protocol 5). Essentially any DNA molecule can therefore be amplified with a bacteriophage promoter at one or both ends of the molecule. The amplified DNA is an efficient template for in vitro transcription reactions. PCR can also be used to introduce convenient restriction sites at the end of the DNA to facilitate linearization of templates and to introduce translation signals at the 5′ end of a DNA fragment to allow the RNA product to be translated efficiently in cell-free protein-synthesizing systems (Browning 1989; Kain et al. 1991).

TABLE 1. Promoter sequences recognized by bacteriophage-encoded RNA polymerases

Bacteriophage	Promoter				
	−15	−10	−5	+1	+5
	|	|	|	|	|
T7	TAATACGACTCACTATAGGGAG A				
T3	AATTAACCCTCACTAAAGGGAGA				
	T				
SP6	ATTTAGG_G GACACTATAGAAG				

The consensus sequences of promoters are recognized by three bacteriophage-encoded RNA polymerases: T7 (Dunn and Studier 1983), T3 (Beck et al. 1989), and SP6 (Brown et al. 1986). All of the bacteriophage promoters share a core sequence that extends from −7 to +1, suggesting that this region has a common role in promoter function. The promoters diverge in the region from −8 to −12, suggesting that promoter-specific contacts are made in this region. By convention, the sequence of the nontemplate strand is shown. (Adapted, with permission, from Jorgensen et al. 1991.)

The affinities of the bacteriophage polymerases for their promoters are rather low ($\sim 10^{-7}$ M^{-1}), and nucleoside triphosphates are required to stabilize transient promoter:enzyme complexes. After a short lag period, RNA synthesis rapidly reaches a rate (for T7 and T3 polymerases) of 200–300 nucleotides/sec at 37°C, almost 10 times faster than that of *E. coli* RNA polymerase, measured under the same conditions. The K_m values of the T3 and T7 RNA polymerases for ATP, UTP, and CTP are between 40 and 100 μM (Oakley et al. 1979). GTP is anomalous, probably because it is used as a chain-initiating nucleotide: RNA chains initiated by T3 RNA polymerase, for example, start with pppGGGA and pppGGGG (for review, see Chamberlin and Ryan 1982). When the concentration of template DNA is 20 nM and the concentration of each of the rNTPs is >50 μM, the rate of RNA synthesis is linear for at least 1 h and is proportional to the amount of enzyme added to the reaction. During the course of the reaction, RNA chains are initiated many times on each molecule of template, and, under optimal conditions, between 10 and 20 mol of full-length transcript are generated per mol of template. However, when a radiolabeled or modified base is used as a precursor, the yield is much lower because reaction conditions are modified to optimize incorporation of the rare component. This is generally done by eliminating the homologous nucleotide from the reaction or by drastically lowering its concentration. However, if the concentration of nucleotide in the reaction drops below the K_m, not only the total yield but also the proportion of full-length RNA chains will drop markedly.

Although all three bacteriophage polymerases share many biochemical properties, each displays considerable preference for its own promoter and does not initiate transcription at other promoters at a substantial rate. The bacteriophage enzymes do not recognize bacterial or plasmid promoters or eukaryotic promoters in cloned DNA sequences. Because efficient signals for termination of RNA chains are also rare, bacteriophage-encoded polymerases are able to synthesize full-length transcripts of almost any DNA that is placed under the control of an appropriate promoter. Because these transcripts are complementary to only one strand of the template, they are excellent for use as probes in virtually any technique involving hybridization, including Southern and northern hybridization, in situ hybridization, and RNase protection assays. In addition, the ability to generate large quantities of long RNAs that are identical in sequence to unstable primary transcripts of mammalian genes has led to the development of in vitro assays for splicing and processing of 3′ termini of eukaryotic mRNAs (e.g., see Green 1991). Synthetic RNAs are also ligands for binding by regulatory proteins such as the tat gene product of HIV (Roy et al. 1990).

Full-length transcripts prepared in vitro are often used as mRNAs in eukaryotic cell-free protein-synthesizing systems. These and other reactions (e.g., in vitro splicing reactions) work efficiently only if the template RNAs are capped at their 5′ termini. The addition of a 5′-capped structure also greatly improves the stability of RNAs injected into oocytes (for references, see Yisraeli and Melton 1989). Capped RNAs can be synthesized in vitro by lowering the concentration of GTP to 50 μM and including a cap analog (such as G[5′]ppp[5′]G) at a 10-fold molar excess in the reaction mixture (Contreras et al. 1982; Konarska et al. 1984). The great majority of transcripts synthesized under these conditions are initiated with a 5′-capped structure. However, once RNA synthesis has begun, the peculiar chemical structure of the cap analog (with two exposed 3′-hydroxyl residues) ensures that no further incorporation of the capping nucleotide can occur.

Finally, the high specificity of bacteriophage RNA polymerases has been exploited in the development of prokaryotic and eukaryotic expression systems. Here, a target cDNA or gene is cloned downstream from a bacteriophage promoter sequence and upstream of an appropriate transcription termination sequence. The resulting recombinant plasmid is then introduced into cells harboring a second plasmid that expresses the bacteriophage RNA polymerase in a regulated fashion. Induction of the polymerase gene results in the transcription of the target DNA on the first plasmid and subsequent abundant expression of its encoded product(s). In *E. coli*, the most widely used binary expression system uses the bacteriophage T7 RNA polymerase and promoter (Tabor and Richardson 1985; Studier and Moffatt 1986; Studier et al. 1990). In eukaryotic cells, the T7 RNA polymerase can be produced by infection with a recombinant vaccinia virus containing the T7 polymerase gene (Fuerst et al. 1986) or by transfection of an expression plasmid.

FACTS AND HINTS

To obtain milligram amounts of RNA from large-scale reactions, adjust the concentration of MgCl$_2$ in the reaction to 6 nM above the total concentration of nucleotides in the reaction (Milligan and Uhlenbeck 1989) and include yeast inorganic pyrophosphatase at a concentration of 5 U/mL (Cunningham and Ofengand 1990). The pyrophosphatase prevents sequestration of Mg^{2+} in the form of magnesium pyrophosphate.

Bacteriophage RNA polymerases will accept biotinylated nucleotides as precursors. However, compared to radiolabeled NTPs, incorporation is less efficient and the products of the reaction contain a higher proportion of truncated RNAs (Grabowski and Sharp 1986; Yisraeli and Melton 1989).

T7 RNA polymerase is strongly inhibited by T7 lysozyme (Moffatt and Studier 1987; Ikeda and Bailey 1992). Coexpression of the T7 lysozyme gene has been used as a method to reduce the activity of T7 polymerase in transformed bacteria (for review, see Studier et al. 1990).

REFERENCES

Beck S, O'Keeffe T, Coull JM, Koster H. 1989. Chemiluminescent detection of DNA: Application for DNA sequencing and hybridization. *Nucleic Acids Res* 17: 5115–5123.

Brown JE, Klement JF, McAllister WT. 1986. Sequences of three promoters for the bacteriophage SP6 RNA polymerase. *Nucleic Acids Res* 14: 3521–3526.

Browning KS. 1989. Transcription and translation of mRNA from polymerase chain reaction-generated DNA. *Amplifications* 3: 15–16.

Butler ET, Chamberlin MJ. 1982. Bacteriophage SP6-specific RNA polymerase. I. Isolation and characterization of the enzyme. *J Biol Chem* 257: 5772–5778.

Chamberlin M, Ryan T. 1982. Bacteriophage DNA-dependent RNA polymerases. In *The enzymes*, 3rd ed. (ed Boyer PD), vol. 15, pp. 87–108. Academic, New York.

Contreras R, Cheroutre H, Degrave W, Fiers W. 1982. Simple, efficient in vitro synthesis of capped RNA useful for direct expression of cloned eukaryotic genes. *Nucleic Acids Res* 10: 6353–6362.

Cunningham PR, Ofengand J. 1990. Use of inorganic pyrophosphatase to improve the yield of in vitro transcription reactions catalyzed by T7 RNA polymerase. *BioTechniques* 9: 713–714.

Davanloo P, Rosenberg AH, Dunn JJ, Studier FW. 1984. Cloning and expression of the gene for bacteriophage T7 RNA polymerase. *Proc Natl Acad Sci* 81: 2035–2039.

Doublie S, Tabor S, Long AM, Richardson CC, Ellenberger T. 1998. Crystal structure of a bacteriophage T7 DNA replication complex at 2.2 Å resolution. *Nature* 391: 251–258.

Dunn JJ, Studier FW. 1983. Complete nucleotide sequence of bacteriophage T7 DNA and the locations of T7 genetic elements. *J Mol Biol* 166: 477–535.

Fuerst TR, Niles EG, Studier FW, Moss B. 1986. Eukaryotic transient-expression system based on recombinant vaccinia virus that synthesizes bacteriophage T7 RNA polymerase. *Proc Natl Acad Sci* 83: 8122–8126.

Grabowski PJ, Sharp PA. 1986. Affinity chromatography of splicing complexes: U2, U5, and U4+U6 small nuclear ribonucleoprotein particles in the spliceosome. *Science* 233: 1294–1299.

Green MR. 1991. Biochemical mechanisms of constitutive and regulated pre-mRNA splicing. *Annu Rev Cell Biol* 7: 559–599.

Green MR, Maniatis T, Melton DA. 1983. Human β-globin pre-mRNA synthesized in vitro is accurately spliced in *Xenopus* oocyte nuclei. *Cell* 32: 681–694.

Ikeda RA, Bailey PA. 1992. Inhibition of T7 RNA polymerase by T7 lysozyme in vitro. *J Biol Chem* 267: 20153–20158.

Jorgensen ED, Durbin RK, Risman SS, McAllister WT. 1991. Specific contacts between the bacteriophage T3, T7, and SP6 RNA polymerases and their promoters. *J Biol Chem* 266: 645–651.

Kain KC, Orlandi PA, Lanar DE. 1991. Universal promoter for gene expression without cloning: Expression-PCR. *BioTechniques* 10: 366–374.

Konarska MM, Padgett RA, Sharp PA. 1984. Recognition of cap structure in splicing in vitro of mRNA precursors. *Cell* 38: 731–736.

Kotani H, Ishizaki Y, Hiraoka N, Obayashi A. 1987. Nucleotide sequence and expression of the cloned gene of bacteriophage SP6 RNA polymerase. *Nucleic Acids Res* 15: 2653–2664.

Loewy ZG, Leary SL, Baum HJ. 1989. Site-directed transcription initiation with a mobile promoter. *Gene* 83: 367–370.

McGraw NJ, Bailey JN, Cleaves GR, Dembinski DR, Gocke CR, Joliffe LK, MacWright RS, McAllister WT. 1985. Sequence and analysis of the gene for bacteriophage T3 RNA polymerase. *Nucleic Acids Res* 13: 6753–6766.

Melton DA, Krieg PA, Rebagliati MR, Maniatis T, Zinn K, Green MR. 1984. Efficient in vitro synthesis of biologically active RNA and RNA hybridization probes from plasmids containing a bacteriophage SP6 promoter. *Nucleic Acids Res* 12: 7035–7056.

Milligan JF, Uhlenbeck OC. 1989. Synthesis of small RNAs using T7 RNA polymerase. *Methods Enzymol* 180: 51–62.

Milligan JF, Groebe DR, Witherell GW, Uhlenbeck OC. 1987. Oligoribonucleotide synthesis using T7 RNA polymerase and synthetic DNA templates. *Nucleic Acids Res* 15: 8783–8798.

Moffatt BA, Studier FW. 1987. T7 lysozyme inhibits transcription by T7 RNA polymerase. *Cell* 49: 221–227.

Moffatt BA, Dunn JJ, Studier FW. 1984. Nucleotide sequence of the gene for bacteriophage T7 RNA polymerase. *J Mol Biol* 173: 265–269.

Morris CE, Klement JF, McAllister WT. 1986. Cloning and expression of the bacteriophage T3 RNA polymerase gene. *Gene* 41: 193–200.

Oakley JL, Strothkamp RE, Sarris AH, Coleman JE. 1979. T7 RNA polymerase: Promoter structure and polymerase binding. *Biochemistry* 18: 528–537.

Roy S, Delling U, Chen CH, Rosen CA, Sonenberg N. 1990. A bulge structure in HIV-1 TAR RNA is required for Tat binding and Tat-mediated *trans*-activation. *Genes Dev* 4: 1365–1373.

Salvo RA, Chakraborty PR, Maitra U. 1973. Studies on T3-induced ribonucleic acid polymerase. IV. Transcription of denatured deoxyribonucleic acid preparations by T3 ribonucleic acid polymerase. *J Biol Chem* 248: 6647–6654.

Studier FW, Moffatt BA. 1986. Use of bacteriophage T7 RNA polymerase to direct selective high-level expression of cloned genes. *J Mol Biol* 189: 113–130.

Studier FW, Rosenberg AH. 1981. Genetic and physical mapping of the late region of bacteriophage T7 DNA by use of cloned fragments of T7 DNA. *J Mol Biol* 153: 503–525.

Studier FW, Rosenberg AH, Dunn JJ, Dubendorff JW. 1990. Use of T7 RNA polymerase to direct expression of cloned genes. *Methods Enzymol* 185: 60–89.

Tabor S, Richardson CC. 1985. A bacteriophage T7 RNA polymerase/promoter system for controlled exclusive expression of specific genes. *Proc Natl Acad Sci* 82: 1074–1078.

Yisraeli JK, Melton DA. 1989. Synthesis of long, capped transcripts in vitro by SP6 and T7 RNA polymerases. *Methods Enzymol* 180: 42–50.

ALKALINE PHOSPHATASE

Several types of alkaline phosphatases (or alkaline phosphomonoesterase) are commonly used in molecular cloning, inlcuding bacterial alkaline phosphatase (BAP) and calf intestinal alkaline phosphatase (CIP, CIAP, or CAP). Similar enzymes isolated from more esoteric cold-blooded organisms (e.g., SAP from shrimp) have become available in recent years and have the advantage of being easier to inactivate than BAP or CIP at the end of dephosphorylation reactions. Alkaline phosphatases, which are all Zn(II) metalloenzymes, catalyze phosphate monoester hydrolysis through the formation of a phosphorylated serine intermediate. They are used for three purposes in molecular cloning:

- to remove phosphate residues from the 5′ termini of nucleic acids before radiolabeling of the resulting 5′-hydroxyl group by transfer of ^{32}P from γ-labeled ATP. This second reaction is catalyzed by bacteriophage T4 polynucleotide kinase (Chaconas and van de Sande 1980).

- to remove 5′-phosphate residues from fragments of DNA to prevent self-ligation. This dephosphorylation reaction is chiefly used to suppress self-ligation of vector molecules and therefore to decrease the number of "empty" clones that are obtained during cloning (Ullrich et al. 1977).

- as reporter enzymes in nonradioactive systems, to detect and localize nucleic acids and proteins. In this case, the alkaline phosphatase is conjugated to a ligand such as streptavidin that specifically interacts with a biotinylated target molecule (Leary et al. 1983).

Dephosphorylation Reactions

Alkaline phosphatases can remove 3′-phosphate groups from a variety of substrate molecules, including 3′-phosphorylated polynucleotides and deoxynucleoside 3′-monophosphates (Reid and Wilson 1971). However, the main use of BAP, CIP, and SAP in molecular cloning is to catalyze the removal of terminal 5′-phosphate residues from ss or dsDNA and RNA. The resulting 5′-hydroxyl termini can no longer take part in ligation reactions but are substrates in radiolabeling reactions catalyzed by polynucleotide kinase (Chaconas and van de Sande 1980).

Although the usefulness of alkaline phosphatases in 5′ labeling of nucleic acids is undisputed, their value in preventing self-ligation is more debatable. There is no doubt that dephosphorylation reduces recircularization of linear plasmid DNA and therefore diminishes the background of transformed bacterial colonies that carry "empty" plasmids (Ullrich et al. 1977; Ish-Horowicz and Burke 1981; Evans et al. 1992). All too frequently, however, there is a parallel decline in the number of colonies that carry the desired recombinant. In addition, some investigators believe that the presence of 5′-hydroxyl groups may lead to an increase in the frequency of rearranged or deleted clones. For these reasons, directional cloning is the preferred method whenever the appropriate restriction sites are available.

Properties of Alkaline Phosphatases

Alkaline phosphatases used in molecular cloning display maximal activity in alkaline Tris buffers (pH 8.0–9.0) in the presence of low concentrations of Zn^{2+} (<1 mM).

- BAP is secreted in monomeric form ($M_r = 47,000$) into the periplasmic space of *E. coli*, where it dimerizes and becomes catalytically active (Bradshaw et al. 1981). At neutral or alkaline pH, dimeric BAP contains up to six Zn^{2+} ions, two of which are essential for enzymatic activity (for review, see Coleman and Gettins 1983). Only one of the two catalytic sites per dimer is active at low concentrations of artificial substrates, whereas both become active at higher concentrations (Heppel et al. 1962; Fife 1967). BAP is a remarkably stable enzyme and is resistant to inactivation by heat and detergents. For this reason, BAP is difficult to remove at the end of dephosphorylation reactions.

- CIP is a dimeric glycoprotein, composed of two 514-residue monomers, that is bound to the plasma membrane by a phosphatidylinositol anchor (Hoffmann-Blume et al. 1991; Weissig et al. 1993) and whose optimal enzymatic activity depends on the concentrations of Mg^{2+} and Zn^{2+}. Some Zn^{2+}, bound at a catalytic site, is required for catalytic activity. Mg^{2+}, which binds at a different site, is an allosteric activator. However, Zn^{2+}, if present in high concentrations, will compete for the Mg^{2+}-binding site and prevent the allosteric activation (Fernley 1971). CIP can be readily digested with proteinase K and/or inactivated by heating (for 30 min to 65°C or for 10–15 min to 75°C) in the presence of 10 mM EGTA. The dephosphorylated DNA can then be purified by extraction with phenol:chloroform.

- SAP is isolated from arctic shrimp and its enzymatic properties are similar to those of CIP. Unlike BAP, SAP is unstable at elevated temperatures and, according to the manufacturers, can be completely inactivated by heating for 15 min to 65°C. However, molecular biology chat sites on the web frequently contain comments from customers who suggest that the enzyme may not be completely inactivated by brief heating. To be fair, about an equal number report that they have no problems in inactivating SAP. However, to be on the safe side, we recommend heating for 20 min to 70°C to ensure complete inactivation of SAP.

Alkaline phosphatases as a group are inhibited by inorganic orthophosphate (Zittle and Della Monica 1950) and chelators of metal ions such as EDTA and EGTA, but not to a significant extent by diisopropylfluorophosphate, a powerful inhibitor of other serine hydrolases (Dabich and Neuhaus 1966). L-phenylalanine is a noncompetitive inhibitor of CIP (Weissig et al. 1993).

REFERENCES

Bradshaw RA, Cancedda F, Ericsson LH, Neumann PA, Piccoli SP, Schlesinger MJ, Shriefer K, Walsh KA. 1981. Amino acid sequence of *Escherichia coli* alkaline phosphatase. *Proc Natl Acad Sci* 78: 3473–3477.

Chaconas G, van de Sande JH. 1980. 5′-^{32}P labeling of RNA and DNA restriction fragments. *Methods Enzymol* 65: 75–85.

Coleman JE, Gettins P. 1983. Alkaline phosphatase, solution structure, and mechanism. *Adv Enzymol Relat Areas Mol Biol* 55: 381–452.

Dabich D, Neuhaus OW. 1966. Purification and properties of bovine synovial fluid alkaline phosphatase. *J Biol Chem* 241: 415–420.

Evans GA, Snider K, Hermanson GG. 1992. Use of cosmids and arrayed clone libraries for genome analysis. *Methods Enzymol* 216: 530–548.

Fernley HN. 1971. Mammalian alkaline phosphatases. In *The enzymes*, 3rd ed. (ed Boyer PD), vol. 4, pp. 417–447. Academic, New York.

Fife WK. 1967. Phosphorylation of alkaline phosphatase (*E. coli*) with *o*- and *p*-nitrophenyl phosphate at pH below 6. *Biochem Biophys Res Commun* 28: 309–317.

Heppel LA, Harkness DR, Hilmoe RJ. 1962. A study of the substrate specificity and other properties of the alkaline phosphatase of *Escherichia coli*. *J Biol Chem* 237: 841–846.

Hoffmann-Blume E, Garcia Marenco MB, Ehle H, Bublitz R, Schulze M, Horn A. 1991. Evidence for glycosylphosphatidylinositol anchoring of intralumenal alkaline phosphatase of the calf intestine. *Eur J Biochem* 199: 305–312.

Ish-Horowicz D, Burke JF. 1981. Rapid and efficient cosmid cloning. *Nucleic Acids Res* 9: 2989–2998.

Leary JJ, Brigati DJ, Ward DC. 1983. Rapid and sensitive colorimetric method for visualizing biotin-labeled DNA probes hybridized to DNA or RNA immobilized on nitrocellulose: Bio-blots. *Proc Natl Acad Sci* 80: 4045–4049.

Reid TW, Wilson IB. 1971. *E coli* alkaline phosphatase In *The enzymes*, 3rd ed. (ed Boyer PD), vol. 4, pp. 373–415. Academic, New York.

Ullrich A, Shine J, Chirgwin J, Pictet R, Tischer E, Rutter WJ, Goodman HM. 1977. Rat insulin genes: Construction of plasmids containing the coding sequences. *Science* 196: 1313–1319.

Weissig H, Schildge A, Hoylaerts MF, Iqbal M, Millán JL. 1993. Cloning and expression of the bovine intestinal alkaline phosphatase gene: Biochemical characterization of the recombinant enzyme. *Biochem J* 290: 503–508.

Zittle CA, Della Monica ES. 1950. Effects of borate and other ions on the alkaline phosphatase of bovine milk and intestinal mucosa. *Arch Biochem Biophys* 26: 112–122.

MELTING TEMPERATURES

Heating a solution of dsDNA unravels the duplexes by disrupting the hydrogen bonds that pin together complementary base pairs. Individual hydrogen bonds are relatively weak ($\sim$5 kcal/mol/bond) and are thus easily disrupted by heat. However, the greater the number of hydrogen bonds holding any two complementary strands of DNA together, the higher the temperature necessary to cause all of those bonds to break, separating the nucleic acid into two strands. This transition from double-stranded (helix) to single-stranded (coil) conformations as a function of temperature can be monitored as an increase in absorbance and is marked by a sharp increase in the extinction coefficient at the temperature at which the conformational transition takes place. The temperature corresponding to the midpoint of the absorbance rise is called the melting temperature (T_m). In structural terms, the T_m is the temperature at which 50% of the base pairs in a duplex have been denatured. This relationship between nucleic acid structure and thermal stability has been experimentally exploited for four decades.

Marmur and Doty (1959, 1962) used this relationship to establish that for a wide range of DNA samples, the T_m is a linear function of the sample's base composition. The slope of the line is 0.41°C per 1% increase in the $(G + C)$ content. For thermal denaturation profiles of DNA samples (with a $G + C$ content between 30% and 75%) run in 0.15 M NaCl and 15 mM sodium citrate, the melting temperature is

$$T_m = 69.3°C + 0.41(\%[G + C]). \tag{1}$$

This equation makes it possible to estimate accurately the base composition of a DNA sample simply by measuring the T_m of the sample. To determine the T_m of DNA samples run under other salt conditions, use the more general equation

$$T_m = 81.5°C + 16.6\log_{10}[M^+] + 0.41(\%[G + C]), \tag{2}$$

where $[M^+]$ is the monovalent cation concentration for $M^+ \leq 0.5$ M.

Molecular Hybridization

The molecular hybridization techniques developed in the 1960s continue to be some of the most useful tools of the molecular biologist. Their simplicity and sensitivity have helped to drive a revolution in our understanding of gene structure, gene expression, and genome organization. These techniques have also been adapted as diagnostic tools in clinical laboratories and pharmaceutical companies.

Establishing optimal conditions for hybridization of a nucleic acid probe and its target sequence requires knowledge of the T_m of the target:probe duplex. Although under some circumstances, the T_m can be experimentally determined (see Protocol 7), it is more common to calculate the T_m. A variety of equations have been derived that can be used to estimate the T_m of a target:probe duplex under a range of conditions.

At the T_m, when the probe and target are both polynucleotides, the molecules contain stretches of native duplex separated by denatured regions. The transition then from helix to coil is intramolecular and is thus independent of polynucleotide concentration (Wetmur 1991). The T_m is a function of base composition, solvent composition, duplex length, and extent of base-pair mismatches (Hall et al. 1980; Wahl et al. 1987; Baldino et al. 1989). For DNA duplexes, Equation 2 has been modified by Wetmur (1991) to permit hybridization in salt up to 1 M NaCl:

$$T_m = 81.5 + 16.6\log_{10}([Na^+]/\{1.0 + 0.7[Na^+]\}) + 0.41(\%[G + C]) - 500/n - P - F, \tag{3}$$

where n is the length of the duplex, P is the temperature correction for the percent mismatch of base pairs (which is typically 1°C per 1% mismatch), and F is 0.63°C per 1% formamide. For RNA duplexes,

$$T_m = 78 + 16.6\log_{10}([Na^+]/\{1.0 + 0.7[Na^+]\}) + 0.7(\%[G + C]) - 500/n - P - F, \tag{4}$$

where F is 0.35°C per 1% formamide. For hybrids of RNA and DNA,

$$T_m = 67 + 16.6 \log_{10}([Na^+]/\{1.0 + 0.7[Na^+]\}) + 0.8(\%[G+C]) - 500/n - P - F, \quad (5)$$

where F is 0.5°C per 1% formamide. Variations on Equations 4 and 5 have been derived by Bodkin and Knudson (1985) and Casey and Davidson (1977), respectively.

When the probe is an oligonucleotide, at the T_m, half of the duplexes have separated. The transition from helix to coil is intermolecular, and thus the T_m is dependent on the oligonucleotide concentration (Wetmur 1991), the oligonucleotide sequence, and the composition of the solvent. Although various permutations of Equations 3–5 are frequently used, the use of %($G + C$) is not sufficient to predict the T_m. A more accurate estimate of T_m can be obtained from a nearest-neighbor model that permits the incorporation of sequence-related thermodynamic data (Breslauer et al. 1986; Freier et al. 1986; Kierzek et al. 1986; Rychlik et al. 1990; Wetmur 1991; Rychlik 1994). One equation derived from such a model is

$$T_m(\text{in degrees Celsius}) = (T^0 \Delta H^0)/(\Delta H^0 - \Delta G^0 + RT^0 \ln[c])$$
$$+ 16.6 \log_{10}([Na^+]/\{1.0 + 0.7[Na^+]\}) - 269.3, \quad (6)$$

where

$$\Delta H^0 = \sum_{nn}(_{nn}\Delta H^0_{nn}) + \Delta H^0_p + \Delta H^0_e,$$

$$\Delta G^0 = \sum_{nn}(N_{nn}\Delta G^0_{nn}) + \Delta G^0_i + \Delta G^0_e,$$

N_{nn} is the number of nearest neighbors (e.g., 13 for a 14-mer), $R = 1.99$ cal/mol (in degrees Kelvin), $T^0 = 298.2°K$, C is the total molar strand concentration, and $[Na^+] \leq 1$ M.

A number of the thermodynamic terms can be approximated: average nearest-neighbor enthalpies,

$$\Delta H^0_{nn} = -8.0 \text{ kcal/mol};$$

average nearest-neighbor free energies,

$$\Delta G^0_{nn} = -1.6 \text{ kcal/mol};$$

initiation term,

$$\Delta G^0_i = +2.2 \text{kcal/mol};$$

average dangling end enthalpy,

$$\Delta H^0_e = -8.0 \text{ kcal/mol/end};$$

average dangling end free energy,

$$\Delta G^0_e = -1 \text{ kcal/mol/end};$$

and average mismatch/loop enthalpy,

$$\Delta H^0_p = -8.0 \text{ kcal/mol/mismatch}.$$

Note that for each mismatch or loop, N_{nn} is reduced by 2. At the opposite end of the complexity spectrum is the equation

$$T_m = 2(A + T) + 4(G + C), \quad (7)$$

where $(A+T)$ is the number of adenines and thymines in the oligonucleotide and $(G + C)$ is the number of guanines and cytosines in the oligonucleotide, which works for short oligonucleotides that perfectly match their target sequence (Suggs et al. 1981). More precisely, this equation predicts the dissociation temperature (T_d) of an oligonucleotide hybridized to a target sequence that is bound to a solid support. For a detailed discussion of T_d, see Wetmur (1991).

The differences in calculated T_m can be significant. For example, the predicted T_m for hybridization, in 0.5 M NaCl, of a 14-bp oligonucleotide probe that perfectly matches a sequence (with a %[$G + C$]=50) in the target DNA will be 59°C using Equation 3, 55°C using Equation 6, and 42°C using Equation 7.

REFERENCES

Baldino FJ, Chesselet MF, Lewis ME. 1989. High resolution in situ hybridization. *Methods Enzymol* 168: 761–777.

Bodkin DK, Knudson DL. 1985. Assessment of sequence relatedness of double-stranded RNA genes by RNA-RNA blot hybridization. *J Virol Methods* 10: 45–52.

Breslauer KJ, Frank R, Blöcker H, Marky LA. 1986. Predicting DNA duplex stability from the base sequence. *Proc Natl Acad Sci* 83: 8893–8897.

Casey J, Davidson N. 1977. Rates of formation and thermal stabilities of RNA:DNA and DNA:DNA duplexes at high concentrations of formamide. *Nucleic Acids Res* 4: 1539–1552.

Freier SM, Kierzek R, Jaeger JA, Sugimoto N, Caruthers MH, Neilson T, Turner DH. 1986. Improved free-energy parameters for predictions of RNA duplex stability. *Proc Natl Acad Sci* 83: 9373–9377.

Hall TJ, Grula JW, Davidson EH, Britten RJ. 1980. Evolution of sea urchin non-repetitive DNA. *J Mol Evol* 16: 95–110.

Kierzek R, Caruthers MH, Longfellow CE, Swinton D, Turner DH, Freier SM. 1986. Polymer-supported RNA synthesis and its application to test the nearest-neighbor model for duplex stability. *Biochemistry* 25: 7840–7846.

Marmur J, Doty P. 1959. Heterogeneity in deoxyribonucleic acids. 1. Dependence on composition of the configurational stability of deoxyribonucleic acids. *Nature* 183: 1427–1429.

Marmur J, Doty P. 1962. Determination of the base composition of deoxyribonucleic acid from its thermal denaturation temperature. *J Mol Biol* 5: 109–118.

Rychlik R. 1994. New algorithm for determining primer efficiency in PCR and sequencing. *J NIH Res* 6: 78.

Rychlik W, Spencer WJ, Rhoads RE. 1990. Optimization of the annealing temperature for DNA amplification in vitro. *Nucleic Acids Res* 18: 6409–6412. (Erratum *Nucleic Acids Res* [1991] 19: 698.)

Suggs SV, Hirose T, Miyake T, Kawashima EH, Johnson MJ, Itakura K, Wallace RB. 1981. Use of synthetic oligodeoxyribonucleotides for the isolation of specific cloned DNA sequences. In *Developmental biology using purified genes* (ed Brown DB, et al.), pp. 683–693. Academic, New York.

Wahl GM, Berger SL, Kimmel AR. 1987. Molecular hybridization of immobilized nucleic acids: Theoretical concepts and practical considerations. *Methods Enzymol* 152: 399–407.

Wetmur JG. 1991. DNA probes: Applications of the principles of nucleic acid hybridization. *Crit Rev Biochem Mol Biol* 26: 227–259.

CHAPTER 14

Methods for In Vitro Mutagenesis

Matteo Forloni, Alex Y. Liu, and Narendra Wajapeyee

Department of Pathology, Yale School of Medicine, New Haven, Connecticut 06520

PROKARYOTIC AND EUKARYOTIC CELLS ARE UNDER CONSTANT exposure to extrinsic and intrinsic stimuli that cause DNA damage, which can lead to mutations if unrepaired (Lombard et al. 2005; Boesch et al. 2011). To survive under these constant genotoxic conditions and still faithfully replicate the genome, prokaryotic and eukaryotic cells have developed specific dedicated DNA repair machineries that repair the DNA damage and slow the accumulation of genetic mutations (Lombard et al. 2005; Boesch et al. 2011). However, not all mutations have an effect on the biological function of a protein. Once a mutation is identified in an organism, functional characterization is still required to understand the effect of the mutation on the function of a specific protein or a noncoding RNA.

Several different methodologies for mutagenesis have been developed to introduce mutations at predetermined sites or regions within mammalian genes. These methods of in vitro mutagenesis have had a transforming effect on the understanding of functions of protein, transcription regulatory elements, and noncoding RNAs and are now integral to molecular biology investigations. In this chapter, we describe some of the most commonly used experimental approaches for mutagenesis and provide detailed protocols for commonly used methods for mutagenesis (see Table 1). The

TABLE 1. Comparison of mutagenesis protocols

Protocol	Template	Method used to enrich for mutant	Characteristics/advantages	Drawbacks
1	Double-stranded DNA	Selective PCR amplification of mutated DNA	A wide spectrum of mutations can be introduced.	Mutations cannot be targeted at specific positions.
2	Double-stranded DNA	Overlap extension	i. Used for the generation of a specific point mutation, insertion, or deletion within a particular DNA sequence of interest ii. Does not require the use of restriction enzymes	The researcher needs to have a specific mutation in mind to implement.
3	Double-stranded DNA	Selection of mutants with DpnI	i. Mutations can be placed precisely in the target gene. ii. 80% of the transformed colonies will contain plasmids with the desired mutation.	Low efficacy with plasmids longer than 7 kb
4	Double-stranded DNA	Altered β-lactamase activity of mutant vector	Like other site-directed mutagenesis methods, targets a predetermined region within a DNA sequence	Requires the presence of antibiotic selection marker in the plasmid
5	Double-stranded DNA	Elimination of a restriction site (USE mutagenesis)	Mutations can be placed precisely in the target gene.	The rate of mutant recovery is low.
6	Double-stranded DNA	Based on SapI digestion	Used to introduce a library of site-specific changes into a specific DNA sequence within a target gene	Requires the presence of SapI digestion site.
7	Single-stranded DNA	Selection against uracil-substituted DNA	A broad spectrum of amino acid changes can be placed at targeted positions.	The mutagenic process takes place during in vitro DNA synthesis and generates plasmid libraries with skewed distributions of variant amino acids at the target region.
8	Double-stranded DNA	Selection of mutants with DpnI	Multiple independent mutations can be introduced by following a single cloning step procedure.	i. This method requires as many rounds of PCR as the target mutation sites to be introduced. ii. When more than three mutations are to be generated, it is quicker to use a commercial kit.
9	Double-stranded DNA	Megaprimer	i. Mutations can be placed precisely in the target DNA sequences. ii. The mutagenesis process can be accomplished by only two rounds of PCR that are carried out sequentially in the same tube.	The yield of mutants has an efficiency of only ∼82%.

nine different methods for performing mutagenesis described here can be used for introduction of mutations based on the requirements of the researcher. Below we summarize the three major types of mutagenesis approaches and their major uses.

HISTORICAL BACKGROUND

The road to developing modern in vitro mutagenesis methods has been paved with Nobel Prizes as progress moved from the level of treating whole organisms to that of targeting specific sites in the DNA molecule. Hermann J. Muller (1890–1967), then at the University of Texas, electrified the scientific world in 1926 when he described evidence of gene mutations and chromosomal changes in *Drosophila* produced by X-irradiation. He announced his results at a scientific meeting in 1926 and then published them in the paper "Artificial Transmutation of the Gene" in *Science* (Muller 1927). The production of mutations by X-irradiation played a major role in the experiments on *Neurospora* published in 1941 by George Beadle and Edward Tatum in which they proposed their one gene–one protein hypothesis (Beadle and Tatum 1941).

H.J. Muller spent some time at Edinburgh University from 1938 to 1940, and while there he worked with Charlotte Auerbach (1899–1994) on substances that could cause mutations. With J.M. Robson in the early 1940s, Auerbach used mustard gas to induce mutations in *Drosophila* (Auerbach and Robson 1947; Stevens et al. 1950). In a later publication, Auerbach related that:

> It seems to me that there may be many ways of affecting the chemical specificity of the gene or the physical integrity of the chromosomes: direct chemical reaction with proteins or nucleic acids; release of energy close to a chromosome; inactivation of enzymes concerned with chromosome metabolism; interference with gene reduplication by competitive analogs, and so on (Auerbach 1951).

The issues raised by Auerbach became approachable after the determination of the structure of DNA in 1953 by James D. Watson and Francis H.C. Crick—a discovery that changed the nature of mutagenesis studies. Mutagenesis techniques became very specific as a result of Michael Smith's (1932–2000) studies of oligonucleotide synthesis and his 1975 sabbatical in Fred Sanger's laboratory sequencing *Escherichia coli* phage φX174. Smith realized that a mutagenic method was needed to target specific bases. His studies showed that small oligonucleotides could form stable duplexes at low temperatures, while work by Clyde A. Hutchison III and Marshall H. Edgell showed that "point mutations could be reverted by annealing mutant phage φX174 with fragments from the complementary strand of wild-type DNA before transfection." Using φX174 DNA as a template, a 120-nucleotide oligomer with a single-nucleotide mismatch as primer, and *E. coli* DNA polymerase, they produced a closed-circular double-stranded DNA with the oligonucleotide incorporated into one strand (Hutchison et al. 1978). At about the same time, the first specific restriction endonucleases were being discovered, making it possible to isolate specific DNA fragments (Smith and Wilcox 1970).

A final key element fell into place in 1983 when Kary Mullis invented the idea of the polymerase chain reaction (Saiki et al. 1985; Mullis et al. 1986). PCR soon became an integral part of in vitro site-directed mutagenesis techniques.

As for the Nobel Prize count, Muller's (Physiology or Medicine) was awarded in 1946, and Smith's in 1993 (Chemistry) was shared with Mullis. Beadle and Tatum were cited in 1958 (Physiology or Medicine), and Watson and Crick in 1962 (Physiology or Medicine).

MUTAGENESIS TERMINOLOGY

There is a plethora of mutagenesis terminology, some of it confusing, with different names for the same approach. The following list clarifies the terminology for the multitude of mutagenesis approaches.

5' Add-On Mutagenesis

A PCR-based method that is useful for adding a new sequence or chemical group to the 5' end of a PCR product.

(Continued)

Alanine Scanning Mutagenesis

Method used to determine the structure–function relationship of a given protein. Alanine is the substitution residue of choice because it eliminates the side chain beyond the β-carbon and yet does not alter the main-chain conformation, nor does it impose extreme electrostatic or steric effects. (See also Scanning Mutagenesis.)

Cassette Mutagenesis

Used for efficient insertion of mutagenic oligodeoxynucleotide cassettes, which allow saturation of a target amino acid codon with multiple mutations.

Chemical Mutagenesis

Exploits the nature of chemical mutagens, like hydroxylamine and N-ethyl-N-nitrosourea, to introduce random mutations into the target DNA sequence.

Circular Mutagenesis

Introduces site-directed mutations into circular DNA molecules of interest by means of mutagenic primer pairs.

Codon Cassette Mutagenesis

Relies on the use of universal mutagenic cassettes to deposit single codons at specific sites in double-stranded DNA.

Deletion Mutagenesis

Used to produce both randomly positioned and targeted deletions over large regions of DNA for constructing deletion mutants.

Directed Evolution

Used in protein engineering to harness the power of natural selection to evolve proteins or RNA with desirable properties not found in nature.

Directed Mutagenesis

See Site-Directed Mutagenesis.

Domain Mutagenesis

Used to introduce multiple mutations into a defined region of cloned DNA.

Insertional Mutagenesis

Used to mutagenize a DNA template by the insertion of one or more bases. These insertions can occur naturally, such as through transposons, or artificially in a laboratory setting.

Linker Scanning Mutagenesis

In linker scanning mutagenesis, a collection of 5′ and 3′ deletants is first created, and the termini are ligated to a linker oligonucleotide. Depending on the DNA sequence of individual deletants, paired combinations are chosen and used to create a new DNA fragment in which the linker sequence precisely replaces a part of the original sequence without altering the spacing of surrounding nucleotides.

Megaprimer PCR-Based Mutagenesis

Two external oligonucleotide primers and one internal mutagenic primer are used in two rounds of PCR to mutagenize a DNA sequence. The first round of PCR is performed using one of the external primers and the mutagenic primer containing the desired mutation. This generates an intermediate PCR product that is then used as a "megaprimer" for the second round of PCR, along with the other external primer. The final PCR product is cloned into appropriate vectors and used in downstream applications.

(Continued)

Misincorporation Mutagenesis

Uses reverse-transcriptases as error-prone polymerases to create mutations.

Mismatch Mutagenesis

Creates specific DNA-based pair mismatches.

Multisite Directed Mutagenesis

Permits mutagenesis at multiple sites simultaneously with only a single oligonucleotide primer per site.

Oligonucleotide-Directed Mutagenesis

Uses a mutagenic oligonucleotide primer to introduce a mutation into a DNA strand. It may or may not involve PCR.

PCR Mutagenesis

Any technique that uses the PCR to generate a mutation in a specific DNA sequence.

PCR Site-Directed Mutagenesis

Uses mutagenic oligonucleotide primers to introduce the desired mutations.

Random Mutagenesis

Random mutations are produced in a specific DNA sequence by, for example, UV irradiation.

Random Scanning Mutagenesis

An oligonucleotide-based method for generating all 19 possible replacements at individual amino acid sites within a protein.

Retroviral Insertional Mutagenesis

Retrovirus particles are used to create insertional mutagenesis.

Saturation Mutagenesis

A form of site-directed mutagenesis, in which one tries to generate all possible (or as close to as possible) mutations at a specific site or narrow region of a gene.

Scannning Mutagenesis

Used to understand structure–function relationships by creating a library of mutants with an alanine or cysteine residue at each position of protein. (See also Alanine Scanning Mutagenesis.)

Sequence Saturation Mutagenesis

A method to create mutations at every single nucleotide position in a given target sequence.

Signature-Tagged Mutagenesis (STM)

A genetic technique used to study gene function. STM can be used to infer the function a particular gene has by observing the effects of mutations on the phenotype. The original and most common use of STM is to discover which genes in a pathogen are involved in virulence in its host, so that better medical treatments can be designed.

Site-Directed Mutagenesis

Also called "site-specific mutagenesis" or "oligonucleotide-directed mutagenesis," this is a technique in which a desired mutation is created at a defined site in a DNA molecule.

(Continued)

Site-Specific Mutagenesis

See Site-Directed Mutagenesis.

Transposon Mutagenesis

Also called "transposition mutagenesis," this is a biological process that allows genes to be transferred into a host organism's chromosome, interrupting or modifying the function of an extant gene.

MUTAGENESIS APPROACHES

Oligonucleotide-Directed Mutagenesis

Oligonucleotide-directed mutagenesis is used to test the role of particular residues in the structure, catalytic activity, and ligand-binding capacity of a protein. In the absence of a three-dimensional (3D) structure, this type of protein engineering relies on informed guesses concerning the structure of the protein and the contribution of individual residues to protein stability and function. A major problem is distinguishing mutations that affect local structures from those that have profound and deleterious effects on the folding or stability of the entire protein. Consider a typical experiment in which several point mutations have been generated at various sites in a gene coding for an enzyme. When the activities of these mutants are assayed, some of them show a reduction in catalytic function and others do not. In the absence of any other data, it is not possible to draw firm conclusions about the structure of the enzyme from this result. There is no way to know whether the substitution of one amino acid for another has affected only the function of the active site or whether it has had more global effects. The problem would remain even if the 3D structure of the wild-type enzyme were known because no algorithms have yet been devised that accurately predict the perturbations in protein structure caused by the substitution, addition, or deletion of amino acid residues. However, these difficulties can be alleviated by developing independent assays for the folding of the protein of interest. Such assays commonly include the ability of the protein to react with monoclonal or polyclonal antibodies that are specific for native or unfolded epitopes, the proper movement and posttranslational modification of the protein within a cell, the retention of catalytic or ligand-binding functions, and the sensitivity or resistance of the mutant protein to digestion with proteases.

If reliable assays are available to confirm that the mutagenized protein is correctly folded, oligonucleotide-directed mutagenesis becomes an analytical technique with both exquisite specificity and extraordinary breadth. Mutations that could never be found in nature can now be placed precisely in the target gene, functions of proteins can be mapped to specific structural domains, undesirable activities of enzymes can be eliminated, and their desirable catalytic and physical properties can be enhanced. In short, oligonucleotide-directed mutagenesis has become the genetic engineer's alchemy.

Saturation Mutagenesis

Saturation mutagenesis is used to generate mutations at many sites in a particular coding sequence. Every effort is made to introduce mutations in an unbiased fashion; preconceptions and knowledge about the functions of individual amino acids in the wild-type sequence are disregarded. The aim is to gather information about the entire "sequence space," that is, about the relationship between the amino acid sequence and the 3D structure of the protein. Saturation mutagenesis is usually performed on small segments of DNA that encode an individual structural domain. At its best, the method can provide catalogs of amino acids or combinations of amino acids that are tolerated within a domain without deleterious effect on structure and function. Studies of the bacteriophage λ repressor, for example, have shown that a large number of combinations of amino acids can satisfy the structural and functional requirements of the hydrophobic core and α-helices of the molecule (Reidhaar-Olson and Sauer 1988; Lim and Sauer 1989).

Scanning Mutagenesis

Alanine-scanning mutagenesis is used to analyze the function(s) of particular amino acid residues on the surface of a protein. The charged residues that normally dapple the surface of proteins are not usually required for structural integrity, but they are generally involved in ligand binding, oligomerization, or catalysis. Systematic replacement of charged amino acids with alanine residues eliminates side chains beyond the β-carbon and disrupts the functional interactions of the amino acids without changing the conformation of the main chain of the protein. Alanine scanning is therefore a powerful method for assigning functions to particular regions of the protein surface (Cunningham and Wells 1989). In an extension of this approach—cysteine-scanning mutagenesis—unpaired cysteine residues are used to replace individual amino acid residues at particular sites in the protein. Unpaired cysteine residues are of average size, uncharged, and hydrophobic. Because they react efficiently with modifying reagents such as *N*-ethylmaleimide, cysteine residues introduced by scanning mutagenesis can be used as biochemical tags to verify the topology of transmembrane proteins and to measure the accessibility of residues to modifying reagents in the aqueous or lipid phases (e.g., see Akabas et al. 1992; Dunten et al. 1993; Kurz et al. 1995; Frillingos and Kaback 1996; He et al. 1996; Frillingos et al. 1997a,b, 1998).

RESEARCH GOALS

In vitro mutagenesis is used specifically to change the base sequence of a segment of DNA. The changes may be localized or general, random or targeted. More general and less specific methods of mutagenesis, such as random mutagenesis using error-prone DNA polymerase (see Protocol 1), are better suited to analysis of regulatory regions of genes, whereas more precise types of mutagenesis, such as overlap extension PCR, site-directed mutagenesis, or, alternatively, megaprimer PCR-based mutagenesis (see Protocols 2, 3, and 7, respectively) are used to understand the contributions of individual amino acids, or groups of amino acids, to the structure and function of a target protein. Both methods—random and targeted—share the virtue of generating mutants in vitro, without phenotypic selection.

During the past decades, several different types of mutagenesis methodologies have been developed that respond to different needs. Although these methods sometimes overlap each other, it is still possible to group the techniques based on the final goal of the research. For example, if the aim of researchers is to analyze the function(s) of particular amino acid residues on the surface of a protein, scanning mutagenesis is well suited for this need. An extension of this approach referred as to "random-scanning mutagenesis" allows researchers to test a broader spectrum of amino acid changes at the targeted positions. For a more comprehensive description of this technique, see Protocol 7.

Alternatively, if the aim is to gather information about the entire "sequence space" (i.e., about the relationship between the amino acid sequence and the 3D structure of the protein), researchers should make use of "saturation mutagenesis." For further details on this technique, see Protocol 6.

If the aim of the project is to test the role of particular residues in the structure, catalytic activity, or ligand-binding capacity of a protein, the methodology of choice should be "oligonucleotide-directed mutagenesis." For a complete description of this methodology, see Protocols 3 and 6. Multisite-directed mutagenesis and megaprimer PCR-based mutagenesis are presented in Protocols 8 and 9.

COMMERCIAL KITS

A list of commercial kits available for the mutagenesis protocols described in this chapter is given in Table 2. These kits simplify the steps for mutagenesis and streamline the whole process. Moreover, each of these kits is reasonably priced and in most cases provides optimized protocols for timely completion of a mutagenesis project.

TABLE 2. Commercial kits available for mutagenesis protocols

Method of mutagenesis	Supplier	Name of kit	Salient features of kit
Random mutagenesis	Clontech	Diversify PCR Random Mutagenesis Kit	Uses *Taq* polymerase for mutagenesis at high magnesium concentration Allows control of mutagenesis rate Allows amplification of large PCR fragment of up to 4 kb Provides opportunity to produce wide mutational diversity
	Agilent	GeneMorph II Random Mutagenesis Kit	Uses a mixture of two enzymes (an error-prone DNA polymerase and a mutated *Taq* DNA polymerase) and is formulated to minimize mutational biases With this kit and using a single buffer condition, one to 16 mutations can be introduced per 1 kb of DNA fragment. The desired mutation rate can be controlled simply by varying the amount of target DNA in the reaction or by varying the number of amplification cycles performed.
Insertion and deletion mutagenesis	Affymetrix	Change-IT Multiple Mutation Site Directed Mutagenesis kit	Creates insertion or deletion Creates deletions as large as 300 bp Single-day method
	Finnzyme	Mutation Generation System Kit	Thousands of insertion clones can be generated from a single kit. Flexibility in mapping mutants of interest Generates insertions of five amino acids in all three reading frames
	Epicentre Biotechnologies	EZ-Tn5 In-Frame Linker Insertion Kit	Allows for generation of random 57-bp insertions that are readable in all three reading frames
Site-directed mutagenesis	Agilent	QuickChange Site-Directed Mutagenesis Kit	Uses DpnI digestion-based protocol (similar to Protocol 3 described in this chapter)
	Affymetrix	Change-IT Multiple Mutation Site Directed Mutagenesis Kit	Creates insertion or deletion Creates deletions as large as 300 bp Single-day method
	New England Biolabs	Phusion Site-Directed Mutagenesis Kit	Simple three-step protocol; Requires 5′-phosphorylated primers
	Life Technologies	GeneArt Site-Directed Mutagenesis System	Combines DNA methylation and amplification steps into a single reaction Eliminates postmutagenesis digestion and purification steps
	Clontech	Transformer Site-Directed Mutagenesis Kit	Highly efficient mutagenesis Uses any double-stranded plasmid No subcloning required
Multisite-directed mutagenesis kit	Affymetrix	Change-IT Multiple Mutation Site Directed Mutagenesis Kit	Creates single or multiple mutations in the plasmid Based on DpnI digestion (see Protocol 3)
	Agilent	QuickChange Lightning Multisite-Directed Mutagenesis Kit	Much faster than other multisite-directed mutagenesis kits (<4 h for completing the mutagenesis and overnight for transformation)
Domain mutagenesis	Agilent	GeneMorph II EZClone Domain Mutagenesis Kit	A specific error-prone DNA polymerase delivers a uniform mutational spectrum. Eliminates the need for restriction sites or subcloning during the mutagenesis process

REFERENCES

Akabas MH, Stauffer DA, Xu M, Karlin A. 1992. Acetylcholine receptor channel structure probed in cysteine-substitution mutants. *Science* **258:** 307–310.

Auerbach C. 1951. Problems in chemical mutagenesis. *Cold Spring Harb Symp Quant Biol* **16:** 199–213.

Auerbach C, Robson JM. 1947. Tests of chemical substances for mutagenic action. *Proc R Soc Edinb Biol* **62:** 284–291.

Beadle GW, Tatum EL. 1941. Genetic control of biochemical reactions in *Neurospora*. *Proc Natl Acad Sci* **27:** 499–506.

Boesch P, Weber-Lotfi F, Ibrahim N, Tarasenko V, Cosset A, Paulus F, Lightowlers RN, Dietrich A. 2011. DNA repair in organelles: Pathways, organization, regulation, relevance in disease and aging. *Biochim Biophys Acta* **1813:** 186–200.

Cunningham BC, Wells JA. 1989. High-resolution epitope mapping of hGH-receptor interactions by alanine-scanning mutagenesis. *Science* **244:** 1081–1085.

Dunten RL, Sahin-Toth M, Kaback HR. 1993. Cysteine scanning mutagenesis of putative helix XI in the lactose permease of *Escherichia coli*. *Biochemistry* **32:** 12644–12650.

Frillingos S, Kaback HR. 1996. Probing the conformation of the lactose permease of *Escherichia coli* by in situ site-directed sulfhydryl modification. *Biochemistry* **35:** 3950–3956.

Frillingos S, Gonzalez A, Kaback HR. 1997a. Cysteine-scanning mutagenesis of helix IV and the adjoining loops in the lactose permease of *Escherichia coli*: Glu126 and Arg144 are essential. *Biochemistry* **36:** 14284–14290.

Frillingos S, Ujwal ML, Sun J, Kaback HR. 1997b. The role of helix VIII in the lactose permease of *Escherichia coli*: I. Cys-scanning mutagenesis. *Protein Sci* **6:** 431–437.

Frillingos S, Sahin-Toth M, Wu J, Kaback HR. 1998. Cys-scanning mutagenesis: A novel approach to structure function relationships in polytopic membrane proteins. *FASEB J* **12:** 1281–1299.

He MM, Sun J, Kaback HR. 1996. Cysteine-scanning mutagenesis of transmembrane domain XII and the flanking periplasmic loop in the lactose permease of *Escherichia coli*. *Biochemistry* **35:** 12909–12914.

Hutchison CA III, Phillips S, Edgell MH, Gillam S, Jahnke P, Smith M. 1978. Mutagenesis at a specific position in a DNA sequence. *J Biol Chem* **253:** 6551–6560.

Kurz LL, Zuhlke RD, Zhang HJ, Joho RH. 1995. Side-chain accessibilities in the pore of a K^+ channel probed by sulfhydryl-specific reagents after cysteine-scanning mutagenesis. *Biophys J* **68:** 900–905.

Lombard DB, Chua KF, Mostoslavsky R, Franco S, Gostissa M, Alt FW. 2005. DNA repair, genome stability, and aging. *Cell* **120:** 497–512.

Muller HJ. 1927. Artificial transmutation of the gene. *Science* **46:** 84–87.

Mullis K, Faloona F, Scharf S, Saiki R, Horn G, Erlich H. 1986. Specific enzymatic amplification of DNA in vitro: The polymerase chain reaction. *Cold Spring Harb Symp Quant Biol* **51:** 263–273.

Saiki RK, Scharf S, Faloona F, Mullis KB, Horn GT, Erlich HA, Arnheim N. 1985. Enzymatic amplification of β-globin genomic sequences and restriction site analysis for diagnosis of sickle cell anemia. *Science* **230:** 1350–1354.

Smith HO, Wilcox KW. 1970. A restriction enzyme from *Hemophilus influenzae*. I. Purification and general properties. *J Mol Biol* **51:** 379–391.

Stevens CM, Mylorie A, Auerbach C, Moser H, Kirk I, Jensen KA, Westergaard M. 1950. Biological action of 'mustard gas' compounds. *Nature* **166:** 1019–1021.

Random Mutagenesis Using Error-Prone DNA Polymerases

"Random mutagenesis" is a technique that allows researchers to develop large libraries of variants of a particular DNA sequence. Once developed, these libraries can then be used for several purposes, including structure–function and directed evolution studies. Random mutagenesis is different from other mutagenesis techniques in that it does not require the researcher to have any prior knowledge about the structural properties of the DNA sequence being targeted, thus allowing for the unbiased discovery of novel or beneficial mutations. For this reason, random mutagenesis is especially useful for protein evolution studies.

The protocol presented below is adapted from Mondon et al. (2010) and is dependent on the hypermutational tendency of the X and Y family of DNA polymerases. Issues regarding DNA polymerases used in this protocol are presented in the Discussion section. Although other methods of random mutagenesis, such as cassette mutagenesis or chemical mutagenesis, do exist, they do not achieve the same mutational spectrum or range as this technique.

This protocol describes mutagenic replication in vitro by a low-fidelity DNA polymerase followed by selective PCR amplification of the newly mutated sequences. The initial mutagenic DNA replication step is accomplished by heat-denaturing the template DNA and annealing primers possessing 5′ extensions that are not complementary to the template. The purpose of the noncomplementary extensions on the primers is to allow for future selection of only the mutant strands. DNA replication is then performed by a low-fidelity DNA polymerase of choice (polymerase β, η, or ι, or any combination of the three). After mutations have been incorporated into the template, the mutagenized strands are then selectively amplified using PCR. Selective amplification of the mutant strands is accomplished by performing a PCR procedure consisting of a first cycle with a low hybridization temperature followed by subsequent selection cycles under higher hybridization temperatures that do not allow amplification of the original unmutagenized template. Figure 1 outlines the major steps of this protocol.

This approach has been used by Mondon et al. (2007) (human DNA polymerases) and by Emond et al. (2008) (human amylosucrase).

MATERIALS

It is essential that you consult the appropriate Material Safety Data Sheets and your institution's Environmental Health and Safety Office for proper handling of equipment and hazardous materials used in this protocol.

Recipes for reagents specific to this protocol, marked <R>, are provided at the end of the protocol. See Appendix 1 for recipes for commonly used stock solutions, buffers, and reagents, marked <A>. Dilute stock solutions to the appropriate concentrations.

Reagents

Because of the length of the Reagents list, the items have been subdivided according to the protocol parts for ease of use.

Cloning of Human DNA Polymerases
Ampicillin (American BioAnalytical)
cDNA for human polymerases (Pol) β, η, and ι (see Step 1)
E. coli TOP10 strain, chemically competent (Life Technologies)

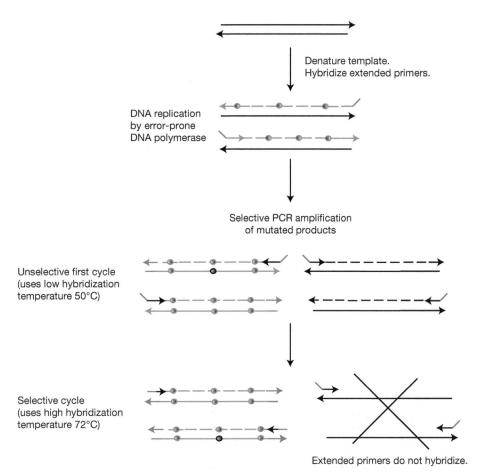

FIGURE 1. Major steps in random mutagenesis by error-prone DNA polymerases. This protocol consists of a single mutagenic replication step followed by selective PCR amplification of the replication products. Primers are designed that carry an extension (green) that is not complementary to the template. After the template has been denatured, primers are allowed to hybridize and replication is performed by a low-fidelity DNA polymerase. The mutant DNA copies (black) are then selectively amplified by a PCR procedure consisting of a first cycle at a low hybridization temperature followed by selective cycles (up to 25) at a high hybridization temperature, disabling template amplification. (Red circles) Sites of random mutations. (Modified from Mondon et al. 2010, with permission from Springer Science+ Business Media.)

Miniprep kit (e.g., QIAprep Spin Miniprep Kit; QIAGEN)
Restriction endonucleases and buffers (New England Biolabs)
Vectors pMG20A and pMG20B
YT medium (2×) <A>

Expression and Purification of Error-Prone Polymerases
Bradford protein assay (Bio-Rad)
Coomassie Blue
Dialysis buffer <R>
E. coli BL21 (DE3) (Stratagene)
Elution buffer <R>
Glycerol (EUROMEDEX)
Isopropyl-β-D-thiogalactopyranoside (IPTG) (1 M stock solution stored at −20°C)
Lysis buffer <R>
Protease inhibitor cocktail (e.g., Roche)
SDS–polyacrylamide gel electrophoresis (SDS-PAGE) gel (40% acrylamide/bis 29:1 solution [Bio-Rad], TEMED [EUROMEDEX], 10% ammonium persulfate [Sigma-Aldrich], 20% SDS [EUROMEDEX])

Washing buffer <R>

YT agar medium (2×) <R>

YT agar medium (2× Amp/1% Glu plates)

> 2× TY agar media containing 100 μg/mL ampicillin and 1% glucose. The antibiotic and glucose are added only when the autoclaved 2× YT agar solution cools to below 50°C.

Preparation of DNA Template

E. coli TOP10 strain, chemically competent (Life Technologies)

LB medium <R>

> Sterilize by autoclaving.

LB/Amp medium <R>

LB/Amp plates <R>

Plasmid vector harboring the gene X to be mutated, pUC18-X (cloned into BamHI, EcoRI restriction sites)

DNA Replication Assay with Error-Prone Polymerases

dNTPs (2.5 mM each deoxyadenosine 5′-triphosphate [dATP], deoxythymidine 5′-triphosphate [dTTP], deoxycytidine 5′-triphosphate [dCTP], and deoxyguanosine 5′-triphosphate [dGTP])

DTT

Ethanol (absolute and 70%, v/v) (Prolabo)

Forward primer MutpUC18_S1 and reverse primer Mutp UC18_R1 (see Fig. 2)

Glycerol (10%)

Human DNA Pol β, Pol η, and Pol ι (purified from Expression and Purification of Error-Prone DNA Polymerases section)

Human DNA Pol β replication buffer <R>

Human DNA Pol η and Pol ι replication buffer <R>

Phenol:chloroform (Sigma-Aldrich)

Plasmid DNA templates (pUC18-X)

Sodium acetate (3 M, pH 5.2)

Water, sterile distilled

Selective Amplification of Replication Products

dNTPs (2.5 mM each deoxyadenosine 5′-triphosphate [dATP], deoxythymidine 5′-triphosphate [dTTP], deoxycytidine 5′-triphosphate [dCTP], and deoxyguanosine 5′-triphosphate [dGTP])

MgCl₂ (50 mM) (Life Technologies)

Milli-Q water, sterile

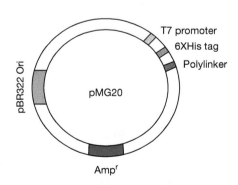

FIGURE 2. **Random mutagenesis.** (*Top*) pMG20A vector. (*Bottom*) Multiple cloning sites (MCS) for pMG20 plasmids. (Modified from Mondon et al. 2010, with permission from Springer Science+Business Media.)

Platinum *Taq* DNA polymerase (Life Technologies)

Platinum *Taq* polymerase buffer (10×) (200 mM Tris-HCl at pH 8.4 and 500 mM KCl) (Life Technologies)

Primers

Forward primer MutpUC18_S1 and reverse primer MutpUC18_R1 (see Step 29)

Cloning of Mutant Libraries

Agarose gel (1%) (1% [w/v] agarose in 1× TAE buffer)

BamHI and EcoRI restriction enzymes, NEB 2 buffer, and 100× BSA solution (New England Biolabs)

E. coli XL1-Blue electrocompetent cells (Stratagene)

SOC medium <R>

T4 DNA ligase and ligation buffer (New England Biolabs)

TAE buffer (50×) <R>

Water, sterile distilled

YT agar medium (2×) <R>

YT (2×)/Amp/1% Glu plates

YT (2×)/Amp/1% Glu/15% glycerol

6 g of Bacto Tryptone, 10 g of yeast extract, and 5 g of NaCl in 1 L of distilled water. Sterilize by autoclaving. The antibiotic, glucose, and glycerol are added only when the autoclaved 2× YT agar solution cools to below 50°C.

Analysis of the Library

LB medium + 100 μg/mL ampicillin

Primers M13(−21) and M13R(−29)

M13(−21): 5'-TGTAAAACGACGGCCAG-3'
M13(−29): 5'-CAGGAAACAGCTATGACC-3'

Equipment

Centrifugal filter devices, Microcon PCR (Millipore)

Centrifuge tubes (50 mL conical; Falcon)

Cold water bath

Dialysis membrane (Visking)

Electroporator 2510 (Eppendorf) and 0.2-cm-gap electroporation cuvettes (Cell Projects UK)

Erlenmeyer flask (1 L), sterile glass (Pyrex)

Homogenizer, ultrasonic (Bandelin SONOPULS HD2200, probe TT13 flat tip)

Incubator shaker, refrigerated rotating (Sartorius)

Microcon PCR spin-column device (Millipore)

MilleGen high-throughput sequence service (see Step 48)

Montage Plasmid Miniprep HTS 96 Kit

MutAnalyse software

Ni–NTA resin (QIAGEN)

PCR thermal cycler (Thermo Cycler PT100 MJ Research)

Plasmid miniprep kit (QIAGEN)

Poly-Prep Chromatography Column, empty (Bio-Rad)

QIAprep Spin Midiprep Kit (QIAGEN)

Spectrophotometer

Speed Vac system (ISS110, Thermo Savant)

Syringe (50 mL) (Terumo)

Syringe filter (0.45 μm) (Luer-Lok)

Tecan Genesis RSP200 platform (or similar robotic system)

UltraClean GelSpin DNA Extraction kit (MO BIO Laboratories)

METHOD

Cloning of Human DNA Polymerases

1. Obtain cDNA for human Pol β (Pubmed GeneID:5434), Pol η (Pubmed GeneID:5429), and Pol ι (Pubmed GeneID: 11201) from a certified source such as Open Biosystems.

2. Subclone the cDNA for Pol β into the pMG20B vector for expression purposes. cDNA for Pol η and Pol ι should be subcloned into the pMG20A vector (see Fig. 2).

3. Set up three separate transformations of chemically competent *E. coli* TOP10 cells with plasmids carrying genes for Pol β, Pol η, and Pol ι DNAs. Grow 5-mL small-scale overnight cultures of individual transformants in $2\times$ YT medium containing 100 µg/mL ampicillin.

4. Use a small-scale "miniprep" method (e.g., QIAprep Spin Miniprep Kit) to isolate plasmid DNA from the small-scale cultures (Step 3).

5. It is recommended that double-strand DNA sequencing be performed on the purified recombinant vectors containing each human DNA polymerase before use.

Expression and Purification of Human Error-Prone Polymerases

6. Transform each of the vectors carrying cloned copies of the cDNAs for human error-prone DNA polymerases into *E. coli* strain BL21(DE3). Plate the transformed cells onto $2\times$ YT agar plates containing 100 µg/mL ampicillin. Incubate the plates overnight at 37°C.

7. Transfer 10 mL of $2\times$ YT containing 100 mg/mL ampicillin into a series of 50-mL Falcon tubes. Pick several well-separated transformants from each plate and grow overnight cultures at 37°C in a shaking incubator.

8. Measure the OD_{600nm} of each overnight culture using a spectrophotometer. Set up three Erlenmeyer flasks, each containing 300 mL of $2\times$ YT and 100 mg/mL of ampicillin. Inoculate each of the large-scale cultures with a volume of one of the starter cultures sufficient to achieve an initial OD_{600} of 0.1. Incubate the cultures at 37°C with rotary shaking at 230 rpm, monitoring the OD_{600nm}.

9. When the OD_{600nm} reaches 0.8, shift each culture to a 15°C cold water bath. Induce the synthesis of the human polymerases by adding 0.2 mM IPTG to each flask and incubating the cultures for 5 h at 15°C with rotary shaking at 230 rpm. Then centrifuge the cultures at 3000*g* for 15 min. Store the resulting cell pellets at -20°C while awaiting purification.

10. Homogeneously resuspend the cell pellet from each expression culture into 20 mL of Lysis buffer containing protease inhibitors (following the manufacturer's instructions).

11. Lyse the cells in a beaker immersed in cold water and ice using an ultrasonic homogenizer. If using the Bandelin SONOPULS HD 2200 (recommended), set the power to 25% and deliver four pulses of 2 min, with cooling between each pulse.

12. Centrifuge the lysates at 16,200*g* for 30 min at 4°C.

13. Filter the supernatants through 0.45-µm filters with 50-mL syringes. Then pass each supernatant through 1-mL Ni–NTA resin columns (50%, v/v) pre-equilibrated with Lysis buffer.

14. Wash the resins with at least 25 mL of Washing buffer after passing the supernatants.

15. Elute the protein in five fractions each of 500 µL of Elution buffer. Evaluate each fraction's protein quantity and quality, respectively, by a Bradford assay (see Chapter 19, Protocol 10) and SDS-PAGE (see Chapter 2, Protocols 3–5), followed by Coomassie Blue staining. Fractions containing the greatest concentration of the error-prone DNA polymerase of the correct protein size should be prioritized.

    See Troubleshooting.

16. Using the results obtained in Step 15, pool the prioritized fractions corresponding to each of the three DNA polymerases separately, and dialyze them separately overnight against a large volume of ice-cold Dialysis buffer.

17. Add glycerol to the dialyzed fractions to a final concentration of 50% v/v glycerol. Measure the final protein quantity by Bradford protein assay. Store the pooled fractions at −80°C.

Preparation of DNA Template

18. Clone gene X to undergo random mutagenesis between the BamHI and EcoRI sites of pUC18-X (see Fig. 3).

    Gene X can be a cDNA coding for a gene or a gene fragment up to 3 kb. In addition, if gene X is cloned using restriction sites other than BamHI and EcoRI, the primers MutpUC18_S1 and MutpUC18_R1 will need to be redesigned with the correct restriction sites incorporated.

19. Transform the pUC18-X vector containing gene X into TOP10 *E. coli* cells, and streak them onto an LB/Amp agar plate.

20. Inoculate a single colony of TOP10 *E. coli* cells harboring the BamHI and EcoRI sites into 5 mL of LB containing 100 μg/mL ampicillin. Incubate the culture with shaking at 200 rpm for 16 h at 37°C.

21. After inoculation, prepare the plasmid DNA using the QIAprep Spin Midiprep Kit according to the manufacturer's instructions. Store the purified pUC18-X vector at −20°C.

DNA Replication Assay with Error-Prone Polymerases

22. The table below shows the expected frequency of mutations generated during in vitro DNA synthesis reactions catalyzed by the three DNA polymerases. The frequency (expressed as mutations per kilobase of DNA synthesized) is affected by the concentrations of dATP and Mn^{2+} in the reaction mixtures.

	Pol β	Pol η	Pol η-ι
Condition A	2–4	ND	ND
Condition B	15–20	ND	ND
Condition E	6–10	ND	ND
Condition N	ND	7–9	10–12

    (ND) Not determined.

    The range of mutation frequencies desired for a particular strand of DNA will determine the type of human polymerase used (Pol β, Pol η, or Pol η-ι) as well as the reaction conditions (see Step 23). Each DNA polymerase has its own mutational properties and generates a unique mutational spectrum. To obtain the most diverse mutational library, it is best to pool the libraries generated by the three DNA polymerases.

23. In an Eppendorf tube, combine 1 μg of the DNA of plasmid pUC18-X, 200 nM of each primer (MutpUC18_S1 and MutpUC18_R1), and 2 μL of 10× Replication buffer. Add dNTPs and

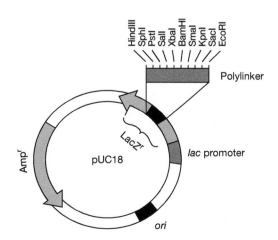

FIGURE 3. **Struture of pUC18.** (Adapted from Griffiths et al. 2000.)

Mn^{2+} in the amounts required to achieve the desired frequency of mutations (see table below). Add H_2O to adjust the final volume of the reactions to 20 μL. This will be the DNA replication mixture.

	dATP (μM)	dATP (μM)	dATP (μM)	dATP (μM)	Mn^{2+} (mM)
Condition A	50	50	100	100	N/A
Condition B	20	100	100	100	0.5
Condition E	20	100	100	100	0.25
Condition N	100	100	100	100	N/A

24. Denature the template DNA by incubation in a water bath for 5 min at 95°C. Immediately transfer the Eppendorf tubes to an ice bath.

25. To each tube, add 4 units of DNA polymerase β, η, or ι in 10 μL of DNA replication buffer to the DNA replication mix.

    One unit is defined as the amount of enzyme required to catalyze the incorporation of 1 nmol of dNTP into an acid-insoluble form in 1 h at 37°C.

26. Incubate the DNA replication mixture for 1 h at 37°C.

27. Purify the mutagenized DNA from the reaction mixture by extraction with phenol:chloroform (see Chapter 1).

28. Precipitate the DNA by adding 0.1× volume of 3 M sodium acetate and 2.5× volume of absolute ethanol. Incubate for at least 2 h at −20°C, and then centrifuge for 20 min at 13,000*g* at 4°C. Wash the pellet two times with 500 μL of 70% ethanol, dry the pellet 3 min in a SpeedVac rotary evaporator, and dissolve the dried pellet in 20 μL of sterile Milli-Q water. Quantify the DNA using a spectrophotometer.

Selective Amplification of Mutagenized DNA Products

29. Perform the selective amplification PCR in a total volume of 30 μL containing 5 ng of purified mutagenized replication product from Step 28, 3 μL of 10× Platinum *Taq* polymerase buffer, 200 μM dNTP, 0.2 μM of primer MutpUC18_S1 and MutpUC18_R1, and 1 U of Platinum *Taq* polymerase. A guideline for the selective PCR cycles is as follows:

Cycle number	Denaturation	Annealing	Polymerization
1 (Denaturation)	2 min at 94°C	N/A	N/A
2 (Low stringency)	2 min at 94°C	10 sec at 58°C	2 min at 72°C
3–28 (High stringency)	20 sec at 94°C	1:30 min at 72°C	Continuation of annealing step

The selective PCR process is based on two different annealing temperatures. The annealing temperature of the first cycle (58°C) allows the hybridization of the primers. Cycles 3–28 are carried out at higher stringency, using an annealing temperature of 72°C.

Primers MutpUC18_S1 and MutpUC18_R1 are designed to anneal to pUC18 at the 5′ end upstream of the BamHI site (27 nucleotides) and at the 3′ end downstream from the EcoRI site, respectively. The sequences of the primers MutpUC18_S1 and MutpUC18_R1 should be as follows:

Primer name	Sequence	Restriction sites
MutpUC18_S1	5′-TCTGACGAGTACTAGCTGCTACATGCA GGTCGACTCTAGAGGATCC-3′	BamHI
MutpUC18_R1	5′-ACAGCTACGTGATACGACTCACACTATG ACCATGATTACGAATTCC-3′	EcoRI
M13 (−21)	5′-TGTAAAACGACGGCCAG-3′	N/A
M13R (−29)	5′-CAGGAAACAGCTATGACC-3′	N/A

30. Purify the PCR products using a Microcon filtration device following the manufacturer's instructions. Quantify the purified DNA using a spectrophotometer.
    See Troubleshooting.

Cloning of Mutant Libraries

31. In separate reactions, digest the totality of the purified PCR products and 2 μg of pUC18 vector with 60 units of BamHI and 60 units of EcoRI.

32. Incubate the reactions for 6 h at 37°C.

33. Purify the digested DNAs (PCR product and pUC18) by electrophoresis through a 1% agarose gel (see Chapter 2, Protocols 1 and 2). Extract the DNA using the UltraClean GelSpin DNA Extraction Kit. Resuspend the purified DNA in 10 μL of water. Measure the concentration of the DNAs by analyzing 1 μL of the DNA solutions on a 1% agarose gel using markers of known size and concentration.

34. Set up a ligation reaction using T4 DNA ligase and a 1:3 molar ratio of vector and insert DNAs.

35. Incubate the reaction overnight at 16°C.

36. Purify the ligation reaction using a Microcon filtration device following the manufacturer's instructions.

37. Add 2 μL of the purified ligation product to 50 μL of XL1-Blue electrocompetent cells. Transfer the mixture to a 0.2-cm-gap electroporation cuvette.

38. Electroporate the cells at 1.8 kV, 25 μF, and 200 W.

39. Add 450 μL of SOC medium to the electrotransformed cells, and incubate the cells with gentle shaking for 1 h at 37°C in an Eppendorf tube.

40. Pool the transformants and plate them on one or two dishes of 2× YT medium containing 100 μg/mL ampicillin and 1% glucose. Plate serial dilutions onto additional plates containing the same medium to determine the size of the library.

41. Incubate the plates overnight at 37°C.

42. After counting colonies on the plate seeded with the diluted stock, scrape the colonies from the plates into 2× YT/Amp/1% Glu medium containing 15% glycerol (v/v), and store the stock at −80°C.

Analysis of the Library

43. Randomly pick 4×96 clones from the dilution plates, and inoculate them into four 96-deep-well plates (from the Montage Plasmid Miniprep HTS 96 Kit) containing LB liquid medium+100 μg/mL ampicillin.

44. Incubate the plates overnight at 37°C under agitation (800 rpm) using an incubator shaker.

45. Isolate the plasmid DNAs using the Montage Plasmid Miniprep HTS 96 Kit on an integrated robotic Tecan Genesis RSP200 platform.
    Similar robotic systems for plasmid minipreps are also available from QIAGEN and other suppliers.

46. Sequence the mutated portions of the DNA using the primers M13 (−21) and M13R (−29) and MilleGen's high-throughput sequencing service (for primer design, see Step 29 and Table 1).

47. Analyze the sequences using MutAnalyse software to determine mutation frequency, library quality (frequency of deletions, insertions, and substitutions), frequency of wild-type sequence, number of mutations per variant, and the distribution of substitutions along the mutagenized DNA.
    See Troubleshooting.

TABLE 1. Primers used for the cloning of DNA Pol β, Pol η, and Pol ι cDNAs and for the construction of random mutagenesis libraries

Primer name	Sequence	Restriction sites
ETAS1	5'-AATAGGATCCATGGCTACTGGACAGGATCG-3'	BamRI
ETAR1	5'-AATAGAATTCCTAATGTGTTAATGGCTTAAAAAATGATTCC-3'	EcoRI
BetaS1	5'-TAGATCATATGAGCAAACGGAAGGCGCCG-3'	NdeI
BetaR1	5'-GACTAAGCTTAGGCCTCATTCGCTCCGGTC-3'	HindIII
IOTAS1	5'-ATATGGATCCATGGAACTGGCGGACGTGGG-3'	BamHI
IOTAR1	5'-TAATAAGCTTTTATTTATGTCCAATGTGGAAATCTGATCC-3'	HindIII
MutpUC18_S1	5'-TCTGACGAGTACTAGCTGCTACATGCAGGTCGACTCTAGAGGATCC-3'	BamHI
MutpUC18_R1	5'-ACAGCTACGTGATACGACTCACACTATGACCATGATTACGAATTCC-3'	EcoRI
M13 (−21)	5'-TGTAAAACGACGGCCAG-3'	
M13R (−29)	5'-CAGGAAACAGCTATGACC-3'	

Reprinted from Mondon et al. 2010, with permission of Springer Science+Business Media.

TROUBLESHOOTING

Problem (Step 15): There are extra proteins in the eluent beside the desired polymerase.
Solution: Block the resin with a protein such as BSA before applying the sample to resin. Wash the column thoroughly before elution.

Problem (Step 15): Low polymerase protein is detectable in the Coomassie Blue assay.
Solution: One possibility is that the polymerases were degraded during the process of purification. Try adding more protease inhibitors. It is also possible that cells were not in the correct stage of log growth for effective induction by IPTG in Step 9.

Problem (Step 30): There is no or little PCR product after selective amplification of mutagenized PCR products.
Solution: First check to see that the primers MutpUC18_S1 and MutpUC18_R1 are capable of driving amplification efficiently. If so, then run a gradient PCR by varying the annealing temperature of the MutpUC18_S1 and MutpUC18_R1 primers.

Problem (Step 47): There is no or little mutagenesis detected upon mutational analysis.
Solution: Try another replication condition in Step 23. In addition, try the initial mutagenesis step using a different polymerase, or try different polymerases in combination.

DISCUSSION

In general, DNA polymerases can be classified into seven different structural families (A, B, C, D, X, Y, and RT) on the basis of structural similarities (Rattray and Strathern 2003). The fidelity of nucleotide incorporation varies according to each DNA polymerase family and is determined by several factors, such as the presence of a $3'-5'$ exonuclease domain. For example, replicative DNA polymerases from families A, B, and C are higher-fidelity enzymes with nucleotide incorporation error frequencies of $\sim 10^{-6}$ (i.e., one error per million nucleotides incorporated).

On the other hand, DNA polymerases of the X and Y structural families are excellent for the purpose of random mutagenesis because of their low fidelity of nucleotide incorporation. For example, DNA Pol β, a well-known member of the X family of DNA polymerases, has an error frequency ranging from 10^{-3} to 10^{-4} (Kunkel 1985). The Y family of polymerases, also known as translesion synthesis (TLS) polymerases, replicate DNA in a distributive manner and lack exonucleolytic proofreading activity. It is no surprise, then, that they show the highest error rates measured among DNA polymerases (10^{-1} to 10^{-3}) (Yang 2005). In humans, three known members of the Y family of polymerases are Pol η, Pol ι, and Pol κ.

RECIPES

It is essential that you consult the appropriate Material Safety Data Sheets and your institution's Environmental Health and Safety Office for proper handling of equipment and hazardous materials used in this protocol.

Dialysis Buffer

Reagent	Final concentration
Tris-HCl (pH 8.0)	40 mM
DTT	2 mM
EDTA	0.2 mM
NaCl	200 mM

Elution Buffer

Reagent	Final concentration
NaH_2PO_4/Na_2HPO_4 (pH 8.0)	50 mM
NaCl	300 mM
Imidazole	250 mM

Human DNA Pol β Replication Buffer

Reagent	Final concentration
Tris-HCl (pH 8.8)	50 mM
$MgCl_2$	10 mM
KCl	100 mM
DTT	1 mM
Glycerol	10%

Human DNA Pol η and Pol ι Replication Buffer

Reagent	Final concentration
Tris-HCl (pH 7.2)	25 mM
DTT	1 mM
$MgCl_2$	5 mM
Glycerol	2.5% (v/v)

LB Medium

Reagent	Quantity (for 1 L)
NaCl	5 g
Bacto Tryptone	10 g
Yeast extract	5 g

Dissolve in 1 L of distilled water; sterilize by autoclaving.

LB/Amp Medium

Prepare LB medium containing 100 μg/mL ampicillin (stock at 50 mg/mL in distilled water).

LB/Amp Plates

Prepare LB medium containing 1.5% (w/v) agar and 100 μg/mL ampicillin (stock at 50 mg/mL in distilled water). Add the antibiotic only when the autoclaved LB-agar solution cools to below 50°C.

Lysis Buffer

Reagent	Final concentration
NaH_2PO_4 (pH 8.0)	50 mM
NaCl	300 mM
Imidazole	10 mM
Triton X-100	0.05%
EDTA	1 mM
DTT	1 mM
Lysozyme	1 mg/mL

SOC Medium

Reagent	Final concentration
Tryptone	2%
Yeast extract	0.5%
NaCl	10 mM
KCl	2.5 mM
$MgCl_2$	10 mM
$MgSO_4$	10 mM
Glucose	20 mM

TAE Buffer (50×)

Reagent	Quantity (for 1 L)
Tris (hydroxymethyl) aminomethane	242 g
Acetic acid	57.1 mL
Na_2EDTA	7.43 g

Dissolve in distilled water and make volume up to 1 L.

Washing Buffer

Reagent	Final concentration
NaH_2PO_4/Na_2HPO_4 (pH 8.0)	50 mM
NaCl	300 mM
Imidazole	20 mM
Triton X-100	0.05%
EDTA	1 mM
DTT	1 mM

YT Agar Medium (2×)

Reagent	Quantity (for 1 L)
Agar	15 g
Bacto Tryptone	16 g
Yeast extract	10 g
NaCl	5 g

Dissolve in 1 L of distilled water; sterilize by autoclaving.

REFERENCES

Emond S, Mondon P, Pizzut-Serin S, Douchy L, Crozet F, Bouayadi K, Kharrat H, Potocki-Veronese G, Monsan P, Remaud-Simeon M. 2008. A novel random mutagenesis approach using human mutagenic DNA polymerases to generate enzyme variant libraries. *Protein Eng Des Sel* **21:** 267–274.

Griffiths AJF, Miller JH, Suzuki DT, Lewontin RC, Gelbart WM. 2000. *An introduction to genetic analysis*, 7th ed. WH Freeman, New York.

Kegler-Ebo DM, Docktor CM, DiMaio D. 1994. Codon cassette mutagenesis: A general method to insert or replace individual codons by using universal mutagenic cassettes. *Nucleic Acids Res* **22:** 1593–1599.

Kunkel TA. 1985. The mutational specificity of DNA polymerase-β during in vitro DNA synthesis. Production of frameshift, base substitution, and deletion mutations. *J Biol Chem* **260:** 5787–5796.

Mondon P, Souyris N, Douchy L, Crozet F, Bouayadi K, Kharrat H. 2007. Method for generation of human hyperdiversified antibody fragment library. *Biotechnol J* **2:** 76–82.

Mondon P, Grand D, Souyris N, Emond S, Bouayadi K, Kharrat H. 2010. Mutagen: A random mutagenesis method providing a complementary diversity generated by human error-prone DNA polymerases. *Methods Mol Biol* **634:** 373–386.

Rattray AJ, Strathern JN. 2003. Error-prone DNA polymerases: When making a mistake is the only way to get ahead. *Annu Rev Genet* **37:** 31–66.

Yang W. 2005. Portraits of a Y-family DNA polymerase. *FEBS Lett* **579:** 868–872.

Creating Insertions or Deletions Using Overlap Extension PCR Mutagenesis

Overlap extension PCR mutagenesis can be used for the generation of a specific point mutation, insertion, or deletion within a particular DNA sequence of interest. Overlap extension PCR mutagenesis requires relatively little preparation compared with other mutagenesis methods and does not require the use of restriction enzymes. Because of its versatility, the method has become widely used. Unlike methods of random mutagenesis, directed mutagenesis requires that the researcher already have a specific mutation in mind to implement.

Traditional overlap extension PCR mutagenesis protocols remain limited in several critical ways, especially when it comes to generating insertions and deletions. For example, traditional protocols require that all sequence alterations be embedded within the primer itself, which makes it difficult to make insertions >30 nucleotides.

This protocol describes an overlap extension PCR mutagenesis method that is more versatile than its predecessors. Using this method, one can essentially make insertions and deletions of any size at any position within a given DNA sequence. To generate an insertion mutation, first prepare an insertion fragment and two flanking fragments by PCR. In the secondary PCR, the insertion fragment is recombined with two flanking fragments derived from the original template. The "chimeric primers" are so named because they are composed of two connected DNA sequences: an 18-nucleotide sequence derived from the insertion cassette and a 9-nucleotide sequence derived from the original template. The 9-nucleotide tail at the 5′ ends of the chimeric primers is critical because it allows the insertion fragment to form hybrids with the flanking DNA sequences. Figure 1A provides an overall schematic of the steps involved in generating an insertion. This method can be used to generate deletions, which is discussed in the latter part of the protocol (Fig. 1B). This protocol is adapted from Lee et al. (2004, 2010).

MATERIALS

It is essential that you consult the appropriate Material Safety Data Sheets and your institution's Environmental Health and Safety Office for proper handling of equipment and hazardous materials used in this protocol.

Recipes for reagents specific to this protocol, marked <R>, are provided at the end of the protocol. See Appendix 1 for recipes for commonly used stock solutions, buffers, and reagents, marked <A>. Dilute stock solutions to the appropriate concentrations.

Reagents

Agarose gel (1%), containing 0.5 µg/mL ethidium bromide
DNA template

Prepare the DNA template using either PEG precipitation or commercial resin kits (such as the QIAprep Spin Miniprep kit from QIAGEN). After purification, the plasmid DNA should be resuspended in 1× Tris-EDTA buffer <R>.

dNTP solution (solution containing four dNTPs, each at a concentration of 2.5 mM)
PCR buffer (10×) <R>
PCR polymerase (2.5 U/µL)

Care should be taken in the choice of Taq DNA. Recommended polymerases are Pfu (Agilent) or some other high-fidelity Taq DNA polymerase.

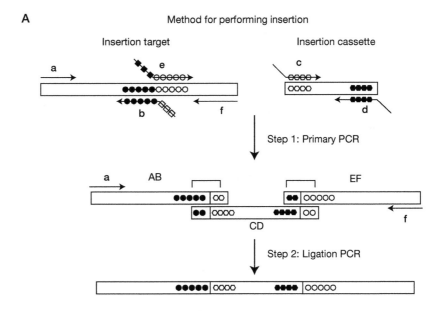

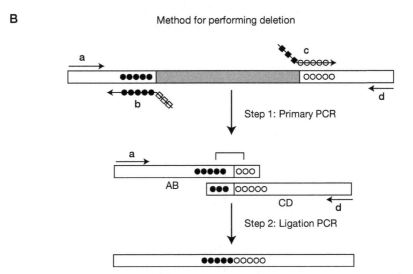

FIGURE 1. **How to generate insertions and deletions using overlap extension PCR.** (*A*) Overlap extension PCR protocol for insertion mutagenesis. The features of an insertion cassette are denoted by hexagons (open or closed), whereas the features of the flanking regions are denoted by circles (open or closed). Chimeric primers of 27 nucleotides (e.g., **b** and **e**) in size are composed of 18 nucleotides derived from the template (circles) and 9 nucleotides from the sequence (hexagons) to be added. Note that each hexagons or circle represents a 3-nucleotide sequence. (Arrow) The 3′ end of the primer. The site where an insertion cassette will be inserted is demarcated by two neighboring circles with different shading (i.e., open or closed). In the first PCR, three PCR products (i.e., AB, CD, and EF) are prepared first by separate reactions with appropriate primer pairs. For example, the DNA fragment AB is a PCR product by primers **a** and **b**, and so forth. Note that the resulting PCR fragments harbor the 18-nucleotide overlap region (bracket) at either one (AB and EF) or both ends (CD). In another reaction, PCR is performed with the outermost flanking primer pair (**a** and **f**) using a mixture of the above three PCR fragments as template. Owing to their terminal complementarities in the overlap region, these products will anneal and subsequently be extended during the first cycle of the second step of PCR. In the second PCR step, the PCR fragments from the first PCR are essentially ligated together. (*B*) The two PCR products that make up the flanking regions of the sequence to be deleted are prepared by using one nonchimeric and one chimeric primer: either **a** and **b** or **d** and **c**, respectively. In the second PCR step, the two products from the first PCR (AB and CD) are used as the template for "a ligation PCR" that contains the outermost primer pair (**a** and **d**). (Gray) The sequence to be deleted; (circles, open or closed) the flanking regions. (Modified from Lee et al. 2004, with permission from *BioTechniques*; and Lee et al. 2010, with permission from Springer Science+Business Media.)

Sterile H$_2$O

Synthetic primers (for design instructions, see the protocol)

Dissolve synthetic nucleotides in water at a concentration of 10 picomol/μL.

Equipment

Gel electrophoresis equipment or QIAquick Gel Extraction kit (QIAGEN)

Microcentrifuge tubes (0.5 mL; thin-walled for PCR amplification)

Micropipettes

NanoDrop machine

Sterile razors (for gel purification)

Thermal cycler

UV light

METHOD

Generation of Insertions

An insertion fragment and two flanking fragments are prepared by PCR, using the appropriate primer pairs. Each of the two flanking fragments contains an overlap region at its 5' end. The insertion fragment has an overlap region at both ends. In the secondary PCR, the insertion and flanking fragments are mixed together, melted, reannealed, and extended to generate the final product. The resulting extended molecule can then be amplified in subsequent PCR cycles using the outermost primer pair (see Fig. 1A).

Primer Design

1. In total, six primers are required for this protocol. Four of these primers are chimeric (primers **b**, **c**, **d**, and **e**) and are used for the primary PCR step. The two other primers (primers **a** and **f**) are nonchimeric and are used for the secondary PCR step.

2. The sequences of the four chimeric primers (primers **b**, **c**, **d**, and **e**) are derived partly from the insertion cassette DNA sequence and partly from the original DNA template (hence the name "chimeric"). To generate the two flanking PCR fragments (i.e., AB and EF), 27-nucleotide-long chimeric primers (primers **b** and **e**) should be designed so that 18 nucleotides of sequence are derived from the template and 9 nucleotides are derived from the insertion cassette (see Fig. 1A). Take care to design the primer with correct polarity: the 18-nucleotide sequence should be placed at the 3' side, whereas the 9-nucleotide sequence is placed at the 5' side of the chimeric primer.

3. Likewise, to generate the insertion fragment (i.e., CD), two more 27-nucleotide chimeric primers (i.e., primers **c** and **d**) are designed such that 18 nucleotides of sequence are derived from the insertion cassette and 9 nucleotides of the sequence are derived from the template. The use of these two primers allows for the generation of two 9-nucleotide terminal sequences on either side of the insertion fragment that are complementary to the two flanking PCR fragments (i.e., AB and EF), allowing for the creation of a hybrid template in the secondary PCR.

4. The two outermost flanking primers (i.e., **a** and **f**) also need to be designed to perform the secondary PCR. The flanking primers may also include restriction sites at their ends to facilitate subsequent cloning of the PCR product in a vector.

Primary PCR

In the primary PCR, three independent PCRs are set up to generate three PCR fragments (AB, EF, and CD) that will be combined to generate the extended product in the secondary PCR (see Fig. 1A). Two of the PCR products generated in the primary PCR step correspond to the flanking regions of the insertion site (i.e., AB and EF), whereas the third PCR product corresponds to the insertion cassette (i.e., CD). The flanking fragments (i.e., AB and EF) are each prepared using one nonchimeric primer and one chimeric primer: primers **a** and **b** or primers **e** and **f**, respectively. As a result, the flanking fragments AB and EF both contain 9 nucleotides corresponding at one end to the original insertion cassette. The insertion fragment (i.e., CD) is prepared with two chimeric primers (i.e., primers **c** and **d**) and has terminal regions of overlap with both of the flanking PCR fragments.

5. Each of the three PCR fragments (AB, CD, and EF) are generated in 50-μL reactions with ∼10–20 ng of PCR template DNA from either the original template DNA to be modified or the insertion cassette, according to the following PCR recipe:

Reagent	Volume added/mass added
PCR template DNA	10–20 ng
PCR buffer (10×)	5 μL
dNTPs	4 μL
Primer 1 (10 μM)	2 μL
Primer 2 (10 μM)	2 μL
Taq DNA polymerase (2.5 U/μL)	0.5 μL
Sterile H_2O	to 50 μL

Different templates and primers are used to generate three PCR fragments.

For Fragment AB:

Original template DNA is mixed with primers **a** and **b**.

For Fragment CD:

Insertion cassette DNA is mixed with primers **c** and **d**.

For Fragment EF:

Original template DNA is mixed with primers **e** and **f**.

6. Run each of the separate PCRs in a thermal cycler using the following PCR program.

Cycle number	Denaturation	Annealing	Polymerization
1	2 min at 98°C		
2–26	10 sec at 98°C	15 sec at 55°C	1 min at 72°C

After the last cycle, incubate each reaction for another 10 min at 72°C to ensure that the final extension step goes to completion.

See Troubleshooting.

7. To purify the PCR products that will be combined in the secondary PCR, load the entire PCR onto a 1% agarose gel containing 0.5 μg/mL ethidium bromide, and perform gel electrophoresis (see Chapter 2). After gel electrophoresis, visualize the PCR product on the gel under UV light to check the efficiency of amplification. A commercial kit (such as the QIAquick Gel Extraction Kit) can be used to purify the PCR product. Before moving on to the next step, quantify the purified DNA using a NanoDrop machine.

Secondary PCR

The purpose of the second PCR is to combine the three PCR fragments generated from the first PCR together to form the final product. The three PCR products (AB, CD, and EF) from the primary PCR are first combined; annealing then occurs because of the complementarity of the terminal oligonucleotides. The resulting hybrid DNA molecules are then extended and amplified by *Taq* polymerase through PCR using the two outermost primers (i.e., **a** and **f**).

8. Set up a 50-μL PCR by combining 10–20 ng of each of the three gel-purified PCR fragments from the primary PCR in a single microcentrifuge tube as follows:

Reagent	Volume/mass added
Gel-purified PCR fragments from primary PCR (AB, CD, and EF)	10–20 ng
PCR buffer (10×)	10 μL
dNTPs	4 μL
Primer **a** (10 μM)	2 μL
Primer **f** (10 μM)	2 μL
Taq DNA polymerase (2.5 U/L)	0.5 μL
Sterile H_2O	to 50 μL

9. Run the PCR in a thermal cycler using the following program.

Cycle number	Denaturation	Annealing	Polymerization
1	2 min at 98°C		
2–26	10 sec at 98°C	15 sec at 55°C	1 min at 72°C

After the last cycle, incubate the reaction mixture for another 10 min at 72°C to let extension go to completion.

> *See Troubleshooting.*

10. Gel-purify the final PCR product as before. The final PCR product can now be cloned into a vector of choice and sequenced for confirmation of mutagenesis.

Generation of Deletions

The strategy used to generate a deletion is similar to the one used to generate an insertion but involves the generation of two PCR fragments in the primary PCR, instead of three as seen in the previous strategy for generating insertions (see Fig. 1B). To generate deletions, PCR is used to generate two fragments that correspond to the flanking regions of the cassette to be deleted. Both of these generated PCR fragments contain a terminal sequence derived from the template on the other side of the sequence to be deleted. These terminal sequences allow for hybridization of the two flanking fragments, which can be subsequently extended to generate a shortened final fragment in the secondary PCR step. In other words, deletion is essentially a ligation or recombination of the two flanking DNA fragments.

Primer Design

11. In total, four primers are required to generate a deletion: two chimeric primers (primers **b** and **c**) and two nonchimeric primers (primers **a** and **d**).

12. The two chimeric primers (i.e., primers **b** and **c**) are designed so that 18 nucleotides of sequence are derived from the last 18 nucleotides of the template adjacent to the deletion region and 9 nucleotides are derived from the nearest 9 nucleotides of the template on the other side of the deletion (Fig. 1B). Be sure to design the primers with the correct polarity.

13. The two outermost flanking primers (i.e., primers **a** and **d**) should be designed in such a way as to allow PCR amplification of the final deletion product. The flanking primers may include restriction sites at their ends to facilitate subsequent cloning of the ligated PCR product.

Primary PCR

In the primary PCR, two independent PCRs are set up to generate fragments AB and CD, which will be hybridized together in the secondary PCR (see Fig. 1B). The two PCR products from this primary PCR correspond to the two flanking regions of the cassette to be deleted. Each flanking fragment is generated using one chimeric primer and one nonchimeric primer.

14. Each of the two PCR fragments (AB, CD) will be generated in a 50-μL reaction with ∼10–20 ng of template DNA.

Reagent	Volume/mass added
Original template DNA	10–20 ng
PCR buffer (10×)	5 μL
dNTPs	4 μL
Primer 1 (10 μM)	2 μL
Primer 2 (10 μM)	2 μL
Taq DNA polymerase (2.5 U/μL)	0.5 μL
Sterile H₂O	to 50 μL

For Fragment AB:

Use primers **a** and **b**.

For Fragment CD:

Use primers **c** and **d**.

15. Run the PCR in a thermal cycler using the following PCR program.

Cycle number	Denaturation	Annealing	Polymerization
1	2 min at 98°C		
2–26	10 sec at 98°C	15 sec at 55°C	1 min at 72°C

After the last cycle, incubate the reaction mixture for another 10 min at 72°C to allow the extension step to go to completion.

See Troubleshooting.

16. Purify the PCR products, which will serve as templates in the next PCR, by loading the entire volume of each PCR onto a 1% agarose gel containing 0.5 μg/mL ethidium bromide and performing gel electrophoresis (see Chapter 2). After gel electrophoresis, visualize the PCR product on a gel under UV light to determine amplification efficiency. A commercial kit (such as the QIAquick Gel Extraction Kit) can be used to purify the PCR product. Before moving on to the next step, quantify the purified DNA using a NanoDrop machine.

Secondary PCR

The purpose of the secondary PCR is to generate the desired deletion by extending the two products of the first PCRs. In the secondary PCR, PCR products AB and CD will be combined; annealing then occurs because of the complementarity of terminal oligonucleotides on the two PCR fragments. The resulting hybrid DNA molecules are then extended and amplified by *Taq* polymerase to generate the desired deletion product.

17. Set up a 50-μL PCR by combining 10–20 ng of the two gel-purified PCR fragments from the primary PCR in a microcentrifuge tube.

Reagent	Volume/mass added
Gel-purified PCR products (fragments AB and CD) from the primary PCR	10–20 ng
PCR buffer (10×)	10 μL
dNTPs	4 μL
Primer **a** (10 μM)	2 μL
Primer **d** (10 μM)	2 μL
Taq DNA polymerase (2.5 U/L)	0.5 μL
Sterile H₂O	to 50 μL

18. Run the PCR in a thermal cycler using the following PCR program.

Cycle number	Denaturation	Annealing	Polymerization
1	2 min at 98°C		
2–26	10 sec at 98°C	15 sec at 55°C	1 min at 72°C

After the last cycle, incubate for another 10 min at 72°C to let extension go to completion.
See Troubleshooting.

19. Purify the deleted product by gel electrophoresis (see Chapter 2). Clone the PCR product into a vector of choice and verify the structure of the deleted product by DNA sequencing.

TROUBLESHOOTING

Problem (Steps 6, 9, 15, and 18): Errors occur in PCR products.
Solution: Amplify PCR products <2.0 kb to minimize errors.

Problem (Steps 6, 9, 15, and 18): One or more mutations are observed in the PCR products.
Solution: Use plasmid DNA that has never been PCR-amplified.

Problem (Steps 6, 9, 15, and 18): Unwanted side products are obtained.
Solution: Several precautions should be taken when performing the PCR. First, the repetitive part of the PCR program should not exceed 25 cycles (in addition to the initial denaturation cycle). Furthermore, it may be necessary to vary the annealing temperature based on the predicted melting temperatures of the primers.

RECIPES

It is essential that you consult the appropriate Material Safety Data Sheets and your institution's Environmental Health and Safety Office for proper handling of equipment and hazardous materials used in this protocol.

PCR Buffer (10×)

Reagent	Final concentration
Tris-HCl (pH 8.3)	100 mM
KCl	500 mM
$MgCl_2$	15 mM
Gelatin	0.1%

Tris-EDTA Buffer (1×)

Reagent	Final concentration
Tris-HCl (pH 8.0)	10 mM
EDTA	1 mM

REFERENCES

Lee J, Lee HJ, Shin MK, Ryu WS. 2004. Versatile PCR-mediated insertion or deletion mutagenesis. *BioTechniques* **36:** 398–400.

Lee J, Shin MK, Ryu DK, Kim S, Ryu WS. 2010. Insertion and deletion mutagenesis by overlap extension PCR. *Methods Mol Biol* **634:** 137–146.

In Vitro Mutagenesis Using Double-Stranded DNA Templates: Selection of Mutants with DpnI

In both this protocol and Protocol 5, two oligonucleotides are used to prime DNA synthesis by a high-fidelity polymerase on a denatured plasmid template. The two oligonucleotides both contain the desired mutation and have the same starting and ending positions on opposite strands of the plasmid DNA. In this protocol, the entire lengths of both strands of the plasmid DNA are amplified in a linear fashion during several rounds of thermal cycling, generating a mutated plasmid containing staggered nicks on opposite strands (see Fig. 1) (Hemsley et al. 1989).

Because of the amount of template DNA used in the amplification reaction, the background of transformed colonies containing wild-type plasmid DNA can be quite high unless steps are taken to enrich for mutant molecules. In this protocol, the products of the linear amplification reaction are treated with the restriction enzyme DpnI, which specifically cleaves fully methylated GMe6ATC

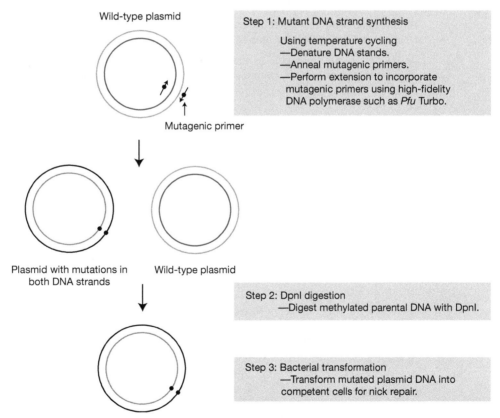

Wild-type plasmid

Step 1: Mutant DNA strand synthesis

Using temperature cycling
—Denature DNA stands.
—Anneal mutagenic primers.
—Perform extension to incorporate mutagenic primers using high-fidelity DNA polymerase such as *Pfu* Turbo.

Mutagenic primer

Plasmid with mutations in both DNA strands

Wild-type plasmid

Step 2: DpnI digestion
—Digest methylated parental DNA with DpnI.

Step 3: Bacterial transformation
—Transform mutated plasmid DNA into competent cells for nick repair.

FIGURE 1. **Site-directed mutagenesis using DpnI and PCR.** In this protocol, a double-stranded DNA (dsDNA) vector containing an insert of interest and two synthetic oligonucleotide primers containing the desired mutation (see Fig. 1 of Protocol 1) are mixed together. Both of the oligonucleotide primers are then extended in opposite directions during PCR. Incorporation of the oligonucleotide primers generates a mutated plasmid containing staggered nicks. Following PCR, the amplified product is treated with DpnI. The DpnI endonuclease is specific for methylated and hemimethylated DNA. Because the parental DNA template is methylated, it is possible to use DpnI cleavage to eliminate parental plasmids and hence to enrich the PCR products for plasmids carrying the desired mutation. Black filled circles represent the mutations incorporated into the site of interest.

sequences (Vovis and Lacks 1977). DpnI will therefore digest the bacterially generated DNA used as template for amplification, but it will not digest DNA synthesized during the course of the reaction in vitro (see the information panel N^6-Methyladenine, Dam Methylase, and Methylation-Sensitive Restriction Enzymes). DpnI-resistant molecules, which are rich in the desired mutants, are recovered by transforming E. coli cells to antibiotic resistance. Depending on the complexity of the mutation and the length of the template DNA, between 15% and 80% of the transformed colonies will contain plasmids with the desired mutation (Weiner et al. 1994). Because the method works well with virtually any plasmid of moderate size (<7 kb), it can be used to introduce mutations directly into full-length cDNAs and eliminates the need for subcloning into specialized vectors.

The key to success with this method, which is often called "circular mutagenesis," lies in the design of the primers and the choice of the appropriate thermostable DNA polymerase, which are described in the Discussion section to this protocol. Commercially available kits for circular mutagenesis include QuickChange (Agilent), which has a plasmid DNA template and mutagenic primers that can be used as a positive control. The kit is especially useful for investigators who are using circular mutagenesis for the first time.

MATERIALS

It is essential that you consult the appropriate Material Safety Data Sheets and your institution's Environmental Health and Safety Office for proper handling of equipment and hazardous materials used in this protocol.

Recipes for reagents specific to this protocol, marked <R>, are provided at the end of the protocol. See Appendix 1 for recipes for commonly used stock solutions, buffers, and reagents, marked <A>. Dilute stock solutions to the appropriate concentrations.

Reagents

Agarose gel (1%), containing 0.5 µg/mL ethidium bromide (see Step 8)
ATP (10 mM)
Bacteriophage T4 DNA ligase (optional)
Bacteriophage T4 polynucleotide kinase and buffer (optional; see Step 10)
Competent E. coli strain with an hsdR17 genotype (e.g., XL1-Blue, XL2-Blue MRF', or DH5α)
DNA ladder (1 kb)
dNTP solution (containing all four dNTPs each at 5 mM)
DpnI restriction endonuclease and buffer
Ethanol (70%), ice cold
Ethidium bromide (5 µg/mL)
Long PCR buffer (10×) (when using mixtures of DNA polymerases) <R>
Mineral oil or paraffin wax (optional; see Step 5)
Mutagenesis buffer (10×) (when using DNA polymerases such as Pfu) <R>
NaOH (1 M)/EDTA (1 mM) (optional)
Oligonucleotide primers

> For advice on the design of oligonucleotide primers, see the introduction to this protocol. The best results are achieved if the oligonucleotide primers are purified by fast-performance liquid chromatography (FPLC) or PAGE to reduce the level of contamination with salts. The purified primers are dissolved in H_2O at a concentration of 20 mM.

Phenol:chloroform (optional; see Step 9)
Plasmid DNA

> The template DNA used for mutagenesis is a circular plasmid containing the gene or cDNA of interest. In general, the shorter the plasmid, the more efficient is the amplification of the target DNA. Plasmids with a total length of <7 kb work well; however, success has been achieved with plasmid templates up to 11.5 kb in length (Gatlin et al. 1995). The plasmid DNA should be dissolved at 1 µg/mL in 1 mM Tris-Cl (pH 7.6) containing a low concentration of EDTA (<0.1 mM; 1× Tris EDTA buffer <R>).

Sodium acetate (3 M, pH 4.8) (optional)

> *This solution is used as a neutralizing agent and therefore has a slightly lower pH than most sodium acetate solutions used in molecular cloning.*

TE buffer (pH 8.0) <R>

Thermostable DNA polymerase (e.g., *Pfu* DNA polymerase)

> *The conditions described in this protocol are optimized for PfuTurbo DNA polymerase. However, they are easily adapted for use with other thermostable polymerases or mixtures of polymerases. Pfu is available from Agilent in three forms: the native enzyme, a recombinant enzyme expressed from a cloned version of the Pfu gene, and a preparation PfuTurbo, which is a formulation of recombinant Pfu DNA polymerase and a novel thermostability factor whose nature is undisclosed but that is said to enhance the yield of amplified product without altering the fidelity of DNA replication. The manufacturer claims that PfuTurbo DNA polymerase is able to amplify DNAs 15 kb in length. In our hands, however, the efficiency of amplification decreases when the length of double-stranded plasmid DNA exceeds 7–8 kb.*

Equipment

Barrier tips for automatic micropipettes
DNA sequencing equipment
Gel electrophoresis equipment
Microcentrifuge tubes (0.5 mL; thin-walled for amplification) or microtiter plates
Positive-displacement pipette
Thermal cycler programmed with desired amplification protocol

> *If the thermal cycler is not equipped with a heated lid, use either mineral oil or paraffin wax to prevent evaporation of liquid from the reaction mixture during PCR.*

Additional Materials

Reagents and equipment for transformation (Step 14; see Chapter 3), hybridization (Step 15; see Chapter 2), and sequencing (Step 16; see Chapter 11)

METHOD

Amplification of the Target DNA with Mutagenic Primers

Steps 1 and 2 are optional (see the note after Step 3).

1. Denature the plasmid DNA template in a reaction containing 1–10 µg of plasmid DNA dissolved in 40 µL of H$_2$O plus 10 µL of 1 M NaOH/1 mM EDTA. Incubate the DNA in this denaturing solution for 15 min at 37°C.

2. Add 5 µL of 3 M sodium acetate (pH 4.8) to neutralize the solution. Precipitate the DNA with 150 µL of ice-cold ethanol.

3. Collect the denatured plasmid DNA by centrifugation for 10 min at 4°C in a microcentrifuge. Carefully decant the ethanolic supernatant, and rinse the pellet with 150 µL of 70% ethanol. Recentrifuge for 2 min, decant the supernatant, and allow the last traces of ethanol to evaporate at room temperature. Resuspend the DNA in 20 µL of H$_2$O.

> *In theory, there is no need to denature the plasmid DNA before it is used as a template in PCR. If the alkaline denaturation step is omitted, superhelical, native, double-stranded DNA will be denatured by heating to 94°C in the first cycle of PCR. So why bother with Steps 1 and 2? The answer probably lies in the state of the plasmid DNA after prolonged denaturation in alkali. During exposure to 0.2 N NaOH, the plasmid DNA collapses into a dense, irreversibly denatured coil that can serve as a template in PCR but has little ability to transform bacteria. The background of colonies containing unmutated wild-type DNA is therefore markedly reduced (Du et al. 1995; Dorrell et al. 1996). In contrast, brief exposure to 95°C in the first cycle of PCR disrupts Watson–Crick base pairs but does not necessarily destroy the transforming capacity of the plasmid molecules. A higher proportion of the transformed colonies will therefore contain unmutated, parental, plasmid molecules. The value of alkaline denaturation does not become apparent until the end of the experiment when the investigator is faced with the task of sifting through colonies to identify those that contain mutated DNA.*

If the efficiency of mutagenesis is low and the selection by DpnI inefficient, the proportion of colonies containing wild-type molecules may be unacceptably high.

4. In sterile, thin-walled 0.5-mL microcentrifuge tubes, set up a series of reaction mixtures containing different amounts (e.g., 5, 10, 25, and 50 ng) of plasmid DNA and a constant amount of each of the two oligonucleotide primers.

Mutagenesis buffer (10×)	5 μL
Template plasmid DNA	5–50 ng
Oligonucleotide primer 1 (20 μL)	1 μL
Oligonucleotide primer 2 (20 μL)	1 μL
dNTP mix	2.5 μL
H_2O	to 50 μL

Add 2.5 units of *PfuTurbo* DNA polymerase to the mixture.

It is important to add the reagents in the order shown to reduce the opportunity for the 3′–5′ exonuclease activity of PfuTurbo to degrade the primers.

5. If the thermal cycler is not fitted with a heated lid, overlay the reaction mixtures with 1 drop (~50 μL) of light mineral oil or a bead of paraffin wax to prevent evaporation of the samples during repeated cycles of heating and cooling. Place the tubes in the thermal cycler.

6. Amplify the nucleic acids using the denaturation, annealing, and polymerization times and temperatures listed in the table below.

Cycle number	Denaturation	Annealing	Polymerization
1 cycle	1 min at 95°C		
2–18 cycles[a]	30 sec at 95°C	1 min at 55°C	2 min per kilobase of plasmid DNA at 68°C
Last cycle	1 min at 94°C	1 min at 55°C	10 min at 72°C

These times are suitable for 50-μL reactions assembled in thin-walled 0.5-mL tubes and incubated in thermal cyclers such as the PerkinElmer 9600 or 9700, Mastercycler (Eppendorf), and PTC 100 (MJ Research). Times and temperatures may need to be adapted to suit other types of equipment and reaction volumes.

[a]For single-base substitutions, use 12 cycles of linear amplification; for substitution of one amino acid with another (usually two or three contiguous base substitutions), use 16 cycles; for insertions and deletions of any size, use 18 cycles.

The rate of DNA synthesis is 1.5–2.0 times slower in amplification reactions catalyzed by *Pfu* than in reactions catalyzed by *Taq*.

A small number of cycles is used together with a large amount of starting template to reduce the introduction of spurious mutations during amplification of the plasmid DNA and gene/cDNA.

7. After amplification of the DNA, place the reactions on ice.

8. Verify that the target DNA was amplified by analyzing 10 μL of each reaction by electrophoresis through a 1% agarose gel containing 0.5 μg/mL ethidium bromide. As standards, load 50 ng of unamplified linearized plasmid DNA and a 1-kb DNA ladder into the outer lanes of the gel.

If the efficiency of amplification is low, set up a series of reactions to optimize the components of the amplification reaction and the cycling parameters.

Ligation and Transformation of Amplified Product

Steps 9–12 are optional and are generally used only when the efficiency of mutagenesis is expected to be low (e.g., when constructing insertions and deletions).

9. Extract the amplified DNAs twice with phenol:chloroform and precipitate with ethanol.

10. Resuspend the DNA pellets in the following:

Bacteriophage T4 polynucleotide kinase buffer (10×)	5 μL
ATP (10 mM)	5 μL
Bacteriophage T4 polynucleotide kinase	5 units
H_2O	to 50 μL

Incubate the reactions for 1 h at 37°C. Inactivate the kinase enzyme by heating for 10 min at 68°C. Extract the phosphorylated DNAs twice with phenol:chloroform, and collect the DNAs by ethanol precipitation.

11. Resuspend the pellets of phosphorylated DNA (~0.9 μg each) in 90 μL of TE buffer. Set up a series of ligation reactions containing the phosphorylated DNAs at concentrations ranging from 0.1 to 1 μg/mL.

Phosphorylated DNA	(10–100 ng)
Bacteriophage T4 DNA ligase buffer (10×)	10 μL
ATP (10 mM)	10 μL
Bacteriophage T4 DNA ligase	4 units
H₂O	to 100 μL

Incubate the reactions for 12–16 h at 16°C.

If the 10× bacteriophage T4 ligase buffer supplied by the manufacturer contains ATP, omit the ATP in the above ligation reactions.

> *Conditions that favor the formation of monomeric circles during ligation are well understood in theory (Collins and Weissman 1984) but are difficult to achieve in practice. The molar concentration of DNA ends must be low to favor the formation of intramolecular circles over concatenates. However, it is difficult to calculate an appropriate concentration when working with amplified DNAs generated by inverse PCR because the proportion of full-length products with undamaged termini is unknown.*

12. Extract the ligated DNAs twice with phenol:chloroform, and collect the DNA by ethanol precipitation. Resuspend each pellet in 45 μL of H₂O. Add 5 μL of 10× DpnI buffer to each tube.

13. Digest the amplified DNAs by adding 10 units of DpnI directly to the remainder of the amplification reactions (Step 7) or to the phosphorylated and ligated DNAs (Step 12). Mix the reagents by pipetting the solution up and down several times, centrifuge the tubes for 5 sec in a microcentrifuge, and then incubate them for 1 h at 37°C.

> *If temperature cycling was performed in the presence of mineral oil or wax, it is important to ensure that the DpnI is added to the aqueous portion of the reaction mixture. Use barrier micropipette tips, and be sure to insert the end of the tip below the mineral oil or wax overlay.*

14. Transform competent *E. coli* cells with 1, 2, and 5 μL of digested DNA (see Chapter 3).

> *Make sure that no mineral oil is transferred from the digestion mixture to the competent cells.*

> *The use of ultracompetent XL2-Blue MRF' E. coli cells (Agilent) and a modified transformation procedure has been reported to facilitate the recovery of mutants (Dorrell et al. 1996). However, in our hands, homemade preparations of highly competent E. coli cells are adequate for most forms of circular mutagenesis.*

15. Prepare plasmid DNA from at least 12 independent transformants. Screen the DNA preparations for mutations by DNA sequencing, by oligonucleotide hybridization (see Chapter 2), or by restriction digestion of small preparations of plasmid DNA if a site was created or destroyed by the introduced mutation, or if an insertion or a deletion was introduced into the template.

> *See Troubleshooting.*

16. Sequence the entire segment of target DNA to verify that the desired mutation has been generated and that no spurious mutations occurred during amplification (see Chapter 11).

> *See Troubleshooting.*

TROUBLESHOOTING

Problem (Step 15): None of the plasmids carry the desired mutation.
Solution: Screen the entire population of transformants by oligonucleotide hybridization to identify rare colonies that carry the desired mutation.

Problem (Step 16): The amplification reactions have worked well, but the yield of mutants is poor.
Solution: DpnI is the likely culprit. Set up a series of reactions to check that the enzyme is capable of digesting 50 ng of parental plasmid to completion in 1× *Pfu* reaction buffer. If necessary, adjust the amount of DpnI used and the time of the digestion. If digestion with DpnI is working efficiently, consider using two-stage amplification reactions (Wang and Malcolm 1999). In the first

stage, two separate asymmetric amplification reactions are set up, each containing just one of the two oligonucleotide primers. The products of these reactions are single-stranded DNAs that can serve as templates for the second-stage linear amplification, which is performed with both oligonucleotides, essentially as described above. The goal of the first-stage reaction is to generate templates to which the mutagenic primers can bind perfectly, without competition from a wild-type duplex.

DISCUSSION

Design of Primers

The two oligonucleotide primers:

- must anneal to the complementary strands of the same target sequence
- must be of equal length (between 25 and 45 bp), with a calculated melting temperature of 78°C or greater. The T_m should be high enough to suppress false priming and low enough to allow complete dissociation of primer–primer hybrids during the denaturation step of the amplification reaction.
- should terminate in a G or C residue
- need not be phosphorylated. This is because *Pfu*, the thermostable DNA polymerase most commonly used to catalyze the amplification reaction, is unable to displace oligonucleotides hybridized to their target sequence, whether or not the oligonucleotides are phosphorylated.
- can generally be used without purification. However, higher efficiencies of mutation, especially with insertions and deletions, are obtained if the primers are purified by either fast performance liquid chromatography (FPLC) or polyacrylamide gel electrophoresis (PAGE).

When a point mutation is to be introduced, one primer carries a wild-type sequence, whereas the other primer carries the desired mutation and has a minimum of 12 bases of correct sequence on each side of the centrally placed mutation. If a deletion is to be introduced, the two primers may both have a wild-type sequence, but they are spaced a distance apart on the template that corresponds to the length of the deletion. Insertion mutations require one primer of wild-type sequence and one with the sequence to be inserted located at its 5′ terminus.

Thermostable DNA Polymerase

Three properties are required of thermostable DNA polymerases used for mutagenesis of denatured plasmid templates:

- an efficient proofreading activity
- a low rate of incorporation of mismatched bases
- a lack of untemplated terminal transferase activity

Taq does not fulfill these criteria and is therefore entirely unsuitable for site-directed mutagenesis (Stemmer and Morris 1992; Watkins et al. 1993). However, a 160:1 mixture of Klentaq (AB Peptides) and *Pfu* polymerase (Agilent) has the appropriate set of properties. A typical mixture contains 0.187 unit of *Pfu* and 33.7 units of Klentaq1 in a total volume of 1.2 μL. Examples of commercially available mixtures include TaqPlus from Agilent and the Expand High Fidelity PCR System from Boehringer-Mannheim.

Single thermostable DNA polymerases that have been used successfully in circular PCR include *Pwo* DNA polymerase (Hidajat and McNicol 1997), rTth DNA polymerase XL (Du et al. 1995; Gatlin et al. 1995), Vent$_R$ DNA polymerase (Hughes and Andrews 1996), and *Pfu* DNA polymerase (see below). These polymerases have one major disadvantage: Relatively high concentrations of oligonucleotide primers are required (1) to counteract the effects of degradation by 3′–5′ exonuclease activity and (2) to ensure that the primers are in vast molar excess (50–100-fold) over the template DNA (see Chapter 7, Table 1). Unfortunately, such high concentrations favor the formation of hybrids between the two complementary oligonucleotide primers, thereby reducing the efficiency of the amplification reaction. Because of the consequent uncertainty about the effective concentration of primers, some investigators perform preliminary experiments to optimize the components of the amplification reaction. Typically, these experiments use agarose gel electrophoresis to measure the yield of linear full-length DNA synthesized in amplification reactions containing a constant amount of template DNA (usually 50 ng) and a range of primer concentrations (e.g., see Parikh and Guengerich 1998). However, in our hands, optimization of primers is unnecessary unless the mutations are complex (more than three single-base changes; deletion or insertion of more than 2 nucleotides).

When using single thermostable DNA polymerases, additional steps may be taken to further minimize the chance of unwanted mutants accumulating during amplification.

- The initial concentrations of dNTPs and Mg^{2+} in the reaction mixture must not exceed 250 μM and 1.5 mM, respectively.

- Because thermostable DNA polymerases such as *Pfu* and *Pwo* require a more alkaline buffer than *Taq*, the pH of the Tris-buffered reaction mixture should be 8.9 (measured at 25°C).

- The number of cycles of linear amplification must be kept to a bare minimum (see Step 6 of this protocol), even if this constraint means using relatively large amounts of template DNA in the reaction mixture.

RECIPES

It is essential that you consult the appropriate Material Safety Data Sheets and your institution's Environmental Health and Safety Office for proper handling of equipment and hazardous materials used in this protocol.

Long PCR Buffer (10×)
Use this with mixtures of DNA polymerases.

Reagent	Final concentration
Tris-Cl (pH 9.0 at room temperature)	500 mM
Ammonium sulfate	160 mM
$MgCl_2$	25 mM
Bovine serum albumin	1.5 mg/mL

Buffers supplied by the manufacturer and suitable for use with thermostable DNA polymerases may be used in place of the above buffer.

Mutagenesis Buffer (10×)
Use this with DNA polymerases such as *Pfu*.

Reagent	Final concentration
KCl	100 mM
Ammonium sulfate	100 mM
Tris (pH 8.9 at room temperature)	200 mM
$MgSO_4$	20 mM
Triton X-100	1%
Bovine serum albumin (nuclease-free)	1 mg/mL

Tris-EDTA (TE) Buffer (1×)

Reagent	Final concentration
Tris-HCl (pH 8.0)	10 mM
EDTA	1 mM

REFERENCES

Collins FS, Weissman SM. 1984. Directional cloning of DNA fragments at a large distance from an initial probe: A circularization method. *Proc Natl Acad Sci* 81: 6812–6816.

Dorrell N, Gyselman VG, Foynes S, Li SR, Wren BW. 1996. Improved efficiency of inverse PCR mutagenesis. *BioTechniques* 21: 604–608.

Du Z, Regier DA, Desrosiers RC. 1995. Improved recombinant PCR mutagenesis procedure that uses alkaline-denatured plasmid template. *BioTechniques* 18: 376–378.

Gatlin J, Campbell LH, Schmidt MG, Arrigo SJ. 1995. Direct-rapid (DR) mutagenesis of large plasmids using PCR. *BioTechniques* 19: 559–564.

Hemsley A, Arnheim N, Toney MD, Cortopassi G, Galas DJ. 1989. A simple method for site-directed mutagenesis using the polymerase chain reaction. *Nucleic Acids Res* 17: 6545–6551.

Hidajat R, McNicol P. 1997. Primer-directed mutagenesis of an intact plasmid by using *Pwo* DNA polymerase in long distance inverse PCR. *BioTechniques* 22: 32–34.

Hughes MJ, Andrews DW. 1996. Creation of deletion, insertion and substitution mutations using a single pair of primers and PCR.

BioTechniques 20: 188–196.

Parikh A, Guengerich FP. 1998. Random mutagenesis by whole-plasmid PCR amplification. *BioTechniques* 24: 428–431.

Stemmer WP, Morris SK. 1992. Enzymatic inverse PCR: A restriction site independent, single-fragment method for high-efficiency, site-directed mutagenesis. *BioTechniques* 13: 214–220.

Vovis GF, Lacks S. 1977. Complementary action of restriction enzymes endo R-DpnI and endo R-DpnII on bacteriophage f1 DNA. *J Mol Biol* 115: 525–538.

Wang W, Malcolm BA. 1999. Two-stage PCR protocol allowing introduction of multiple mutations, deletions and insertions using QuikChange Site-Directed Mutagenesis. *BioTechniques* 26: 680–682.

Watkins BA, Davis AE, Cocchi F, Reitz MSJ. 1993. A rapid method for site-specific mutagenesis using larger plasmids as templates. *BioTechniques* 15: 700–704.

Weiner MP, Costa GL, Schoettlin W, Cline J, Mathur E, Bauer JC. 1994. Site-directed mutagenesis of double-stranded DNA by the polymerase chain reaction. *Gene* 151: 119–123.

Altered β-Lactamase Selection Approach for Site-Directed Mutagenesis

Many protocols exist to perform site-directed mutagenesis, and here we present one of the more commonly used ones—site-directed mutagenesis by altered β-lactamase selection. β-Lactamase is an enzyme that cleaves ampicillin, rendering it impotent to bacteria. Certain mutations in the active site of β-lactamase can alter the substrate specificity of the enzyme and allow it to have increased hydrolytic activity for the cephalosporin family of antibiotics, a property not shared by wild-type lactamases (Cantu et al. 1996). *E. coli* cells carrying the β-lactamase triple mutant G238S:E240:R241G show increased resistance to cefotaxime and ceftriaxone, two cephalosporins, compared with wild-type cells. This protocol takes advantage of this property to select for plasmids that have undergone site-directed mutagenesis.

Specifically, a double-stranded plasmid DNA template containing a cloned target gene (gene X) to be mutagenized and the β-lactamase gene is first alkaline-denatured. Next, two oligonucleotide primers are simultaneously annealed to the template. The first oligonucleotide—an antibiotic selection oligonucleotide—encodes residue changes in the β-lactamase gene that confers increased resistance to cephalosporins. The second oligonucleotide is a mutagenic oligonucleotide that (1) anneals to gene X and (2) encodes the mutations to be incorporated into gene X. The mutagenic oligonucleotide hybridizes to the same strand as the antibiotic selection oligonucleotide. Extension of the two oligonucleotides by DNA polymerase creates a heteroduplex, which links the altered β-lactamase gene with the mutagenized gene X. The DNA is then transformed into BMH71-18 *mutS*, a repair-deficient *E. coli* host. This strain is used to propagate the mutant plasmid while preventing selection against the desired mutation by cellular machinery. Plasmids prepared from the BMH71-18 *mutS* cells are used to transform cells of strain JM109. Finally, the transformed JM109 cells are plated onto LB plates supplemented with a cocktail of antibodies that includes cephalosporins. In this way, clones harboring plasmids with the mutagenized gene linked to the altered-specificity β-lactamase can be selected and isolated. Figure 1 shows the major steps of this protocol. The protocol presented here is adapted from Andrews and Lesley (2002).

MATERIALS

It is essential that you consult the appropriate Material Safety Data Sheets and your institution's Environmental Health and Safety Office for proper handling of equipment and hazardous materials used in this protocol.

Recipes for reagents specific to this protocol, marked <R>, are provided at the end of the protocol. See Appendix 1 for recipes for commonly used stock solutions, buffers, and reagents, marked <A>. Dilute stock solutions to the appropriate concentrations.

Reagents

▲ Store *E. coli* BMH 71-18 *mutS* and JM109 strains at –70°C until use. Store all other components at –20°C until use. The antibiotic used for selection should not undergo more than five freeze–thaw cycles.

Agarose gel (1%) (see Chapter 2)
Ammonium acetate (2 M, pH 4.6)

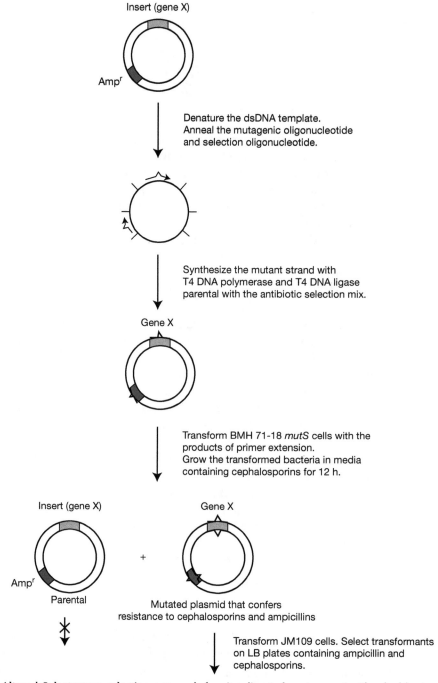

FIGURE 1. Altered β-lactamase selection approach for site-directed mutagenesis. The double-stranded plasmid DNA (dsDNA) template is first alkali-denatured. Two oligonucleotides are simultaneously annealed to the template. The first encodes nucleotide changes to the β-lactamase gene that confer increased resistance to cephalosporins. The second oligonucleotide encodes changes to gene X. The second oligonucleotide, which carries the mutation(s) destined for gene X, hybridizes the same strand of the plasmid DNA as the antibiotic selection oligonucleotide. Extension of the two oligonucleotides by DNA polymerase creates a heteroduplex, which links the antibiotic-resistance gene to the mutagenized gene. The plasmid DNA is then transformed into a repair-deficient strain of *E. coli* (e.g., BMH71-18 *mutS*), followed by clonal separation in a repair-competent strain of *E. coli*. (Modified from Lewis and Thompson 1990, with permission of Oxford University Press.)

Ampicillin solution (100 mg/mL; filter-sterilized)

Annealing buffer (10×) <R>

Antibiotic selection mixture (5 mg/mL ampicillin, 25 μg/mL cefotaxime, 25 μg/mL ceftriaxone [Sigma-Aldrich], 100 mM potassium phosphate at pH 6.0)

Chloroform:isoamyl alcohol (24:1) <A>

dsDNA template plasmid

The candidate gene to be mutagenized (gene X) should be cloned into the pGEM-11Zf(+) plasmid (Promega).

E. coli BMH 71-18 *mutS* strain

Several other E. coli *strains are compatible with BMH 81-18 mutS, including BJ5183, JM107, SK1590, DH1, JM108, SK2287, DH5α, JM109, TB1, JM103, MC1061, TG1, JM105, MM294, XL1-Blue, JM106, Q358, and Y1088. The BMH 71-18 mutS strain is chosen because it suppresses in vivo mismatch repair.*

E. coli JM109 strain

Ethanol (70% and 100%)

LB medium <R>

LB plates with ampicillin <R>

NaOH (2 M)/EDTA (2 mM)

pGEM-11Zf(+) (Promega)

Plasmid purification kit for conventional miniprep (QIAprep Spin Miniprep Kit)

Primers:

Antibiotic selection oligonucleotide primer

The sequence is GATAAATCTGGAGCC<u>***TCCAAGGG****T*</u>*GGGTCTCGCGGT; substitutions are boldface and underlined.*

Mutagenic oligonucleotide primer (phosphorylated)

See Step 2 for synthesis information.

Synthesis buffer (10×) <R>

T4 DNA ligase (5 U/μL) (NEB)

T4 DNA polymerase (10 U/μL) (NEB)

TE buffer <R>

Equipment

Agarose gel electrophoresis equipment

Cell culture plates

Heating block

Incubator, shaking

Polypropylene tubes (17×100 mm), sterile

Vacuum dryer

Water bath (42°C)

METHOD

Preparing the Double-Stranded Template Plasmid

1. Clone gene X to be mutagenized into the pGEM-11Zf(+) plasmid (Promega), which carries the TEM-1 β-lactamase gene. Bacteria transformed by the plasmid are resistant to ampicillin.

Mutagenic and Antibiotic Selection Oligonucleotides Preparation

2. Synthesize a mutagenic oligonucleotide primer containing the mutation of interest.

The stability of a hybrid formed between an oligonucleotide and its complementary DNA strand depends on a number of factors, including the length and G+C content of the oligonucleotide and the annealing conditions. For more details, see Chapter 13. As a general rule, mutagenic oligonucleotides are designed with a mismatch base located in the center of a 17–20-base oligonucleotide (see other protocols in this chapter for details). To generate mutations requiring two or more mismatches, oligonucleotides of 25 bases or longer are needed so that there are 12–15 perfectly matched oligonucleotides on either side of the mismatch.

Then synthesize the antibiotic selection oligonucleotide primer (sequence GATAAATCTG-GAGCC**TCCAAGGGT**GGGTCTCGCGGT; substitutions to the wild-type sequence of the β-lactamase gene are in boldface and underlined).

Remember that the mutagenic oligonucleotide and the antibiotic selection nucleotide must be complementary to the same strand of plasmid DNA to achieve coupling of the mutation to the antibiotic selection.

Denaturation of the dsDNA Template

3. Use alkali to denature 2 μg of the double-stranded DNA plasmid carrying gene X. This reaction generates enough denatured DNA for 10 mutagenesis reactions. Incubate the reaction mixture for 5 min at room temperature.

Reagent	Volume added
dsDNA template	0.5 pmol (2 μg)
NaOH (2 M), EDTA (2 mM)	2 μL
Sterile, deionized water	to a final volume of 20 μL

In general, (nanograms of dsDNA) = (pmoles of dsDNA) × 0.66 × N, where N is the length of the dsDNA in bases.

4. Precipitate the DNA by adding 2 μL of 2 M ammonium acetate (pH 4.6) and 75 μL of 100% ethanol (4°C) to the reaction mixture.

5. Incubate the mixture for 30 min at −70°C.

6. Collect the DNA by centrifugation at top speed (∼13,000 rpm) in a microcentrifuge for 15 min at 4°C.

7. Drain and wash the pellet with 200 μL of 70% ethanol (4°C). Centrifuge again as in Step 6. Dry the DNA pellet under vacuum.

8. Resuspend the pellet in 100 μL of TE buffer (pH 8.0).

Annealing the Mutagenic Primers

9. Prepare the primer annealing reaction as follows:

Reagent	Volume added
Denatured template DNA (from Step 8)	10 μL (0.05 pmol)
Selection oligonucleotide (2.9 ng/μL), phosphorylated (from Step 1)	1 μL (0.25 pmol)
Mutagenic oligonucleotide, phosphorylated (from Step 1)	1.25 pmol
Annealing buffer (10×)	2 μL
Sterile, deionized water	to a final volume of 20 μL

Heat the annealing reaction to an annealing temperature for 5 min at 75°C, and allow it to cool slowly to 37°C. Slow cooling helps to minimize nonspecific annealing of both the mutagenic and antibiotic selection oligonucleotides. It is recommended that the reaction be cooled at a rate of ∼1.5°C/min.

Optimal annealing temperatures and times may depend on the composition of the mutagenic oligonucleotide used. It may be necessary to try several annealing temperatures in order to optimize the mutagenesis efficiency of your particular oligonucleotide.

See Troubleshooting.

Mutant Strand Synthesis and Ligation

10. After the annealing reaction mixture has cooled to 37°C, centrifuge the reaction mixture briefly to collect the contents at the bottom of the microcentrifuge tube. Then add the following components in the order listed to assemble the reaction mixture for synthesis and ligation of the mutant strand.

Reagent	Volume added
Sterile, deionized water	5 μL
Synthesis buffer (10×)	3 μL
T4 DNA polymerase (5–10 U)	1 μL
T4 DNA ligase (1–3 U)	1 μL
Final volume	30 μL

T4 DNA polymerase is used in this reaction because it is a high-fidelity DNA polymerase and does not displace the hybridized oligonucleotide. As a result, multiple site-directed mutations can be introduced in one reaction simply by annealing additional mutagenic oligonucleotides at different locations in the DNA insert.

11. Incubate the reaction for 90 min at 37°C to allow the mutant strand synthesis and ligation to occur. The mutant DNA is now ready for transformation of BMH 71-18 *mutS* strain competent cells.

    The BMH71-18 mutS *strain is used because it suppresses in vivo mismatch repair.*

Transformation of BMH 71-18 *mutS* Strain Competent Cells

12. Prechill a sterile 17-mm × 100-mm polypropylene tube on ice.

13. Thaw 100 μL of competent *E. coli* BMH71-18 *mutS* cells, and place them into the sterile polypropylene tube on ice. Add 1.5 μL of the mutagenesis reaction mixture from Step 11 to the competent cells. Mix gently.

14. Incubate cells on ice with the DNA for 10 min.

15. Heat-shock the cells for 45–50 sec in a 42°C water bath.

16. Place the heat-shocked cells back on ice for 2 min.

17. Add 900 μL of room-temperature LB medium (without antibiotic) to the BMH71-*18 mutS* cells (from Steps 13–16) and incubate the culture with shaking for 60 min at 37°C.

18. Prepare overnight cultures by adding 4 mL of LB medium to the cells from Step 17, and then add 100 μL of ampicillin (100 mg/mL) to the culture.

19. Incubate the culture overnight at 37°C with vigorous shaking.

    See Troubleshooting.

20. Perform a plasmid DNA miniprep following conventional protocols (see Chapter 1) to extract and purify the plasmid DNA.

Transformation into JM109 Cells and Clonal Segregation

21. Before beginning the transformation procedure, pour molten LB agar containing 7.5 mL/L of antibiotic selection mix (with cephalosporins) and 1 mL/L of 100 mg/mL ampicillin solution into cell culture plates.

22. Prechill sterile 17-mm × 100-mm polypropylene tubes on ice.

23. Place 100 μL of competent JM109 cells on ice, and add 1.5 μL of the plasmids purified from the BMH71-18 *mutS* cells from Step 20 to the competent JM109 cells. Mix gently.

24. Incubate the cells on ice for 10 min.

25. Heat-shock the cells for 45–50 sec in a 42°C water bath.

26. Place the tubes on ice for 2 min.

27. Add 900 μL of room-temperature LB medium (without antibiotics) to the transformed JM109 cells and incubate with shaking for 60 min at 37°C.

28. Plate 100 μL of the transformation reaction from Step 27 onto the LB agar plates containing the antibiotic selection mix and ampicillin (prepared in Step 21). As a control, also plate the cells on LB agar plates without the antibiotic selection mix. Incubate the plates for 12–14 h at 37°C.

29. After the incubation period, screen the white colonies on the plate. It is best to screen at least five colonies by direct sequencing.

TROUBLESHOOTING

Problem (Step 9): The template is degraded.
Solution: Components should be added in the order listed because addition of polymerase without dNTPs can trigger exonuclease activity.

Problem (Step 19): No growth is seen in the BMH *mutS* overnight culture.
Solution: Decrease the amount of antibiotic in the culture. Use only high-efficiency competent cells.

RECIPES

It is essential that you consult the appropriate Material Safety Data Sheets and your institution's Environmental Health and Safety Office for proper handling of equipment and hazardous materials used in this protocol.

Annealing Buffer (10×)

Reagent	Final concentration
Tris-HCl (pH 7.5)	200 mM
$MgCl_2$	100 mM
NaCl	500 mM

LB Medium

Reagent	Quantity (for 1 L)
Bacto Tryptone	10 g
Bacto-yeast extract	5 g
NaCl	10 g

Dissolve in 1 L of distilled water. Adjust pH to 7.5 with 5 N NaOH. Sterilize by autoclaving.

LB Plates with Ampicillin

Make in the same way as LB medium, except add 15 g/L of Bacto Agar before autoclaving. After autoclaving, add 125 μg of ampicillin/mL of medium before pouring.

Synthesis Buffer (10×)

Reagent	Final concentration
Tris-HCl (pH 7.5)	100 mM
DNTPs	5 mM
ATP	10 mM
DTT	20 mM

Tris-EDTA (TE) Buffer (1×)

Reagent	Final concentration
Tris-HCl (pH 8.0)	10 mM
EDTA	1 mM

REFERENCES

Andrews CA, Lesley SA. 2002. Site-directed mutagenesis using altered β-lactamase specificity. *Methods Mol Biol* **182:** 7–17.

Cantu C III, Huang W, Palzkill T. 1996. Selection and characterization of amino acid substitutions at residues 237–240 of TEM-1 β-lactamase with altered substrate specificity for aztreonam and ceftazidime. *J Biol Chem* **271:** 22538–22545.

Lewis MK, Thompson DV. 1990. Efficient site directed in vitro mutagenesis using ampicillin selection. *Nucleic Acids Res* **18:** 3439–3443.

Oligonucleotide-Directed Mutagenesis by Elimination of a Unique Restriction Site (USE Mutagenesis)

In this method, two oligonucleotide primers are hybridized to the same strand of a denatured double-stranded recombinant plasmid. One primer (the mutagenic primer) introduces the desired mutation into the target sequences, whereas the second primer carries a mutation that destroys a unique restriction site (called "unique site elimination," or USE) in the plasmid (see Fig. 1). Both primers are elongated in a reaction catalyzed by bacteriophage T4 or T7 DNA polymerase. Nicks in the strand of newly synthesized DNA are sealed with bacteriophage T4 DNA ligase. The product of the first part of the method is a heteroduplex plasmid consisting of a wild-type parental strand and a new full-length strand that carries the desired mutation but no longer contains the unique restriction site. The population of plasmids therefore consists of (1) wild-type molecules that, for one reason or another, were never used as templates for DNA synthesis primed by the two oligonucleotide primers and (2) heteroduplex molecules that have lost the unique restriction site and gained the desired mutation.

In the second phase of the method, the mixed population is incubated with the restriction enzyme that cleaves the unique site. The wild-type molecules are linearized, and the mutated plasmids are resistant to digestion. The mixture of circular heteroduplex DNA and linear wild-type DNA is then used to transform a strain of *E. coli* that is deficient in repair of mismatched bases. Because linear DNA transforms 10–1000-fold less efficiently than circular DNA (Conley and Saunders 1984), many of the wild-type molecules are unable to reestablish themselves in *E. coli* cells. The circular heteroduplex molecules, however, begin to replicate. Because the mismatched bases are not repaired, the first round of replication generates a wild-type plasmid that carries the original restriction site and a mutated plasmid that does not. DNA from the first set of transformants is recovered, digested once more with the same restriction enzyme to linearize the wild-type molecules, and then used to transform a standard laboratory strain of *E. coli*. This biochemical selection can be sufficiently powerful to ensure that a high proportion of the resulting transformants carry the desired mutation (Deng and Nickoloff 1992; Zhu 1996).

UNIQUE SITE ELIMINATION (USE)

The theoretical maximum yield of mutants in the standard form of unique site elimination (USE) described in this protocol is 50%. However, in most laboratories, the frequency with which mutant plasmids are recovered varies between 5% and 30%, depending on the complexity of the mutation and the efficiency of cleavage with the restriction enzyme. If necessary, the rate of mutant recovery can be improved significantly by combining USE with the Kunkel method of selection against uracil-containing template DNAs (Markvardsen et al. 1995) or by increasing the concentration of the mutagenic primer to 10:1 in favor of the mutagenic primer.

With the USE method, a gene may, in principle, be mutated in any double-stranded circular vector provided the vector contains a unique restriction site and a selectable marker (e.g., a gene conferring antibiotic resistance). Several companies sell selection primers and kits containing a range of the components required for USE mutagenesis (e.g., the Transformer Site-Directed Mutagenesis Kit sold by Clontech). These kits can be used successfully with many of the commonly used vectors; however, see the Troubleshooting section at the end of this protocol.

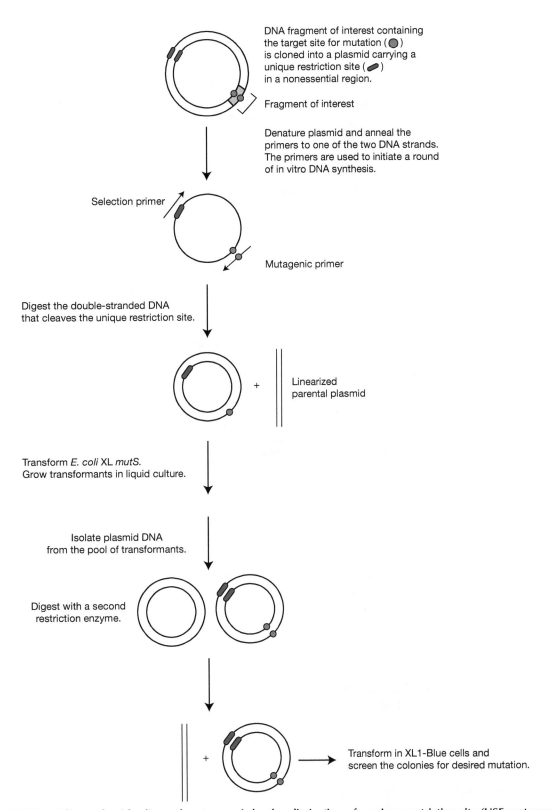

FIGURE 1. Oligonucleotide-directed mutagenesis by the elimination of a unique restriction site (USE mutagenesis). The plasmid of interest is amplified using a selection primer and a mutagenic primer. The PCR product is used for transformation as shown. This protocol generates site-directed mutants with an efficiency of ~90%. (Adapted from Braman et al. 2000.)

MATERIALS

It is essential that you consult the appropriate Material Safety Data Sheets and your institution's Environmental Health and Safety Office for proper handling of equipment and hazardous materials used in this protocol.

Recipes for reagents specific to this protocol, marked <R>, are provided at the end of the protocol. See Appendix 1 for recipes for commonly used stock solutions, buffers, and reagents, marked <A>. Dilute stock solutions to the appropriate concentrations.

Reagents

Agarose gel (1%), containing 0.5 μg/mL ethidium bromide

Annealing buffer (10×) <R>

Bacteriophage T4 DNA ligase

Bacteriophage T4 DNA polymerase or Sequenase

> *The native DNA polymerase encoded by bacteriophage T4 is unable to displace the oligonucleotide primers from the template DNA (Nossal 1974; Kunkel 1985; Bebenek and Kunkel 1989; Schena 1989).*

E. coli strain with a *mutS* genotype (e.g., BMH 71-18), made competent for transformation

E. coli strain with a *mut*$^+$ phenotype, made competent for transformation (see Step 14)

LB agar plates <R>

LB medium containing the appropriate antibiotic <R>

Plasmid DNA

> *Closed-circular plasmid DNA, purified either by chromatography through a commercial resin or by the alkaline lysis method (see Chapter 1, Protocol 1 or 2).*

Primers (mutagenic and selection)

> *Both mutagenic and selection primers must anneal to the same strand of the target DNA, and the 5' end of each primer must be phosphorylated. The mutagenic primer should be designed with the engineered mutations in the middle and flanked on each side by 10–15 bases that pair perfectly with the template DNA. The oligonucleotide primers should be purified by FPLC or PAGE (see Chapter 2, Protocol 3) before use.*

Restriction enzyme that cleaves the plasmid at a single site

Synthesis buffer (10×) <R>

Additional Materials

Reagents and equipment for transformation of *E. coli* (see Steps 7 and 14; see Chapter 3)

Reagents and equipment for minipreparation of plasmid DNA (see Steps 10 and 16; see Chapter 1, Protocols 1 and 2)

Equipment

Cell culture equipment

Gel electrophoresis equipment

Water baths (boiling and preset to 70°C and to the appropriate temperature for restriction endonuclease digestion)

METHOD

Synthesis of the Mutant DNA Strand

1. Mix the following components in a microcentrifuge tube:

Annealing buffer (10×)	2 μL
Plasmid DNA	0.025–0.25 pmol
Selection primer	25 pmol
Mutagenic primer	25 pmol
H$_2$O	to 20 μL

Incubate the tube in a boiling water bath for 5 min.

> *The protocol will tolerate a wide range of concentrations of primers and plasmid DNA. If necessary, the optimal stoichiometry of the components can be determined empirically.*

2. Immediately chill the tube for 5 min in ice. Centrifuge the tube for 5 sec in a microcentrifuge to deposit the fluid at the base of the tube.

3. To the tube of annealed primers and plasmid, add:

Synthesis buffer (10×)	3 μL
Bacteriophage T4 DNA polymerase (2–4 units/μL)	1 μL
Bacteriophage T4 DNA ligase (4–6 units/μL)	1 μL
H₂O	5 μL

Mix the reagents well by gentle pipetting up and down. Centrifuge the tube for 5 sec in a microcentrifuge to deposit the fluid at the base of the tube. Incubate the reaction for 1–2 h at 37°C.

4. Stop the reaction by heating the tube for at least 5 min at 70°C in a water bath to inactivate the enzymes. Store the tube on the bench to allow it to cool to room temperature.

Primary Selection by Restriction Endonuclease Digestion

5. Adjust the NaCl concentration of the reaction to a level that is optimal for the selected unique site restriction endonuclease. Use the 10× annealing buffer, a stock of NaCl, or the 10× buffer supplied with the restriction enzyme.

> *The concentration of NaCl in the synthesis/ligation mixture is 37.5 mM in a total volume of 30 μL.*

> *If the restriction digestion needs less or no NaCl, the DNA mixture in the synthesis/ligation buffer can be precipitated with ethanol or passed through a spun column and resuspended in the appropriate restriction enzyme buffer.*

6. Add 20 units of the selective restriction endonuclease to the reaction mixture. Incubate the reaction for at least 1 h at the appropriate digestion temperature.

> ▲ *The volume of enzymes added to the reactions (including polymerase and ligase) should not exceed 10% of the total reaction volume. Adjust the reaction volume accordingly.*

First Transformation and Enrichment for Mutant Plasmids

7. Transform a *mutS E. coli* strain such as BMH 71-18 with the plasmid DNAs contained in the digestion mixture, using one of the transformation procedures described in Chapter 3.

8. Spread 10, 50, and 250 μL of the transformation mixture onto LB agar plates containing the appropriate antibiotic. Incubate the plates overnight at 37°C. Then perform Step 9 while the plates are incubating.

> *These plates are used to determine the number of primary transformants, which should be between 100 and 300 colonies per 50 μL of transformation mixture plated.*

9. Add the remaining transformation mixture to 3 mL of LB medium containing the appropriate antibiotic. Incubate the culture overnight at 37°C with shaking.

10. The next day, prepare plasmid DNA from ~2.5 mL of the overnight culture (see Chapter 1).

11. Digest the plasmid DNA prepared in Step 10 with the selective restriction enzyme as follows:

Plasmid DNA	500 ng
Restriction enzyme buffer (10×)	2 μL
Unique site restriction endonuclease	20 units
H₂O	to 20 μL

Incubate the reaction for 2 h at the appropriate temperature.

12. Add an additional 10 units of the restriction enzyme, and incubate for at least 1 h further.

13. Assess the extent of digestion by running 5–10 μL of the plasmid DNA on a 1% agarose gel containing 0.5 μg/mL ethidium bromide (see Chapter 2, Protocols 1 and 2).

> *Linearized plasmid DNA will run as a discrete band; undigested (circular) DNA will run as two bands, corresponding to the relaxed circular form and the supercoiled form. However, because the parental plasmid makes up a greater part of the total plasmid pool, the bands corresponding to the undigested mutant plasmids may be quite faint compared with the digested parental plasmid band.*

Final Transformation

14. Transform competent $mutS^+$ *E. coli* cells with either 2–4 μL of the digested plasmid DNA (~50–100 ng) (from Step 12) for transformation of chemically treated competent cells or 1 μL of plasmid DNA diluted fivefold with sterile H_2O (~5 ng) for transformation by electroporation (see Chapter 3, Protocol 4).

15. Spread 10, 50, and 250 μL of the transformation mixture onto LB agar plates containing the appropriate antibiotic. Incubate the plates overnight at 37°C.

16. The next day, prepare minipreparations of plasmid DNA (see Chapter 1) from at least 12 independent transformants. Screen the preparations by restriction endonuclease digestion and agarose gel electrophoresis to identify plasmids that are resistant to cleavage by the selective restriction enzyme.

17. Use DNA sequencing to confirm that the plasmids contain the desired mutation.

TROUBLESHOOTING

Problem: The rate of mutant recovery is low.
Solution: Try one or more of the following:

- Combine USE with the Kunkel method of selection against uracil-containing template DNAs (Markvardsen et al. 1995).

- Increase the concentration of the mutagenic primer so that the molar ratio of the two primers is 10:1 in favor of the mutagenic primer (Hutchinson and Allen 1997).

- Incubate the extension reaction catalyzed by T4 DNA polymerase at 42°C rather than 37°C (Pharmacia Instruction Booklet).

Problem: Vectors of the pBluescript family have been reported to yield mutants with very low efficiency, perhaps because they contain a knotty region of secondary structure that impedes the progress of the DNA polymerase during the extension of primers in vitro.
Solution: This problem can be avoided completely by using a plasmid vector that is not a member of the pBluescript family or by using a thermostable DNA polymerase and DNA ligase and incubating the extension reaction at high temperatures (Wong and Komaromy 1995).

RECIPES

It is essential that you consult the appropriate Material Safety Data Sheets and your institution's Environmental Health and Safety Office for proper handling of equipment and hazardous materials used in this protocol.

Annealing Buffer (10×)

Reagent	Final concentration
Tris-HCl (pH 7.5)	200 mM
MgCl₂	100 mM
NaCl	500 mM

LB Medium

Reagent	Quantity (for 1 L)
Bacto Tryptone	10 g
Bacto-yeast extract	5 g
NaCl	10 g

Dissolve in 1 L of distilled water. Adjust pH to 7.5 with 5 N NaOH. Sterilize by autoclaving.

LB Plates with Ampicillin

Make in the same way as LB medium, except add 15 g/L Bacto Agar before autoclaving. After autoclaving, add 125 μg of ampicillin/mL of medium before pouring.

Synthesis Buffer (10×)

Reagent	Final concentration
Tris-HCl (pH 7.5)	100 mM
dNTPs (dTTP, dATP, dCTP, and dGTP)	5 mM each
ATP	10 mM
DTT	20 mM

REFERENCES

Bebenek K, Kunkel TA. 1989. The use of native T7 DNA polymerase for site-directed mutagenesis. *Nucleic Acids Res* 17: 5408.

Braman J, Papworth C, Greener A. 2000. Site-directed mutagenesis using double-stranded DNA templates. In *The nucleic acid protocols handbook* (ed Rapley R), pp. 835–844. Humana, Totowa, NJ.

Conley EC, Saunders JR. 1984. Recombination-dependent recircularization of linearized pBR322 plasmid DNA following transformation of *Escherichia coli. Mol Gen Genet* 194: 211–218.

Deng WP, Nickoloff JA. 1992. Site-directed mutagenesis of virtually any plasmid by eliminating a unique site. *Anal Biochem* 200: 81–88.

Hutchinson MJ, Allen JM. 1997. Improved efficiency of two-primer mutagenesis. *Elsevier Trends Journals Technical Tips Online (#40072)*.

Kunkel TA. 1985. Rapid and efficient site-specific mutagenesis without phenotypic selection. *Proc Natl Acad Sci* 82: 488–492.

Markvardsen P, Lassen SF, Borchert V, Clausen IG. 1995. Uracil-USE, an improved method for site-directed mutagenesis on double-stranded plasmid DNA. *BioTechniques* 18: 370–372.

Nossal NG. 1974. DNA synthesis on a double-stranded DNA template by the T4 bacteriophage DNA polymerase and the T4 gene 32 DNA unwinding protein. *J Biol Chem* 249: 5668–5676.

Schena M. 1989. High efficiency oligonucleotide-directed mutagenesis using Sequenase. *Comments (US Biochemical Corp)* 15: 23.

Wong F, Komaromy M. 1995. Site-directed mutagenesis using thermostable enzymes. *BioTechniques* 18: 1034–1038.

Zhu L. 1996. In vitro site-directed mutagenesis using the unique restriction site elimination (USE) method. *Methods Mol Biol* 57: 13–29.

Saturation Mutagenesis by Codon Cassette Insertion

Saturation mutagenesis by cassette insertion introduces a library of site-specific changes into a specific DNA sequence within a target gene and is especially useful for analyzing the effect of specific residues on the structure and function of a protein. In general, cassette insertion mutagenesis involves excising a target region by restriction enzyme cleavage at two flanking sites and inserting a custom oligonucleotide containing the desired mutation in place of the excised fragment. This approach is effective for generating a single substitution at a single site, but it is not practical if the goal is to generate a library of mutants containing all possible amino acid substitutions at a single site because of the costs involved in synthesizing a complete canon of oligonucleotides.

This protocol overcomes many of these issues (Fig. 1). Here a set of 11 universal oligodeoxyribonucleotide cassettes is used to generate mutations. The major advantage of this method is that a single set of mutagenic codon cassettes can be used to insert codons encoding all possible

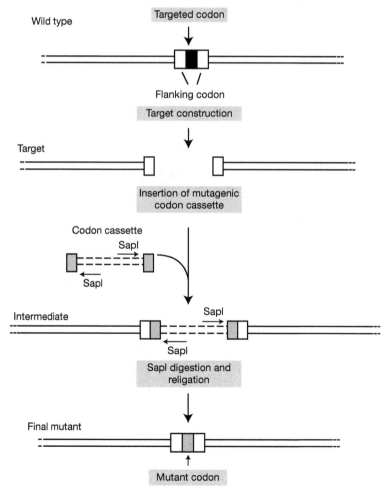

FIGURE 1. **Saturation mutagenesis by cassette insertion.** (Dark box) The codon targeted for mutagenesis; (open boxes) the codons immediately adjacent to the targeted codon "flanking codons." (Horizontal arrows) The sites of SapI cleavage. (Modified from Kegler-Ebo et al. 1996, with permission from Springer Science+Business Media.)

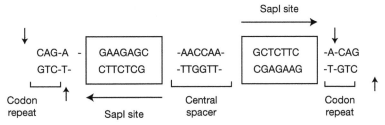

FIGURE 2. Structure of a universal codon cassette. (Boxes) The SapI recognition sequences; (horizontal arrows) point toward the cleavage sites; (vertical arrows) sites of SapI cleavage. The direct terminal codon repeat and central spacer between the two SapI recognition sequences are indicated. The cassette shown introduces CAG (glutamine) and CUG (leucine) codons. (Redrawn from Kegler-Ebo et al. 1996, with permission from Springer Science+Business Media.)

amino acids at any predetermined site within a gene. Each of the 11 cassettes contains two recognition sequences for SapI, a restriction enzyme that cleaves outside of its recognition sequence (as shown in Fig. 2). The recognition sequences for SapI are arranged in opposite orientations and are separated by a central spacer. At the end of each cassette is a 3-bp direct repeat, positioned such that the sites of SapI cleavage bracket each repeat. Cleavage by SapI will result in the generation of three base-cohesive single-stranded ends on the end of the cassette. These three base-cohesive single-stranded ends can then be ligated together to regenerate the original 3-bp direct repeat, while excising the central spacer. It is this 3-bp repeat sequence that is ultimately incorporated into the template. By substituting the 3-bp direct repeats in the universal cassette shown in Figure 2 (5'-CAG...CAG-3') with other sequences, one can essentially generate all possible amino acid substitutions. Table 1 shows the series of 11 universal oligonucleotide cassettes needed for this protocol and the substitutions they generate.

The overall scheme of steps needed to generate codon substitutions is shown in Figure 1. The blunt-ended mutagenic cassettes shown in Figure 2 are inserted into the target molecule by ligation at the blunt end sites, generating an intermediate molecule, which is then cleaved by SapI. Finally, the two single-stranded tails generated by SapI are ligated together to generate the mutagenic molecule. This essentially transfers a copy of the 3-base repeat from the cassette into the target molecule, which generates the final mutation. The protocol presented here is adapted from Kegler-Ebo et al. (1994, 1996).

MATERIALS

It is essential that you consult the appropriate Material Safety Data Sheets and your institution's Environmental Health and Safety Office for proper handling of equipment and hazardous materials used in this protocol.

Recipes for reagents specific to this protocol, marked <R>, are provided at the end of the protocol. See Appendix 1 for recipes for commonly used stock solutions, buffers, and reagents, marked <A>. Dilute stock solutions to the appropriate concentrations.

Reagents

Ammonium acetate (3.75 M, pH 5.0)
dNTP mix for Klenow reaction (10×)
  500 µM each in TE buffer; store at –20°C.

E. coli, DH5α strain
Ethanol (100 and 70%)
  Store at –20°C.

Klenow fragment of DNA polymerase I (5 U/µL)
Klenow reaction buffer (10×) <R>
Ligase reaction buffer (10×) <R>

TABLE 1. Universal codon cassette design

Mutagenic cassette[a]	Inserted amino acid[b]
ATG...ATG	Methionine
TAC...TAC	Histidine
TGG...TGG	Tryptophan
ACC...ACC	Proline
CAG...CAG	Glutamine
GTC...GTC	Leucine
GAC...GAC	Aspartic acid
CTG...CTG	Valine
AAC...AAC	Asparagine
TTG...TTG	Valine
TAT...TAT	Tyrosine
ATA...ATA	Isoleucine
GGC...GGC	Glycine
CCG...CCG	Alanine
AAA...AAA	Lysine
TTT...TTT	Phenylalanine
TTC...TTC	Phenylalanine
AAG...AAG	Glutamic acid
AGA...AGA	Arginine
TCT...TCT	Serine
ACA...ACA	Threonine
TGT...TGT	Cysteine

Reprinted from Kegler-Ebo et al. 1994, by permission of Oxford University Press.

[a]Schematic version of a series of universal codon cassettes and their direct terminal repeats are shown. The top strain is written in the 5′ to 3′ direction and the bottom strand in the 3′ to 5′ direction.

[b]The amino acids inserted by the cassettes in the first column. For each cassette, the top amino acid is introduced if the cassette is inserted in the orientation shown in the first column, and the bottom amino acid is introduced if the cassette is inserted in the opposite orientation.

Phenol:chloroform:isoamyl alcohol (PCA solution; 25:24:1 mixture of TE-saturated phenol, chloroform, and isoamyl alcohol)
    Store at 4°C.

RNase A, from bovine pancreas

SapI reaction buffer (10×) <R>

SapI restriction endonuclease (1 U/μL)

Sodium acetate (3 M, pH 5.2)

Sodium hydroxide (NaOH, 1 M)

Target molecule
    See Generation of the Blunt-Ended Target Molecule below.

T4 DNA ligase (400 U/μL)

TE buffer <R>

tRNA (10 mg/mL yeast tRNA; Sigma-Aldrich)
    Store at −20°C.

Universal codon cassettes
Obtain from New England Biolabs, and dilute to 10 ng/μL in TE buffer <R>.

Equipment

Gel electrophoresis equipment
QIAGEN Spin Miniprep Kit
Spectrophotometer
Thermal cycler
Vortex machine

METHOD

Generation of the Blunt-Ended Target Molecule

To perform codon cassette mutagenesis, a target molecule must be constructed that (1) does not contain any endogenous SapI cleavage sites and (2) is linear and blunt-ended as a result of a double-strand break at the desired site of mutagenesis. To generate the target molecule, the following outline and considerations must be taken into account.

- A pretarget molecule must be constructed with restriction enzyme cleavage sites so that digestion generates blunt ends at the correct position for codon substitution or insertion.
- The pretarget molecule must not have any endogenous SapI sites.

Any endogenous SapI sites within the pretarget molecule must be removed before performing codon cassette mutagenesis. Although the SapI recognition sequence is 7 bp long and occurs rarely in DNA sequences, it is important to check for the presence of these sites and amend them using either subcloning techniques or standard mutagenesis techniques (i.e., overlap mutagenesis).

When designing the pretarget molecule, it is important to develop an advance strategy for generating the blunt ends at the correct sites. A general method for constructing the pretarget molecule involves arranging SapI sites around the codon to be replaced so that cleavage by SapI and subsequent filling in of the single-stranded terminal ends by a Klenow fragment results in blunt ends at the appropriate position. This approach can be used at any amino acid sequence because the SapI sites are removed and the wild-type codons flanking the site of mutagenesis are regenerated.

1. Mix 2 μL of 10× SapI reaction buffer, 5 μL (5 μg) of pretarget molecule, 5 μL of SapI, and H₂O up to 20 μL. Incubate the mixture for 4 h at 37°C.

2. Add 30 μL of TE buffer to the reaction mixture. Next, add 50 μL of phenol:chloroform:isoamyl alcohol (PCA) solution and vortex. Centrifuge briefly in a microcentrifuge. Transfer the upper layer, which contains the DNA, to a new Eppendorf tube. Add 5 μL of sodium acetate to the tube, vortex, add 100 μL of ethanol, vortex, and store the tube for 2 h at −20°C or for 15 min on dry ice. Once again, collect the precipitate of DNA by centrifugation, wash the pellet with 70% ethanol, air-dry for 5 min, and dissolve the pellet in 20 μL of TE buffer. Quantify the DNA on a spectrophotometer at 260 nm.

3. To 16 μL (4 g) of the precipitated SapI-digested DNA, add 1 μL of 10× Klenow reaction buffer, 1 μL of 10× dNTP mix, and 1 μL of Klenow polymerase. Incubate the mixture for 15 min at 22°C.

4. Purify the DNA as described in Step 2. Dissolve the DNA pellet in 20 μL of TE.

Insertion of Codon Cassettes to Generate Intermediate Molecules

5. After the target molecule with the appropriate blunt ends has been prepared, set up 11 ligation reactions. In each reaction, the molar ratio of the blunt-end target molecule to the codon cassette should be ~1:25. Set up each reaction as follows and incubate overnight at 16 h:

Reagent	Volume/mass added
Ligase reaction buffer (10×)	1 μL
Blunt-end target molecule	1 μL
Double-stranded codon cassette	2.5 μL
H₂O	4.5 μL
T4 DNA ligase	1 μL

6. Add 40 μL of TE buffer and 2 μL of the yeast tRNA carrier to each reaction, and then extract once again, as described in Step 2. Ethanol-precipitate the DNA, centrifuge, and dissolve the pellet in 10 μL of TE buffer.

7. Transform competent DH5α *E. coli* cells separately (see Chapter 3) with 2.5 μL of solution from each of the 11 reaction mixtures from Step 6. Streak the cells onto LB agar plates containing the appropriate antibiotic.

8. Prepare miniprep DNA from individual colonies using a commercial kit. Dissolve the DNA in 20 μL of TE buffer (pH 8.0) containing 100 μg/mL RNase A. Store the DNA at −20°C.

DNA Sequencing to Confirm Insertion of Codon

9. Add 4 μL of 1 M NaOH to 16 μL of miniprep DNA (from Step 8) to denature the DNA. Incubate for 10 min at room temperature.

10. Add 2.5 μL of ammonium acetate and vortex. Precipitate the DNA by adding 100 μL of ethanol, and chill the DNA on ice for 20 min at −70°C or 2 h at −20°C.

11. Pellet the DNA using a microcentrifuge, and discard the supernatant. Wash the pellet once with cold 70% ethanol and dry the pellet.

12. Dissolve the pellet in 7 μL of water, and proceed with primer annealing and sequencing as described in the Sequenase Version 2.0 Manual (*United States Biochemical [1994] Step-by-Step Protocols for DNA Sequencing with Sequenase Version 2.0 T7 DNA Polymerase*, 8th ed.).

Resolution of the Intermediate Molecules to Generate Final Product

13. To generate the final mutations, set up a separate digestion reaction for each individual molecule. Mix the following in a test tube. Incubate for at least 4 h at 37°C.

Reagent	Volume/mass added
SapI reaction buffer (10×)	2 μL
Miniprep DNA of the intermediate molecule (from Step 6)	5 μL
H₂O	11 μL
SapI	2 μL

14. Use gel electrophoresis to confirm that digestion is complete for each sample (see Chapter 2).

15. To circularize the digested intermediate molecule, add:

Reagent	Volume/mass added
T4 DNA ligase buffer (10×)	5 μL
SapI-digested intermediate DNA (from Step 13)	10 μL
H₂O	34 μL
T4 DNA ligase	1 μL

Incubate for at least 1 h at room temperature.

16. Extract the solution with phenol:chloroform:isoamyl alcohol and precipitate the DNA with ethanol. Dissolve the pellet in 10 μL of TE buffer. This solution contains the final mutagenized product.

17. Transform competent *E. coli* cells separately with 2–4 μL of DNA solution from each reaction and streak onto LB agar plates containing the appropriate selection antibiotic. Incubate the plates overnight. The next day, colonies can then be picked and grown in liquid culture. Minipreparations of the plasmid DNAs can be digested with SapI to confirm incorporation of the cassette. If cassettes have been incorporated, no linearization should be observed.

    See Troubleshooting.

18. To detect transformants that lack cassettes, screen colonies by colony hybridization. For more information, see Kegler-Ebo (1994) for the colony hybridization method. Alternatively, one could screen clones using SapI digestion (plasmids that have incorporated a cassette should be resistant to SapI digestion).

TROUBLESHOOTING

Problem (Step 17): No transformants are detected upon plating.
Solution: Extend the length of SapI digestion (Step 13) to ensure that the reaction goes to completion.

Problem (Step 17): SapI digestion linearizes plasmids.
Solution: If the SapI digestion of the codon cassette in Step 13 was successful, the plasmid should be resistant to digestion. If this is not the case, then check to make sure that codon cassettes were constructed correctly and that the ligation reaction in Step 5 is working.

RECIPES

It is essential that you consult the appropriate Material Safety Data Sheets and your institution's Environmental Health and Safety Office for proper handling of equipment and hazardous materials used in this protocol.

Klenow Reaction Buffer (10×)

Reagent	Final concentration
Tris-HCl (pH 7.5)	100 mM
MgCl$_2$	50 mM
Dithiothreitol	75 mM

Ligase Reaction Buffer (10×)

Reagent	Final concentration
Tris-HCl (pH 7.8)	500 mM
MgCl$_2$	100 mM
Dithiothreitol	100 mM
ATP	10 mM
Bovine serum albumin	250 μg/mL

SapI Reaction Buffer (10×)

Reagent	Final concentration
Potassium acetate	500 mM
Tris-acetate	200 mM
Magnesium acetate	100 mM
Dithiothreitol	10 mM

Adjust pH to 7.9.

TE Buffer

Reagent	Final concentration
Tris-HCl (pH 8.0)	10 mM
EDTA	1 mM

REFERENCES

Kegler-Ebo DM, Docktor CM, DiMaio D. 1994. Codon cassette mutagenesis: A general method to insert or replace individual codons by using universal mutagenic cassettes. *Nucleic Acids Res* 22: 1593–1599.

Kegler-Ebo DM, Polack GW, DiMaio D. 1996. Use of codon cassette mutagenesis for saturation mutagenesis. *Methods Mol Biol* 57: 297–310.

Random Scanning Mutagenesis

In vitro oligonucleotide and PCR-based mutagenesis is generally used for altering the nucleotide sequence of genes to study their functional importance and the products they encode (Hutchison et al. 1978; Botstein and Shortle 1985; Kunkel 1985; Higuchi et al. 1988). A thorough approach to this problem is to systematically change each successive amino acid residue in the protein to alanine (i.e., alanine-scanning mutagenesis) or to a limited number of alternative amino acids (Cunningham and Wells 1989). Although these strategies can provide useful information, it is sometimes desirable to test a broader spectrum of amino acid changes at the targeted positions. Recently, Smith and colleagues developed an approach called "random scanning mutagenesis" to examine the functional importance of individual amino acid residues in the conserved structural motif of human immunodeficiency virus (HIV) reverse transcriptase, and this protocol is adapted from their method (Smith et al. 2004, 2006). This strategy is an oligonucleotide-based method for generating all 19 possible replacements at individual amino acid sites within a protein. Figure 1 illustrates the major steps of this protocol.

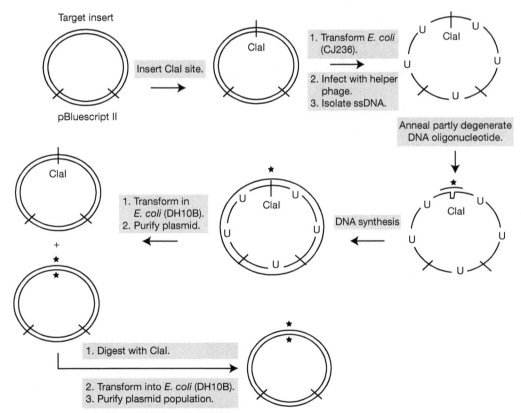

FIGURE 1. **Random scanning mutagenesis approach.** In this example, a ClaI restriction site is introduced into the target region. The mutated construct is used as the template to produce single-stranded, uracil-containing DNA. Following in vitro DNA synthesis using a pool of mutagenic oligonucleotide primers, the plasmids are transformed into competent dUTPase⁺ UNG⁺ E. coli cells to select against the uracil-containing template DNA. To further enrich the mutant population, plasmids are digested with ClaI and reintroduced into an E. coli host. The resulting plasmid library is greatly enriched for amino acid–altering mutations at the target codon site. (Stars) Mutated sites. (Modified from Smith 2010, with permission from Springer Science+Business Media.)

MATERIALS

It is essential that you consult the appropriate Material Safety Data Sheets and your institution's Environmental Health and Safety Office for proper handling of equipment and hazardous materials used in this protocol.

Recipes for reagents specific to this protocol, marked <R>, are provided at the end of the protocol. See Appendix 1 for recipes for commonly used stock solutions, buffers, and reagents, marked <A>. Dilute stock solutions to the appropriate concentrations.

Reagents

Ampicillin (100 mg/mL)
Annealing buffer (10×) <R>
Chloramphenicol (15 µg/mL)
E. coli strains DH10B (Life Technologies) and CJ236 (Takara Bio Inc.)
Ethanol (100%)
Gene fragment or open reading frame of interest
Kanamycin (70 µg/mL)
LB-ampicillin (100 µg/mL) agar plates
LB-ampicillin-chloramphenicol agar plates
LB medium <R>
pBluescript II KS(−) (Agilent)
PEG/NaOAc solution (20% [w/v] polyethylene glycol 8000 in 2.5 M sodium acetate)
Phenol:chloroform:isoamyl alcohol (25:24:1)
Plasmid DNA purification kit
RNase A (25 mg/mL), DNase-free
SOC medium <A>
Synthesis buffer (10×) <R>
T4 DNA ligase (New England Biolabs)
T7 DNA polymerase (New England Biolabs)
VCSM13 interference-resistant helper phage (Agilent)
Yeast tryptone broth (YT broth) (2×) <R>

Equipment

Centrifuge
DNA sequencing equipment
Incubator
Petri dishes
Spectrophotometer
Thermal cycler
Water bath or thermoblock

METHOD

1. Introduce a unique restriction enzyme site into the DNA fragment of interest by means of either oligonucleotide-directed mutagenesis (see Protocol 5) or by using a PCR-based strategy (see Protocol 2).

   Disrupting the polymerase reading frame is important to avoid expression of the target protein.

Production of Uracil-Containing ssDNA

2. Transform the modified plasmid into chemically competent CJ236 *E. coli* host, which is deficient in deoxyuridine triphosphatase (dUTPase) and uracil DNA glycosylase (UDG) activity (see Chapter 3). As a result, uracil will be incorporated into the plasmid DNA.

DESIGN OF THE MUTAGENIC PRIMERS

The mutagenic oligonucleotides should be long enough to ensure that, when annealed to the ssDNA template, the primers are bound by 20–25 complementary base pairs at both termini. Moreover, it is of fundamental importance that the oligonucleotides carry a 5′ phosphate group. This can be achieved by either chemical synthesis of the primers or by incubating the oligonucleotides with T4 polynucleotide kinase. (See Chapter 13.)

Other Important Considerations

- Because this protocol requires single-stranded DNA (ssDNA), the plasmid must contain an f1 origin of replication. Examples of plasmids producing sufficient amounts of ssDNA upon helper phage infection are pBluescript II KS(−), pGEM-3Zf, and pTZ18U.

- The engineered restriction site should be located at the target codon if one single codon is targeted. Otherwise, if more than one codon is being randomized, the restriction site should be placed in the middle of the target region.

- The mutagenic process that takes place during in vitro DNA synthesis produces plasmid libraries with an uneven distribution of variant amino acids at the target region. This is due to both the intrinsic asymmetry of the genetic code and different annealing capacity of the oligonucleotide primers in the pools. To overcome this problem, synthesize oligonucleotides encoding for less-represented amino acids separately, and mix them into the pools at defined ratios.

- Incubate the phage-infected cultures for no more than 12 h, otherwise the result will be large contamination of *E. coli* genomic DNA and RNA in the final preparation of ssDNA.

3. Recover bacteria by incubating them in SOC medium for 1 h at 37°C.

4. Spread the bacteria onto LB-ampicillin-chloramphenicol plates, and incubate the plates overnight at 37°C.

5. Pick a colony from the plate, and inoculate it into 2 mL of 2× YT medium containing ampicillin (50–100 μg/mL) and chloramphenicol (15 μg/mL). Incubate the plate for 12 h at 37°C.

6. Transfer the 2-mL mixture to a flask containing 300 mL of 2× YT with ampicillin (50–100 μg/mL) and chloramphenicol (15 μg/mL). Add VCSM13 helper phage at a final concentration of 2×10^8 pfu/mL.

7. Incubate the flask for 2 h at 37°C, and then add kanamycin (70 μg/mL) and continue the incubation for an additional 12 h.

8. Centrifuge at 7000g for 15 min at 4°C. Carefully decant the supernatant, which contains the bacteriophage particles, into fresh tubes.

9. Transfer the supernatant into new centrifuge tubes and repeat the centrifugation step.

10. Transfer 80% of the supernatant to new centrifuge tubes and add 1/4 volume of PEG/NaOAc solution to each tube. Mix the solution and store it on ice for 1 h or overnight. The phage particles will precipitate during this step.

11. Centrifuge the solution at 9000g for 25 min at 4°C.

12. Discard the supernatant, and resuspend the pellets (containing phage particles) in 2 mL of TE buffer (pH 8).

13. Divide the phage suspension into two tubes (1 mL each), and add 2 μL of 25 mg/mL DNase-free RNase A to each tube. Incubate the tubes for 1 h at 37°C.

14. Centrifuge the tubes at 15,000g for 5 min. This step is important to eliminate any cells or cell debris. Transfer the supernatants to new tubes, and repeat the precipitation of the phage particles by adding 250 μL of PEG/NaOAc to each tube and placing the tubes on ice for 15 min. Then centrifuge them at 14,000 rpm for 5 min at 4°C.

15. Discard the supernatant, and resuspend the pellet in 300 μL of TE buffer (pH 8).

16. Extract the ssDNA with phenol:chloroform (1:1) followed by phenol:chloroform:isoamyl alcohol (25:24:1).

17. Precipitate the ssDNA by adding 1/10 volume of 3 M NaOAc (pH 5) and 2 volumes of 100% ethanol and incubating on ice for 30 min. Centrifuge the tubes at 14,000 rpm for 15 min.

18. Discard the supernatant, allow the DNA pellets to dry, and then resuspend them in 500 μL of TE buffer (pH 8). Determine the DNA concentration by using a spectrometer. A sample of pure DNA has a 260/280 ratio of 1.8.

Mutagenesis Reaction

19. Perform the following reaction on ice and include a negative control (no primer) reaction. To each tube, add 0.6 pmol of ssDNA template. Then, add 6.6 pmol of the random oligonucleotide pool, and bring the volume to 9 μL with distilled water.

20. Add 1 μL of 10× annealing buffer to each tube. Anneal the primers to the ssDNA template by adding the tubes to a bath, previously warmed at 80°C, and allowing the water to cool slowly to 30°C. Finally, place the tubes on ice.

21. Add 1 μL of 10× Synthesis buffer, 6 Weiss units of T4 DNA ligase, and 0.5 units of T7 DNA polymerase. Mix by pipetting and incubate for 2 h at 37°C.

22. Transform 1 μL of each reaction into the *E. coli* strain. The host strain must have both deoxyuridine triphosphatase (dUTPase) and uracil DNA glycosylase (UDG) activities to select against uracil-containing plasmids.

23. Recover the cells by incubating them for 1 h at 37°C in SOC medium. Then, spread 100 μL of cells onto LB-ampicillin agar plates, followed by incubation overnight at 37°C. In addition, transfer 200 μL of the cells to a flask containing 100 mL of LB-ampicillin (50–100 mg/mL), and incubate the cultures overnight at 37°C.

24. Centrifuge the cell cultures from Step 23 at 6000*g* for 15 min at 4°C.

25. Count the number of colonies on each LB-ampicillin plate from Step 23. If the mutagenic primer–containing reactions give rise to at least threefold more colonies compared with the control reaction (no primer), purify plasmid DNA from the pellets prepared in Step 24 (see Chapter 1, or use a commercial kit). Store the DNA at −20°C.

Restriction Enzyme Digestion

26. Digest DNA libraries from Step 25 with an enzyme that recognizes the engineered restriction sequence.

27. Transform an aliquot of each digest into the *E. coli* strain (see Chapter 3), and proceed as described in Step 23.

28. Characterize individual clones from plasmid libraries by DNA sequencing and use them for subsequent functional assays.

RECIPES

It is essential that you consult the appropriate Material Safety Data Sheets and your institution's Environmental Health and Safety Office for proper handling of equipment and hazardous materials used in this protocol.

Annealing Buffer (10×)

Reagent	Final concentration
Tris-HCl (pH 7.4)	200 mM
MgCl$_2$	20 mM
NaCl	500 mM

Autoclave and store at −20°C.

LB Medium

Reagent	Quantity (for 1 L)
Bacto Tryptone	10 g
Bacto-yeast extract	5 g
NaCl	10 g

Dissolve in 1 L of distilled water. Adjust pH to 7.5 with 5 N NaOH. Sterilize by autoclaving.

Synthesis Buffer (10×)

Reagent	Final concentration
Tris-HCl (pH 7.4)	100 mM
dNTPs (dTTP, dATP, dCTP, and dGTP)	5 mM each
ATP	10 mM
$MgCl_2$	50 mM
DTT	20 mM

Divide into aliquots and store at −80°C.

Yeast Tryptone (YT) Broth (2×)

Reagent	Quantity (for 1 L)
Bacto Tryptone	16 g
Bacto-yeast extract	10 g
NaCl	5 g

Dissolve in 1 L of distilled water. Adjust pH to 7.0 with 5 N NaOH. Sterilize by autoclaving.

REFERENCES

Botstein D, Shortle D. 1985. Strategies and applications of in vitro mutagenesis. *Science* **229:** 1193–1201.

Cunningham BC, Wells JA. 1989. High-resolution epitope mapping of hGH-receptor interactions by alanine-scanning mutagenesis. *Science* **244:** 1081–1085.

Higuchi R, Krummel B, Saiki RK. 1988. A general method of in vitro preparation and specific mutagenesis of DNA fragments: Study of protein and DNA interactions. *Nucleic Acids Res* **16:** 7351–7367.

Hutchison CA III, Phillips S, Edgell MH, Gillam S, Jahnke P, Smith M. 1978. Mutagenesis at a specific position in a DNA sequence. *J Biol Chem* **253:** 6551–6560.

Kunkel TA. 1985. The mutational specificity of DNA polymerase-β during in vitro DNA synthesis. Production of frameshift, base substitution, and deletion mutations. *J Biol Chem* **260:** 5787–5796.

Smith RA. 2010. Random-scanning mutagenesis. *Methods Mol Biol* **634:** 387–397.

Smith RA, Anderson DJ, Preston BD. 2004. Purifying selection masks the mutational flexibility of HIV-1 reverse transcriptase. *J Biol Chem* **279:** 26726–26734.

Smith RA, Anderson DJ, Preston BD. 2006. Hypersusceptibility to substrate analogs conferred by mutations in human immunodeficiency virus type 1 reverse transcriptase. *J Virol* **80:** 7169–7178.

Multisite-Directed Mutagenesis

Most of the PCR-based methods of mutagenesis currently in use are direct descendants of techniques originally described in the late 1980s, soon after the introduction of thermostable DNA polymerases to PCR (Higuchi et al. 1988; Ho et al. 1989; Vallette et al. 1989). PCR-based site-directed mutagenesis (SDM) methods are widely used in molecular biology to selectively change the sequence of genes and to gain insight into their function. The early protocols allowed for the introduction of only one or two point mutations at a time (see Protocol 2), but more recently, several protocols for mutagenesis have been developed that create multiple independent mutations with high efficiency (Mikaelian and Sergeant 1992; Ishii et al. 1998; Kim and Maas 2000), and several commercial kits are now available. Most of them are not suitable for both general multisite-directed and close-proximity mutagenesis. Some methods require phosphorylated primers at the 5' end, and others require more rounds of PCR than the number of mutation sites to be introduced.

Recently, Tian et al. (2010) developed an effective approach that introduces multiple independent mutations in a procedure that has only one cloning step, and the protocol presented here is adapted from their method. This method suits for both general and close-proximity mutagenesis and includes a simple and rapid procedure that combines PCR, DpnI digestion, and overlap extension. The key point of this approach is the use of overlap extension to form a circular DNA plasmid with mutations without the need for phosphorylated primers or ligase reactions. Essentially, during the first round of PCR, the new DNA is synthesized with nicks between the 3' ends of the synthesized DNA and the 5' ends of the first pair of primers (see Fig. 1). During successive rounds of PCR, a new pair of mutagenic oligonucleotides leads to the synthesis of two DNA segments that anneal together with the overlap sequence inside the two primers. This new mutated molecule also contains nicks but at different positions compared with those formed in the first round of PCR that had been "repaired" by overlap extension. Mutations can be introduced successfully by this method. Finally, the circular DNA is transformed into *E. coli* cells, where the nicks are ligated into a circular plasmid. One important requirement is that the parental plasmid carrying the target gene needs to be methylated by Dam methyltransferase or purified from Dam$^+$ *E. coli* (i.e., DH5a).

ADVANTAGES AND DISADVANTAGES

Major Advantages of This Method

- The targeted mutations can be close together or far apart.
- The method requires only one cloning step.
- The overall cost of this method is lower compared with that of the commercial kits.
- Primer phosphorylation or DNA ligation is not required.

Major Drawbacks of This Method

- The efficiency for three-site mutation in a 4-kb plasmid is only ~36.4%, lower than that for a commercial kit (>50% efficiency for the QuikChange Multi-Site-Directed Mutagenesis Kit; Agilent) (Tian et al. 2010).
- This method requires as many rounds of PCR as the target mutation sites to be introduced.
- The time required to create more than three mutations by this method is greater than the time required by commercial kits.

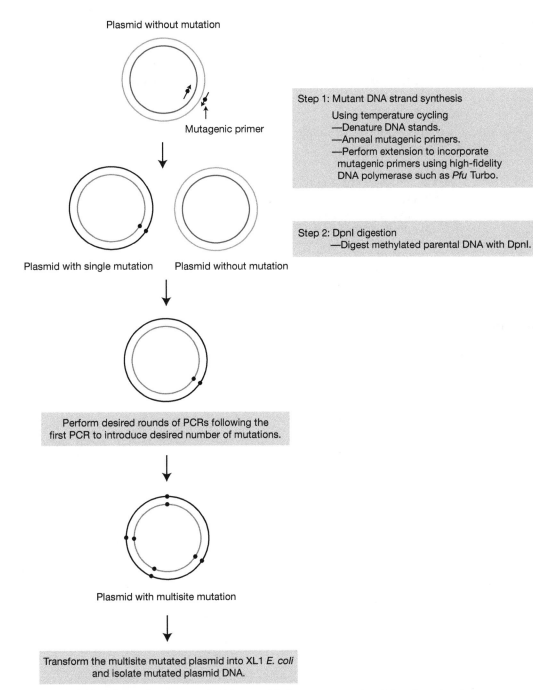

FIGURE 1. **Multisite-directed mutagenesis.** Primers with the target mutation site and overlapping sequence generate the first circular DNA molecule. The mutated DNA is enriched by digesting the parental template DNA with DpnI and used for subsequent rounds of PCR with different mutagenic primers. Finally, the DNA carrying the desired mutations is transformed into competent bacteria, isolated, and sequenced. (Black circle) The mutation. (Adapted from Tian et al. 2010, with permission from Elsevier.)

MATERIALS

It is essential that you consult the appropriate Material Safety Data Sheets and your institution's Environmental Health and Safety Office for proper handling of equipment and hazardous materials used in this protocol.

Recipes for reagents specific to this protocol, marked <R>, are provided at the end of the protocol. See Appendix 1 for recipes for commonly used stock solutions, buffers, and reagents, marked <A>. Dilute stock solutions to the appropriate concentrations.

Reagents

Agarose gel electrophoresis reagents
Antibiotics appropriate for bacterial strain
DNA template containing cloned gene(s) to be mutated
dNTPs mix
DpnI (10 U/μL)
E. coli XL-Blue 1 competent cells
LB-agar plates <R>
LB medium <R>
PCR buffer (according to DNA polymerase used)
Primers, forward and reverse mutagenic

Both of the mutagenic primers must contain the desired mutation and anneal to the same sequence on opposite strands of the plasmid. They may be designed to contain a point mutation, insertion, or deletion, as desired. The primers can be 25–45 bases in length and should have at least 10–15 bases of complementarity on both sites of the mutation. This is especially important when multiple base mismatches are present.

Thermostable DNA polymerase

Equipment

Centrifuge
Gel electrophoresis equipment
Incubator
Petri dishes
Plasmid DNA purification kit
Thermal cycler
Water bath or thermal block

METHOD

1. Perform a PCR with the following reagents (see also Chapter 7):

Forward and reverse mutagenic primers	Final concentration of 0.4 μM
dNTPs mixture	Final concentration of 0.2 mM of each base
DNA template	100 ng

   Perform the PCR in a thermal cycler under the following conditions:

Number of cycles	Denaturation	Annealing	Polymerization
1	5 min at 94°C		
30–40	30 sec at 94°C	30 sec at 55°C–62°C	2 min at 68°C
1			5 min at 68°C

2. Add 1 μL of DpnI restriction enzyme (10 U/μL) directly to each amplification reaction.

3. Gently and thoroughly mix each reaction mixture by pipetting the solution up and down several times. Centrifuge the reaction mixtures in a microcentrifuge, and then immediately incubate each reaction for 1 h at 37°C to digest the parental DNA.

4. Perform another PCR using the product from the first PCR and a different pair of mutagenic primers to introduce further mutation sites. The procedure is the same as that described in Step 1.

5. After all of the desired mutation sites are introduced, use 10 μL of the final PCR product to transform 50–100 μL of *E. coli* XL-Blue1 competent cells (see Chapter 3). Incubate the cells on ice for 15–30 min, and then heat-shock them for 90 sec at 42°C. Allow the bacteria to

recover in 1 mL of LB medium for 1 h at 37°C. Plate the transformed *E. coli* cells onto LB-agar plates containing appropriate antibiotics, and incubate the plates overnight at 37°C.

6. Pick 12 colonies at random, and inoculate them overnight in 2–3 mL of LB medium plus appropriate antibiotics.

7. Extract and purify plasmid DNAs using a commercial kit. Sequence tha plasmids to find some that contain all the desired mutations.

RECIPES

It is essential that you consult the appropriate Material Safety Data Sheets and your institution's Environmental Health and Safety Office for proper handling of equipment and hazardous materials used in this protocol.

LB-Agar Plates
Make in the same way as LB medium, except add 15 g/L Bacto Agar before autoclaving. After autoclaving, add 125 μg of appropriate antibiotic/mL of medium before pouring.

LB Medium

Reagent	Quantity (for 1 L)
Bacto Tryptone	10 g
Bacto-yeast extract	5 g
NaCl	10 g

Dissolve in 1 L of distilled water. Adjust pH to 7.5 with 5 N NaOH. Sterilize by autoclaving.

REFERENCES

Higuchi R, Krummel B, Saiki RK. 1988. A general method of in vitro preparation and specific mutagenesis of DNA fragments: Study of protein and DNA interactions. *Nucleic Acids Res* 16: 7351–7367.

Ho SN, Hunt HD, Horton RM, Pullen JK, Pease LR. 1989. Site-directed mutagenesis by overlap extension using the polymerase chain reaction. *Gene* 77: 51–59.

Ishii TM, Zerr P, Xia XM, Bond CT, Maylie J, Adelman JP. 1998. Site-directed mutagenesis. *Methods Enzymol* 293: 53–71.

Kim YG, Maas S. 2000. Multiple site mutagenesis with high targeting efficiency in one cloning step. *BioTechniques* 28: 196–198.

Mikaelian I, Sergeant A. 1992. A general and fast method to generate multiple site directed mutations. *Nucleic Acids Res* 20: 376.

Tian J, Liu Q, Dong S, Qiao X, Ni J. 2010. A new method for multi-site-directed mutagenesis. *Anal Biochem* 406: 83–85.

Vallette F, Mege E, Reiss A, Adesnik M. 1989. Construction of mutant and chimeric genes using the polymerase chain reaction. *Nucleic Acids Res* 17: 723–733.

Megaprimer PCR-Based Mutagenesis

The megaprimer method is a really simple and versatile approach that can be adopted to create a single mutation in a specific target region as well as to create site-specific insertions, deletions, and gene fusions (see Fig. 1). This method uses three oligonucleotide primers, two rounds of PCR, and a DNA template containing the gene to be mutated. As shown in Figure 1, where A and B are the flanking primers and M represents the primer carrying the desired mutation, the first round of PCR generates a fragment with the desired mutation introduced by using one of the flanking primers (A) and the mutant primer (M). This amplified fragment—the megaprimer—is used in the second PCR along with the remaining external primer (B) to amplify a longer region of the template DNA. The final product is purified and can be cloned into an appropriate vector. By designing flanking primers with universal restriction site sequences, compatible with the vector of choice, it is possible to create different mutant clones by changing only the mutant primer. Recently, this approach has been improved by the use of forward and reverse flanking primers with significantly different melting temperatures. This allows researchers to perform both PCR steps in a single tube (Ke and Madison 1997). Here, we describe the protocol adapted from Brons-Poulsen et al. (1998, 2002). This protocol has been successfully applied on templates with either low or high G+C content to amplify megaprimers 71–800 bp in length and final products ranging from 400 to 2500 bp. Moreover, the investigators were able to introduce single-base-pair mutations as well as 24-bp deletions and substitutions.

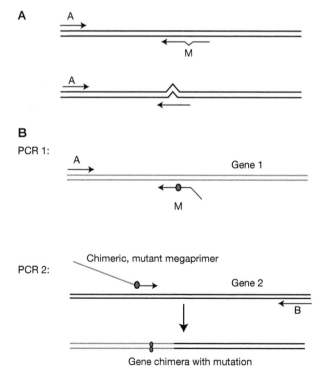

FIGURE 1. **Procedure for insertion, deletion, and gene fusion mutagenesis using the megaprimer approach.** For insertion (*A, upper*) and deletion (*A, lower*) mutagenesis, only the first round of PCR is shown; the subsequent second PCR will use the megaprimer produced during the first PCR, primer B, and the same template as for the first PCR, as described above. For the gene fusion (*B*) the first and the second PCRs use two different genes, (gray and black lines, respectively). The M primer, used during the first PCR (PCR 1), must have a sequence complementary to the second gene that will allow the megaprimer to anneal to the DNA template during the second round of PCR (PCR 2). The blue dot on the M primer represents the desired mutation to be introduced into the chimeric gene. (Adapted from Barik 1997, with permission from Springer Science+ Business Media.)

MATERIALS

It is essential that you consult the appropriate Material Safety Data Sheets and your institution's Environmental Health and Safety Office for proper handling of equipment and hazardous materials used in this protocol.

Recipes for reagents specific to this protocol, marked <R>, are provided at the end of the protocol. See Appendix 1 for recipes for commonly used stock solutions, buffers, and reagents, marked <A>. Dilute stock solutions to the appropriate concentrations.

Reagents

Agarose gel
DNA template containing the cloned gene to be mutated
dNTPs mix
Oligonucleotide primers A, B, and M (see box below on Primer Design)
PCR buffer (supplied with enzyme)
Reagents for template and PCR product purification
> *Several commercially available kits can be useful, but a classic extraction with phenol:chloroform also works well.*

Thermostable DNA polymerase

Equipment

Equipment for agarose gel electrophoresis of PCR products (see Chapter 2, Protocols 1 and 2)
Gel extraction kit
Long-wave UV transilluminator
Thermal cycler

PRIMER DESIGN

When designing primers, be aware that the PCR product arising from the second PCR must differ in length from the megaprimers. This ensures appropriate separation of the two fragments during gel electrophoresis. Second, the mutation in the M primer should reside in the center of the primer. Mutations close to the 5′ end can impair the efficiency of the second round of mutation. A mutation located at the 3′ end can suppress the yield of the megaprimer. A mutation placed in the middle of a primer can affect the melting temperature. Therefore, to ensure efficient annealing, the length of the primer should be adjusted. Finally, both A and B primers should include specific restriction sites so that the final product can be cloned in the vector of choice.

The same considerations apply when designing primers for insertion, deletion, and gene fusion. In addition, gene fusion can be combined with mutagenesis by incorporating the desired mutations in the M primer.

METHOD

Megaprimer Synthesis

1. In a sterile 0.5-mL microcentrifuge tube, mix the following reagents for the first round of PCR:

Amplification buffer (10×)	10 µL
Plasmid template DNA	50–200 ng
dNTPs mix	200 µM (of each of the four dNTPs)
Mutagenic/flanking primers	100 pmol (of each primer)
Thermostable DNA polymerase	2.5 U
Water	to 100 µL

The cycling protocol is:

Number of cycles	Denaturation	Annealing	Polymerization
1	2 min at 94°C		
10–25	20–45 sec at 94°C	30–120 sec at 50°C–60°C	30–90 sec at 72°C
1			10 min at 72°C

2. Purify the product obtained from the PCR by agarose gel electrophoresis (see Chapter 2, Protocols 1 and 2).

3. Cut out the band containing the megaprimer and purify the DNA using a gel extraction kit.

Second PCR with Megaprimer

The second round of PCR is performed by adding the megaprimer from the first PCR to the DNA template, along with primer B. The crucial point of this step is the primer concentration. Although the reasons for this observation remain elusive, they probably relate to the secondary and tertiary structures formed when single-stranded DNA of considerable size reaches critical concentrations. Some investigators have found that megaprimer concentrations higher than 0.01 µM inhibit the reaction. In contrast, other studies show that a megaprimer concentration between 0.02 and 0.04 µM is advantageous, compared with concentrations of 0.01 µM and below. With certain megaprimers, a complete inhibition of the reaction may be observed. In this case, the concentration of the megaprimer should be decreased to an absolute minimum. Note that both primers should be equal in concentration, to avoid asymmetric amplification, resulting in a single-stranded product.

4. In a sterile 0.5-mL microcentrifuge tube, mix the following reagents for the last round of PCR:

Amplification buffer (10×)	10 µL
Plasmid template DNA	50–200 ng
dNTPs	200 µM (of each of the four dNTPs)
Megaprimer/flanking primers	4 pmol (of each primer)
Thermostable DNA polymerase	2.5 U
Water	to 100 µL

Use the following PCR conditions for the synthesis of the final mutated product:

Number of cycles	Denaturation	Annealing	Polymerization
1	2 min at 94°C		
10–25	20–45 sec at 94°C	120 sec at 50°C–60°C	30–90 sec at 72°C
1	10 min at 72°C		

5. Analyze the PCR product on an agarose gel, and recover the band of higher-molecular-weight DNA.

6. Purify the fragment using a gel extraction kit, and clone it into the desired vector (see Chapter 3). If the yield is poor, reamplify the mutant product using primers A and B and standard PCR.

REFERENCES

Barik S. 1997. Mutagenesis and gene fusion by megaprimer PCR. *Methods Mol Biol* 67: 173–182.

Brons-Poulsen J, Petersen NE, Horder M, Kristiansen K. 1998. An improved PCR-based method for site directed mutagenesis using megaprimers. *Mol Cell Probes* 12: 345–348.

Brons-Poulsen J, Nohr J, Larsen LK. 2002. Megaprimer method for polymerase chain reaction-mediated generation of specific mutations in DNA. *Methods Mol Biol* 182: 71–76.

Ke SH, Madison EL. 1997. Rapid and efficient site-directed mutagenesis by single-tube 'megaprimer' PCR method. *Nucleic Acids Res* 25: 3371–3372.

DOMAIN MUTAGENESIS

Domain mutagenesis is used to introduce multiple mutations into a defined region of cloned DNA. One of the most powerful tools exploited for introducing random mutations in protein domains or promoter elements is the error-prone PCR method (see Protocol 1). The GeneMorph II EZClone Domain Mutagenesis Kit (Agilent) offers a simple and rapid system to perform targeted random mutagenesis on protein domains. The kit requires a double-stranded DNA vector containing the gene region to be mutated, the megaprimers generated during the mutagenesis reaction, and a digestion step with the restriction enzyme DpnI to eliminate the unmutated parental plasmid DNA. Finally, the mutated library is transformed into competent *E. coli* cells. The Mutazyme II DNA polymerase included in the kit introduces a more uniform mutational spectrum with minimal mutational bias compared with other error-prone PCR enzymes so that mutations at As and Ts occur at the same frequency as at Gs and Cs (Fig. 1).

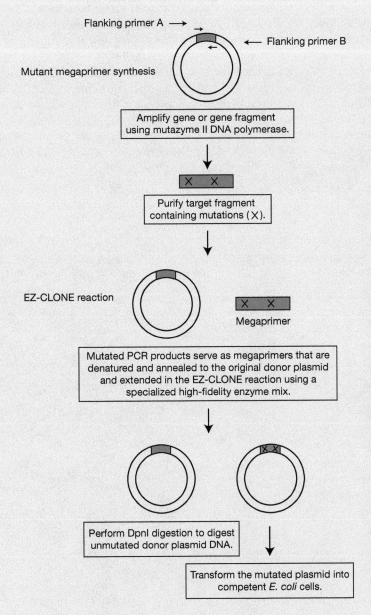

FIGURE 1. **Domain mutagenesis.** The first round of PCR (mutant megaprimer synthesis) is performed by using a primer pair external to the target domain along with the error-prone Mutazyme II DNA polymerase. The mutated PCR products are purified and serve as a megaprimer in the second round of PCR (EZ-CLONE reaction). This step uses the same template as in the first PCR but exploits a high-fidelity DNA polymerase. Finally, the mutated vector is enriched by depletion of the parental template DNA with DpnI and amplified by transformation into *E. coli* cells. (Medium blue rectangle) Target domain; (black cross) mutated base pair. (Modified from Fig. 1 of Protocol 1 in the GeneMorph II EZ-CLONE Domain Mutagenesis Kit Instruction Manual, Agilent Technologies, Stratagene Products Division. © Agilent Technologies, Inc. Reproduced with permission, courtesy of Agilent Technologies, Inc.)

HIGH-THROUGHPUT SITE-DIRECTED MUTAGENESIS OF PLASMID DNA

This information panel outlines an efficient method for high-throughput in vitro site-directed mutagenesis. This protocol can be used to generate insertions, deletions, and substitutions of up to 21 nucleotides from any source. Unlike many other methods, this protocol does not require specialized vectors, host strains, or restriction sites. Only a pair of mutagenic oligonucleotide primers is necessary to generate the desired mutation. A commercial kit for performing this—the GeneTailor Site-Directed Mutagenesis System—is available from Life Technologies.

This method is a simple three-step protocol and can be easily scaled up to 96 mutations in a given DNA sequence. First, it requires designing forward and reverse primers that are at least 30 nucleotides in length, not including the mutation site on the mutagenic primer. The forward and reverse primers should have an overlapping region at the 5′ ends of 15–20 nucleotides to efficiently end-join the mutagenesis products. The mutation site should be located on only one of the primers and can be up to 21 bases. After designing the primers, the template plasmid should be methylated. To do so, the template plasmid is incubated with DNA methylase in methylation reaction buffer for 1 h at 37°C. DNA methylase methylates cytosine residues within a specific sequence throughout the double-stranded DNA. As a result, the methylated DNA template is made vulnerable to Mcr and Mrr restriction systems in E. coli, as described by Bandaru et al. (1995) and Wyszynski et al. (1994).

Next, the methylated plasmid is used for the mutagenesis reaction, which incorporates both the mutagenic primer and the complementary primer, together with thermostable polymerase, dNTPs, MgSO$_4$, and PCR buffer. The reaction mixture is placed in a thermocycler for 20 rounds of cycling. The PCR product is linear, double-stranded DNA containing the mutation of interest. Finally, the mutagenesis mixture is transformed into a wild-type E. coli cell. The host cell circularizes the linear mutated DNA, and endogenous host McrBC endonuclease digests the original methylated template, leaving only the unmethylated mutated product. The mutated plasmid can then be purified from the E. coli cell through conventional methods.

REFERENCES

Bandaru B, Wyszynski M, Bhagwat AS. 1995. HpaII methyltransferase is mutagenic in *Escherichia coli*. *J Bacteriol* 177: 2950–2952.

Wyszynski M, Gabbara S, Bhagwat AS. 1994. Cytosine deaminations catalyzed by DNA cytosine methyltransferases are unlikely to be the major cause of mutational hot spots at sites of cytosine methylation in *Escherichia coli*. *Proc Natl Acad Sci* 91: 1574–1578.

N⁶-METHYLADENINE, DAM METHYLASE, AND METHYLATION-SENSITIVE RESTRICTION ENZYMES

Methylation of Adenine Residues by Dam Methylase

In *E. coli* cells, adenine residues embedded in the sequence $^{5'}$...GATC...$^{3'}$ carry a methyl group attached to the N^6 atom (Hattman et al. 1978). More than 99% of these modified adenine bases, which are found on both strands of the palindromic recognition sequence, are formed by action of DNA adenine methylase (Dam), a single-subunit nucleotide-independent (type II) DNA methyltransferase that transfers a methyl group from *S*-adenosylmethionine to adenine residues in the recognition sequence (Geier and Modrich 1979; for reviews, see Marinus 1987; Palmer and Marinus 1994).

The recognition sites of several restriction enzymes (including PvuI, BamHI, BclI, BglII, XhoII, MboI, and Sau3AI) contain the sequence $^{5'}$...GATC...$^{3'}$, as do a proportion of the sites recognized by ClaI (about one site in four), XbaI (one site in 16), MboII (one site in 16), TaqI (one site in 16), and HphI (one site in 16). The transfer of a methyl group to the N^6 atom of adenine places a bulky alkyl substituent in the major groove of B-form DNA and completely prevents cleavage in vitro by some restriction enzymes (e.g., MboI) (Dreiseikelmann et al. 1979). Other restriction enzymes will at best cleave a subset of their recognition sites. For example, ClaI recognizes the sequence $^{5'}$-ATCGAT-$^{3'}$. If this sequence is preceded by G, followed by C, or both, either or both of the A residues will be methylated and the site will be protected from cleavage. However, methylation does not endow the sequence GATC with absolute immunity from cleavage by any and all restriction enzymes. For example, the restriction enzyme Sau3AI, an isoschizomer of MboI, cleaves ...GATC... sequences regardless of their state of adenine methylation, whereas the enzyme DpnI specifically cleaves fully methylated ...G^{me6}ATC... sequences (Lacks and Greenberg 1975, 1977; Geier and Modrich 1979). DNAs modified by Dam methylase in vitro also remain sensitive to restriction by the *E. coli* Mcr and Mrr systems. Mammalian DNA is not methylated at the N^6 position of adenine and therefore can be cleaved to completion by restriction enzymes that are sensitive to Dam methylation of prokaryotic DNA.

Lists of restriction enzymes whose pattern of cleavage is affected by Dam methylation have been assembled by Kessler and Manta (1990) and by McClelland and Nelson (1988); additional information may be found in the brochures of most commercial suppliers of enzymes and in a database of restriction and modification enzymes (REBASE) accessible via the Internet at http://rebase.neb.com/rebase/rebase.html.

When it is necessary to cleave prokaryotic DNA at every possible site with restriction enzymes that are sensitive to Dam methylation, the DNA must be isolated from strains of *E. coli* that are Dam⁻ (Marinus 1973; Backman 1980; Roberts et al. 1980; McClelland and Nelson 1988). These strains (e.g., GM2163 is available from New England Biolabs and JM110 from the ATCC) show a quirky phenotype, including elevated rates of spontaneous mutation and recombination, increased sensitivity to UV irradiation, high rates of recombination, and increased rates of induction of lysogenic bacteriophages (for discussion, see Marinus 1987; Palmer and Marinus 1994). Dam⁻ strains are generally less robust than wild-type K-12 strains because a deficiency of Dam impairs the ability of the mismatch repair system to correct errors in the progeny strand of newly replicated DNA (Lu et al. 1983; Pukkila et al. 1983; for reviews, see Modrich 1989; Palmer and Marinus 1994). The absence of Dam methylation therefore leads to increased rates of spontaneous mutation. Dam⁻ strains may also show aberrant regulation of certain classes of genes and a reduced efficiency of initiation of DNA replication. Because of these problems, Dam⁻ strains should not be grown for long periods in continuous culture or stored on plates or stab cultures for long periods. Instead, these strains should be stored in small aliquots at −70°C. In some *E. coli* strains, maintenance of the *dam* mutation is enhanced by growing the cells in the presence of chloramphenicol or kanamycin. In these strains, the *dam* gene was inactivated by insertion of the transposon Tn9, which encodes chloramphenicol resistance (Marinus et al. 1983), or by replacement of part of the gene with a DNA fragment carrying a kan^r marker to inactivate the *dam* gene (Parker and Marinus 1988). A list of commonly used Dam⁻ strains has been published by Palmer and Marinus (1994).

Unmethylated and Hemimethylated Adenine Residues in GATC

In some strains of *E. coli*, a small fraction (~0.2% or less) of the estimated 18,000 $^{5'}$...GATC...$^{3'}$ sequences are unmethylated (Ringquist and Smith 1992). These unmodified sites appear to be preferentially located near the origin of replication or in regions of genomic DNA that are protected from Dam methylation in vivo by DNA-binding proteins (e.g., cAMP receptor protein) (Ringquist and Smith 1992; Wang and Church 1992; Hale et al. 1994).

The unique origin of replication in the chromosome of *E. coli* (oriC) is highly enriched in GATC sequences. The time of initiation of DNA synthesis is controlled by several factors, including the state of methylation of these sequences (for review, see Crooke 1995). Punctual and timely initiation requires that these *oriC*-proximal GATC sequences be fully methylated by the Dam nuclease. Semiconservative replication generates hemimethylated GATC sequences, which become bound to the bacterial membrane, where they are no longer accessible to Dam methylase. Before the next round of DNA synthesis is initiated, the *oriC* region is released from the membrane, and the local GATC sites once again become fully methylated. How these processes are coordinated is not fully understood.

REFERENCES

Backman K. 1980. A cautionary note on the use of certain restriction endonucleases with methylated substrates. *Gene* 11: 169–171.

Crooke E. 1995. Regulation of chromosomal replication in *E. coli*: Sequestration and beyond. *Cell* 82: 877–880.

Dreiseikelmann B, Eichenlaub R, Wackernagel W. 1979. The effect of differential methylation by *Escherichia coli* of plasmid DNA and phage T7 and λ DNA on the cleavage by restriction endonuclease MboI from *Moraxella bovis*. *Biochim Biophys Acta* 562: 418–428.

Geier GE, Modrich P. 1979. Recognition sequence of the *dam* methylase of *Escherichia coli* K12 and mode of cleavage of *Dpn* I endonuclease. *J Biol Chem* 254: 1408–1413.

Hale WB, van der Woude MW, Low DA. 1994. Analysis of nonmethylated GATC sites in the *Escherichia coli* chromosome and identification of sites that are differentially methylated in response to environmental stimuli. *J Bacteriol* 176: 3438–3441.

Hattman S, van Ormondt H, de Waard A. 1978. Sequence specificity of the wild-type (*dam*⁺) and mutant (*dam*ʰ) forms of bacteriophage T2 DNA adenine methylase. *J Mol Biol* 119: 361–376.

Kessler C, Manta V. 1990. Specificity of restriction endonucleases and DNA modification methyltransferases (review, edition 3). *Gene* 92: 1–248.

Lacks S, Greenberg B. 1975. A deoxyribonuclease of *Diplococcus pneumoniae* specific for methylated DNA. *J Biol Chem* 250: 4060–4066.

Lacks S, Greenberg B. 1977. Complementary specificity of restriction endonucleases of *Diplococcus pneumoniae* with respect to DNA methylation. *J Mol Biol* 114: 153–168.

Lu AL, Clark S, Modrich P. 1983. Methyl-directed repair of DNA base-pair mismatches in vitro. *Proc Natl Acad Sci* 80: 4639–4643.

Marinus MG. 1973. Location of DNA methylation genes on the *Escherichia coli* K-12 genetic map. *Mol Gen Genet* 127: 47–55.

Marinus MG. 1987. Methylation of DNA. In Escherichia coli *and* Salmonella typhimurium: *Cellular and molecular biology* (ed Neidhardt FC, et al.), Vol. 1, pp. 697–702. American Society for Microbiology, Washington, DC.

Marinus MG, Carraway M, Frey AZ, Brown L, Arraj JA. 1983. Insertion mutations in the *dam* gene of *Escherichia coli* K-12. *Mol Gen Genet* 192: 288–289.

McClelland M, Nelson M. 1988. The effect of site-specific DNA methylation on restriction endonucleases and DNA modification methyltransferases—A review. *Gene* 74: 291–304.

Modrich P. 1989. Methyl-directed DNA mismatch correction. *J Biol Chem* 264: 6597–6600.

Palmer BR, Marinus MG. 1994. The *dam* and *dcm* strains of *Escherichia coli*—A review. *Gene* 143: 1–12.

Parker B, Marinus MG. 1988. A simple and rapid method to obtain substitution mutations in *Escherichia coli*: Isolation of a *dam* deletion/insertion mutation. *Gene* 73: 531–535.

Pukkila PJ, Peterson J, Herman G, Modrich P, Meselson M. 1983. Effects of high levels of DNA adenine methylation on methyl-directed mismatch repair in *Escherichia coli*. *Genetics* 104: 571–582.

Ringquist S, Smith CL. 1992. The *Escherichia coli* chromosome contains specific, unmethylated *dam* and *dcm* sites. *Proc Natl Acad Sci* 89: 4539–3543.

Roberts TM, Swanberg SL, Poteete A, Riedel G, Backman K. 1980. A plasmid cloning vehicle allowing a positive selection for inserted fragments. *Gene* 12: 123–127.

Wang MX, Church GM. 1992. A whole genome approach to in vivo DNA–protein interactions in *E. coli*. *Nature* 360: 606–610.

WWW RESOURCE

REBASE, a database of restriction and modification enzymes
http://rebase.neb.com/rebase/rebase.html

CHAPTER 15

Introducing Genes into Cultured Mammalian Cells

Priti Kumar[1], Arvindhan Nagarajan[2], and Pradeep D. Uchil[3]

[1]*Department of Internal Medicine/Section for Infectious Diseases, Yale School of Medicine, New Haven, Connecticut 06511 (corresponding author)*

[2]*Department of Pathology, Yale School of Medicine, New Haven, Connecticut 06520*

[3]*Section of Microbial Pathogenesis, Yale School of Medicine, New Haven, Connecticut 06536*

STRATEGIES FOR THE DELIVERY OF GENES INTO EUKARYOTIC cells fall into three categories: transfection by biochemical methods, transfection by physical methods, and virus-mediated transduction. This chapter deals with the first two categories; virus-mediated transduction is described in Chapter 16.

The choice of transfection method is determined by several experimental factors:

- whether gene expression or protein production is the desired end result
- the cell type to be used (adherent vs. suspension, culture-adapted vs. primary cells)
- the ability of the cell line to survive the stress of transfection
- the type of assay to be used for screening
- the culture conditions
- the nucleic acid to be transfected (DNA, RNA, siRNA, or oligonucleotides)
- the efficiency required of the system
- whether the application requires low, medium, or high throughput

Commercial transfection reagents are routinely tested for their efficacy with multiple cell types, different nucleic acids, and a variety of specific applications. Before starting, we recommend consulting the transfection reagent selection guide provided on many manufacturers' websites, and if time and resources permit, comparing the efficiencies of different transfection reagents.

Biochemical methods of transfection, including calcium-phosphate-mediated and diethylaminoethyl (DEAE)–dextran-mediated transfection, have been used for more than 45 years to deliver nucleic acids into cultured cells. The work of Graham and van der Eb (1973) on transformation of mammalian cells by viral DNAs in the presence of calcium phosphate laid the foundation for the biochemical transformation of genetically marked mouse cells by cloned DNAs (Maitland and McDougall 1977; Wigler et al. 1977), for the transient expression of cloned genes in a variety of mammalian cells (e.g., see Gorman et al. 1983b), and for the isolation and identification of cellular oncogenes, tumor-suppressing genes, and other single-copy mammalian genes (e.g., see Wigler et al. 1978; Perucho and Wigler 1981; Weinberg 1985; Friend et al. 1988). Nowadays, cationic/neutral lipid (liposome) reagents are the most popular reagents for better delivery of genes into a wide range of cell types. Here, the chemical agent forms a complex with the DNA, neutralizing or masking the negative charge (and sometimes even rendering a positive charge) on the DNA molecule, thus enabling interaction with the negatively charged cell membrane, which facilitates uptake.

In addition to biochemical methods, electroporation—a physical transfection method—is widely used. Electroporation is a process in which brief electrical pulses create transient pores in the plasma membrane that allow nucleic acids to enter the cellular cytoplasm. Nucleofection, an improvement of the electroporation technology (developed by the company Amaxa), permits the introduction of nucleic acids directly into the nucleus using a combination of optimized electrical parameters with cell-type-specific reagents. Finally, "optical transfection" exploits the ability of light to create small transient pores in the plasma membrane of mammalian cells (see the information panel Optical Transfection).

TRANSIENT VERSUS STABLE TRANSFECTION

Two different approaches are used to transfer DNA into eukaryotic cells: transient transfection and stable transfection. In transient transfection, recombinant DNA introduced into a recipient cell line generates a temporary but high level of expression of the target gene. The transfected DNA does not necessarily become integrated into the host chromosome. Transient transfection is the method of choice when large numbers of samples are to be analyzed within a short period of time. Typically, the cells are harvested 1 and 4 d after transfection, and the resulting cell lysates are assayed for expression of the target gene.

Stable or permanent transfection is used to establish clonal cell lines in which the transfected target gene is integrated into chromosomal DNA, from where it directs the synthesis of the target protein, usually in moderate amounts. In general (depending on the cell type), the formation of stably transfected cells occurs with an efficiency one to two orders of magnitude lower than the efficiency of transient transfection. Isolation of rare stable transformants from a background of nontransfected cells is facilitated by the use of a selectable genetic marker. The marker may be present on the recombinant plasmid carrying the target gene, or it may be carried on a separate vector and introduced, together with the recombinant plasmid, into the desired cell line by cotransfection (for further details, see the information panels Cotransformation, and Selective Agents for Stable Transformation). In general, the methods described below are suitable for use both in transient transfection assays and for the generation of stable transfectants.

TRANSFECTION METHODS

In the past, cloned DNA was introduced into cultured eukaryotic cells chiefly by biochemical methods using either calcium phosphate or DEAE-dextran. Lipid reagents are now preferred because of the high efficiency of transfection that can be obtained and because of the ability of this class of reagents to mediate transfection of all types of nucleic acids into a wide range of cell types. Additional advantages include their ease of use, reproducibility, low toxicity, and ability to mediate effective transfection of suspension cultures. Physical methods, such as electroporation and nucleofection, are options to consider with cell lines that are resistant to transfection by other means. A brief summary of transfection methods is given in Table 1.

TRANSFECTION CONTROLS

All transfection experiments should include controls to test individual reagents and plasmid DNA preparations, and to test for toxicity of the gene or construct being introduced.

Controls for Transient Expression

Negative Controls

In transient transfection experiments, one or two dishes of cells should be transfected with the carrier DNA and/or buffer used to dilute the test plasmid or gene. Typically, salmon sperm DNA or another inert carrier such as the empty vector used to construct the recombinant is transfected into

TABLE 1. Transfection methods

Method	Expression		Cell toxicity	Cell types	Comments	Pros	Cons
	Transient	Stable					
Lipid-mediated; Protocol 1	Yes	Yes	Varies	Adherent cells, primary cell lines, suspension cultures	Cationic lipids are used to bind negatively charged DNA molecules, sometimes creating artificial membrane vesicles (liposomes). The resulting stable cationic complexes adhere to and fuse with the negatively charged cell membrane (Felgner et al. 1987, 1994).	Very low immunogenicity and cytotoxicity. They can be used with large plasmids. Low safety risk, simple handling procedure, single protocol for all nucleic acids (DNA, mRNA, siRNA, etc.), rapid procedure, comparatively low costs, and adaptable to high-throughput systems.	Low to variable efficiency in primary and suspension cells
Calcium-phosphate-mediated; Protocols 2 and 3	Yes	Yes	No	Adherent cells (CHO, 293); suspension cultures	Calcium phosphate forms an insoluble coprecipitate with DNA, which attaches to the cell surface and is absorbed by endocytosis or phagocytosis (Graham and van der Eb 1973).	Low cytotoxicity, can be used with large plasmids, low safety risk, simple handling procedure, very inexpensive	Reagent consistency is critical to achieve transfection efficiency; small changes in pH (± 0.1) can greatly affect transfection efficiency. The size and quality of the precipitate are crucial for efficient transfection. Sometimes can prove to be immunogenic. The protocol may require tweaking for different nucleic acids and may result in variable efficiencies, and variable efficiency for adherent and suspension cells. Ineffective for primary cells. Complexes cannot be made in culture medium because of the high concentration of phosphate in the medium.
DEAE-dextran-mediated; Protocol 4	Yes	No	Yes	BSC-1, CV-1, and COS	Positively charged DEAE-dextran binds to negatively charged—phosphate groups of DNA, forming aggregates that bind to the negatively charged plasma membrane. Uptake into the cell is believed to be mediated by endocytosis, which is potentiated by osmotic shock (Vaheri and Pagano 1965).	Can be used with large plasmids, low safety risk, simple handling procedure, inexpensive	Can prove immunogenic and cytotoxic. Can only be used for transient transfections. The protocol may require tweaking for different nucleic acids with highly variable efficiencies and variable efficiency for adherent and suspension cells. Ineffective for primary cells.

Method				Description	Advantages	Disadvantages
Electroporation; Protocol 5	Yes	Yes	Yes	Many; including cells refractory to lipofection	Application of brief high-voltage electrical pulses to a variety of mammalian and plant cells leads to the formation of nanometer-sized pores in the plasma membrane (Neumann et al. 1982; Zimmermann 1982). DNA is taken directly into the cell cytoplasm either through these pores or as a consequence of the redistribution of membrane components that accompanies the closure of the pores.	Can be used with large plasmids, low safety risk, fairly simple handling procedure, single protocol for all nucleic acids (DNA, mRNA, siRNA, etc.), rapid procedure, nonchemical method that does not alter the biological structure or function of cells	Can be highly cytotoxic, poor efficiencies for suspension and primary cells, moderate costs, requires specialized equipment. Experimental settings vary based on the cell type used.
Nucleoporation; discussed in Protocol 5	Yes	Yes	Varies	Many; including cells refractory to lipofection	Based on the physical method of electroporation. Nucleofection uses a combination of electrical parameters, generated by a device called a Nucleofector, with cell-type-specific reagents to transfer nucleic acids directly into the cell nucleus and the cytoplasm (Greiner et al. 2004; Johnson et al. 2005).	Fairly low cytotoxicity for many cell types; can be used with large plasmids; low safety risk; good efficiencies with adherent, suspension, and primary cell types; fairly simple handling procedure; single protocol for all nucleic acids (DNA, mRNA, siRNA, etc.); rapid procedure	Moderate costs, requires specialized equipment. Experimental settings vary for each cell type and require specialized buffers provided by a single manufacturer.

adherent cells in the absence of the test gene. After transfection, the cultured cells should not detach from the dish nor become rounded and glassy in appearance.

Positive Controls

One or two dishes of cells are transfected with a plasmid encoding a readily assayed gene product such as chloramphenicol acetyl transferase, luciferase (Chapter 17, Protocols 2 and 3), *Escherichia coli* β-galactosidase (Chapter 17, Protocol 1), or fluorescent proteins such as green fluorescent protein (GFP) and its derivatives (Chapter 17, Protocol 4), whose expression is driven by a panspecific promoter such as the human cytomegalovirus immediate early gene region promoter and its associated enhancer. Tracer plasmids of this kind are available from commercial suppliers who sell kits containing the enzymes and reagents needed for detection of the encoded protein. Because the endogenous levels of these reporter activities are typically low, the increase in enzyme activity or fluorescence provides a direct indication of the efficiency of the transfection and the quality of the reagents used for a particular experiment. This control is especially important when comparing results of transfection experiments performed at different times. Cotransfecting the reporter plasmid with the test plasmid or genomic DNA also provides a control for nonspecific toxicity in the overall transfection process. For more information on these reporter systems, see Chapter 17 and the information panels within.

Controls for Stable Expression

Negative Controls

One or two dishes should be transfected with an inert nucleic acid such as salmon sperm DNA, in the absence of the selectable marker. After culturing for 2–3 wk in the presence of the appropriate selective agent (e.g., G418, hygromycin, mycophenolic acid), no colonies should be visible.

Positive Controls

One or two dishes of cells should be transfected with the empty plasmid encoding the selectable marker. The number of viable colonies detected at the end of a 2–3-wk selection period is a measure of the efficiency of the transfection/selection process. A similar number of colonies should be present on dishes into which both the selectable marker and the test plasmid or gene were introduced. A marked discrepancy in the number of colonies on these two sets of dishes can be an indication of a toxic gene product (or, in rare instances, of a gene product that enhances survival of the transfected cells). If a particular cDNA or gene proves toxic to recipient cells, consider the use of a regulated promoter such as metallothionein (a Zn^{2+}- or Cd^{2+}-responsive DNA), the mouse mammary tumor virus long terminal repeat promoter (a glucocorticoid-responsive DNA), or a tetracycline-regulated promoter (Chapter 17, Protocol 5) (Gossen and Bujard 1992; Gossen et al. 1995; Shockett et al. 1995).

OPTIMIZATION AND SPECIAL CONSIDERATIONS

Irrespective of the method used to introduce DNA into cells, the efficiency of transient or stable transfection is determined largely by the cell type that is used (see Table 1). Different lines of cultured cells vary by several orders of magnitude in their ability to take up and express exogenously added DNA. In addition, a method that works well for one type of cultured cell may be useless for another. Many of the protocols described in this chapter have been optimized for the standard lines of cultured cells. When using more exotic lines of cells, it is important to compare the efficiencies of several different methods. Protocols 1–5 describe commonly used transfection techniques as well as methods that have been successful with cell lines that are resistant to transfection by standard techniques. Commercial kits are available that provide collections of reagents for many types of transfections (Table 2).

TABLE 2. Commercial kits and reagents for transfection

Manufacturer	Website address	Method or reagents	Kit/product
Bio-Rad	www.bio-rad.com	Nonliposomal cationic lipid	Transfectin
Clontech	www.clontech.com	Nonliposomal polymer	Xfect
GE Healthcare Lifesciences	www.gelifesciences.com	DEAE-dextran	CellPhect Transfection Kit
Life Technologies	www.invitrogen.com	Calcium phosphate	Calcium phosphate transfection kit
		Liposomal cationic lipid	Lipofectamine 2000
		Liposomal cationic lipid	Lipofectamine LTX
Mirus Bio	www.mirusbio.com	Nonliposomal cationic lipid	*TransIT*-2020
		Nonliposomal cationic lipid	*TransIT*-LT1
Polyplus Transfection	www.polyplus-transfection.com	Nonlipid based	jetPRIME, jetPEI
Promega	www.promega.com	Calcium phosphate	ProFection Mammalian Transfection System
		Nonliposomal cationic lipid	FuGENE HD Transfection System
		Nonliposomal cationic lipid	TransFast Transfection Reagent
		Nonliposomal cationic lipid	FuGENE 6 Transfection Reagent
QIAGEN	www.qiagen.com	Nonliposomal cationic lipid	Effectene Transfection Reagent
		Nonliposomal cationic lipid	Attractene Transfection Reagent
		Activated dendrimer	SuperFect Transfection Reagent
		Activated dendrimer	PolyFect Transfection Reagent
Roche Applied Science	www.roche-applied-science.com	Liposomal cationic lipid	DOTAP Transfection Reagent
Sigma-Aldrich	www.sigmaaldrich.com	Calcium phosphate	Calcium phosphate transfection
		DEAE-dextran	DEAE-dextran transfection kit
		Nonliposomal cationic lipid	Escort, DOTAP, DOPE

ASSESSING CELL VIABILITY IN TRANSFECTED CELL LINES

In determining the best option for transfecting a particular cell line, it may be worthwhile to assess the effects of transfection on cell viability. Assessment of cellular health requires accurate quantitative and qualitative measurements over a period of time. Assays are available to measure a variety of different markers that indicate the number of dead cells (cytotoxicity assay), the number of live cells (viability assay), the total number of cells, or the mechanism of cell death (e.g., apoptosis vs. necrosis). The DNA content, the presence or absence of enzyme activity within the cell, the amount of ATP within cells, membrane integrity, and metabolic activity are all indicators of cellular integrity. Assays (e.g., ATP quantitation or LDH-release assays) that generate a measurable signal quickly enough to provide information representing a snapshot in time have the advantage of speed over assays that require several hours of incubation to develop a signal (e.g., MTT or resazurin). The latter are useful for measurements in unsynchronized and heterogeneous cell populations that respond over the course of several hours or days.

Detection sensitivity, a parameter of great significance when measuring the toxicity of drugs, is of limited concern when optimizing transfection reagents, because one measures a change in the viability of a large population of cells. In addition, assaying a large number of samples requires a high-throughput method that uses a simple, stable, nonradioactive reagent that is compatible with robotic handling. The method must also be relatively inexpensive, not labor intensive, and capable of determining the kinetics of cell growth in multiple cell types without interfering with basic cellular functions. It is for these reasons that when dealing with a large number of samples, the classical Trypan Blue exclusion test, used for estimating the number of dead cells, is less desirable than the many other easy-to-perform assays that provide an accurate quantitative and qualitative indicator of cell viability.

Protocols 6, 7, and 8 detail the use of three reagents: alamarBlue, lactate dehydrogenase (LDH), and MTT [3-(4,5-dimethylthiazol-2-yl)-2,5-diphenyltetrazolium bromide, a yellow tetrazole], respectively. The alamarBlue and MTT assays measure cell viability directly, whereas the LDH assay measures cell death and is hence a cytotoxicity assay. In addition, although all three assays provide time-dependent quantitation of the cell death mechanism (i.e., apoptosis vs. necrosis), only the LDH assay can distinguish between the two mechanisms in single-time-point measurements. The LDH assay measures the release of a cytoplasmic enzyme, a process that occurs only when necrosis causes loss of cellular membrane integrity. Because all three assays can be adapted for high-throughput measurements, they have largely replaced older radioactive assays such as incorporation of [^{3}H] thymidine into cellular DNA.

ACKNOWLEDGMENTS

The authors thank Michelle Collins of Bio-Rad, Kevin Smoker of Life Technologies, and Daniela Bruell of Lonza Cologne AG for technical and editorial support.

REFERENCES

Felgner PL, Gadek TR, Holm M, Roman R, Chan HW, Wenz M, Northrop JP, Ringold GM, Danielsen M. 1987. Lipofection: A highly efficient, lipid-mediated DNA-transfection procedure. *Proc Natl Acad Sci* **84**: 7413–7417.

Felgner JH, Kumar R, Sridhar CN, Wheeler CJ, Tsai YJ, Border R, Ramsey P, Martin M, Felgner PL. 1994. Enhanced gene delivery and mechanism studies with a novel series of cationic lipid formulations. *J Biol Chem* **269**: 2550–2561.

Friend SH, Dryja TP, Weinberg RA. 1988. Oncogenes and tumor-suppressing genes. *N Engl J Med* **318**: 618–622.

Gorman C, Padmanabhan R, Howard BH. 1983. High efficiency DNA-mediated transformation of primate cells. *Science* **221**: 551–553.

Gossen M, Bujard H. 1992. Tight control of gene expression in mammalian cells by tetracycline-responsive promoters. *Proc Natl Acad Sci* **89**: 5547–5551.

Gossen M, Freundlieb S, Bender G, Müller G, Hillen W, Bujard H. 1995. Transcriptional activation by tetracyclines in mammalian cells. *Science* **268**: 1766–1769.

Graham FL, van der Eb AJ. 1973. A new technique for the assay of infectivity of human adenovirus 5 DNA. *Virology* **52**: 456–467.

Greiner J, Wiehe J, Wiesneth M, Zwaka TP, Prill T, Schwarz K, Bienek-Ziolkowski M, Schmitt M, Döhner H, Hombach V, et al. 2004. Transient genetic labeling of human CD34 positive hematopoietic stem cells using nucleofection. *Transfus Med Hemother* **31**: 136–141.

Johnson BD, Gershan JA, Natalia N, Zujewski H, Weber JJ, Yan X, Orentas RJ. 2005. Neuroblastoma cells transiently transfected to simultaneously express the co-stimulatory molecules CD54, CD80, CD86, and CD137L generate antitumor immunity in mice. *J Immunother* **28**: 449–460.

Maitland NJ, McDougall JK. 1977. Biochemical transformation of mouse cells by fragments of herpes simplex DNA. *Cell* **11**: 233–241.

Neumann E, Schaefer-Ridder M, Wang Y, Hofschneider PH. 1982. Gene transfer into mouse lyoma cells by electroporation in high electric fields. *EMBO J* **1**: 841–845.

Perucho M, Wigler M. 1981. Linkage and expression of foreign DNA in cultured animal cells. *Cold Spring Harb Symp Quant Biol* **45**: 829–838.

Shockett P, Difilippantonio M, Hellman N, Schatz D. 1995. A modified tetracycline-regulated system provides autoregulatory, inducible gene expression in cultured cells and transgenic mice. *Proc Natl Acad Sci* **92**: 6522–6526.

Vaheri A, Pagano JS. 1965. Infectious poliovirus RNA: A sensitive method of assay. *Virology* **27**: 434–436.

Weinberg RA. 1985. Oncogenes and the molecular basis of cancer. *Harvey Lect* **80**: 29–136.

Wigler M, Pellicer A, Silverstein S, Axel R. 1978. Biochemical transfer of single-copy eukaryotic genes using total cellular DNA as a donor. *Cell* **14**: 725–731.

Wigler M, Silverstein S, Lee LS, Pellicer A, Cheng VC, Axel R. 1977. Transfer of purified herpes virus thymidine kinase gene to cultured mouse cells. *Cell* **11**: 223–232.

Zimmermann U. 1982. Electric field-mediated fusion and related electrical phenomena. *Biochim Biophys Acta* **694**: 227–277.

DNA Transfection Mediated by Cationic Lipid Reagents

Liposomal transfection reagents vary in their ability to transfect cell lines efficiently. Some are generalists, whereas others are best used with specific cell types. The nonliposomal FuGENE 6 and the cationic liposomal Lipofectamine 2000 are examples of reagents that can successfully transfect most adherent and suspension cell types (including several primary and hard-to-transfect cell types) with negligible toxicity and a minimal number of manipulations. Importantly, both reagents can be used to transfect cells in the presence of serum, minimizing the number of manipulations during the transfection procedure. These two reagents are used in the main protocol below, which has been modified from its original form after obtaining permission from the manufacturers—Roche Applied Science and Life Technologies. We also provide an alternative protocol that uses the cationic lipid reagents Lipofectin and Transfectam.

Because a large number of variables affect the efficiency of lipofection, we suggest that the conditions outlined in the following protocol be used as a starting point for systematic optimization of the system (for further details, see the information panel Lipofection). Once a positive signal has been obtained with a plasmid carrying a standard reporter gene, each of the parameters discussed in the information panel Lipofection may be changed systematically to obtain the maximal ratio of signal to background and to minimize variability between replicate assays. From these results, optimal protocols can be developed to assay the expression of the genes of interest.

MATERIALS

It is essential that you consult the appropriate Material Safety Data Sheets and your institution's Environmental Health and Safety Office for proper handling of equipment and hazardous materials used in this protocol.

Reagents

Cell culture growth medium (complete, serum-free, and [optional] selective)
Exponentially growing cultures of mammalian cells
FuGENE 6[1] (Promega Corporation)
Giemsa stain (10%)
  The Giemsa stain should be freshly prepared in phosphate-buffered saline or H_2O and filtered through Whatman No. 1 filter paper before use.

Lipofectamine 2000[2] (Life Technologies)
Lipofection reagents
Methanol, ice-cold
Plasmid DNA
  If performing lipofection for the first time or if using an unfamiliar cell line, obtain an expression plasmid encoding E. coli β-galactosidase or green fluorescent protein (see the information panels β-Galactosidase, and Fluorescent Proteins in Chapter 17). These can be purchased from Addgene (a nonprofit plasmid repository) or several commercial manufacturers (e.g., pCMV-SPORT-β-gal, Life Technologies, or pEGFP-F, Clontech) (see Figs. 1 and 2).

---

[1] https://www.promega.com/de-de/aboutus/press-releases/2011/20110202-fugene6/.

[2] http://www.invitrogen.com/site/us/en/home/Global/trademark-information/life-technologies-trademarks-list.html.

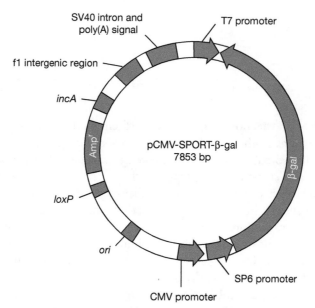

FIGURE 1. pCMV-SPORT-β-gal. pCMV-SPORT-β-gal is a reporter vector that may be used to monitor transfection efficiency. It carries the *E. coli* gene encoding β-galactosidase preceded by the CMV (cytomegalovirus) immediate early promoter, which drives high levels of transcription in mammalian cells. The SV40 polyadenylation signal downstream from the β-galactosidase sequence directs proper processing of the 3′ end of the mRNA in eukaryotic cells. (Reproduced, with permission, from Life Technologies Corporation.)

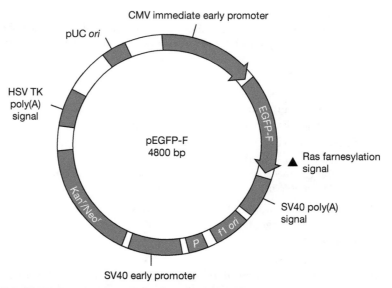

FIGURE 2. pEGFP-F. pEGFP-F is a reporter vector that may be used both to monitor transfection efficiency and as a cotransfection marker. The vector encodes a modified form of the green fluorescent protein, a farnesylated enhanced GFP (EGFP-F) that remains bound to the plasma membrane in both living and in fixed cells. The EGFP-F coding sequence is preceded by the CMV (cytomegalovirus) promoter, which drives high levels of transcription, and is followed by the SV40 polyadenylation signal to direct proper processing of the 3′ end of the mRNA in eukaryotic cells. The plasmid carries (i) sequences that allow replication in prokaryotic (pUC ori) as well as eukaryotic (SV40 ori) cells and (ii) markers that facilitate selection for the plasmid in prokaryotic (kanamycin) cells as well as eukaryotic (neomycin) cells. The presence of EGFP-F can be detected by fluorescence microscopy. (Adapted, with permission, from Clontech.)

TABLE 1. Dimensions of dishes used for cell culture

Size of plate	Growth area (cm^2)	Relative area[a]	Recommended volume
96 wells	0.32	0.04×	200 µL
24 wells	1.88	0.25×	500 µL
12 wells	3.83	0.5×	1.0 mL
Six wells	9.4	1.2×	2.0 mL
35 mm	8.0	1.0×	2.0 mL
60 mm	21	2.6×	5.0 mL
10 cm	55	7.0×	10.0 mL
Flask	25	3.0×	5.0 mL
Flask	75	9.0×	12.0 mL

[a]Relative area is expressed as a factor of the growth area of a 35-mm culture plate.

Purify closed-circular plasmid DNAs using a transfection-grade plasmid DNA preparation kit that preferably removes bacterial endotoxins during the purification procedure (e.g., EndoFree Plasmid Maxi Kit from QIAGEN). Alternatively, DNA may be purified by column chromatography or ethidium bromide-CsCl gradient centrifugation. Dissolve the DNAs in sterile Tris-EDTA buffer or sterile H_2O at 0.2–2 µg/µL. Determine DNA purity using a 260 nm/280 nm ratio; the ratio should be 1.8.

Linear DNA, that is, DNA linearized with restriction enzymes, is sometimes preferred, particularly if the aim is to generate stable transfectants (see the information panel Linearizing Plasmids before Transfection).

Equipment

Microcentrifuge tubes, polypropylene (sterile)

Tissue culture multiwell plates (six well) or culture dishes (35 mm)

This protocol is designed for cells grown in six-well tissue culture plates or 35-mm culture dishes. If multiwell plates, flasks, or dishes of a different diameter are used, scale the cell density and plating volume according to Table 1. The corresponding starting volume of the transfection reagent and the amount of DNA can be scaled according to the manufacturer's instructions.

METHOD

Growing Cells for Transfection

1. Twenty-four hours before lipofection, harvest exponentially growing mammalian cells by trypsinization, and replate them in six-well or 35-mm dishes at a density appropriate for the desired transfection reagent (see the table below). Add 2 mL of growth medium, and incubate the cultures for 20–24 h at 37°C in a humidified incubator with an atmosphere of 5%–7% CO_2.

   If the cells are grown for <12 h before transfection, they may not be well anchored to the substratum and are likely to detach during exposure to lipid.

   For suspension cells, use freshly passaged cells at the appropriate concentration as indicated in the table below (use 2 mL in a 35-mm culture dish or six-well plate).

Transfection reagent	Adherent cell density (cells/well)	Suspension cell concentration (cells/mL)	Confluency at lipofection
FuGENE 6	1×10^5 to 3×10^5	5×10^4 to 1×10^6	50%–80%
Lipofectamine 2000	2×10^5 to 8×10^5	2×10^5 to 3×10^6	90%–95%
DOTMA	1×10^5 to 3×10^5	5×10^4 to 1×10^6	75%
DOGS	1×10^5 to 3×10^5	5×10^4 to 1×10^6	75%

Preparation of FuGENE 6–DNA Complexes

For initial optimization, use reagent (in microliters) to DNA (in micrograms) ratios of 3:1, 3:2, and 6:1. These ratios provide good transfection efficiencies for commonly used adherent and suspension cells.

2. Bring the FuGENE 6 reagent to room temperature and mix before use by vortexing for 1 sec or by inverting the bottle. Dilute the reagent into serum-free medium (without antibiotics or fungicides). Label three small sterile tubes: "3:1," "3:2," and "6:1." Pipette 97 μL of serum-free medium into the first two tubes and 94 μL into the last tube. Pipette the FuGENE 6 directly into the medium without touching the walls of the plastic tube: 3 μL of FuGENE 6 into each of the first two tubes, and 6 μL into the tube labeled "6:1." Vortex for 1 sec or flick the tube to mix. Incubate for 5 min at room temperature.

   ▲ *The reagent–DNA complex must be prepared in serum-free medium, even if the cells are transfected in the presence of serum. The order and manner of addition of components to form the transfection complex is critical. Serum-free medium must be pipetted first. Undiluted FuGENE 6 should not come into contact with any plastic surfaces (such as the walls of the tube that contains the serum-free medium) other than pipette tips.*

3. Add 50 μL of DNA to the diluted transfection reagent from Step 2. Add 1 μg of plasmid DNA into each of the tubes labeled 3:1 and 6:1, and 2 μg of DNA (also in a total of 50 μL) into the tube labeled 3:2.

4. Tap the tubes or vortex for 1 sec to mix the contents and incubate for 15–45 min at room temperature.

5. Proceed to Step 10.

Preparation of Lipofectamine 2000–DNA Complexes

For initial optimization, prepare complexes using a Lipofectamine 2000 (in microliters) to DNA (in micrograms) ratio of 2:1–3:1.

6. Mix Lipofectamine 2000 gently, then dilute 10 μL in 250 μL of serum-free medium. Incubate for 5 min at room temperature.

   ▲ *Proceed to Step 8 within 25 min.*

7. Dilute 4 μg of DNA into 250 μL of serum-free medium. Mix gently.

8. Combine the diluted Lipofectamine 2000 with the diluted DNA (total volume = 500 μL). Mix gently and incubate for 20 min at room temperature (the solution may appear cloudy).

   Complexes are stable for 6 h at room temperature.

9. Proceed to Step 10.

Transfection of Cells

10. Add the complexes dropwise to cells in the existing growth medium in the six-well plate or 35-mm dish (from Step 1). Swirl the plate or dish to ensure distribution over the entire surface. Return the cells to a 37°C humidified incubator with an atmosphere of 5%–7% CO_2.

    There is no need to remove and replace with fresh growth medium. However, the growth medium may be changed after 4–6 h if Lipofectamine 2000 causes any kind of toxicity.

11. Analyze cell viability at 6–24 h after transfection by a Trypan Blue exclusion test (see Chapter 17, Protocol 3), or quantify cell viability using cytotoxicity tests incorporating alamarBlue, lactate dehydrogenase, or MTT (see Protocols 6, 7, or 8, respectively).

    See Troubleshooting.

12. If the objective is stable transformation of the cells, proceed to Step 13. To assay for transient transfection, examine the cells 24–96 h after lipofection using one of the following assays:

    i. If a plasmid DNA expressing *E. coli* β-galactosidase was used, follow the steps outlined in Chapter 17, Protocol 1 to measure enzyme activity in cell lysates. Alternatively, perform a histochemical staining assay as detailed in the Additional Protocol Histochemical Staining of Cell Monolayers for β-Galactosidase.

    ii. If a fluorescence protein expression vector was used, examine the cells with a microscope under appropriate illumination conditions. (Green fluorescent protein expression can be observed at 450–490 nm.) Alternatively, a small portion of the cells can be analyzed by flow cytometry to obtain an estimate of transfection efficiency and viability as described in the information panel Fluorescent Proteins in Chapter 17.

    iii. For other gene products, newly synthesized protein may be analyzed by radioimmunoassay, by immunoblotting, by immunoprecipitation following in vivo metabolic labeling, or by assays of enzymatic activity in cell extracts.

> *To minimize the effects of dish-to-dish variation in transfection efficiency, it is best to (1) transfect several dishes with each construct, (2) trypsinize the cells after 24 h of incubation, (3) pool the cells, and (4) replate them on several dishes.*
>
> *See Troubleshooting.*

13. Isolate stable transfectants.

    i. Incubate the cells for 48–72 h in complete medium to allow time for expression of the transferred gene(s).

    ii. Trypsinize the cells, and replate them in the appropriate selective medium.

    iii. Change the medium every 2–4 d for 2–3 wk to remove the debris of dead cells and to allow colonies of resistant cells to grow.

    iv. Clone individual colonies, and propagate them for assay (for methods, see Jakoby and Pastan 1979 or Spector et al. 1998).

    v. Obtain a permanent record of the numbers of colonies by fixing the remaining cells with ice-cold methanol for 15 min followed by staining with 10% Giemsa for 15 min at room temperature before rinsing in tap water.

TROUBLESHOOTING

Problem (Step 11): Cells sicken.
Solution: The antibiotic used for selection is the cause. Consider the following:

- Wait to add the selection antibiotic until 24–48 h after the transfection procedure.

- Use a lower (or a range of) concentration of the selection antibiotic.

Problem (Step 11): Cells sicken.
Solution: The expressed protein is cytotoxic or is produced at levels that are too high. Consider the following:

- Analyze cytotoxicity by preparing experimental controls: untransfected cells, cells exposed to DNA alone without a transfection reagent, and cells exposed to transfection reagent alone. Compare the transfected cells with the experimental construct to the wells containing these experimental controls.

- Consider repeating the experiment with a secreted reporter gene assay such as secreted alkaline phosphatase or human growth hormone. Secreting cells should show little or no evidence of cytotoxicity.

Problem (Step 11): Cells sicken.
Solution: The cultures are contaminated with mycoplasma. Consider the following:

- Use a commercial mycoplasma detection kit to determine if the culture is contaminated.

- Treat the cells with an antibiotic like BM-Cyclin (Roche Applied Sciences) to eliminate the mycoplasma, or start over with fresh clean cultures.

Problem (Step 11): Cells sicken.

Solution: If none of the solutions above are effective, then consider the following:

- DNA containing higher endotoxin levels may cause cytotoxicity to sensitive cell lines such as Huh-7 and primary cells. Use an endotoxin removal kit or restart the experiment using a kit that incorporates this step for generating the plasmid.

- In rare cases, the transfection reagent may be toxic to the cell line used. Try repeating transfection in the presence of fetal bovine serum, under reduced exposure time to the transfection reagent, at a higher cell density, and with varying ratios of lipid:DNA.

Problem (Step 12): Transfection efficiency is low.

Solution: The nucleic acids may be of poor quality or are insufficient in quantity. Consider the following:

- Use only high-quality plasmid preparations and at the recommended concentration.

- Perform a control transfection experiment with a commercially available transfection-grade plasmid preparation containing a marker gene like GFP.

- DNA containing higher endotoxin levels may cause cytotoxicity to sensitive cell lines like Huh-7 and primary cells. Use an endotoxin removal kit, or start over with a kit that incorporates this step for generating the plasmid.

Problem (Step 12): Transfection efficiency is low.

Solution: The transfection reagents were not stored as recommended. Consider the following:

- Store the reagent in the original container from the manufacturer and do not aliquot it.

- Do not freeze lipid transfection reagents.

Problem (Step 12): Transfection efficiency is low.

Solution: The lipid-to-DNA ratio is suboptimal. It may be necessary to empirically determine the ratio of the lipid transfection reagent to the DNA used for transfection. This can be performed in a multiwell plate and then scaled up accordingly.

Problem (Step 12): Transfection efficiency is low.

Solution: If none of the solutions above are effective, then consider the following:

- Some lipofection reagents, particularly FuGENE 6, are highly sensitive to contact with plastic. Make sure to pipette the reagent directly into the medium.

- In some cases, serum and other additives in the medium can inhibit complex formation. Prepare complexes in medium that does not contain additives (e.g., serum, antibiotics, growth enhancers).

Transfection Using DOTMA and DOGS

The following protocol details the use of DOTMA (Lipofectin[1]) and DOGS (Transfectam[2]) for transfecting cells in 60-mm tissue culture dishes and varies from the protocols outlined above mainly in the fact that transfections need to be performed in the absence of serum (see the box DOTMA and DOGS).

ADDITIONAL MATERIALS

It is essential that you consult the appropriate Material Safety Data Sheets and your institution's Environmental Health and Safety Office for proper handling of equipment and hazardous materials used in this protocol.

Reagents

Lipofection reagents
    Lipofectin (Life Technologies)
    Transfectam (Promega)

NaCl (300 mM)
    Use as the diluent for DOGS.

Sodium citrate (20 mM, pH 5.5), containing 150 mM NaCl
    Use instead of sterile H_2O as the diluent for the plasmid DNA if DOGS is the lipofection reagent (Kichler et al. 1998).

Equipment

Test tubes, polypropylene or polystyrene
    See note to Step 2.

Tissue culture dishes (60 mm)

METHOD

1. Prepare cells in 60-mm dishes as described in Step 1 of Protocol 1.

2. For each 60-mm dish of cultured cells to be transfected, dilute 1–10 μg of plasmid DNA into 100 μL of sterile deionized H_2O (if using Lipofectin) or 20 mM sodium citrate containing 150 mM NaCl (pH 5.5) (if using Transfectam) in a polystyrene or polypropylene test tube. In a separate tube, dilute 2–50 μL of the lipid solution to a final volume of 100 μL with sterile deionized H_2O or 300 mM NaCl. Incubate the tubes for 10 min at room temperature.

   ▲ *When transfecting with Lipofectin, use polystyrene test tubes; do not use polypropylene tubes because the cationic lipid DOTMA binds nonspecifically to polypropylene.*

---

[1] http://www.invitrogen.com/site/us/en/home/Global/trademark-information/life-technologies-trademarks-list.html.

[2] http://www.promega.com/aboutus/corporate/trademarks/.

DOTMA AND DOGS

Although it is now commonplace to use commercially available transfection reagents composed of a proprietary blend of lipids, it is also possible to prepare a homebrew of lipid-based transfection reagents; two are presented here.

- Lipofectin (N-[1-(2,3-dioleoyloxy)propyl]-N,N,N-trimethylammonium chloride; DOTMA) (Fig. 3). This monocationic lipid mixed with a helper lipid is usually purchased at a concentration of 1 mg/mL. DOTMA can also be synthesized with the help of an organic chemist (Felgner et al. 1987). If synthesized in-house, dissolve 10 mg each of dried DOTMA and the helper lipid dioleoyl phosphatidylethanolamine (DOPE; purchased from Sigma-Aldrich) in 2 mL of sterile deionized H_2O in a polystyrene tube (do not use polypropylene tubes). Sonicate the turbid solution to form liposomes before diluting to a final concentration of 1 mg/mL. Store the solution at 4°C.

- Transfectam (spermine-5-carboxy-glycinedioctadecyl-amide; DOGS) (Fig. 3). Prepare a stock solution of the cationic lipid DOGS as follows: Dissolve 1 mg of polyamine in 40 μL of 96% (v/v) ethanol for 5 min at room temperature with frequent solute vortexing. Add 360 μL of sterile H_2O, and store the solution at 4°C. Vortex the solution just before use. Polyamines, such as DOGS, do not require the use of polystyrene tubes; polypropylene tubes (i.e., standard microcentrifuge tubes) can be safely used with these reagents.

Although in-house preparation of these two lipid reagents may prove to be cost effective (Loeffler and Behr 1993), the disadvantages include higher cell toxicity, the ability to transfect only a restricted number of cell types, and the need for a large number of manipulations during transfection, because efficiencies are severely compromised in the presence of serum.

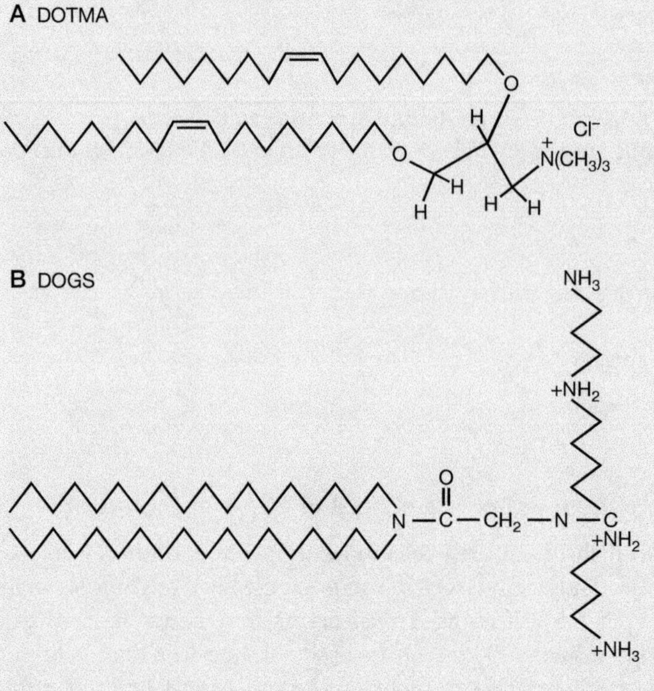

FIGURE 3. Structures of lipids used in lipofection.

3. Add the lipid solution to the DNA, and mix the solution by pipetting up and down several times. Incubate the mixture for 10 min at room temperature.

4. While the DNA–lipid solution is incubating, wash the cells to be transfected three times with serum-free medium. After the third rinse, add 0.5 mL of serum-free medium to each 60-mm dish, and return the washed cells to a 37°C humidified incubator with an atmosphere of 5%–7% CO_2.

   ▲ *The cells are rinsed free of serum before the addition of the lipid–DNA liposomes. In some cases, serum is a very effective inhibitor of the transfection process (Felgner and Holm 1989). Similarly, extracellular matrix components such as sulfated proteoglycans can also inhibit lipofection, presumably by binding the DNA–lipid complexes and preventing their interaction with the plasma membranes of the recipient cells.*

5. Add 900 μL of serum-free medium to each tube. Mix the solution by pipetting up and down several times. Incubate for an additional 10 min at room temperature.

6. Transfer each tube of DNA–lipid–medium solution to a 60-mm dish of cells. Incubate the cells for 1–24 h at 37°C in a humidified incubator with an atmosphere of 5%–7% CO_2.

7. After the cells have been exposed to the DNA for the appropriate time, wash them three times with serum-free medium. Feed the cells with complete medium and return them to the incubator. Analyze for viability and transfection efficiency and/or proceed to make stable transfectants as detailed in Protocol 1, Steps 11–13.

Histochemical Staining of Cell Monolayers for β-Galactosidase

There are several methods for assaying the success of transient transfections. If a plasmid expressing *E. coli* β-galactosidase was used, then this histochemical staining procedure is simple to perform and yields dependable results. The following method, designed for cells growing in 60-mm culture dishes, was adapted from Sanes et al. (1986). Kits that contain all of the necessary reagents for immunohistochemical detection of β-galactosidase are available from several manufacturers.

ADDITIONAL MATERIALS

It is essential that you consult the appropriate Material Safety Data Sheets and your institution's Environmental Health and Safety Office for proper handling of equipment and hazardous materials used in this protocol.

Recipes for reagents specific to this protocol, marked <R>, are provided at the end of the protocol. See Appendix 1 for recipes for commonly used stock solutions, buffers, and reagents, marked <A>. Dilute stock solutions to the appropriate concentrations.

Reagents

Cell fixative <R>
Histochemical stain <R>
Phosphate-buffered saline <A>

METHOD

1. Wash the transfected cells twice with 2–3 mL of phosphate-buffered saline at room temperature.

2. Add 5 mL of cell fixative to the cells.

3. Wash the cells once with phosphate-buffered saline.

4. Add 3–5 mL of histochemical stain to the cells.

5. Incubate the cells for 14–24 h at 37°C.

6. Wash the cell monolayer several times with phosphate-buffered saline.

7. Cover the cell monolayer with a small amount of phosphate-buffered saline, and examine it under a light microscope.

    Cells that have expressed the β-galactosidase expression vector should be a brilliant blue. The transfection frequency can be estimated by counting the relative numbers of stained and unstained cells.

RECIPES

It is essential that you consult the appropriate Material Safety Data Sheets and your institution's Environmental Health and Safety Office for proper handling of equipment and hazardous materials used in this protocol.

Cell Fixative

Reagent	Final concentration
Formaldehyde	2% (v/v)
Glutaraldehyde	0.2% (v/v)
Phosphate-buffered saline	$1\times$

Prepare the cell fixative solution in a chemical fume hood, and store it at room temperature.

Histochemical Stain

Reagent	Final concentration
Potassium ferricyanide ($K_3Fe[CN]_6$)	5 mM
Potassium ferrocyanide ($K_4Fe[CN]_6$)	5 mM
$MgCl_2$	2 mM
Phosphate-buffered saline	$1\times$
X-Gal (5-bromo-4-chloro-3-indolyl-β-galactoside)	1 mg/mL

REFERENCES

Felgner PL, Holm M. 1989. Cationic liposome-mediated transfection. *Focus* **11:** 21–25.

Felgner PL, Gadek TR, Holm M, Roman R, Chan HW, Wenz M, Northrop JP, Ringold GM, Danielsen M. 1987. Lipofection: A highly efficient, lipid-mediated DNA-transfection procedure. *Proc Natl Acad Sci* **84:** 7413–7417.

Jakoby WB, Pastan IH., eds. 1979. *Cell culture.* Methods in Enzymology, Vol. 58. Academic, New York.

Kichler A, Zauner W, Ogris M, Wagner E. 1998. Influence of the DNA complexation medium on the transfection efficiency of lipospermine/ DNA particles. *Gene Ther* **5:** 855–860.

Loeffler J-P, Behr J-P. 1993. Gene transfer into primary and established mammalian cell lines with lipopolyamine-coated DNA. *Methods Enzymol* **217:** 599–618.

Sanes J, Rubinstein JL, Nicolas JF. 1986. Use of recombinant retroviruses to study post-implantation cell lineages in mouse embryos. *EMBO J* **5:** 3133–3142.

Spector DL, Goldman RD, Leinwand LA. 1998. *Cells: A laboratory manual,* Vol. 2: *Light microscopy and cell structure.* Cold Spring Harbor Laboratory Press, Cold Spring Harbor, NY.

Calcium-Phosphate-Mediated Transfection of Eukaryotic Cells with Plasmid DNAs

This protocol, which describes a calcium-phosphate-mediated transfection method for use with plasmid DNAs and adherent cells, was modified from Jordan et al. (1996), who rigorously optimized calcium-phosphate-based transfection methods for Chinese hamster ovary cells and human embryonic kidney 293 cells. Following this protocol are variations on the basic method.

- For high-efficiency generation of stable transfectants, see the alternative method at the end of this protocol.

- For use with high-molecular-weight genomic DNAs, see Protocol 3, and for use with adherent cells that have been released from the substratum with trypsin, see the alternative method at the end of Protocol 3.

- For use with non-adherent cells, see the alternative method at the end of Protocol 3.

INCREASING TRANSFECTION EFFICIENCY

The uptake of DNA by cells in culture is markedly enhanced when the nucleic acid is presented as a coprecipitate of calcium phosphate and DNA. Graham and van der Eb (1973) initially described this method; their work laid the foundation for the introduction of cloned DNAs into mammalian cells and led directly to reliable methods for both stable transformation of cells and transient expression of cloned DNAs. For further details on the procedure, see the information panel Transfection of Mammalian Cells with Calcium Phosphate–DNA Coprecipitates.

Since the publication of the original method, increases in the efficiency of the procedure have been achieved by incorporating additional steps, such as a glycerol shock (Parker and Stark 1979) and/or a chloroquine treatment (Luthman and Magnusson 1983) in the transfection protocol. Treatment with sodium butyrate has been shown to enhance the expression of plasmids that contain the SV40 early promoter/enhancer in simian and human cells (Gorman et al. 1983a,b). Transfection kits, which frequently include these and other modifications to the original protocol, are available from several companies (see Table 2 in the chapter introduction).

MATERIALS

It is essential that you consult the appropriate Material Safety Data Sheets and your institution's Environmental Health and Safety Office for proper handling of equipment and hazardous materials used in this protocol.

Recipes for reagents specific to this protocol, marked <R>, are provided at the end of the protocol. See Appendix 1 for recipes for commonly used stock solutions, buffers, and reagents, marked <A>. Dilute stock solutions to the appropriate concentrations.

Reagents

Calcium chloride ($CaCl_2$) (2.5 M) <A>
Cell culture growth medium (complete and [optional] selective)
Chloroquine diphosphate (100 mM) (optional) <R>
Giemsa stain (10%)

> The Giemsa stain should be freshly prepared in phosphate-buffered saline or H_2O and filtered through Whatman No. 1 filter paper before use.

Glycerol (15%, v/v) in 1× HEPES-buffered saline (optional)

Add 15% (v/v) autoclaved glycerol to filter-sterilized HEPES-buffered saline solution just before use. See Step 5.

HEPES-buffered saline (2×) <A>

Mammalian cells, exponentially growing cultures

Methanol, ice-cold

Phosphate-buffered saline <A>

The solution should be sterilized by filtration before use and stored at room temperature.

Plasmid DNA

Dissolve the DNA in 0.1× TE (pH 7.6) at a concentration of 25 µg/mL; 50 µL of plasmid solution is required per milliliter of medium. A DNA solution with an A_{260}/A_{280} ratio of 1.8 or greater is desirable. The plasmid DNA to be transfected should be free of protein, RNA, chemical and preferably endotoxin contamination, and can be purified using any standard DNA purification kit. Alternatively, DNA may be purified by column chromatography or ethidium bromide-CsCl gradient centrifugation. If the starting amount of plasmid DNA is limiting, then add carrier DNA to adjust the final concentration to 25 µg/ mL. Eukaryotic carrier DNA prepared in the laboratory usually gives higher transfection efficiencies than commercially available DNA such as calf thymus or salmon sperm DNA. Sterilize the carrier DNA before use by ethanol precipitation or extraction with chloroform.

Sodium butyrate (500 mM) (optional) <R>

TE (0.1×, pH 7.6) <A>

Equipment

Tissue culture dishes (60 mm) or plates (12 well)

METHOD

This protocol is designed for cells grown in 60-mm culture dishes or 12-well plates. If other multi-well plates, flasks, or dishes of a different diameter are used, scale the cell density and reagent volumes accordingly (see Table 1 of Protocol 1).

1. Twenty-four hours before transfection, harvest exponentially growing cells by trypsinization, and replate them so that cells will be at 30%–60% confluency on the day of transfection. A general guideline is to maintain a density of 1×10^5 to 4×10^5 cells/cm^2 in 60-mm tissue culture dishes or 12-well plates in the appropriate complete medium. Incubate the cultures for 20–24 h at 37°C in a humidified incubator with an atmosphere of 5%–7% CO_2. Change the medium 1–3 h before transfection.

   To obtain optimum transfection frequencies, use exponentially growing cells. Cell lines used for transfection should never be allowed to grow to >80% confluency.

2. Prepare the calcium phosphate–DNA coprecipitate as follows:

   i. Combine 100 µL of 2.5 M $CaCl_2$ with 25 µg of plasmid DNA in a sterile 5-mL plastic tube. If necessary, bring the final volume to 1 mL with 0.1× TE (pH 7.6).

   ii. Add this 2× $CaCl_2$/DNA mix dropwise with a pipette to an equal volume of 2× HEPES-buffered saline at room temperature, mixing gently during the addition. When DNA addition is complete, the solution should appear slightly opaque because of the formation of a fine calcium phosphate–DNA coprecipitate.

   iii. Incubate the solution for 30 min at room temperature.

   The precipitation reaction mixture can be doubled or quadrupled in volume if a larger number of cells are to be transfected. Normally, 0.1 mL of the calcium phosphate–DNA coprecipitate is added per 1 mL of medium in the culture dish, well, or flask. However, because efficient DNA coprecipitation occurs at smaller volumes of the reaction mix, the recommended DNA concentration while transfecting larger dishes is 0.2–1 mg/mL.

3. Immediately transfer the calcium phosphate–DNA suspension into the medium above the cell monolayer. Use 0.1 mL of suspension for each 1 mL of medium in a well or 60-mm dish. Rock the plate gently to mix the medium, which will become yellow-orange and turbid. Perform this step as quickly as possible because the efficiency of transfection declines rapidly once the DNA precipitate is formed. If the cells will be treated with chloroquine, glycerol, and/or sodium butyrate, proceed directly to Step 5.

> *In some instances, higher transfection frequencies are achieved by first removing the medium and then directly adding the calcium phosphate–DNA suspension to the exposed cells. Incubate the cells for 15 min at room temperature, and then add medium to the dish.*

4. Transfected cells that will not be treated with transfection facilitators should be incubated at 37°C in a humidified incubator with an atmosphere of 5%–7% CO_2. After 2–6 h of incubation, remove the medium and DNA precipitate by aspiration. Add 5 mL of warmed (37°C) complete growth medium, and return the cells to the incubator for 1–6 d. Proceed to Step 6 to assay for transient expression of the transfected DNA, or proceed directly to Step 7 if the objective is stable transformation of the cells.

5. The uptake of DNA can be increased by treatment of the cells with chloroquine in the presence of the calcium phosphate–DNA coprecipitate, or exposure to glycerol and sodium butyrate following removal of the coprecipitate solution from the medium.

Treatment of Cells with Chloroquine

Chloroquine is a weak base that is postulated to act by inhibiting the intracellular degradation of the DNA by lysosomal hydrolases (Luthman and Magnusson 1983). The concentration of chloroquine added to the growth medium and the time of treatment are limited by the sensitivity of the cells to the toxic effect of the drug. The optimal concentration of chloroquine for the cell type used should be determined empirically (see the information panel Chloroquine Diphosphate).

  i. Dilute 100 mM chloroquine diphosphate 1:1000 directly into the medium either before or after the addition of the calcium phosphate–DNA coprecipitate to the cells.

 ii. Incubate the cells for 3–5 h at 37°C in a humidified incubator with an atmosphere of 5%–7% CO_2.

> *Most cells can survive in the presence of chloroquine for 3–5 h. Cells often develop a vesicularized appearance during treatment with chloroquine.*

iii. After the treatment with DNA and chloroquine, remove the medium, wash the cells with phosphate-buffered saline, and add 5 mL of warmed complete growth medium. Return the cells to the incubator for 1–6 d. Proceed to Step 6 to assay for transient expression of the transfected DNA, or proceed directly to Step 7 if the objective is stable transformation of the cells.

Treatment of Cells with Glycerol

This procedure may be used following treatment with chloroquine. Because cells vary widely in their sensitivity to the toxic effects of glycerol, each cell type must be tested in advance to determine the optimum time (30 sec to 3 min) of treatment.

  i. After cells have been exposed for 2–6 h to the calcium phosphate–DNA coprecipitate in growth medium (± chloroquine), remove the medium by aspiration, and wash the monolayer once with phosphate-buffered saline.

 ii. Add 1.5 mL of 15% glycerol in 1× HEPES-buffered saline to each monolayer, and incubate the cells for the predetermined optimum length of time at 37°C.

iii. Remove the glycerol by aspiration, and wash the monolayers once with phosphate-buffered saline.

iv. Add 5 mL of warmed complete growth medium, and incubate the cells for 1–6 d. Proceed to Step 6 to assay for transient expression of the transfected DNA, or proceed directly to Step 7 if the objective is stable transformation of the cells.

Treatment of Cells with Sodium Butyrate

The mechanism through which sodium butyrate acts is not known with certainty; however, the compound is an inhibitor of histone deacetylation (Lea and Randolph 1998), which suggests that treatment may lead to histone hyperacetylation and a chromatin structure on the incoming plasmid DNA that is predisposed to transcription (Workman and Kingston 1998).

i. Following the glycerol shock, dilute 500 mM sodium butyrate directly into the growth medium (Treatment of Cells with Glycerol, Step 5.iv). Different concentrations of sodium butyrate are used depending on the cell type. For example:

CV-1	10 mM
NIH-3T3	7 mM
HeLa	5 mM
CHO	2 mM

The correct amount for other cell lines that may be transfected should be determined empirically.

ii. Incubate the cells for 1–6 d. Proceed to Step 6 to assay for transient expression of the transfected DNA, or proceed directly to Step 7 if the objective is stable transformation of the cells.

6. To assay the transfected cells for transient expression of the introduced DNA, harvest the cells 1–6 d after transfection. Analyze RNA or DNA using hybridization. Analyze newly synthesized protein by radioimmunoassay, by immunoblotting, by immunoprecipitation following in vivo metabolic labeling, or by assays of enzymatic activity in cell extracts.

To minimize the effects of dish-to-dish variation in transfection efficiency, it is best to (1) transfect several dishes with each construct, (2) trypsinize the cells after 24 h of incubation, (3) pool the cells, and (4) replate them on several dishes.

See Troubleshooting.

7. To isolate stable transfectants, do the following.

i. Incubate the cells for 24–48 h in nonselective medium to allow time for expression of the transferred gene(s).

ii. Either trypsinize and replate the cells in the appropriate selective medium, or add the selective medium directly to the cells without further manipulation.

iii. Change the selective medium with care every 2–4 d for 2–3 wk to remove the debris of dead cells and to allow colonies of resistant cells to grow.

iv. Clone individual colonies, and propagate them for assay (for methods, see Jakoby and Pastan 1979 or Spector et al. 1998).

v. Obtain a permanent record of the numbers of colonies by fixing the remaining cells with ice-cold methanol for 15 min followed by staining with 10% Giemsa for 15 min at room temperature before rinsing in tap water.

The dilution at which the transfected cells should be replated to yield well-separated colonies is determined by the efficiency of stable transformation, which can vary over several orders of magnitude (e.g., see Spandidos and Wilkie 1984). The efficiency is dependent on the recipient cell type (significant differences have been observed even between different clones or different passage numbers of the same cell line) (Corsaro and Pearson 1981; Van Pel et al. 1985), the nature of the introduced gene and the efficacy of the transcriptional control signals associated with it, and the amount of DNA used in the transfection.

TROUBLESHOOTING

Problem (Step 6): Transfection efficiency is low, or there is no transfection at all.

Solution: The precipitate may be poor because the 2× HBS solution is no longer at the appropriate pH. The optimum pH range for calcium phosphate transfection is extremely narrow (between 7.05 and 7.12) (Graham and van der Eb 1973). The pH of the solution can change during storage, and an old solution of 2× HBS solution may not work well. Check the pH and use freshly made 2× HBS of the appropriate pH.

Problem (Step 6): Transfection efficiency is low, or there is no transfection at all.

Solution: The precipitate may be poor because over time the 2.5 M $CaCl_2$ solution can deteriorate. Prepare a fresh solution or use a new aliquot from a frozen stock.

Problem (Step 6): Transfection efficiency is low, or there is no transfection at all.

Solution: The precipitate may be poor because complex formation was not carried out at room temperature. The $CaCl_2$–DNA and 2× HBS solutions should be at room temperature (22°C–25°C) when they are mixed.

Problem (Step 6): Transfection efficiency is low, or there is no transfection at all.

Solution: The precipitate may be poor because the steps for forming the complex were not followed closely. The addition of $CaCl_2$–DNA to the 2× HBS solution must be performed dropwise and with continuous mixing. The precipitate must be added dropwise and distributed quickly around the dish to the medium bathing the cells, and the medium should be mixed thoroughly at the end of the addition to distribute the precipitate evenly and avoid localized acidification of cells.

Problem (Step 6): Transfection efficiency is low, or there is no transfection at all.

Solution: The precipitate may be poor because too much time elapsed during critical steps. The dropwise addition of $CaCl_2$–DNA to the 2× HBS solution should be performed in under a minute followed by an incubation time of no more than 30 min. Longer incubations can result in the formation of fewer but larger precipitates that reduce transfection efficiency.

Problem (Step 6): Transfection efficiency is low, or there is no transfection at all.

Solution: If the pH of the medium turns acidic during transfection, this will result in an extremely heavy precipitate, and the medium will turn orange. Care should be taken to maintain a pH of 7.2–7.4 and CO_2 concentrations in the incubator as listed in the protocols. Incubator and medium conditions that are fine for routinely growing cells may not suffice for calcium phosphate transfection.

Problem (Step 6): Transfection efficiency is low, or there is no transfection at all.

Solution: The nucleic acids may be of poor quality or are insufficient in quantity. Consider the following.

- Use only high-quality plasmid preparations and at the recommended concentration.

- Perform a control transfection experiment with a commercially available transfection-grade plasmid preparation containing a marker gene like GFP.

- DNA containing higher endotoxin levels may cause cytotoxicity to sensitive cell lines like Huh-7 and primary cells. Use an endotoxin removal kit, or start over with a kit that incorporates this step for generating the plasmid.

- Low amounts of DNA (<1 μg) may be supplemented with sheared carrier DNA such as salmon or herring sperm DNA; however, effects on transfection efficiencies may vary.

Problem (Step 6): Transfection efficiency is low, or there is no transfection at all.
Solution: The cell density may be too high or too low at the time of transfection. It is critical that the cell density be at 30%–60% confluency on the day of transfection.

Problem (Step 6): Cells sicken.
Solution: The antibiotic used for selection is the cause. Consider the following.

- Wait to add the selection antibiotic until 24–48 h after the transfection procedure.

- Use a lower (or a range of) concentration of the selection antibiotic.

Problem (Step 6): Cells sicken.
Solution: The expressed protein is cytotoxic or is produced at levels that are too high. Consider the following.

- Analyze cytotoxicity by preparing experimental controls: untransfected cells, cells exposed to DNA alone without a transfection reagent, and cells exposed to transfection reagent alone. Compare the transfected cells with the experimental construct to the wells containing these experimental controls.

- Consider repeating the experiment with a secreted reporter gene assay such as secreted alkaline phosphatase or human growth hormone. Secreting cells should show little or no evidence of cytotoxicity.

Problem (Step 6): Cells sicken.
Solution: The cultures are contaminated with mycoplasma. Consider the following.

- Use a commercial mycoplasma detection kit to determine if the culture is contaminated.

- Treat the cells with an antibiotic like BM-Cyclin (Roche Applied Sciences) to eliminate the mycoplasma, or start over with fresh clean cultures.

Problem (Step 6): Cells sicken.
Solution: If none of the solutions above are effective, then consider the following.

- DNA containing higher endotoxin levels may cause cytotoxicity to sensitive cell lines like Huh-7 and primary cells. Use an endotoxin removal kit, or start over with a kit that incorporates this step for generating the plasmid.

High-Efficiency Calcium-Phosphate-Mediated Transfection of Eukaryotic Cells with Plasmid DNAs

A modification of the classical calcium phosphate transfection method that greatly enhances the efficiency of the procedure was developed by Hiroto Okayama and colleagues (Chen and Okayama 1987, 1988). Their method works particularly well when stable transfectants are to be isolated using supercoiled plasmid DNAs and differs from the classical procedure in that the calcium phosphate–DNA coprecipitate is allowed to form in the tissue culture medium during prolonged incubation (15–24 h) under controlled conditions of pH (6.96) and reduced CO_2 tension (2%–4%) (see the box Variables Affecting the Efficiency of Transfection).

Chen and Okayama (1987) reported that this method could be used for transient analysis of gene expression and that the simultaneous introduction of two or more plasmids reduced the overall efficiency of transfection. The overall frequency was still much higher than that obtained with other calcium phosphate methods. When cotransfecting with a selectable marker, it is usually necessary to optimize the system using mixtures containing different ratios of plasmids carrying the selectable marker or the gene of interest (e.g., 1:2, 1:5, and 1:10).

VARIABLES AFFECTING THE EFFICIENCY OF TRANSFECTION

Variables that affect transfection include the purity, form, and amount of the DNA; the pH of the $2\times$ BES buffer; and the concentration of CO_2 in the incubator.

Impure plasmid DNAs transfect poorly because of the inhibitory effects of bacterial contaminants. For this reason, the best results are obtained with DNA obtained using any of the commercial DNA purification kits. Scrupulously clean plasmid DNA purified through specialized chromatography resins or two rounds of CsCl centrifugation is preferred. If necessary, the plasmid can be further purified by phenol: chloroform extraction in the presence of 1% (w/v) SDS.

- Linear DNAs yield very low transformation frequencies, perhaps because the slow formation of the calcium phosphate–DNA coprecipitate leaves the DNA exposed for long periods of time to cell nucleases.

- The nature of the precipitate is affected by the amount of DNA used. A transition (visible under the microscope) from a coarse precipitate to a fine precipitate occurs at the optimal DNA concentration (usually 2–3 μg/mL in the growth medium). The optimum DNA concentration encompasses a narrow range and should be determined empirically for individual cell lines.

- The slow formation of the calcium phosphate–DNA coprecipitate requires a slightly acidic pH and incubation in an atmosphere containing low concentrations of CO_2. The pH curve is very sharp, with a clearly defined optimum at 6.96, whereas the CO_2 concentration is optimal between 2% and 4%.

- The $CaCl_2$–DNA and $2\times$ HBS formation solutions should be at room temperature (22°C–25°C) when they are mixed. Higher or lower temperatures for precipitate formation can lead to decreased transfection efficiency.

ADDITIONAL MATERIALS

It is essential that you consult the appropriate Material Safety Data Sheets and your institution's Environmental Health and Safety Office for proper handling of equipment and hazardous materials used in this protocol.

Recipes for reagents specific to this protocol, marked <R>, are provided at the end of the protocol. See Appendix 1 for recipes for commonly used stock solutions, buffers, and reagents, marked <A>. Dilute stock solutions to the appropriate concentrations.

Reagents

BES-buffered saline (BBS) (2×) <R>

$CaCl_2$ (0.25 M)

Dissolve 1.1 g of $CaCl_2 \cdot 6H_2O$ in 20 mL of distilled H_2O. Sterilize the solution by passing it through a 0.22-μm filter. Store the filtrate in 1-mL aliquots at −20°C.

Superhelical plasmid (1 μg/μL in 0.1× TE at pH 7.6)

Equipment

Tissue culture dishes (90 mm)

METHOD

This protocol is designed for cells grown in 90-mm culture dishes. If other multiwell plates, flasks, or dishes of a different diameter are used, scale the cell density and reagent volumes accordingly (see Table 1 of Protocol 1).

1. Twenty-four hours before transfection, harvest exponentially growing cells by trypsinization, and replate aliquots of 5×10^5 cells onto 90-mm tissue culture dishes. Add 10 mL of complete growth medium, and incubate the cultures overnight at 37°C in a humidified incubator with an atmosphere of 5%–7% CO_2.

2. Mix 20–30 μg of superhelical plasmid DNA with 0.5 mL of 0.25 M $CaCl_2$. Add 0.5 mL of 2× BES-buffered saline (BBS), and incubate the mixture for 10–20 min at room temperature.

   Do not expect a visible precipitate to form during this time.

3. Add the $CaCl_2$–DNA–BBS solution dropwise to the dishes of cells, swirling gently to mix well. Incubate the cultures for 15–24 h at 37°C in a humidified incubator in an atmosphere of 2%–4% CO_2.

4. Remove the medium by aspiration, and rinse the cells twice with medium. Add 10 mL of non-selective medium, and incubate the cultures for 18–24 h at 37°C in a humidified incubator in an atmosphere of 5% CO_2.

5. Following 18–24 h of incubation in nonselective medium, to allow expression of the transfected gene(s) to occur, trypsinize and replate the cells in the appropriate selective medium. Change the selective medium with care every 2–4 d for 2–3 wk to remove the debris of dead cells and to allow colonies of resistant cells to grow.

   The dilution at which the transfected cells should be replated to yield well-separated colonies is determined by the efficiency of stable transformation, which can vary over several orders of magnitude (e.g., see Spandidos and Wilkie 1984). The efficiency is dependent on the recipient cell type (significant differences have been observed even between different clones or different passage numbers of the same cell line) (Corsaro and Pearson 1981; Van Pel et al. 1985), the nature of the introduced gene and the efficacy of the transcriptional control signals associated with it, and the amount of donor DNA used in the transfection.

6. Clone individual colonies and propagate for assay (for methods, see Jakoby and Pastan 1979 or Spector et al. 1998).

   A permanent record of the number of colonies may be obtained by fixing the remaining cells with ice-cold methanol for 15 min followed by staining with 10% Giemsa for 15 min at room temperature before rinsing in tap water. The Giemsa stain should be freshly prepared in phosphate-buffered saline or H_2O and filtered through Whatman No. 1 filter paper before use.

   See Troubleshooting.

TROUBLESHOOTING

Problem (Step 6): There is no or low transfection efficiency.

Solution: Transfection with $2\times$ BBS is highly dependent on the pH of the BBS and the percentage of CO_2 in the incubator during precipitate formation. Consider the following.

- Perform pilot experiments with $2\times$ BBS buffers of varying pH to obtain a pH curve. The optimal pH is within a very narrow range (6.95–6.98). Once the optimal buffer is found, use it as a reference to prepare buffer stocks.

- Be sure to mix $2\times$ BBS and 2.5 M $CaCl_2$ thoroughly before adding to the DNA. Crystal formation upon the addition of calcium chloride indicates incorrect calcium chloride concentration, and the transfection must be repeated.

- The first overnight incubation should be ideally at 3% CO_2, but a variation of 1% is acceptable. After overnight incubation, the culture medium should be alkaline (pH 7.6). Measure the CO_2 levels of the incubator with a Fyrite device before incubating cells.

Problem (Step 6): There is no or low transfection efficiency.

Solution: The form and amount of DNA used might be incorrect. Only superhelical plasmid DNA yields high efficiencies with this protocol using $2\times$ BBS.

RECIPES

It is essential that you consult the appropriate Material Safety Data Sheets and your institution's Environmental Health and Safety Office for proper handling of equipment and hazardous materials used in this protocol.

BES-Buffered Saline (BBS) (2×)

Reagent	Quantity (for 100 mL)	Final concentration
BES (*N,N*-bis[2-hydroxyethyl]-2-aminoethanesulfonic acid)	1.07 g	50 mM
NaCl	1.6 g	280 mM
$Na_2HPO_4 \cdot 2H_2O$	0.027 g	1.5 mM

Dissolve the reagents in 90 mL of distilled H_2O. Adjust the pH of the solution to 6.96 with HCl at room temperature, and then adjust the volume to 100 mL with distilled H_2O. Sterilize the solution by passing it through a 0.22-μm filter. Store the filtrate in aliquots at −20°C.

Chloroquine Diphosphate (100 mM)

Chloroquine diphosphate	52 mg
Deionized, distilled H_2O	1 mL

Dissolve 52 mg of chloroquine diphosphate in 1 mL of deionized, distilled H_2O. Sterilize the solution by passing it through a 0.22-μm filter. Store the filtrate in foil-wrapped tubes at −20°C. See Step 5.

Sodium Butyrate (500 mM)

In a chemical fume hood, bring an aliquot of stock butyric acid solution to a pH of 7.0 with 10 N NaOH. Sterilize the solution by passing it through a 0.22-μm filter. Store in 1-mL aliquots at −20°C. See Step 5.

REFERENCES

Chen C, Okayama H. 1987. High-efficiency transformation of mammalian cells by plasmid DNA. *Mol Cell Biol* **7**: 2745–2752.

Chen C, Okayama H. 1988. Calcium phosphate-mediated gene transfer: A highly efficient transfection system for stably transforming cells with plasmid DNA. *BioTechniques* **6**: 632–638.

Corsaro CM, Pearson ML. 1981. Enhancing the efficiency of DNA-mediated gene transfer in mammalian cells. *Somat Cell Mol Genet* **7**: 603–616.

Gorman CM, Howard BH, Reeves R. 1983a. Expression of recombinant plasmids in mammalian cells is enhanced by sodium butyrate. *Nucleic Acids Res* **11**: 7631–7648.

Gorman C, Padmanabhan R, Howard BH. 1983b. High efficiency DNA-mediated transformation of primate cells. *Science* **221**: 551–553.

Graham FL, van der Eb AJ. 1973. A new technique for the assay of infectivity of human adenovirus 5 DNA. *Virology* **52**: 456–467.

Jakoby WB, Pastan IH, eds. 1979. *Cell culture*. Methods in Enzymology, Vol. 58. Academic, New York.

Jordan M, Schallhorn A, Wurm FW. 1996. Transfecting mammalian cells: Optimization of critical parameters affecting calcium-phosphate precipitate formation. *Nucleic Acids Res* **24**: 596–601.

Lea MA, Randolph VM. 1998. Induction of reporter gene expression by inhibitors of histone deacetylase. *Anticancer Res* **18**: 2717–2722.

Luthman H, Magnusson G. 1983. High efficiency polyoma DNA transfection of chloroquine treated cells. *Nucleic Acids Res* **11**: 1295–1308.

Parker BA, Stark GR. 1979. Regulation of simian virus 40 transcription: Sensitive analysis of the RNA species present early in infections by virus or viral DNA. *J Virol* **31**: 360–369.

Spandidos DA, Wilkie NM. 1984. Expression of exogenous DNA in mammalian cells. In *Transcription and translation: A practical approach* (ed Hames BD, Higgins SJ), pp. 1–48. IRL, Oxford.

Spector DL, Goldman RD, Leinwand LA. 1998. *Cells: A laboratory manual*, Vol. 2: *Light microscopy and cell structure*. Cold Spring Harbor Laboratory Press, Cold Spring Harbor, NY.

Van Pel A, De Plaen E, Boon T. 1985. Selection of highly transfectable variant from mouse mastocytoma P815. *Somat Cell Mol Genet* **11**: 467–475.

Workman JL, Kingston RE. 1998. Alteration of nucleosome structure as a mechanism of transcriptional regulation. *Annu Rev Biochem* **67**: 545–579.

Calcium-Phosphate-Mediated Transfection of Cells with High-Molecular-Weight Genomic DNA

Mammalian genes have been successfully isolated by transfecting cultured mammalian cells with genomic DNA, followed by selection for the gene of interest. This includes dominant cellular onco-genes, genes that encode cell surface molecules, and, as selection/identification strategies and techniques have improved, genes that encode intracellular proteins. Target genes are recovered from the chromosomal DNA of stably transfected cells by virtue of their species-specific repetitive DNA elements or by linkage to cotransfected plasmid DNAs.

The method outlined below is a modification of the calcium phosphate procedure described by Graham and Van der Eb (1973), using high-molecular-weight genomic DNA instead of plasmid DNA. The procedure works especially well to generate stable lines of cells carrying transfected genes that complement mutations in the hosts' chromosomal genes (Sege et al. 1984; Kingsley et al. 1986).

This protocol was supplied by P. Reddy (Amgen Inc.) and M. Krieger (Massachusetts Institute of Technology).

MATERIALS

It is essential that you consult the appropriate Material Safety Data Sheets and your institution's Environmental Health and Safety Office for proper handling of equipment and hazardous materials used in this protocol.

Recipes for reagents specific to this protocol, marked <R>, are provided at the end of the protocol. See Appendix 1 for recipes for commonly used stock solutions, buffers, and reagents, marked <A>. Dilute stock solutions to the appropriate concentrations.

Reagents

$CaCl_2$ (2 M) <A>

Sterilize by filtration, and store frozen as 5-mL aliquots.

Cell culture growth medium (complete and selective)
Exponentially growing cultures of mammalian cells
Genomic DNA

*Prepare high-molecular-weight DNA in TE from appropriate cells as described in **Chapter 1, Protocol 12 or 13**. Dilute the DNA to 100 μg/mL in TE (pH 7.6). Approximately 20–25 μg of genomic DNA is required to transfect each 90-mm plate of cultured cells. The genomic DNA must be sheared to a size range of 45–60 kb before using it to transfect cells (see Steps 2 and 3). The appropriate conditions for shearing the genomic DNA are best determined in preliminary experiments as follows. Shear 2-mL aliquots of high-molecular-weight DNA by passing each aliquot through a 22-gauge needle for a different number of times (e.g., three, four, five, or six times). Examine the DNA by electrophoresis on a 0.6% (w/v) agarose gel followed by staining with either ethidium bromide or SYBR Gold. As markers, use monomeric and dimeric forms of linear bacteriophage λ DNA. To optimize the remaining steps, sheared DNA of the proper size should then be taken through Step 9 of the protocol using dishes without cells.*

Giemsa stain (10%)

The Giemsa stain should be freshly prepared in phosphate-buffered saline or H_2O and filtered through Whatman No. 1 filter paper before use.

Glycerol (15%, v/v) in 1× HEPES-buffered saline

Add 15% (v/v) autoclaved glycerol to filter-sterilized HEPES-buffered saline solution just before use.

HEPES-buffered saline <A>
Isopropanol
Methanol, ice-cold
NaCl (3 M)
   Sterilize by filtration and store at room temperature.
Plasmid with selectable marker (optional; see notes to Steps 3 and 12)

Equipment

Polyethylene tubes (12 mL)
Shepherd's Crook
   Siliconized glass Pasteur pipette containing a hook at the end.
Tissue culture dishes (90 mm)

METHOD

1. On Day 1 of the experiment, plate exponentially growing cells (e.g., CHO cells) at a density of 5×10^5 cells per 90-mm culture dish in appropriate growth medium containing serum. Incubate the cultures for ~16 h at 37°C in a humidified incubator with an atmosphere of 5% CO_2.

2. On Day 2, shear an appropriate amount of high-molecular-weight DNA into fragments ranging in size from 45 kb to 60 kb, by passing it through a 22-gauge needle for the predetermined number of times (see the note to the Genomic DNA entry above in Materials).
   Cells should be transfected with 20–25 µg of genomic DNA per 90-mm dish.

3. Precipitate the sheared DNA by adding 0.1 volume of 3 M NaCl and 1 volume of isopropanol. Collect the DNA on a Shepherd's Crook. Drain the precipitate briefly against the side of the tube, and transfer it to a second tube containing HEPES-buffered saline (1 mL per 12–15 µg of DNA). Redissolve the DNA by gentle rotation for 2 h at 37°C. Make sure that all the DNA has dissolved before proceeding.
   When cotransfecting with a selectable marker (see the note in Step 12), add to the genomic DNA a sterile solution of the appropriate plasmid to a final concentration of 0.5 µg/mL.

4. Transfer 3-mL aliquots of sheared genomic DNA into 12-mL polyethylene tubes (one aliquot per two dishes to be transfected).
   The number of dishes required and transfectants obtained will vary from one cell line to another and depending on the efficiency of the selection method. As a guide, about 15–20 dishes of CHO cells must be transfected to obtain three to 10 stable transformants.

5. To form the calcium phosphate–DNA coprecipitate, gently vortex an aliquot of sheared genomic DNA, and add 120 µL of 2 M $CaCl_2$ in a dropwise fashion. Incubate the tube for 15–20 min at room temperature.
   The solution should turn hazy, but it should not form visible clumps of precipitate.

6. Aspirate the medium from two dishes of cells (from Step 1), and gently add 1.5 mL of the calcium phosphate–DNA coprecipitate to each dish. Carefully rotate the dishes to swirl the medium and spread the precipitate over the monolayer of cells. Incubate the cells for 20 min at room temperature, rotating the dishes once during the incubation.

7. Gently add 10 mL of warmed (37°C) growth medium to each dish, and incubate them for 6 h at 37°C in a humidified incubator with an atmosphere of 5% CO_2.

8. Repeat Steps 5–7 until all of the dishes of cells contain the calcium phosphate–DNA precipitate.

9. After 6 h of incubation, examine each dish under a light microscope. A "peppery" precipitate should be seen adhering to the cells. The precipitate should be neither too fine nor clumpy.

Experience will dictate how a "peppery" precipitate looks under the microscope. Terminate the experiment with cells if a very fine or clumpy precipitate is visualized at this step. The failure to form a peppery precipitate at this step or a hazy solution at Step 5 could be due to the use of a HEPES-buffered saline solution of improper pH, an overly long incubation at Step 5, or a suboptimal concentration of $CaCl_2$ or DNA.

10. In most cases, treatment with glycerol at this step will enhance the transfection frequency. To shock the cells with glycerol:

    i. Aspirate the medium containing the calcium phosphate–DNA coprecipitate.

    ii. To each dish of cells, add 3 mL of 15% glycerol in 1× HEPES-buffered saline that has been warmed to 37°C. Incubate for no longer than 3 min at room temperature.

       It is important that the glycerol in the HEPES-buffered saline not be left in contact with the cells for too long. The optimum time period usually spans a narrow range and varies from one cell line to another and from one laboratory to the next. For these reasons, treat only a few dishes at a time, and take into account the length of time to aspirate the glycerol in the HEPES-buffered saline. Do not exceed the optimum incubation period. Seconds can count!

    iii. Aspirate the glycerol in the HEPES-buffered saline, and rapidly wash the dishes twice with 10 mL of warmed growth medium.

    iv. Add 10 mL of warmed growth medium, and incubate the cultures for 12–15 h at 37°C in a humidified incubator with an atmosphere of 5% CO_2.

11. Replace the medium with 10 mL of fresh growth medium. Continue the incubation overnight at 37°C in a humidified incubator with an atmosphere of 5% CO_2.

12. Microscopic examination of cells at this point (Day 4) should reveal a normal morphology. Cells can be trypsinized and replated in selective medium on Day 4. Continue the incubation for 2–3 wk to allow growth of complemented and/or resistant colonies. Change the medium every 2–3 d.

    The length of the selection period, the cell density of replating, and the selection conditions all depend on the mutation or gene being complemented or selected. Optimum cell density for replating at Step 12 usually varies between 2.5×10^5 and 1×10^6 cells per 90-mm dish. Determine this parameter empirically by plating different numbers of cells without transfection and applying the selection procedure. For logistical reasons, use the highest density that still allows efficient cell killing.

    Cotransfection (e.g., with a plasmid conferring G418 resistance) can be used to distinguish between transfectants and revertants. Because the reversion frequency for some mutant cell lines can be as high as 10^{-6} (i.e., one per 1 million cells plated), false positives can be a problem. The transfection frequency is usually 2×10^{-7}, and the cotransfection frequency is $\sim 10^{-8}$. The use of a selection (e.g., G418 resistance) in conjunction with the mutation/gene selection should eliminate false positives. For further details, see the information panel Cotransformation.

13. Clone individual colonies and propagate them for assay (for methods, see Jakoby and Pastan 1979 and Spector et al. 1998).

14. Obtain a permanent record of the numbers of colonies by fixing the remaining cells with ice-cold methanol for 15 min followed by staining with 10% Giemsa for 15 min at room temperature before rinsing in tap water.

Calcium-Phosphate-Mediated Transfection of Adherent Cells

This protocol can be used with all types of adherent cells, but is particularly useful for polarized epithelial cells, which do not efficiently take up material by endocytosis through the apical plasma membrane. To improve transfection efficiency, adherent cells are trypsinized and collected by centrifugation. The cells are resuspended in the calcium phosphate–DNA coprecipitate and then plated again on tissue culture dishes.

ADDITIONAL MATERIALS

It is essential that you consult the appropriate Material Safety Data Sheets and your institution's Environmental Health and Safety Office for proper handling of equipment and hazardous materials used in this protocol.

Recipes for reagents specific to this protocol, marked <R>, are provided at the end of the protocol. See Appendix 1 for recipes for commonly used stock solutions, buffers, and reagents, marked <A>. Dilute stock solutions to the appropriate concentrations.

Reagents

Phosphate-buffered saline (PBS) <A>

Equipment

Rotor, Sorvall H1000B or equivalent

METHOD

1. Harvest exponentially growing adherent cells by trypsinization. Resuspend the cells in growth medium containing serum, and centrifuge aliquots containing $\sim 10^6$ cells at 800g (2000 rpm in a Sorvall H1000B rotor) for 5 min at 4°C. Discard the supernatants.

2. Form the calcium phosphate–DNA coprecipitate as described in Protocol 2, Step 2 if plasmid DNA is used or as described in Protocol 3, Step 5 if genomic DNA is used.

    Note that the coprecipitate with plasmid DNA takes only ~ 5 min to prepare, whereas the coprecipitate containing genomic DNA takes ~ 25 min to prepare. Execute the initial two steps of this protocol so that cells and coprecipitate are ready at the same time.

3. Resuspend each aliquot of 10^6 cells in 0.5 mL of the calcium phosphate–DNA suspension, and incubate for 15 min at room temperature.

    This protocol can be easily modified to accommodate greater numbers of cells. For example, Chu and Sharp (1981) used 10^8 cells in 2 mL of calcium phosphate–DNA suspension containing 25 μg of DNA. In this case, after 15 min, dilute the mixture with 40 mL of complete growth medium supplemented with 0.05× HEPES-buffered saline and 6.25 mM CaCl$_2$. Plate the cells at a density of 5×10^7 cells per 150-mm dish.

4. To each aliquot, add 4.5 mL of warmed growth medium (with or without chloroquine; see Protocol 2, Step 5), and plate the entire suspension (~ 5 mL) in a single 90-mm tissue culture dish.

Incubate the cells for up to 24 h at 37°C in a humidified incubator with an atmosphere of 5%–7% CO_2.

5. Some types of cells may be further treated with glycerol and sodium butyrate to facilitate transfection. Follow the procedures in Protocol 2, Step 5.

6. Thereafter, assay the cells for transient expression or place in the appropriate selective medium for the isolation of stable transformants (see Protocol 2, Steps 6 and 7).

Calcium-Phosphate-Mediated Transfection of Cells Growing in Suspension

A few cell lines grown as suspension cultures (e.g., HeLa cells) can be transfected using the modified calcium phosphate procedure described in this protocol. However, most lines of cells grown in suspension are resistant to calcium-phosphate-mediated transfection methods. Intransigent cell lines are best transfected using electroporation (Protocol 5) or lipofection (Protocol 1).

ADDITIONAL MATERIALS

It is essential that you consult the appropriate Material Safety Data Sheets and your institution's Environmental Health and Safety Office for proper handling of equipment and hazardous materials used in this protocol.

Equipment

Rotor, Sorvall H1000B or equivalent

METHOD

1. Collect cells from an exponentially growing suspension culture by centrifugation at 800g (2000 rpm in a Sorvall H1000B rotor) for 5 min at 4°C. Discard the supernatant, and resuspend the cell pellet in 20 volumes of ice-cold PBS. Divide the suspension into aliquots containing 1×10^7 cells each. Recover the washed cells by centrifugation as before, and again discard the supernatant.

2. Form the calcium phosphate–DNA coprecipitate as described in Protocol 2, Step 2 if plasmid DNA is used for transfection or as described in Protocol 3, Step 5 if genomic DNA is used.

   Note that preparation of the coprecipitate with plasmid DNA takes only ~5 min to prepare, whereas the coprecipitate containing genomic DNA takes ~25 min to prepare. Execute the initial two steps of this protocol so that cells and coprecipitate are ready at the same time.

3. Gently resuspend 1×10^7 cells in 1 mL of calcium phosphate–DNA suspension (containing ~20 μg of DNA), and allow the suspension to stand for 20 min at room temperature.

4. Add 10 mL of complete growth medium (with or without chloroquine; see Protocol 2, Step 5) to a tube of cells, and plate the entire suspension in a single 90-mm tissue culture dish. Incubate the cells for 6–24 h at 37°C in a humidified incubator with an atmosphere of 5%–7% CO_2.

5. (Optional; for cells known to survive a glycerol shock) At 4–6 h after beginning Step 4, perform the following (otherwise, proceed to Step 6):

   i. Collect the cells by centrifugation at 800g (2000 rpm in a Sorvall H1000B rotor) for 5 min at room temperature, and wash them once with PBS.

   ii. Resuspend the washed cells in 1 mL of 15% glycerol in 1× HEPES-buffered saline, and incubate the cells for 30 sec to 3 min at 37°C.

   See the note to Protocol 3, Step 10.ii.

iii. Dilute the suspension with 10 mL of PBS, and recover the cells by centrifugation as described in Step 5.i. Wash the cells once in PBS.

iv. Resuspend the washed cells in 10 mL of complete growth medium, and plate them in a 90-mm tissue culture dish. Incubate the culture for 48 h at 37°C in a humidified incubator with an atmosphere of 5%–7% CO_2.

6. Recover the cells by centrifugation at 800g (2000 rpm in a Sorvall H1000B rotor) for 5 min at room temperature, and wash them once with PBS.

7. Resuspend the cells in 10 mL of complete growth medium warmed to 37°C. Return the cells to the incubator for 48 h before assaying for transient expression of transfected genes (Protocol 2, Step 6) or replating the cells in selective medium for isolation of stable transformants (Protocol 2, Step 7).

REFERENCES

Chu G, Sharp PA. 1981. SV40 DNA transfection of cells in suspension: Analysis of efficiency of transcription and translation of T-antigen. *Gene* **13:** 197–202.

Graham FL, van der Eb AJ. 1973. A new technique for the assay of infectivity of human adenovirus 5 DNA. *Virology* **52:** 456–467.

Jakoby WB, Pastan IH, eds. 1979. *Cell culture.* Methods in Enzymology, Vol. 58. Academic, New York.

Kingsley DM, Sege RD, Kozarsky KF, Krieger M. 1986. DNA-mediated transfer of a human gene required for low-density lipoprotein receptor expression and for multiple Golgi processing pathways. *Mol Cell Biol* **6:** 2734–2737.

Sege RD, Kozarsky K, Nelson DL, Krieger M. 1984. Expression and regulation of human low-density lipoprotein receptors in Chinese hamster ovary cells. *Nature* **307:** 742–745.

Spector DL, Goldman RD, Leinwand LA. 1998. *Cells: A laboratory manual,* Vol. 2: *Light microscopy and cell structure.* Cold Spring Harbor Laboratory Press, Cold Spring Harbor, NY.

Transfection Mediated by DEAE-Dextran: High-Efficiency Method

Transfection mediated by DEAE-dextran differs from calcium phosphate coprecipitation in three important respects. First, it is used for transient expression of cloned genes and not for stable transformation of cells (Gluzman 1981). Second, it works very efficiently with lines of cells such as BSC-1, CV-1, and COS but is unsatisfactory with many other types of cells. Third, smaller amounts of DNA are used for transfection with DEAE-dextran than with calcium phosphate coprecipitation. Maximal transfection efficiency of 10^5 simian cells is achieved with 0.1–1.0 µg of supercoiled plasmid DNA; larger amounts of DNA (>2–3 µg) can be inhibitory. In contrast to transfection mediated by calcium phosphate, where high concentrations of DNA are required to promote the formation of a coprecipitate, carrier DNA is rarely used with the DEAE-dextran transfection method. For additional details on the procedure, see the information panel DEAE-Dextran Transfection.

Here, we describe two variations on the classical DEAE-dextran transfection procedure. The first involves a brief exposure of cells to a high concentration of DEAE-dextran and yields higher transfection frequencies but elevated cellular toxicity. The second (see Alternative Protocol Transfection Mediated by DEAE-Dextran: Increased Cell Viability) involves a longer exposure of cells to a lower concentration of DEAE-dextran, which produces lower transfection frequencies but increased cell survival.

TRANSFECTION OF COS CELLS

The DEAE-dextran procedure is most often used to transfect simian COS cells. These cells, which express the SV40 large T antigen, were developed by Yasha Gluzman (1981). (For an account of the origins of COS cells, see Witkowski et al. 2008.) Introduction of the SV40 origin of replication, typically by use of the SV40 early region promoter–enhancer/origin to express the gene or cDNA of interest, results in the amplification of the origin-containing plasmid to a very high copy number (Gluzman 1981). This amplification, in turn, produces a high level of expression of the transfected cDNA or gene but severely stresses and eventually kills cells that take up the plasmid. COS cells are thus usually used as transient transfection hosts and analyzed 48–72 h posttransfection.

The efficiency of DEAE-dextran-mediated transfection of COS cells is very high, often approaching 50% of the cells on a dish. For this reason, COS cells are frequently used in expression cloning. The high efficiency of transfection also allows multiple plasmids to be introduced simultaneously into the cells. For example, entire intermediary metabolism pathways can be reconstituted in COS cells by introducing expression plasmids encoding individual enzymes in the pathway (Zuber et al. 1988).

MATERIALS

It is essential that you consult the appropriate Material Safety Data Sheets and your institution's Environmental Health and Safety Office for proper handling of equipment and hazardous materials used in this protocol.

Recipes for reagents specific to this protocol, marked <R>, are provided at the end of the protocol. See Appendix 1 for recipes for commonly used stock solutions, buffers, and reagents, marked <A>. Dilute stock solutions to the appropriate concentrations.

Reagents

Cell culture growth medium (complete and serum-free)

Chloroquine diphosphate (100 mM) <R>

DEAE-dextran (50 mg/mL) <R>

DEAE-dextran transfection kits

Several manufacturers sell kits that provide all of the materials listed in this protocol (e.g., ProFection Mammalian Transfection System from Promega). These kits are somewhat expensive, but they serve as a useful source of control reagents when performing DEAE-dextran transfection experiments for the first time.

Exponentially growing cultures of mammalian cells

Phosphate-buffered saline (PBS) <A>

Sterilize the solution by filtration before use, and store it at room temperature.

Plasmid DNA

Purify closed-circular plasmid DNAs using a transfection-grade plasmid DNA preparation kit that preferably removes bacterial endotoxins during the purification procedure (e.g., EndoFree Plasmid Maxi Kit from QIAGEN). Alternatively, DNA may be purified by column chromatography or ethidium bromide-CsCl gradient centrifugation. Dissolve the DNAs in sterile Tris-EDTA buffer or sterile H_2O at 0.2–2 µg/µL. Determine DNA purity using a 260 nm/280 nm ratio; the ratio should be 1.8.

Tris-buffered saline with dextrose (TBS-D) <R>

Equipment

Tissue culture dishes (60 mm or 35 mm)

METHOD

This protocol is designed for cells grown in 60-mm or 35-mm culture dishes. If multiwell plates, flasks, or dishes of a different diameter are used, scale the cell density and reagent volumes accordingly (see Table 1 of Protocol 1).

1. Twenty-four hours before transfection, harvest exponentially growing cells by trypsinization, and transfer them to 60-mm tissue culture dishes at a density of 10^5 cells/dish (or 35-mm dishes at a density of 5×10^4 cells/dish). Add 5 mL (or 3 mL for 35-mm dish) of complete growth medium, and incubate the cultures for 20–24 h at 37°C in a humidified incubator with an atmosphere of 5%–7% CO_2.

   The cells should be ~75% confluent at the time of transfection. If the cells are grown for <12 h before transfection, they will not be well anchored to the substratum and are more likely to detach during exposure to DEAE-dextran.

2. Prepare the DNA/DEAE-dextran/TBS-D solution by mixing 0.1–4 µg of supercoiled or circular plasmid DNA into 1 mg/mL DEAE-dextran in TBS-D.

   0.25 mL of the solution is required for each 60-mm dish; 0.15 mL is required for each 35-mm dish.

   The amount of DNA required to achieve maximal levels of transient expression depends on the exact nature of the construct and should be determined in preliminary experiments. If the construct carries a replicon that will function in the transfected cells (e.g., the SV40 early region promoter/origin of replication), 100–200 ng of DNA per 10^5 cells should be sufficient; if no replicon is present, larger amounts of DNA may be required (up to 1 µg per 10^5 cells).

3. Remove the medium from the cell culture dishes by aspiration, and wash the monolayers twice with warmed (37°C) PBS and once with warmed TBS-D.

4. Add the DNA/DEAE-dextran/TBS-D solution (250 µL per 60-mm dish, 150 µL per 35-mm dish). Rock the dishes gently to spread the solution evenly across the monolayer of cells. Return the cultures to the incubator for 30–90 min (the time will depend on the sensitivity of each batch of cells to the DNA/DEAE-dextran/TBS-D solution). At 15–20-min intervals, remove the dishes from the incubator, swirl them gently, and check the appearance of the cells under

the microscope. If the cells are still firmly attached to the substratum, continue the incubation. Stop the incubation when the cells begin to shrink and round up.

5. Remove the DNA/DEAE-dextran/TBS-D solution by aspiration. Gently wash the monolayers once with warmed TBS-D and then once with warmed PBS, taking care not to dislodge the transfected cells.

6. Add 5 mL (per 60-mm dish) or 3 mL (per 35-mm dish) of warmed medium supplemented with serum and chloroquine (100 μM final concentration), and incubate the cultures for 3–5 h at 37°C in a humidified incubator with an atmosphere of 5%–7% CO_2.

   The efficiency of transfection is increased severalfold by treatment with chloroquine, which may act by inhibiting the degradation of the DNA by lysosomal hydrolases (Luthman and Magnusson 1983). Note, however, that the cytotoxic effects of a combination of DEAE-dextran and chloroquine can be severe. It is therefore important to perform preliminary experiments to determine the maximum permissible length of exposure to chloroquine after treatment of cells with DEAE-dextran. (For further details, see the information panel Chloroquine Diphosphate.)

7. Remove the medium by aspiration, and wash the monolayers three times with serum-free medium. Add to the cells 5 mL (per 60-mm dish) or 3 mL (per 35-mm dish) of medium supplemented with serum, and incubate the cultures for 36–60 h at 37°C in a humidified incubator with an atmosphere of 5%–7% CO_2 before assaying for transient expression of the transfected DNA.

   The time of incubation should be optimized for the particular cell line and construct under study.

8. To assay the transfected cells for transient expression of the introduced DNA, harvest the cells 36–60 h after transfection. Analyze RNA or DNA using hybridization. Analyze newly synthesized protein by radioimmunoassay, by immunoblotting, by immunoprecipitation following in vivo metabolic labeling, or by assays of enzymatic activity in cell extracts.

   To minimize the effects of dish-to-dish variation in transfection efficiency, it is best to (i) transfect several dishes with each construct, (ii) trypsinize the cells after 24 h of incubation, (iii) pool the cells, and (iv) replate them on several dishes.

Transfection Mediated by DEAE-Dextran: Increased Cell Viability

In contrast to the DEAE-dextran method described in Protocol 4, this alternative protocol uses a lower concentration of DEAE-dextran (250 μg/mL) that remains in contact with the cells for longer periods of time (up to 8 h). Although transfection frequencies are not as high as those obtained in the presence of elevated DEAE-dextran concentrations, the use of reduced levels of DEAE-dextran is associated with less cell toxicity.

ADDITIONAL MATERIALS

It is essential that you consult the appropriate Material Safety Data Sheets and your institution's Environmental Health and Safety Office for proper handling of equipment and hazardous materials used in this protocol.

Reagents

Dulbecco's modified Eagle's medium
> *This is standard DMEM buffered with $NaHCO_3$ and supplemented with serum.*

Dulbecco's modified Eagle's medium buffered with HEPES (HEPES-buffered DMEM)
> *This is DMEM lacking $NaHCO_3$ but containing 10 mM HEPES (pH 7.15). No serum should be added to this reagent.*

METHOD

1. Twenty-four hours before transfection, harvest exponentially growing cells by trypsinization, and transfer them to 60-mm tissue culture dishes with 10^5 cells/dish (or 35-mm dishes with 5×10^4 cells/dish). Add 5 mL (or 3 mL for 35-mm dish) of complete growth medium, and incubate the cultures for 20–24 h at 37°C in a humidified incubator with an atmosphere of 5%–7% CO_2.

   > *The cells should be ~75% confluent at the time of transfection. If the cells are grown for <12 h before transfection, they will not be well anchored to the substratum and are more likely to detach during exposure to DEAE-dextran.*

2. Mix 0.1–1 μg of supercoiled or circular plasmid DNA and 250 μg of DEAE-dextran per 1 mL of HEPES-buffered DMEM. The resulting solution will be used at 500 μL per 60-mm dish or 250 μL per 35-mm dish.

   > *The amount of DNA required to achieve maximal levels of transient expression depends on the exact nature of the construct and should be determined in preliminary experiments. If the construct carries a replicon that will function in the transfected cells (e.g., the SV40 early region promoter/origin of replication), 100–200 ng of DNA per 10^5 cells should be sufficient; if no replicon is present, larger amounts of DNA may be required (up to 1 μg per 10^5 cells).*

3. Remove the medium from the cell culture dishes by aspiration, and wash the monolayers twice with warmed (37°C) HEPES-buffered DMEM.

4. Add the DNA/DEAE-dextran/DMEM solution to the cells (500 μL per 60-mm dish, 250 μL per 35-mm dish), and return the cells to the incubator for up to 8 h. Gently rock the dishes every 2 h to ensure even exposure to the DNA/DEAE-dextran/DMEM solution.

   > *The efficiency of transfection is increased severalfold by concurrent treatment of the cells with chloroquine diphosphate. If used, add the drug (100 μM final concentration) to the DNA/DEAE-*

dextran solution just before it is applied to the cells. Because chloroquine is toxic to the cells, the time of incubation must then be limited to 3–5 h.

A simple variation on this step is reported to double the transfection frequency obtained with COS cells (Gonzales and Joly 1995): Plate the cells in small culture flasks with screw caps at the beginning of the experiment and tightly screw the cap after addition of the DNA/DEAE-dextran/DMEM solution in Step 4. Continue the incubation for 8 h, during which time the medium alkalinizes slowly, owing to the metabolism of the small amount of CO_2 remaining in the flask. This change, which is marked by the gradual deepening in color from crimson to burgundy of the Phenol Red indicator in the medium, may stimulate transfection in a manner similar to the use of a reduced CO_2 atmosphere within the incubator (see the Alternative Protocol High-Efficiency Calcium-Phosphate-Mediated Transfection of Eukaryotic Cells with Plasmid DNAs in Protocol 2).

5. Remove the DNA/DEAE-dextran/DMEM solution from the cells by aspiration, and gently wash the cell monolayers twice with warmed (37°C) HEPES-buffered DMEM. Take care not to dislodge the transfected cells.

6. Wash the cells once with warmed DMEM (buffered with $NaHCO_3$, not HEPES) supplemented with serum. Add to the cells 5 mL (per 60-mm dish) or 3 mL (per 35-mm dish) of complete growth medium, and incubate the cultures for 36–60 h at 37°C in a humidified incubator with an atmosphere of 5%–7% CO_2 before assaying for transient expression of the transfected DNA.

7. To assay the transfected cells for transient expression of the introduced DNA, harvest the cells 36–60 h after transfection. Analyze RNA or DNA using hybridization. Analyze newly synthesized protein by radioimmunoassay, by immunoblotting, by immunoprecipitation following in vivo metabolic labeling, or by assays of enzymatic activity in cell extracts.

To minimize the effects of dish-to-dish variation in transfection efficiency, it is best to (i) transfect several dishes with each construct, (ii) trypsinize the cells after 24 h of incubation, (iii) pool them, and (iv) replate them on several dishes.

RECIPES

It is essential that you consult the appropriate Material Safety Data Sheets and your institution's Environmental Health and Safety Office for proper handling of equipment and hazardous materials used in this protocol.

Chloroquine Diphosphate (100 mM)

Chloroquine diphosphate	60 mg
H_2O, deionized, distilled	1 mL

Dissolve 60 mg of chloroquine diphosphate in 1 mL of deionized, distilled H_2O. Sterilize the solution by passing it through a 0.22-μm filter. Store the filtrate in foil-wrapped tubes at −20°C.

See the information panel Chloroquine Diphosphate.

DEAE-Dextran (50 mg/mL)

DEAE-dextran ($M_r = 500,000$)	100 mg
H_2O, distilled	2 mL

Dissolve 100 mg of DEAE-dextran ($M_r = 500,000$) in 2 mL of distilled H_2O. Sterilize the solution by autoclaving for 20 min at 15 psi (1.05 kg/cm^2) on liquid cycle. Autoclaving also assists dissolution of the polymer.

The molecular weight of the DEAE-dextran originally used for transfection was $>2 \times 10^6$ (McCutchan and Pagano 1968). Although this material is no longer available commercially, it is still occasionally found in chemical storerooms. The older batches of higher-molecular-weight DEAE-dextran are more efficient for transfection than the lower-molecular-weight polymers currently available.

Tris-Buffered Saline with Dextrose (TBS-D)

Immediately before use, add 20% (w/v) dextrose (prepared in H$_2$O and sterilized by autoclaving or filtration) to the TBS solution. The final dextrose concentration should be 0.1% (v/v).

REFERENCES

Gluzman Y. 1981. SV40-transformed simian cells support the replication of early SV40 mutants. *Cell* 23: 175–182.

Gonzales AL, Joly E. 1995. A simple procedure to increase efficiency of DEAE-dextran transfection of COS cells. *Trends Genet* 11: 216–217.

Luthman H, Magnusson G. 1983. High efficiency polyoma DNA transfection of chloroquine treated cells. *Nucleic Acids Res* 11: 1295–1308.

McCutchan JH, Pagano JS. 1968. Enhancement of the infectivity of simian virus 40 deoxyribonucleic acid with diethyl aminoethyl-dextran. *J Natl Cancer Inst* 41: 351–357.

Witkowski JA, Gann A, Sambrook JA, eds. 2008. *Life illuminated: Selected papers from Cold Spring Harbor*, Vol. 2, 1972–1994, pp. 125–128. Cold Spring Harbor Laboratory Press, Cold Spring Harbor, NY.

Zuber MX, Mason JI, Simpson ER, Waterman MR. 1988. Simultaneous transfection of COS-1 cells with mitochondrial and microsomal steroid hydroxylases: Incorporation of a steroidogenic pathway into non-steroidogenic cells. *Proc Natl Acad Sci* 85: 699–703.

DNA Transfection by Electroporation

Electroporation, which uses pulsed electrical fields, can be used to introduce DNA into a variety of animal cells (Neumann et al. 1982; Wong and Neumann 1982; Potter et al. 1984; Sugden et al. 1985; Toneguzzo et al. 1986; Tur-Kaspa et al. 1986), plant cells (Fromm et al. 1985, 1986; Ecker and Davis 1986), and bacteria. Electroporation works well with cell lines that are refractory to other transformation techniques, such as lipofection and calcium phosphate–DNA coprecipitation. But, as with other transfection methods, the optimal conditions for electroporating DNA into untested cell lines must be determined empirically.

Several different electroporation instruments are available commercially, and manufacturers supply detailed protocols for their use with specific cell types and guidelines for optimization with others. The following method describes the conditions for electroporating mammalian cell lines using the Gene Pulser Xcell[1] Electroporation System (Bio-Rad). When working with a new experimental system, begin with an optimization protocol using incremental voltage steps to determine the best electroporation conditions (see the box Electroporation Efficiency). Optimization can be greatly facilitated by using an electroporator that can handle multiwell electroporation plates. For further information on the history, mechanism, and optimization of electroporation, see the information panel Electroporation. For a description of an electroporation technique that transfers DNA directly into the nucleus, see the box DNA Transfection by Nucleofection.

ELECTROPORATION EFFICIENCY

The efficiency of transfection by electroporation is influenced by several factors as described below.

- *Strength of the applied electric field.* At low voltage, the plasma membranes of cultured cells are not sufficiently altered to allow passage of DNA molecules; at higher voltage, the cells are irreversibly damaged. For most lines of mammalian cells, the maximal level of transient expression is reached when voltages between 250 V/cm and 500 V/cm are applied. Typically, between 20% and 50% of the cells survive this treatment (as measured by exclusion of Trypan Blue) (Patterson 1979; Baum et al. 1994).

- *Duration of the electric pulse.* Usually, a single electric pulse is applied to cells. The duration, field shape, and strength of the pulse are determined by the capacitance of the power supply and the dimensions of the cuvette. Most electroporation devices grant the investigator control over the characteristics of the pulse. The optimal length of the electric pulse required for electroporation is 20–100 msec. The electroporator can also offer the choice of varying the electrical pulse shape between exponential decay and square wave (for pulse wave characteristics, see the information panel Electroporation).

- *Temperature.* Some investigators have reported that maximal levels of transient expression are obtained when the cells are maintained at room temperature during electroporation (Chu et al. 1987). Others have obtained better results with cells maintained at 0°C (Reiss et al. 1986). These differences may be cell type specific or dependent on the amount of heat generated during electroporation when large electrical voltages (>1000 V/cm) and extended electric pulses (>100 msec) tend to produce more heat. The efficiency of transient expression is increased if the cells are incubated for 1–2 min in the electroporation chamber after exposure to the electric pulse (Rabussay et al. 1987).

(Continued)

---

[1] http://www.bio-rad.com/evportal/evolutionPortal.portal?_nfpb=true&_pageLabel=trademarks_page&country=US&lang=en&javascriptDisabled=true.

- *Conformation and concentration of DNA.* Although both linear and circular DNAs can be transfected by electroporation, higher levels of both transient expression and stable transformation are obtained when linear DNA is used (Neumann et al. 1982; Potter et al. 1984; Toneguzzo et al. 1986). Effective transfection has been obtained with concentrations of DNA ranging from 1 μg/mL to 40 μg/mL. The purity of the plasmid is also a prominent factor that contributes to the efficiency of electroporation. For most practical purposes, plasmid DNA purified by the many commercial kits (preferably those that also use a step to remove bacterial endotoxins) yields satisfactory results.

- *Ionic composition of the medium.* The efficiency of transfection is severalfold higher when the cells are suspended in buffered salt solutions (e.g., HEPES-buffered saline) rather than in buffered solutions containing nonionic substances such as mannitol or sucrose (Rabussay et al. 1987). Some companies also market proprietary buffers recommended for use with any electroporator for any nucleic acid and mammalian cell line, for example, the Gene Pulser Electroporation Buffer (Bio-Rad), or Ingenio Electroporation Solution (Mirus Bio).

- *Cell physiology.* The best transfections are achieved with actively growing, healthy cells that are free of contamination and preferably of low passage number.

DNA TRANSFECTION BY NUCLEOFECTION

"Nucleofection"[2] is a specialized form of electroporation that delivers nucleic acids through the nuclear membrane and directly into the nucleus. Because DNA or RNA is transferred directly into the nucleus, cell division is not required for the incorporation of nucleic acid into the cell. Thus, nucleofection permits comparatively efficient transfection of nondividing primary cells such as neurons, as well as difficult-to-transfect cell lines. Unlike other prevalent transfection methods, the same conditions are used for the nucleofection of any nucleic acid; however, the electrical parameters and buffer solutions are specific to the cell type being transfected.

The proprietary electrical conditions are preloaded in the nucleofection device, and proprietary cell-specific solutions are required. As with other transfection technologies, the cell conditions (confluency, passage number, etc.), the amount and quality of DNA, and the manner in which cells are handled all influence transfection efficiency. Protocols that have been optimized for more than 500 cell types are available from the manufacturer of the nucleofection instruments and reagents (see http://www.lonzabio.com/resources/product-instructions/protocols/). In addition, a cell line optimization kit and optimization kits for primary cells are available to standardize transfection according to one's experimental setup.

MATERIALS

It is essential that you consult the appropriate Material Safety Data Sheets and your institution's Environmental Health and Safety Office for proper handling of equipment and hazardous materials used in this protocol.

Recipes for reagents specific to this protocol, marked <R>, are provided at the end of the protocol. See Appendix 1 for recipes for commonly used stock solutions, buffers, and reagents, marked <A>. Dilute stock solutions to the appropriate concentrations.

Reagents

Carrier DNA (10 mg/mL; e.g., sonicated salmon sperm DNA) (optional)
Cell culture growth medium (complete and [optional] selective)
Electroporation buffer

> *The electroporation buffer needs to be optimized based on the cell type or on the manufacturer's recommendations. Any of the following buffers could be used: phosphate-buffered saline <A>. HEPES-buffered saline (HBS) <A>; phosphate-buffered sucrose; and HEPES-buffered sucrose. For all preset protocols programmed in the Gene Pulser Xcell System, the manufacturer recommends the use of Opti-MEM or growth media without serum as electroporation buffer.*

---

[2] http://www.lonzabio.com/meta/legal/.

Exponentially growing cell cultures
> *For best results, passage cells 1–2 d before electroporation, and harvest cells at a density of 60%–80% confluency on the day of electroporation.*

Linearized or circular plasmid DNA (1–5 μg/μL in sterile deionized H_2O)

Phosphate-buffered saline (PBS) <A>

Trypsin

Equipment

Electroporation cuvettes

Electroporation instrument
> *This protocol presumes the use of the Gene Pulser Xcell System with ShockPod[3] chamber for eukaryotic cells (Bio-Rad; catalog no. 165-2661; operational at 100–240 V).*

Hemocytometer

Tissue culture dish or multiwell plate

Tissue culture flasks (75 cm^2)

METHOD

1. Prepare cells for electroporation.
   For Adherent Cells

   i. One day before electroporation, use trypsin to release adherent cells, and transfer the cells into 75-cm^2 flasks with fresh growth medium at a density sufficient to obtain 50%–70% confluency on the day of electroporation. (For most cell lines, this translates to $\sim 1 \times 10^5$ to 10×10^5 cells per electroporation.)

   ii. On the day of the experiment, aspirate the growth medium, rinse cells with PBS, and use trypsin to release adherent cells. Remove an aliquot of trypsinized cell suspension, and, using a hemocytometer, count the cells to determine the cell density. Centrifuge the cell suspension at 500g for 5 min at room temperature.

   iii. Resuspend the cell pellet in the appropriate electroporation buffer at a density of 1×10^6 to 5×10^6 cells/mL. Gently pipette cells to obtain a single suspension.

   For Suspension Cells

   i. One day before electroporation, dilute the cells into fresh growth medium in a 75-cm^2 flask to obtain a mid-log phase of growth on the day of electroporation confluency ($\sim 0.5 \times 10^6$ to 4×10^6 cells/mL). Count cells and collect them by centrifugation as described above.

   ii. Resuspend the cell pellet at a density of 1×10^6 to 5×10^6 cells/mL in the appropriate electroporation buffer. Gently pipette cells up and down to obtain a homogeneous suspension.

Electroporation

2. Set the parameters on the electroporation instrument. Depending on the cell type, choose either a preset protocol or an optimization protocol. For an exponential protocol (i.e., exponential decay pulse), a typical capacitance value is 1050 μF, and the voltage ranges from 200 to 350 V. Generally a starting voltage of 260 V with 50-V increments works for most cells. Use an infinite internal resistance value.

3. Add 10–50 μg of plasmid DNA to the electroporation cuvette. If needed, carrier DNA (e.g., salmon sperm DNA) may be added to bring the total amount of DNA to 120 μg.

---

[3] http://www.bio-rad.com/evportal/evolutionPortal.portal?_nfpb=true&_pageLabel=trademarks_page&country=US&lang=en&javascriptDisabled=true.

TABLE 1. Recommendations for electroporation using the Bio-Rad ShockPod

Cuvette (cm)	Cell concentration (cells/mL)	Cell volume (µL)	Growth conditions following electroporation
0.2[a]	1×10^6	100	48-well plate with 0.5 mL of growth medium
0.2	5×10^6	200	Six-well plate with 2 mL of growth medium
0.4	2.5×10^6	400	Six-well plate with 2 mL growth medium

Reproduced from Bio-Rad Laboratories, Inc.
[a]Not recommended for eukaryotic cells.

4. Add cells to the cuvette, and gently mix the cells and DNA by pipetting up and down. For suggestions on cell concentrations and volumes, see Table 1.

   ▲ *Do not introduce air bubbles into the suspension during the mixing step.*

5. Place the cuvette into the electroporator, close the lid, and apply one electric pulse.

   Record the actual pulse time and time constant for each cuvette to facilitate comparisons between experiments.

6. Immediately add 0.5 mL of growth medium to the cuvette, and transfer the electroporated cells to an appropriately sized tissue culture dish or multiwell plate. If desired, rinse out 0.2- or 0.4-cm cuvettes with growth medium, and add the wash to the cells on the plate or dish.

7. Repeat Steps 5 and 6 for each sample and each voltage increment. Record the actual pulse time and time constant for each cuvette to facilitate comparisons between experiments. For mammalian cells, conditions that result in pulse times or time constants of 10–40 msec at field strengths of 400–900 V/cm are optimal. Rock the plate gently to distribute the cells evenly throughout the well or dish.

8. Transfer the dish or plate to a humidified incubator at 37°C with an atmosphere of 5%–7% CO_2. Incubate the cells for 6–24 h.

9. Determine cell viability using any of the methods outlined in Protocols 6, 7, or 8.

10. To isolate stable transfectants, follow Step 13 of Protocol 1.

11. For transient expression, examine the cells 24–96 h after electroporation using one of the assays described in Step 12 of Protocol 1.

REFERENCES

Baum C, Forster P, Hegewisch-Becker S, Harbers K. 1994. An optimized electroporation protocol applicable to a wide range of cell lines. *Bio-Techniques* 17: 1058–1062.

Chu G, Hayakawa H, Berg P. 1987. Electroporation for the efficient transfection of mammalian cells with DNA. *Nucleic Acids Res* 15: 1311–1326.

Ecker JR, Davis RW. 1986. Inhibition of gene expression in plant cells by expression of antisense RNA. *Proc Natl Acad Sci* 83: 5372–5376.

Fromm M, Taylor LP, Walbot V. 1985. Expression of genes transferred into monocot and dicot plant cells by electroporation. *Proc Natl Acad Sci* 82: 5824–5828.

Fromm M, Taylor LP, Walbot V. 1986. Stable transformation of maize after gene transfer by electroporation. *Nature* 319: 791–793.

Neumann E, Schaefer-Ridder M, Wang Y, Hofschneider PH. 1982. Gene transfer into mouse lyoma cells by electroporation in high electric fields. *EMBO J* 1: 841–845.

Patterson MK Jr. 1979. Measurement of growth and viability of cells in culture. *Methods Enzymol* 58: 141–152.

Potter H, Weir L, Leder P. 1984. Enhancer-dependent expression of human κ immunoglobulin genes introduced into mouse pre-B lymphocytes by electroporation. *Proc Natl Acad Sci* 81: 7161–7165.

Rabussay D, Uher L, Bates G, Piastuch W. 1987. Electroporation of mammalian and plant cells. *Focus* 9: 1–3.

Reiss M, Jastreboff MM, Bertino JR, Narayanan R. 1986. DNA-mediated gene transfer into epidermal cells using electroporation. *Biochem Biophys Res Commun* 137: 244–249.

Sugden B, Marsh K, Yates J. 1985. A vector that replicates as a plasmid and can be efficiently selected in B-lymphoblasts transformed by Epstein-Barr virus. *Mol Cell Biol* 5: 410–413.

Toneguzzo F, Hayday AC, Keating A. 1986. Electric field-mediated DNA transfer: Transient and stable gene expression in human and mouse lymphoid cells. *Mol Cell Biol* 6: 703–706.

Tur-Kaspa R, Teicher L, Levine BJ, Skoultchi AI, Shafritz DA. 1986. Use of electroporation to introduce biologically active foreign genes into primary rat hepatocytes. *Mol Cell Biol* 6: 716–718.

Wong T-K, Neumann E. 1982. Electric field mediated gene transfer. *Biochem Biophys Res Commun* 107: 584–587.

Analysis of Cell Viability by the alamarBlue Assay

The water-soluble dye alamarBlue (which is the registered product name for the chemical resazurin [see the box Resazurin]) has been used for quantifying in vitro viability of multiple cell types (Fields and Lancaster 1993; Ahmed et al. 1994). Because the reagent is extremely stable and nontoxic, it can be used to continuously monitor cells over time (Ahmed et al. 1994). It is mainly for this reason that the alamarBlue assay is considered superior to classical tests for cell viability, such as the MTT test (Protocol 8). In a comparative study of cell viability after exposure to each of 117 different toxic molecules, the alamarBlue assay was slightly more sensitive than the MTT assay for most compounds (Hamid et al. 2004).

The following protocol describes viability measurements for cell cultures in a 96-well tissue culture plate. The assay can be modified to accommodate larger plates; however, for a preliminary analysis of transfection reagents and parameters of the transfection protocol on cell viability, a 96-well plate format is the most cost effective.

RESAZURIN

Resazurin is cell permeable and virtually nonfluorescent. Upon entering cells, resazurin is reduced to resorufin through the activity of cellular redox enzymes (Gonzales and Tarloff 2001) by accepting electrons from NADPH, $FADH_2$, $FMNH_2$, NADH, and the cytochromes (Fig. 1) (O'Brien et al. 2000). This redox reaction is accompanied by a shift in color from indigo blue to a bright fluorescent red, which diffuses out of the cells into the culture medium, where it can be easily measured by colorimetry or fluorometry at 590 nm. Viable cells continuously convert resazurin to resorufin, thereby generating a quantitative measure of cell viability.

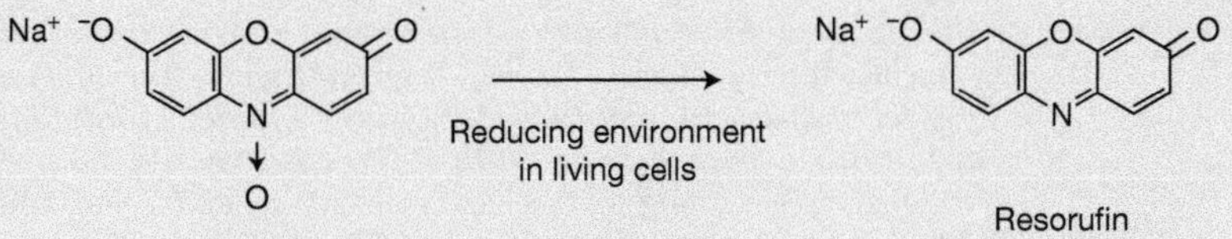

FIGURE 1. Conversion of resazurin to resorufin by redox reactions in viable cells.

MATERIALS

It is essential that you consult the appropriate Material Safety Data Sheets and your institution's Environmental Health and Safety Office for proper handling of equipment and hazardous materials used in this protocol.

Recipes for reagents specific to this protocol, marked <R>, are provided at the end of the protocol. See Appendix 1 for recipes for commonly used stock solutions, buffers, and reagents, marked <A>. Dilute stock solutions to the appropriate concentrations.

Reagents

alamarBlue (Life Technologies)

alamarBlue is available commercially as a sterile 10× solution. Because the resazurin dye in the alamarBlue reagent and the resorufin produced are both light sensitive, protect them from prolonged exposure to light.

Cultured mammalian cells that have been transfected
SDS (3%) (optional; see Step 5)

Equipment

Fluorometer with plate reading capability

The readout of the reaction is a measure of the reduced resorufin, which has an excitation/emission maximum of 570/590 nm. This can be measured as absorbance or fluorescence. Fluorescence measurements are more sensitive than absorbance measurements and hence are the preferred method of detection.

Plates (96 well), standard flat-bottom wells
Plates (96 well), tissue-culture treated

METHOD

1. Plate 50–50,000 cells/well in a 96-well plate. For each set of conditions, perform the assay in triplicate.

   Include a control well with untreated cells and a control well with no cells. In addition, wells containing the transfection reagents but no cells, as well as a positive control for cytotoxicity can be included if desired.

   Viability assays are generally performed after a genotoxic stress. As per the requirements of the experiment, the toxic agent can be retained in the cell culture medium during the assay or it can be removed before the assay. Toxic agents that interfere with the function of mitochondria can yield misleading results because a bulk of the resazurin reduction takes place in the mitochondrion.

2. Measure the culture medium volume in a well. Add $0.1\times$ volume of alamarBlue to the well.

   If present in the cell culture medium, the pH indicator Phenol Red does not interfere with the assay.

3. Incubate for 1–4 h at 37°C in a tissue culture incubator.

   The ability of different cell types to reduce resazurin to resorufin varies depending on the metabolic capacity of the cell line and the length of incubation time with the reagent. For most applications a 1–4-h incubation is adequate. Although a 1-h incubation time is enough for 5000–50,000 cells, lower cell numbers ranging from 50 to 5000 cells may need a longer incubation time of up to 24 h. For optimizing screening assays, the number of cells/well and the length of the incubation period should be determined empirically.

4. Transfer 100 µL of the cell culture supernatant to a flat-bottomed, 96-well plate. Proceed with either Step 5 or Step 6.

5. If the reading is not performed immediately, stop and stabilize the reaction by adding 50 µL of 3% SDS per 100 µL of original culture volume. Store the plate at ambient temperature for up to 24 h before proceeding, provided that the contents are protected from light and covered to prevent evaporation.

6. Read the plate at an excitation wavelength of 560 nm and an emission wavelength of 590 nm. For a fixed-wavelength plate reader, an excitation range of 540–570 nm and an emission range of 580–610 nm can be used.

   If the readout exceeds the detection limit of the fluorometer, dilute the supernatant 10-fold to 100-fold with cell culture medium or PBS, and repeat the reading.

   alamarBlue measurements can also be read as absorbance on a spectrophotometer at 570 nm, with a 600-nm reading as reference. However, fluorescence measurements are far more sensitive than absorbance measurements.

7. Subtract the blank value (i.e., well with no cells) from each reading, and plot the adjusted values. The readings are directly proportional to the metabolic activity and, in turn, the viability of the cells.

REFERENCES

Ahmed SA, Gogal RM Jr, Walsh JE. 1994. A new rapid and simple non-radioactive assay to monitor and determine the proliferation of lymphocytes: An alternative to [³H]thymidine incorporation assay. *J Immunol Methods* **170:** 211–224.

Fields RD, Lancaster MV. 1993. Dual-attribute continuous monitoring of cell proliferation/cytotoxicity. *Am Biotechnol Lab* **11:** 48–50.

Gonzalez RJ, Tarloff JB. 2001. Evaluation of hepatic subcellular fractions for Alamar blue and MTT reductase activity. *Toxicol In Vitro* **15:** 257–259.

Hamid R, Rotshteyn Y, Rabadi L, Parikh R, Bullock P. 2004. Comparison of alamar blue and MTT assays for high through-put screening. *Toxicol In Vitro* **18:** 703–710.

O'Brien J, Wilson I, Orton T, Pognan F. 2000. Investigation of the Alamar Blue (resazurin) fluorescent dye for the assessment of mammalian cell cytotoxicity. *Eur J Biochem* **267:** 5421–5426.

Analysis of Cell Viability by the Lactate Dehydrogenase Assay

A common method for determining cytotoxicity is based on measuring the activity of cytoplasmic enzymes released by damaged cells. Lactate dehydrogenase (LDH) is a stable cytoplasmic enzyme that is found in all cells. There are five different isoforms of lactate dehydrogenase present in different tissue types, and they differ in their quantity, specificity, and kinetics of enzyme action. But all of the different enzyme isoforms catalyze the common reaction of interconverting pyruvate to lactate accompanied by the oxidation of NADH to NAD^+ (Fig. 1).

LDH is rapidly released into the cell culture supernatant when the plasma membrane is damaged, a key feature of cells undergoing apoptosis, necrosis, and other forms of cellular damage. LDH activity can be easily quantified by using the NADH produced during the conversion of lactate to pyruvate to reduce a second compound in a coupled reaction into a product with properties that are easily quantitated. This protocol measures the reduction of a yellow tetrazolium salt, INT, by NADH into a red, water-soluble formazan-class dye by absorbance at 492 nm. The amount of formazan is directly proportional to the amount of LDH in the culture, which is, in turn, directly proportional to the number of dead or damaged cells. Several companies market kits that use INT (2-*p*-iodophenyl-3-*p*-nitrophenyl tetrazolium chloride) as a substrate. Promega markets a kit that uses resazurin as a substrate for the coupled LDH assay, which can be measured as described in Protocol 6.

MATERIALS

It is essential that you consult the appropriate Material Safety Data Sheets and your institution's Environmental Health and Safety Office for proper handling of equipment and hazardous materials used in this protocol.

Recipes for reagents specific to this protocol, marked <R>, are provided at the end of the protocol. See Appendix 1 for recipes for commonly used stock solutions, buffers, and reagents, marked <A>. Dilute stock solutions to the appropriate concentrations.

Reagents

Cultured mammalian cells that have been transfected
> *The assay must be performed in triplicates. For an LDH assay, a set of wells with cells not exposed to the toxic agent needs to be set aside for estimating the total amount of LDH.*

LDH assay substrate solution <R>
LDH standards (optional; see Step 8)
> *Purified LDH is available commercially. Prepare a standard solution by dissolving LDH in cell culture medium, and dilute it to desired concentrations of 0.2–2.0 U/mL.*

Lysis solution (9% [v/v] Triton X-100, for measurement of total cellular LDH)
Stop solution (50% dimethylformamide and 20% SDS at pH 4.7)

> *Hydrochloric acid (1 N) can be used to stop the reaction, but DMF/SDS solution is better when used with medium containing Phenol Red, because it neutralizes background absorption.*

Equipment

Microtiter-plate spectrophotometer
Plates (96 well), standard flat-bottom wells
Plates (96 well), either V or round bottom

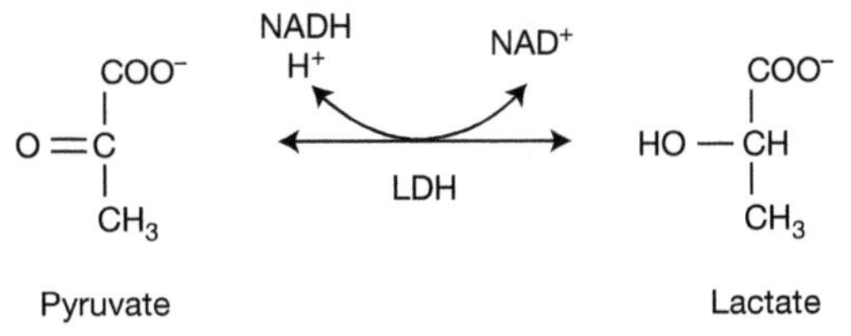

FIGURE 1. **The reversible reaction catalyzed by lactate dehydrogenase.**

IMPORTANT CONSIDERATIONS

Two factors in the cell culture medium can influence the background of LDH assays—Phenol Red and serum. The absorbance value of a culture medium control is used to normalize the values obtained from other samples. Background absorbance from Phenol Red can also be eliminated by using a Phenol Red–free medium. Serum contains a significant level of LDH activity. Human AB serum is relatively low in LDH activity, whereas calf serum is relatively high. Heat-inactivated serum has a much lower concentration of active LDH. In general, decreasing the serum concentration to 5% will significantly reduce background without affecting cell viability. Certain detergents (e.g., SDS and cetrimide) can inhibit LDH activity.

When performing the cytotoxicity assay, seed additional wells in order to perform the following controls.

1. *Spontaneous LDH release.* This corrects for spontaneous release of LDH from cells. Seed wells with cells that have not been transfected.

2. *Culture medium background.* This corrects for LDH activity contributed by serum in culture medium and the varying amounts of Phenol Red in the culture medium. Seed wells with culture medium alone.

3. *Maximum LDH release.* This yields an estimate of the maximum LDH activity that could be expected if all cells in the well were killed under the assay conditions employed. For this, wells with untreated target cells are lysed with lysis solution and measured as described in Steps 1–8.

METHOD

Optimization of Target Cell Number (Total LDH Release Assay)

Because different target cell types (YAC-1, K562, Daudi, etc.) contain different amounts of LDH, it is valuable to perform a preliminary experiment using the cell type under investigation in order to determine the optimum number of target cells needed to ensure an adequate signal-to-noise ratio.

1. Prepare target cell dilutions (0, 5000, 10,000, and 20,000 cells/100 µL), and add 100 µL to each well in a V- or round-bottom, 96-well plate. Perform the assay in triplicate.

2. Add 15 µL of lysis solution to each well. Centrifuge the plate at 250g for 4 min.

3. Transfer 50 µL of supernatant to a 96-well, flat-bottom enzymatic assay plate.

4. Add 50 µL of LDH assay substrate to the medium. Cover the plate with foil or a small opaque box to protect it from light. Incubate for 15–30 min at 37°C.

   To include measurements for the preparation of a standard curve, see Step 8.

5. Add 100 µL of Stop solution.

6. Ensure that there are no bubbles in the wells. Measure the absorbance at 490 nm within 1 h of adding the Stop solution. Set the background absorbance at 690 nm, and subtract this value from the primary wavelength measurement (490 nm).

7. Determine the concentration of target cells yielding absorbance values at least two times the background absorbance of the medium control.

8. (Optional) Add 50 µL of assay substrate to 50 µL of different LDH standards. Incubate for 15 min and measure absorbance as above. A standard curve is prepared with the values obtained and is used to compare the enzyme activity of test samples.

Cytotoxicity Assay

9. Transfer 50 μL of cell culture supernatant into a 96-well plate. With suspension cell cultures, centrifuge the cells at 250*g* for 4 min, and then transfer 50 μL of the supernatant.

10. Add 50 μL of LDH assay substrate to the medium. Cover the plate with foil or a small opaque box to protect it from light. Incubate it for 15–30 min at 37°C.

11. Add 100 μL of Stop solution.

12. Ensure that there are no bubbles in the wells. Measure the absorbance at 490 nm within 1 h of adding the Stop solution. Set the background absorbance at 690 nm, and subtract this value from the primary wavelength measurement (490 nm).

13. Determine the percent of cell death (% cytotoxicity) using the following equation:

$$\% \text{ Cytotoxicity} = \frac{\text{Experimental LDH release (OD}_{490})}{\text{Maximum LDH release (OD}_{490})}.$$

14. (Optional) Add 50 μL of assay substrate to 50 μL of different LDH standards. Incubate for 15 min and measure the absorbance as above. A standard curve is prepared with the values obtained and is used to compare the enzyme activity of test samples.

DISCUSSION

The total LDH release assay provides a measure of the lactate dehydrogenase activity present in the cytoplasm of intact cells. Cell numbers can therefore be quantitated when the cells are lysed to release the LDH present inside. The number of cells present will be directly proportional to the absorbance values measured at 490 nm, which represent total LDH activity. The resulting data can be plotted with absorbance at 490-nm values along the ordinate and cell number along the abscissa.

RECIPES

It is essential that you consult the appropriate Material Safety Data Sheets and your institution's Environmental Health and Safety Office for proper handling of equipment and hazardous materials used in this protocol.

INT Solution
Dissolve 2-*p*-iodophenyl-3-*p*-nitrophenyl tetrazolium chloride (INT) in PBS to a final concentration of 100 mM.

LDH Assay Substrate Solution

L-(+)-lactic acid	0.054 M
β-NAD$^+$	1.3 mM
2-*p*-Iodophenyl-3-*p*-nitrophenyl tetrazolium chloride (INT)	0.66 mM
1-Methoxyphenazine methosulfate (MPMS)	0.28 mM
Tris-HCl buffer (pH 8.2)	0.2 M

Dissolve the ingredients in 0.2 M Tris-HCl buffer (pH 8.2). One milliliter of assay substrate is needed for 20 reactions.

Over time, competitive inhibitors of LDH can form in phosphate solutions of NADH. Prepare the LDH assay substrate solution fresh from stock solutions each time.

MPMS Solution
Dissolve 1-methoxyphenazine methosulfate (MPMS) in PBS to a concentration of 100 mM. Store at 4°C for up to 1 mo.

Analysis of Cell Viability by the MTT Assay

Among viability assays that depend on the conversion of substrate to chromogenic product by live cells, the MTT assay developed by Mossman (1983) is still among one of the most versatile and popular assays. The MTT assay involves the conversion of the water-soluble yellow dye MTT [3-(4,5-dimethylthiazol-2-yl)-2,5-diphenyltetrazolium bromide] to an insoluble purple formazan by the action of mitochondrial reductase (Fig. 1). Formazan is then solubilized and the concentration determined by optical density at 570 nm. The result is a sensitive assay with excellent linearity up to $\sim 10^6$ cells per well. As with the alamarBlue assay, small changes in metabolic activity can generate large changes in MTT, allowing one to detect cell stress upon exposure to a toxic agent in the absence of direct cell death. The assay has been standardized for adherent or nonadherent cells grown in multiple wells. The protocol uses a standard 96-well plate. This can be scaled up, however, to suit a different plate format. Plate 500–10,000 cells per well in a 96-well plate. The assay has good linearity up to 10^6 cells.

MATERIALS

It is essential that you consult the appropriate Material Safety Data Sheets and your institution's Environmental Health and Safety Office for proper handling of equipment and hazardous materials used in this protocol.

Recipes for reagents specific to this protocol, marked <R>, are provided at the end of the protocol. See Appendix 1 for recipes for commonly used stock solutions, buffers, and reagents, marked <A>. Dilute stock solutions to the appropriate concentrations.

Reagents

Cultured mammalian cells transfected

Most mammalian cell types reduce the tetrazolium salt sufficiently to perform MTT assays accurately at low cell numbers. However, it may be necessary to increase the cell number to 1×10^5 to $5 \times 10^5/mL$ when using blood cells in order to obtain a significant 570-nm absorbance reading (Chen et al. 1990). The optimum number of cells for the assay may also be determined empirically by performing the assay with varying numbers of cells in triplicate wells.

DMSO
MTT stock solution <R>
SDS-HCl solution (10%) <R>
Tissue culture medium

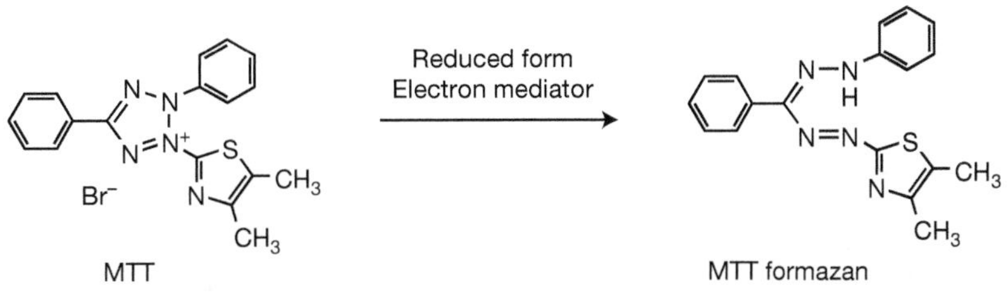

FIGURE 1. Reduction of MTT to MTT formazan.

Equipment

Plates (96 well), standard flat-bottom wells
Plates (96 well), either V or round bottom
Spectrophotometer with multiwell tissue culture plate reading ability

METHOD

Labeling Cells

1. For adherent cells, remove the medium, and replace it with 100 μL of fresh culture medium. For nonadherent cells, centrifuge the microplate, pellet the cells, carefully remove as much medium as possible, and replace it with 100 μL of fresh medium.

   Phenol Red indicator in cell culture medium can interfere with the MTT reaction. It may be necessary to grow the cells in an indicator-free medium or replace it with an indicator-free medium just before the addition of MTT reagent.

2. Add 10 μL of the 12 mM MTT stock solution to each well. Include a negative control of 10 μL of the MTT stock solution added to 100 μL of medium alone.

3. Incubate the plate for 4 h at 37°C. At high cell densities (>100,000 cells per well), the incubation time can be shortened to 2 h. Solubilize the formazan produced using either SDS-HCl or DMSO.

   Several companies market the MTT assay kit in its original format. Some kits (e.g., Promega's Cell-Titer 96 Aqueous) use MTS instead of MTT as a substrate. The MTS tetrazolium is similar to the MTT tetrazolium, with the advantage that the formazan product of MTS reduction is soluble in cell culture medium and does not require a solubilization step in the assay protocol.

Solubilizing Formazan with SDS-HCl

4. Add 100 μL of SDS-HCl solution to each well and mix thoroughly using the pipette. This dissolves the formazan produced.

5. Incubate the microplate for 4–18 h at 37°C in a humidified chamber depending on the amount of formazan product formed. Longer incubations will decrease the sensitivity of the assay.

6. Mix each sample again using a pipette, and read the absorbance at 570 nm.

Solubilizing Formazan with DMSO

A quicker and more popular solubilization reagent than SDS-HCl is DMSO.

7. Remove all but 25 μL of medium from the wells.

   For nonadherent cells, it may be necessary to first centrifuge the plates to sediment the cells.

8. Add 50 μL of DMSO to each well and mix thoroughly with a pipette.

9. Incubate the plates for 10 min at 37°C.

10. Mix each sample again, and read the absorbance at 540 nm.

DISCUSSION

To calculate the percentage of dead cells, use a positive control of 100% lysed cells. Cells are lysed by freeze–thaw and pipetting before addition of the MTT. The average of the absorbance values for the positive control (100% lysed cells) is used as a blank value and subtracted from all other absorbance values to yield the corrected absorbance values. Plot the corrected absorbance values 570 nm (on the ordinate) versus concentration of cytotoxic agent (abscissa, log scale) as shown in Figure 2, and determine the IC_{50} value by locating the abscissa value corresponding to one-half the maximum

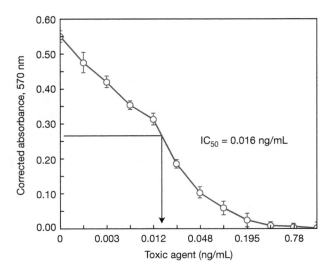

FIGURE 2. **A sample plot for determining the IC$_{50}$ of a toxic agent.** (Based on an assay for TNF-α effects on L929 cells, modified with permission from Promega Corp.)

absorbance value (IC$_{50}$ is the inhibitory concentration of cytotoxic agent necessary to kill one-half of the cell population).

RECIPES

It is essential that you consult the appropriate Material Safety Data Sheets and your institution's Environmental Health and Safety Office for proper handling of equipment and hazardous materials used in this protocol.

MTT Stock Solution (12 mM)

MTT	5 mg
PBS	1 mL

Prepare a 12 mM stock solution by dissolving 5 mg of MTT in 1 mL of PBS. This stock solution will be enough for 100 reactions (10 μL/reaction). Once prepared, the MTT solution can be stored for 4 wk at 4°C protected from light.

SDS-HCl Solution (10%)

SDS	1 g
HCl (0.01 N)	10 mL

This solution is needed for dissolving the insoluble formazan product formed at the end of the reaction. Prepare by dissolving 1 g of SDS in 10 mL of 0.01 N HCl. Mix the solution by vortexing until the SDS completely dissolves. One milliliter is sufficient for 100 reactions (10 μL/ reaction). Alternatively, the formazan product can be dissolved in dimethyl sulfoxide (DMSO).

REFERENCES

Chen C-H, Campbell PA, Newman LS. 1990. MTT colorimetric assay detects mitogen responses of spleen but not blood lymphocytes. *Int Arch Allergy Appl Immunol* **93:** 249–255.

Mossman T. 1983. Rapid colorimetric assay for cellular growth and survivals: Application to proliferation and cytotoxicity assays. *J Immunol Methods* **65:** 55–63.

OPTICAL TRANSFECTION

Light in the form of a focused laser can create small transient pores in the plasma membrane of mammalian cells through which plasmid DNA and other macromolecules can enter (Tsukakoshi et al. 1984). Based on this observation, various methods have been developed that use different forms of light to introduce macromolecules into cultured cells. The methods have been called optical transfection, optoporation, opto-injection, and laserfection; however, the term "optical transfection" is commonly used to refer to all of these techniques (for a general review, see Yao et al. 2008). Although optical transfection methods have been classified based on the nature of the laser source or the mode of macromolecule entry, a more generalized and acceptable classification is either (1) transfection with laser alone or (2) transfection by a combination of chemical and laser.

The earliest optical transfection methods used lasers as the light source. In the first report, by Tsukakoshi et al. (1984), a nanosecond-pulsed laser at 355 nm was focused to a 0.5-μm-diameter spot on the plasma membrane of normal rat kidney cells suspended in a solution containing plasmid DNA. Only cells within this area were found to have taken up plasmid DNA. Tao et al. (1987) used a 355-nm yttrium–aluminum garnet laser with light pulse energies in the range of 23–67 J (and focused the light to a spot of ~2 μm in diameter, enabling transfection efficiencies 100-fold higher than calcium-phosphate transfection efficiencies). Nanosecond lasers have been largely replaced by femtosecond lasers with an emission at 800 nm and a laser power of 50–225 mW (Tirlapur and Konig 2002; Zeira et al. 2003; Stevenson et al. 2010). These lasers, with near-infrared wavelength emissions at 800 or 1064 nm and femtosecond pulses, are able to transfect cells with reduced associated cytotoxicity (Brown et al. 2008; Yao et al. 2008).

The mechanism by which macromolecules enter into cells during optical transfection appears to be either through a small transient hole that is made in the cell membrane by the interaction of the laser beam with the membrane, or indirectly through the creation of shock waves, in which the focused laser beam interacts with an absorptive medium, frequently polyimide, rather than with the cell. A mechanical stress wave is then produced as a result of rapid heating of the absorbing medium. This shock wave interacts with nearby cells, producing temporary alterations in the plasma membrane. The exact nature by which shock waves transiently permeabilize membranes remains unknown. However, the method can be efficiently used for adherent cells.

Another optical transfection method, termed "photochemical internalization" (PCI), uses lasers to activate photosensitive chemicals that have been localized to the membranes of endocytic vesicles. Shortly following the uptake of the desired macromolecule by endocytosis, laser light is applied, which activates the photosensitizers, inducing the formation of reactive oxygen species and causing the endosomal membranes to rupture, releasing the vesicle's contents (Berg et al. 1999; Niemz 2003). An improvement on this method has been the conjugation of light-absorbing particles (such as gold nanoparticles) to cells followed by laser irradiation of the cells contained in medium with plasmid DNA (Pitsillides et al. 2003).

The maximum transfection efficiency achieved using optical transfection appears to be comparable to other transfection methods. The major hurdles to using these light-based methods, however, are the high cost of the instrumentation required and the inability to transfect large numbers of cells simultaneously. The greatest potential of optical transfection lies in its ability to target single cells, an application that other transfection procedures fail to deliver. Techniques currently available for manipulating single cells include microinjection of virions or nanoparticle capsules optically tweezed into cells (Ashkin and Dziedzic 1987; Sun and Chiu 2004). Optical transfection provides a simpler alternative to these procedures. For a review focusing on optical transfection of single cells, see Stevenson et al. (2010).

The application of optical transfection to address various biological questions at the single-cell level is now being explored. For instance, optical transfection methods were used to show that the injection of mRNA into the dendrite of a neuron resulted in a different outcome from that resulting from the injection of mRNA into the neuronal cell body (Barrett et al. 2006). Using optical transfection methods, Sul et al. (2009) showed that the entire transcriptome from one cell type can be transferred into another cell type, leading to reprogramming of the recipient cells. Major advances in overcoming the engineering challenges faced by optical transfection have led to better adaptation of the technology to answer some of biology's enduring questions. As the workbench shrinks to the size of a microchip and studies are performed at the level of a single cell, optical transfection will likely emerge as one of the major tools.

REFERENCES

Ashkin A, Dziedzic JM. 1987. Optical trapping and manipulation of viruses and bacteria. *Science* 235: 1517–1520.

Barrett LE, Sul JY, Takano H, Van Bockstaele EJ, Haydon PG, Eberwine JH. 2006. Region-directed phototransfection reveals the functional significance of a dendritically synthesized transcription factor. *Nat Methods* 3: 455–460.

Berg K, Weyergang A, Prasmickaite L, Bonsted A, Høgset A, Strand MT, Wagner E, Selbo PK. 1999. Photochemical internalization: A novel technology for delivery of macromolecules into cytosol. *Cancer Res* 59: 1180–1183.

Brown CT, Stevenson DJ, Tsampoula X, McDougall C, Lagatsky AA, Sibbett W, Gunn-Moore FJ, Dholakia K. 2008. Enhanced operation of femtosecond lasers and applications in cell transfection. *J Biophotonics* 1: 183–199.

Niemz MH. 2003. *Laser–tissue interactions: Fundamentals and applications* Springer-Verlag, Berlin.

Pitsillides CM, Joe EK, Wei X, Anderson RR, Lin CP. 2003. Selective cell targeting with light-absorbing microparticles and nanoparticles. *Biophys J* 84: 4023–4032.

Stevenson DJ, Gunn-Moore FJ, Campbell P, Dholakia K. 2010. Single cell optical transfection. *J R Soc Interface* 7: 863–871.

Sul JY, Wu CW, Zeng F, Jochems J, Lee MT, Kim TK, Peritz T, Buckley P, Cappelleri DJ, Maronski M, et al. 2009. Transcriptome transfer produces a predictable cellular phenotype. *Proc Natl Acad Sci* 106: 7624–7629.

Sun B, Chiu DT. 2004. Synthesis, loading, and application of individual nanocapsules for probing single-cell signaling. *Langmuir* 20: 4614–4620.

Tao W, Wilkinson J, Stanbridge EJ, Berns MW. 1987. Direct gene transfer into human cultured cells facilitated by laser micropuncture of the cell membrane. *Proc Natl Acad Sci* 84: 4180–4184.

Tirlapur UK, Konig K. 2002. Femtosecond near-infrared laser pulses as a versatile non-invasive tool for intra-tissue nanoprocessing in plants without compromising viability. *Plant J* 31: 365–374.

Tsukakoshi M, Kurata S, Nomiya Y, Ikawa Y, Kasuya T. 1984. A novel method of DNA transfection by laser microbeam cell surgery. *Appl Phys B* 35: 135–140.

Yao CP, Zhang ZX, Rahmanzadeh R, Huettmann G. 2008. Laser-based gene transfection and gene therapy. *IEEE Trans Nanobioscience* 7: 111–119.

Zeira E, Manevitch A, Khatchatouriants A, Pappo O, Hyam E, Darash-Yahana M, Tavor E, Honigman A, Lewis A, Galun E. 2003. Femtosecond infrared laser—An efficient and safe in vivo gene delivery system for prolonged expression. *Mol Ther* 8: 342–350.

COTRANSFORMATION

Analysis of function and expression of transfected genes may require stable integration of the transfected DNA into a host chromosome. After entering the cell, some of the transfected nucleic acid gains entry into the nucleus from the cytoplasm during the process of cell division. Depending on the cell type, up to 80% of a population of cells will then express the transfected gene in a transient fashion. At some point within the first few hours after transfection, the incoming DNA undergoes a series of nonhomologous intermolecular recombination and ligation events to form a large concatemeric structure that eventually integrates into a cellular chromosome. Each transformed cell usually contains only one of these packages, which can exceed 2 Mb in size (Perucho et al. 1980). Stable cell lines can then be isolated that carry integrated copies of the transfected DNA. Transformation rates vary widely from cell type to cell type. In the best cases, approximately one cell in 10^3 in the original transfected population stably expresses a gene(s) carried by the transfected DNA.

Because uptake, integration, and expression of DNA are relatively rare events, stable transformants are usually isolated by selection of cells that have acquired a new phenotype. Typically, this phenotype is conferred by the presence in the transfection mixture of a gene encoding antibiotic resistance. Cells transformed for a genetic marker present on one piece of DNA frequently express another genetic marker that was originally carried on a separate DNA molecule. Therefore, cells that stably express selectable (e.g., antibiotic-resistant) markers are also likely to have incorporated other DNA sequences present among the carrier DNA. This phenomenon, in which physically unlinked genes are assembled into a single integrated array and expressed in the same transformed cell, is known as "cotransformation."

The first gene to be used extensively for selection in mammalian cells was a viral (herpes simplex) gene encoding thymidine kinase (TK) (Wigler et al. 1977). Although many mammalian cell lines express thymidine kinase, several TK$^-$ lines were created by selection for growth in the presence of 5-bromodeoxyuridine (BrdU). When transfected and stably integrated into the host genome of cell lines lacking thymidine kinase, the viral gene confers the TK$^+$ phenotype, thereby allowing growth in the presence of aminopterin (for a discussion of the basis for selection, see the information panel Selective Agents for Stable Transformation). Thereafter, this strategy was used to introduce foreign DNA into mammalian cells by cotransfection with a plasmid encoding the *tk* gene (Perucho et al. 1980; Robins et al. 1981). The difficulties or additional efforts involved in creating *tk$^-$* mutants promoted the search for other selection schemes. Other possibilities for selection in cotransformation studies were explored, leading to the development of vectors that express bacterial proteins that confer drug resistance in mammalian cell lines. These selectable markers include, for example, aminoglycoside phosphotransferase (resistance to G418 or neomycin), hygromycin B phosphotransferase (resistance to hygromycin B), xanthine-guanine phosphoribosyltransferase (resistance to mycophenolic acid and aminopterin), puromycin-*N*-acetyl transferase (resistance to puromycin), and blasticidin deaminase (resistance to Blasticidin S). All have been used with considerable success to establish stably transformed lines of mammalian cells.

In addition to providing a means to introduce exogenous genes into mammalian cells in a stable manner, it is often desirable to increase the stringency of the selective conditions in order to obtain higher levels of expression of the transfected genes. This enhancement in expression can be achieved as a result of increase in copy number, or coamplification, of the target gene and the gene conferring the resistance. A target gene that has been cotransfected with (and become integrated near) a particular marker gene is highly likely to undergo amplification with the marker under selection. For example, amplification of the dihydrofolate reductase (*dhfr*) gene resulting from exposure to increasing levels of methotrexate has been used successfully to overexpress cotransfected foreign genes (Schimke 1984). Similarly, the gene encoding adenosine deaminase (ADA) can be amplified through the stepwise increase in concentrations of 2′-deoxycoformycin (dCF) (Kaufman et al. 1986; for further details on gene amplification, see Stark and Wahl 1984). For further details on the basis for selection, as well as, selective conditions required for these systems, see the information panel Selective Agents for Stable Transformation.

Optimizing the Selection of Stable Transfectants

The selectable markers required to generate mammalian cell lines stably expressing the gene of interest could be located either on the gene expression plasmid or on a separate plasmid that is cotransfected with the gene expression plasmid. When present on the same plasmid, there is a greater likelihood of producing drug-resistant stable transfectants that express the gene of interest. However, the cotransfection approach is also an effective method. When using cotransfection, it is advisable to transfect a plasmid mixture containing five to 10 parts gene expression plasmid and one part selectable marker plasmid. These ratios help ensure that the stably transfected cells will express both the gene of interest and the selectable marker.

REFERENCES

Kaufman RJ, Murtha P, Ingolia DE, Yeung CY, Kellems RE. 1986. Selection and amplification of heterologous genes encoding adenosine deaminase in mammalian cells. *Proc Natl Acad Sci* **83**: 3136–3140.

Perucho M, Hanahan D, Wigler M. 1980. Genetic and physical linkage of exogenous sequences in transformed cells. *Cell* **22**: 309–317.

Robins DM, Ripley S, Henderson AS, Axel R. 1981. Transforming DNA integrates into the host cell chromosome. *Cell* **23**: 29–39.

Schimke RT. 1984. Gene amplification in cultured animal cells. *Cell* **37**: 705–713.

Stark G, Wahl G. 1984. Gene amplification. *Annu Rev Biochem* **53**: 447–491.

Wigler M, Silverstein S, Lee LS, Pellicer A, Cheng VC, Axel R. 1977. Transfer of purified herpes virus thymidine kinase gene to cultured mouse cells. *Cell* **11**: 223–232.

SELECTIVE AGENTS FOR STABLE TRANSFORMATION

Resistance to antibiotics has proven to be effective in selecting cotransformants and, in some cases, as a driver for gene amplification.

Aminopterin

Mode of Action

Thymidine kinase catalyzes a reaction in an alternative pathway for the synthesis of dTTP from thymidine. The enzyme is not required under normal conditions of growth because cells typically synthesize dTTP from dCDP. However, cells grown in the presence of aminopterin (an analog of dihydrofolate) are unable to use the usual pathway for synthesizing dTTP and thus require thymidine kinase to make use of the alternative pathway.

Selective Conditions

Cell lines lacking endogenous thymidine kinase activity are grown in a complete medium supplemented with 100 μM hypoxanthine, 0.4 μM aminopterin, 16 μM thymidine, and 3 μM glycine (HAT medium).

G418

Mode of Action

This aminoglycoside antibiotic, similar in structure to neomycin, gentamycin, and kanamycin, is the most commonly used selective agent in permanent transfection experiments. G418 and its relatives block protein synthesis through interference with ribosomal functions. The bacterial enzyme aminoglycoside phosphotransferase, carried on the transposon sequence Tn5, converts G418 to a nontoxic form.

Selective Conditions

Because each eukaryotic cell line shows a different sensitivity to this antibiotic (and some are completely resistant to it), the optimum amount required to kill nontransfected cells must be established empirically for each new cell line or strain used for permanent transfection. This optimum is established by determining a killing curve for the cell line of interest. In this type of experiment, a plasmid conferring resistance to G418 (e.g., pSV2neo or pSV3neo) (Southern and Berg 1982) is transfected into the cells, and plates of transfected cells are subjected to different concentrations of G418. After a 2–3-wk selection period, the concentration of G418 giving rise to the largest number of viable colonies is determined by visual inspection or, better, by counting the colonies after staining with Giemsa or Gentian violet.

Commercial preparations of G418 vary in their concentration of active antibiotic, with the average purity being ~50%. For this reason, each batch of G418 should be titrated before use in tissue culture. Despite this variation, the amount of G418 used to obtain optimal numbers of transfected colonies is constant for well-characterized cell lines. Table 1 gives the optimum G418 concentration ranges for use with several commonly used cell lines.

TABLE 1. Selective G418 concentration ranges

Cell line or organisms	G418 concentration (μg/mL)
Chinese hamster ovary cells	700–800
Madin–Darby canine kidney cells	500
Human epidermoid A431 cells	400
Simian CV-1 cells	500
Dictyostelium	10–35
Plant	10
Yeast	125–500

Hygromycin B

Mode of Action

Hygromycin B is an aminocyclitol antibiotic produced by *Streptomyces hygroscopicus* (Pittenger et al. 1953). Hygromycin B inhibits protein synthesis in both prokaryotes and eukaryotes by interfering with translocation (Cabañas et al. 1978; González et al. 1978) and causing mistranslation in vivo and in vitro (Singh et al. 1979).

A bacterial plasmid-borne gene encoding a 341-amino-acid hygromycin B phosphotransferase (Rao et al. 1983) that inactivates the antibiotic has been identified and sequenced (Gritz and Davies 1983). This gene has been used as a selectable marker in *E. coli*, and chimeric genes constructed with the appropriate promoters act as dominant selectable markers in *Saccharomyces cerevisiae* (Gritz and Davies 1983; Kaster et al. 1984), mammalian cells (Santerre et al. 1984; Sugden et al. 1985), and plants (van den Elzen 1985; Waldron et al. 1985).

Selective Conditions

The concentrations of antibiotic required to inhibit growth of various organisms are presented in Table 2.

Methotrexate (MTX)

Mode of Action

An analog of dihydrofolate, methotrexate is a powerful inhibitor of dihydrofolate reductase (DHFR), an enzyme required for purine biosynthesis. Increasing levels of methotrexate can result in amplification of the gene encoding DHFR with concomitant increase in its levels of expression. The system is therefore extremely effective for amplification of cotransfected genes (Simonsen and Levinson 1983).

Selective Conditions

The medium is typically supplemented with 0.01–300 μM methotrexate.

Mycophenolic Acid

Mode of Action

Mycophenolic acid, a weak dibasic acid with antibiotic properties, specifically inhibits inosinate (IMP) dehydrogenase, an enzyme of mammalian cells that converts IMP to xanthine monophosphate (XMP). This block in the biosynthetic pathway of guanosine monophosphate (GMP) can be relieved by supplying cells with xanthine and a functional *E. coli gpt* gene, which encodes an enzyme, xanthine–guanine phosphoribosyltransferase, that converts xanthine to XMP. *E. coli gpt* can therefore be used in the presence of mycophenolic acid as a dominant selectable marker for cotransformation of mammalian cells of any type (Mulligan and Berg 1981a,b). The selection can be made more efficient by the addition of aminopterin, which blocks the endogenous pathway of purine biosynthesis (for further details, see Gorman et al. 1983).

Selective Conditions

The concentration of antibiotic required to inhibit growth in mammalian cells is ~25 μg/mL.

TABLE 2. Selective concentrations of Hygromycin B

Organism	Inhibitory concentration of Hygromycin B	Reference(s)
Escherichia coli	200 μg/mL	Gritz and Davies 1983
Saccharomyces cerevisiae	200 μg/mL	Gritz and Davies 1983
Mammalian cell lines	12–400 μg/mL (depending on the cell line)	Sugden et al. 1985; Palmer et al. 1987

TABLE 3. Selective concentrations of Blasticidin S

Cell line	Concentration (µg/mL)	Reference(s)
HeLa (human cervical cancer)	10–20	Kanada et al. 2008; Cheng et al. 2008
HEK293 (human embryonic kidney)	5–10	Oka et al. 2008; Jiang et al. 2008
Hep G2 (human hepatocellular carcinoma)	5	Maxson et al. 2009
MCF-7 (human breast cancer)	5	Denger et al. 2008
A549 (lung cancer)	10	Huang et al. 2008

Puromycin

Mode of Action

Puromycin, acting as an analog of aminoacyl tRNA, inhibits protein synthesis by causing premature chain termination. The antibiotic becomes acetylated, and thereby inactivated, by the action of puromycin-N-actyl transferase (de la Luna et al. 1988).

Selective Conditions

The concentrations of antibiotic required to inhibit growth of mammalian cell lines is typically in the range of 0.5–10 µg/mL; many transformed cell lines are effectively selected at 2 µg/mL.

Blasticidin S

Mode of Action

Blasticidin S is a peptidyl nucleoside that inhibits protein synthesis in both prokaryotes and eukaryotes by blocking the release of peptides from the ribosome following aminoacyl-tRNA formation and, thus, inhibiting peptide-bond formation (Izumi et al. 1991; Kimura et al. 1994). Blasticidin S is used to select transfected cells carrying the resistance genes *bsr* from *Bacillus cereus* or *BSD* from *Aspergillus terreus*, which encode blasticidin deaminase. The differences in CpG dinucleotide representation in bacterial versus mammalian species sometimes results in the silencing of *bsr* gene expression through methylation of the DNA. To avoid this, a functional reduced-CpG (14 to 4) *bsr* gene is incorporated in eukaryotic expression vectors. This synthetic gene also uses codon optimization to incorporate nucleotides encoding codons preferentially used in mammalian cells to ensure high-level expression.

Selective Conditions

The working concentration of Blasticidin S for mammalian cell lines varies from 3 to 50 µg/mL, and the optimal concentration has to be titered for each cell line. After treatment, cell death occurs rapidly, allowing selection of transfected cells in as little as 7 d posttransfection. Table 3 lists suggested working conditions for selection with some commonly used mammalian cells.

REFERENCES

Cabañas MJ, Vázquez D, Modolell J. 1978. Dual interference of hygromycin B with ribosomal translocation and with aminoacyl-tRNA recognition. *Eur J Biochem* 87: 21–27.

Cheng N, He R, Tian J, Ye PP, Ye RD. 2008. Cutting edge: TLR2 is a functional receptor for acute-phase serum amyloid A. *J Immunol* 181: 22–26.

de la Luna S, Soria I, Pulido D, Ortin J, Jiminez A. 1988. Efficient transformation of mammalian cells with constructs containing puromycin-resistance marker. *Gene* 62: 121–126.

Denger S, Bähr-Ivacevic T, Brand H, Reid G, Blake J, Seifert M, Lin CY, May K, Benes V, Liu ET, Gannon F. 2008. Transcriptome profiling of estrogen-regulated genes in human primary osteoblasts reveals an osteoblast-specific regulation of the insulin-like growth factor binding protein 4 gene. *Mol Endocrinol* 22: 361–379.

González A, Jiménez A, Vázquez D, Davies JE, Schindler D. 1978. Studies on the mode of action of hygromycin B, an inhibitor of translocation in eukaryotes. *Biochim Biophys Acta* 521: 459–469.

Gorman C, Padmanabhan R, Howard BH. 1983. High efficiency DNA-mediated transformation of primate cells. *Science* 221: 551–553.

Gritz L, Davies J. 1983. Plasmid-encoded hygromycin-B resistance: The sequence of hygromycin-B-phosphotransferase and its expression in *E. coli* and *S. cerevisiae*. *Gene* 25: 179–188.

Huang G, Eisenberg R, Yan M, Monti S, Lawrence E, Fu P, Walbroehl J, Löwenberg E, Golub T, Merchan J, et al., 2008. 15-Hydroxy-prostaglandin dehydrogenase is a target of hepatocyte nuclear factor 3β and a tumor suppressor in lung cancer. *Cancer Res* 68: 5040–5048.

Izumi M, Miyazawa H, Kamakura T, Yamaguchi I, Endo T, Hanaoka F. 1991. Blasticidin S-resistance gene (*bsr*): A novel selectable marker for mammalian cells. *Exp Cell Res* **197**: 229–233.

Jiang D, Guo H, Xu C, Chang J, Gu B, Wang L, Block TM, Guo JT. 2008. Identification of three interferon-inducible cellular enzymes that inhibit the replication of hepatitis C virus. *J Virol* **82**: 1665–1678.

Kanada M, Nagasaki A, Uyeda TQ. 2008. Novel functions of Ect2 in polar lamellipodia formation and polarity maintenance during "contractile ring-independent" cytokinesis in adherent cells. *Mol Biol Cell* **19**: 8–16.

Kaster KR, Burgett S, Ingolia TD. 1984. Hygromycin B resistance as dominant selectable marker in yeast. *Curr Genet* **8**: 353–358.

Kimura M, Takatsuki A, Yamaguchi I. 1994. Blasticidin S deaminase gene from *Aspergillus terreus* (BSD): A new drug resistance gene for transfection of mammalian cells. *Biochim Biophys Acta* **1219**: 653–659.

Maxson JE, Enns CA, Zhang AS. 2009. Processing of hemojuvelin requires retrograde trafficking to the Golgi in HepG2 cells. *Blood* **113**: 1786–1793.

Mulligan RC, Berg P. 1981a. Selection for animal cells that express the *E. coli* gene coding for xanthine-guanine phosphoribosyltransferase. *Proc Natl Acad Sci* **78**: 2072–2076.

Mulligan RC, Berg P. 1981b. Factors governing the expression of a bacterial gene in mammalian cells. *Mol Cell Biol* **1**: 449–459.

Oka T, Mazack V, Sudol M. 2008. Mst2 and Lats kinases regulate apoptotic function of Yes kinase-associated protein (YAP). *J Biol Chem* **283**: 27534–27546.

Palmer TD, Hock RA, Osborne WRA, Miller AD. 1987. Efficient retrovirus-mediated transfer and expression of a human adenosine deaminase gene in diploid skin fibroblasts from an adenine-deficient human. *Proc Natl Acad Sci* **84**: 1055–1059.

Pittenger RC, Wolfe RN, Hoehn MM, Marks PN, Dailey WA, McGuire JM. 1953. Hygromycin. I. Preliminary studies on the production and biological activity of a new antibiotic. *Antibiot Chemotherl* **3**: 1268–1278.

Rao R, Allen N, Hobbs J, Alborn W, Kirst H, Paschal J. 1983. Genetic and enzymatic basis of hygromycin B resistance in *Escherichia coli*. *Antimicrob Agents Chemother* **24**: 689–695.

Santerre R, Allen N, Hobbs J, Rao R, Schmidt R. 1984. Expression of prokaryotic genes for hygromycin B and G418 resistance as dominant-selection markers in mouse. *Gene* **30**: 147–156.

Simonsen CC, Levinson AD. 1983. Isolation and expression of an altered mouse dihydrofolate reductase cDNA. *Proc Natl Acad Sci* **80**: 2495–2499.

Singh A, Ursic D, Davies J. 1979. Phenotypic suppression and misreading in *S. cerevisiae*. *Nature* **277**: 146–148.

Southern PJ, Berg P. 1982. Transformation of mammalian cells to antibiotic resistance with a bacterial gene under control of the SV40 early region promoter. *J Mol Appl Genet* **1**: 327–341.

Sugden B, Marsh K, Yates J. 1985. A vector that replicates as a plasmid and can be efficiently selected in B-lymphoblasts transformed by Epstein-Barr virus. *Mol Cell Biol* **5**: 410–413.

van den Elzen P, Townsend J, Lee K, Bedbrook J. 1985. A chimaeric hygromycin resistance gene as a selectable marker in plant cells. *Plant Mol Biol* **5**: 299–302.

Waldron C, Murphy EB, Roberts JL, Gustafson GD, Armour SL, Malcolm SK. 1985. Resistance to hygromycin B: A new market for plant transformation studies. *Plant Mol Biol* **5**: 103–108.

LIPOFECTION

Lipofection is the generic name of a set of techniques used to introduce exogenous DNAs into cultured mammalian cells. Many variants of the basic method have been developed, but they all adhere to the same general principle: The DNA to be transfected is coated by a lipid, which either interacts directly with the plasma membrane of the cell (Bangham 1992) or is taken into the cell by non-receptor-mediated endocytosis (Zhou and Huang 1994; Zabner et al. 1995), presumably as a prelude to membrane fusion in endosomes (Pinnaduwage et al. 1989; Leventis and Silvius 1990; Rose et al. 1991). However, as with other transfection techniques, only a small percentage of liposomes deliver their cargo of DNA into the nucleus (Tseng et al. 1997). As judged from microscopy, most of the DNA remains associated with the membrane compartments of the cell, where it is unavailable for transport into the cytoplasm and subsequent movement to the nucleus, which is necessary for transcription and expression of cloned genes (Zabner et al. 1995). Nevertheless, when working at its best, lipofection can deliver DNA into cells more efficiently than precipitation with polycations such as calcium phosphate and at lower cost than electroporation.

Like other transfection techniques, lipofection is not universally successful: The efficiency of both transient expression and stable transformation by exogenously added genes varies widely from cell line to cell line. Different types of cells may show a range of quantitative responses to the same lipofection protocol. Different protocols used on the same cell line may generate results that span an extensive range. However, lipofection works very well in many situations in which standard methods are notoriously inefficient, for example, transfection of primary cultures or cultures of differentiated cells (e.g., see Thompson et al. 1999) or introduction of very high-molecular-weight DNA into standard cell lines (e.g., see Strauss 1996). Lipofection is therefore the method of choice for introducing genes into differentiated cells in vitro and is the technique of first resort when older methods of transfection are inadequate.

The Chemistry of Lipofection

There are two general classes of liposomal transfection reagents: those that are anionic and those that are cationic. Transfection with anionic liposomes, which was first used in the late 1970s to deliver DNA and RNA to cells in a biologically active form, requires that the DNA be trapped in the internal aqueous space of large artificial lamellar liposomes (for reviews of this early work, see Fraley and Papahadjopoulos 1981, 1982; Fraley et al. 1981; Straubinger and Papahadjopoulos 1983). However, the technique in its basic form never entered widespread use, perhaps because of its time-consuming nature and problems with reproducibility by investigators who were not expert in lipid chemistry.

The lipofection techniques in common use today stem from a seminal discovery by Peter Felgner that cationic lipids react spontaneously with DNA to form a unilamellar shell that can fuse with cell membranes (Felgner et al. 1987; Felgner and Ringold 1989). The formation of DNA–lipid complexes is due to ionic interactions between the head group of the lipid, which carries a strongly positive charge, and the negatively charged phosphate groups on the DNA with a concomitant neutralization of charge (see Fig. 3 in Protocol 1, Alternative Protocol Transfection Using DOTMA and DOGS).

The first-generation cationic lipids were monocationic double-chain amphiphiles with a positively charged quaternary amino head group (Duzgunes et al. 1989), linked to the lipid backbone by ether or ester linkages. Such monocationic lipids suffer from two major problems: They are toxic to many types of mammalian and insect cells, and their ability to promote transfection is restricted to a small range of cell lines (Felgner et al. 1987; Felgner and Ringold 1989). The later generations of cationic lipids are polycationic, have a far wider host range, and are considerably less toxic than their predecessors (for review, see Gao and Huang 1993). In most cases, preparations of cationic lipids used for transfection consist of a mixture of synthetic cationic lipid and a fusogenic lipid (phosphatidylethanolamine or DOPE). Depending on the composition of the lipid mixture, the DNA to be transfected becomes incorporated either into multilamellar structures composed of alternating layers of lipid bilayer and hydrated DNA or into hexagonal columns arranged in a honeycomb structure (Labat-Moleur et al. 1996; Koltover et al. 1998). Each column or tube in the honeycomb consists of a central core of hydrated DNA molecules and a surrounding hexagonal shell of lipid monolayers. Experiments with model systems suggest that honeycomb arrangements of this type deliver DNA across lipid bilayers more efficiently than multilamellar structures do.

A Plethora of Lipofectants Is Available

The same properties of lipids that facilitate the formation of transfection-competent structures with DNA also bring unwanted side effects. Chief among these are a generalized toxicity, which is manifested by cells rounding up and detaching from the dish. In addition, lipofection is vulnerable to interference by fats and lipoproteins in serum and by charged components of the extracellular matrix such as chondroitin sulfate (Felgner and Holm 1989). Systematic modifications of the cationic and neutral lipids have been made in an effort to overcome these drawbacks (e.g., see Behr et al. 1989; Felgner et al. 1994), resulting in a wealth of effective lipofection reagents. Unfortunately, these reagents, many of which are commercially available, work with varying efficiencies with different types of cells. Although few head-to-head comparisons of efficiency are available, the companies that market these reagents for lipofection provide useful bibliographies and lists of cell lines that can be efficiently transfected with the help of their particular products. The toxicity of these compounds varies from cell line to cell line, as does inter alia, the optimal ratio of cationic lipids:DNA and the amount of cationic lipid that can be added to a given number of cells (e.g., see Felgner et al. 1987; Ho et al. 1991; Ponder et al. 1991; Farhood et al. 1992; Harrison et al. 1995).

Optimizing Lipofection

In addition to the properties and chemical composition of the cationic and neutral lipids, several other variables affect the efficiency of lipofection, including the following.

- *Cell physiology and initial density of the cell culture.* Cell cultures that need to be transfected should not be allowed to remain confluent for more than 24 h. Because cell cultures evolve over time in the laboratory, transfection efficiencies could be compromised because of changes in cell behavior. Restarting the culture from frozen stocks can help recover transfection activity. Cell monolayers should be in mid-log phase and should be between 40% and 75% confluent during transfection.

- *Medium and serum used to grow the cells.* The increase in cell permeability caused by cationic lipid reagents results in higher concentrations of antibiotic delivered inside the cells. This leads to increased cytotoxicity and lower transfection efficiency. Therefore, use of antibiotics in transfection medium is not advisable. Serum can be present during transfection as long as the DNA–cationic lipid reagent complexes are formulated in the absence of serum. Some serum proteins interfere with complex formation.

- *Purity of the DNA preparation.* The 260:280-nm ratio of the DNA preparation should be 1.8. With some lipids, it may be advisable to dissolve the DNA in water rather than buffers containing EDTA. Plasmid preparations used for lipofection should be free of bacterial lipopolysaccharides and should preferably be purified by chromatography on anion exchange resins or by CsCl–ethidium bromide equilibrium density centrifugation.

- *Amount of DNA added.* Depending on the concentration of the sequences of interest, as little as 50 ng and as much as 40 μg of DNA might be required to obtain maximum signal from a reporter gene.

- *Time of exposure of cells to the cationic lipid–DNA complex.* This varies from 0.1 to 24 h.

All of these variables must be optimized in order to establish optimum transfection frequencies for a target cell line.

REFERENCES

Bangham AD. 1992. Lipsomes: Realizing their promise. *Hosp Pract* 27: 51–62.

Behr J-P, Demeneix B, Loeffler J-P, Perez-Mutul J. 1989. Efficient gene transfer into mammalian primary endocrine cells with lipopolyamine-coated DNA. *Proc Natl Acad Sci* 86: 6982–6986.

Duzgunes N, Goldstein JA, Friend DS, Felgner PL. 1989. Fusion of liposomes containing a novel cationic lipid, N-[2,3-(dioleyloxy)propyl]-N,N,N-trimethylammonium: Induction by multivalent anions and asymmetric fusion with acidic phospholipid vesicles. *Biochemistry* 28: 9179–9184.

Farhood H, Bottega R, Epand RM, Huang L. 1992. Effect of cationic cholesterol derivatives on gene transfer and protein kinase C activity. *Biochim Biophys Acta* 1111: 239–246.

Felgner PL, Holm M. 1989. Cationic liposome-mediated transfection. *Focus* 11: 21–25.

Felgner PL, Ringold G. 1989. Cationic liposome-mediated transfection. *Nature* 337: 387–388.

Felgner PL, Gadek TR, Holm M, Roman R, Chan HW, Wenz M, Northrop JP, Ringold GM, Danielsen M. 1987. Lipofection: A highly efficient, lipid-mediated DNA-transfection procedure. *Proc Natl Acad Sci* 84: 7413–7417.

Felgner JH, Kumar R, Sridhar CN, Wheeler CJ, Tsai YJ, Border R, Ramsey P, Martin M, Felgner PL. 1994. Enhanced gene delivery and mechanism studies with a novel series of cationic lipid formulations. *J Biol Chem* 269: 2550–2561.

Fraley R, Papahadjopoulos D. 1981. New generation liposomes: The engineering of an efficient vehicle for intracellular delivery of nucleic acids. *Trends Biochem Sci* 6: 77–80.

Fraley R, Papahadjopoulos D. 1982. Liposomes: The development of a new carrier system for introducing nucleic acid into plant and animal cells. *Curr Top Microbiol* 96: 171–191.

Fraley R, Straubinger RM, Rule G, Springer EL, Papahadjopoulos D. 1981. Liposome-mediated delivery of deoxyribonucleic acid to cells: Enhanced efficiency of delivery related to lipid composition and incubation conditions. *Biochemistry* 20: 6978–6987.

Gao X, Huang L. 1993. Cationic liposomes and polymers for gene transfer. *Liposome Res* 3: 17–30.

Harrison GS, Wang Y, Tomczak J, Hogan C, Shpall EJ, Curiel TJ, Felgner PL. 1995. Optimization of gene transfer using catioinic lipids in cell lines and primary human CD4+ and CD34+ hematopoietic cells. *BioTechniques* 19: 816–823.

Ho W-Z, Gonczol E, Srinivasan A, Douglas SD, Plotkin SA. 1991. Minitransfection: A simple, fast technique for transfections. *J Virol Methods* 32: 79–88.

Koltover I, Salditt T, Radler JO, Safinya CR. 1998. An inverted hexagonal phase of cationic liposome–DNA complexes related to DNA release and delivery. *Science* 281: 78–81.

Labat-Moleur F, Steffan A-M, Brisson C, Perron H, Feugeas O, Furstenberger P, Oberling F, Brambilla E, Behr J-P. 1996. An electron microscopy study into the mechanism of gene transfer with lipopolyamines. *Gene Ther* 3: 1010–1017.

Leventis R, Silvius JR. 1990. Interactions of mammalian cells with lipid dispersions containing novel metabolizable cationic amphiphiles. *Biochim Biophys Acta* 1023: 124–132.

Pinnaduwage P, Schmitt L, Huang L. 1989. Use of a quaternary ammonium detergent in liposome mediated DNA transfection of mouse L-cells. *Biochim Biophys Acta* 985: 33–37.

Ponder KP, Dunbar RP, Wilson DR, Darlington GJ, Woo SLC. 1991. Evaluation of relative promoter strength in primary hepatocytes using optimized lipofection. *Hum Gene Ther* 2: 41–52.

Rose JK, Buonocore L, Whitt MA. 1991. A new cationic liposome reagent mediating nearly quantitative transfection of animal cells. *BioTechniques* 10: 520–525.

Straubinger RM, Papahadjopoulos D. 1983. Liposomes as carriers for intracellular delivery of nucleic acids. *Methods Enzymol* 101: 512–527.

Strauss WM. 1996. Transfection of mammalian cells by lipofection. *Methods Mol Biol* 54: 307–327.

Thompson CD, Frazier-Jessen MR, Rawat R, Nordan RP, Brown RT. 1999. Evaluation of methods for transient transfection of a murine macrophage cell line, RAW 264.7. *BioTechniques* 27: 824–832.

Tseng W-C, Haselton FR, Giorgio TD. 1997. Transfection by cationic liposomes using simultaneous single cell measurements of plasmid delivery and transgene expression. *J Biol Chem* 272: 25641–25647.

Zabner J, Fasbender AJ, Moninger T, Poellinger KA, Welsh MJ. 1995. Cellular and molecular barriers to gene transfer by a cationic lipid. *J Biol Chem* 270: 18997–19007.

Zhou X, Huang L. 1994. DNA transfection mediated by cationic liposomes containing lipopolylysine: Characterization and mechanism of action. *Biochim Biophys Acta* 1189: 195–203.

LINEARIZING PLASMIDS BEFORE TRANSFECTION

A stable cell line is generated when the transfected plasmid undergoes integration into a chromosome by nonhomologous recombination. The recombination site is random and could be within any region of the plasmid, including the gene expression or selectable marker cassettes. Recombination within any of these critical regions can be detrimental. To increase the frequency of recombination in the nonessential region, the plasmid can be linearized using a restriction enzyme that maps within a nonessential region, such as the bacterial replicon or bacterial marker gene. Linearization creates DNA ends that have a higher frequency of recombination over internal site(s), thus promoting integration in the nonessential plasmid regions.

TRANSFECTION OF MAMMALIAN CELLS WITH CALCIUM PHOSPHATE–DNA COPRECIPITATES

DNA can be introduced into many lines of cultured mammalian cells as a coprecipitate with calcium phosphate. After entering the cell by endocytosis, some of the coprecipitate escapes from endosomes or lysosomes and enters the cytoplasm, from where it is transferred to the nucleus. This method works best in cell lines that are (1) highly transformed and (2) adherent, like HeLa, NIH-3T3, CV-1, 293T, and CHO, in which transient transfection efficiencies between 50% and 100% are obtained depending on the cell line. Transformed cell lines that carry integrated copies of the transfected DNA can also be selected, although at a much lower frequency. Transformation rates vary widely from cell type to cell type. In the best cases, about one cell in 10^3 permanently expresses a selectable marker(s) carried by the transfected DNA.

Calcium-phosphate-mediated DNA transfection was developed by Frank Graham and Alex van der Eb (1973) as a method to introduce adenovirus and SV40 DNA into adherent cultured cells. Graham and van der Eb worked out optimal conditions for the formation of calcium phosphate–DNA coprecipitates and for subsequent exposure of cells to the coprecipitate. Their work laid the foundation for the biochemical transformation of genetically marked mouse cells by cloned DNAs (Maitland and McDougall 1977; Wigler et al. 1977); for the transient expression of cloned genes in a variety of mammalian cells (e.g., see Gorman 1985); and for the isolation and identification of cellular oncogenes, tumor-suppressing genes, and other single-copy mammalian genes (e.g., see Wigler et al. 1978; Perucho and Wigler 1981; Weinberg 1985; Friend et al. 1988). However, Graham and van der Eb never profited financially from their discovery. That was left to Wigler, Axel, and their colleagues, who in 1983 were awarded a lucrative patent for cotransformation of unlinked segments of DNA by the calcium phosphate method (see the information panel Cotransformation).

Published procedures differ widely in the manner in which calcium phosphate–DNA coprecipitates are formed before their addition to cells. Some methods advise against anything but the gentlest agitation and suggest, for example, that air bubbled gently from an electric pipetting device should be used to mix the DNA and the buffered solution of calcium phosphate. Other methods advocate slow mixing during addition of the DNA solution, followed by gentle vortexing. Whatever technique is chosen, the aim should be to avoid creation of coarse precipitates that are endocytosed and processed inefficiently by cells. In addition to the speed of mixing, the following are other factors that affect the efficiency of transfection.

- *Size and concentration of the DNA.* The inclusion of high-molecular-weight genomic DNA in the coprecipitate increases the efficiency of transformation by small DNAs (e.g., plasmids) (e.g., see Chen and Okayama 1987). Soon after transfection, the small DNAs integrate into the carrier DNA, often forming an array of head-to-tail tandems. This assemblage subsequently integrates into the chromosome of the transfected cell (Perucho and Wigler 1981).

- *Exact pH of the buffer and the concentration of calcium and phosphate ions* (Jordan et al. 1996). Some investigators make up several batches of HEPES-buffered saline over the pH range 6.90–7.15 and test each batch for the quality of the calcium phosphate–DNA precipitates and for the efficiency of transformation.

- *Use of facilitators.* Increases in the efficiency of transient expression and transformation can be achieved by exposing cells to glycerol (Parker and Stark 1979), chloroquine (Luthman and Magnusson 1983), commercially available "transfection maximizers" (e.g., see Zhang and Kain 1996), or certain inhibitors of cysteine proteases (Coonrod et al. 1997). In general, these agents are toxic to cells, and their effects on viability and transfection efficiently vary from one type of cell to another. For example, chloroquine, an amine that prevents acidification of endosomes and lysosomes and inhibits lysosomal protease cathepsin B (Wibo and Poole 1974), improves the transfection efficiency of some types of cells and decreases the efficiency of others (Chang 1994). The optimal time, length, and intensity of treatment with facilitators must therefore be determined empirically for each cell line.

The level of transient expression is determined chiefly by the intensity of transcription from the promoter and its associated *cis*-acting control elements. In specific cases, it may be possible to increase the level of expression by exposing the transfected cells to hormones, heavy metals, or other substances that activate the appropriate cellular transcription factors. In addition, expression of genes carried on plasmids that contain the SV40 enhancer can be enhanced by treating transfected simian and human cells with sodium butyrate (Gorman et al. 1983a,b). Transfection kits, which frequently include these and other modifications to the original protocol, are available from several companies (see Table 2 in the chapter introduction).

DNA transfected as a calcium phosphate coprecipitate or with DEAE-dextran as a facilitator is mutated at a high frequency (~1% per gene) in all mammalian cells examined (Calos et al. 1983; Lebkowski et al. 1984). This effect is confined to the transfected sequences and does not affect the chromosomal DNA of the host cell (Razzaque et al. 1983). The mutations, which are predominantly base substitutions and deletions, appear to occur shortly after the transfecting DNA arrives in the nucleus (Lebkowski et al. 1984). However, replication of the incoming DNA is not necessary. Because almost all of the base substitutions occur at G:C base pairs, it seems likely that the major premutational events are hydrolysis of the sugar base glycosyl bond of deoxyguanosine residues and deamination of cytosine residues. Both of these reactions occur readily at acid pH and would take place as the incoming DNA passes through endosomes, which maintain a pH of ~5 (de Duve et al. 1974). Linear DNA is especially prone to deletions (Razzaque et al. 1983; Miller et al. 1984), presumably because it serves as an attractive substrate for exonucleases. Although these mutation rates are extraordinarily high, they have little relevance to transient expression of transfected genes unless the gene of interest is large and/or has a very high content of G+C. Most of the work on mutation rates was performed with LacI, which is encoded by a 750-bp segment of DNA. A gene that is 10 kb in length might therefore be expected to suffer a mutation rate of 12% or more depending on its content of G+C.

REFERENCES

Calos M, Lebkowski JS, Botchan MR. 1983. High mutation frequency in DNA transfected into mammalian cells. *Proc Natl Acad Sci* **80**: 3015–3019.

Chang PL. 1994. Calcium phosphate-mediated DNA transfection. In *Gene therapeutics: Methods and applications of direct gene transfer* (ed Wolfe JA, Crow JF), pp. 157–179. Birkhäuser, Boston.

Chen C, Okayama H. 1987. High-efficiency transformation of mammalian cells by plasmid DNA. *Mol Cell Biol* **7**: 2745–2752.

Coonrod A, Li F-Q, Horwitz M. 1997. On the mechanism of DNA transfection: Efficient gene transfer without viruses. *Gene Ther* **4**: 1313–1321.

de Duve C, de Barsy T, Poole B, Tronet A, Tulkens P, Van Hoof E. 1974. Lysosomotropic agents. *Biochem Pharmacol* **23**: 2495–2531.

Friend SH, Dryja TP, Weinberg RA. 1988. Oncogenes and tumor-suppressing genes. *N Engl J Med* **318**: 618–622.

Gorman C. 1985. High efficiency gene transfer into mammalian cells. In *DNA cloning: A practical approach* (ed Glover D), Vol. 2, pp. 143–190. IRL, Oxford.

Gorman CM, Howard BH, Reeves R. 1983a. Expression of recombinant plasmids in mammalian cells is enhanced by sodium butyrate. *Nucleic Acids Res* **11**: 7631–7648.

Gorman C, Padmanabhan R, Howard BH. 1983b. High efficiency DNA-mediated transformation of primate cells. *Science* **221**: 551–553.

Graham FL, van der Eb AJ. 1973. A new technique for the assay of infectivity of human adenovirus 5 DNA. *Virology* **52**: 456–467.

Jordan M, Schallhorn A, Wurm FW. 1996. Transfecting mammalian cells: Optimization of critical parameters affecting calcium-phosphate precipitate formation. *Nucleic Acids Res* **24**: 596–601.

Lebkowski JS, DuBridge RB, Antell EA, Greisen KS, Calos MP. 1984. Transfected DNA is mutated in monkey, mouse, and human cells. *Mol Cell Biol* **4**: 1951–1960.

Luthman H, Magnusson G. 1983. High efficiency polyoma DNA transfection of chloroquine treated cells. *Nucleic Acids Res* **11**: 1295–1308.

Maitland NJ, McDougall JK. 1977. Biochemical transformation of mouse cells by fragments of herpes simplex DNA. *Cell* **11**: 233–241.

Miller JH, Lebkowski JS, Greisen KS, Calos MP. 1984. Specificity of mutations induced in transfected DNA by mammalian cells. *EMBO J* **3**: 3117–3121.

Parker BA, Stark GR. 1979. Regulation of simian virus 40 transcription: Sensitive analysis of the RNA species present early in infections by virus or viral DNA. *J Virol* **31**: 360–369.

Perucho M, Wigler M. 1981. Linkage and expression of foreign DNA in cultured animal cells. *Cold Spring Harb Symp Quant Biol* **45**: 829–838.

Razzaque A, Mizusawa H, Seidman MM. 1983. Rearrangement and mutagenesis of a shuttle vector plasmid after passage in mammalian cells. *Proc Natl Acad Sci* **80**: 3010–3014.

Weinberg RA. 1985. Oncogenes and the molecular basis of cancer. *Harvey Lect* **80**: 29–136.

Wibo M, Poole B. 1974. Protein degradation in cultured cells. II. The uptake of chloroquine by rat fibroblasts and the inhibition of cellular protein degradation and cathepsin B1. *J Cell Biol* **63**: 430–440.

Wigler M, Pellicer A, Silverstein S, Axel R. 1978. Biochemical transfer of single-copy eukaryotic genes using total cellular DNA as a donor. *Cell* **14**: 725–731.

Wigler M, Silverstein S, Lee LS, Pellicer A, Cheng VC, Axel R. 1977. Transfer of purified herpes virus thymidine kinase gene to cultured mouse cells. *Cell* **11**: 223–232.

Zhang G, Kain SR. 1996. Transfection maximizer increases the efficiency of calcium phosphate transfections with mammalian cells. *BioTechniques* **21**: 940–945.

CHLOROQUINE DIPHOSPHATE

Chloroquine (F.W.=519.5), an amine that prevents acidification of endosomes and lysosomes and inhibits lysosomal protease cathepsin B (Wibo and Poole 1974), increases the efficiency of transfection of some types of cells and decreases the efficiency of others (Chang 1994). By inhibiting acidification of lysosomes, chloroquine may prevent or delay the degradation of transfecting DNA by lysosomal hydrolases (Luthman and Magnusson 1983). Unfortunately, the beneficial effects of chloroquine are modest and do not extend to all cell lines. Because the balance between the benefits and disadvantages of chloroquine varies so widely from cell line to cell line, there is simply no way to predict whether the drug will lead to a useful increase in transfection frequency in a particular circumstance. However, if low frequencies of transfection are a problem, it is certainly worth exploring whether chloroquine can help. The optimal time, length, and intensity of treatment must be determined empirically for each cell line. Typically, however, cells will be exposed to chloroquine diphosphate at a final concentration of 100 μM for 3–5 h either before, during, or after the cells are exposed to a calcium phosphate–DNA coprecipitate, or during exposure of cells to a mixture of DNA and DEAE-dextran. In the presence of chloroquine, the cells develop a vesicularized appearance. After the treatment, the cells are washed with phosphate-buffered saline and medium and then incubated for 24–60 h before assaying for expression of the transfected DNA. Chloroquine diphosphate is prepared as a 100 mM stock solution (52 mg/mL in H_2O), which should be sterilized by filtration and stored in foil-covered tubes at $-20°C$.

REFERENCES

Chang PL. 1994. Calcium phosphate-mediated DNA transfection. In *Gene therapeutics: Methods and applications of direct gene transfer* (ed Wolfe JA, Crow JF), pp. 157–179. Birkhäuser, Boston.

Luthman H, Magnusson G. 1983. High efficiency polyoma DNA transfection of chloroquine treated cells. *Nucleic Acids Res* 11: 1295–1308.

Wibo M, Poole B. 1974. Protein degradation in cultured cells. II. The uptake of chloroquine by rat fibroblasts and the inhibition of cellular protein degradation and cathepsin B1. *J Cell Biol* 63: 430–440.

DEAE-DEXTRAN TRANSFECTION

The first transfection methods, developed in the late 1950s, used hyperosmotic and polycationic proteins to promote entry of DNA into cells (for review, see Felgner 1990). The results were erratic, and the efficiency of transfection was, at best, very poor. The situation improved dramatically in the mid 1960s when DEAE-dextran (diethylaminoethyl-dextran) was used to introduce poliovirus RNA (Pagano and Vaheri 1965) and SV40 and polyomavirus DNAs (McCutchan and Pagano 1968; Warden and Thorne 1968) into cells. The procedure, with slight modifications, continues to be widely used for transfection of cultured cells with viral genomes and recombinant plasmids. Although the mechanism of action of DEAE-dextran is not understood in detail, it seems likely that the high-molecular-weight, positively charged polymer serves as a bridge between the negatively charged nucleic acid and the negatively charged surface of the cell (Lieber et al. 1987; Holter et al. 1989). After the DEAE-dextran/DNA complexes have been internalized by endocytosis (Ryser 1967; Yang and Yang 1997), the DNA somehow escapes from the increasingly acidic endosomes and is transported by unknown mechanisms across the cytoplasm and into the nucleus.

Since the method was introduced more than 20 years ago, many variants of DEAE-dextran transfection have been described. In most cases, the cells are exposed to a preformed mixture of DNA and high-molecular-weight DEAE-dextran (M.W. > 500,000). However, a modified procedure has been described in which the cells are exposed first to DEAE-dextran and then to DNA (al-Moslih and Dubes 1973; Holter et al. 1989). All of these methods seek to maximize the uptake of DNA and to minimize the cytotoxic effects of DEAE-dextran. The following are among the variables that influence the efficiency of transfection.

- *Concentration of DEAE-dextran used and length of time cells are exposed to it.* It is possible to use either a relatively high concentration of DEAE-dextran (1 mg/mL) for short periods (30 min to 1.5 h) or a lower concentration (250 μg/mL) for longer periods of time (up to 8 h). The first of these transfection procedures is the more efficient, but it involves monitoring the cells for early signs of distress when they are exposed to the DNA/DEAE-dextran mixture. The second technique is less demanding and more reliable, but slightly less efficient. However, it can be combined with shock treatments (see below) that can raise the efficiency of transfection to very high levels.

- *Use of facilitators such as DMSO, chloroquine, or glycerol.* The efficiency of transient expression of genes introduced by DEAE transfection is increased ~50-fold if cells are exposed to DMSO, glycerol, polyethyleneimine, or other substances such as Starburst dendrimers that perturb osmosis and increase the efficiency of endocytosis (Lopata et al. 1984; Sussman and Milman 1984; Kukowska-Latello et al. 1996; Zauner et al. 1996; Godbey et al. 1999). A similar increase in efficiency of transfection of some lines of cultured cells may be obtained by exposing the transfected cells to chloroquine, which prevents acidification of endosomes and promotes early release of DNA into the cytoplasm (Luthman and Magnusson 1983). In the best cases, 80% of the cells in a transfected population can express foreign genes when DEAE-dextran and facilitators are used in combination (e.g., see Kluxen and Lübbert 1993). However, the efficiency of DNA transfection using DEAE-dextran with a facilitator varies greatly from cell line to cell line. Conditions that are optimal for one cell line may not work at all for another. To obtain consistently high efficiencies of transformation with a particular cell line, the following factors should be standardized:

  - the density of cells and their state of growth

  - the amount of transfecting DNA

  - the concentration and molecular weight of DEAE-dextran

  - the length of time the cells are exposed to DNA

  - whether the DEAE-dextran and DNA are added to the cells simultaneously or sequentially (al-Moslih and Dubes 1973; Holter et al. 1989)

  - length and temperature of the posttransfection facilitation and the concentration of the facilitating agent

- whether the cells are transfected while growing on a solid support or are first removed from the solid support and transfected in suspension (Golub et al. 1989)

For publications that analyze the effects of some or all of these conditions on transfection efficiency, see Holter et al. (1989), Fregeau and Bleackley (1991), Kluxen and Lübbert (1993), and Luo and Saltzman (1999).

In addition to its use as a primary agent for transfection, DEAE-dextran can also be used as an adjuvant to enhance the efficiency of electroporation. Although the effects appear to vary from one cell line to another, the combination of electroporation and DEAE-dextran in some cases can improve the efficiency of transfection by a factor of 10–100 (Gauss and Leiber 1992).

DNA transfected into cells by the DEAE-dextran method is prone to mutation. This is particularly true of sequences cloned in vectors that can replicate in transfected mammalian cells. For example, when the *E. coli lacI* gene, cloned in a plasmid containing an SV40 origin of replication, was introduced into COS-7 cells, allowed to replicate for several generations, and then returned to *E. coli*, mutations occurred at a frequency of one to several percent (Calos et al. 1983). Stunning in their variety, mutations induced during replication in mammalian cells include deletions, insertions, and base substitutions (Razzaque et al. 1983; Lebkowski et al. 1984; Ashman and Davidson 1985). These mutations are thought to arise as a consequence of damage caused by the action of degradative enzymes and low pH in the lysosomes and also perhaps by the lack of a complete chromatin structure after the transfecting DNA enters the nucleus (Miller et al. 1984; Reeves et al. 1985).

REFERENCES

al-Moslih MI, Dubes GR. 1973. The kinetics of DEAE-dextran induced cell sensitization to transfection. *J Gen Virol* 18: 189–193.

Ashman CR, Davidson RL. 1985. High spontaneous mutation frequency of BPV shuttle vector. *Somat Cell Mol Genet* 11: 499–504.

Calos M, Lebkowski JS, Botchan MR. 1983. High mutation frequency in DNA transfected into mammalian cells. *Proc Natl Acad Sci* 80: 3015–3019.

Felgner PL. 1990. Particulate systems and polymers for in vitro and in vivo delivery of polynucleotides. *Adv Drug Delivery Rev* 5: 163–187.

Fregeau CJ, Bleackley RC. 1991. Factors influencing transient expression in cytotoxic T cells following DEAE dextran-mediated gene transfer. *Somat Cell Mol Genet* 17: 239–257.

Gauss GH, Lieber MR. 1992. DEAE-dextran enhances electroporation of mammalian cells. *Nucleic Acids Res* 20: 6739–6740.

Godbey W, Wu K, Hirasaki G, Mikos A. 1999. Improved packing of poly-(ethyleneimine)/DNA complexes increases transfection efficiency. *Gene Ther* 6: 1380–1388.

Golub EI, Kim H, Volsky DJ. 1989. Transfection of DNA into adherent cells by DEAE-dextran/DMSO method increases dramatically if the cells are removed from surface and treated in suspension. *Nucleic Acid Res* 17: 4902.

Holter W, Fordis CM, Howard BH. 1989. Efficient gene transfer by sequential treatment of mammalian cells with DEAE-dextran and deoxyribonucleic acid. *Exp Cell Res* 184: 546–551.

Kluxen FW, Lübbert H. 1993. Maximal expression of recombinant cDNAs in COS cells for use in expression cloning. *Anal Biochem* 208: 352–356.

Kukowska-Latello JF, Bielinska AU, Johnson J, Spindler R, Tomalia DA, Baker JR Jr. 1996. Efficient transfer of genetic material into mammalian cells using Starburst polyamidoamine dendrimers. *Proc Natl Acad Sci* 3: p4897–4902.

Lebkowski JS, DuBridge RB, Antell EA, Greisen KS, Calos MP. 1984. Transfected DNA is mutated in monkey, mouse, and human cells. *Mol Cell Biol* 4: 1951–1960.

Lieber MR, Hesse JE, Mizuuchi K, Gellert M. 1987. Developmental stage specificity of the lymphoid V(D)J recombination activity. *Genes Dev* 1: 751–761.

Lopata MA, Cleveland DW, Sollner-Webb B. 1984. High level expression of a chloramphenicol acetyl transferase gene by DEAE-dextran mediated DNA transfection coupled with a dimethyl sulfoxide or glycerol shock treatment. *Nucleic Acids Res* 12: 5707–5717.

Luo DL, Saltzman WM. 1999. Synthetic DNA delivery systems. *Nat Biotechnol* 18: 33–37.

Luthman H, Magnusson G. 1983. High efficiency polyoma DNA transfection of chloroquine treated cells. *Nucleic Acids Res* 11: 1295–1308.

McCutchan JH, Pagano JS. 1968. Enhancement of the infectivity of simian virus 40 deoxyribonucleic acid with diethyl aminoethyl-dextran. *J Natl Cancer Inst* 41: 351–357.

Miller JH, Lebkowski JS, Greisen KS, Calos MP. 1984. Specificity of mutations induced in transfected DNA by mammalian cells. *EMBO J* 3: 3117–3121.

Pagano JS, Vaheri A. 1965. Enhancement of infectivity of poliovirus RNA with diethyl-aminoethyl-dextran (DEAE-D). *Arch Gesamte Virusforsch* 17: 456–464.

Razzaque A, Mizusawa H, Seidman MM. 1983. Rearrangement and mutagenesis of a shuttle vector plasmid after passage in mammalian cells. *Proc Natl Acad Sci* 80: 3010–3014.

Reeves R, Gorman CM, Howard BH. 1985. Minichromosome assembly of non-integrated plasmid DNA transfected into mammalian cells. *Nucleic Acids Res* 13: 3599–3615.

Ryser HJ-P. 1967. A membrane effect of basic polymers dependent on molecular size. *Nature* 215: 934–936.

Sussman DJ, Milman G. 1984. Short-term, high efficiency expression of transfected DNA. *Mol Cell Biol* 4: 1641–1643.

Warden D, Thorne HV. 1968. The infectivity of polyoma virus DNA for mouse embryo cells in the presence of diethylaminoethyl-dextran. *J Gen Virol* 3: 371–377.

Yang YW, Yang JC. 1997. Studies of DEAE-dextran-mediated gene transfer. *Biotechnol Appl Biochem* 25: 47–51.

Zauner W, Kichler A, Schmidt W, Sinski A, Wagner E. 1996. Glycerol enhancement of ligand-polylysine/DNA transfection. *BioTechniques* 20: 905–913.

ELECTROPORATION

Nucleic acids do not enter cells under their own power; they require assistance in crossing physical barriers at the cell boundary and in reaching an intracellular site where they can be expressed and/or replicated. Exposure of many types of cells to an electrical discharge reversibly destabilizes their membranes and transiently induces the formation of aqueous pathways or membrane pores (Neumann and Rosenheck 1972; Neumann et al. 1982; Wong and Neumann 1982; for reviews, see Zimmermann 1982; Andreason and Evans 1988; Tsong 1991; Weaver 1993) that potentiate the entry of DNA molecules (Neumann et al. 1982). This method, which is known as electroporation, has been developed into a rapid, simple, and efficient technique for introducing DNA into a wide variety of cells, including bacteria, yeasts, plant cells, and a large number of cultured mammalian cell lines. The chief practical advantages of electroporation are that it can be applied to a wide variety of cells, both prokaryotic and eukaryotic, and that it is extremely simple to perform.

The Mechanism of Electroporation

Because the changes in membrane structure that accompany electroporation cannot be visualized in real time by microscopy, our understanding of the mechanism is based on evidence that is both patchy and circumstantial. The following model (Weaver 1993) has been used for many years to provide a plausible account of the sequence of events that are initiated by increasing the transmembrane voltage from its physiological value of ~0.1 to 0.5–1.0 V. Figure 1 shows the following sequence of events.

- The onset of electroporation causes a membrane dimple followed by formation of transient hydrophobic pores whose diameter fluctuates from a minimum of 2 nm to a maximum of several nanometers.

- Some of the larger *hydrophobic* pores are converted to *hydrophilic* pores because the energy needed to form an aqueous pore is reduced as the transmembrane voltage is increased and the energy required

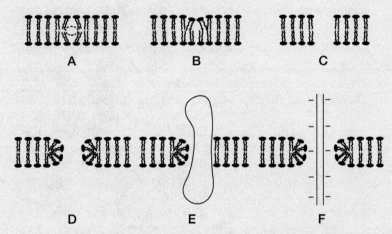

FIGURE 1. **Hypothetical structures for transient and metastable membrane conformations believed to be relevant to electroporation.** (A) Fredd volume fluctuation; (B) aqueous protrusion or "dimple"; (C,D) hydrophobic pores usually regarded as the "primary pores" through which ions and molecules pass; (E) composite pore with "foot in the door" charged macromolecule inserted into a hydrophilic pore. The transient aqueous pore model assumes that transitions from A→B→C or D occur with increasing frequency as U is increased. Type F may form by entry of a tethered macromolecule, while the transmembrane voltage is significantly elevated, and then persist after U has decayed to a small value through pore conduction. It is emphasized that these hypothetical structures have not been directly observed and that support for them derives from the interpretation of a variety of experiments involving electrical, optical, mechanical, and molecular transport behavior. (Redrawn from Weaver 1993, with permission from John Wiley & Sons, Inc.)

to maintain the circumference of a large hydrophilic pore is significantly lower than that required to maintain a large hydrophobic pore. For this reason, hydrophilic pores have an extended half-life and may be further stabilized by attachment to underlying cytoskeletal elements. The generation of such long-lived metastable pores allows small ions and molecules to enter and leave the cell long after the transmembrane voltage has returned to low values (Rosenheck et al. 1975; Zimmermann et al. 1976; Lindner et al. 1977). The detailed mechanism by which molecules pass through hydrophilic pores is not known but may include electrophoresis (Chermodnick et al. 1990), electroendo-osmosis, diffusion, and endocytosis (Weaver and Barnett 1992). Reclosing of the pores appears to be a stochastic process that can be delayed by keeping the cells at 0°C. While the pores remain open, up to 0.5 pg of DNA can enter the cell (Bertling et al. 1987). Size seems to be no impediment because DNA molecules up to 150 kb in pass through the pores (Knutson and Yee 1987). Because the DNA enters directly into the cytoplasm, it is not exposed to acid conditions in endosomes and lysosomes. This route may explain why the rate of mutation in DNA introduced to cells by electroporation is apparently very low (Drinkwater and Kleindienst 1986; Bertling et al. 1987) compared with DNA transfected as calcium phosphate coprecipitates (e.g., see Calos et al. 1983). Despite intensive research, nobody has been able to see or provide other evidence of these pores. The alternative term "electropermeabilization" referring to changes that may contribute to make the cell membrane reversibly permeable, without "holes," has been put forth to explain the mechanism of DNA uptake into cells.

Electropermeabilization might result from the following.

- *The electrocompressive forces generated by an electric field associated with the transmembrane potential difference.* These electrocompressive forces bring the two lipid layers closer together than the distance that allows them to remain parallel to each other, which could disrupt the ordered stacking of the lipids that comprise the plasma membrane.

- *Changes in the lipid polar head orientation*, detected by ^{31}P magnetic resonance changes (Lopez et al. 1988).

- *The penetration of water in the lipid layer (hydration of the membrane) resulting from the two aforementioned structural changes of the lipids.* This mechanism appears plausible based on simulations of membranes exposed to high transmembrane potential differences, using validated molecular dynamics programs (Tieleman et al. 2003; Tarek 2005; Vernier and Ziegler 2007) that also reveal the influence of the lipid composition on the stability and the electropermeabilization of the membranes.

- Possible changes in the structure of transmembrane proteins (Teissie et al. 2005).

A good discussion of the mechanisms of cell membrane electroporation and electropermeabilization can be found in the review by Teissie et al. (2005).

For *E. coli*, electroporation is currently the most efficient method available for transformation with plasmids. In excess of 80% of the cells in a culture can be transformed to ampicillin resistance by this method, and efficiencies of transformation approaching the theoretical maximum of one transformant per molecule of plasmid DNA have been reported (Smith et al. 1990). However, the number of transformants obtained is marker dependent. When pBR322, which carries genes conferring resistance to two antibiotics (ampicillin and tetracycline), is introduced into *E. coli* by electroporation, the number of tetracycline-resistant transformants is ~100-fold less than the number of ampicillin-resistant transformants (Steele et al. 1994). This effect is not seen when the plasmid is introduced into the bacteria by the calcium chloride method. One possible explanation is that electroporation damages or changes the bacterial membrane so that it can no longer interact efficiently with the tetracycline-resistance protein.

Typically, between 50% and 70% of cells exposed to high electric field strengths are killed. The lethal effects, which vary in intensity from one cell type to another, are not due to heating or electrolysis and are independent of the current density and energy input. Instead, cell killing is dependent on field strength and the total time of treatment (Sale and Hamilton 1967). The most likely cause of cell killing is the inability to restore membrane structure and barrier function, leading to rupture of cell membranes, a rapid loss of ionic balance, and massive efflux of cellular components. Nevertheless, this process is still termed "reversible electroporation" and should not be confused with "irreversible electroporation," in which cell death is the intended end point. Irreversible electroporation is used in medical applications like atrial heart ablation (Lavee et al. 2007) or tumor ablation by electrical nonthermal means (Edd et al. 2006; Al-Sakere et al. 2007).

Electrical Conditions Required for Electroporation

Electroporation of almost all mammalian cells is induced when the transmembrane voltage, $\Delta U(t)$, is increased to 0.5–1.0 V for durations of microseconds to milliseconds. This translates to an electric field strength of $\sim$7.5–15.0 kV/cm. Because this value is constant and is independent of the biochemical nature of the cell membrane, it seems likely that variations in the efficiency of electroporation from cell line to cell line are due to differences in the rate and efficiency of membrane recovery at the end of the pulse.

The transmembrane voltage, $\Delta U(t)$, induced by electric fields varies in direct proportion to the diameter of the cell that is the target for transfection (Knutson and Yee 1987). Electroporation of mammalian cells, for example, requires smaller electric fields ($<$10 kV/cm) than does electroporation of yeasts or bacteria (12.5–16.5 kV/cm). Most of the commercial suppliers of electroporation machines provide literature describing the approximate voltages required for transfection of specific types of cells in their particular apparatus.

Three important characteristics of the pulse affect the efficiency of electroporation: the length of the pulse, its field strength, and its shape. Most of the commercial electroporation machines use capacitative discharge to produce controlled pulses whose length is mainly determined by the value of the capacitor and the conductivity of the medium. Thus, the time constant of the pulse can be altered by switching capacitors according to the manufacturer's instructions or by changing the ionic strength of the medium. When the charge from the capacitor is directed to a sample placed between two electrodes, the voltage across the electrodes rises rapidly to a peak (V_0) and declines over time (t) according to

$$V_t = V_0[e^{-F(t,T)}],$$

where t is the time constant, which is equal to the time over which the voltage declines to $\sim$37% of peak value. T (measured in seconds) is also equal to the product of the resistance (R, measured in ohms) and the capacitance (C, in farads):

$$T = RC.$$

From this equation, it follows that (1) a larger capacitor requires more time to discharge through a medium of a given resistance, and (2) a capacitor of a given size discharges more slowly as the resistance of the medium increases.

The field strength (E) of the pulse varies in direct proportion to the applied voltage (V) and in inverse proportion to the distance (d) between the two electrodes, which is usually determined by the size of the cuvette through which the pulse travels:

$$E = F(V, d).$$

Most manufacturers provide cuvettes in three sizes, where the interelectrode distances are, respectively, 0.1 cm, 0.2 cm, and 0.4 cm. When 1000 V is discharged into these cuvettes, E_0 is 10,000 V/cm in the 0.1-cm cuvette, 5000 V/cm in the 0.2-cm cuvette, and 2500 V/cm in the 0.4-cm cuvette. Newer machines like the Gene Pulser MXcell (Bio-Rad) also allow high-throughput electroporation using lower cell numbers in a plate well format (12, 24, or 96 well).

The shape of the pulse is determined by the design of the electroporation device (Fig. 2). The waveform produced by most commercial machines is simply the exponential decay pattern of a discharging capacitor. They are thus defined by only two parameters, the field strength E and the time constant T, as explained above. In some types of electroporation apparatuses, square waves can be generated by rapidly increasing the voltage, maintaining it at the desired level for a specified time (pulse width), and then rapidly reducing the voltage to 0. Thus they can be defined by the voltage delivered, the length of each pulse, the number of pulses, and the length of interval between the pulses. Square pulses can be grouped into two general categories: very high field strength of very short duration (typically 8 kV/cm for 5.4 msec) (Neumann et al. 1982), and low field strength of medium to long duration (e.g., $<$2 kV/cm for $>$10 msec) (e.g., see Potter et al. 1984). Although square waves have been reported to offer increased

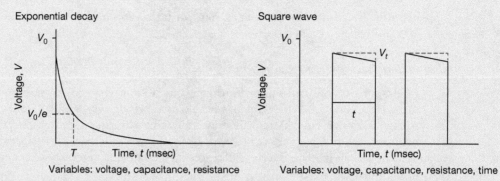

FIGURE 2. Electrical pulses used for electroporation. Exponential decay pulse: When a capacitor charged to a voltage V_0 is discharged into cells, the voltage applied to the cells reduces exponentially with time. The time required for V_0 to drop to V_0/e is the time constant T, which denotes the pulse length. Square wave pulse: When the pulse from a capacitor is abruptly stopped after discharge, a square wave pulse is generated. The pulse length is the time for which the cells are subjected to the discharge. A slight drop in voltage occurs during the pulse length in all instruments. This is called the "pulse droop" and is measured as a percentage of V_0.

viability and efficiency for particularly sensitive cell lines and for in vivo applications (Hewapathirane and Haas 2008; Jordan et al. 2008), the exponential wave forms produced by most commercial electroporation machines are satisfactory for most applications.

Electroporation of mammalian cells is usually performed in a buffered saline solution or culture medium with pulse lengths of 10–40 msec and strengths of 400–900 V/cm.

Optimizing Conditions for Electroporation

A major advantage of electroporation over other methods of transfection is that it works for a very wide variety of mammalian cells, including those that are difficult to transfect by other means (e.g., see Potter et al. 1984; Tur-Kaspa et al. 1986; Chu et al. 1987). However, despite its advantages, electroporation is not always the most efficient way to introduce DNA into a particular cell line. For example, most cell lines like 293T and Vero are easily transfected by lipofection. To find out whether electroporation is a useful method of transfection for a particular cell line, it is important to use a range of field strengths and pulse lengths/types and thereby to establish conditions that generate the maximum numbers of transfectants. Such conditions have been reported for more than 50 types of mammalian cells, and it is sometimes possible to save a lot of work by simply reading the relevant literature. Most of the companies that sell electroporation devices produce excellent up-to-date lists of papers in which electroporation has been used for transfection. These bibliographies are often the easiest way to gain access to information about the properties of a particular cell line or its close relatives. However, because of variation in properties between different cultivars of the same cell line, it is important for investigators to confirm that the conditions described in the literature are optimal for cells grown in their laboratory. Newer electoporators also provide both the exponential decay and square wave pulse formats for optimizing the best conditions for DNA delivery into target cells.

Because transfection and cell killing are independently determined by field strength (Chu et al. 1987), it is best to expose aliquots of cells to electric fields of increasing strength with time constants between 50 and 200 msec. For each field strength, measure (1) the number of cells that express a transfected reporter gene (10–40 μg/mL linearized plasmid DNA in the electroporation buffer) and (2) the proportion of cells that survive exposure to the electric field. Plating efficiency is a more accurate measure of cell survival than staining with vital dyes because, after electroporation, cells can remain permeable to vital dyes such as Trypan Blue for an hour or two. The following are other variables that have been reported to affect the efficiency of electroporation.

- *The temperature of the cells before, during, and after electroporation* (e.g., Potter et al. 1984; Chu et al. 1987). Usually, electroporation is performed on cells that have been prechilled to 0°C. The cells are held at 0°C after electroporation (to maintain the pores in an open position) and are diluted into warm medium for plating (Rabussay et al. 1987).

- *The concentration and conformation of the DNA* (e.g., Neumann et al. 1982; Potter et al. 1984; Toneguzzo and Keating 1986). Linear DNA is preferred for stable transformation; circular DNA is for transient transfection. Preparations containing DNA at a concentration of 1–80 μg/mL are optimal.

- *The state of the cells.* The best results are obtained with cell cultures in the mid-log phase of growth that are actively dividing.

Nucleofection

A specialized form of electroporation, under standardized conditions nucleofection permits the introduction of nucleic acids directly into the nucleus as well as into the cytoplasm of the host cell. Although a direct comparison between electroporation and nucleofection has not been made, nucleofection appears to be much more efficient in direct nuclear transfer of the substrate as determined by early expression kinetics of nucleofected genes like GFP (30 min postnucleofection) and detection of at least 3 log copies of plasmid DNA in the nucleus by quantitative PCR 4 h after the procedure (Greiner et al. 2004; Johnson et al. 2005). The high operating costs (both the device and specialty reagents) have limited the use of nucleofection to primary cells such as stem cells, neurons, T-cells, macrophages, and other mammalian cells that are refractory to conventional lipid-based transfection methods. However, other commercial vendors (e.g., the Ingenio from Mirus Bio) now market less expensive cell-specific reagents for nucleofection that are compatible with the Amaxa Nucleofector.

As is true for conventional electroporation, nucleofection results in a high rate of cell death and requires optimization for best results.

Optimization of Nucleofection

The optimization process involves subjecting cells to various combinations of electrical parameters (termed "programs" in the nucleofector device) and different nucleofector solutions. The condition that results in the highest transfection efficiencies with the lowest amount of cell death is chosen for the rest of the studies. Important considerations are as follows.

- Excessive cell death during the nucleofection procedure can be minimized by using cells that are maintained under the best possible growth conditions and preferably at a low passage number.

- Cells should be handled gently, and procedures such as centrifugation should be limited and performed at the lowest possible speeds both before and after nucleofection.

- Warm up only the necessary amount of medium (rather than the entire bottle) to obtain consistent growth of cells and reproducible data.

- Cell cultures should be free of mycoplasma contamination because it drastically impairs transfection efficiency.

- DNA used for transfection should be of high purity with an OD_{260} to OD_{280} ratio of ≥ 1.8 and should be freshly prepared to avoid using nicked DNA.

REFERENCES

Al-Sakere B, Bernat C, Andre F, Connault E, Opolon P, Davalos RV, Mir LM. 2007. A study of the immunological response to tumor ablation with irreversible electroporation. *Technol Cancer Res Treat* 6: 301–306.

Andreason GL, Evans GA. 1988. Introduction and expression of DNA molecules in eukaryotic cells by electroporation. *BioTechniques* 6: 650–660.

Bertling W, Hunger-Bertling K, Cline M-J. 1987. Intranuclear uptake and persistence of biologically active DNA after electroporation of mammalian cells. *J Biochem Biophys Methods* 14: 223–232.

Calos M, Lebkowski JS, Botchan MR. 1983. High mutation frequency in DNA transfected into mammalian cells. *Proc Natl Acad Sci* 80: 3015–3019.

Chermodnick LV, Sokolov AV, Budker VG. 1990. Electrostimulated uptake of DNA by liposomes. *Biochim Biophys Acta* 1024: 179–183.

Chu G, Hayakawa H, Berg P. 1987. Electroporation for the efficient transfection of mammalian cells with DNA. *Nucleic Acids Res* 15: 1311–1326.

Drinkwater NR, Kleindienst DK. 1986. Chemically induced mutagenesis in a shuttle vector with a low-background mutant frequency. *Proc Natl Acad Sci* 83: 3402–3406.

Edd JF, Horowitz L, Davalos RV, Mir LM, Rubinsky B. 2006. In vivo results of a new focal tissue ablation technique: Irreversible electroporation. *IEEE Trans Biomed Eng* 53: 1409–1415.

Greiner J, Wiehe J, Wiesneth M, Zwaka TP, Prill T, Schwarz K, Bienek-Ziolkowski M, Schmitt M, Döhner H, Hombach V, et al. 2004. Transient genetic labeling of human CD34 positive hematopoietic stem cells using nucleofection. *Transfus Med Hemother* 31: 136–141.

Hewapathirane DS, Haas K. 2008. Single cell electroporation in vivo within the intact developing brain. *J Vis Exp* 17: 705.

Johnson BD, Gershan JA, Natalia N, Zujewski H, Weber JJ, Yan X, Orentas RJ. 2005. Neuroblastoma cells transiently transfected to simultaneously express the co-stimulatory molecules CD54, CD80, CD86, and CD137L generate antitumor immunity in mice. *J Immunother* 28: 449–460.

Jordan ET, Collins M, Terefe J, Ugozzoli L, Rubio T. 2008. Optimizing electroporation conditions in primary and other difficult-to-transfect cells. *J Biomol Tech* 19: 328–334.

Knutson JC, Yee D. 1987. Electroporation: Parameters affecting transfer of DNA into mammalian cells. *Anal Biochem* 164: 44–52.

Lavee J, Onik G, Mikus P, Rubinsky B. 2007. A novel nonthermal energy source for surgical epicardial atrial ablation: Irreversible electroporation. *Heart Surg Forum* 10: E162–E167.

Lindner P, Neumann E, Rosenheck K. 1977. Kinetics of permeability changes induced by electric impulses in chromaffin granules. *J Membr Biol* 32: 231–254.

Lopez A, Rols MP, Teissie J. 1988. ^{31}P NMR analysis of membrane phospholipid organization in viable, reversibly electropermeabilized Chinese hamster ovary cells. *Biochemistry* 27: 1222–1228.

Neumann E, Rosenheck K. 1972. Permeability changes induced by electric impulses in vesicular membranes. *J Membr Biol* 10: 279–290.

Neumann E, Schaefer-Ridder M, Wang Y, Hofschneider PH. 1982. Gene transfer into mouse lyoma cells by electroporation in high electric fields. *EMBO J* 1: 841–845.

Potter H, Weir L, Leder P. 1984. Enhancer-dependent expression of human κ immunoglobulin genes introduced into mouse pre-B lymphocytes by electroporation. *Proc Natl Acad Sci* 81: 7161–7165.

Rabussay D, Uher L, Bates G, Piastuch W. 1987. Electroporation of mammalian and plant cells. *Focus* 9: 1–3.

Rosenheck K, Lindner P, Pecht I. 1975. Effect of electric fields on light-scattering and fluorescence of chromaffin granules. *J Membr Biol* 12: 1–12.

Sale AJH, Hamilton WA. 1967. Effects of high electric fields on microorganisms. I. Killing of bacteria and yeasts. *Biochim Biophys Acta* 148: 781–788.

Smith M, Jessee J, Landers T, Jordan J. 1990. High efficiency bacterial electroporation 1×10^{10} *E. coli* transformants per microgram. *Focus* 12: 38–40.

Steele C, Zhang S, Shillitoe EJ. 1994. Effect of different antibiotics on efficiency of transformation of bacteria by electroporation. *BioTechniques* 17: 360–365.

Tarek M. 2005. Membrane electroporation: A molecular dynamics simulation. *Biophys J* 88: 4045–4053.

Teissie J, Golzio M, Rols MP. 2005. Mechanisms of cell membrane electropermeabilization: A minireview of our present (lack of ?) knowledge. *Biochim Biophys Acta* 1724: 270–280.

Tieleman DP, Leontiadou H, Mark AE, Marrink SJ. 2003. Simulation of pore formation in lipid bilayers by mechanical stress and electric fields. *J Am Chem Soc* 125: 6382–6383.

Toneguzzo F, Keating A. 1986. Stable expression of selectable genes introduced into human hematopoietic stem cells by electric field-mediated DNA transfer. *Proc Natl Acad Sci* 83: 3496–3499.

Tsong TY. 1991. Electroporation of cell membranes. *Biophys J* 60: 297–306.

Tur-Kaspa R, Teicher L, Levine BJ, Skoultchi AI, Shafritz DA. 1986. Use of electroporation to introduce biologically active foreign genes into primary rat hepatocytes. *Mol Cell Biol* 6: 716–718.

Vernier PT, Ziegler MJ. 2007. Nanosecond field alignment of head group and water dipoles in electroporating phospholipid bilayers. *J Phys Chem B* 111: 12993–12996.

Weaver JC. 1993. Electroporation: A general phenomenon for manipulating cells and tissues. *J Cell Biochem* 51: 426–435.

Weaver JC, Barnett A. 1992. Progress towards a theoretical model of electroporation mechanism: Membrane behavior and molecular transport. In *Guide to electroporation and electrofusion* (ed Chang DC, et al.), pp. 91–117. Academic, San Diego.

Wong T-K, Neumann E. 1982. Electric field mediated gene transfer. *Biochem Biophys Res Commun* 107: 584–587.

Zimmermann U. 1982. Electric field-mediated fusion and related electrical phenomena. *Biochim Biophys Acta* 694: 227–277.

Zimmermann U, Riemann F, Pilwat G. 1976. Enzyme loading of electrically homogeneous human red blood cell ghosts prepared by dielectric breakdown. *Biochim Biophys Acta* 436: 460–474.

CHAPTER 16

Introducing Genes into Mammalian Cells: Viral Vectors

Guangping Gao and Miguel Sena-Esteves

University of Massachusetts Medical School, Worcester, Massachusetts 01605

MUCH OF THE EARLY EXCITEMENT ABOUT MOLECULAR CLONING WAS fueled by a belief that cloned copies of mammalian coding sequences could be inserted into prokaryotic vectors, expressed to high levels in a bacterial host, and the expressed proteins purified in an active form and used not only in research projects but also therapeutically. Because of the wealth of knowledge of its physiology and genetics, its fast growth rate, and ease of handling, the host organism of first choice was *Escherichia coli*. The spectacular demonstration that *E. coli* could be used to clone and express small eukaryotic proteins, such as human growth hormone (Goedell et al. 1979) and rat insulin (Ullrich et al. 1977), provided significant impetus for the rapid commercialization of DNA cloning technology in the late 1970s and early 1980s. Many of the early recombinant DNA companies were floated on the optimistic assertion that any difficulties encountered using *E. coli* to synthesize large quantities of biologically active mammalian proteins would be minor and that any problems could be solved reasonably quickly. However, this initial bubble of confidence dissolved when many eukaryotic proteins expressed in *E. coli* were found to be biologically inactive, denatured, and/or aggregated into insoluble inclusion bodies. Other eukaryotic proteins could be expressed only at very low levels, whereas a few either were not expressed at all or were toxic to their bacterial hosts. Purification of inclusion bodies and in vitro refolding of proteins offered a potential solution to this problem. But the conditions for refolding varied from protein to protein and the efficiency was generally low even after optimization.

These problems result from the inability of eukaryotic proteins expressed in *E. coli* to form the same folds, domains, and three-dimensional structures as are present in their natural hosts. As a rule, only small (<23 kDa) globular cytoplasmic proteins with a high content of charged amino acids and few contiguous hydrophobic residues can be expressed in soluble form in *E. coli*. In some cases, the solubility of the expressed protein can be improved by the addition of a hydrophilic tag (Dyson et al. 2004) or, more recently, by genetic selection (Lim et al. 2009). However, the majority of eukaryotic proteins are much larger than 23 kDa, have complex folds, and consist of several subunits. By the early 1980s, it had become clear that cloned cDNAs encoding mammalian proteins were best expressed in host cells that carry out the appropriate posttranslational modifications and are equipped with the specific chaperones, partner-binding proteins, and cellular trafficking systems required for accurate, rapid folding and assembly of nascent eukaryotic polypeptides.

The past 30 years have seen the development of increasingly sophisticated systems to express genes (or more often, complementary DNAs [cDNAs]) in cultured mammalian cells. These systems are of two types: (1) those that involve transient or stable expression of transfected DNA (see Chapter 15) and (2) those that involve the use of viral expression vectors, which take advantage of the multiplicity of mechanisms that viruses use to gain entry into mammalian cells and once there to harness the cellular machinery to express their cargo of genetic material.

Over the years, many different viral vector systems have been developed to take advantage of the specific biological properties and tropisms of a large number of mammalian viruses. Some systems have not yet progressed beyond an early exploratory stage; others have been designed to solve specific biological problems; and a few require detailed knowledge of, and sometimes affection for, the biology of the parental virus.

In their natural state, the virions of some vectors contain DNA as their genetic material, and others contain RNA. But whatever the natural state of the vector's genome, recombinants are constructed (1) using standard in vitro techniques to introduce an expression cassette into a cloned DNA copy of the vector's genome and (2) by recombination in cells transfected with the vector and an expression cassette. If the foreign DNA replaces viral sequences that are not essential for viral growth, the recombinant DNA can be transfected into an appropriate cell type and the recombinant virions can be harvested a few days later. If the segment of foreign DNA replaces one or more viral genes that are essential for growth, the missing functions must be provided in *trans*—either from a helper virus or, in some cases, by a cell that carries an integrated copy of the essential gene(s).

FACTORS TO CONSIDER WHEN CHOOSING A VIRAL VECTOR

Mammalian viruses display an extraordinary variety of genetic structures and biological behaviors. As a result, researchers wanting to introduce and/or express genes in mammalian cells have many options, and the choice among them is not always clear. Factors to consider when selecting a vector include the following.

The Purpose of the Experiment

The requirements made of a viral vector depend on the goals of the experiment. If the goal is to produce a biologically active protein, major factors to consider include the following:

1. The amounts of protein that will be required and the yields that can be expected from the various available vectors

2. The size of the expression cassette, which must fit comfortably within the space available in the vector's genome

3. The assays that will be used to monitor protein production and measure the biological activity of the protein

4. Whether purification of the protein will be required. This becomes an important consideration if a major goal is to express and assay a large number of mutant forms of the target protein.

5. The scalability and estimated cost of the project (in both time and materials). In some cases (e.g., when assaying the activity of a series of mutant proteins), the cheapest and fastest option may be to use transfection systems rather than a viral vector. And, depending on the experience and skills available, the best option may be to buy a commercial viral expression kit, should a suitable one be available.

If the goal of the experiment is to stably express a transgene in cultured mammalian cells, the best option is to use a retrovirus or lentivirus vector, because they integrate into the genome without killing the host cells.

The Design of the Expression Cassette to Be Inserted into the Vector

Mammalian viruses have become a fruitful source of functional elements (promoters, splicing signals, transcriptional terminators, poly(A) addition sites, etc.) that can be used to construct synthetic cassettes containing the elements required for expression of proteins at different levels and in different types of host cells. The various control elements required for construction of expression cassettes are available as precloned modules from a number of the large commercial suppliers, many of whom also sell prefabricated, empty cassettes.

Most of the vectors designed for protein production use strong promoters cloned from viral genomes. These promoters are much more compact and far more powerful than promoters from eukaryotic genes, and many of them are indifferent to the species and tissue of origin of the host cell. Viral promoters commonly used in expression cassettes include the cytomegalovirus immediate-early promoter (CME-1E), the SV40 early promoter (SV40-E), and the long terminal repeats (LTRs) of murine and avian retroviruses. Other control elements commonly used in expression cassettes include a poly(A) addition site (usually from SV40) and an internal ribosome entry site (IRES) that allows two open reading frames to be translated from a single messenger RNA (mRNA) (for review, see de Felipe 2002).

For cell-type or tissue-specific expression, the best option is to use a promoter and associated *cis*-acting elements from a gene that is expressed strongly and specifically in the target tissue of cell type. For a useful catalog of tissue-specific promoters, see Papadakis et al. (2004). The information panel at the end of this chapter, Basic Elements in Viral Vectors, shows diagrams of commonly used expression cassettes.

Whether the Capacity of the Vector Is Sufficient to Accommodate the Segment of Foreign DNA

Many viruses used as vectors (e.g., adenoviruses and SV40) have strict genome size requirements for effective packaging into viral particles. Trying to circumvent packaging limits that have been established and optimized over long periods of evolutionary time is futile.

Whether the Appropriate Cell Lines Are Available

No single cell line exists that is permissive for all viral vectors. For example, many retroviruses exhibit strong species specificity and show a marked preference for dividing cells. On the other hand, adenoviruses and adeno-associated viruses (AAVs) can infect a variety of mammalian cell types with high efficiency, in sharp contrast to SV40, which has a very strong preference for the cells of its natural simian hosts. (For more details, see descriptions of the individual vector systems below.)

To increase the amount of space available for insertion of an expression cassette, genes necessary for replication or other viral functions may be deleted from the vector's genome. Recombinants constructed in such vectors require the essential functions to be supplied in *trans*, either from a helper cell, a helper virus, or from cotransfection of plasmids encoding those functions. For further information about helper-dependent vectors and cells that supply functions in *trans*, see the sections below describing adenovirus vectors and retrovirus vectors.

Time Constraints and Skills

Generating a stock of recombinant virus is a lengthy process that involves choosing an appropriate vector, designing and constructing a modular expression cassette, inserting the cassette into the backbone of the vector, transfecting cells, harvesting the recombinant virus, generating and purifying a high-titer stock, and then confirming the genetic structure of the recombinant structure (best done by DNA sequencing). The entire process requires a wide range of skills and takes a minimum of a few weeks, even for people with experience in animal virology. First-time users of animal virus vectors might want to consider using one of the commercial kits that are available from several companies for the construction, purification, and titration of recombinants constructed in adenovirus vectors, lentivirus vectors, and adeno-associated viral vectors.

THE MAJOR TYPES OF VIRUSES CURRENTLY USED AS VECTORS

The power of mammalian virus vectors was amply demonstrated in the early 1980s when SV40 was used to express high levels of the hemagglutinin gene of influenza virus (Gething and Sambrook 1981). The nascent subunits of the hemagglutinin were synthesized in large quantities, translocated into the endoplasmic reticulum, glycosylated, assembled into mature trimers, and transported efficiently to the cell surface. Since then, mammalian viruses of almost every type have been used as vehicles to express a wide variety of proteins. Some of them, including SV40, whose genome is too small to accommodate transgenes larger than a couple of kilobases, have limited applicability; others are still works in progress. But during the past decade, robust vectors have been developed from four types of viruses (adenoviruses, AAVs, lentiviruses, and retroviruses) that can satisfy most experimental needs. These four vectors differ dramatically in their biology and modus operandi. Adenoviruses and AAVs infect the host cells efficiently as episomal forms and express large quantities of the protein(s) encoded by their cargo of transgenes. Lentiviruses and retroviruses, in contrast, integrate the DNA form of their genomes into the host cell genome. These vectors, which do not kill the host cell, are used for stable transduction of mammalian cells. Although commercial kits have mitigated much of the grunt work involved in using these vectors, prior experience in culturing animal cells and in handling viral stocks remains a prerequisite.

Recombinant Adenovirus Vectors

The molecular biology of adenoviruses is known in great detail: The viruses are easy to handle and grow to high titer (10^9–10^{10} plaque-forming units [pfu]/mL of medium), they infect both dividing and nondividing cells, replication of the virus does not involve integration of the viral DNA into the cell's genome, and the viral particles can comfortably accommodate ~36 kb of DNA. With such attractive qualities, it is not surprising that adenoviruses have become vectors of choice for both florid expression of eukaryotic proteins and experimental forms of gene therapy (for reviews, see Hitt and Graham 2000; McConnell and Imperiale 2004; Palmer and Ng 2008).

The more than 50 serologically different adenoviruses that infect humans can be classified into six species, A–F, based on antigenic relationship, oncogenicity, DNA homology, G/C content, and the patterns of the cleavage by restriction enzymes, and other properties. Virtually all adenovirus vectors are based on two closely related serotypes, adenovirus types 2 and 5, that belong to species C (Fig. 1). The molecular biology of these serotypes was worked out in the 1970s and 1980s after it became possible to dissect their genomes with restriction enzymes and hence to map biological functions and transcripts to specific sectors of the ~36-kb linear viral DNA. The DNA sequences of both of these viruses are available at GenBank.

First-Generation Recombinant Adenovirus Vectors

Particles of adenovirus 2 and 5 can, at most, accommodate ~37 kb of double-stranded DNA (dsDNA). The E3 region (78.5–84.3 map units) is not required for viral growth in cultured cells. E3-deleted mutants are viable and can accept DNA segments of ~4.4 kb in length, which is large enough to accommodate a small expression cassette but too small for most cloning projects. However, the cloning capacity of type C adenoviruses can be increased by deleting both the nonessential E3 region and the essential E1A region, which codes for a set of immediate-early proteins that activate transcription of the three other early regions (E2–E4; see Fig. 1). Mutants carrying deletions of E1 grow poorly or not at all in standard lines of cultured cells, rendering the virus replication defective. However, the E1 region becomes nonessential when the virus is propagated in the 293 or 911 helper lines of human cells that have been engineered to express the E1 region in *trans*. E1/E3 double-deleted recombinant adenoviruses can accommodate transgenes that are ~6.5 kb in length—enough for an expression cassette of average size. Although recombinants constructed in E1/E3 deletion mutants do not grow quite as well as wild-type adenoviruses, they can usually be propagated in permissive cell lines to a titer of 10^8–10^9 pfu/mL. The viral particles of E1/E3 deletion mutants are sufficiently robust to withstand the rigors of purification by standard methods (e.g., centrifugation through cesium chloride [CsCl] density gradients [Gerard 1995]). E1/E3 double mutants have become the adenovirus vectors of choice for production of many recombinant proteins because of their ease of handling and reasonable cloning capacity. Many first-generation

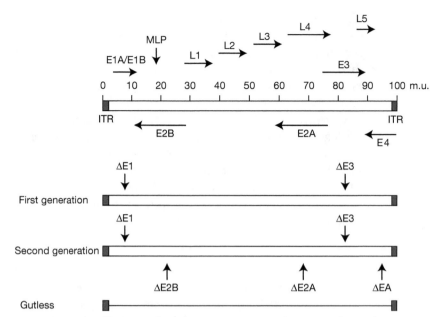

FIGURE 1. Transcription map of human adenovirus serotypes 2 and 5. (*Top*) The viral genome (~36 kb) is represented as a horizontal line with map units (m.u.). Transcription units are shown as arrows indicating direction of transcription. There are four early transcription units (E1–E4) and five families of late RNAs (L1–L5), which are the alternately spliced products of a common precursor expressed from a single major late promoter (MLP). The termini of the double-stranded genome consist of 103-bp inverted repeats (ITR) that are involved in viral DNA replication. Signals for encapsidation of the viral DNA during assembly of viral particles are located 190–380 nucleotides from the left-hand end of the viral genome. (*Bottom*) Diagrams of the structures of three generations of adenovirus vectors (first, second, and gutless). (Downward arrows) Viral regions that are deleted in first- and second-generation vectors. All second-generation vectors lack early regions E1 and E3; some vectors also lack early regions E2A and E4. The only viral DNA sequences retained in gutless adenovirus vectors are the ITR and packaging signals. (Redrawn from Wu et al. 2001, with permission from Bentham Science Publishers Ltd.)

vectors that differ from one another in detail have been described in the academic literature over the years; others are available commercially (for review, see Danthinne and Imperiale 2000).

Second-Generation and Third-Generation (Gutless) Recombinant Adenovirus Vectors

Second-generation vectors lack three (E1, E2, and E3 or E1, E3, and E4) of the viral regions that are expressed early after infection (see, e.g., Gao et al. 1996; Amalfitano et al. 1998; Lusky et al. 1998; Moorhead et al. 1999). Although these mutants have a greater carrying capacity than first-generation vectors, they retain the ability to induce inflammatory and cytotoxic responses that lead to the elimination of the virally transduced cells from the body. To minimize this problem, recombinant adenovirus vectors that lack all viral genes have been generated. These vectors express transgenes in vivo to a high level for extended periods of time and have a greatly expanded cloning capacity (~36 kb) and a greatly reduced ability to stimulate the immune system (Schiedner et al. 1998; Kochanek et al. 2001). However, production of helper-virus-free gutless vectors is more challenging as compared to that of earlier generations of adenovirus vectors. For further information about the construction and use of these gutless adenovirus vectors, see Parks et al. 1996; Kochanek et al. 2001; Alba et al. 2005; Palmer and Ng 2008.

Expression cassettes can be cloned into the DNA of first-generation adenovirus vectors by ligating the expression cassette directly into plasmids (see, e.g., He et al. 1998; Mizuguchi et al. 2001) or cosmids (Giampaoli et al. 2002) carrying a full-length copy of an E1-deleted vector technique (see Protocol 1).

Expression cassettes can also be introduced into the backbone of an adenovirus vector by recombination in either mammalian cells or *E. coli*. In mammalian cells, recombinants are generated in cells that express an integrated copy of the adenoviral E1 region and are permissive to viral growth (see Table 1). These helper cells are cotransfected with (1) the genomic DNA of an adenovirus

TABLE 1. Examples of helper cell lines that support the growth of adenovirus E1 mutants

Name of cell line	Description	Reference(s)
293	Generated by transforming normal human embryonic kidney cells with sheared adenovirus 5 DNA. The transformed cells contain 4.5 kb from the left-hand end of the adenovirus 5 genome integrated into human chromosome 19 of the HEK cells.	Graham et al. 1977; Louis et al. 1997
911	Generated by transforming human embryonic retinoblasts with a plasmid containing sequences from the left-hand end of adenovirus 5 (bp 79–5789 of the Ad5 genome). Claimed to be better than 293 cells (faster plaque formation [4–5 d rather than 5–8 d] and higher virus yields).	Fallaux et al. 1996
N52.E6	Generated by transforming primary human amniocytes with adenovirus 5 E1 DNA.	Schiedner et al. 2000
PER.C6	Contains the E1A and E1B genes of adenovirus 5 under the control of the human phosphoglycerokinase promoter. Yields of superinfecting adenoviruses are equivalent to the yields from 911 cells.	Fallaux et al. 1998

E1/E3 deletion and (2) a plasmid carrying an expression cassette flanked on both sides by tracts of DNA homologous to viral DNA sequences immediately upstream of and downstream from the E1 deletion. Homologous recombination between the plasmid and the viral DNA transfected into the helper cell line generates a recombinant in which the E1 region is replaced by the expression cassette (see Fig. 2).

Homologous recombination between the E1 sequences in the expression plasmid and the vector DNA results in formation of a recombinant with the expression cassette integrated into the E1 region of the vector. The recombinant virus particles are then plated on helper cells and the viral DNAs in individual plaques are screened by polymerase chain reaction (PCR) and/or cleavage with restriction enzymes. However, wild-type adenoviruses, generated by recombination between the vector and viral DNA sequences integrated into the genome of the helper cells, may also be present in the viral yield. Removing contaminating wild-type adenoviruses from the recombinant stock requires at least two rounds of plaque purification and screening. The frequency of wild-type contamination can also be reduced, but not entirely eliminated, by using cell lines such as 911 that carry a smaller segment of the adenoviral genome than 293 cells.

Constructing, purifying, and generating a high-titer stock of an adenovirus recombinant, by either direct in vitro insertion of an expression cassette into a second-generation vector or in vivo recombination, is a lengthy process that, even in experienced hands, may take several weeks to complete. If only one or two recombinants are required, it may be more effective, both for time required and for cost, to use a local core facility or a commercial kit. Commercial recombination systems are

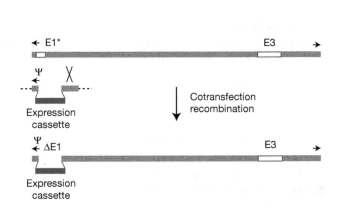

FIGURE 2. **Construction of first-generation recombinant Ad vectors.** E1-complementing cells (e.g., 293 cells) are cotransfected with a shuttle plasmid containing a transgene expression cassette flanked by sequences derived from the "left" end of the adenovirus genome together with genomic adenovirus DNA. The E1 region ("E1") of the genomic DNA, which can be either viral or plasmid derived, should be modified to reduce infectivity of the parental genome. Homologous recombination between the shuttle and genomic sequences results in a first-generation vector with the expression cassette replacing E1. (Gray bars) Adenovirus sequences, (blue bars) heterologous DNA, (white bars) sequences of the adenovirus genome that are not required for the construction of first-generation vectors in E1-complementing cell lines. (Redrawn from Hitt and Graham 2000, with permission from Elsevier.)

available from Clontech (AdenoX system, http://www.clontech.com/images/pt/PT3674-1.pdf) and Qbiogene (AdEasy system, http://www.qbiogene.com/products/adenovirus/adeasy.shtml).

If your laboratory will have an ongoing need to produce adenovirus recombinants, the best option in the long term may be to establish an advanced in vitro recombination system or ligation-based direct cloning system for reduced wild-type contamination, using, for example, the *E. coli* or cre-loxP recombination systems (Chartier et al. 1996; He et al. 1998; Luo et al. 2007; Reddy et al. 2007) or standard molecular cloning techniques to directly insert transgene expression cassettes into the deleted E1 locus in a clone of infectious recombinant adenovirus genome (Mizuguchi et al. 1998, 2001; Gao et al. 2003). The latter method is described in detail in Protocol 1 of this chapter.

Adeno-Associated Viral Vectors

At first sight, AAVs seem to have few of the qualities required of a good vector. Their single-stranded DNA genomes are small (4.7 kb), and their replication depends on coinfection of cells with a helper virus (e.g., adenovirus, herpes simplex virus, papilloma virus). In the absence of helper functions—for example, those supplied by adenoviral proteins encoded by early regions E1, E2a, VA RNA, and E4—the wild-type AAV genome enters a state of latency by integrating at a single specific locus (19q13.3-qter) in the genome of cultured human cells. Recombinant AAV vectors express transgenes at high level in target tissues for extended periods of time, in some instances remaining unchanged for >10 yr. Other advantageous qualities include a lack of pathogenicity, an ability to infect a broad range of cell types (both proliferating and quiescent), and a capacity to be rescued from an integrated state by superinfection with a helper virus or by transfection with plasmid vectors that supply helper functions in *trans* (for reviews, see Büning et al. 2008; Daya and Berns 2008).

The majority of AAV recombinants are generated in AAV serotype 2, whose linear genome, which is 4675 nucleotides in length (Srivastava et al. 1983), consists of a long, single-stranded coding region bracketed between identical but inverted 145-bp terminal repeats. Each repeat can fold into a T-shaped, inward-facing, hairpin-like structure that contains information required to prime synthesis of a complementary-strand DNA. The product is a monomer-length duplex, one end of which is covalently closed. This cross-linked structure creates a replication origin that, after activation by the viral Rep protein, generates a rolling hairpin structure, which allows the replication fork to shuttle up and down the length of the viral genome. Progeny genomes are excised by the introduction of specific nicks at the closed circular terminus of each full-length duplex copy of the viral DNA. Both sense and antisense single-stranded viral DNAs are packaged efficiently into viral particles. For further information, please see Cotmore and Tattersall (1996).

The coding region of the viral genome contains two open reading frames (*rep* and *cap*). *rep* encodes four proteins: (1) Rep 78, (2) its spliced variant Rep 40, (3) Rep 68, and (4) its spliced variant Rep 52 (see Fig. 3). *cap* codes for VP1, VP, and V3, the three viral capsid proteins, which differ in amino acid sequence only at their amino termini. Differences among the amino acid sequences of the capsid proteins account for the specific tissue tropisms displayed by different AAV serotypes and define the epitopes recognized by the host's immune system.

Construction of rAAV vectors is carried out in standard plasmids by simple molecular cloning methods. rAAV vectors do not encode the Rep proteins necessary for integration and, as a result, long-term transgene expression depends on the persistence of extrachromosomal concatamers of vector genomes. For a discussion of in vivo applications of AAV vectors, see the In Vivo Expression section.

Retroviruses

Retroviruses have a long history as vectors for introducing and expressing ectopic genes in cultured mammalian cells and as exploratory vehicles for gene therapy. In retroviral particles, the viral genome is complexed with a nucleocapsid protein within a lipid envelope that is studded by viral glycoprotein protein. Interactions between the viral glycoprotein and receptors on the cell surface

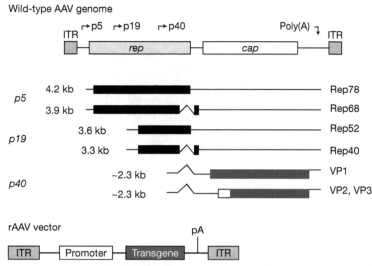

FIGURE 3. **Structure of adeno-associated virus (AAV) vectors.** The wild-type AAV consists of the viral genes *rep* and *cap* coding for the different rep (Rep78, Rep68, Rep52, Rep42) and cap (VP1, VP2, VP3) proteins, the AAV promoters (p5, p19, p40), the polyadenylation site (pA), and the inverted terminal repeats (ITR). In rAAV vectors, the viral *rep* and *cap* genes are replaced by a transgene cassette carrying the promoter, the transgene, and the pA site. (Redrawn from Walther and Stein 2000, with permission from Wolters Kluwer Pharma Solutions, Inc.)

determine the tropism of the virus. The viral genome consists of two identical copies of single-stranded RNA (ssRNA) that are 8–10 kb in length. The viral RNA replicates through a DNA intermediate that becomes stably integrated into the genome of infected cells via a recombination reaction catalyzed by the virally coded integrase. Production of progeny viral particles requires transcription of the integrated viral DNA by host-encoded RNA polymerase II (Pol II). The resulting mRNA is processed, transported, and translated using machinery translation of the host cell. Progeny viral particles are assembled at, and bud through, the cell surface.

All retroviral genomes contain at least three genes (see Fig. 4). In murine retroviruses, these are the *gag* gene that encodes the Gag polyprotein that is cleaved into the viral matrix, capsid, and nucleocapsid proteins; the *pol* gene that encodes a protease, reverse transcriptase, and integrase; and the *env* gene that encodes a polyprotein that is cleaved by the viral protease into a surface glycoprotein (gp70) and a transmembrane protein (p15E).

The regions of the viral genome that regulate DNA synthesis and transcription are clustered in the two LTRs and consist of three functional regions, as shown in Table 2. The region downstream of the 5′-LTR contains a binding site for the transfer RNA that serves as a primer for initiation of DNA synthesis by reverse transcriptase. Finally, the sequences needed for packaging of the viral RNA are located between the primer-binding site and the *gag* open reading frame.

Most retroviral vectors in current use are derived from the Moloney strain of murine leukemia virus (Mo-MLV), which is amphotropic and has the ability to infect both murine and human cells. Recombinants are generated by replacing the viral *gag*, *pol*, and *env* genes in a cloned copy of the vector genome with an appropriately designed transgene. The maximum size of the transgene that can be efficiently cloned in a typical retroviral vector is ~6.5 kb. Expression of the transgene is driven either by the natural Mo-MLV control elements in the upstream LTR or by an interval promoter (e.g., the CMV promoter) cloned along with the transgene.

To generate a stock of recombinant virus, the DNA of the retroviral recombinant constructed in vitro is transfected into a packaging cell line that expresses the three viral genes necessary for particle formation and replication: *gag*, *pol*, and *env*. The backbone of the vector provides the signals (ψ) required in *cis* for packaging of the recombinant genomes into viral particles. The viral Env protein expressed by the packaging cell line determines the host range of the progeny viral particles. For example, the EcoPack 2-293 cell line sold by Clontech generates ecotropic recombinant viruses

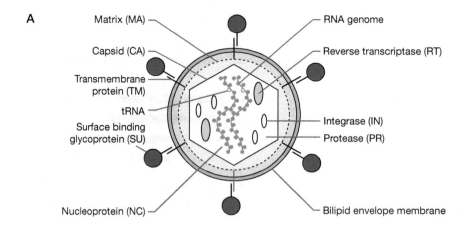

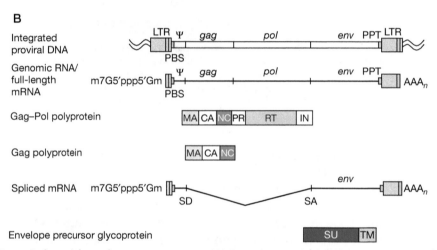

FIGURE 4. Retroviral particle and genome structure. (*A*) Retrovirus particle showing approximate location of its components using standardized two-letter nomenclature for retroviral proteins. (*B*) Genome organization and gene expression pattern of a simple retrovirus showing the structure of an integrated provirus linked to flanking host cellular DNA at the termini of its LTR sequences (U3-R-U5) and the full-length RNA that serves as genomic RNA and as mRNA for translation of the *gag* and *pol* open reading frames (ORFs) into polyproteins. *env* mRNA is generated by splicing and encodes an Env precursor glycoprotein. (LTR) (U3-R-U5) for proviral DNA, derived from R-U5 downstream from 5′ cap and U3-R upstream of 3′ poly(A) in genomic RNA; (PBS) primer binding site; (Ψ) packaging signal; (PPT) polypurine tract; (SD) splice donor site; (SA) splice acceptor site. (Redrawn from Pedersen and Duch 2003, with permission from John Wiley & Sons Ltd.)

that can infect only rodent cells. In contrast, other cell lines, such as AmphoPack-293 cells and RetroPack PT67 cells (both Clontech), generate amphotropic or dual-tropic viruses, respectively, with the ability to infect a broad range of mammalian cell types. If necessary, the host range of the recombinant virus can be extended still further by producing the recombinant virus in a cell line (e.g., GP2-293 [Clontech]) that expresses the G glycoprotein of the vesicular stomatitis virus

TABLE 2. Regions of the viral genome that regulate DNA synthesis and transcription are clustered in the two LTRs that consist of three functional regions

Region of LTR	Functions
U3 (unique-3′)	Transcriptional promoters and enhancers
R (repeat)	Reverse transcription and replication; polyadenylation
U5 (unique-5′)	Initiation of reverse transcription

LTR, long terminal repeat.

(VSV), which mediates viral entry through lipid binding and plasma membrane fusion rather than by attachment to a specific cell-surface receptor.

The virus stocks generated in helper cells are used to transduce actively dividing cultures of target cells that are of interest to the researcher. In contrast to lentiviruses (see below), recombinant murine retroviruses can transduce only actively dividing cells. This is because the murine viral preintegration complex cannot enter the nucleus of a nondividing cell. Once the recombinant provirus penetrates the nucleus, the viral genome integrates into the cellular genome, enabling long-term expression of the transgene. Each transduced cell carries the transgene at a different chromosomal location because integration is not site- or sequence-specific. The realization that integration of recombinant murine viral genomes at chromosomal sites close to cellular oncogenes (Montini et al. 2009) could lead to proliferation of clonal populations of T cells in patients treated by gene therapy for X-linked severe combined immunodeficiency (SCID-XI; Hacein-Bey et al. 2001) is a considerable obstacle to the therapeutic use of recombinant murine retroviruses.

Lentiviruses, a genus of the Retroviridae family, are more complicated than murine retroviruses. In addition to the three canonical genes (*gag, pol,* and *env*), their genome encodes at least six additional proteins (tat, rev, vpr, vpu, nef, and vif) that perform a variety of regulatory functions involved in viral pathogenicity. However, lentiviral vectors derived from the human immunodeficiency virus type 1 (HIV-1) have two advantages over murine retrovirus vectors: (1) Lentiviruses stably integrate the viral genome into the genomes of both dividing and nondividing cells (Naldini et al. 1996a,b) and (2) the genotoxic potential of lentiviruses appears to be considerably lower than that of murine retrovirus (Montini et al. 2009).

Recombinants constructed in lentivirus vectors can be used as simple expression plasmids for transient transfection of cells in culture or they can be packaged into high-titer viral stocks and used to genetically modify cells in culture or in vivo. In addition, lentiviruses can be used to generate transgenic animals by transduction of murine embryonic stem (ES) cells in culture (Pfeifer et al. 2002), or by injection of vector stocks into the perivitelline space of single-cell mouse embryos (Lois et al. 2002).

Exceptionally useful libraries to probe gene function and regulation have been constructed in lentivirus vectors:

- the RNAi Consortium small hairpin RNA (shRNA) library, where shRNAs expressed from a U6 promoter target most mouse and human genes (commercialized by Sigma-Aldrich and Open Biosystems)

- a microRNA (miRNA)-adapted shRNA (shRNAmir) library, where shRNAs are expressed in the context of a miRNA precursor (shRNAmir) from a RNA Pol II promoter (Open Biosystems)

- a library of miRNA precursors (Open Biosystems) or cDNAs (Open Biosystems; GeneCopoeia)

The great advantage of lentivirus-based libraries is that they dramatically accelerate the ability to perform genome-wide, high-throughput screens of primary cultures of cells such as neurons that might not be susceptible to transfection by traditional methods. Importantly, lentivirus vector transduction does not appear to significantly change the gene expression pattern of target cells (Cassani et al. 2009).

Examples of lentiviral expression systems that have been designed for specific purposes include:

- bicistronic vectors, in which the transgene is coexpressed with drug-selection genes or fluorescent proteins (or both) to allow for selection and easy identification of transduced cells (for review, see de Felipe 2002)

- polycistronic vectors that encode the four transcription factors necessary to generate induced pluripotent stem (iPS) cells from somatic cells along with green fluorescent protein (GFP) (Carey et al. 2009) (Cell Bio-Systems)

- vectors that allow for drug-regulated gene expression of cDNAs (Pluta et al. 2005), shRNAs (Szulc et al. 2006; Wiznerowicz et al. 2006) (Open Bio-Systems; Addgene), and shRNAmirs (Stegmeier et al. 2005) (Open Bio-Systems)

Many of these lentivirus vectors are available through companies such as Life Technologies, Open Biosystems, Cell Biolabs, and GeneCopoeia, or at Addgene.org, where plasmids are deposited for distribution to other academic institutions. Figure 5 shows an example of a basic lentivirus backbone deposited at Addgene that includes all of the elements necessary to achieve highly efficient transgene expression in cultured cells or in vivo. The vector shown, pRRLSIN.cPPT.PGK-GFP.WPRE, was among the first to incorporate all of the design features that enhance transduction and safety of lentivirus vectors (Follenzi et al. 2000). Many other lentivirus vectors carrying a variety of different promoters are also available through Addgene.

Lentivirus vectors can be used to generate mammalian cell lines that produce high levels of recombinant mammalian proteins. Because lentivirus vectors stably transduce target cells, enhanced levels of recombinant protein can be achieved by sequential rounds of transduction and selection. Using a bicistronic vector in which GFP (or any other fluorescent marker protein) and the transgene of interest are placed under the control of the same promoter, the brightest cells can be selected by fluorescent cell sorting and reinfected until the required level of production of the target protein is reached.

For many laboratories, lentivirus vectors are the best choice as mammalian vectors. They can be manipulated using standard recombinant DNA techniques, they can be used as simple expression

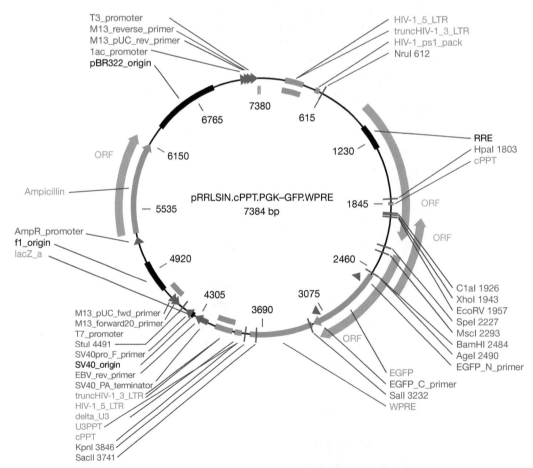

FIGURE 5. Vector pRRLSIN.cPPT.PGK-GFP.WPRE. Example of a lentiviral vector genome that includes all of the elements needed to achieve highly efficient transgene expression in cultured cells or in vivo. (Plasmid 12252 redrawn from Addgene, http://www.addgene.org/pgvec1?f=c&cmd=findpl&identifier=12252, and Follenzi et al. 2000.)

plasmids, and the production of vector stocks is straightforward and quick, requiring only a simple cotransfection of 293 cells with helper plasmids and harvesting of the recombinant virus in the supernantant a few days later. Thus, lentivirus vectors may be the best system to rapidly express and screen large numbers of mutant recombinant proteins.

IN VIVO EXPRESSION

Despite the rapidly increasing use of lentivirus for genetic modification of cells in culture and in vivo, these vectors still have some limitations that include their total transgene capacity (~8 kb) and restricted transduction after systemic gene delivery in vivo. Vector systems with considerably larger transgene capacity (e.g., high-capacity adenovirus vectors [Schiedner et al. 1998] and herpes simplex 1 amplicons [up to ~150 kb] [Geller and Breakefield 1988]), have been used to deliver entire genomic loci to target cells to achieve regulated gene expression (Wade-Martins et al. 2001). However, these vectors are extremely difficult to produce at high titers and rarely achieve long-term expression of target genes.

rAAV vectors share many of the attributes of lentivirus vectors; AAV plasmids can be easily manipulated by standard recombinant DNA techniques and they can be used as conventional expression plasmids. However, the most useful properties of AAV vectors are their exceptional efficiency and duration of transgene expression in vivo, which, in many cases, appears to last for the life span of the experimental animal. In rhesus macaques that received an intramuscular injection of an AAV vector encoding a primate transgene, expression remained unabated 11 years later (G Gao, pers. comm.). Moreover, the in vivo gene-delivery efficiency of AAV vectors is such that a single intravenous infusion in adult animals is sufficient to achieve nearly complete transduction of liver, heart, and skeletal muscle as well as transduction in the central nervous system (CNS) (Gregorevic et al. 2004; Wang et al. 2005; Inagaki et al. 2006; Duque et al. 2009; Foust et al. 2009; Hester et al. 2009). Despite these extraordinary properties, AAV vectors are not the best choice for transduction of dividing cells in culture because they do not integrate into the target cell genome and are slowly lost over time.

ADENOVIRUS VECTORS

Recombinant adenoviruses are generated by ligating the gene of interest into a plasmid carrying a replication-defective adenovirus genome. This method was developed by Mizuguchi and Kay (1998) as a simple and efficient ligation-based platform, and it is commercially available under the trade name Adeno-X Expression System (Clontech). Gao and colleagues (2003) further modified this system with a convenient green–white selection feature that facilitates the isolation of transgene-positive recombinant adenovirus clones. The ligation-based direct cloning system is considerably faster and more efficient than traditional methods for generating recombinant adenovirus via homologous recombination. Direct ligation, although quicker and more efficient, is technically more challenging, requiring proficient skills to manipulate the large sizes of plasmids (>36 kb). Homologous recombination usually generates a mixture of recombinants, including replication-competent adenovirus (RCA), removal of which requires several time-consuming rounds of plaque purification and characterization.

Production of a recombinant adenovirus simply requires linearization of the infectious clone by digestion with the endonuclease PacI, followed by transfection into human embryonic kidney 293 cells or other packaging cell lines (see Table 1) with the necessary ability to complement viral genes deleted during construction of the vector backbone. Figure 6 depicts the entire process for generating a recombinant adenovirus vector using direct cloning with the green–white selection method. The process starts with cloning the gene of interest into the shuttle plasmid pShuttle-*pk-GFP* (pSh-*pkGFP*), followed by another cloning step to transfer the transgene expression cassette from the shuttle plasmid into the adenovirus clone pAd-*pkGFP*. Both pSh-*pkGFP* and pAd-*pkGFP*

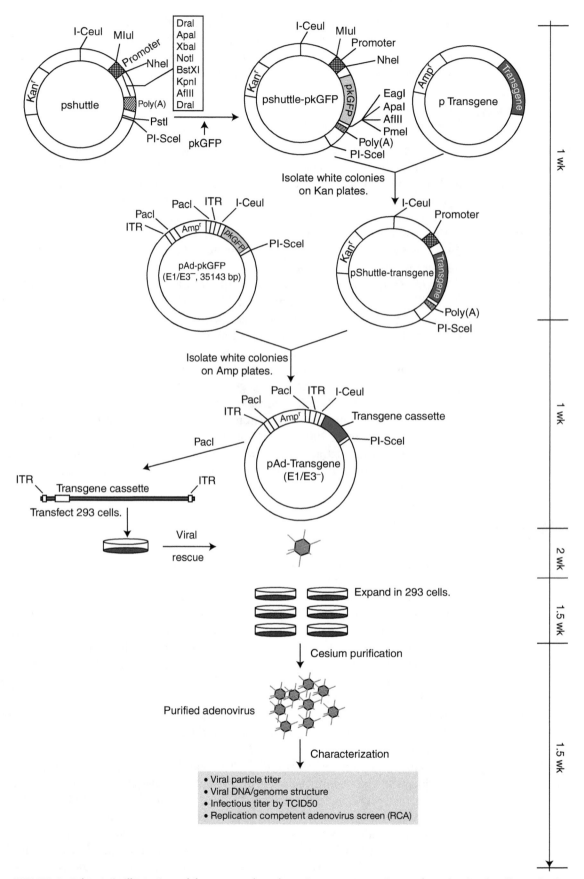

FIGURE 6. Schematic illustration of the process for adenovirus vector creation and production by direct cloning (Protocol 1).

plasmids carry a prokaryotic green fluorescent protein (GFP) expression cassette for easy selection of transgene-positive/GFP-negative bacterial colonies. The next steps are to linearize the recombinant adenovirus plasmid by PacI digestion to expose the inverted terminal repeats (ITRs), followed by transfection into 293 cells for rescue and expansion. Finally, purified high-titer recombinant adenovirus stocks are prepared by cesium chloride (CsCl) gradient sedimentation, followed by detailed characterization of vector genome structure, vector particle titer, and infectious titer, as well as testing for the presence of RCA.

Protocol 1 provides step-by-step instructions for creating recombinant adenovirus by this method. Protocol 2 describes the process of virus rescue and expansion, and Protocol 3 describes the purification of adenovirus vectors by gradient sedimentation. The final set of protocols (Protocols 4–6) is a series of supporting protocols for detailed characterization of recombinant adenovirus stocks.

Strategic Planning

Selection of Vector Backbone and Corresponding Packaging Cell Line

Several critical factors need to be considered in choosing the backbone, including the intended target (e.g., cells in culture or tissues/organs in intact animals), the impact of vector immunogenicity on the outcome of gene transfer, vector toxicity, transgene size, and desired level of transgene expression (Table 3). For simple gene-transfer experiments in cell culture where a high level of transient transgene expression is desired, a backbone with an E1 deletion only or with an E1/E3 double deletion is ideal (see the information panel Basic Elements in Viral Vectors). However, for systemic administration of an adenovirus recombinant, where liver toxicity and strong T-cell responses to viral and transgene products expressed in target cells may limit the interpretation of gene-transfer data, or when extended transgene expression is necessary, an E1/E4 double-deleted vector backbone may be a more appropriate choice.

For applications where large transgene capacity is necessary, two types of backbones can be used. One is an E1/E3/E4 triple-deletion vector backbone with a transgene capacity of up to 8 kb; however, E1/E4 double-complementing cell lines, such as 10-3 cells, are required for recombinants. Because constitutive high-level expression of E4 in 293 cells is toxic, 10-3 cells carry an E4 orf6 expression cassette driven by the metallothionein promoter, which is inducible by heavy metals (Table 4). Alternatively, the transgene capacity of an E1/E3-deleted vector backbone can be increased by deleting most of E4, except for orf6. Because E4 orf6 is sufficient to provide the major

TABLE 3. Comparison of major characteristics of four commonly used RNA and DNA viral vectors

Vector	Transgene capacity	Titer	Scale-up	Efficiency/ stability	Tropism and primary use	Host responsible	Genetic fate (genotoxicity)	Biosafety
Retro-	Up to 7 kb	10^6–10^8 IU/mL	Difficult	Low/stable	Dividing cells in vitro and ex vivo	Low immunogenicity and toxicity	Integrated[a]	BSL-II
Lenti-	Up to 7.5 kb	10^6–10^{10} IU/mL	Difficult	Moderate–high/stable	Dividing/nondividing cells, envelope/pseudotype-dependent tissue tropism in vitro and in vivo	Low immunogenicity and toxicity	Integrated[a]	BSL-II
Adeno-	Up to 35 kb	10^{12}–10^{13} virus particles/mL	Scalable	High/transient	Broad in vitro and in vivo	Highly immunogenic and toxic	Episomal	BSL-II
AAV	Up to 4.5 kb	10^{12}–10^{13} GC/mlL	Scalable	High/stable	Broad in vivo	Low immunogenicity and toxicity	Episomal	BSL-I

[a]See the information panel Lentivirus Vectors.

TABLE 4. Choices of backbones for clones of adenovirus vectors

Vector backbone	Viral gene expression	Transgene capacity/ expression	Packaging cell line	Vector yield	Innate (cytokine)	T cell Viral	T cell Transgene	Vector toxicity
					Immunogenicity			
ΔE1	Detectable	4 kb/strong	293 cells	High	IL6$^\uparrow$ and IL10$^\uparrow$	Strong	Strong	High
ΔE1+E3	Same	6 kb/strong	293 cells	Same	Same	Same	Same	Same
ΔE1+E3+E4	Diminished	8 kb/reduced	10-3 cells	Reduced	Same	Diminished	Diminished	Diminished

A $\uparrow$ represents an increase in expression.

functions of the E4 gene, an adenovirus backbone with E1/E3 full and E4 partial deletions can be rescued and grown in regular 293 cells and have a transgene capacity of up to 7 kb (Table 4).

Selection of Promoter for Transgene Expression

For most vector backbones listed in Table 4, strong viral promoters such as the cytomegalovirus (CMV) immediate-early promoter, drive high-level transgene expression both in culture and in vivo. However, the CMV promoter appears to shut down over time after systemic infusion of E4-deleted adenovirus recombinants in animals (Armentano et al. 1997). In this case, strong constitutive cellular promoters should be considered in the vector design. Additionally, when adenovirus vectors carry genes that regulate the cell cycle, or are cytotoxic, introduction of a mechanism to regulate transgene expression during vector production should be considered. Otherwise, expression during production could lead to loss of transfected or infected cells, reducing vector yields significantly or preventing virus rescue and infection entirely (Bruder et al. 2000).

ADENO-ASSOCIATED VIRUS VECTORS

To date, at least five different methods have been developed for the production of recombinant adeno-associated virus (rAAV) vectors (Zhang et al. 2009). These include (1) helper-free triple trans-fection of 293 cells, (2) infection of a stable rep/cap cell line with Ad-AAV hybrid, (3) infection of an rAAV producer cell line with wild-type adenovirus helper, (4) co-infection of 293 cells with two recombinant herpes simplex virus (rHSV) vectors, and (5) a recombinant baculovirus-based system. The method of helper-free triple transfection of 293 cells is the method most commonly used for production of rAAV vectors in the laboratory setting (Grieger et al. 2006). The process described in this chapter is simple and versatile and can be implemented in most laboratories. The simplicity and flexibility of this method facilitates the simultaneous production of rAAVs carrying different transgenes with the same or different capsids. Two drawbacks include difficulties in scaling up and the potential emergence of low levels of replication-competent AAV (rcAAV) particles in the vector preparations.

An overview of the rAAV production process using the helper-free triple-transfection method in 293 cells is summarized in Figure 7. In this production system, all genetic components that are necessary for the packaging of rAAV particles in 293 cells are provided by three plasmids: (1) the vector plasmid or pCis carries a transgene cassette of interest that is flanked by ITRs, the only viral elements from AAV (<4% of the AAV genome), (2) the packaging plasmid or pTrans expressing AAV Rep and Cap proteins for rescue and packaging of vector genomes, and (3) an adenovirus helper plasmid supplying all required helper functions from E2a, E4, and VA RNA genes except for the E1 gene, which is *trans*-complemented by E1 expression in 293 cells.

The production process of AAV recombinants starts with the generation of a pCis plasmid carrying an expression cassette for the transgene of interest. This is accomplished by subcloning the gene of interest, which could be a complementary DNA (cDNA), small hairpin RNA (shRNA),

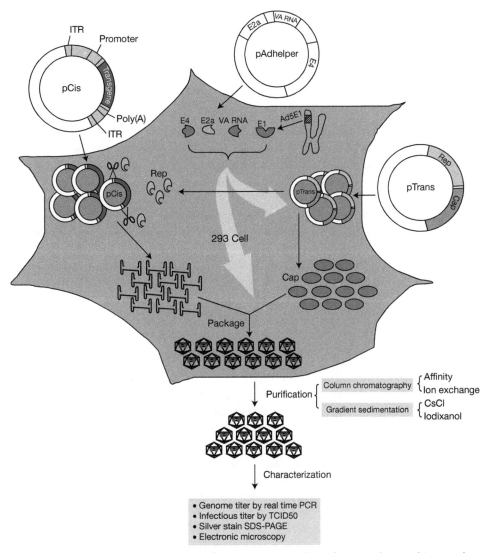

FIGURE 7. **Schematic illustration of the process for production and purification of recombinant adeno-associated virus by 293 cell–triple-transfection method (Protocol 7).**

or artificial miRNA shuttle, into the pCis vector plasmid containing transcription elements of choice (promoter and polyadenylation signal) flanked by AAV ITRs (Fig. 7). Most AAV vectors carry ITR elements derived from AAV2. The second step is to select the packaging plasmid (pTrans) that expresses the AAV serotype capsid of interest. The third step is to perform triple transfection of the pCis, pTrans, and pAd helper plasmids into 293 cells for AAV vector genome rescue and packaging (Protocol 7). This is followed by purification of rAAV particles from the crude lysate of transfected cells.

As shown in Figure 7, the methods used in the purification of rAAVs can be divided into three major categories: (1) gradient sedimentation, (2) column chromatography, and (3) combination of gradient centrifugation with column chromatography.

Gradient Sedimentation

For the first category of purification methods, either cesium chloride (CsCl) or iodixanol formed gradients can be used to purify rAAVs by ultracentrifugation (Protocol 8 or 9) (Grieger et al. 2006). The major advantages of gradient centrifugation–based methods are the ability to separate virions carrying vector genomes from empty viral particles and their applicability to all serotypes.

CsCl gradient centrifugation is the most commonly used method for laboratory purification of AAV vectors; however, this method is time consuming and difficult to scale up. In addition, it requires an additional step to remove toxic chemicals used to generate the gradient and reformulate the vector stock in a physiological buffer.

Column Chromatography–Based Methods

The second category of purification methods is column chromatography–based and includes affinity chromatography (Protocol 10) and ion-exchange chromatography (Protocol 11) (Grieger et al. 2006). Overall, the column chromatography–based methods are faster and more efficient in removing cellular and viral impurities. However, they also have shortfalls, including in developing a generic purification protocol for all serotypes and the inability to separate empty particles from vector preparations. An effective strategy to enrich the fully packaged virions in column-purified rAAV preparations is to combine gradient centrifugation with column chromatography, as described in Protocol 11.

Finally, the genome of the purified rAAV should be quantitated by real-time PCR (Protocol 12); the infectivity of the rAAV should be measured by infection of a rep-cap cell line and real-time qPCR (Protocol 13). The morphology of the rAAV can be confirmed by electron microscopy (Protocol 14), and its purity estimated by SDS-PAGE and silver staining (Protocol 15).

Strategic Planning

Transgene Expression Cassette Design

The maximum size of the genome that can be efficiently packaged in most AAV capsids is 4.7–4.8 kb. This constraint on vector size dictates a minimalistic approach to the design of transgene expression cassettes in rAAVs, which should include only essential transcriptional accessories (e.g., promoter and polyadenylation signals of minimal size). The choice of promoter should be guided by, for example, whether ubiquitous or tissue-specific expression is desired (Le Bec and Douar 2006). In the former case, hybrid promoters such as CBA (also known as CAG, or CB), which is composed of the cytomegalovirus (CMV) enhancer fused to the chicken β-actin promoter, mediates robust stable expression both in culture and in vivo in most cell types and tissues targeted by AAV vectors. Strong viral promoters such as the CMV immediate-early gene promoter works well for rAAV-mediated gene transduction in most target tissues except for liver, where CMV-directed transgene expression rapidly shuts off. Incorporation of tissue- or cell-type-specific promoters is reasonably effective in restricting expression to the desired target. Alternatively, tissue-specific expression can be achieved by incorporating into the vector mRNA target sequences that are recognized by miRNAs differentially expressed in target (absent) and nontarget (present) tissues. Finally, pharmacologically regulated rAAV-mediated transgene expression can be realized by incorporating the transcription regulator(s) and transgene cassettes separately into two vector genomes, packaging them individually, and co-injecting both vectors into the target tissue (Rivera et al. 1999; Ye et al. 1999).

Serotype Selection

Recombinant AAVs carrying the same transgenic genome (same AAV2 ITR-flanked expression cassette for a particular transgene), but with capsids derived from different AAV serotypes/strains, display dramatically different transduction properties in vivo (Gao et al. 2005). The AAV capsid determines the cell/tissue tropism of AAV vectors and other aspects of their biology. The choice of AAV capsid for a particular application should be guided by the ability to transduce the intended target tissue using a particular delivery route (Gao et al. 2005). Table 5 characterizes the major AAV serotype vectors with regard to their cellular receptor(s), optimal target tissues, and relative transduction efficiency in culture (293 cells) and in vivo. Production of AAV vectors with different capsids is accomplished by (1) *trans*-encapsidation of the same vector genome flanked by AAV2 ITRs, (2) capsid protein expressed from a chimeric packaging plasmid carrying the *rep* gene from AAV2

TABLE 5. Target tissue–specific AAV capsid selection

Capsid	Receptor	Optimal target tissues	Transduction (293 cells/in vivo)
AAV1	N-linked sialic acid	Skeletal muscle, CNS	Moderate/good
AAV2	HSPG	Skeletal muscle, CNS	Good/poor
	αVβ5 integrin		
	FGFR1		
	Laminin		
AAV4	O-linked sialic acid	CNS, eye/RPE	Poor/moderate
AAV5	N-linked sialic acid PDGFR	CNS, lung, eye/RPE/photoreceptor	Poor/moderate
AAV6	N-linked sialic acid	Skeletal muscle, cardium	Poor/good
AAV7	Unknown	Skeletal muscle, pancreas, liver	Poor/good
AAV8	Laminin	Liver, skeletal muscle, pancreas	Poor/good
AAV9	Laminin	Liver, lung, skeletal muscle, cardium, CNS (via both local and transvascular delivery)	Poor/good
rh.10	Unknown	Lung, CNS	Poor/good

AAV, adeno-associated virus; HSPG, heparan sulfate proteoglycan; FGFR, fibroblast growth factor receptor; RPE, retinal pigment epithelium; PDGFR, platelet-derived growth factor receptor; CNS, central nervous system.

(which is necessary to replicate AAV2 ITR-flanked genomes during packaging), and (3) the *cap* gene from an AAV serotype/strain of choice.

One paradoxical property of AAV recombinants is that they are generally inefficient for gene delivery to cells in culture but display remarkable transduction efficiency in vivo. The transduction efficiency of these vectors in culture, regardless of capsid type, is considerably (several orders of magnitude) lower than adenovirus or lentivirus vectors at comparable MOI values. Interestingly, AAV2 recombinants are generally the most effective for gene transfer in cell culture but are among the least efficient for many in vivo gene-delivery applications.

RETROVIRUS AND LENTIVIRUS VECTORS

Retrovirus vectors were among the first gene-transfer vehicles to gain wide acceptance in the scientific community as a tool for engineering cells in culture with genes of interest. Most retrovirus vectors currently in use are derived from Moloney murine leukemia virus (Mo-MLV), and numerous design improvements have been implemented over the years. The fastest and most efficient method to generate retrovirus vector stocks is by transient transfection of 293-based packaging cell lines such as Phoenix-ECO and Phoenix-AMPHO (Dr. Gary P. Nolan, Stanford University) or other commercially available cell lines such as Plat-A, Plat-E, and Plat-GP (Cell Biolabs); AmphoPak-293; EcoPak 2-293; and RetroPack PT67 (Clontech). For pseudotyping retrovirus vectors with other envelope proteins such as vesicular stomatitis virus glycoprotein (VSV-G), cell lines can be used that express Mo-MLV *gag-pol* genes only (Plat-GP cells, Cell Biolabs; GP2-293 cells, Clontech) cotransfected with the vector plasmid and an expression plasmid encoding the envelope of interest (e.g., see pVSV-G below). An alternative is triple transient transfection of 293 cells with the retrovirus vector plasmid and two expression plasmids separately encoding Mo-MLV *gag-pol* genes (e.g., pUMVC plasmid; Plasmid 8449 at Addgene.org) and an envelope protein (similar to the approach for production of lentivirus vectors, see below). The protocols for production and titration of retrovirus vector stocks by transient transfection of both 293-based packaging cell lines and 293 cells are the same as those used for lentivirus vectors.

The most efficient method for producing high-titer lentivirus vectors is by transient cotransfection of human embryonic kidney 293T cells with transfer vector plasmid, packaging plasmid(s), and an envelope expression plasmid. Lentiviruses are surprisingly tolerant of envelope proteins

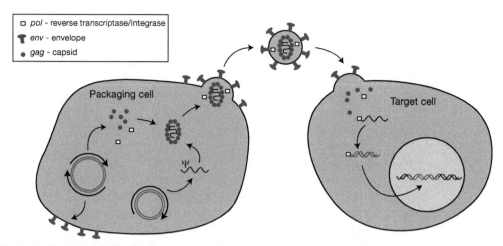

□ *pol* - reverse transcriptase/integrase

♆ *env* - envelope

● *gag* - capsid

Packaging cell

Target cell

FIGURE 8. Packaging and infection by a lentiviral vector. During production, the lentivirus vector plasmid generates genomic RNA that is transported to the cytoplasm, and virion assembly takes place at the cellular membrane with structural proteins encoded by the helper plasmid (*left* panel). Upon entry into a target cell, the lentivirus genome is reverse-transcribed in a pre-integration complex, which is transported into the nucleus of dividing and nondividing cells. Once in the cell nucleus, the viral integrase (part of the pre-integration complex) mediates integration of the vector genome into the host cell genome. Integration into the host genome is not necessary for lentivirus vector–mediated transgene expression as integrase-deficient vectors are capable of mediating robust expression.

derived from other viruses, a process known as pseudotyping. In consequence, VSV-G is the most commonly used envelope for producing recombinant lentivirus vectors. This is mainly due to the broad tropism of VSV-G pseudotyped lentivirus vectors, which can transduce a wide range of cell types from many different organisms both in culture and in vivo. Moreover, these vectors can be concentrated by ultracentrifugation to titers in excess of 10^{10} TU/mL, which are useful for gene-transfer experiments in vivo (Fig. 8). The protocol included in this chapter describes the production of HIV-1–derived lentivirus vectors pseudotyped with a VSV-G envelope because these vectors have become the most widely used vectors of their class. Lentivirus vectors are highly effective for gene transfer both to dividing and to nondividing cells. Thus, they have largely replaced Mo-MLV–based retrovirus vectors, which require actively dividing cells as the tools of choice for gene function and regulation studies. Additional gene-transfer systems have been developed based on other lentiviruses such as equine infectious anemia virus (EIAV), simian immunodeficiency virus (SIV), and feline immunodeficiency virus (FIV). The method described here can be easily adapted for production of any retrovirus/lentivirus vectors and would differ only in the type of transfer vector and plasmid encoding helper functions.

After the production method (Protocol 16), we describe different approaches to determine the titer of a lentivirus vector (Protocol 17) and, finally, a method to screen stocks for replication-competent lentiviruses (RCLs) (Protocol 18). These protocols can also be adapted for other retrovirus/lentivirus vector systems.

Strategic Planning

Retrovirus Vector Design

Retrovirus vectors can accommodate up to 7–8 kb of foreign sequences. The Babe-Puro retrovirus vector (Morgenstern and Land 1990) is a prototypical classical design still in use. In this vector, the gene of interest is expressed from the Mo-MLV LTR promoter, and an internal immediate-early SV40 promoter drives expression of a drug resistance marker (puromycin, neomycin, or hygromycin) (Fig. 9). These vectors incorporate several modifications that reduce the risk of generating replication-competent retroviruses (RCRs) during packaging. Later versions of retrovirus vectors (Fig. 9B) carry internal mammalian or viral promoters in the context of a self-inactivating (SIN) design to prevent interference between the 5′-LTR and the internal promoter. During reverse

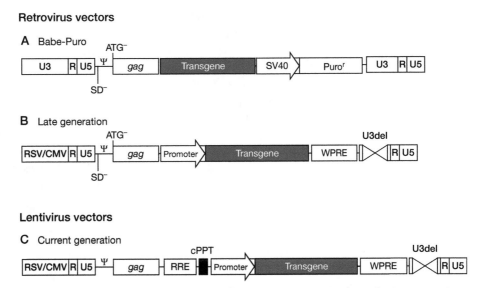

FIGURE 9. Basic structure of retrovirus and HIV-1 lentivirus vectors. (*A*) The Babe-Puro retrovirus vector is a classical design still in use today. The vector was optimized from previous generations to reduce the risk of generating replication-competent retroviruses (RCRs) during packaging. The gene of interest is expressed from the Moloney murine virus LTR promoter. The vector also carries a drug-resistance gene such as puromycin (other versions carry neomycin or hygromycin genes) under the immediate-early SV40 promoter to allow for selection of retrovirus vector–transduced cells in culture. (*B*) Later generations of retrovirus vectors carry internal mammalian or viral promoters in the form of a self-activating (SIN) design in which the promoter elements (U3 region) in the 3′ LTR are deleted, resulting in the inactivation of both LTR elements in the integrated provirus genome. Many of these vectors also carry the woodchuck hepatitis virus posttranslational regulatory element (WPRE) to increase transgene expression levels and may also carry a hybrid 5′ LTR composed of the CMV or RSV promoter in place of the native viral U3 promoter elements. (*C*) The design of lentivirus vectors is identical to late-generation retrovirus vectors. Most HIV-1–based lentivirus vectors are self-activating and carry a chimeric 5′ LTR in which the CMV and RSV promoters replace the HIV-1 native U3 promoter elements, making packaging of these vectors Tat independent and thus compatible with third-generation packaging systems. In addition to the packaging signal (Ψ) immediately following the 5′ LTR, all lentivirus vectors carry a portion of the *gag* gene. In addition, all vectors carry the Rev-responsive element (RRE) necessary for nuclear export of the vector RNA genome for packaging into virions. Inclusion of the central polypurine tract (cPPT) dramatically enhances transduction efficiency. WPRE is also commonly incorporated into lentivirus vectors. The choice of internal promoter to drive transgene expression is dictated by the specific experimental application.

transcription of retroviruses, the U3 region present only at the 3′ end of the genomic RNA is duplicated to the 5′-LTR in the provirus. As a consequence, deletion of the U3 enhancer elements in the 3′-LTR of a retrovirus (or lentivirus) vector plasmid leads to transcriptionally inactive 5′- and 3′-LTRs in the integrated provirus. It is interesting to note that the transcriptional status of the 3′-LTR in Mo-MLV retrovirus vectors is a key determinant of their genotoxicity (Montini et al. 2009). To reduce the risk of generating RCRs by recombination, the U3 promoter in the 5′ LTR has been replaced with another viral promoter such as the CMV immediate-early promoter or the Rous sarcoma virus (RSV) promoter. In addition, incorporation of posttranscriptional regulatory elements at the 3′ end of the expression cassette, such as the woodchuck posttranscriptional regulatory element (WPRE) or the constitutive transport element (CTE) from the Mason–Pfizer monkey virus, enhances transgene expression levels quite considerably. One of the drawbacks of Mo-MLV retrovirus vectors is their inability to transduce (infect) nondividing cells. This is because their preintegration complex can only gain access to the host cell genome after loss of nuclear membrane integrity during mitosis. Most of the design features first developed for retrovirus vectors have also been adopted for lentivirus vectors.

Lentivirus Vector Design

Lentivirus vectors can accommodate up to 7.5–8 kb of transgenic sequences and, as such, are compatible with an enormous variety of designs to accomplish different goals. There are, however, a number of basic components that have been shown to enhance safety, gene transfer, and expression efficiency (Fig. 9): (1) The U3 promoter region in the 5′-LTR has been replaced with the RSV or CMV promoter, rendering packaging of these vectors independent of Tat expression. (2) Deletion of U3 promoter elements in the 3′-LTR renders the LTR element transcriptionally inactive after integration into the host cell genome. These vectors are known as SIN, and recent evidence indicates that this is a critical feature that significantly reduces genotoxicity (see the information panel Lentivirus Vectors). (3) Incorporation of a second polypurine tract in the vector genome, central polypurine tract, or central flap is critical for nuclear transport of the preintegration complex and dramatically increases transduction efficiency (Follenzi et al. 2000; Sirven et al. 2000). (4) Incorporation of a post-transcriptional regulatory element derived from the WPRE enhances transgene expression (Zufferey et al. 1999). Numerous lentivirus vector designs have been developed to achieve the following:

- tissue-specific expression by incorporation of tissue-specific promoters and/or miRNA targets to eliminate off-target effects (Frecha et al. 2008)

- expression of multiple proteins from a single vector using IRES, bidirectional promoters, or multiple open reading frames separated by self-cleaving 2A peptides (Carey et al. 2009)

- shRNA and miRNA expression

- generation of cDNA, shRNA (Moffat et al. 2006), and miRNA (Open Biosystems, Inc.) libraries

- drug-regulated gene expression

Envelope Selection

HIV-1–derived lentivirus vectors are permissive for incorporation of different envelope proteins in the virion. VSV-G pseudotyped lentivirus vectors appear to be highly effective for in vivo and ex vivo gene delivery. Lentivirus vectors pseudotyped with envelope proteins derived from other viruses or engineered to target specific cell-surface receptors have also been successfully used to increase transduction efficiency of specific cell types and to facilitate targeted transduction, respectively (Cockrell and Kafri 2007).

ACKNOWLEDGMENTS

The authors would like to acknowledge the technical assistance of Qin Su and Ran He of the University of Massachusetts Medical School, and Julio Sanmiguel and Michael Korn of the University of Pennsylvania in the development of this chapter.

REFERENCES

Alba R, Bosch A, Chillon M. 2005. Gutless adenovirus: Last-generation adenovirus for gene therapy. *Gene Ther* 12: S18–S27.

Amalfitano A, Hauser MA, Hu H, Serra D, Begy CR, Chamberlain JS. 1998. Production and characterization of improved adenovirus vectors with the E1, E2b, and E3 genes deleted. *J Virol* 72: 926–933.

Armentano D, Zabner J, Sacks C, Sookdeo CC, Smith MP, St George JA, Wadsworth SC, Smith AE, Gregory RJ. 1997. Effect of the E4 region on the persistence of transgene expression from adenovirus vectors. *J Virol* 71: 2408–2416.

Bruder JT, Appiah A, Kirkman WM III, Chen P, Tian J, Reddy D, Brough DE, Lizonova A, Kovesdi I. 2000. Improved production of adenovirus vectors expressing apoptotic transgenes. *Hum Gene Ther* 11: 139–149.

Büning H, Perabo L, Coutelle O, Quadt-Humme S, Hallek M. 2008. Recent developments in adeno-associated virus vector technology. *J Gene Med* 10: 717–733.

Carey BW, Markoulaki S, Hanna J, Saha K, Gao Q, Mitalipova M, Jaenisch R. 2009. Reprogramming of murine and human somatic cells using a single polycistronic vector. *Proc Natl Acad Sci* 106: 157–162.

Cassani B, Montini E, Maruggi G, Ambrosi A, Mirolo M, Selleri S, Biral E, Frugnoli I, Hernandez-Trujillo V, Di Serio C, et al. 2009. Integration of retroviral vectors induces minor changes in the transcriptional activity of T cells from ADA-SCID patients treated with gene therapy. *Blood* 114: 3546–3556.

Chartier C, Degryse E, Gantzer M, Dieterle A, Pavirani A, Mehtali M. 1996. Efficient generation of recombinant adenovirus vectors by homologous recombination in *Escherichia coli. J Virol* 70: 4805–4810.

Cockrell AS, Kafri T. 2007. Gene delivery by lentivirus vectors. *Mol Biotechnol* 36: 184–204.

Cotmore SF, Tattersall P. 1996. Parvovirus DNA replication. In *DNA replication in eukaryotic cells*, p. 799–813. Cold Spring Harbor Laboratory Press, Cold Spring Harbor, NY.

Danthinne X, Imperiale MJ. 2000. Production of first generation adenovirus vectors: A review. *Gene Ther* 7: 1707–1714.

Daya S, Berns KI. 2008. Gene therapy using adeno-associated virus vectors. *Clin Microbiol Rev* 21: 583–593.

de Felipe P. 2002. Polycistronic viral vectors. *Curr Gene Ther* 2: 355–378.

Duque S, Joussemet B, Riviere C, Marais T, Dubreil L, Douar AM, Fyfe J, Moullier P, Colle MA, Barkats M. 2009. Intravenous administration of self-complementary AAV9 enables transgene delivery to adult motor neurons. *Mol Ther* 17: 1187–1196.

Dyson MR, Shadbolt SP, Vincent KJ, Perera RL, McCafferty J. 2004. Production of soluble mammalian proteins in *Escherichia coli*: Identification of protein features that correlate with successful expression. *BMC Biotechnol* 4: 32.

Fallaux FJ, Kranenburg O, Cramer SJ, Houweling A, Van Ormondt H, Hoeben RC, Van Der Eb AJ. 1996. Characterization of 911: A new helper cell line for the titration and propagation of early region 1-deleted adenoviral vectors. *Hum Gene Ther* 7: 215–222.

Fallaux FJ, Bout A, van der Velde I, van den Wollenberg DJ, Hehir KM, Keegan J, Auger C, Cramer SJ, van Ormondt H, van der Eb AJ, et al. 1998. New helper cells and matched early region 1-deleted adenovirus vectors prevent generation of replication-competent adenoviruses. *Hum Gene Ther* 9: 1909–1917.

Follenzi A, Ailles LE, Bakovic S, Geuna M, Naldini L. 2000. Gene transfer by lentiviral vectors is limited by nuclear translocation and rescued by HIV-1 pol sequences. *Nat Genet* 25: 217–222.

Foust KD, Nurre E, Montgomery CL, Hernandez A, Chan CM, Kaspar BK. 2009. Intravascular AAV9 preferentially targets neonatal neurons and adult astrocytes. *Nat Biotechnol* 27: 59–65.

Frecha C, Szecsi J, Cosset FL, Verhoeyen E. 2008. Strategies for targeting lentiviral vectors. *Curr Gene Ther* 8: 449–460.

Gao GP, Yang Y, Wilson JM. 1996. Biology of adenovirus vectors with E1 and E4 deletions for liver-directed gene therapy. *J Virol* 70: 8934–8943.

Gao G, Zhou X, Alvira MR, Tran P, Marsh J, Lynd K, Xiao W, Wilson JM. 2003. High throughput creation of recombinant adenovirus vectors=by direct cloning, green-white selection and I-Sce I-mediated rescue of circular adenovirus plasmids in 293 cells. *Gene Ther* 10: 1926–1930.

Gao G, Vandenberghe LH, Wilson JM. 2005. New recombinant serotypes of AAV vectors. *Curr Gene Ther* 5: 285–297.

Geller AI, Breakefield XO. 1988. A defective HSV-1 vector expresses *Escherichia coli* β-galactosidase in cultured peripheral neurons. *Science* 241: 1667–1669.

Gerard RD. 1995. Adenovirus vectors. In *DNA cloning: Mammalian systems* (ed Glover BD, Hames BD), pp. 285–330. Oxford University Press, Oxford.

Gething MJ, Sambrook J. 1981. Cell-surface expression of influenza haemagglutinin from a cloned DNA copy of the RNA gene. *Nature* 293: 620–625.

Giampaoli S, Nicolaus G, Delmastro P, Cortese R. 2002. Adeno-cosmid cloning vectors for regulated gene expression. *J Gene Med* 4: 490–497.

Goeddel DV, Heyneker HL, Hozumi T, Arentzen R, Itakura K, Yansura DG, Ross MJ, Miozzari G, Crea R, Seeburg PH. 1979. Direct expression in *Escherichia coli* of a DNA sequence coding for human growth hormone. *Nature* 281: 544–548.

Graham FL, Smiley J, Russell WC, Nairn R. 1977. Characteristics of a human cell line transformed by DNA from human adenovirus type 5. *J Gen Virol* 36: 59–74.

Gregorevic P, Blankinship MJ, Allen JM, Crawford RW, Meuse L, Miller DG, Russell DW, Chamberlain JS. 2004. Systemic delivery of genes to striated muscles using adeno-associated viral vectors. *Nat Med* 10: 828–834.

Grieger JC, Choi VW, Samulski RJ. 2006. Production and characterization of adeno-associated viral vectors. *Nature Protocols* 1: 1412–1428.

Hacein-Bey S, Gross F, Nusbaum P, Yvon E, Fischer A, Cavazzana-Calvo M. 2001. Gene therapy of X-linked sever combined immunologic deficiency (SCID-X1). *Pathol Biol (Paris)* 49: 57–66.

He TC, Zhou S, da Costa LT, Yu J, Kinzler KW, Vogelstein B. 1998. A simplified system for generating recombinant adenoviruses. *Proc Natl Acad Sci* 95: 2509–2514.

Hester ME, Foust KD, Kaspar RW, Kaspar BK. 2009. AAV as a gene transfer vector for the treatment of neurological disorders: Novel treatment thoughts for ALS. *Curr Gene Ther* 9: 428–433.

Hitt MM, Graham FL. 2000. Adenovirus vectors for human gene therapy. *Adv Virus Res* 55: 479–505.

Inagaki K, Fuess S, Storm TA, Gibson GA, McTiernan CF, Kay MA, Nakai H. 2006. Robust systemic transduction with AAV9 vectors in mice: Efficient global cardiac gene transfer superior to that of AAV8. *Mol Ther* 14: 45–53.

Kochanek S, Schiedner G, Volpers C. 2001. High-capacity "gutless" adenoviral vectors. *Curr Opin Mol Ther* 3: 454–463.

Le Bec C, Douar AM. 2006. Gene therapy progress and prospects–vectorology: Design and production of expression cassettes in AAV vectors. *Gene Ther* 13: 805–813.

Lim HK, Mansell TJ, Linderman SW, Fisher AC, Dyson MR, DeLisa MP. 2009. Mining mammalian genomes for folding competent proteins using Tat-dependent genetic selection in *Escherichia coli*. *Protein Sci* 18: 2537–2549.

Lois C, Hong EJ, Pease S, Brown EJ, Baltimore D. 2002. Germline transmission and tissue-specific expression of transgenes delivered by lentiviral vectors. *Science* 295: 868–872.

Louis N, Evelegh C, Graham FL. 1997. Cloning and sequencing of the cellular-viral junctions from the human adenovirus type 5 transformed 293 cell line. *Virology* 233: 423–429.

Luo J, Deng ZL, Luo X, Tang N, Song WX, Chen J, Sharff KA, Luu HH, Haydon RC, Kinzler KW, et al. 2007. A protocol for rapid generation of recombinant adenoviruses using the AdEasy system. *Nature Protocols* 2: 1236–1247.

Lusky M, Christ M, Rittner K, Dieterle A, Dreyer D, Mourot B, Schultz H, Stoeckel F, Pavirani A, Mehtali M. 1998. In vitro and in vivo biology of recombinant adenovirus vectors with E1, E1/E2A, or E1/E4 deleted. *J Virol* 72: 2022–2032.

McConnell MJ, Imperiale MJ. 2004. Biology of adenovirus and its use as a vector for gene therapy. *Hum Gene Ther* 15: 1022–1033.

Mizuguchi H, Kay MA. 1998. Efficient construction of a recombinant adenovirus vector by an improved in vitro ligation method. *Hum Gene Ther* 9: 2577–2583.

Mizuguchi H, Kay MA, Hayakawa T. 2001. Approaches for generating recombinant adenovirus vectors. *Adv Drug Deliv Rev* 52: 165–176.

Moffat J, Grueneberg DA, Yang X, Kim SY, Kloepfer AM, Hinkle G, Piqani B, Eisenhaure TM, Luo B, Grenier JK, et al. 2006. A lentiviral RNAi library for human and mouse genes applied to an arrayed viral high-content screen. *Cell* 124: 1283–1298.

Montini E, Cesana D, Schmidt M, Sanvito F, Bartholomae CC, Ranzani M, Benedicenti F, Sergi LS, Ambrosi A, Ponzoni M, et al. 2009. The genotoxic potential of retroviral vectors is strongly modulated by vector design and integration site selection in a mouse model of HSC gene therapy. *J Clin Invest* 119: 964–975.

Moorhead JW, Clayton GH, Smith RL, Schaack J. 1999. A replication-incompetent adenovirus vector with the preterminal protein gene deleted efficiently transduces mouse ears. *J Virol* 73: 1046–1053.

Naldini L, Blomer U, Gage FH, Trono D, Verma IM. 1996a. Efficient transfer, integration, and sustained long-term expression of the transgene in adult rat brains injected with a lentiviral vector. *Proc Natl Acad Sci* 93: 11382–11388.

Naldini L, Blomer U, Gallay P, Ory D, Mulligan R, Gage FH, Verma IM, Trono D 1996b. In vivo gene delivery and stable transduction of nondividing cells by a lentiviral vector. *Science* 272: 263–267.

Palmer DJ, Ng P. 2008. Methods for the production of first generation adenoviral vectors. *Methods Mol Biol* 43: 55–78.

Papadakis ED, Nicklin SA, Baker AH, White SJ. 2004. Promoters and control elements: Designing expression cassettes for gene therapy. *Curr Gene Ther* 4: 89–113.

Parks RJ, Chen L, Anton M, Sankar U, Rudnicki MA, Graham FL. 1996. A helper-dependent adenovirus vector system: Removal of helper virus by Cre-mediated excision of the viral packaging signal. *Proc Natl Acad Sci* 93: 13565–13570.

Pedersen FS, Duch M. 2003. Retroviral replication. In *Encyclopedia of life sciences*. Wiley & Sons, Chichester. doi: 101038/npgels0000430 (http://onlinelibrarywileycom/doi/101002/9780470015902a0000430 pub3/otherversions).

Pfeifer A, Ikawa M, Dayn Y, Verma IM 2002. Transgenesis by lentiviral vectors: Lack of gene silencing in mammalian embryonic stem cells and preimplantation embryos. *Proc Natl Acad Sci* 99: 2140–2145.

Pluta K, Luce MJ, Bao L, Agha-Mohammadi S, Reiser J. 2005. Tight control of transgene expression by lentivirus vectors containing second-generation tetracycline-responsive promoters. *J Gene Med* 7: 803–817.

Reddy PS, Ganesh S, Hawkins L, Idamakanti N. 2007. Generation of recombinant adenovirus using the *Escherichia coli* BJ5183 recombination system. *Methods Mol Med* 130: 61–68.

Rivera VM, Ye X, Courage NL, Sachar J, Cerasoli F Jr, Wilson JM, Gilman M. 1999. Long-term regulated expression of growth hormone in mice after intramuscular gene transfer. *Proc Natl Acad Sci* 96: 8657–8662.

Schiedner G, Morral N, Parks RJ, Wu Y, Koopmans SC, Langston C, Graham FL, Beaudet AL, Kochanek S. 1998. Genomic DNA transfer with a high-capacity adenovirus vector results in improved in vivo gene expression and decreased toxicity. *Nat Genet* 18: 180–183.

Schiedner G, Hertel S, Kochanek S 2000. Efficient transformation of primary human amniocytes by E1 functions of Ad5: Generation of new cell lines for adenoviral vector production. *Hum Gene Ther* 11: 2105–2116.

Sirven A, Pflumio F, Zennou V, Titeux M, Vainchenker W, Coulombel L, Dubart-Kupperschmitt A, Charneau P. 2000. The human immunodeficiency virus type-1 central DNA flap is a crucial determinant for lentiviral vector nuclear import and gene transduction of human hematopoietic stem cells. *Blood* 96: 4103–4110.

Srivastava A, Lusby EW, Berns KI. 1983. Nucleotide sequence and organization of the adeno-associated virus 2 genome. *J Virol* 45: 555–564.

Stegmeier F, Hu G, Rickles RJ, Hannon GJ, Elledge SJ. 2005. A lentiviral microRNA-based system for single-copy polymerase II-regulated RNA interference in mammalian cells. *Proc Natl Acad Sci* 102: 13212–13217.

Szulc J, Wiznerowicz M, Sauvain MO, Trono D, Aebischer P. 2006. A versatile tool for conditional gene expression and knockdown. *Nat Meth* 3: 109–116.

Ullrich A, Shine J, Chirgwin J, Pictet R, Tischer E, Rutter WJ, Goodman HM. 1977. Rat insulin genes: Construction of plasmids containing the coding sequences. *Science* 196: 1313–1319.

Wade-Martins R, Smith ER, Tyminski E, Chiocca EA, Saeki Y. 2001. An infectious transfer and expression system for genomic DNA loci in human and mouse cells. *Nat Biotechnol* 19: 1067–1070.

Walther W, Stein U. 2000. Viral vectors for gene transfer. *Drugs* 60: 249–271.

Wang Z, Zhu T, Qiao C, Zhou L, Wang B, Zhang J, Chen C, Li J, Xiao X. 2005. Adeno-associated virus serotype 8 efficiently delivers genes to muscle and heart. *Nat Biotechnol* 23: 321–328.

Wiznerowicz M, Szulc J, Trono D. 2006. Tuning silence: Conditional systems for RNA interference. *Nat Meth* 3: 682–688.

Wu Q, Moyana T, Xiang J. 2001. Cancer gene therapy by adenovirus-mediated gene transfer. *Curr Gene Ther* 1: 101–122.

Ye X, Rivers VM, Zoltick P, Cerasoli F Jr, Schnell MA, Gao G, Hughes JV, Gilman M, Wilson JM. 1999. Regulated delivery of therapeutic proteins after in vivo somatic cell gene transfer. *Science* 283: 88–91.

Zhang H, Xie J, Xie Q, Wilson JM, Gao G. 2009. Adenovirus-adeno-associated virus hybrid for large-scale recombinant adeno-associated virus production. *Hum Gene Ther* 20: 922–929.

Zufferey R, Donello JE, Trono D, Hope TJ. 1999. Woodchuck hepatitis virus posttranscriptional regulatory element enhances expression of transgenes delivered by retroviral vectors. *J Virol* 73: 2886–2892.

Construction of Recombinant Adenovirus Genomes by Direct Cloning

This protocol describes how to generate an infectious adenovirus vector by direct ligation and cloning. This is the first step in the production of a recombinant adenovirus vector. The protocol begins with a convenient and efficient double-selection procedure based on antibiotic resistance and identification of GFP-negative bacterial colonies (Fig. 6 in the chapter introduction). In this protocol, the prokaryotic expression cassette for GFP in the shuttle plasmid pSh-*pkGFP* is replaced with the transgene. The resulting pShuttle–transgene plasmid is then used to clone the transgene expression cassette into pAd-*pkGFP*, using the same double-selection procedure (Fig. 6 in the chapter introduction).

Biosafety Consideration

According to the NIH Guidelines for Research Involving Recombinant DNA Molecules (April 2000), all types of wild-type and replication-competent human adenoviruses are classified as risk group 2 of biohazard agents. The human disease associated with this group of biohazard agents is usually treatable and preventable and is rarely serious. All work with adenovirus vectors should be conducted at Biosafety Level 2 (BL2) with the approval by the Institutional Biosafety Committee of the home institution.

This protocol was contributed by Xiangyang Zhou (Vaccine Research Center, Wistar Institute, Philadelphia, Pennsylvania).

MATERIALS

It is essential that you consult the appropriate Material Safety Data Sheets and your institution's Environmental Health and Safety Office for proper handling of equipment and hazardous materials used in this protocol.

Recipes for reagents specific to this protocol, marked <R>, are provided at the end of the protocol. See Appendix 1 for recipes for commonly used stock solutions, buffers, and reagents, marked <A>. Dilute stock solutions to the appropriate concentrations.

Reagents

Agar plates
Antibiotics (ampicillin, kanamycin)
ATP (20 mM)
Calf intestinal alkaline phosphatase
DH5a *E. coli*, chemically competent (Life Technologies)
Gene of interest in plasmid or as PCR fragment
Glycerol (20%)
I-CeuI (5 U/μL)
KCM buffer (5×) <R>
LB medium <A>
Low melting agarose (LMA) (1%)
PI-SceI (1 U/μL) and 10× buffer of PI-SceI
Plasmid DNA miniprep and midiprep purification kits (QIAGEN)
pSh-*pkGFP* and pAd-*pkGFP* plasmids (available at the University of Pennsylvania, Penn Vector Core, vector@mail.med.upenn.edu)

Restriction enzymes and restriction enzyme buffers (10×)
Stbl2 *E. coli*, chemically competent (Life Technologies)
T4 DNA ligase (5 U/μL) with ligation buffer (included with purchased enzyme)

Equipment

Bacterial incubator
Electrophoresis apparatus
Heat block (65°C)
Ice bucket
Microcentrifuge
Orbital shaking incubator
Polypropylene round-bottomed tube (14 mL)
Refrigerated tabletop centrifuge
UV microscope
Water bath (37°C)

METHOD

Cloning the Gene of Interest into pShuttle Plasmid

pShuttle is a generic shuttle plasmid used to transfer the expression cassette carrying the gene of interest into the adenovirus vector backbone. Steps 1–9 describe how to clone the gene of interest into the pSh-*pkGFP* plasmid using the double-selection process (Fig. 6 in the chapter introduction). Currently available pShuttle or pSh-*pkGFP* constructs contain a CMV-promoter-driven expression cassette only. However, Figure 6 in the chapter introduction also illustrates those restriction enzyme sites that are suitable for either cloning the transgene of interest behind the CMV promoter or replacing it with a promoter of choice for transgene expression.

1. In sterile 0.5-mL microcentrifuge tubes, digest the transgene source vector and pSh-*pkGFP* vector separately using the following conditions:

DNA	1–2 μg
Restriction enzyme A	0.5 μL
Restriction enzyme B	0.5 μL
Restriction enzyme buffer (10×)	2.0 μL

   Add H₂O to 20 mL and mix well. Incubate for 1 h in a water bath at 37°C.

2. Analyze 2 μL of the digest by electrophoresis through a 1% low-melting agarose gel to confirm the accuracy of the plasmid maps.

3. Proceed with dephosphorylation of the pShuttle backbone using calf intestinal alkaline phosphatase following the manufacturer's instructions.

4. Separate the remaining 18 μL of the digests by preparative gel electrophoresis through a 1% LMA gel. Cut gel slices around the DNA fragments corresponding to the transgene insert and the shuttle backbone and place them into separate 1.5-mL microcentrifuge tubes.

5. Incubate the microcentrifuge tubes containing gel slices for 3–5 min on a heat block set to 65°C. Once both gel slices are melted, prepare the ligation reaction as follows:

Shuttle backbone DNA	The total volume for both DNAs is 10 μL (up to 200 ng) with
Transgene insert DNA	a typical molecular ratio of 1:3 for backbone:insert.
T4 DNA ligase (5 U/μL)	1 μL
Ligation buffer (5×)	4 μL
ATP (20 mM)	1 μL
H₂O	4 μL

   Incubate the ligation mixture overnight (16 h) at 16°C.

6. On the following day, incubate the ligation reaction for 3–5 min at 65°C on a heating block. Add 40 μL of 5× KCM buffer and 140 μL of nuclease-free water to the 20-μL ligation reaction and mix well by pipetting. Place the ligation reactions on ice for 3–5 min before transformation. Prechill one 14-mL polypropylene round-bottomed tube on ice for each transformation reaction.

7. Thaw chemically competent DH5α *E. coli* on ice. Add 50 μL of competent cells and 25 μL of diluted ligation reaction to the bottom of the prechilled polypropylene round-bottomed tube. Mix the reaction by gently swirling the tube. Incubate the transformation reaction on ice for at least 30 min. Refer to the competent cell manufacturer's instructions for details regarding the remaining steps of transformation. Plate the cells on LB agar plates with 50 μg/mL kanamycin and incubate overnight at 37°C.

8. Place the LB plates on a UV microscope for evaluation through the fluorescein isothiocyanate (FITC) channel at 510-nm wavelength. The bacterial colonies representing transformants of the pShuttle transgene are colorless (white) when illuminated by UV light; colonies transformed with the original pSh-*pkGFP* plasmid will fluoresce green. Pick several white colonies to grow in 2 mL of LB medium containing kanamycin (50 mg/L) for 18 h at 37°C with agitation (220 rpm).

9. Prepare plasmid DNA from the bacterial minicultures using a commercially available miniprep kit. Perform diagnostic restriction enzyme digestions to identify and confirm the structure of pShuttle–transgene-positive clones.

Creation of Adenoviral Plasmid Vector Expressing the Transgene of Interest

In the pShuttle–transgene plasmid, the transgene expression cassette is flanked by two rare restriction sites, CeuI and PI-SceI, that are used to clone the expression cassette into pAd-*pkGFP*. The available vector backbones are different versions of the pAd-*pkGFP* plasmid with E1-, E1/E3- (the version illustrated in Fig. 6 in the chapter introduction, 35,143 bp in size), E1/E4-, or E1/E3/E4-deleted adenovirus genomes, where a GFP expression cassette driven by a prokaryotic promoter is inserted into the E1-deletion locus and flanked by the same CeuI and PI-SceI restriction sites.

10. Prepare the insert and vector backbone DNA fragments by digesting the pShuttle–transgene (for insert) and pAd-pkGFP (for backbone) plasmids as follows:

DNA	1–2 μg
I-CeuI (5 U/mL)	1.0 μL
PI-SceI (1 U/mL)	2.0 μL
PI-SceI buffer (10×)	4.0 μL
H$_2$O	add to 40 μL

Incubate the digest reactions for 2–3 h at 37°C.

11. Repeat Steps 2–8 with the following changes.

   i. Use chemically competent Stbl2 *E. coli* for the transformation at 37°C.

   ii. Centrifuge the transformed cells at 1000 rpm for 5 min at 4°C. Then, remove 800 μL of the supernatant, resuspend the bacteria in the remaining 200 μL of medium, and plate the entire resuspended pellet on a single LB plate. Incubate the plate at 37°C.

   iii. Grow all bacterial cultures in LB medium at 30°C.

   iv. Use ampicillin as the antibiotic for selection of transformants.

   v. By 16–18 h after transformation, 50–100 colonies should be visible on the plate, of which up to 90% are white. At least 50% of the white colonies should carry the adenovirus vector containing the transgene, as described in Gao et al. (2003).
   If molecular cloning of recombinant adenovirus is unsuccessful, see Troubleshooting.

12. Prepare minipreps of DNAs in the white colonies and confirm the structure of the adenovirus recombinant by restriction enzyme analysis. In addition to the enzymes that cut specifically in the transgene cassettes, restriction enzymes such as BglII, XhoI, and HindIII that cut multiple times in the different regions of the adenovirus DNA should be chosen for diagnostic digestions.

13. Prepare midiprep plasmid DNA of the correct clone using a QIAGEN midiprep kit. As a backup, store aliquots of the confirmed clone in medium containing 20% glycerol at −80°C.

TROUBLESHOOTING

Problem (Step 11): Few or no white colonies are on the LB plate. The clones do not contain a transgene cassette or their patterns of cleavage by restriction enzymes are incorrect.

Solution: The adenovirus vector is a large plasmid (>36 kb) containing the entire adenoviral vector genome and unstable repetitive sequences of ITRs. Particular care is required when growing and manipulating large plasmids, especially those that contain unstable repetitive elements (the adenovirus LTRs). Consider the following adjustments to the protocol.

- Add 50 mL of the diluted ligation reaction to 100 mL of Stbl 2–competent cells for transformation. Alternatively, use Stbl 4 cells and the electroporation method for transformation.

- Reduce the total amount of vector and backbone DNA to 100 ng but keep the molar ratio of backbone:insert at 1:3.

- Grow minicultures and midicultures of the clones of recombinant adenovirus plasmids at a reduced agitation speed (160–180 rpm) at 32°C.

- Do not expose DNA to UV light. When recovering DNA from the agarose gel, cut a small strip from the gel containing the markers and a small sliver of agar containing the plasmid digest. Use this to mark the location of the desired DNA fragments and then cut the main part of the well without exposing it to UV light.

DISCUSSION

The major advantage of this direct cloning method for creation of a recombinant adenovirus vector is its efficiency. However, depending on the researcher's experience in manipulating large plasmids, subcloning the transgene from the pShuttle to the adenovirus vector backbone may become the rate-limiting step. Strategies such as solid-phase ligation in low-melting-temperature gel, antibiotic and green–white double selection, growth of bacterial cultures at 32°C, and use of low agitation speed can significantly enhance the cloning efficiency and success rate.

RECIPE

It is essential that you consult the appropriate Material Safety Data Sheets and your institution's Environmental Health and Safety Office for proper handling of equipment and hazardous materials used in this protocol.

KCM Buffer (5×)

Reagent	Quantity (for 0.1 L)	Final concentration
KCl (1 M)	50 mL	500 mM
CaCl$_2$ (2.5 M)	6 mL	150 mM
MgCl$_2$ (1 M)	25 mL	250 mM
H$_2$O	19 mL	

Mix well, sterilize by passing through a 0.22-μm filter, and store at room temperature.

REFERENCES

Gao G, Zhou X, Alvira MR, Tran P, Marsh J, Lynd K, Xiao W, Wilson JM. 2003. High throughput creation of recombinant adenovirus vectors by direct cloning, green-white selection and I-Sce I-mediated rescue of circular adenovirus plasmids in 293 cells. *Gene Ther* 10: 1926–1930.

NIH Guidelines for Research Involving Recombinant DNA Molecules. (April) 2000. http://oba.od.nih.gov/rdna/nih_guidelines_oba.html.

Release of the Cloned Recombinant Adenovirus Genome for Rescue and Expansion

After generating the recombinant adenovirus plasmid by direct cloning, the viral genome must be released from the plasmid for virus rescue and expansion in corresponding packaging cell lines. This protocol describes how to process and rescue a recombinant adenovirus and grow the rescued virus to large scale.

Biosafety Consideration

According to the NIH Guidelines for Research Involving Recombinant DNA Molecules (April 2000), all types of wild-type and replication-competent human adenoviruses are classified as risk group 2 of biohazard agents. The human disease associated with this group of biohazard agents is usually treatable and preventable and is rarely serious. All work with adenovirus vectors should be conducted at Biosafety Level 2 (BL2) with the approval by the Institutional Biosafety Committee of the home institution.

MATERIALS

It is essential that you consult the appropriate Material Safety Data Sheets and your institution's Environmental Health and Safety Office for proper handling of equipment and hazardous materials used in this protocol.

Recipes for reagents specific to this protocol, marked <R>, are provided at the end of the protocol. See Appendix 1 for recipes for commonly used stock solutions, buffers, and reagents, marked <A>. Dilute stock solutions to the appropriate concentrations.

Reagents

Agarose gel (1%)
$CaCl_2$ (2 M)
Calcium Phosphate Transfection Kit (Promega ProFection Mammalian Transfection System)
10-3 cells (available from University of Pennsylvania, Penn Vector Core) or other commercially available cells
Complete growth medium <R>
Dulbecco's minimal essential medium (DMEM)
Dulbecco's phosphate-buffered saline (D-PBS) (purchased)
Fetal bovine serum (FBS)
HBS buffer (2×) <R>
HEK-293 cells (American Type Culture Selection [ATCC], catalog no. CRL-1573)
Lipofectamine (Life Technologies)
PacI restriction enzyme and buffer
Penicillin–streptomycin (P/S) solution (100×)
Recombinant adenovirus plasmid (from Protocol 1)
Tris (10 mM, pH 8.0)
Trypsin–EDTA (0.05%)
$ZnSO_4$ (see Step 4.iv)

Equipment

Biosafety Level II tissue culture hood
Conical centrifuge bottles, sterile (150 and 500 mL)
Conical centrifuge tubes (15 and 50 mL)
Dry-ice/ethanol bath
Freezer (−80°C)
Gel electrophoresis apparatus
Humidified cell culture incubator (37°C, 5% CO_2)
Microcentrifuge
Pasteur pipette attached to vacuum flask
Polystyrene round-bottomed tubes with snap caps (5 mL)
T-25 culture flasks
Tabletop centrifuge
Vortex machine
Water bath (37°C)

METHOD

Preparation of DNA from the Directly Cloned Recombinant Adenovirus Plasmid for Use in Transfection

▲ *It is critical to practice aseptic technique during the following preparation steps.*

1. Digest the recombinant adenovirus plasmid with PacI endonuclease to release the intact adeno-virus vector genome. The PacI sites flank both 5′ and 3′ ITRs located at the ends of the adeno-virus vector genome. Set up the reaction as follows:

Recombinant adenovirus plasmid DNA	5 mg
PacI	2 μL
PacI buffer (10×)	5 μL
H_2O	to 50 μL

Incubate the digest reaction for 60 min at 37°C.

2. Analyze 5 μL of the reaction by electrophoresis on a 1% agarose gel to confirm successful release of the vector genome from the plasmid backbone.

Transfection of Packaging Cells with Linearized Adenovirus Vector DNA for Rescue

3. One day before transfection, seed early- or medium-passage HEK-293 cells (passage 31–60 for E1-deleted, E1/E3-deleted, and E1/E3-deleted+E4 orf6 adenovirus vectors) or 10-3 cells (for E1/E4- or E1/E3/E4-deleted adenovirus) in T-25 tissue culture flasks at 2×10^6 cells per flask.

4. On the next day, when cell confluency reaches ∼50%–70%, proceed with one of the following transfection methods.

For Transfection Using Lipofectamine (see also Chapter 15)

▲ *No serum or antibiotics should be used.*

i. For each transfection, label two 5-mL round-bottomed polystyrene tubes with snap caps as A and B. Add 5 mg of PacI-digested recombinant adenovirus plasmid DNA to tube A and adjust the total volume to 300 mL with DMEM. Add 32 mL of Lipofectamine and 268 mL of DMEM to tube B. Gently mix the contents of each tube separately by swirling the tubes. Do not vortex. Combine the solutions of tube A and B using a wide-bore pipette. Gently mix them by swirling. Do not vortex. Let the mixture stand for 45 min at room temperature to form DNA–liposome complexes.

   ii. At 5 min before the end of the incubation, gently rinse the cells once with DMEM prewarmed to 37°C and then add 3 mL of DMEM. Add the transfection mixture to the cells dropwise, rock the flask gently, and return the cells to the incubator. Three hours later, add 300 mL of FBS and incubate the cultures overnight.

   iii. On the following day, replace the transfection medium with fresh Complete growth medium and continue to incubate the transfected cells. Add 1 mL of fresh Complete growth medium to the flask every 3 d and examine the cells for cytopathic effect (CPE).

> *CPE reflects the morphological changes of the cultured cells as infection proceeds. Such distinctive morphological changes result from the accumulation of newly produced virus progeny. CPE is different from necrosis; CPE begins with rounding of adhered cells followed by gradual detachment of the rounded cells from the plate. When the cells detach from the plate, they may form "grape-like" clusters that float in the growth medium.*

   iv. To rescue an adenovirus vector with a complete deletion of region E4, add 175 mM of $ZnSO_4$ to the growth medium of 10-3 cells (in Step 4.iii).

> *The addition of $ZnSO_4$ is necessary to induce the metallothionein promoter that drives E4-orf6 expression in 10-3 cells (Gao et al. 1996).*

For Calcium Phosphate Transfection

   i. Replace the growth medium in the T-25 culture flask with 3 mL of fresh Complete growth medium prewarmed to 37°C for at least 3 h before transfection.

   ii. Thaw all reagents in the calcium phosphate transfection kit at room temperature. Label as tubes A and B two polystyrene 5-mL tubes with snap caps. Aliquot 300 mL of $2\times$ HBS buffer into tube A and prepare 300 mL of DNA mix in tube B by adding 5 mg of PacI-linearized clone of recombinant adenovirus plasmid DNA, 37.5 mL of 2 M $CaCl_2$, and the required volume of sterile H_2O. Mix the transfection solution by adding DNA mix in tube B into the $2\times$ HBS in tube A dropwise while vigorously vortexing tube A. Incubate the transfection cocktail for 20 min at room temperature to form DNA−calcium phosphate precipitates. The solution should become translucent. After the incubation period, slowly add the transfection cocktail to the T-25 culture flask. Gently rock the flask to evenly distribute the DNA−calcium precipitates over the entire monolayer. Return the cells to the incubator.

   iii. The next morning, wash cells once gently with DMEM prewarmed to 37°C and then add fresh Complete growth medium and continue the incubation. Add 1 mL of fresh growth medium to the flask every 3 d and examine the cells for CPE.

   iv. To rescue an E4-deleted adenovirus vector, add 175 mM $ZnSO_4$ to the Complete growth medium of 10-3 cells.

> *Addition of $ZnSO_4$ is necessary to induce the metallothionein promoter that drives E4-orf6 expression in 10-3 cells.*

5. Once 90% of the cell monolayer shows CPE, dislodge any remaining cells by gently tapping the flask on the work surface of the cell culture biosafety hood or pipetting the medium against the flask growth surface. Transfer the cell suspension to a 15-mL conical centrifuge tube and store it at −80°C for virus expansion.

> *If the transfected cells do not show any sign of CPE 2 wk after the transfection, see Troubleshooting.*

Expansion of Adenovirus Vector for Large-Scale Production

6. Seed 7×10^6 HEK-293 cells onto each of two 150-mm plates 1 d before viral infection.

   i. To prepare a crude cell lysate, thaw the cell suspension from Step 5 in a 37°C water bath. Repeat the freeze−thaw cycle twice more in a dry-ice/ethanol bath and a 37°C water bath. Shake the tube several times after each thaw to ensure that cells do not settle.

ii. Centrifuge the cells at 3200 rpm in a tabletop centrifuge for 10 min at 4°C.

iii. Remove the supernatant and add it directly to two 70%–80% confluent 150-mm plates of HEK-293 cells. Return the cells to the incubator and monitor CPE on a daily basis.

Usually, CPE will become noticeable 24 h after infection and will be fully evident within 2–3 d.

7. Harvest the cells when 90% of the cells show CPE, as before, and centrifuge at 3200 rpm for 10 min at 4°C. Discard the supernatant and resuspend the cell pellet in 2 mL of 10 mM Tris, pH 8.0. After three cycles of freezing/thawing, centrifuge the crude cell lysate at 3200 rpm for 10 min at 4°C and collect the supernatant for infection of eight plates of HEK-293 cells, as in Step 6.

8. At ∼40–45 h after infection, or when 90% of the infected cells show CPE, harvest the cells as before and centrifuge at 3200 rpm for 10 min at 4°C. Resuspend the cell pellet in 8 mL of 10 mM Tris, pH 8.0, and store at −80°C.

i. For expansion of the E4-deleted adenovirus vectors, use 10-3 cells and supplement the Complete growth medium with 175 mM ZnSO$_4$ when sending the cells to induce expression of the E4 orf6 necessary for replication of this vector.

If the crude viral lysate harvested from the transfection/rescue step fails to expand after 7 d, see Troubleshooting.

Large-Scale Production of Adenovirus Vector

9. Seed HEK-293 (or 10-3) cells 1 d before infection. At the time of infection, cells should be ∼70%–80% confluent. Seeding the cells at a 1:3 split ratio from a 100% confluent plate works well for this purpose. Prepare 40 150-mm plates of HEK-293 cells with 20 mL of Complete growth medium per plate.

For large-scale production of E4-deleted adenovirus vectors, use 10-3 cells and supplement the growth medium with 175 mM ZnSO$_4$.

10. Use the procedure described in Step 6 to generate a crude supernatant from the cell suspension generated in Step 8. To infect 40 plates of cells, transfer 8 mL of crude supernatant into a 150-mL sterile bottle containing 112 mL of Complete growth medium and mix well by swirling.

11. Remove the tissue culture plates (maximum of 12 plates at a time) from the incubator. Add 3 mL of infectious medium from Step 10 to each plate slowly and gently from the side of the plate to avoid dislodging cells from the plate. Repeat this procedure until all plates have been infected. Periodically, swirl the bottle containing the inoculum. Return the plates to the incubator.

12. Approximately 40–45 h postinfection, or when 90% of the cells show CPE, remove the tissue culture plates from the incubator and place them in the biosafety cabinet. Using a 25-mL pipette, rinse off any cells that remain attached to the tissue culture plate after gentle agitation (with the lid on). Transfer equal volumes of the suspensions from the 40 plates to two 500-mL sterile conical centrifuge bottles.

13. Pellet the cells by centrifuging at 3200 rpm in a refrigerated tabletop centrifuge for 15 min at 4°C.

14. Remove the supernatants using a sterile Pasteur pipette connected to a vacuum flask. Resuspend the cell pellets in 0.5 mL of 10 mM Tris, pH 8.0, per plate. For example, for a bottle containing the cell pellet from 20 plates, resuspend the pellet in 10 mL of 10 mM Tris, pH 8.0. Resuspend the cell pellet by gently mixing the cell suspension with a 10-mL pipette.

15. Transfer the suspension into a sterile 50-mL conical centrifuge tube. Store at −80°C for future purification.

Purified Virus as Seed Virus for Large-Scale Infection

16. Thaw the frozen seed lot of adenovirus stock in an ice bucket. To infect 40 150-mm plates, aliquot enough adenovirus seed stock into a 150-mL sterile bottle containing 120 mL of growth medium so that the multiplicity of infection (MOI) is 2000 adenovirus vector particles per cell; for example, (# of plates) $\times$ (2000 vector particles/cell) $\times$ (2×10^7 cells/plate) = particles of virus needed. Thoroughly mix the solution of virus/medium by gently swirling the bottle.

17. Start the procedure from Step 10 as above.

TROUBLESHOOTING

Problem (Step 5): Transfected HEK-293 cells do not show any sign of CPE 2 wk after transfection.
Solutions: Repeat the transfection using 7.5 mg of linearized recombinant adenovirus plasmid DNA. Use a different transfection method.

- Extend the observation time up to 4 wk after transfection.

- If the transgene is known to be cytotoxic or cytostatic, introduce a gene regulation mechanism into the transgene expression cassette.

Problem (Step 8): The crude viral lysate fails to expand in HEK-293 (or 10-3 cells) cells.
Solution: There are several possible causes for this problem, including cytotoxicity associated with transgene expression, which leads to cell death instead of viral CPE; an oversized transgene cassette; the nature and numbers of early gene deletions in the vector backbone (i.e., E1 deletion only, E1/E3 deletions, or E1/E3/E4 deletions); levels of complementing gene expression in the cell line (e.g., 10-3 cells); and nonsynchronized viral infection caused by a low MOI. In the last case, when the cells are infected at low MOIs, in order to reach full CPE, serial passage of the virus stock may be required. For vectors that have a slow and delayed infection process, it can be difficult to distinguish cell death from viral CPE. The following strategies should be considered for overcoming this problem.

- Construct an inducible transgene expression cassette to regulate expression of the toxic transgene in 293 cells.

- Slow down the expansion process. Start the expansion from the crude lysate of the original transfection/rescue to one 150-mm plate of 293 cells to 3 plates, 12 plates, and finally 40 plates of 293 cells.

- In the expansion process, if the cells do not reach full CPE in 72 h, prepare and clarify the crude lysate from the entire infection (not from the cell pellet) for the next stage of infection.

DISCUSSION

As shown in Figure 6 in the chapter introduction, for nontoxic and nonoversized transgenes, rescue of virus from a recombinant adenovirus plasmid and expansion of the rescued infectious viral vector for large-scale production should be accomplished in ~3 wk. Conversely, for recombinant adenovirus plasmids oversized or toxic transgenes, viral vector rescue and expansion steps may take up to 12 wk.

RECIPES

Complete Growth Medium

Reagent	Quantity (for 1 L)	Final concentration
DMEM	890 mL	
FBS	100 mL	10%
Penicillin/ streptomycin solution (100×)	10 mL	1×

Store at 4°C.

HBS Buffer (2×)

Reagent	Quantity (for 1 L)	Final concentration
NaCl	16.4 g	280 mM
HEPES ($C_8H_{18}N_2O_4S$)	11.9 g	50 mM
$Na_2HPO_4 \cdot 7H_2O$	0.38 g	1.42 mM
H_2O	To 1 L	

Adjust the pH to 7.05 with 10 M NaOH. Sterilize by passing through a 0.22-μm filter and store at room temperature.

REFERENCES

Gao GP, Yang Y, Wilson JM. 1996. Biology of adenovirus vectors with E1 and E4 deletions for liver-directed gene therapy. *J Virol* 70: 8934–8943.

NIH Guidelines for Research Involving Recombinant DNA Molecules. (April) 2000. http://oba.od.nih.gov/rdna/nih_guidelines_oba.html.

Purification of the Recombinant Adenovirus by Cesium Chloride Gradient Centrifugation

Cesium chloride gradient centrifugation is the most widely used method for purification of recombinant adenovirus. This protocol describes the entire process, from the preparation and clarification of crude viral lysate to the formulation and storage of purified virus.

Biosafety Consideration

According to the NIH Guidelines for Research Involving Recombinant DNA Molecules (April 2000), all types of wild-type and replication-competent human adenoviruses are classified as risk group 2 of biohazard agents. The human disease associated with this group of biohazard agents is usually treatable and preventable and is rarely serious. All work with adenovirus vectors should be conducted at Biosafety Level 2 (BL2) with the approval by the Institutional Biosafety Committee of the home institution.

MATERIALS

It is essential that you consult the appropriate Material Safety Data Sheets and your institution's Environmental Health and Safety Office for proper handling of equipment and hazardous materials used in this protocol.

Recipes for reagents specific to this protocol, marked <R>, are provided at the end of the protocol. See Appendix 1 for recipes for commonly used stock solutions, buffers, and reagents, marked <A>. Dilute stock solutions to the appropriate concentrations.

Reagents

Adenovirus cell suspension (from Protocol 2)
Bleach (10%)
Dulbecco's phosphate-buffered saline (D-PBS)
Ethanol (70%)
Glycerol (autoclave sterilized)
Heavy CsCl solution (H-CsCl) <R>
Light CsCl solution (L-CsCl) <R>
Milli-Q H_2O, autoclaved
Phosphate-buffered saline
Tris-HCl (10 mM, pH 8.0) <R>

Equipment

Beaker, plastic (4 L)
Cell culture biosafety cabinet
Conical centrifuge tube (15 mL)
Cryovials, sterile
Dry-ice/ethanol bath
Magnetic stirring plate and magnetic stir bar
Needles, sterile, disposable (21 and 18 gauge)
Ring stand

Slide-A-Lyzer dialysis cassettes (10,000 molecular weight cut-off [MWCO], Pierce)
Syringes, sterile, disposable (3 and 5 mL)
Ultracentrifuge (SW28 and SW41 rotors; Beckman)
Ultracentrifuge tubes (SW28 and SW41; Beckman)
UV spectrophotometer
Water bath (37°C)
Wet ice

METHOD

Preparation and Clarification of Viral Lysate

1. Remove the 50-mL conical centrifuge tube containing the adenovirus-infected cell suspension from the −80°C freezer and thaw in a 37°C water bath. Repeat the freeze–thaw cycle twice in a 37°C water bath and dry-ice/ethanol bath. Clarify the crude lysate of cell debris by centrifugation at 4000 rpm for 20 min at 4°C.

2. Remove the lysate tube from the centrifuge and place it in a cell culture biosafety cabinet. Transfer the supernatant to another sterile 50-mL conical tube. Place the tube on ice. Resuspend the cell lysate pellet with 0.5 mL of 10 mL Tris-HCl, pH 8.0, per plate.

3. Repeat the freeze–thaw cycle once more. Centrifuge the cell debris as in Step 1 and combine the supernatant with that collected after the first centrifugation. Place the tube on ice. Using a sterile pipette, measure the total volume of supernatant in the tube (∼36 mL). If the final volume is <36 mL, add 10 mL of Tris-HCl, pH 8.0, to make up the difference.

Preparation of CsCl Gradient and Ultracentrifugation

4. Using a 10-mL sterile pipette, add 9 mL of L-CsCl solution to each of two SW28 ultracentrifuge tubes. Then, fill a 10-mL sterile pipette with 9 mL of H-CsCl solution and lower it to the bottom of the tube through the L-CsCl layer. Slowly dispense the 9 mL of H-CsCl solution under the layer of L-CsCl in each SW28 centrifuge tube. Then, carefully and slowly, overlay 18 mL of the adenovirus vector supernatant into each centrifuge tube on top of the two-layered CsCl gradient.

   Add the adenovirus supernatant to the side of the tube to avoid disturbing the gradient.

5. Very carefully, so as not to disturb the gradient, place the SW28 tubes into the rotor buckets and balance them. Attach the buckets to a SW28 rotor and centrifuge at 20,000 rpm for at least 2 h at 4°C.

   Be sure to align the buckets numerically in the appropriate position on the rotor.

   To pair and balance a bucket containing viral lysate, fill an empty bucket with the tube containing equal volumes of 10 mL of Tris-HCl, pH 8.0, and the two layers of CsCl.

6. After centrifugation, carefully remove one bucket from the rotor. Remove the tube from one bucket and clamp it to a ring stand. Wipe the outside of the tube with 70% ethanol. Visually inspect the tube. Two distinct bands should be observed in the middle of the centrifuge tube. The top band usually contains either empty or incomplete virions and the bottom band contains intact infectious virions (Fig. 1A).

7. Puncture the tube just below the bottom band with an 18- or 21-gauge needle attached to a 5-mL syringe. Turn the needle bevel up and very slowly withdraw the plunger to collect the band. Be careful to collect only the viral band and a minimum of the cesium chloride solution (Fig. 1A).

   Depending on the size of the bands in each tube, 1–4 mL of virus per tube is recovered.

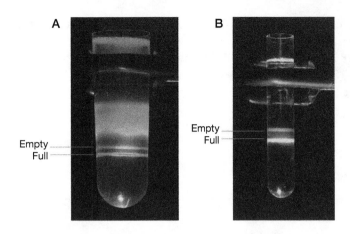

FIGURE 1. Photographs of adenovirus vector bands after the first (*A*) and second (*B*) round of CsCl gradient purification (Protocol 3).

8. After removing the needle, transfer the viral solution into a sterile 15-mL conical tube on ice. Discard the waste remaining in the centrifuge tube into a beaker containing 10% bleach for disinfection.

9. Repeat Steps 5–8 with the other centrifuged tubes. Dilute the collected viral suspension with 10 mM Tris-HCl, pH 8.0, to a final volume of 3.5 mL for each SW41 ultracentrifuge tube to be set up as described below.

10. Using a 5-mL pipette, add 3.5 mL of L-CsCl solution to SW41 ultracentrifuge tubes. Underlay it with 3.5 mL of H-CsCl as described in Step 4. Then, carefully overlay 3.5 mL of viral suspension onto the top of each of the two-layered CsCl gradients. Place the SW41 tubes into rotor buckets and balance them. Attach the buckets to the SW41 rotor and centrifuge at 20,000 rpm for 15–18 h at 4°C.

11. After centrifugation, repeat Steps 5–8 using a 3-mL syringe to collect the viral vector band (Fig. 1B).

    After the second centrifugation, only one band should be visible. If two bands are visible, some of the bands of empty virus particles might have been harvested after the first centrifugation. In this case, take the lower band only (Fig. 1B).

    Depending on the size of the bands, it may be necessary to extract 0.5–2.0 mL of viral vector per tube.

12. Use a 5-mL syringe with a 18-gauge × $1\frac{1}{2}$-inch needle to transfer the virus into 0.5–3-mL or 3–12-mL Slide-A-Lyzer Dialysis Cassettes (10,000 MWCO), depending on the volume of the collected viral vector. Dialyze the viral vector against 3 L of cold PBS in a 4-L plastic beaker and a magnetic stir bar. Place the beaker on a magnetic stirring plate at 4°C and stir slowly. Replace with 3 L of fresh cold PBS every 2–3 h for a period of 12 h.

13. Use another 5-mL syringe with an 18-gauge × $1\frac{1}{2}$-inch needle to carefully remove the desalted virus from the dialysis cassette to a sterile 15-mL conical centrifuge tube and mix well by pipetting.

14. Measure virus particle titer by transferring 10 mL of the virus into a sterile Eppendorf tube containing 90 mL of sterile water. Mix the virus dilution thoroughly by pipetting. Measure the sample optical density (OD) at 260 and 280 nm on a UV spectrophotometer, using water as the blank. Calculate the particle concentration of the virus using the following formula:

$$OD_{260} \times \text{dilution factor} \times 10^{12} = \text{virus particles/mL.}$$

The reliable range for OD_{260} reading is 0.1–1. If the readings are outside of this range, adjust dilution ratios accordingly. The vector OD_{260}/OD_{280} ratio can serve as a reference for the quality of virus packaging. The OD_{260}/OD_{280} ratio of a good quality preparation should be 1.20–1.40.

    Viral yields from 40-plate infections range from 0.2×10^{13} to 2×10^{13} virus particles. If the yield is $<2 \times 10^{12}$ virus particles, see Troubleshooting.

15. Add sterile glycerol to the virus preparation to a final concentration of 10% and mix gently by pipetting. Repeat Step 14 to measure the final virus particle concentration. Aliquot the viral vector into sterile cryovials. Immediately store the cryovials at −80°C.

TROUBLESHOOTING

Problem (Step 14): There is poor yield of purified recombinant adenovirus.

Solution: The most common cause of this problem is the lack of synchronization in the step of large-scale infection. In other words, because the cells are infected at low MOIs, in order to reach full CPE stage, subsequent secondary and even tertiary infections are required of the noninfected cells after completion of the localized primary infection and release of newly replicated virus into the infection medium. The best solution to this problem is to slow down the viral expansion process and to accumulate enough infectious viruses in the crude viral lysate to accomplish a synchronized and productive infection at the 40-plate infection stage, which is usually indicated by observation of CPE in 90% of cells no more than 48 h after infection.

DISCUSSION

Purification by CsCl gradient ultracentrifugation is relatively straightforward and fast, as long as precautions are taken in creating well-segregated multilayer step gradients. Purified and concentrated adenovirus suspension should have an opaque appearance with a light bluish tint. After formulating in 10% glycerol/PBS, keep adenovirus at −80°C and avoid repeated freeze–thaw cycles. Three cycles of freeze–thaw can result in up to a 90% reduction in adenovirus infectivity. Nonetheless, for repeated uses and short-term storage (up to 2 wk) of adenovirus vectors for cell culture applications, an effective strategy to avoid the detrimental effect of freeze–thaw cycles on the infectivity of adenovirus is to add more glycerol to the purified adenovirus to a final concentration of 40% and to keep the virus in a −20°C freezer.

For gene-transfer applications in cultured cells, the recommended MOIs should be in the range of 1000–10,000 virus particles per cell (∼10–100 infectious units per cell) for most cell types that do not support viral replication. This could also be influenced by some intrinsic properties of the target cells such as expression level of the adenovirus receptors (e.g., integrins and Coxsackie adenovirus receptor [CAR]) on the cell surface and cell density at the time of infection (the OD should be 60%–70% confluency). For in vivo gene-transfer applications, doses depend on the purpose of gene transfer, the intended target tissues, and routes of vector administration. For instance, recombinant adenoviruses delivered by intravenous injection primarily target the liver; the recommended dose for this application is 5×10^{12} virus particles/kg or 10^{11} virus particles per mouse. Adenoviruses at a dose of 5×10^{10} virus particles per mouse can achieve efficient gene transfer in the lung after intratracheal or intranasal administration. The suggested dose for gene-transfer applications in a mouse by a single-site intramuscular injection is 5×10^{10} virus particles per mouse.

Because of the strong immunogenicity of adenovirus vectors, transgene expression usually reaches its peak within a week; afterward, transgene expression declines as transduced cells are cleared (eliminated) by transgene and viral-protein-specific cytolytic T-cell responses. When using adenovirus vectors to study gene function, include a control vector without the transgene of interest to control for cellular responses elicited by adenovirus infection.

RECIPES

It is essential that you consult the appropriate Material Safety Data Sheets and your institution's Environmental Health and Safety Office for proper handling of equipment and hazardous materials used in this protocol.

Heavy CsCl Solution (H-CsCl, density=1.45 g/mL)
Dissolve 442.3 g of biological-grade cesium chloride (CsCl) in 578 mL of 10 mM Tris-Cl (pH 8.0), sterilize by passing through a 0.22-μm filter, and store at room temperature.

Light CsCl Solution (L-CsCl, density=1.25 g/mL)
Dissolve 223.9 g of biological-grade cesium chloride (CsCl) in 776 mL of 10 mM Tris-Cl, pH 8.0, sterilize by passing through a 0.22-μm filter, and store at room temperature.

Tris-Cl (10 mM, pH 8.0)
Mix 10 mL of 1 M Tris-Cl (pH 8.0) with 990 mL of H$_2$O, sterilize by passing through a 0.22-μm filter, and store at room temperature.

REFERENCE

NIH Guidelines for Research Involving Recombinant DNA Molecules. (April) 2000. http://oba.od.nih.gov/rdna/nih_guidelines_oba.html.

Characterization of the Purified Recombinant Adenovirus for Viral Genome Structure by Restriction Enzyme Digestions

The easiest way to confirm the structure and identity of genomic DNA isolated from purified adeno-viral recombinants is restriction enzyme digestion and gel electrophoresis. This analysis entails com-paring the restriction patterns of the adenoviral vector DNA with that plasmid that was used to initiate the entire rescue and expansion process. The integrity of the viral backbone and the presence of both the transgene and viral ITRs are assessed.

Biosafety Consideration

According to the NIH Guidelines for Research Involving Recombinant DNA Molecules (April 2000), all types of wild-type and replication-competent human adenoviruses are classified as risk group 2 of biohazard agents. The human disease associated with this group of biohazard agents is usually treatable and preventable and is rarely serious. All work with adenovirus vectors should be conducted at Biosafety Level 2 (BL2) with the approval by the Institutional Biosafety Committee of the home institution.

MATERIALS

It is essential that you consult the appropriate Material Safety Data Sheets and your institution's Environmental Health and Safety Office for proper handling of equipment and hazardous materials used in this protocol.

Recipes for reagents specific to this protocol, marked <R>, are provided at the end of the protocol. See Appendix 1 for recipes for commonly used stock solutions, buffers, and reagents, marked <A>. Dilute stock solutions to the appropriate concentrations.

Reagents

Agarose
DNA marker ladder (1 kb)
EDTA (0.5 M, pH 8.0)
Ethanol (70% and 100%)
Ethidium bromide (0.5 mg/mL)
  This is a hazardous substance; review the Material Data Safety Sheet before handling.

Isopropanol
Phenol:chloroform:isoamyl alcohol solution (25:24:1, v/v)
Plasmid DNA of the matched adenovirus vector clone (created in the research lab and used for viral
  vector rescue and production)
Pronase solution (2×) <R>
Restriction enzymes and digestion buffers
RNase A (20 mg/mL)
SDS (10%)
Sodium acetate (3 M, pH 5.2)
Sterile water

TAE running buffer <A>
TE buffer (1×, pH 8.0) <A>
Tris-HCl (1 M, pH 7.6)

Equipment

Gel electrophoresis apparatus
Imaging station or Polaroid camera
Microcentrifuge
Refrigerated tabletop centrifuge
Spectrophotometer
Vortex machine
Water bath (37°C)

METHOD

Extraction of Recombinant Adenovirus Genomic DNA from the Purified Virus

1. Pipette the appropriate volume of purified adenovirus vector containing 0.5×10^{12} to 1×10^{12} vector particles into a 1.5-mL microcentrifuge tube. Add an equal volume of 2× pronase solution. Mix well by inverting and incubate the digestion reaction for at least 4 h or overnight in a 37°C water bath.

    An alternative is to add an equal volume of sterile TE buffer to the vial of virus and then an equal volume of 2× pronase to the virus plus TE buffer; this dilutes the virus and allows maximum digestion of the virus.

2. Add an equal volume of phenol:chloroform:isoamyl alcohol solution to the digest. Mix well by repeatedly inverting the tube for at least 1 min. Centrifuge at 13,000 rpm in a microcentrifuge for 10 min at room temperature.

3. Remove the top (aqueous) layer (be sure to avoid the interface) and transfer to a new 1.5-mL microcentrifuge tube. Add 10% of the volume of 3 M sodium acetate, pH 5.2. Then, add an equal volume of isopropanol. Mix well by inverting the tube several times and incubate for at least 20 min at room temperature. Centrifuge at 13,000 rpm in a refrigerated microcentrifuge for 20 min at 4°C.

4. Decant the isopropanol from the pellet by slowly inverting the tube. Rinse the pellet by adding 1 mL of ice-cold 70% ethanol to the tube and then vortex the tube briefly for 5 sec. Centrifuge at 13,000 rpm for 5 min at 4°C.

5. Repeat Step 4 once. Discard the 70% ethanol by slowly inverting the tube. Pulse-centrifuge the tube and remove any remaining liquid using a micropipette (keep track of the pellet at all times). Allow the pellet to air-dry for 1 min and resuspend the vector DNA in 50 mL of TE buffer, pH 8.0, with 20 mg/mL of RNase A. Leave the tube overnight at room temperature. Determine the DNA concentration in a spectrophotometer after a 1:20 dilution.

Restriction Enzyme Analysis of Viral Vector Genome

6. On the basis of the electronic map for the recombinant adenovirus plasmid created in Protocol 1, identify suitable restriction enzymes that can be used for the digestion. The enzymes chosen should not only cut in the transgene cassette but should also cleave the vector genome several times, generating fragments mostly in the range of 0.5 to 5 kb that ideally should not comigrate with one another. Some suitable enzymes include BglII, HindIII, and XhoI. Two separate digests (which can either be single- or double-enzyme digests) need to be performed on both of the viral DNA and the parental plasmid DNA samples.

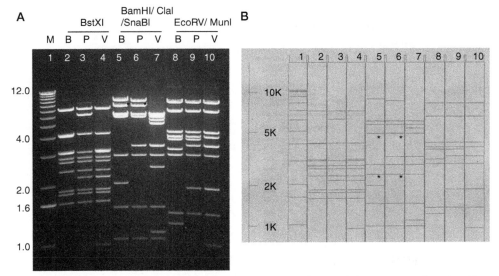

FIGURE 1. Characterization of a recombinant adenovirus genome by restriction enzyme digestion (Protocol 4). One microgram of the DNAs of a plasmid carrying an empty adenovirus vector, a plasmid carrying an adenovirus recombinant, and the empty adenoviral vector were cleaved with different restriction enzymes and analyzed by agarose gel electrophoresis. (*A*) Ethidium bromide–stained gel. M, 1 kb molecular weight ladder; B, plasmid carrying recombinant constructed in EGFP vector; P, plasmid carrying empty adenovirus vector; V, viral vector DNA. (*B*) Predicted restriction patterns by Vector NTI DNA software (Life Technologies). (Asterisk) Dam methylation of ClaI sites results in a shift of the expected 4871- and 2409-bp bands to a 7280-bp single band in the ClaI digest of plasmid carrying an empty adenovirus vector (Protocol 4).

7. Digest 0.75 µg of viral DNA or 0.5 µg of the plasmid DNA in a volume of 10 mL 1× restriction buffer with 1 mL of enzyme of choice. Incubate the mixture for at least 2 h at the temperature required for the enzyme.

8. Pulse-centrifuge the digest and carry out the gel electrophoresis of all of the samples together with a 1-kb DNA marker ladder on a 0.8% agarose gel in 1× TAE running buffer containing 0.5 mg/mL ethidium bromide at a constant voltage of 4–5 V/cm (100–110 V for a 10-cm-long gel). (See Chapter 2.) Photograph the gel at two stages: When the marker dye (1) reaches one-fourth to one-fifth of the way down the gel and (2) has moved past the end of the gel. This allows visualization of small fragments below 0.5 kb and separation of larger fragments in the 2–10-kb range.

9. Compare the electronically generated pattern of DNA fragments with the DNA bands on the gel. Pay close attention to determine whether all internal fragments of the digested viral vector genome (i.e., not including the ITR-containing fragments) match those of the vector plasmid. In other words, with the exception of fragments containing ITR elements, any additional, missing, or incorrectly sized fragments seen in the vector DNA (i.e., not predicted by the map) should also be present in the digest of the plasmid carrying the empty adenovirus vector. Search carefully for any signs of rearrangements of the viral genome after rescue and growth (Fig. 1).

DISCUSSION

The genome of a recombinant adenovirus vector is characterized using standard molecular biology techniques such as DNA extraction, restriction enzyme digestion, and gel electrophoresis; however, close attention must be paid to possible deletions and rearrangements in the viral vector genome, particularly in the transgene expression cassette. If high-level expression of the transgene interferes with adenovirus biology, vector genomes that carry deletions or rearrangements in the transgene cassette, or those that carry mutations that result in diminished transgene expression, have a growth advantage that allows them to become dominant over the course of repeated cycles of viral growth. Therefore, this analysis of genome structure is particularly important for vectors that have undergone repeated amplifications.

RECIPE

It is essential that you consult the appropriate Material Safety Data Sheets and your institution's Environmental Health and Safety Office for proper handling of equipment and hazardous materials used in this protocol.

Pronase Solution (2×)

Reagent	Quantity (for 20 mL)	Final concentration
Pronase (Roche)	40 mg	0.2%
Tris-Cl (1 M, pH 7.6)	2 mL	100 mM
EDTA (0.5 M, pH 8.0)	0.08 mL	2 mM
SDS (10%)	2 mL	1%
H$_2$O	15.9 mL	

Mix the solution well by votexing and incubate for 45 min at 37°C for self-digestion and activation. Aliquot the ready-to-use 2× pronase solution into 1.5-mL Eppendorf tubes and store at −20°C.

REFERENCE

NIH Guidelines for Research Involving Recombinant DNA Molecules. (April) 2000. http://oba.od.nih.gov/rdna/nih_guidelines_oba.html.

Measuring the Infectious Titer of Recombinant Adenovirus Using TCID$_{50}$ End-Point Dilution and qPCR

Traditionally, adenovirus and recombinant adenovirus infectious titers have been measured by plaque assay, in which the cells are infected with serially diluted adenovirus stock and then overlaid with agar; a plaque will form as the result of a single infectious event (Lawrence and Ginsberg 1967). Although this method gives a quantitative readout (number of plaques corrected for the dilution), there can be issues with sensitivity and reproducibility, especially when adenovirus serotypes are used that infect standard cell lines with poor efficiency.

An alternative approach is to plate serial dilutions of the cells growing in the wells of a 96-well tissue culture plate and determine the dilution at which 50% of the wells are infected. This ancient and reliable technique known as the "tissue culture infection dose 50%" (TCID$_{50}$) end-point dilution method has been used for titering a number of viruses, especially those that do not readily form plaques (Heldt et al. 2006). Usually, infected wells are determined by direct examination for CPE or cell viability. However, by combining a 96-well TCID$_{50}$ format and the power of quantitative PCR (qPCR) for detection, a large increase in sensitivity—in our hands 10-fold, with a range of both transgenes and adenovirus serotypes—can be achieved.

The following protocol uses a 96-well TCID$_{50}$ format, in conjunction with qPCR for sensitive and quantitative positive-well calling, to determine infectious titer of adenovirus vectors.

Biosafety Consideration

According to the NIH Guidelines for Research Involving Recombinant DNA Molecules (April 2000), all types of wild-type and replication-competent human adenoviruses are classified as risk group 2 of biohazard agents. The human disease associated with this group of biohazard agents is usually treatable and preventable and is rarely serious. All work with adenovirus vectors should be conducted at Biosafety Level 2 (BL2) with the approval by the Institutional Biosafety Committee of the home institution.

This protocol was contributed by Martin Lock, Michael Korn, and James Wilson (Gene Therapy Program, University of Pennsylvania, Philadelphia).

MATERIALS

It is essential that you consult the appropriate Material Safety Data Sheets and your institution's Environmental Health and Safety Office for proper handling of equipment and hazardous materials used in this protocol.

Recipes for reagents specific to this protocol, marked <R>, are provided at the end of the protocol. See Appendix 1 for recipes for commonly used stock solutions, buffers, and reagents, marked <A>. Dilute stock solutions to the appropriate concentrations.

Reagents

Adenovirus test vector (created by the method described in Protocol 1)
Complete growth medium (DMEM+10% FBS+1% P/S) <R>
Deoxycholate (DOC) (10%) <R>
DNA extraction solution <R>
HEK-293 cells

PCR standards (set of eight, $10–10^8$ copies)

> See the Additional Protocol Preparation of DNA Standard for qPCR after Protocol 5, for preparation of PCR standards.

Phosphate-buffered saline (PBS)

Proteinase K (1.2 mg/mL)

Proteinase K buffer (10×) <R>

qPCR master mix <R>

qPCR mix (2×)

> qPCR mix containing, buffer, nucleotide triphosphates, and thermostable polymerase is available from various sources (e.g., Applied Biosystems, catalog no. 4326614).

qPCR Primer/Probe set, targeting the E2a region of the vector genome

> qPCR primer and probe design has been described extensively in the published literature. Helpful information is available on the Applied Biosystems website (www.appliedbiosystems.com). See also Chapters 7 and 9.

Serum-free medium (DMEM+1% P/S) <R>

Sterile H_2O

Trypan Blue dye

Trypsin–EDTA solution (0.25%)

Tween solution <R>

Equipment

Adhesive AirPore Tape Sheets plate-sealing film (QIAGEN, catalog no. 120001)

Cell culture biosafety cabinet

Centrifuge, with 96-well plate

Conical tubes (15 and 50 mL)

Hemocytometer

Humidified cell culture incubator (37°C, 5% CO_2)

Hybridization oven

Micropipettes (20 and 200 mL)

Microscope

PCR plate (96 well) and optical-grade plate-sealing film

Pipettes (8 or 12 channel)

qPCR machine (96 well)

Reservoirs, sterile (12 well)

Solution reservoir for multichannel pipettes, sterile

Sterile cryovials (nine per test vector)

Tabletop centrifuge, with swinging plate holder

Vortex machine

Water bath (37°C)

METHOD

Plating Cells

1. Warm Complete growth medium, PBS, and trypsin in a water bath to 37°C.

   Perform all subsequent steps in the cell culture biosafety cabinet using sterile technique.

2. Aspirate the medium from a confluent monolayer of HEK-293 cells.

   HEK-293 cells are human embryonic kidney cells transformed with the E1 region of adenovirus and are in widespread use. Stocks of HEK-293 cells with different passage histories often show different susceptibility to adenovirus infection. Early passage numbers are preferred. If no in-house stock is kept, the cells are available from the American Type Culture Collection (ATCC), catalog no. CRL1573.

3. Wash the cells with 10 mL of PBS, aspirate the PBS, and then trypsinize the cells with 2 mL of trypsin for 5 min at 37°C. Tap the flasks gently to remove the cells and then add 8 mL of Complete growth medium. Transfer the cell suspension to a 50-mL conical tube. Repeat the trypsinization for additional flasks and pool the cells into the same 50-mL conical tube. Mix the cell suspension thoroughly and then count the living cells using a hemocytometer and Trypan Blue dye.

4. Dilute the cell suspension with Complete growth medium to a concentration of 8×10^5 cells/mL. Vortex the diluted cell suspension and pour it into a sterile plastic reservoir. Using a 12-channel pipette, pipette 50 mL of cell suspension into each well of a 96-well plate. Incubate the plates overnight in a 37°C incubator.

   When pipetting the cell suspension, hold the plate at an angle and dispense the cell suspension along the side of the wells to minimize the formation of air bubbles in the medium.

Infection

5. Prepare all vector dilutions in serum-free medium. Thaw the vector on ice and pipette to mix. Eight serial dilutions in a total of 1 mL are made in cryovials for each vector. For example, the following dilution scheme can be used:

   $1 \times 10^{-2} = 990$ mL diluent + 10 mL stock
   $1 \times 10^{-4} = 990$ mL diluent + 10 mL previous dilution
   $1 \times 10^{-6} = 990$ μL diluent + 10 μL previous dilution
   $1 \times 10^{-7} = 900$ μL diluent + 100 μL previous dilution
   $1 \times 10^{-8} = 900$ μL diluent + 100 μL previous dilution
   $1 \times 10^{-9} = 900$ μL diluent + 100 μL previous dilution
   $1 \times 10^{-10} = 900$ μL diluent + 100 μL previous dilution
   $1 \times 10^{-11} = 900$ μL diluent + 100 μL previous dilution

   The dilution range plated should be determined empirically for each vector and vector serotype.

6. Infect the HEK-293 cells in the 96-well plates with the diluted vector.

   Infection schemes similar to those shown in Figure 1 are typically used in our laboratories but can be adapted as required by the researcher. A 12-well dilution reservoir is used in conjunction with a multichannel pipette to simplify the infection process.

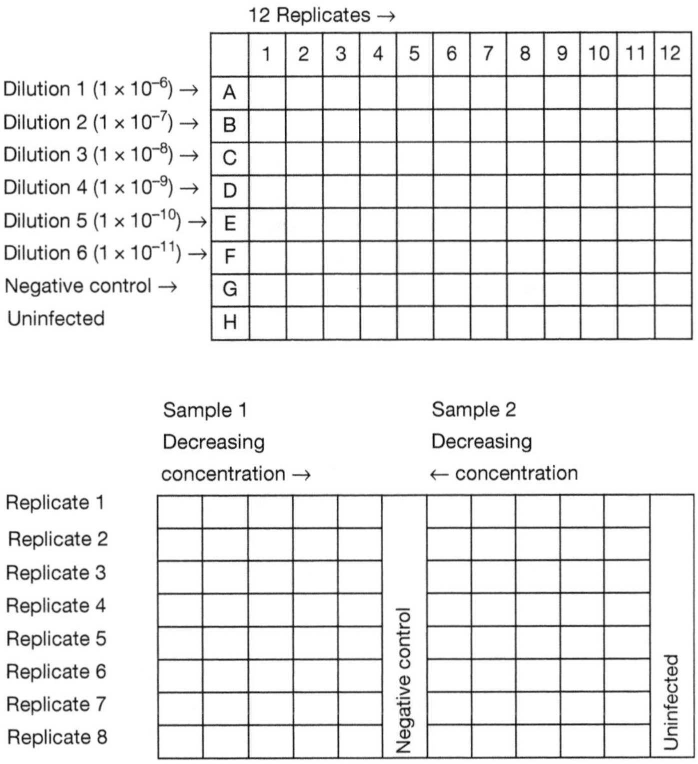

FIGURE 1. **Plate layouts for infection setup in the adenovirus vector infectious titer determination assay.** The first two dilutions made (i.e., 1×10^{-2} and 1×10^{-4}) will not be used.

7. Add the relevant dilution to each well of the 12-well reservoir and align the reservoir and the 96-well plate such that the vector dilution is in line with the row or column of wells to which the dilution will be added.

   Take care to open the cryovial of vector pointing away from the multiwell basin to minimize cross-contamination.

8. Aspirate the Complete growth medium from the cells in the 96-well plate. Use a 200-µL unplugged plastic pipette tip on the end of a 2-mL aspiration pipette and hold the plate at an angle while aspirating from the side of the well.

9. Pipette 50 µL of a virus dilution to the first column or row (see layouts in Fig. 1) of the 96-well plate, again ensuring that the liquid is added to the side of the well. Repeat this maneuver for all other vector dilutions. Add plain diluent to negative control wells.

   When pipetting the vector dilutions, enter the plate from the side adjacent to the appropriate well such that pipette tips cross only those wells designated for that dilution.

10. Cover the plate with the lid, label the plate, and then incubate it for 2 h in a 37°C incubator.

11. Warm Complete growth medium to 37°C in a water bath and then pipette 1 mL of the medium into each well of a 12-well reservoir, using a new 12-well reservoir for each plate of infected cells. Use a 12-channel pipette to add 50 µL of Complete growth medium to each well of the 96-well plate containing infected cells. Avoid contact with the infection medium.

    The plate should be placed in the opposite orientation such that the Complete growth medium is added to the opposite wall of the well to which the vector dilution was added.

12. Seal the plate with adhesive AirPore film, cover the plate with a lid, and incubate the plate for 72 h in a 37°C, CO_2 incubator

    The adhesive AirPore film is used to minimize cross contamination by aerosols during the 3-d incubation period.

DNA Extraction

13. Prepare enough DNA extraction solution for all plates by combining the reagents listed in the recipe.

14. Centrifuge the 96-well plate at 1500 rpm for 30 sec in a tabletop centrifuge with a swinging plate holder to deposit all liquid on the bottom of the wells.

15. Carefully remove the adhesive AirPore film from the plate (peel from the bottom row upward) and inspect the cells under a microscope and score each well for the presence or absence of CPE. Add 100 µL of DNA extraction solution to each well. Mix the reaction by pipetting five times while exercising caution to avoid cross-contaminating adjacent wells.

16. Seal the 96-well plate with adhesive sealing film and incubate it in a hybridization oven at 37°C for 1 h, 55°C for 2 h, and then 95°C for 30 min (to heat-inactivate the proteinase K). The plate containing the DNA samples can be used immediately for PCR or may be stored covered with adhesive sealing film for up to 7 d at 4°C.

Real-Time PCR

17. Working in a dedicated PCR facility or under a chemical fume hood, set up the PCR plate in the same format as the infection plate but include wells for quantification standards and a "no-template" control (water). Figure 2 depicts an example of a PCR setup.

18. For each plate, make a qPCR master mix for 100 reactions as shown in the recipes section at the end of the protocol.

    The qPCR master mix contains PCR primers and a Taqman probe specific to an element within the adenovirus vector. These reagents are usually designed to target the adenovirus DNA-binding protein (E2A gene) of a particular adenovirus serotype so that they can be used for multiple vectors made with this serotype. Several PCR mixes containing the other essentials of the PCR (nucleotide

		1	2	3	4	5	6	7	8	9	10	11	12
Dilution 1 (1×10^{-6}) →	A												
Dilution 2 (1×10^{-7}) →	B												
Dilution 3 (1×10^{-8}) →	C												
Dilution 4 (1×10^{-9}) →	D												
Dilution 5 (1×10^{-10}) →	E												
Dilution 6 (1×10^{-11}) →	F												
Negative control →	G												
Standards →	H	10^8	10^7	10^6	10^5	10^4	10^3	10^2	10^1	H_2O			

FIGURE 2. **Representative layout of the qPCR plate setup for the adenovirus vector infectious titer determination assay.**

*triphosphates, buffer, thermostable enzyme) are available from various commercial vendors at a 2×
concentration.*

19. Combine the appropriate volumes of reagents in a 15-mL conical tube, mix well, and then add
45 μL of the qPCR master mix to each well of the 96-well PCR plate. Pipette 5 μL of each standard into the qPCR master mix in the appropriate wells of row H of the 96-well PCR plate. Mix
by pipetting 10 times. Add 2.5 μL of water to the rest of the wells, with the exception of the
no-template control, into which 5 μL of water should be added.

    *Exercise extreme caution when adding the standards to minimize cross-contamination of other
    wells. Ensure that the pipette crosses only those wells designated for the appropriate standard.*

20. Add 2.5 μL of cell lysate (prepared in Step 16) to the qPCR master mix in the corresponding
wells of the 96-well PCR plate.

    Align the plates and enter the PCR plate from the side adjacent to the appropriate wells so that pipette tips cross only those wells designated for that sample dilution.

21. Cover the PCR plate with an optical-grade plastic adhesive sealing film and ensure that a
tight seal is formed. Perform PCRs in a real-time qPCR machine following the manufacturer's
directions.

    *Typical cycling times are an initial denaturation period of 10 min at 95°C followed by 40 cycles for 15
    sec at 95°C/1 min at 60°C.*

Data Collection

22. Refer to the appropriate instrument user guide for instructions on analyzing the data. The general process involves viewing the amplification plots and then setting the baseline and threshold
values. During early PCR cycles, use the background signal in all wells to determine the baseline
fluorescence. Place the threshold in the region of exponential amplification. For ABI 7500 or
similar machines, the baseline can be routinely set as cycles 3–12 and the threshold adjusted
so that the 10^8 standard reads 13.09 Ct. Alternatively, auto baseline and auto Ct programs
included in the instrument software can be used. The important point is to use the same
method consistently.

    See Troubleshooting.

Data Analysis

23. First calculate the vector input value at each dilution by multiplying the input vector concentration (viral genomes/mL) by the dilution factor and divide the result by 20 to obtain genomes
per 50 μL, the volume of the inoculum. For example,

$$\text{Vector input concentration} = 1 \times 10^{12} \text{ genomes/mL,}$$

$$\text{Dilution} = 1 \times 10^{-8},$$

$$\text{Input vector genome copies} = (1 \times 10^{12} \text{ genomes/mL} \times 1 \times 10^{-8})/20 = 500.$$

24. Multiply the copy numbers detected at a dilution by a dilution factor of 80 (the total dilution of sample in all of the manipulations described) to obtain a total number of vector genome copies in the assay well. Subtract the number of input vector genomes at a particular dilution from the total vector copy number for each well at that dilution; repeat this operation for all dilutions.

    For instance, genome copies detected at 1×10^{-8} dilution by PCR $= 50,000$; thus, genome copies in the assay plate $= (50,000 \times 80) - 500 = 3.99 \times 10^6$.

    The input vector is subtracted so that only the vector that is replicating actively, and thus has infected the cell, is measured. The mathematical operations described here can be simplified by transferring the data to a Microsoft Excel spreadsheet.

25. Next, determine the number of positive replicates at each dilution. For the purposes of this assay, the quantifiable detection limit of the qPCR is set at 10 vector copies. Multiplying by the dilution factor of 80, this translates to 800 vector copies in the assay well. Thus, any input-subtracted copy number <800 is considered background, and input-subtracted vector copy numbers >800 are designated positive for infectious particles as measured by replicated vector DNA.

    See Troubleshooting.

26. The calculations described allow one to transform the data into a typical "hits per dilution" format, which is the basis of calculating a tissue culture infectious dose at which 50% of the wells are infected ($TCID_{50}$). In our laboratories, we use the Spearman–Karber formula to perform this calculation:

$$\log TCID_{50} = \sum_{i=1}^{K-1} (P_{i+1} - P_i) \cdot \frac{1}{2} (X_{i+1} - X_i),$$

where K is the number of dosages (or dilutions), X_i is the log of dosage level at the ith dose (dilution), and P_i is the proportion of wells with adenovirus DNA replication at the ith dose (dilution).

The dosage level here refers to the log of the dilution plated, not the number of genome copies. Genome copies are used solely to quantitatively call whether a well has scored a hit (replicated adenovirus DNA), as described in Steps 24 and 25. The log $TCID_{50}$ from which the infectious titer is derived is calculated from the log dosages X_i (log dilutions) and the percent of hits (P_i) using the Spearman–Karber formula as indicated above. P_i is the fraction of wells at a given dilution (i) that contain replicated DNA according to the definition in Steps 24 and 25.

This formula thus converts the fraction of wells that contain replicated adenovirus DNA (P_i) at a given dilution (i) to log $TCID_{50}$. For example,

Log dilution (X)	Hit percentage (P)
$X_1 = -8$	$P_1 = 0$
$X_2 = -9$	$P_2 = 1/12$
$X_3 = -6$	$P_3 = 8/12$
$X_4 = -5$	$P_4 = 1$

For $K = 4$, expanding the formula, where $i = 1$ to 3 ($K - 1$), gives

$$\text{Log } TCID_{50} = (P_2 - P_1)*(1/2)(X_2 + X_1) \ldots . i = 1$$
$$+ (P_3 - P_2)*(1/2)(X_3 + X_2) \ldots . i = 2$$
$$+ (P_4 - P_3)*(1/2)(X_4 + X_3) \ldots . i = 3.$$

Substituting the values above, the calculated log $TCID_{50}$ here is -6.25.

Titer in infectious units (IU) is then determined as follows:

$$\text{TCID}_{50} \ \text{IU/mL} = \frac{10^{-(\log \text{TCID}_{50})}}{A},$$

where A is the amount of inoculum (in milliliters). A convenient computer program to calculate TCID$_{50}$ IU using this formula has been written (Lynn 1992).

Note that 0.7 TCID$_{50}$ IU is equivalent to 1 IU as determined by plaque assay.

TROUBLESHOOTING

Problem (Step 22): Negative control wells have a positive signal.

Solution: Take steps to minimize cross-contamination between wells. Vials containing vector dilutions should be opened pointing away from the infection and PCR plates. Track pipette tips containing a vector dilution only over those wells that will receive the same dilution. Centrifuge the film-covered plates before opening them to ensure that condensation on the sealing film is deposited in the well. Remove the sealing film from the plate in the direction from the least to the most concentrated vector dilution. Arrange the negative control wells in the plate such that they are adjacent to the highest vector dilution.

Problem (Step 25): All vector dilutions give a positive PCR signal; there is no apparent decrease in copy number with increasing dilution.

Solution: Ensure that the correct dilution range has been chosen. The two most likely causes for this problem are (1) cross-contamination (see above) and (2) insufficient dilution of the vector. It is recommended that, when titrating a vector (or particular vector serotype) for the first time, an extended range of dilutions be tested on several plates. Thereafter, the appropriate range of six dilutions can be selected and used.

Inclusion of a validation vector in the assay for which the titer is known will be helpful in eliminating more general problems, such as a cross-contaminated qPCR master mix or primer stock. Inclusion of the no-template qPCR control is also important for pinpointing this type of problem.

Problem (Step 25): No positive signal can be seen in the vector dilution wells.

Solution: Ensure that the correct dilution range and qPCR primer–probe set are chosen. It is possible that the vector dilutions plated are too dilute. We recommend that an extended range of dilutions be tested on several plates when titrating a vector (or particular vector serotype) for the first time. A validation vector, preferably with the same or similar vector genome as the test vector, will again be useful to ensure that a more general problem does not exist (e.g., a degraded primer or probe). The target sequence in the vector should be aligned with the primer–probe sequences to detect any possible mismatches.

DISCUSSION

This method measures the infectious titer of adenovirus vectors with high sensitivity. In addition, the 96-well format introduces ease of handling and the possibility of adapting the technique to automation. One unwanted consequence of the sensitivity of qPCR quantitation is the detection of low-level cross-contamination with adenovirus genome copies, which may lead to false-positive infected-well calling. However, with proper handling and the precautions outlined in the sections above, this phenomenon can be minimized and excellent interassay reproducibility is possible.

RECIPES

It is essential that you consult the appropriate Material Safety Data Sheets and your institution's Environmental Health and Safety Office for proper handling of equipment and hazardous materials used in this protocol.

Complete Growth Medium

Reagent	Quantity (for 1 L)	Final concentration
DMEM	890 mL	
FBS	100 mL	10%
P/S solution (100×)	10 mL	1×

Store at 4°C.

Deoxycholate (DOC) (10%)

Reagent	Quantity (for 100 mL)	Final concentration
Sodium deoxycholate	10 g	10%, w/v
H_2O	to 100 mL	

Dissolve 10 g of deoxycholate in 85 mL of dH_2O, bring the volume to 100 mL, and sterilize by passing through a 0.22-μm filter. Store at room temperature.

DNA Extraction Solution

Reagent	Quantity (for 10 mL)	Final concentration
DOC (10%)	0.5 mL	0.25%
Tween solution (10%)	0.9 mL	0.45%
Proteinase K buffer (10×)	2 mL	2×
Proteinase K (10 mg/mL)	0.6 mL	0.3 mg/mL
H_2O	6 mL	

Do not store; make fresh immediately before use. Final concentrations take into account the twofold dilution incurred when adding the extraction solution to the tissue culture wells.

Proteinase K Buffer

Reagent concentration	Quantity (for 100 mL)	Final concentration
Tris-HCl (1 M, pH 8)	1 mL	10 mM
EDTA (500 mM)	2 mL	10 mM
SDS (10%)	10 mL	1%
H_2O	87 mL	

Sterilize by passing through a 0.22-μm filter. Store at −20°C.

qPCR Master Mix

Reagent	Stock	1 Reaction (μL)	100 Reactions (μL)	Final concentration
qPCR mix (2×)	2×	25 μL	2500 μL	1×
Forward primer	10 μM	1 μL	100 μL	200 nM
Reverse primer	10 μM	1 μL	100 μL	200 nM
Probe	10 μM	0.5 μL	50 μL	100 nM
Water	NA	17.5 μL	1750 μL	NA
Total		45 μL	4500 μL	

Serum-Free Medium

Reagent	Quantity (for 1 L)	Final concentration
DMEM	965 mL	
HEPES (1 M, pH 8.0)	25 mL	25 mM
P/S solution (100×)	10 mL	1×

Store at 4°C.

Tween Solution

Reagent concentration	Quantity (for 50 mL)	Final concentration
Tween 20 (50%)	10 mL	10%, w/v
HEPES (1 M, pH 8)	1 mL	20 mM
H_2O	39 mL	

Make a 50% solution of Tween 20 by diluting 1:1 with water. Combine 10 mL of 50% Tween 20, 1 mL of 1 M HEPES, and 39 mL of water to make a 50-mL stock solution. Sterilize by passing through a 0.22-μm filter. Store at 4°C.

Preparation of a DNA Standard for qPCR

Biosafety Consideration

According to the NIH Guidelines for Research Involving Recombinant DNA Molecules (April 2000), all types of wild-type and replication-competent human adenoviruses are classified as risk group 2 of biohazard agents The human disease associated with this group of biohazard agents is usually treatable and preventable and is rarely serious. All work with adenovirus vectors should be conducted at Biosafety Level 2 (BL2) with the approval by the Institutional Biosafety Committee of the home institution.

MATERIALS

It is essential that you consult the appropriate Material Safety Data Sheets and your institution's Environmental Health and Safety Office for proper handling of equipment and hazardous materials used in this protocol.

Reagents

DNA purification kit (e.g., QIAGEN, catalog no. 28104)
Nuclease-free water
PCR buffer (10×; Applied Biosystems, catalog no. N8080189)
Plasmid standard
Restriction enzyme with buffer
Sheared salmon sperm (SSS) DNA

Equipment

Biosafety cabinet
Spectrophotometer
Vortex mixer

METHOD

Use a biosafety cabinet to prevent potential contamination from the laboratory environment. Real-time PCR is extremely sensitive to contamination.

1. Linearize 40 μg of plasmid DNA used as the standard in four 100-μL reactions with a restriction enzyme that cuts outside the intended PCR target. Purify the linearized plasmid from each reaction using a suitable DNA purification kit.
   Do not perform a phenol:chloroform:isoamyl purification of the digest because any contamination with phenol will affect the accuracy of the spectrophotometric reading.

2. Dilute the linearized plasmid 1:20 with water and determine the concentration by spectrophotometry. Convert the concentration readout to grams per liter (g/L).
   Ensure the reading falls within the linear range of the instrument. If necessary, use a more concentrated set of samples. Read against a buffer blank and apply background correction at 320 nm.

3. Calculate the formula weight (F.W.) of the standard plasmid: F.W. = Plasmid size (in base pairs) × 662 g/mol.bp. Calculate the molar concentration (M) of the linearized plasmid. M = mol/L = (mass [in grams]/F.W.)/1 L.

4. Determine the copy number per microliter of the linearized plasmid based on molar concentration. 1 M is equivalent to 6.02×10^{23} copies. Dilute the linearized plasmid in $1 \times$ PCR buffer and 2 ng/μL SSS DNA so that the final concentration is 2×10^{11} copies per 100 μL (1×10^{10} copies/5 μL) in $1 \times$ PCR buffer/2 ng/μL SSS DNA.

   Salmon sperm DNA is included as a blocking agent to prevent plasmid loss caused by nonspecific binding to surfaces.

5. Perform a 10-fold serial dilution of the stock standard (1×10^{10} copies/5 μL) using $1 \times$ PCR buffer/2 ng/μL SSS DNA as the diluent. Use the 1×10^8 copies/5 μL through 10 copies/ 5 μL dilutions as the test standards.

   All standards should be aliquoted and stored at $-20°C$. Repeated freeze–thawing should be avoided.

REFERENCES

Lawrence WC, Ginsberg HS. 1967. Intracellular uncoating of type 5 adenovirus deoxyribonucleic acid. *J Virol* **1**: 851–867.

Lynn DE. 1992. A BASIC computer program for analyzing endpoint assays. *BioTechniques* **12**: 880–881.

NIH Guidelines for Research Involving Recombinant DNA Molecules. (April) 2000. http://oba.od.nih.gov/rdna/nih_guidelines_oba.html.

Detection Assay for Replication-Competent Adenovirus by Concentration Passage and Real-Time qPCR

Replication-defective adenovirus vectors are deficient in early viral gene *E1* and are generally grown in cell lines such as HEK-293, which contain and express an integrated copy of *E1* (Louis et al. 1997) for the purpose of complementing the defective adenoviral genome. Although the host cell E1 gene is required for complementation, its presence can also be problematic because recombination between vector and host cell genomes may result in the reacquisition of *E1* by the vector and the formation of replication-competent adenoviruses (RCAs). This problem has long been recognized, and approaches, such as the minimization of the sequence homology between the adenovirus genes inserted into the host chromosome DNA and viral sequences remaining in the 5′ end of the vector genome, have been taken to reduce recombination and RCA formation (Fallaux et al. 1998). Nevertheless, in the clinical setting it is important to ensure that the vector dose given to patients contains no more than one RCA infectious particle per patient dose (NIH 2001). Historically, this has been done by plating the vector stock on a cell line that does not contain the adenovirus E1 gene and then visually scoring the RCA cytopathic effect (CPE) (Hehir et al. 1996). The sensitivity of the assay is established by spiking known amounts of an RCA surrogate (e.g., wild-type human adenovirus type 5 [HuAd5] for vectors based on this serotype) into the vector stock and assaying the spiked samples alongside the test vector. Reliance on the detection of CPE results in low sensitivity and is generally only sufficient to screen for relatively high amounts of RCAs. More recently, qPCR directed at the adenovirus E1 gene has been used to enhance sensitivity and also shorten the assay time (Schalk et al. 2007). The sensitivity of the assay can also be affected by a phenomenon known as "interference," whereby a low level of RCAs is outcompeted by large excesses of vector for occupancy of available cell-surface viral receptors and subsequent replication pathways. An excess of vector can also lead to toxicity and cell death. Thus, the initial ratio of virus to cells (or MOI) must be kept to levels at which toxicity and interference are minimized but are still practical in terms of the tissue culture burden incurred.

Sensitivity can often be enhanced by biological amplification of the RCAs with serial passage. Here, we describe an extension of this technique, termed "concentration passage," in which RCA replicated during the first plating of the vector is collected and concentrated onto 1/10th of the original number of cells. This significantly increases the chances of detecting the RCAs. Combining this approach with the use of qPCR for sensitive detection of the RCA E1 gene, we are able to reach levels of sensitivity of 1 IU of RCAs in 10^{11} vector particles. The protocol described here is tailored for HuAd5 vectors using wild-type HuAd5 as the RCA surrogate. However, we have also adapted this technique with similar sensitivity to vectors based on other adenovirus serotypes. If other adenovirus serotypes are assayed, careful consideration should be given to the appropriate RCA surrogate. Strictly speaking, if the vector is propagated in HEK-293 or similar cell lines, the RCA surrogate should be a hybrid virus containing the HuAd5 E1 gene.

Biosafety Consideration

According to the NIH Guidelines for Research Involving Recombinant DNA Molecules (April 2000), all types of wild-type and replication-competent human adenoviruses are classified as risk group 2 of biohazard agents. The human disease associated with this group of biohazard agents is usually treatable and preventable and is rarely serious. All work with adenovirus vectors should be conducted at Biosafety Level 2 (BL2) with the approval by the Institutional Biosafety Committee of the home institution.

This protocol was contributed by Martin Lock, Mauricio Alvira, and James Wilson (Gene Therapy Program, University of Pennsylvania, Philadelphia).

MATERIALS

It is essential that you consult the appropriate Material Safety Data Sheets and your institution's Environmental Health and Safety Office for proper handling of equipment and hazardous materials used in this protocol.

Recipes for reagents specific to this protocol, marked <R>, are provided at the end of the protocol. See Appendix 1 for recipes for commonly used stock solutions, buffers, and reagents, marked <A>. Dilute stock solutions to the appropriate concentrations.

Reagents

A549 cells (160 150-mm plates)
Bleach 10%
Complete growth medium <R>
Dry ice
Dulbecco's phosphate-buffered saline solution (D-PBS)
Ethanol (70% and 100%)
F-12/K medium
Fetal bovine serum (FBS)
HEPES (1 M, pH 8.0)
Infection medium <R>
Master mix for nonspiked reactions <R>
Master mix for spiked reactions <R>
PCR standards (eight, $10-10^8$ copies)
  See the Additional Protocol Preparation of a DNA Standard for qPCR for preparation of PCR standards.

Penicillin–streptomycin (P/S)
QIAamp DNA Mini Kit (QIAGEN, catalog no. 51104)
qPCR mix (2×)
  qPCR mix containing buffer, nucleotide triphosphates, and thermostable polymerase is available from various sources (e.g., Applied Biosystems, catalog no. 4326614).

qPCR primer/probe set: HuAd5
  Forward primer   AGATACACCCGGTGGTCCC
  Reverse Primer   CGACGCCCACCAACTCTC
  Probe   6FAM-CTGTGCCCCATTAAACCAGTTGCCG-TAMRA
Spike DNA (H5 E1 plasmid standard)
Sterile water
Supplementation medium <R>
Test vector
Trypan Blue solution (0.4%)
Trypsin–EDTA solution (1×, 0.25%)
Wild-type HuAd5 adenovirus at a titer of 1×10^{12} particles/mL or greater
  Aliquots of HuAd5 adenovirus are available for purchase from Penn Vector Core (vector@mail.med.upenn.edu).

Equipment

ABI 7500 Fast Real-Time PCR system machine or equivalent
Centrifuge, with swinging plate holder
Falcon tube (50 mL)
Hemocytometer

Humidified cell culture incubator (37°C, 5% CO_2)

Laminar flow biosafety cabinet with UV light

> *The biosafety cabinet should be approved for work with Biosafety Level 2 (BSL-2) pathogens.*

Micropipettes (20, 200, and 1000 μL)

PCR plate, 96 well, and optical-grade adhesive plate-sealing film

Pipettes and pipette tips, assorted, disposable, and sterile

qPCR machine, 96 well

Refrigerated tabletop centrifuge

Screw-cap tubes, sterile

Sterile filter units (0.2 μM)

T-225 tissue culture flasks

Tissue culture plates (150 mm)

Vortex mixer

Water bath (37°C)

METHOD

The cell culture work is divided into two phases. In the first, 5×10^{11} particles of the test vector, the same amount of the vector spiked with 5 IU of the RCA surrogate (spiked control plates), and 5 IU of the RCA surrogate alone (positive control plates) are each inoculated into 50 plates of A549 cells. Fifty plates are used at this point because only 1×10^{10} particles can be added per plate (MOI = 3300 particles/cell $\approx$ 330 $TCID_{50}$ IU/cell $\approx$ 33 pfu/cell) without toxicity. In the second phase, the total lysate from the first 50 plates is used to infect five plates of A549 cells. This operation is referred to as a "concentration passage."

This assay requires the use of wild-type adenovirus; therefore, all work up to Step 25 must be performed in a BSL-2–approved biosafety cabinet. It is advisable that handling the wild-type adenovirus be the last thing done in the biosafety cabinet on the day that the assay is set. It is good practice to aliquot the medium first, using a separate medium tube for each set of plates and discarding the remainder. Thoroughly disinfect the biosafety cabinet with bleach solution and then ethanol immediately after handling wild-type adenovirus. Keep the UV light on for at least 1 h. Do not transport positive control plates at the same time as any other plates. Treat all plastics exposed to wild-type adenovirus with bleach solution before they are removed from the biosafety cabinet. Take any spill from a plate seriously and clean thoroughly with bleach solution.

Cell Culture: Phase I

For this phase, 160 150-mm plates of A549 cells are required for each test vector. The seed density for 150-mm plate is 3×10^6 A549 cells plated the day before the assay.

Preparation of Cells

1. Maintain A549 cells in T-225 flasks and passage them as required using Complete medium. Cells may come to confluence before passaging.

2. Using a 1× trypsin–EDTA solution, trypsinize 23 confluent T-225 flasks of A549 cells into 230 mL of Complete medium.

3. Obtain a cell count by diluting the cells 1:2 in Trypan Blue solution and counting the four corner grids of a hemocytometer.

4. Prepare a cell suspension with 1.5×10^5 cells/mL in Complete growth medium and add 20 mL per 150-mm plate (the total number of A549 cells per plate is 3×10^6). Record the number of cells added per plate and incubate overnight at 37°C, 5% CO_2.

Vector/Virus Dilution and Plating

Note the A549 cells' confluency; cells should be 80%–90% confluent before proceeding.

5. Groups of plates will be infected with the following vector/surrogate RCA combinations:

Group A, test sample: 5×10^{11} particles of test vector alone (50 plates)

Group B, spiked sample: 5×10^{11} particles of test vector + 5 IU of RCA surrogate (wtHuAd5) (50 plates)

Group C, positive control: 5 IU of RCA surrogate (wtHuAd5) (50 plates)

Group D, negative control: medium alone (10 plates)

Titers of the RCA surrogate ($TCID_{50}$ IU/mL) are determined using the infectious titer assay described in Protocol 5. Note that 1 IU = 0.7 $TCID_{50}$ IU. The number of vector particles is set by OD_{260} determination using an extinction coefficient of 1×10^{12} particles/OD unit. Vector particle titers are used here because this measure is frequently used to set the dose for experimental animals and patients.

6. Clean the cell culture biosafety cabinet working surfaces with bleach solution and then with ethanol. Pass the bleach solution through the aspirator hose. Lower the front glass panel and keep the UV light on for at least 10–15 min.

7. Prepare the amount of infection medium needed (160 plates $\times$ 4 mL/150-mm plate $\times$ 1.2 = 768 mL) and filter through a 0.2-μm sterile filter unit.

Serial Dilution of the RCA Surrogate

The following is an example of a dilution scheme using wtHuAd5 as the RCA surrogate (4.42×10^{10} IU/mL) and a HuAd5 test vector (6.5×10^{12} pt/mL).

8. Set up the following dilutions of HuAd5 in screw-cap tubes using Infection medium:

Dilution	Sample	Infection medium
4.42×10^{8} IU/mL (10^{-2})	40 μL of stock	3.96 mL
4.42×10^{7} IU/mL (10^{-3})	1 mL of 10^{-2}	9 mL
4.42×10^{6} IU/mL (10^{-4})	1 mL of 10^{-3}	9 mL
4.42×10^{5} IU/mL (10^{-5})	1 mL of 10^{-4}	9 mL
4.42×10^{4} IU/mL (10^{-6})	1 mL of 10^{-5}	9 mL
4.42×10^{3} IU/mL (10^{-7})	1 mL of 10^{-6}	9 mL
4.42×10^{2} IU/mL (10^{-8})	1 mL of 10^{-7}	9 mL
4.42×10^{1} IU/mL (10^{-9})	1 mL of 10^{-8}	9 mL
4.42 IU/mL (diluted stock)	1 mL of 10^{-9}	9 mL

Prepare enough inoculum for 60 plates of each for group A–C in Step 5 (60 plates of inoculum are prepared rather than 50 plates to allow for loss during pipetting). This corresponds to 6 IU wtHuAd5 and/or 6×10^{11} test vector particles per 60 plates in 240 mL of Infection medium. In this example, the following amounts are required:

wtHuAd5 diluted stock: 6 IU/4.42 = 1.357 mL

HuAd5 Vector concentrated stock: 6.5×10^{12}/1,000 = 6×10^{11}/92.3 μL

9. Prepare each inoculum as follows:

Group A
92.3 μL of HuAd5 vector concentrated stock
239.9 mL of Infection medium

Group B
1.357 mL of wtHuAd5 diluted stock
92.3 μL of HuAd5 vector concentrated stock
238.63 mL of Infection medium

Group C
1.357 mL of wtHuAd5 diluted stock
238.64 mL of Infection medium

Group D
40 mL of Infection medium

10. Aspirate the medium from each group of plates separately and add 4 mL of the appropriate inoculum per plate in the same order as indicated in Step 9.

The combination of absolute minimum volume and maximum infection time is key for the detection of extremely low levels of RCAs by the concentration passage method.

It is very important to have the plates exactly level and to stack them in groups of only three.

11. Rock the plates well to distribute the medium and return them to the 37°C incubator in stacks of three. Use a separate shelf for each group and take great care to ensure that each stack is level.

12. After the infection process is completed, clean the hood as described in Step 6 and keep the UV light on for 1 h.

13. At 24 h postinfection, add 15 mL of filtered Supplementation medium to each plate. Incubate the plates until day 8 postinfection.

Make sure to use different pipettes and media aliquots for each group of plates.

Cell Culture: Phase II (Day 7 Postinfection)

14. Set up five new 150-mm plates of A549 cells for each of groups A–C infected in Step 9 and one plate for group D according to the instructions in Steps 1–4.

Concentration Passage (Day 8 Postinfection)

At this stage, the plates can be examined and scored for signs of CPE, if desired. The procedure outlined below passages replication-competent virus onto new cells and concentrates the assay from 50 plates to 5 plates per 5×10^{11} vector particles.

15. Wash each plate with 5 mL of D-PBS, add 1 mL of $1 \times$ trypsin–EDTA solution (0.25%), and incubate for 5 min at room temperature. Make sure that all cells have lifted from the plate.

16. Add 4 mL of Infection medium to each plate. Within each group, pool the cell suspension from 10 plates into a 50-mL falcon tube.

Be sure to use different pipettes, D-PBS, and medium aliquots for each group of 10 plates.

17. Centrifuge the tubes at 1500 rpm in a benchtop centrifuge for 20 min at 4°C and aspirate the supernatant, leaving ∼5 mL over the cell pellet. Pool the cell pellets from the same group and subject them to three cycles of freeze–thaw (dry ice/37°C). Clear the lysates by centrifuging at 2000 rpm in a refrigerated tabletop centrifuge for 10 min at 4°C.

18. Aspirate the medium from the 150-mm A549 plates seeded on day 7 postinfection and infect 5 plates per 50-plate pool using the ∼25 mL of cleared lysate prepared in Step 17 (5 mL per plate).

19. Incubate the infected plates overnight in a 37°C incubator, ensuring that the plates are level. Clean the hood according to Step 6 and keep the UV light on for 1 h.

20. At 24 h post concentration passage infection, add 15 mL of filtered Supplementation medium to each plate and return them to the incubator. Incubate the plates to day 15 postinfection (i.e., post phase I infection) at 37°C.

Be sure to use different pipettes and medium aliquots for each group of plates.

DNA Harvest (Day 15 Postinfection)

Extract total DNA from the infected cells using the QIAGEN QIAamp DNA Mini Kit according to kit instructions.

21. Wash each plate with 5 mL of D-PBS, add 1 mL of $1\times$ trypsin–EDTA solution (0.25%), and incubate for 5 min at room temperature. Ensure that all cells have lifted.

22. Add 5 mL of Complete growth medium to each plate and transfer 1.5 mL of the cell suspension to a microcentrifuge tube.

23. Pellet the cells at 300g in a refrigerated microcentrifuge for 5 min at 4°C and resuspend the pellet in 200 μL of PBS (tubes may be stored at −80°C).

24. For each 200-μL cell suspension (five cell suspensions per groups A–C and one for group D), follow the instructions in the QIAGEN QIAamp DNA Mini Kit manual. Pool the DNA eluate from five columns for each of groups A–C.

25. Dilute 20 μL of the pooled eluate and the negative control 1:5 with dH$_2$O and measure the OD$_{260/280}$.

E1 qPCR

For detection of RCA, 0.5–1 μg of extracted DNA is included in real-time PCRs using primer–probe sets directed against the HuAd5 E1 gene (wtHuAd5 RCA surrogate). HuAd5 E1–containing plasmids are used as standards.

26. Working in a dedicated PCR facility or chemical fume hood, set up the PCR plate according to the format in Figure 1 and the instructions below. The following samples are included on the plate.

    i. No template control (NTC): six replicates

    ii. Spiked controls (SC): one replicate for each group of tissue culture plates

    The spiked controls consist of plasmid DNA spiked into each sample and allows for the measurement of PCR inhibition by the sample.

    iii. Quantification standards (Std) in duplicates, 10–10^8 copies of DNA standards

    See the Additional Protocol Preparation of a DNA Standard for qPCR at the end of Protocol 5.

    iv. Samples (Spl) in duplicates: 1 μg of DNA per PCR

27. Prepare two master mix solutions:

    i. Master mix for nonspiked reactions (number of reactions = 6 NTC+16 Std+8 Spl = 30)

    Include two extra reactions to allow for pipetting losses and then calculate the volume of reagents according to the master mix for nonspiked reactions recipe at the end of this protocol. Combine the appropriate volumes of reagents in a 15-mL conical tube and vortex to mix.

	1	2	3	4	5	6	7	8	9	10	11	12
A	NTC	NTC	NTC	NTC	NTC	NTC	Std 10^8	Std 10^8	Std 10^7	Std 10^7	Std 10^6	Std 10^6
B	Std 10^5	Std 10^5	Std 10^4	Std 10^4	Std 10^3	Std 10^3	Std 10^2	Std 10^2	Std 10^1	Std 10^1	Spl A1	Spl A2
C	Spl B1	Spl B2	Spl C1	Spl C2	Spl D1	Spl D2	SC A	SC B	SC C	SC D		
D												
E												
F												
G												
H												

FIGURE 1. **Representative layout of the qPCR plate setup in the sensitive assay for detection of RCA in adenovirus vector preparations.**

ii. Master mix for spiked reactions (control samples) (number of reactions = number of samples = 4)

> *Calculate reagent volumes for the number of reactions and include two extra reactions to adjust for pipetting losses according to the master mix for spiked reactions recipe at the end of this protocol. Combine the appropriate volumes of reagents in a 15-mL conical tube and vortex to mix.*

28. Aliquot 45 µL of the two master mixes to the appropriate wells of the PCR plate as shown in Figure 1 and then add 5 µL of the appropriate DNA (100–200 ng/µL) to the standard wells, test wells, and spiked wells. Mix the wells by pipetting several times.

> *Vortex the spiked master mix between each aliquot for greater consistency. Exercise extreme caution when adding the standards to minimize cross-contamination of other wells. Enter the plate from the bottom or side and ensure that the pipette crosses only those wells designated for the appropriate standard.*

29. Cover the PCR plate with an optical-grade plastic adhesive plate cover, ensure that a tight seal is formed, and then quickly centrifuge the plate to ensure all reagents are mixed at the bottom of the wells. Perform the PCRs in an ABI 7500 Fast Real-Time PCR system machine (or similar), following the manufacturer's directions.

> *Typical cycling times are an initial denaturation period of 95°C for 10 min followed by 40 cycles of 95°C for 15 sec/60°C for 1 min.*

Data Collection

30. Refer to the appropriate instrument user guide for instructions for analyzing the data. The general process for this analysis involves viewing the amplification plots and then setting the baseline and threshold values. During early PCR cycles, use the background signal in all wells to determine the baseline fluorescence. Place the threshold in the region of exponential amplification. For ABI 7500 machines or similar, the baseline can be routinely set as cycles 3–12 and the threshold adjusted such that the 10^8 standard reads 13.09 Ct. Alternatively, auto baseline and auto ct programs included in the instrument software can be used. The important point is to consistently use the same method. (See also Chapter 9.)

Data Analysis

31. Before the test sample results can be interpreted, it is first necessary to ensure that all controls give results in the expected range. Importantly, the negative control wells (group D) should give only background levels of signal, close to the no-template control (C_T of 35 or greater). The group D DNA spiked control well should give a value close to 1×10^5 copies, indicating that the extracted DNA sample is not inhibiting the PCR. Plot the C_T values of the plasmid standards against copy number. The calculated slope should be ~ -3.3, giving an amplification efficiency of 95%–100%. Once these conditions are verified, the wtHuAd5 spiked test sample (group B) should be compared with the positive control sample (group C) to assess the degree of interference of the vector with the RCA surrogate replication. Ideally, signal obtained for groups B and C should be similar.

Finally, the test vector signal should be examined. Assuming that the controls display the expected results and the copy number in the test wells is close to the background levels exhibited by the negative and no-template controls, the sample can be assumed to contain less than 1 IU of RCA per 1×10^{11} vector particles. Again, assuming that all controls give expected results, a positive signal substantially greater than background levels indicates RCA contamination.

> *See Troubleshooting.*

TROUBLESHOOTING

Problem (Step 31): The negative control group gives a positive signal.

Solution: Take steps to minimize cross contamination during both assay setup and PCR. Contamination of the negative control group with RCA surrogate presents the possibility that a positive signal obtained in the test sample is due to cross-contamination and the result thus invalidates the assay. Groups of plates, pelleted cells, and extracted DNAs should always be handled in the group order A, D, B, and C, as delineated in Step 5. Positive and spiked controls should be separated from test and negative controls—if not in a different incubator, then at least on a separate shelf with the test and negative controls on the top shelf. New pipettes should be used for each group, and RCA surrogate–contaminated pipettes should be soaked in a 10% solution of bleach inside the biosafety cabinet. Empty plates should also be soaked in bleach solution for 5 min before discarding.

The second area of possible cross-contamination is at the qPCR stage. Again, plate the samples in the order given above and take care not to move pipette tips containing RCA DNA or plasmid standards across wells containing negative or test samples.

Problem (Step 31): The group D spiked DNA PCR control wells give values more than fivefold lower than 1×10^5 copies.

Solution: Dilute the sample to overcome the inhibition. It is possible that trace contaminants in the purified DNA will interfere with the PCR, but they may be diluted to levels at which this is no longer an issue.

Problem (Step 31): The RCA surrogate–spiked vector control (group B) gives a much lower signal than the RCA surrogate positive control (group C).

Solution: Reduce the amount of vector added per plate and include additional plates such that the same dose can be screened. If the spiked vector control signal is much lower than the positive control signal, it is likely that the adenoviral vector is interfering with the replication of the RCA surrogate. Ideally, the optimal amount of vector to add per plate should be established empirically before commencing the assay, especially when different adenovirus serotypes are being used.

DISCUSSION

Properly performed, this assay should allow the detection of 1 IU of wtHuAd5 in a dose of 1×10^{11} HuAd5 vector particles. The sensitivity of the assay is derived from both biological amplification and the use of qPCR for detection. Like all RCA assays, large numbers of cells must be grown in the initial infection phase to accommodate the high vector dose without toxicity or interference; however, in the second passage, the number of cells is reduced 10-fold. This concentration passage is useful for the reduction of labor and material cost but also serves to enhance detection.

At first glance, it may appear that RCA content of a vector preparation might be directly assessed by simply quantifying the amount of E1 gene in the purified vector stock. However, this approach is usually not considered, mainly because it cannot be proven that this DNA is derived from infectious RCA and may in fact represent contaminating host cell E1 DNA. In developing the method described here, we found that an excess of exogenous E1 DNA spiked into the vector was diluted and removed during the first passage and was undetectable in DNA extracted and quantified by qPCR. Hence, the assay format distinguishes noninfectious host cell E1 DNA from true RCA and minimizes the possibility of a false-positive result.

The copy numbers obtained in the assay presented here are not directly quantitative because substantial amplification of any initial RCA contamination will have occurred. If desired, a semiquantitative measure of the degree of contamination can be obtained by including in the assay additional spiked controls with increasing IU values of the RCA surrogate. The test vector signal can then be compared to these additional controls and a range of RCA contamination can be determined.

RECIPES

It is essential that you consult the appropriate Material Safety Data Sheets and your institution's Environmental Health and Safety Office for proper handling of equipment and hazardous materials used in this protocol.

Complete Growth Medium (10%)

Reagent	Quantity (for 1 L)	Final concentration
F-12/K media	890 mL	
FBS	100 mL	10%
P/S solution (100×)	10 mL	1×

Store at 4°C.

Infection Medium (2%)

Reagent	Quantity (for 1 L)	Final concentration
DMEM	945 mL	
FBS	20 mL	2%
HEPES (1 M, pH 8.0)	25 mL	25 mM
P/S solution (100×)	10 mL	1×

Store at 4°C.

Master Mix for Nonspiked Reactions

Reagent	Stock	1 Reaction	32 Reactions	Reaction
qPCR mix (2×)	2×	25 μL	800 μL	1×
Forward primer	3 μM	5.0 μL	160 μL	300 nM
Reverse primer	3 μM	5.0 μL	160 μL	300 nM
Probe	2 μM	5.0 μL	160 μL	200 nM
Water		5.0 μL	160 μL	
Total		45 μL	1440 μL	

Master Mix for Spiked Reactions

Reagent	Stock	1 Reaction	4 Reactions	Reaction
qPCR mix (2×)	2×	25 μL	100 μL	1×
Forward primer	3 μM	5.0 μL	20 μL	300 nM
Reverse primer	3 μM	5.0 μL	20 μL	300 nM
Probe	2 μM	5.0 μL	20 μL	200 nM
Water		4.0 μL	20 μL	
Spiked DNA	10^5 copies/μL	1 μL	4 μL	10^5 copies
Total		45 μL	184 μL	

Supplementation Medium (20%)

Reagent	Quantity (for 1 L)	Final concentration
F-12/K medium	790 mL	
FBS	200 mL	20%
P/S solution (100×)	10 mL	1×

Store at 4°C.

REFERENCES

Fallaux FJ, Bout A, van der Velde I, van den Wollenberg DJ, Hehir KM, Keegan J, Auger C, Cramer SJ, van Ormondt H, van der Eb AJ, et al. 1998. New helper cells and matched early region 1-deleted adenovirus vectors prevent generation of replication-competent adenoviruses. *Hum Gene Ther* **9:** 1909–1917.

Hehir KM, Armentano D, Cardoza LM, Choquette TL, Berthelette PB, White GA, Couture LA, Everton MB, Keegan J, Martin JM, et al. 1996. Molecular characterization of replication-competent variants of adenovirus vectors and genome modifications to prevent their occurrence. *J Virol* **70:** 8459–8467.

Louis N, Evelegh C, Graham FL. 1997. Cloning and sequencing of the cellular-viral junctions from the human adenovirus type 5 transformed 293 cell line. *Virology* **233:** 423–429.

NIH. 2001. Guidance for human somatic cell therapy and gene therapy. *Hum Gene Ther* **12:** 303–314.

NIH Guidelines for Research Involving Recombinant DNA Molecules. (April) 2000. http://oba.od.nih.gov/rdna/nih_guidelines_oba.html.

Schalk JA, de Vries CG, Orzechowski TJ, Rots MG. 2007. A rapid and sensitive assay for detection of replication-competent adenoviruses by a combination of microcarrier cell culture and quantitative PCR. *J Virol Methods* **145:** 89–95.

Production of rAAVs by Transient Transfection

The most commonly used method for production of rAAVs in research laboratories is by transient triple transfection of 293 cells with AAV *cis* and *trans* plasmids and an adenovirus helper plasmid. This protocol describes the processes required to prepare the transfected cell suspension for virus purification by the various methods described in subsequent protocols.

Biosafety Consideration

According to NIH Guidelines for Research Involving Recombinant DNA Molecules (April 2000), both wild-type AAVs and recombinant AAV constructs are not associated with disease in healthy adult humans, if the transgene does not encode either a potentially tumorigenic gene product or a toxic molecule and is produced in the absence of a helper virus. All work involving AAV vectors can be conducted at Biosafety Level 1 (BL1) with the approval of the Institutional Biosafety Committee at the home institution.

MATERIALS

It is essential that you consult the appropriate Material Safety Data Sheets and your institution's Environmental Health and Safety Office for proper handling of equipment and hazardous materials used in this protocol.

Recipes for reagents specific to this protocol, marked <R>, are provided at the end of the protocol. See Appendix 1 for recipes for commonly used stock solutions, buffers, and reagents, marked <A>. Dilute stock solutions to the appropriate concentrations.

Reagents

$CaCl_2$ solution (2.5 M) <R>
cis plasmid containing the gene of interest
Complete growth medium <R>
ΔF6 Adeno helper plasmid (available at the Penn Vector Core, vector@mail.med.upenn.edu)
Dulbecco's minimal essential medium (DMEM)
Dulbecco's phosphate-buffered saline solution (D-PBS)
Fetal bovine serum (FBS)
HBS solution (2×) <R>
Milli-Q water, sterile
Penicillin–streptomycin (P/S)
trans plasmids of different serotypes (Gao et al. 2002, 2003, 2004) (available at Penn Vector Core, vector@mailmed.upenn.edu)
Tris (50 mM, pH 7.4), $MgCl_2$ (1 mM) buffer (see Step 8.i)
Tris (50 mM), NaCl (150 mM) buffer (see Step 8.ii)

Equipment

Aspirating pipettes, sterile tissue culture plates (150 mm)
Cell scrapers, sterile

Centrifuge bottles, sterile (125 and 500 mL)
Conical tubes, sterile (50 mL)
Humidified cell culture incubator (37°C, 5% CO_2)
Pipettes, sterile and disposable
Receiver, sterile (125 mL)
Refrigerated tabletop centrifuge
Vortex machine
Water bath (37°C)

METHOD

Preparation of 293 Cells

1. Prepare Complete growth medium. A volume of 20 mL of medium is needed for each 150-mm plate. For large-scale rAAV production, prepare 800 mL of Complete growth medium for 40 plates of 293 cells.

2. Seed the cells 1 d before transfection at 7×10^6 293 cells/150-mm plate. The cells should be ~70%–80% confluent at the time of transfection.

   293 cells are suitable for AAV vector production until passages 100–130 (293 cells obtained from ATCC are usually around passage 40).

3. On the day of transfection, replace the medium 2 h before transfection with fresh Complete growth medium. Carefully add 20 mL of medium per plate without disturbing the cell monolayer.

   Maintaining an intact cell monolayer is critical to the efficiency of transfection. See Troubleshooting.

Preparation of Transfection Cocktail

4. Prepare the following DNA mix in a sterile 125-mL bottle:

Milli-Q water, sterile	to make a final volume of 54 mL
$CaCl_2$ (2.5 M)	5.2 mL
pΔF6 adeno helper plasmid	1040 μg
trans plasmid	520 μg
cis plasmid	520 μg

   These numbers are for transfection of 40 plates.

5. Prepare four 50-mL conical tubes each containing 12.5 mL of 2× HBS solution. To make the transfection cocktail, add 12.5 mL of the DNA mixture from Step 4 dropwise to each 2× HBS tube while vortexing it. Incubate the transfection cocktail for 5 min at room temperature.

   After 5 min of incubation, the transfection cocktail should show a uniform whitish cloudiness. If it is not cloudy at all or has large precipitates, see Troubleshooting.

Transfection of 293 Cells

6. Add 2.5 mL of the transfection cocktail dropwise to each tissue culture plate from Step 2. Gently rock the plate to distribute the cocktail evenly over the entire cell monolayer. Place the transfected plates back in the 37°C, 5% CO_2 incubator. Sixteen h after transfection, aspirate the transfection medium from the plates and replace with fresh Complete growth medium.

   When changing the medium, take no more than 12 plates out of the incubator at a time and add the medium slowly from the side of the plate without disturbing the cell monolayer.

Harvesting of Transfected Cells

7. At ~72 h after transfection, scrape the cells into the medium using a sterile cell scraper. Transfer the cell suspension from all plates to two 500-mL sterile centrifuge bottles using a 25-mL pipette.

8. Centrifuge cells at 4000 rpm in a refrigerated tabletop centrifuge for 20 min at 4°C. Decant the supernatant and process the cell pellets in the following ways, depending on the method to be used for AAV vector purification.

For Cesium Chloride Gradient Purification

i. Resuspend the cell pellet in each of the 500-mL centrifuge bottles in ~14 mL of 50 mM Tris (pH 7.4), 1 mM $MgCl_2$ buffer. Transfer the cell suspensions to a sterile 50-mL conical centrifuge tube.

ii. Use 7 mL of the same buffer to rinse the 500-mL bottles.

iii. Combine the rinsing buffer with the rest of cell suspension in the 50-mL conical centrifuge tube and then store at ~80°C until time of processing.

For Iodixanol Gradient Purification

i. Resuspend the cell pellet in each of the 500-mL centrifuge bottles in a total of 40 mL of 50 mM Tris (pH 8.4), 150 mM NaCl.

ii. Combine the cell suspensions in a sterile 50-mL conical centrifuge tube and store at −80°C until time of processing.

For Heparin Column Purification

i. Resuspend the cell pellets in a total of 110 mL of DMEM and aliquot into four 50-mL conical tubes. Freeze the cell suspension at −80°C until time of processing.

TROUBLESHOOTING

Problem (Step 3): The cells come off of the plates while changing the growth medium.
Solution: Take no more than 12 plates out of the incubator at a time for medium change. When removing the medium from plates, leave a few milliliters of old medium to keep the cells covered before adding 20 mL of fresh medium.

Problem (Step 5): The transfection cocktail is not cloudy at all or has large precipitates.
Solution:

- Ensure that the pH value of 2× HBS is 7.05.

- Check that 2.5 M $CaCl_2$ has been added to the transfection cocktail.

- Before preparation of the cocktail, take the transfection reagents out of refrigerator and allow to set for 30 min at room temperature.

DISCUSSION

The efficiency of the transfection process determines the overall rAAV yield of this production method. One of the key components to achieve high transfection efficiencies (>70%) is careful preparation and testing of 2× HBS solution with the correct pH. When making this solution, ensure that the pH meter is properly calibrated. Each batch of the transfection reagents should be tested for transfection efficiency in 293 cells using an enhanced green fluorescent protein (EGFP)-expressing plasmid before their use for rAAV production. Quality and lot-to-lot variation of commercial FBS could also have a significant impact on transfection efficiency. We strongly recommend testing different lots and manufacturers of FBS for their effect on transfection by calcium phosphate precipitation. In addition, the pH of the culture medium in the transfected cell plates is an influential factor in the efficiency of transfection. It is best to take no more than 12 plates out of the incubator right before the transfection and to return the plates to the incubator immediately after adding the transfection cocktail.

The structure of the pCis (vector) plasmid is also critical for the efficient packaging of rAAV genomes. The ITR sequences in the pCis plasmids serve not only as the recognition and cleavage sites for Rep proteins to release the viral genome from the plasmid backbone but also as the replication origin and packaging signal. Therefore, ensuring the integrity of the ITR sequences in the pCis plasmids is a key step in rAAV production. The easiest and most effective way to confirm the integrity of ITRs is to analyze the pCis plasmid by restriction digestion with SmaI (multiple recognition sites are present in the ITRs), alone or in combination with some other single- or double-cutting enzymes near the ITRs.

RECIPES

It is essential that you consult the appropriate Material Safety Data Sheets and your institution's Environmental Health and Safety Office for proper handling of equipment and hazardous materials used in this protocol.

CaCl₂ Solution (2.5 M)

Dissolve 183.74 g of calcium chloride ($CaCl_2$) in 500 mL of H_2O. Sterilize by passing through a 0.22-μm filter and store at room temperature.

Complete Growth Medium

Reagent	Quantity (for 1 L)	Final concentration
DMEM	890 mL	
FBS	100 mL	10%
P/S solution (100×)	10 mL	1×

HBS Solution (2×)

Reagent	Quantity (for 1 L)	Final concentration
NaCl	16.4 g	280 mM
HEPES ($C_8H_{18}N_2O_4S$)	11.9 g	50 mM
$Na_2HPO_4 \cdot 7H_2O$	0.38 g	1.42 mM
H_2O	To 1 L	

Adjust pH to 7.05 with 10 M NaOH. Sterilize by passing through a 0.22-μm filter and store at room temperature.

REFERENCES

Gao GP, Alvira MR, Wang L, Calcedo R, Johnston J, Wilson JM. 2002. Novel adeno-associated viruses from rhesus monkeys as vectors for human gene therapy. *Proc Natl Acad Sci* 99: 11854–11859.

Gao G, Alvira MR, Somanathan S, Lu Y, Vandenberghe LH, Rux JJ, Calcedo R, Sanmiguel J, Abbas Z, Wilson JM. 2003. Adeno-associated viruses undergo substantial evolution in primates during natural infections. *Proc Natl Acad Sci* 100: 6081–6086.

Gao G, Vandenberghe LH, Alvira MR, Lu Y, Calcedo R, Zhou X, Wilson JM. 2004. Clades of adeno-associated viruses are widely disseminated in human tissues. *J Virol* 78: 6381–6388.

NIH Guidelines for Research Involving Recombinant DNA Molecules. (April) 2000. http://oba.od.nih.gov/rdna/nih_guidelines_oba.html.

Purification of rAAVs by Cesium Chloride Gradient Sedimentation

Centrifugation to equilibrium in cesium chloride gradients has been used for more than 40 years to purify viruses. The application of high G-forces for a long period of time to a solution of CsCl generates a density gradient that allows separation of empty, partially packaged, and fully packaged viral particles from cellular debris, proteins, and nucleic acids in the crude viral lysate on the basis of their buoyant densities. This protocol describes the use of CsCl gradients to purify AAV vectors from crude viral lysates.

Biosafety Consideration

According to NIH Guidelines for Research Involving Recombinant DNA Molecules (April 2000), both wild-type AAVs and recombinant AAV constructs are not associated with disease in healthy adult humans, if the transgene does not encode either a potentially tumorigenic gene product or a toxic molecule and is produced in the absence of a helper virus. All work involving AAV vectors can be conducted at Biosafety Level 1 (BL1) with the approval of the Institutional Biosafety Committee at the home institution.

MATERIALS

It is essential that you consult the appropriate Material Safety Data Sheets and your institution's Environmental Health and Safety Office for proper handling of equipment and hazardous materials used in this protocol.

Recipes for reagents specific to this protocol, marked <R>, are provided at the end of the protocol. See Appendix 1 for recipes for commonly used stock solutions, buffers, and reagents, marked <A>. Dilute stock solutions to the appropriate concentrations.

Reagents

Benzonase (EMD Chemicals, Gibbstown, NJ)
Cell resuspension buffer for CsCl gradient purification <R>
Cell suspension, frozen (from Protocol 7)
CsCl, ultrapure Deoxycholic acid, 10%
D-Sorbitol (5%)/D-PBS (pH 7.6) <R>
Dulbecco's phosphate-buffered saline (D-PBS)
Ethanol (70%)
Glycerol, sterile (100%)
Heavy CsCl (H-CsCl) solution for rAAV purification <R>
Light CsCl (L-CsCl) solution for rAAV purification <R>
Tris (50 mM, pH 7.4), MgCl$_2$ (1 mM) buffer

Equipment

Conical centrifuge tube (15 mL)
Glass beaker (1 and 4 L), autoclaved
Glass stir bar, autoclaved

Ice-water bath
Magnetic stirring plate
Needles (16 and 18 gauge x $1\frac{1}{2}$ inch)
Quick-Seal tubes, for 70.1 Ti rotor
Refractometer (Milton Roy)
Refrigerated tabletop centrifuge
Slide-A-Lyzer cassette (0.5–3 mL or 3–12 mL)
Sonicator
SW28 centrifuge tubes
SW28 rotor
Syringe (3 and 10 mL)
70.1 Ti rotor (Beckman)
Tube topper, cordless (Beckman Coulter, Fullerton, CA)
Ultracentrifuge (Beckman)
Water bath (37°C)

METHOD

Processing of Transfected Cell Suspension

1. Thaw the 50-mL conical centrifuge tube with 35 mL of frozen cell suspension (from Protocol 7) for 10 min in a 37°C water bath. Sonicate the cell lysate at 25% output for 1 min in an ice-water bath. Repeat twice, with a 2-min interval between each sonication. Always keep the tube on ice.

2. Add 150 μL of Benzonase to the sonicated cell lysate to a final concentration of 100 U/mL. Invert the tube gently to mix. Incubate the samples for 20 min at 37°C, inverting the tube every 5 min. Then, add 1.25 mL of 10% deoxycholic acid; mix by gently inverting the tube and extend the incubation for a further 10 min at 37°C. Then, immediately place the tube on ice for 10–20 min. Clarify the lysate by centrifugation at 4000 rpm in a refrigerated tabletop centrifuge for 30 min at 4°C. Transfer the supernatant to a sterile 50-mL conical tube.

   Benzonase is a genetically engineered endonuclease from the bacterium Serratia marcescens. *It has no proteolytic activity but it is used to completely degrade all forms of DNA and RNA (single-stranded, double-stranded, linear, and circular) to 5′-monophosphate-terminated oligonucleotides that are 2–5 bases in length. It is ideal for removal of nucleic acids (of both cellular and plasmid origins) from crude viral lysates for viscosity reduction and improvement of virus purification.*

   Sodium deoxycholate, the sodium salt of deoxycholic acid, is used here as a biological detergent to help lyse the 293 cells and free the rAAV virions from cellular and membrane components with which the virions are usually associated.

First Cesium Chloride Gradient Centrifugation

3. Add 0.454 g of ultrapure CsCl for every milliliter of clarified lysate. Mix well by gentle inversion. The final volume of the viral lysate should be ~40 mL.

4. Add 9 mL of L-CsCl solution to two SW28 centrifuge tubes. Fill a 10-mL pipette with 9 mL of H-CsCl solution and carefully lower the pipette to the bottom of the tube through the L-CsCl layer. Dispense the H-CsCl solution slowly to prevent mixing and to generate a sharp interface between the two solutions. Slowly add 18–20 mL of clarified lysate (from Step 2) on top of the two-layered CsCl gradient in each SW28 centrifuge tube using a 10-mL pipette. Again, take care to avoid any mixing of the lysate and L-CsCl layer.

5. Very carefully, so as not to disturb the gradient, place the tubes in the rotor buckets and balance them. Attach the buckets on the SW28 rotor and centrifuge at 25,000 rpm in an ultracentrifuge for 18–20 h at 15°C.

6. To collect gradient fractions, set up a set of 16 sterile 1.5-mL microcentrifuge tubes for each SW28 centrifuge tube.

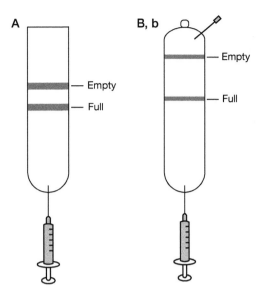

FIGURE 1. CsCl gradients after the first (*A*), second (*B*), and third (*b*) rounds of CsCl gradient centrifugation.

7. Remove a centrifuge tube from its bucket without disturbing the gradient. Secure the SW28 centrifuge tube to a holder. Carefully insert an 18-gauge × 1½-inch needle at the bottom of the tube without disturbing the gradient (Fig. 1A), collect the first 11 mL from the tube into a 15-mL conical tube, and discard it appropriately.

8. Collect the next 16 mL in 1-mL fractions into the 1.5-mL microcentrifuge tubes. Label each tube with the fraction number. Take 5 µL of each fraction and read the refraction index using a refractometer. The fractions containing rAAV virions have a refractive index from 1.3650 to 1.3760. Discard the fractions that are outside of this range.

9. Repeat Steps 7 and 8 with the second SW28 centrifuge tube.

Second Cesium Chloride Gradient Centrifugation

10. Combine all fractions with the appropriate refractive index and transfer them into two 70.1 Ti Quick-Seal tubes using a 10-mL syringe with an 18-gauge × 1½-inch needle. Fill the remaining space in each tube with L-CsCl solution that contains 1.4 g/mL CsCl. Seal the tubes using a cordless tube topper. Centrifuge at 60,000 rpm for 20–24 h at 15°C.

11. To collect fractions from the second centrifugation, set up a set of 13 sterile 1.5-mL microcentrifuge tubes for each gradient. Carefully remove the centrifuge tubes from the rotor without disturbing the gradient. Secure the first tube on a tube holder. Insert a 16-gauge needle at the top of the tube and then an 18-gauge × 1½-inch needle at the bottom of the tube (Fig. 1B).

12. Collect 13 mL in 1-mL fractions from the tube into 1.5-mL microcentrifuge tubes. Label each tube with the fraction number. Take 5 µL of each fraction and read the refraction index using a refractometer. The fractions containing rAAV virions have a refractive index from 1.3650 to 1.3760.

13. Repeat Steps 11 and 12 for the second gradient.

Third Cesium Chloride Gradient Centrifugation

14. Combine the rAAV-containing fractions collected from the two 70.1 Ti tubes and transfer them into one new 70.1 Ti Quick-Seal tube using an 10-mL syringe with an 18-gauge needle. Fill the remaining space in the tube with an L-CsCl solution that contains 1.4 g/mL CsCl, seal the tube, and centrifuge as in Step 10.

15. Repeat Step 11 and collect 13 mL in 0.5-mL fractions in a total of 26 sterile 1.5-mL microcentrifuge tubes on a tube rack. Read the refractive indexes as in Step 12. After this final centrifugation, pool the fractions with refractive indexes in the range of 1.3670–1.3740 containing rAAV virions.

Desalting, Formulation, and Storage

16. Use a 5-mL syringe fitted with a 18-gauge × $1\frac{1}{2}$-inch needle to transfer the rAAV collected from Step 15 into a 0.5–3-mL or 3–12-mL Slide-A-Lyzer cassette, depending on the volume of collected vector. Dialyze in a 4-L plastic beaker containing 3 L of cold D-PBS, pH 7.6, with slow continuous stirring on a magnetic stirring plate at 4°C. Replace with 3 L of fresh cold PBS three times, at 3-h intervals. After the third time, leave for overnight dialysis. Use cold 5% sorbitol/PBS, pH 7.6, for the final overnight dialysis. The virus preparation will be used or stored in the dialysis buffer.

17. Use a 5-mL syringe fitted with an 18-gauge × $1\frac{1}{2}$-inch needle to carefully transfer the desalted virus from the dialysis cassette to a sterile 15-mL conical centrifuge tube. Take 20-μL aliquots from the conical tube for genome copy titration (Protocol 12), infectivity titration (Protocol 13), electron microscopic analysis (Protocol 14), and silver staining of SDS-PAGE (Protocol 15). Store the remaining rAAV at 4°C temporarily until the completion of vector characterization.

    Depending on the concentration of viral capsid and viral particles in the preparation, a whitish precipitate may appear, which is the result of aggregation of viral particles. Aggregation is undesirable because it causes a reduction in the efficiency of gene delivery, particularly systemic delivery. See Troubleshooting for strategies to prevent aggregation of rAAV vectors.

18. Aliquot the rAAV in cryovials and store at −80°C.

TROUBLESHOOTING

Problem (Step 17): Whitish precipitates appear in the purified rAAV preparations.
Solutions:

- To reduce the risk of aggregation before dialysis of the CsCl gradient fractions, it is helpful to perform a quick SDS-PAGE silver staining analysis on the sample to estimate the concentration of virus particles in the sample by SDS-PAGE (see Protocol 15). If the concentration is greater than 2×10^{13} to 3×10^{13} virus particles/mL, add an equal volume of PBS for 1:1 dilution and then proceed with dialysis.

- If the genome copy titer of a purified and desalted vector preparation is higher than 1×10^{13} genomic copies (GC)/mL, dilute the vector to 1×10^{13} GC/mL before freezing. High-titer vector stocks may sometimes become aggregated after storage in the freezer.

- Once the virions become aggregated, it is very difficult to resuspend the virus without affecting its integrity. Aggregation has significant impact on the transducing activity of the virus in systemic delivery. See Discussion for more details.

DISCUSSION

It is well known that AAV2 vectors and recombinants tend to aggregate at high concentrations, which has a negative impact on transduction efficiency, biodistribution, and immunological profiles of rAAV2 vectors. The factors that contribute to aggregation, and formulations that prevent it, have been thoroughly investigated (Wright et al. 2005). Storage of AAV2 vectors and recombinants in high-ionic-strength buffers prevents aggregation. Because the biological properties of any rAAV are determined by the amino acids that are exposed on the surface of the viral capsid, it seems reasonable to speculate that different rAAV capsids have different surface composition and possibly different overall charge. These changes are likely to lead to differences in solubility and susceptibility to aggregation, but for rAAV capsids other than rAAV2, this remains to be investigated. Nonetheless, as a general precaution, it is best to store other serotypes of rAAVs at a virus particle concentration or GC titer no higher than 2×10^{13} to 3×10^{13} (virus particles/mL or GC/mL) unless they are in formulations that can prevent aggregation.

Depending on the titer and purity, rAAV preparations purified by CsCl gradient centrifugation should be colorless solutions. In contrast to adenovirus recombinants, purified AAV recombinants are much more resistant to temperature and pH changes. In fact, if an adenovirus is used to provide helper functions for rAAV production, an effective way to inactivate infectious adenovirus in rAAV preparations is to treat the preparations for 0.5–1 h at 56°C. Although this heating process leads to rapid loss of adenovirus infectivity, it has little impact on the infectivity of rAAV vectors. For long-term storage, rAAVs should be kept at −80°C. Once thawed, it is best to store rAAVs where it is stable for several weeks at 4°C.

RECIPES

It is essential that you consult the appropriate Material Safety Data Sheets and your institution's Environmental Health and Safety Office for proper handling of equipment and hazardous materials used in this protocol.

Cell Resuspension Buffer for CsCl Gradient Purification

Reagent	Quantity (for 1 L)	Final concentration
Tris-Cl (1 M, pH 7.4)	50 mL	50 mM
MgCl$_2$ (1 M)	1 mL	1 mM
H$_2$O	to 1 L	

Mix well, sterilize by passing through a 0.22-μm filter, and store at room temperature.

D-Sorbitol/D-PBS (pH 7.6) (5%)

Dissolve 25 g of D-sorbitol in 5 L of D-PBS, pH 7.6. Sterilize by passing through a 0.22-μm filter and store at 4°C.

Heavy CsCl for AAV Purification (Density = 1.61 g/mL)

Dissolve 672.1 g of biological grade CsCl in 672.9 mL of 10 mM Tris-Cl (pH 8.0), sterilize by passing through a 0.22-μm filter, and store at room temperature.

Light CsCl for rAAV Purification (Density = 1.41 g/mL)

Dissolve 513.89 g of biological grade CsCl in 786 mL of 10 mM Tris-Cl (pH 8.0), sterilize by passing through a 0.22-μm filter, and store at room temperature.

REFERENCES

NIH Guidelines for Research Involving Recombinant DNA Molecules. (April) 2000. http://oba.od.nih.gov/rdna/nih_guidelines_oba.html.

Wright JF, Le T, Prado J, Bahr-Davidson J, Smith PH, Zhen Z, Sommer JM, Pierce GF, Qu G. 2005. Identification of factors that contribute to recombinant AAV2 particle aggregation and methods to prevent its occurrence during vector purification and formulation. *Mol Ther* 12: 171–178.

Purification of rAAVs by Iodixanol Gradient Centrifugation

This is a simple method for rapid preparation of rAAV stocks, which can be used for in vivo gene delivery. Infusion of these vectors into the brain appears to be well tolerated without significant side effects (Hermens et al. 1999). The purity of these vectors is considerably lower than that obtained by either CsCl gradient centrifugation or by combination of iodixanol gradient ultracentrifugation followed by column chromatography.

Biosafety Consideration

According to NIH Guidelines for Research Involving Recombinant DNA Molecules (April 2000), both wild-type AAVs and recombinant AAV constructs are not associated with disease in healthy adult humans, if the transgene does not encode either a potentially tumorigenic gene product or a toxic molecule and is produced in the absence of a helper virus. All work involving AAV vectors can be conducted at Biosafety Level 1 (BL1) with the approval of the Institutional Biosafety Committee at the home institution.

MATERIALS

It is essential that you consult the appropriate Material Safety Data Sheets and your institution's Environmental Health and Safety Office for proper handling of equipment and hazardous materials used in this protocol.

Recipes for reagents specific to this protocol, marked <R>, are provided at the end of the protocol. See Appendix 1 for recipes for commonly used stock solutions, buffers, and reagents, marked <A>. Dilute stock solutions to the appropriate concentrations.

Reagents

Benzonase (EMD Chemicals, Gibbstown, NJ)
Cell suspension, frozen (from Protocol 7)
Dulbecco's phosphate-buffered saline (D-PBS)
Ethanol (7%)
Iodixanol solution (15%), (OptiPrep; Sigma-Aldrich) <R>
Iodixanol solution (25%), (OptiPrep; Sigma-Aldrich) <R>
Iodixanol solution (40%), (OptiPrep; Sigma-Aldrich) <R>
Iodixanol solution (60%), (OptiPrep; Sigma-Aldrich) <R>
Phenol red (0.5%) <R>

Equipment

Amicon Ultra-15 100K concentrators
Beckman ultracentrifuge
Conical tubes (15 and 50 mL)
Dry-ice/ethanol bath
Glass micropipettes
Needles (16 and 18 gauge $\times 1\frac{1}{2}$ inch)
Quick-Seal centrifuge tubes (25×89 mm)
Refrigerated tabletop centrifuge

Syringe, plastic disposable (5 and 10 mL)
70 Ti rotor
Water bath (37°C)

METHOD

Processing of Transfected Cell Suspension

1. Thaw the 50-mL conical tube with 40 mL of frozen cell suspension (from Protocol 7) in a 37°C water bath. Invert the tube several times to mix. Perform two additional freeze–thaw cycles using a dry-ice/ethanol bath and a 37°C water bath.

2. Add Benzonase to a final concentration of 50 U/mL. Incubate for 30 min at 37°C.

   Benzoase is an endonuclease that is used to degrade all forms of DNA and RNA (single-stranded, double-stranded, linear, and circular) in cell lysates.

3. Centrifuge at 4000g in a refrigerated tabletop centrifuge for 20 min at 4°C. Transfer the supernatant to a new 50-mL conical tube and discard the cell pellet.

Centrifugation in Discontinuous Iodixanol Gradient

4. Using a 20-mL syringe with an 18-gauge × 1½-inch needle, load 12 mL of supernatant into each Quick-Seal centrifuge tube.

5. Insert a glass micropipette into the tube and attach to one end of the provided tubing.

6. Fill a 10-mL plastic disposable syringe with 9 mL of 15% iodixanol solution. Attach the syringe to the other end of the tubing and introduce the iodixanol solution under the supernatant slowly to generate a sharp interface without mixing of the two layers.

   Make sure that no bubbles are present in the tubing or syringe because they will disturb the iodixanol step gradient.

7. Once finished with the infusion, detach the syringe, fill it with the next iodixanol solution, and repeat the process. Infuse the solutions necessary to make the iodixanol step gradient in the following amounts and order (each solution is colored differently with Phenol red to facilitate their identification in the gradient) (Fig. 1):

15% Iodixanol solution (orange)	9 mL
25% Iodixanol solution (red)	6 mL
40% Iodixanol solution (transparent)	5 mL
60% Iodixanol solution (light yellow)	5 mL

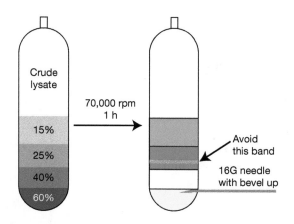

FIGURE 1. Iodixanol gradients.

8. Once all gradient solutions are loaded, carefully add D-PBS to fill the tube according to the manufacturer's instructions.

9. Seal the tubes and centrifuge at 70,000 rpm in a Beckman 70 Ti rotor for 1 h at 20°C.

10. Carefully remove the tubes from the rotor to avoid disturbing the gradient. Puncture the top of the tube near the neck with an 18-gauge × 1½-inch needle. Using a 5-mL disposable plastic syringe attached to an 18-gauge × 1½-inch needle, puncture the tube 2–3 mm below the 40%–60% iodixanol interface (see Fig. 1). Turn the bevel up and withdraw 4 mL at the 40%–60% interface, avoiding the 25%–40% interface. Transfer the withdrawn solution to a new 15-mL conical tube.

    The white band near the bottom of the 25% layer does not correspond to the rAAV.

11. Store the 15-mL conical tube containing rAAV at 4°C for use the next day or frozen at −20°C.

12. The rAAV-containing fraction withdrawn in Step 10 can be processed to remove the iodixanol as follows and used or further purified by anion-exchange chromatography (Protocol 11).

Iodixanol Removal and Concentration of rAAV

13. Fill two Amicon Ultra-15 100K concentrators with 70% ethanol and leave for 20–30 min at room temperature.

14. Decant the 70% ethanol. Fill the chamber with D-PBS and centrifuge at 2600g in a tabletop centrifuge for 10 min at room temperature. Remove the D-PBS from the bottom chamber and repeat the procedure.

15. Add 4 volumes of D-PBS to the iodixanol-containing vector fraction (1:5 dilution) to reduce the viscosity of the solution. Pipette 15 mL into each concentrator and centrifuge at 2600g in a refrigerated tabletop centrifuge for 20–30 min at 4°C. If necessary, extend the centrifugation time to bring the volume down to <1 mL. Repeat the procedure until all vector-containing solution has been put through the concentrators.

    See Troubleshooting.

16. Wash with D-PBS three times. Add 14–15 mL of D-PBS and mix by gently pipetting. Centrifuge as before until the solution volume in each concentrator is 200–300 μL.

17. At the end of the three cycles of desalting/concentration, collect the rAAV by centrifugation according to the manufacturer's instructions. Aliquot and store at −80°C.

TROUBLESHOOTING

Problem (Step 15): The solution does not go through the centrifugal device.
Solution: Dilute further the iodixanol-containing vector fraction with D-PBS and concentrate in the same manner. Alternatively, increase the centrifugation time.

DISCUSSION

This method of rAAV purification yields consistent results across a number of different rAAV serotypes. During removal of iodixanol using the centrifugal device, the concentration of vector increases considerably, which could lead to capsid aggregation. This can affect dramatically the in vivo gene-delivery properties of AAV vectors. AAV vectors purified by this method can be used for in vivo gene delivery to the central nervous system without significant detrimental effects. However, the presence of contaminating cellular proteins could affect transduction efficiency and/or the immunological properties of the vectors.

RECIPES

It is essential that you consult the appropriate Material Safety Data Sheets and your institution's Environmental Health and Safety Office for proper handling of equipment and hazardous materials used in this protocol.

Iodixanol Solution (15%)

Reagent	Quantity (for 100 mL)	Final concentration
Iodixanol stock	25 mL	15%, w/v
D-PBS (10×)	10 mL	1×
NaCl (5 M)	20 mL	1 M
MgCl$_2$ (1 M)	100 μL	1 mM
KCl (1 M)	250 μL	2.5 mM
Phenol red (0.5%)	150 μL	
H$_2$O	To 100 mL	

Iodixanol Solution (25%)

Reagent	Quantity (for 100 mL)	Final concentration
Iodixanol stock	41.66 mL	25%, w/v
D-PBS (10×)	10 mL	1×
MgCl$_2$ (1 M)	100 μL	1 mM
KCl (1 M)	250 μL	2.5 mM
Phenol red (0.5%)	200 μL	
H$_2$O	To 100 mL	

Iodixanol Solution (40%)

Reagent	Quantity (for 100 mL)	Final concentration
Iodixanol stock	66.8 mL	40%, w/v
PBS (10×)	10 mL	1×
MgCl$_2$ (1 M)	100 μL	1 mM
KCl (1 M)	250 μL	2.5 mM
H$_2$O	To 100 mL	

Iodixanol Solution (60%)

Reagent	Quantity (for 100 mL)	Final concentration
Iodixanol stock	100 mL	~60%, w/v
MgCl$_2$ (1 M)	100 μL	1 mM
KCl (1 M)	250 μL	2.5 mM
Phenol red (0.5%)	50 μL	
H$_2$O	To 100 mL	

Phenol Red (0.5%, w/v)
Add 0.25 g of phenol red to 50 mL of 50% ethanol.

REFERENCES

Hermens WT, ter Brake O, Dijkhuizen PA, Sonnemans MA, Grimm D, Kleinschmidt JA, Verhaagen J. 1999. Purification of recombinant adeno-associated virus by iodixanol gradient ultracentrifugation allows rapid and reproducible preparation of vector stocks for gene transfer in the nervous system. *Hum Gene Ther* **10:** 1885–1891.

NIH Guidelines for Research Involving Recombinant DNA Molecules. (April) 2000. http://oba.od.nih.gov/rdna/nih_guidelines_oba.html.

Purification of rAAV2s by Heparin Column Affinity Chromatography

This protocol describes a simple single-step column purification (SSCP) of rAAV2 by gravity flow based on its affinity to heparin, without ultracentrifugation. This method reproducibly yields preparations of AAV2 vectors with high titers, infectivity, and purity (Auricchio et al. 2001).

Biosafety Consideration

According to NIH Guidelines for Research Involving Recombinant DNA Molecules (April 2000), both wild-type AAVs and recombinant AAV constructs are not associated with disease in healthy adult humans, if the transgene does not encode either a potentially tumorigenic gene product or a toxic molecule and is produced in the absence of a helper virus. All work involving AAV vectors can be conducted at Biosafety Level 1 (BL1) with the approval of the Institutional Biosafety Committee at the home institution.

This protocol was contributed by Alberto Auricchio (Department of Pediatrics, "Federico II" University and Telethon Institute of Genetics and Medicine, Napoli, Italy).

MATERIALS

It is essential that you consult the appropriate Material Safety Data Sheets and your institution's Environmental Health and Safety Office for proper handling of equipment and hazardous materials used in this protocol.

Recipes for reagents specific to this protocol, marked <R>, are provided at the end of the protocol. See Appendix 1 for recipes for commonly used stock solutions, buffers, and reagents, marked <A>. Dilute stock solutions to the appropriate concentrations.

Reagents

Cell suspension, frozen (from Protocol 7)
DNase I from bovine pancreas, grade II (20 mg/mL)
DMEM, high glucose
Elution buffer <R>
Glycerol, sterile
Heparin-agarose, type I, saline suspension (Sigma-Aldrich)
Phosphate-buffered saline (PBS; 1×, pH 7.4)
RNase A from bovine pancreas, grade II (20 mg/mL)
Sodium deoxycholate (10%, w/v)
Washing buffer <R>

Equipment

Clamp holder, right angle, and extension clamp
Conical centrifuge bottle (500 mL)
Dry-ice/ethanol bath
Liquid chromatography columns, Luer-Lok (bed volume 98 mL, 2.5 cm × 20 cm)
Magnetic stirrer

Refrigerated tabletop centrifuge
Sample diffusion disk
Slide-A-Lyzer dialysis cassettes (10,000 MWCO, 0.5–3 mL)
Syringe filters (5 and 0.8 μm)
Stopcock, three-way Luer-Lok
Water bath (37°C)

METHOD

Preparation of Crude Cell Lysates

1. Remove the 50-mL conical centrifuge tubes from Protocol 7 containing transfected cell suspension in DMEM (serum free) from of the −80°C freezer and thaw in a 37°C water bath. Perform two rounds of freeze–thawing in a dry-ice/ethanol bath and a 37°C water bath. Transfer the crude viral lysate to a 500-mL conical centrifuge bottle.

   The use of two rounds (instead of three) of freezing and thawing is critical. A third round decreases the yield of infectious virus.

2. Add 250 μL of DNase (20 mg/mL; 10,000 U) and 250 μL of RNase (20 mg/mL). Incubate for 30 min at 37°C. Centrifuge at 1900g for 10 min at room temperature and carefully pipette the supernatant into a new 500-mL conical centrifuge bottle.

   Removal of cellular debris before adding sodium deoxycholate increases virus recovery and purity.

3. Add sodium deoxycholate (10%, w/v) to the clarified lysate to a final concentration of 0.5%. Mix and incubate for 10 min at 37°C. Filter the lysate through a 5-μm-pore-size filter using a 50-mL syringe first, then a 0.8-μm-pore-size filter.

   Addition of sodium deoxycholate to the lysate is another critical step; without it, the virus does not bind to heparin. Sodium deoxycholate is a biological detergent that helps to lyse the 293 cells and free rAAV virions from cellular and membrane components that associate with the virions and interfere with their binding to heparin.

Affinity Column Chromatography

4. Assemble the liquid chromatography column and Luer-Lok and place it in the clamp holder. (Columns may be reused if purifying the same virus, but a new diffusion disc should be used each time.) Add 8 mL of heparin-agarose slurry per column, making sure that the heparin-agarose slurry is thoroughly mixed. Open the Luer-Lok completely and allow the liquid to flow through into the waste tray.

5. Place a diffusion disc gently on the top of the heparin-agarose bed in the column. Equilibrate the column with 24 mL of PBS buffer, open the Luer-Lok completely, and then close it after all liquid has flowed through.

6. Gently apply the filtered viral lysate to the column and open the Luer-Lok so that the rate of flow is 1 drop/sec. Continue to load the viral lysate as needed until all of the liquid has flowed through the column and then close the Luer-Lok.

7. Apply 40 mL of Washing buffer per column and open the Luer-Lok so that the rate of flow is 1 drop/sec. Close the Luer-Lok after all of the buffer has washed through the column.

8. Apply 5 mL of Elution buffer per column and resuspend the heparin-agarose in elution buffer. Open the Luer-Lok again and allow the Elution buffer to drop into a 15-mL conical centrifuge tube.

Desalting, Formulation, and Storage

9. Centrifuge the tube at 300g for 3 min and transfer the supernatant into a Slide-A-Lyzer dialysis cassette. Dialyze in 2 L of cold PBS overnight at 4°C. The following morning, change the PBS

and allow the sample to dialyze for an additional 3–4 hr. Collect the rAAV vector and add sterile glycerol to a final concentration of 5%. Mix well by pipetting and aliquot into cryovials for storage at −80°C.

DISCUSSION

This purification method has several advantages over traditional CsCl gradient ultracentrifugation: (1) It does not require more technical skill or equipment than does purification of plasmid DNA in a commercially available column; (2) virus can be purified within a half day, reducing the total production time by at least 3 d compared with the most commonly used CsCl purification method; (3) affinity purification of rAAV2 allows the recovery of virus without the major cellular contaminants that are present when a physical separation method such as CsCl density centrifugation is used, although it is potentially more enriched in empty capsids; (4) rAAV2 is unstable in CsCl, but this method yields highly infectious virus with ratios of genome copies to transducing units of up to six. Obviously, as in other chromatography purification methods, the heparin affinity column purification process cannot separate the empty particles from fully packaged virions.

RECIPES

It is essential that you consult the appropriate Material Safety Data Sheets and your institution's Environmental Health and Safety Office for proper handling of equipment and hazardous materials used in this protocol.

Elution Buffer (0.4 M NaCl PBS)

Reagent	Quantity (for 1 L)	Final concentration
NaCl	23.38 g	0.4 M

Adjust the volume to 1 L with PBS. Filter the solution through a nitrocellulose filter (0.45-μm pore size) and store at room temperature.

Washing Buffer (0.1 M NaCl PBS)

Reagent	Quantity (for 1 L)	Final concentration
NaCl	5.84 g	0.1 M

Adjust the volume to 1 L with PBS. Filter the solution through a nitrocellulose filter (0.45-μm pore size) and store at room temperature.

REFERENCES

Auricchio A, Hildinger M, O'Connor E, Gao G, Wilson JM. 2001. Isolation of highly infectious and pure Adeno-Associated Virus type 2 vectors with a single-step gravity-flow column. *Hum Gene Ther* 12: 71–76.

NIH Guidelines for Research Involving Recombinant DNA Molecules. (April) 2000. http://oba.od.nih.gov/rdna/nih_guidelines_oba.html.

Enrichment of Fully Packaged Virions in Column-Purified rAAV Preparations by Iodixanol Gradient Centrifugation Followed by Anion-Exchange Column Chromatography

This rapid and efficient method to prepare highly purified rAAVs, first described by Zolotukhin et al. (2002), is based on binding of negatively charged rAAV capsids to an anion-exchange resin that is pH dependent. The isoelectric point (pI) of different rAAV capsids may be slightly different, and as such, the pH of Buffer A may need to be optimized for new serotypes (capsids) to ensure an overall negative charge and thus efficient binding to the anion-exchange resin. Because rAAV capsids are stable in a wide range of pH values, cation-exchange chromatography can also be used. In fact, tandem cation-anion–exchange chromatography has been used for purification of rAAVs without an ultracentrifugation step (Debelak et al. 2000). However, it is important to understand that chromatographic methods are not suitable for complete elimination of empty capsids from rAAV vector preparations. The elution buffer composition can be optimized for the type of ion used and pH to maximize the separation of empty from full capsids (Qu et al. 2007). The conditions described here are effective for purification of rAAV1, 2, 5, rh8, and 8 vectors (Debelak et al. 2000; Gao et al. 2000; Zolotukhin et al. 2002; Qu et al. 2007).

Biosafety Consideration

According to NIH Guidelines for Research Involving Recombinant DNA Molecules (April 2000), both wild-type AAVs and recombinant AAV constructs are not associated with disease in healthy adult humans, if the transgene does not encode either a potentially tumorigenic gene product or a toxic molecule and is produced in the absence of a helper virus. All work involving AAV vectors can be conducted at Biosafety Level 1 (BL1) with the approval of the Institutional Biosafety Committee at the home institution.

MATERIALS

It is essential that you consult the appropriate Material Safety Data Sheets and your institution's Environmental Health and Safety Office for proper handling of equipment and hazardous materials used in this protocol.

Recipes for reagents specific to this protocol, marked <R>, are provided at the end of the protocol. See Appendix 1 for recipes for commonly used stock solutions, buffers, and reagents, marked <A>. Dilute stock solutions to the appropriate concentrations.

Reagents

Buffer A <R>
Buffer B <R>
Phosphate-buffered saline (PBS)

Equipment

Conical tube (50 mL)
Fast protein liquid chromatography (FPLC) system
HiTrap Q anion-exchange chromatography columns (5 mL) or equivalent

Luer-Lok disposable plastic syringe
Refrigerated tabletop centrifuge
Slide-A-Lyzer cassette (10,000 MWCO) or Amicon Ultra-15 100K concentrator
Syringe filter (0.22 μm)
Tubes, round bottomed (15 or 5 mL)

METHOD

1. Perform the iodixanol gradient centrifugation described in Protocol 9. Transfer all rAAV-containing fractions (~12 mL) to a 50-mL conical tube and bring the volume to 50 mL with Buffer A.

2. Attach the 5-mL HiTrap Q column to an FPLC system and load a fraction collector with 5- or 15-mL round-bottomed disposable tubes.

3. Set up a program to:

   i. equilibrate the column with 25 mL of Buffer A at a flow rate of 5 mL/min

   ii. inject the sample (50 mL) into the column at 2 mL/min

   iii. wash the column with 50 mL of Buffer A at 5 mL/min

   iv. elute with a 50-mL gradient of 0%–100% Buffer B (15–500 mM NaCl) over 50 mL at 2 mL/min with collection of 2-mL fractions

      If using this method for the first time, or for a new serotype, set up the system to collect fractions as soon as the rAAV sample is injected into the column.

4. Run the program. Store the fractions at 4°C.

   HiTrap Q columns can be reused for purification of the same rAAV. Run 100% Buffer B at 5 mL/min (or at a speed that does not exceed the maximum recommended pressure for these columns), followed by 100% Buffer A. Run the storage buffer (according to the manufacturer's instructions) and store at 4°C. To avoid cross contamination, do not use the same column for purification of different rAAV.

5. Screen fractions for rAAV by PCR. Add 1 μL of each fraction to 100 μL of water in a microcentrifuge tube. Use 1 μL of dilution for PCR amplification using a primer set specific for the genome of the particular AAV being purified. A typical chromatogram for an rAAV1 vector and the location of positive fractions are shown in Figure 1.

   See Troubleshooting.

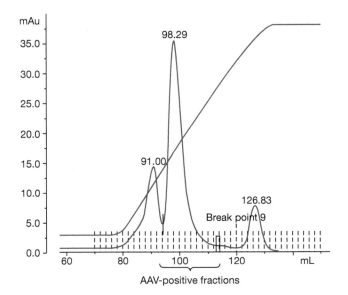

FIGURE 1. Typical chromatogram for an rAAV1 vector purified by iodixanol followed by anion-exchange column chromatography.

6. Pool all rAAV-positive fractions into a single tube. Attach a 0.22-μm syringe filter to a 20-mL Luer-Lok disposable plastic syringe and filter the rAAV solution into a sterile 5-mL conical tube.

7. Exchange the buffer by dialysis against sterile PBS at 4°C using a Slide-A-Lyzer cassette (10,000 MWCO) and multiple changes of buffer, as described in Protocol 8, Step 14 and Protocol 10, Step 9.

8. Alternatively, use a concentrator to exchange the buffer to PBS using the conditions described in Protocol 9.

9. Remove the rAAV stock from the Slide-A-Lyzer cassette or centrifugal concentrator, aliquot, and store at −80°C.

TROUBLESHOOTING

Problem (Step 5): No elution peak is detectable on the chromatogram.
Solutions:

- An FPLC system measures multiple parameters simultaneously, including absorbance at one or more wavelengths (most commonly used is at 280 nm), conductance, temperature, pressure, and pH (depending on the system). Iodixanol absorbs very strongly at 280 nm, causing the system to automatically reset the absorbance scale in order to represent that absorbance. In this extended scale, the absorbance of rAAV preparation proteins becomes almost indistinguishable from the baseline. To correct for this, reset the scale of the chromatogram in the analysis software to a maximum of 50–100 mAU. Elution peaks should become apparent.

- Check the conductance profile during the entire procedure. Make sure that before the elution starts, the conductance is low (∼4 mS/cm). rAAVs elute at 10–15 mS/cm.

- Check pH of Buffers A and B.

- When using this method for purification of a new rAAV serotype/capsid, test binding to the anion-exchange resin with Buffers A and B of different pH values.

DISCUSSION

This is an efficient method to generate highly purified rAAV preparations with high titers and very good transduction profiles in culture and in vivo. In addition, it can be completed in a considerably shorter period than the CsCl gradient ultracentrifugation method (Protocol 8). However, this chromatography-based method does not eliminate empty capsids from rAAV preparations, and it has to be tested and optimized for production of each new rAAV serotype.

RECIPES

It is essential that you consult the appropriate Material Safety Data Sheets and your institution's Environmental Health and Safety Office for proper handling of equipment and hazardous materials used in this protocol.

Buffer A (pH 8.5)

Reagent	Quantity (for 1 L)	Final concentration
Tris-HCl (1 M, pH 8.5)	20 mL	20 mM
NaCl (5 M)	3 mL	15 mM
H_2O	to 1 L	

Adjust the pH of the final solution to 8.5. Pass through a 0.22-μm filter and de-gas for 10 min before using in the FPLC system.

Buffer B (pH 8.5)

Reagent	Quantity (for 1 L)	Final concentration
Tris-HCl (1 M, pH 8.5)	20 mL	20 mM
NaCl (5 M)	100 mL	500 mM
H_2O	to 1 L	

Adjust the pH of the final solution to 8.5. Pass through 0.22-μm filter and de-gas for 10 min before using in the FPLC system.

REFERENCES

Debelak D, Fisher J, Iuliano S, Sesholtz D, Sloane DL, Atkinson EM. 2000. Cation-exchange high-performance liquid chromatography of recombinant adeno-associated virus type 2. *J Chromatogr* **740:** 195–202.

Gao G, Qu G, Burnham MS, Huang J, Chirmule N, Joshi B, Yu QC, Marsh JA, Conceicao CM, Wilson JM. 2000. Purification of recombinant adeno-associated virus vectors by column chromatography and its performance in vivo. *Hum Gene Ther* **11:** 2079–2091.

NIH Guidelines for Research Involving Recombinant DNA Molecules. (April) 2000. http://oba.od.nih.gov/rdna/nih_guidelines_oba.html.

Qu G, Bahr-Davidson J, Prado J, Tai A, Cataniag F, McDonnell J, Zhou J, Hauck B, Luna J, Sommer JM, et al. 2007. Separation of adeno-associated virus type 2 empty particles from genome containing vectors by anion-exchange column chromatography. *J Virol Methods* **140:** 183–192.

Zolotukhin S, Potter M, Zolotukhin I, Sakai Y, Loiler S, Fraites TJ Jr, Chiodo VA, Phillipsberg T, Muzyczka N, Hauswirth WW, et al. 2002. Production and purification of serotype 1, 2, and 5 recombinant adeno-associated viral vectors. *Methods* **2:** 158–167.

Titration of rAAV Genome Copy Number Using Real-Time qPCR

This protocol is used to determine the concentration of DNase-resistant vector genomes (i.e., packaged in the capsid) in purified rAAV preparations. The protocol begins with treatment of the vector stock with DNase I to eliminate unencapsidated AAV DNA or contaminating plasmid DNA. This is followed by a heat treatment to heat-inactivate DNase I, to disrupt the viral capsid, and to release the packaged vector genomes for quantification by real-time PCR using a set of standards (linearized pCis plasmid used for vector production) containing known copy numbers. To accomplish high-throughput titration, the primer and probe sets used in real-time PCR are usually designed to target common elements present in most rAAV genomes, such as promoters and poly(A) signals. This strategy significantly reduces the number of PCRs, controls, and turnaround time. Several important controls should be included in the assay as follows: The first two controls should have a known copy number of the rAAV genome plasmid treated or not treated with DNase I. This control tests the effectiveness of DNase treatment. To control for potential cross-contamination between samples during the preparation process, a blank control containing nuclease-free water only should be processed and tested in parallel. A validation vector sample with a known titer should be included in every assay to monitor interassay variability. Finally, for the PCR run, a no-template control (NTC) is included to indicate cross-contamination during PCR setup.

Biosafety Consideration

According to NIH Guidelines for Research Involving Recombinant DNA Molecules (April 2000), both wild-type AAVs and recombinant AAV constructs are not associated with disease in healthy adult humans, if the transgene does not encode either a potentially tumorigenic gene product or a toxic molecule and is produced in the absence of a helper virus. All work involving AAV vectors can be conducted at Biosafety Level 1 (BL1) with the approval of the Institutional Biosafety Committee at the home institution.

MATERIALS

It is essential that you consult the appropriate Material Safety Data Sheets and your institution's Environmental Health and Safety Office for proper handling of equipment and hazardous materials used in this protocol.

Recipes for reagents specific to this protocol, marked <R>, are provided at the end of the protocol. See Appendix 1 for recipes for commonly used stock solutions, buffers, and reagents, marked <A>. Dilute stock solutions to the appropriate concentrations.

Reagents

DNase I and DNase buffer
DNA standard set (see the Additional Protocol Preparation of a DNA Standard for qPCR after Protocol 5)
6FAM fluorescent probe working stock solution (2 μM)
Forward primer working stock solution (9 μM)
GeneAmp PCR buffer (10×; Applied Biosystems, CA)
Nuclease-free water

rAAV genome standard set and NTC
Reverse primer working stock solution (9 μM)
Sample dilution buffer <R>
Sheared salmon sperm (SSS) DNA (Applied Biosystems)
TaqMan Universal PCR Mix, No UNG (uracil *N*-glycosylase) (Applied Biosystems)
Test rAAV and rAAV validation sample

Equipment

Cap-It tool and Ultra Clear cap strips
Microcentrifuge tubes (0.65 and 1.7 mL) and tube rack
PCR-dedicated hood
Real-time PCR station from any manufacturer
Vortex machine

METHOD

Sample Preparation

1. Set up 0.65-mL microcentrifuge tubes on a tube rack and label the tubes as 10-fold dilutions for each vector preparation to be assayed, validation vector sample, (+) and (−) DNase treatment controls, and sample preparation blank control.

2. Prepare the DNase digestion reaction master mix in a 1.7-mL microcentrifuge tube as follows:

Reagent	Volume per sample
Nuclease-free H$_2$O	38 μL
DNase buffer (10×)	5 μL
DNase I, RNase-free	2 μL (20 U)

   Calculate the total volume of each reagent to add to the reaction master mix by multiplying the volume per sample (above) by the total number of samples (vector dilutions and controls). Before adding DNase I to the reaction mixture, add 2 μL of nuclease-free H$_2$O and 43 μL of DNase digestion reaction master mix without DNase I to bring the volume of the (−) DNase control to 45 μL total volume. Mix well by vortexing. Next, add DNase I to the remaining DNase digestion reaction master mix (minus one volume, i.e., 2 μL), vortex, and add 45 μL of this reaction mixture to all other tubes.

3. In a PCR-dedicated hood, add 5 μL of the rAAV preparation to be assayed and controls to dedicated tubes in the following order:

   i. Control plasmid (5×10^7 GC/μL) to (+) DNase control tube

   ii. Test rAAV

   iii. Nuclease-free water to sample preparation blank control

   iv. Control plasmid (5×10^7 GC/μL) to (−) DNase control tube

   v. rAAV validation sample

   Incubate for 30 min at 37°C.

4. Meanwhile, set up microcentrifuge tubes in a tube rack to prepare the following 10-fold serial dilutions:

Test vector	100- to 10,000-fold
Sample preparation blank control	100-fold
(+) DNase treatment control	100-fold
(−) DNase control	100- to 1000-fold
Validation vector sample	Extent of dilution depends on the concentration of the vector (e.g., 1000-fold dilution for the validation vector with a titer in the range of 10^{12} GC/mL).

5. Thaw the sample dilution buffer on ice. Dispense 45 µL of this buffer to the serial dilution tubes in the second microcentrifuge tube rack that is set up in Step 4. Keep these tubes at room temperature and proceed to the next step below.

Real-Time PCR Setup

6. Thaw the primers and probe in advance of sample preparation. The TaqMan Universal PCR Mix, No UNG should be kept at 4°C at all times. Thaw the set of DNA standards immediately before use.

7. While DNase digestion is in progress, and after dispensing the Sample dilution buffer into the dilution tubes, prepare the real-time PCR master mix as follows:

Reagent	Volume per reaction
TaqMan Universal PCR Mix (2×)	25 µL
Forward primer (9 µM)	5 µL
Reverse primer (9 µM)	5 µL
6FAM fluorescent probe (2 µM)	5 µL
Nuclease-free water	5 µL
Test vector samples	5 µL

8. Multiply the volumes of each reagent per reaction (with the exception of the test vector samples sample DNA) by three (for triplicate sets) and by the total number of samples. Include an additional two volumes to compensate for pipetting errors. Prepare the real-time PCR master mix in a 15-mL conical tube. Mix by vortexing and keep on ice.

   To minimize cross-contamination, while dispensing reagents or samples, keep all containers and microcentrifuge tubes closed at all times when not in use. In addition, keep pipette tip rack lids closed when not in use.

9. Dispense 45 µL of real-time PCR master mix into triplicate wells for each of NTC, standards ($1-10^8$ copies), test vectors, plasmid DNA without DNase I (two dilutions), plasmid DNA treated with DNase I, sample preparation blank control, and validation vector sample in a 96-well plate. Keep the plate on ice.

10. After the DNase digestion is complete, prepare 10-fold serial dilutions of DNase I–treated test vectors, validation vector sample, and controls using the microcentrifuge tubes prepared in Step 4 in a PCR-dedicated hood.

11. Dispense 5 µL each of nuclease-free water for NTC, standards, test vectors, controls, and validation sample at appropriate dilutions to triplicate wells in 96-well plates, with each well containing 45 µL of PCR master mix in the following order:

    i. standards ($1-10^8$ genome copies in 10-fold serial dilutions in ascending order)

    ii. DNase I–treated test vectors (10,000-fold dilution)

    iii. plasmid control without DNase I (1000-fold dilution)

    iv. plasmid control without DNase I (100-fold dilution)

    v. DNase I–treated plasmid control (100-fold dilution)

    vi. sample preparation blank control (100-fold dilution)

    vii. DNase I–treated rAAV validation control

    viii. nuclease-free water for NTC

12. Seal the 96-well plate tightly using the Ultra Clear cap strips and the Cap-It tool (or film) and place the plate in the thermocycler block. Start the PCR run with the following PCR program:

    i. one cycle of 2 min at 50°C

    ii. 10 min at 95°C to release the viral DNA from the capsid

      iii. 40 cycles of 15 sec at 95°C

      iv. 1 min at 60°C

13. After the PCR run, analyze the data following the manufacturer's instructions. Calculate the final vector genome copy titer after and conversion factors and the single-stranded nature of the AAV genome using the following formula:

Average copy number from PCR run ÷ 5 μL

    × 2 (to account for the single-stranded nature of most rAAVs)

    × dilution factor (10-fold for DNase treatment × 10, 000-fold subsequent dilution)

    × 1000 (μL/mL) = GC titer/mL.

TROUBLESHOOTING AND DISCUSSION

The problems and issues that usually arise with this assay are very similar to those of the qPCR steps in Protocol 5 for titration of adenovirus vector infectivity. See the corresponding sections in Protocol 5 for the solutions.

RECIPE

It is essential that you consult the appropriate Material Safety Data Sheets and your institution's Environmental Health and Safety Office for proper handling of equipment and hazardous materials used in this protocol.

Sample Dilution Buffer (1× PCR Buffer/20 ng/μL SSS DNA)

Reagent	Quantity for 10 mL	Final concentration
GeneAmp PCR buffer (10×)	1 mL	1×
SSS DNA (1 mg/mL)	0.2 mL	0.02 mg/mL
Nuclease-free water	8.8 mL	

Mix well, aliquot into 1.5-mL Eppendorf tubes, and store at −20°C.

REFERENCE

NIH Guidelines for Research Involving Recombinant DNA Molecules. (April) 2000. http://oba.od.nih.gov/rdna/nih_guidelines_oba.html.

Sensitive Determination of Infectious Titer of rAAVs Using TCID$_{50}$ End-Point Dilution and qPCR

AAV recombinants are currently the vector of choice for many gene therapy applications. As experimental therapies progress to clinical trials, the need to characterize rAAVs accurately and reproducibly increases. Accurate determination of rAAV infectious titer is important for determining the activity of each vector lot and for ensuring lot-to-lot consistency. Early assays to determine rAAV infectious titer, such as the infectious center assay (Salvetti et al. 1998) and a hybridization-based 96-well TCID$_{50}$ assay (Atkinson et al. 1998), suffer from high background levels, subjective calling of infectious events, and low sensitivity.

Much like the adenovirus infectivity assay described in Protocol 5, the power of real-time qPCR can be combined with the 96-well TCID$_{50}$ format to achieve a large increase in sensitivity and the detection of single infectious rAAV events (Zen et al. 2004). The following protocol developed in our laboratory adapts the 96-well TCID$_{50}$ format and qPCR detection described in Protocol 5 to the determination of rAAV infectious titer.

Only those reagents and methods that differ from Protocol 5 are described in detail below. Identical steps in the method are noted.

Biosafety Consideration

According to NIH Guidelines for Research Involving Recombinant DNA Molecules (April 2000), both wild-type AAVs and recombinant AAV constructs are not associated with disease in healthy adult humans, if the transgene does not encode either a potentially tumorigenic gene product or a toxic molecule and is produced in the absence of a helper virus. All work involving AAV vectors can be conducted at Biosafety Level 1 (BL1) with the approval of the Institutional Biosafety Committee at the home institution.

This protocol was contributed by Martin Lock and James Wilson (Gene Therapy Program, University of Pennsylvania, Philadelphia).

MATERIALS

It is essential that you consult the appropriate Material Safety Data Sheets and your institution's Environmental Health and Safety Office for proper handling of equipment and hazardous materials used in this protocol.

Recipes for reagents specific to this protocol, marked <R>, are provided at the end of the protocol. See Appendix 1 for recipes for commonly used stock solutions, buffers, and reagents, marked <A>. Dilute stock solutions to the appropriate concentrations.

See Protocol 5 for the complete list of reagents, equipment, and recipes required. Only the reagents required for rAAV titration that differ from those required in Protocol 5 are listed here.

Reagents

Human adenovirus 5 (HuAd5) at a titer of 1×10^{12} particles/mL or greater
Aliquots of HuAd5 are available for purchase from Penn Vector Core (vector@mail.med.upenn.edu).
rAAV test vector
RC32 cells (ATCC, CRL-2972)

RC32 cells, transformed HeLa cells expressing AAV2 rep and cap genes (Chadeuf et al. 2000), are available directly from American Type Culture Collection (www.atcc.org). Alternatively, B50 cells, another HeLa-based AAV2 rep/cap cell line (Gao et al. 1998), can be obtained from the Penn Vector Core (vector@mail.med.upenn.edu). These cell lines are fully interchangeable in our hands.

METHOD

Many steps in the method for the determination of rAAV infectious titer are identical to those described in the protocol for measuring the titer of an adenovirus vector (Protocol 5). The steps that differ are described in detail below.

Plating Cells

1. Warm Complete medium, PBS, and trypsin in a water bath to 37°C. Perform all subsequent steps in the biosafety cabinet using sterile technique.

2. Aspirate the medium from a confluent monolayer of RC32 cells.

   The presence of the AAV2 rep and cap gene in cells co-infected with adenovirus and the rAAV leads to the amplification of the rAAV genome, packaging of the genome into new capsids, and a second cycle of vector infection. This results in increased sensitivity of the assay.

3. Plate the RC32 cells as described in Protocol 5, Steps 3 and 4.

Infection

4. The following morning, prepare diluent solution containing HuAd5 in serum-free medium (SFM) at a concentration of 3.2×10^8 adenovirus particles/mL. Prepare 15 mL of diluent solution per 96-well plate.

5. Prepare the rAAV dilutions. Thaw the vector on ice and mix it by pipetting. For each vector, prepare eight 1-mL serial dilutions in cryovials. Determine empirically the dilution range plated for each vector or vector serotype. For example, use the following dilution schema:

   1×10^{-2} = 990 μL of diluent + 10 μL of stock
   1×10^{-4} = 990 μL of diluent + 10 μL of previous dilution
   1×10^{-6} = 990 μL of diluent + 10 μL of previous dilution
   1×10^{-7} = 900 μL of diluent + 100 μL of previous dilution
   1×10^{-8} = 900 μL of diluent + 100 μL of previous dilution
   1×10^{-9} = 900 μL of diluent + 100 μL of previous dilution
   1×10^{-10} = 900 μL of diluent + 100 μL of previous dilution
   1×10^{-11} = 900 μL of diluent + 100 μL of previous dilution

6–12. Follow the procedures described in the Protocol 5, Steps 6–12.

DNA Extraction

13–16. Perform DNA extraction from the infected cells in 96-well plates exactly as described in Protocol 5, Steps 13–16.

Real-Time PCR

17. Working in a dedicated PCR facility or chemical fume hood, set up the PCR plate in the same format as the infection plate but include wells for quantification standards and an NTC (water). The PCR setup is the same as that depicted in Figure 2 in Protocol 5.

18. For each plate, make a master mix for 100 reactions (see Protocol 5, qPCR master mix recipe).

   The master mix contains PCR primers and a TaqMan probe specific to an element within the rAAV expression cassette. For convenience, these reagents usually are designed to target a polyadenylation signal or promoter sequence so that they can be used for multiple vectors. Several PCR mixes containing the other essentials of the PCR (nucleotide triphosphates, buffer, and thermostable enzyme) are available from commercial vendors at a 2× concentration.

19–21. Perform as described in Protocol 5, Steps 19–21.

Data Collection

22. Perform the data collection as described in Protocol 5, Step 22.

Data Analysis

23–26. Perform data analysis as indicated in Protocol 5, Steps 23–26.

TROUBLESHOOTING

See the Troubleshooting section for Protocol 5.

DISCUSSION

This method measures the infectious titer of rAAVs with greater sensitivity than early infectious titer methods (in our hands, five- to 10-fold greater) and features nonsubjective and quantitative calling of positive wells. In addition, the 96-well format simplifies and speeds up the assay and opens the way to adapting the technique to automation. One unwanted consequence of the sensitivity of qPCR quantitation is the detection of low-level cross-contamination with AAV genome copies, which may lead to false-positive calling of rAAV-infected wells; however, with proper handling and the precautions outlined in Protocol 5, this phenomenon can be minimized, and excellent interassay reproducibility is possible.

It is worthwhile to note that the in vitro infectivity measurements between different lots of the same serotype vector are useful in assessing the relative potency of a particular vector preparation. However, this information does not predict activity between different serotypes for in vivo transduction. In general, the early generations of rAAVs, such as rAAV2, transduce cells in culture very well but usually underperform in vivo. On the other hand, vectors derived from newly isolated primate AAVs are very inefficient for transduction of cells in culture but are highly efficient for in vivo gene-delivery applications.

REFERENCES

Atkinson EM, Debelak DJ, Hart LA, Reynolds TC. 1998. A high-throughput hybridization method for titer determination of viruses and gene therapy vectors. *Nucleic Acids Res* **26:** 2821–2823.

Chadeuf G, Favre D, Tessier J, Provost N, Nony P, Kleinschmidt J, Moullier P, Salvetti A. 2000. Efficient recombinant adeno-associated virus production by a stable rep-cap HeLa cell line correlates with adenovirus-induced amplification of the integrated rep-cap genome. *J Gene Med* **2:** 260–268.

Gao GP, Qu G, Faust LZ, Engdahl RK, Xiao W, Hughes JV, Zoltick PW, Wilson JM. 1998. High-titer adeno-associated viral vectors from a Rep/Cap cell line and hybrid shuttle virus. *Hum Gene Ther* **9:** 2353–2362.

NIH Guidelines for Research Involving Recombinant DNA Molecules. (April) 2000. http://oba.od.nih.gov/rdna/nih_guidelines_oba.html.

Salvetti A, Oreve S, Chadeuf G, Favre D, Cherel Y, Champion-Arnaud P, David-Ameline J, Moullier P. 1998. Factors influencing recombinant adeno-associated virus production. *Hum Gene Ther* **9:** 695–706.

Zen Z, Espinoza Y, Bleu T, Sommer JM, Wright JF. 2004. Infectious titer assay for adeno-associated virus vectors with sensitivity sufficient to detect single infectious events. *Hum Gene Ther* **15:** 709–715.

Analysis of rAAV Sample Morphology Using Negative Staining and High-Resolution Electron Microscopy

Negative staining is a simple and rapid method for studying the morphology and ultrastructure of small particulate specimens (e.g., viruses, bacteria, cell fragments, and isolated macromolecules such as proteins and nucleic acids). Samples stained in this manner show great structural integrity because the stains used not only delineate the ultrastructure but also act as a fixative, thus protecting samples from irradiation damage by the electron beam. Negative staining also reduces the surface-tension forces at the air–liquid interface, thus reducing shrinkage and specimen collapse. These effects are promoted by the heavy metal salts used in negative stains. Although negative staining is one of the oldest techniques for studying the ultrastructure of particulate samples at the level of electron microscopy, it has never been surpassed by any other technique for determining the surface structures, size, and shapes of specimens such as viruses, bacteria, and macromolecules, with resolution down to the 2-nm level (Hayat 2000). Because it is a simple and rapid method, negative staining combined with high-resolution transmission electron microscopy is a very effective method for determining the morphology and relative purity of rAAVs. Images taken at 40,000 to 50,000 diameters in magnification can be quickly scanned to differentiate empty versus fully packaged virions. For a research-grade laboratory preparation of rAAVs, a full-to-empty ratio of at least 20% may be acceptable; however, the higher the ratio of full to empty, the better the result for in vivo applications of rAAVs.

One technique of negative staining involves allowing particles or fragments of cells to settle onto a support film, then applying a drop of metal salt solution to the adherent particulate specimen. The stain penetrates the interstices of the particles to bring out detail. In this situation, the preparation dries rapidly. The dissolved substance precipitates out of solution in an amorphous condition at the 0.1-nm level, and it is deposited over the support film and exposed surface of the specimen. The theoretical requirements of a good negative staining are a substance (1) of high density to provide high contrast, (2) at high solubility so that the stain does not come out of solution prematurely but does so only at the final stage of drying, (3) of high melting point and boiling point so that the material does not evaporate at high temperatures induced by the electron beam, and (4) in which the precipitate should be essentially amorphous down to the limit of resolution.

Biosafety Consideration

According to NIH Guidelines for Research Involving Recombinant DNA Molecules (April 2000), both wild-type AAVs and recombinant AAV constructs are not associated with disease in healthy adult humans, if the transgene does not encode either a potentially tumorigenic gene product or a toxic molecule and is produced in the absence of a helper virus. All work involving AAV vectors can be conducted at Biosafety Level 1 (BL1) with the approval of the Institutional Biosafety Committee at the home institution.

This protocol was contributed by Gregory Hendricks (Electronic Microscopy Core, University of Massachusetts Medical School, Worcester).

MATERIALS

It is essential that you consult the appropriate Material Safety Data Sheets and your institution's Environmental Health and Safety Office for proper handling of equipment and hazardous materials used in this protocol.

Recipes for reagents specific to this protocol, marked <R>, are provided at the end of the protocol. See Appendix 1 for recipes for commonly used stock solutions, buffers, and reagents, marked <A>. Dilute stock solutions to the appropriate concentrations.

Reagents

rAAV sample
Uranyl acetate (1%, w/v) <R>

Equipment

Filter paper
Formvar support films, carbon stabilized, on 200-mesh copper grids
Humidity chamber, controlled (60% relative humidity)
Transmission electron microscope

METHOD

1. Prepare all spreads on freshly prepared carbon-stabilized Formvar support films on 200-mesh copper grids.

2. Place 5 μL of rAAV preparation in solution onto the Formvar support films and allow it to stand for 30 sec.

   Sometimes, excessive sugar content in the buffer used to stabilize viruses can result in artifacts in negative staining of high-titer preparations. See Troubleshooting for a solution to minimize the artifacts.

3. Remove the excess liquid with filter paper and negatively stain the sample by running 6 drops of 1% uranyl acetate over the grid to fix and contrast the spread virus particles.

4. Remove excess stain with filter paper and dry the sample in a controlled humidity chamber (60% relative humidity). Examine the samples with a transmission electron microscope and record the images at magnifications that allow the fine structure of the virus to be seen (Fig. 1A).

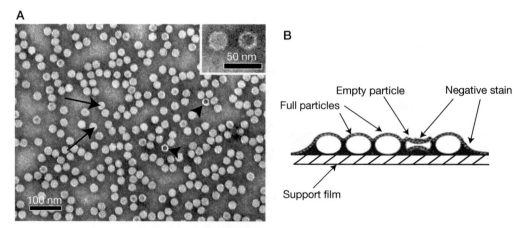

FIGURE 1. Transmission electron microscopy of negative-stained AAV recombinants. (*A*) rAAV particles prepared as described in Protocol 14, spread on a freshly prepared carbon-coated Formvar support film, and stained with 1% uranyl acetate. The large field of virus particles was taken at 92,000× (*inset at upper right* is taken at 190,000×). Arrowheads point to two empty viral particles in a field of full particles (arrows). (*Inset*) A full viral particle (*left*) next to an empty particle. The hexagonal shape of the AAV particles is clearly evident. (*B*) Illustration of the empty rAAV particles exhibiting as donut-like shapes formed by the accumulation of the uranyl acetate stain on the dimples.

TROUBLESHOOTING

Problem (Step 2): False-negative staining occurs that is caused by excessive sugar in high-titer viral samples formulated in a sugar-containing buffer.

Solution: Rinse the support film with a few drops of distilled water, remove the excess water with filter paper, and proceed with negative staining.

DISCUSSION

The negative-staining method is based on the principle that there is no reaction between the stain and the specimen. This is accomplished by adjusting the pH of the stain so that attraction between the stain and specimen is negligible. Many factors affect the appearance of negatively stained specimen preparations. The shapes and sizes of the viral particles are influenced by both the stain that is used and its mode of application. Prefixation with aldehydes often allows internal components to be seen; however, the presence of excess, unreacted aldehyde groups can cause artifacts in the staining process due to reaction between the metal ions in the stain and the very reactive, electronegative oxygen of the polar formyl groups. Excess sugars in buffer solutions used to purify and stabilize many viral preparations can react with the metal salts as a result of the formation of aldehyde residues on the sugar molecules, causing crystallization of the stain-sugar complex (Hayat 2000). As explained earlier, the interaction of the electron beam with the sample generates contrast in the image. In negative-stain microscopy, the electron beam primarily interacts with the stain. When stain is added to a sample, the stain surrounds the sample but is excluded from the volume occupied by the sample; hence the use of the term "negative stain" (Fig. 1B).

It is worthwhile to note that the viral particles with the dimpled (dark) center in Figure 1A are the empty particles. Because of the way the negative-staining procedure works, the empty particles with the dark center would have a well-like pool of stain collected on the capsid surface following the application and drying of the heavy metal staining solution. This suggests that the structure of empty and full capsids is different. Figure 1A illustrates the "empty" viral particle, an intact viral particle, and the thick buildup of stain at the base of the particles, and Figure 1B shows the very thin layer of stain that covers the surface of all of the particles.

RECIPE

It is essential that you consult the appropriate Material Safety Data Sheets and your institution's Environmental Health and Safety Office for proper handling of equipment and hazardous materials used in this protocol.

Uranyl Acetate (1%, w/v)

Add 10 mL of ddH$_2$O to a 15-mL conical centrifuge tube containing 10 mg of uranyl acetate powder. Cover the tube with foil and rotate it for several days in a cold room until fully dissolved. Filter through a 0.22-μm syringe filter that has be prerinsed well with ddH$_2$O. Filtered stain can be used for >1 yr if stored at 4°C in a foil-wrapped tube.

REFERENCES

Hayat MA. 2000. *Principles and techniques of electron microscopy: Biological applications*, pp. 367–399. Cambridge University Press, Cambridge.

NIH Guidelines for Research Involving Recombinant DNA Molecules. (April) 2000. http://oba.od.nih.gov/rdna/nih_guidelines_oba.html.

Analysis of rAAV Purity Using Silver-Stained SDS-PAGE

AAV virions are built from three major capsid proteins, VP1, VP2, and VP3, at a ratio of 1:1:18. On a silver-stained SDS–polyacrylamide gel, VP1, VP2, and VP3 should be the only visible bands in a highly purified rAAV preparation, migrating at approximately 87, 73, and 62 kDa, respectively. Silver-stained SDS-PAGE analysis of rAAV preparations aims to scrutinize the test vectors for purity and reveal whether any other cellular or transgene protein contaminants are present. Nonetheless, this method cannot differentiate empty virions from fully packaged virions, which can be best accomplished by examination by electron microscopy of negative-stained virus samples (see Protocol 14).

This protocol describes how SDS-PAGE and silver staining can be used to determine the purity of an rAAV preparation. In addition, using a highly purified rAAV preparation whose particle titer is known, this assay can be used to derive a semiquantitative estimate of the particle concentration of a test vector (Fig. 1). Note that virus particle titer estimation based on a single loading of a test vector is sometimes unreliable. See Troubleshooting for an alternative strategy.

Biosafety Consideration

According to NIH Guidelines for Research Involving Recombinant DNA Molecules (April 2000), both wild-type AAVs and recombinant AAV constructs are not associated with disease in healthy adult humans, if the transgene does not encode either a potentially tumorigenic gene product

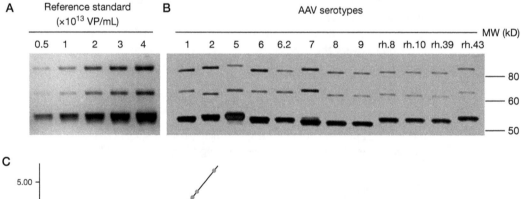

FIGURE 1. Silver-stained SDS-PAGE analysis of CsCl gradient–purified AAV recombinants of 12 different serotypes. (A) 0.5, 1, 2, 3, and 4 µL each of an AAV2 vector with a concentration of 1×10^{13} virus particles/mL were loaded in the corresponding lane as reference standards. (B) Approximately 1.5×10^{10} virus particles each of AAVs 1, 2, 5, 6, 6.2, 7, 8, 9, rh.8, rh.10, rh.39, and rh.43 were loaded in the corresponding lanes. (C) Linear regression plot to compare and interpret the standards (black dots) against unknown samples (blue dots). The regression equation is Conc = 8.45E-006 * Vol − 0.762, where $R^2 = 0.999521$.

or a toxic molecule and is produced in the absence of a helper virus. All work involving AAV vectors can be conducted at Biosafety Level 1 (BL1) with the approval of the Institutional Biosafety Committee at the home institution.

MATERIALS

It is essential that you consult the appropriate Material Safety Data Sheets and your institution's Environmental Health and Safety Office for proper handling of equipment and hazardous materials used in this protocol.

Reagents

Novex 15-well Tris-glycine (10%) gels (Life Technologies) or equivalent
Novex Tris-glycine SDS running buffer ($10\times$; Life Technologies)
NuPAGE Reducing Agent ($10\times$; Life Technologies)
Protein molecular mass marker in the range of 20–220 kDa
Reference standard (1×10^{13} virus particles/mL of rAAV)

> *The reference standard is selected from rAAV preparations whose genome concentrations are either known or can be estimated. These preparations are first titrated for GC titer per milliliter and then subjected to analysis by electron microscopy and silver-stained SDS-PAGE. The vector preparation to be used as the reference standard has a GC titer $\geq 1\times10^{13}$ GC/mL, no more than 10% empty particles, and only VP1, VP2, and VP3 bands.*

SilverXpress Silver Staining Kit (Life Technologies) or equivalent
Test rAAV
Tris-glycine SDS sample buffer ($2\times$; Life Technologies)

Equipment

Gel electrophoresis apparatus
Gel Station (Universal Hood II, Bio-Rad), and Quantity One 1-D analysis software
Microcentrifuge tubes (0.5 and 0.65 mL)

METHOD

1. Prepare a $1\times$ gel running buffer from Novex Tris-glycine SDS running buffer $10\times$ stock.

2. Prepare the reference standard and test rAAV as follows.

   i. Reference standard with 1×10^{13} virus particles/mL of rAAV

Reference standard	0.5, 1, 2, 3, and 4 μL each in 0.65-mL microcentrifuge tubes
Tris-glycine SDS sample buffer ($2\times$)	5 μL
NuPAGE Reducing Agent ($10\times$)	1 μL
dH$_2$O	to a final volume of 10 μL

   The use of NuPAGE Reducing Agent in the sample preparation helps to disassociate and stabilize VP1, VP2, and VP3 viral capsid proteins.

   ii. Test rAAV

Test rAAV	2 and 6 μL each in 0.5-mL microcentrifuge tubes
Tris-glycine SDS sample buffer ($2\times$)	7.5 μL
NuPAGE Reducing Agent ($10\times$)	1.5 μL
dH$_2$O	to a final volume of 15 μL

   The use of NuPAGE Reducing Agent in the sample preparation helps to disassociate and stabilize VP1, VP2, and VP3 viral capsid proteins.

3. Prepare an appropriate amount of protein molecular mass standard according to the manufacturer's instructions.

4. Mix the samples well, heat for 5 min at 95°C, and pulse-centrifuge.

5. Set up the gel apparatus. Load 10 μL of reference standards and 12 μL of test samples into the wells. Run the gel at 125 V for 1.5–2 h.

   See Troubleshooting.

6. Perform silver staining of the gel using a silver staining kit according to the manufacturer's instructions.

7. Scan the silver-stained gel on a Gel Station (e.g., Universal Hood II, Bio-Rad) and perform semiquantitative analysis of the test vectors (Fig. 1) using Quantity One 1-D analysis software (Bio-Rad), which is based on a linear regression comparison and interpolation of the standards against the unknown samples (Fig. 1C).

TROUBLESHOOTING

Problem (Step 5): Semiquantitative estimation of virus particle titer in a test vector is not reliable on the basis of single loading.

Solution: Prepare duplicates of the test vector for the analysis. Load the reference standard dilutions in the middle wells and one set each of the unknown samples on the left and right sides of the reference set.

REFERENCE

NIH Guidelines for Research Involving Recombinant DNA Molecules. (April) 2000. http://oba.od.nih.gov/rdna/nih_guidelines_oba.html.

Production of High-Titer Retrovirus and Lentivirus Vectors

The most commonly used method for production of retrovirus and lentivirus vectors in research laboratories is of transient transfection of 293T cells with a lentivirus vector plasmid, packaging genome plasmid(s), and an envelope expression plasmid. High-titer stocks are obtained by concentration of vector supernatants by ultracentrifugation.

Biosafety Considerations

National Institutes of Health (NIH) Guidelines for Research Involving Recombinant DNA Molecules do not directly address work with lentivirus vectors. Therefore, the Recombinant DNA Advisory Committee (RAC) has issued guidelines on how to conduct a risk assessment for lentivirus vector research. The major risks associated with lentivirus vector work are the potential for generation of replication-competent lentiviruses (RCLs) and oncogenesis. In the laboratory setting, BL2 or enhanced BL2 containment procedures are adequate for most applications not involving transgenes with high oncogenic potential. A risk assessment by the local Institutional Biosafety Committee (IBC) based on the lentivirus vector design and packaging system, nature of transgene insert, vector titer and total amount of vector produced, inherent biological containment of the animal host, and negative RCL testing will determine the appropriate procedures. Review and approval by the local IBC is mandatory before initiating any work with lentivirus vectors.

MATERIALS

It is essential that you consult the appropriate Material Safety Data Sheets and your institution's Environmental Health and Safety Office for proper handling of equipment and hazardous materials used in this protocol.

Recipes for reagents specific to this protocol, marked <R>, are provided at the end of the protocol. See Appendix 1 for recipes for commonly used stock solutions, buffers, and reagents, marked <A>. Dilute stock solutions to the appropriate concentrations.

Reagents

CaCl$_2$ solution (2 M) <R>
Complete growth medium <R>
DMEM
Ethanol (or isopropanol) (70%)
Fetal bovine serum (FBS)
HBS solution (2×) <R>
HEPES (2.5 mM, pH 7.3)
OptiMEM+1% penicillin/streptomycin (P/S)
OptiMEM without P/S
pCMVΔR8.91 plasmid
> *This is obtainable from Dr. Didier Trono (Swiss Institutes of Technology [EPFL], Lausanne, Switzerland); other second- and third-generation packaging plasmids can be found at Addgene.com.*

Penicillin/streptomycin (P/S) solution (100×)
Phosphate-buffered saline (PBS)
pUMVC plasmid (Addgene plasmid no. 8449)
> *This plasmid, which encodes Mo-MLV gag-pol genes, can be used to produce Mo-MLV retrovirus vectors by triple transient transfection of 293T cells.*

pVSV-G plasmid (Addgene plasmid no. 8454)

This plasmid is available from Clontech; comparable VSV-G envelope expression plasmids can be found at Addgene.com.

Spray bottle
293T cells
Transfer lentivirus vector plasmid
Trypsin–EDTA (0.05%)

Equipment

Benchtop centrifuge
Biosafety cabinet
Bleach (14%)
Cell culture incubator (37°C)
Conical tubes, sterile (15 mL and 50 mL)
Culture dishes (150 mm)
Hemocytometer
Humidified cell culture incubator (37°C, 5% CO_2)
Ice and ice bucket
Micropipettes with sterile barrier tips (20, 200, and 1000 μL)
PVDF (polyvinylidene fluoride) Durapore filtration device (150 mL, 0.45-μM pore size) and vacuum tube
SW32 or SW28 rotor
Tube holding rack
Tubes, sterile (0.65 mL)
Ultracentrifuge with SW32 or SW28 rotor (Beckman)
Ultra-Clear round-bottomed tubes, disposable (25 × 89 mm; Beckman)
UV light source
Vortex machine

METHOD

Plating 293T Cells for Transfection

1. Grow 293T cells in 20 mL of Complete growth medium in 150-mm dishes at 37°C in a humidified cell culture incubator with a 5% CO_2 atmosphere. Split cells 1:10 every 3 or 4 d.

2. One day before transfection, remove the Complete growth medium, wash the cells with 10 mL of PBS, remove the PBS, and add 5 mL 0.05% trypsin–EDTA. Leave the plate for 5–10 min at room temperature. Rock the plate back and forth until the cells dislodge from the bottom of the plate. Add 5 mL of Complete growth medium to neutralize the trypsin and then pipette up and down vigorously by squirting the trypsin–EDTA solution against the bottom of the plate to generate a single-cell suspension. Transfer the cell suspension to a 50-mL conical tube.

3. Use a hemocytometer to estimate the concentration of cells in the suspension. A confluent 150-mm dish of 293T cells will yield $\sim 60 \times 10^6$ to 80×10^6 cells.

4. Plate 21×10^6 293T cells per 150-mm dish in 20 mL of Complete growth medium, prewarmed to 37°C. Distribute the cells uniformly throughout the dish by gently rocking the dish(es) back and forth and then sideways. Then, place the dishes in a 37°C cell culture incubator overnight. Prepare a total of five dishes.

   When preparing large numbers of dishes, calculate the volume of cell suspension necessary for each 150-mm dish and then prepare a cell suspension master mix with a final cell concentration of 4.2×10^6 cells/mL. Add 15 mL of Complete growth medium to all dishes and then 5 mL of cell suspension master mix to all dishes. Distribute the cells uniformly in the dishes, as above.

Cotransfection of 293T Cells with Lentivirus Vector and Helper Plasmids

5. On day 1 at 2–4 h before transfection, replace the culture medium of all dishes with 20 mL of fresh Complete growth medium, prewarmed to 37°C.

6. Prepare the following two tubes, which will contain enough DNA to transfect five 150-mm dishes. Use sterile barrier tips to reduce the risk of contamination.

   i. Tube 1: DNA mix (in a sterile 15-mL conical tube)

Transfer lentivirus vector plasmid	90 μg
pCMVΔR8.91 plasmid	90 μg
pVSV-G plasmid	60 μg
CaCl$_2$ (2 M)	486 μg
HEPES (2.5 mM)	to a total volume of 3.9 mL

   ii. Tube 2: HBS solution (in a sterile 50-mL conical tube)

HBS (2×)	3.9 mL

7. Spray a vortex machine thoroughly with 70% ethanol, wipe it with paper towels, and place it in a biosafety cell culture cabinet.

8. Vortex tube 2 at high speed, while adding the content of tube 1 dropwise over a period of 1–2 min. Incubate the tube for 20–30 min at room temperature.

   For transfection of 10 or more dishes, tube 2 should be a conical tube larger than 50 mL to avoid spilling transfection solution during the mixing process.

9. Add 1.56 mL of the transfection solution from Step 8 to each 150-mm dish (prepared in Step 5). Distribute the transfection solution uniformly throughout the dishes by gently rocking the dishes back and forth, and then sideways, before returning them to the 37°C cell culture incubator. Incubate overnight.

10. Early in the morning on day 2, remove the medium and wash the cells twice with 15 mL of DMEM, prewarmed to 37°C. Remove as much of the washing medium as possible.

    Use great care during the washes to avoid dislodging the 293T cells from the dishes. Add medium to the dishes by dispensing it against the dish wall and not directly onto the cells. Do not use PBS for washing cells because it will cause cell detachment from the bottom of the dish.

    See Troubleshooting.

11. To the cell dishes, add 15 mL of OptiMEM supplemented with 1% P/S, prewarmed to 37°C. Return the dishes to the 37°C cell culture incubator.

    It is important that the 37°C incubator and shelves be properly leveled to ensure an even distribution of medium in the dishes. Given the relatively small volume, tilted shelves (or incubator) can lead to areas of the dishes drying out with corresponding cell death, resulting in overall lower vector yields.

    See Troubleshooting.

Harvesting Lentivirus Vector Supernatants

12. During the morning of day 4, transfer the culture medium from all dishes to two 50-mL conical tubes. Cap the tubes tightly and spray them with 70% ethanol before removing from the biosafety cabinet.

    See Troubleshooting.

13. Centrifuge the 50-mL conical tubes at 500g for 10 min at 4°C to remove any cells.

14. Transfer the supernatants to a 150-mL, 0.45-μm-pore PVDF filtration device and connect it to a vacuum line. Collect the virus filtrate (∼70 mL) and place it on ice.

15. The titer of the lentivirus vector supernatant at this point should be between 10^6 and 10^8 transducing units (TU)/mL, depending on the vector and transgene. If such titers are sufficient for the desired application, aliquot the lentivirus vector supernatant and store at −80°C; otherwise, continue immediately to the concentration step.

    Repeated cycles of freezing and thawing dramatically decrease the titer of lentivirus vector stocks. Aliquot sizes should be such that each aliquot is used only once.

Concentration of Lentivirus Vector Supernatants by Ultracentrifugation

16. Open the SW32 buckets and place them in a tube rack. Invert the caps to ensure that the inside will be exposed to UV light. Irradiate the buckets and caps with UV light for 10–20 min in the biosafety cabinet. Alternatively, treat the buckets and caps with 70% ethanol for 20–30 min. Remove the ethanol and let the buckets and caps air-dry.

17. Place two 25×89-mm ultracentrifuge tubes in a rack and fill them with 70% ethanol. Incubate the tubes for 20–30 min.

18. Decant the 70% ethanol. Remove as much ethanol as possible using a sterile Pasteur pipette connected to the vacuum line. Wash tubes twice with sterile PBS.

19. Distribute the ~70 mL of filtered lentivirus supernatant from Step 16 equally between the buckets and close them before removing from the biosafety cabinet. Place the buckets for 10 min on ice. In addition, the night before concentrating lentivirus vector supernatants, place the SW32 (or SW28) rotor at 4°C. Otherwise, it will take a considerable amount of time for the ultracentrifuge to cool down the large rotor mass.

20. Centrifuge the tubes at 28,000 rpm in a SW32 or SW28 rotor for 75 min at 4°C.

21. Remove the buckets from the rotor and place them on ice. Spray the buckets with 70% ethanol before transferring them to the biosafety cabinet. Open them in the biosafety cabinet, use clean forceps to remove the tubes from the buckets, and decant the supernatants to a container. Add 1/6 volume of 14% bleach to the container and store for 1 h before discarding. Invert the ultracentrifuge tubes on a paper towel and remove as much medium from the tube walls as possible using a sterile Pasteur pipette connected to the vacuum line. Avoid touching the round bottom of the tube.

22. Using a 200-μL micropipette with a sterile barrier tip, add 50–100 μL of OptiMEM (without P/S) to the center of the round bottom of each tube and let the tubes stand for 5–10 min at room temperature. Alternatively, the lentivirus vector can be resuspended in 50–100 μL of PBS. In this case, pipette gently up and down to resuspend the lentivirus vector virions, avoiding the formation of bubbles. Another option is to store the tubes for 2 h on ice in a biosafety cabinet, followed by gentle pipetting.

    The lentivirus vector pellets are transparent and very difficult to identify at the bottom of the tubes. However, after careful resuspension, the concentrated lentivirus vector stocks will be slightly more turbid than plain OptiMEM.

23. Transfer the lentivirus suspension to a single sterile 0.65-mL tube and mix. Prepare 10-μL aliquots and store at −80°C.

TROUBLESHOOTING

Problem: There is difficulty in growing the pCMVdR8.91 plasmid.
Solution: Grow this plasmid in HB101 *E. coli.*

Problem: The lentivirus-encoded transgene is not expressed in the target cells, or the anticipated biological effect is not observed.
Solutions:

- Determine whether the issue is at the level of production or titration/infection. Use a lentivirus encoding GFP to perform these tests. Over time, monitor GFP expression in transfected 293T cells, which should be >90% GFP positive by day 4. Perform test infections by exposing naïve 293T cells to increasing amounts of supernatant and analyze GFP expression by days 2–3 postinfection using a fluorescence microscope. Use the protocol provided for titration by limiting dilution (see Protocol 17). If transfection efficiency is as indicated above, but there is no evidence of transduction, obtain a test supernatant of a lentivirus vector encoding GFP from a company or academic core facility and perform test infections. If these work, this indicates that there is an issue with one of the plasmids used for cotransfection of 293T cells. Ideally, the researcher should produce a large amount of packaging genome(s) and envelope expression plasmids before initiating regular production of lentivirus vectors so that any changes in packaging efficiency can be identified with one or two variables, such as the structure of the lentivirus vector plasmid and potential toxicity from overexpression of the transgene in 293T cells during packaging.

- Confirm the identity of all plasmids by analyzing their patterns of cleavage by several informative restriction enzymes.

- Test lentivirus production using a commercial packaging mix.

- Test different methods/kits for plasmid production.

Problem: Low titers of virus are produced.

Solution: The transfection efficiency using the calcium phosphate precipitation method is highly dependent on a number of parameters, and the transfection efficiency should be optimized in preliminary experiments using a lentivirus plasmid encoding GFP under a strong promoter functional in 293T cells (e.g., CMV immediate-early promoter). Achieving transfection efficiencies above 70% at day 3 posttransfection is critical for production of lentivirus vector stocks with titers above 1×10^7 TU/mL. The following parameters affect transfection efficiency.

- *pH of the 2× HBS solution.* Before initiating production of lentivirus vectors, this buffer composition can be optimized for the specific conditions by testing the buffer to optimize the pH of 2× HBS solution by testing batches adjusted to pH values ranging from 6.95 to 7.05 at room temperature. It is critically important to use an accurate pH meter properly calibrated with reference solutions. Alternatively, use a commercially available kit.

- *Fetal bovine serum.* The source and lot of fetal bovine serum used during transfection can dramatically affect transfection efficiency. We recommend testing different lots and vendors for their ability to support high transfection efficiencies.

- *Cell density.* If the conditions indicated in this protocol do not yield the desired transfection efficiency, test different cell densities.

- *Plasmid quality.* DNA quality is critical for transfection efficiency. Column-purified plasmids appear to perform comparably to plasmids purified by double CsCl ultracentrifugation. However, if plasmids have been purified by column chromatography, some kits can leave a residue of resin, which can interfere significantly with transfection efficiency. After resuspending alcohol-pelleted plasmids in TE buffer, centrifuge the tube at 15,000g for 5 min at room temperature in a tabletop microcentrifuge to pellet the resin. Remove the supernatant without disturbing the resin.

Problem (Steps 10–12): There is an unusual amount of cell death/detachment during lentivirus vector production.

Solutions:

- Packaging of lentivirus vectors with VSV-G envelope results in some degree of cell fusion and some cell detachment, especially on day 4. Cell detachment takes place mostly as a result of culturing 293T cells in serum-free medium. Therefore, it is important to exercise particular care in pipetting during medium exchanges to avoid detaching cells from the plate.

- Check whether a specific sector of the plates is particularly affected by cell loss. This suggests that the incubator and/or shelves are not level, resulting in uneven distribution of medium. Level the incubator and/or shelves or add additional OptiMEM medium to the plates (20 mL instead of 15 mL) during production.

- Check whether this problem persists during production of a lentivirus vector encoding GFP. In this case, increase the cell density and compare the reagents made in the laboratory with a commercially available kit. If the problem continues, test a different method or kit for producing the plasmids.

- Overexpression of a particular transgene could be toxic to transfected 293T cells. If this is the case, consider using a drug-regulated lentivirus vector.

Problem: Lentivirus vector pseudotyped with VSV-G envelope transduces 293T cells efficiently but does not appear to transduce target cells.
Solutions:

- Test the functionality of the promoter in the lentivirus vector in the target cells by transfection with an identical lentivirus vector plasmid encoding GFP. If expression of GFP is poor, try to find a promoter that is more active in the target cells.

- Consult the literature and identify an envelope that has been used to pseudotype lentivirus vectors for efficient transduction of the target cells. Obtain the corresponding envelope expression plasmid and repeat the experiments. Many envelope expression plasmids are available at Addgene.com. Note that other lentivirus pseudotypes are less stable than VSV-G during ultracentrifugation (i.e., there is considerable loss in functional lentivirus vectors in the process). Alternative methods can be used, such as slower centrifugation for longer times. In addition, production of other lentivirus pseudotypes may require some optimization to achieve high titers. A good starting point is to vary the amount of envelope expression plasmid in the transfection mix. Finally, alternative titration protocols may need to be used because some envelopes may be inefficient for transduction of 293T cells (or other cells commonly used for titration), resulting in underestimation of titer.

Problem: High-titer lentivirus stocks are toxic to a specific type of target cells (e.g., primary cultures of neurons).
Solutions:

- Perform the ultracentrifugation step with a 4 mL of 20% sucrose cushion (20% sucrose, 100 mM NaCl, 20 mM HEPES [pH 7.4], and 1 mM EDTA) underlying the lentivirus vector supernatants.

- Add an additional step of purification by anion-exchange chromatography using Mustang Q disks, as described by Kutner et al. (2009).

DISCUSSION

This is a highly effective and reproducible protocol for production of lentivirus vectors with titers of 10^7–10^8 TU/mL for unconcentrated supernatants and 10^9–10^{10} TU/mL for vectors concentrated by ultracentrifugation. Because the transient transfection protocol is so robust, it can be used with virtually any lentivirus and with many constructs simultaneously. Stocks with a known titer can be produced within 1 wk. One important aspect concerning lentivirus vector stocks is that they should not be subjected to repeated cycles of freezing and thawing because this will cause a marked loss in titer. Lentivirus vector aliquots should be of single-use size and stored at −80°C. No information is currently available on the long-term stability of lentivirus vector stocks stored at −80°C.

Transduction titers determined on 293T cells can be somewhat misleading because these cells are easily transduced with VSV-G pseudotyped lentivirus vectors. A good starting range of multiplicities of infection (MOI; number of TU per target cell) to test in a new cell line is 5, 10, 20, 50, and 100. It is important to reduce the volume of infection to the minimum level possible to increase the chances of transduction. Transduction can be enhanced by the presence of polybrene (4–16 μg/mL) or protamine sulfate (5 μg/mL) in the culture medium. There are other commercially available transduction enhancement agents that can be tested. However, some cell lines or primary cells are adversely affected by these agents. For instance, mouse primary neuronal cultures are highly sensitive to polybrene, and transduction should only be done in its absence. For this particular application, MOI values of 10, 50, and 100 can yield close to 100% transduction of these primary cultures. Despite the broad tropism of the VSV-G envelope, many cell types are quite refractory to transduction with lentivirus vectors pseudotyped with this envelope. Examples include human mesenchymal stem cells and quiescent T cells, which can be efficiently transduced with lentivirus vectors pseudotyped with RD114 (Zhang et al. 2004), and measles virus glycoproteins (Frecha et al. 2008), respectively. For transduction of cells that prove to be difficult to transduce with VSV-G pseudotyped lentivirus vectors, it is worth testing the efficiency of additional pseudotypes produced by incorporation of other envelope glycoproteins.

RECIPES

It is essential that you consult the appropriate Material Safety Data Sheets and your institution's Environmental Health and Safety Office for proper handling of equipment and hazardous materials used in this protocol.

CaCl₂ Solution (2 M)
Dissolve 147.02 g of calcium chloride dihydrate ($CaCl_2 \cdot 2H_2O$) in 500 mL of H_2O. Sterilize by passing through a 0.22-μm filter and store at room temperature.

Complete Growth Medium

Reagent	Quantity (for 1 L)	Final concentration
DMEM	890 mL	
FBS	100 mL	10%
P/S solution (100×)	10 mL	1×

HBS (2×)

Reagent	Quantity (1 L)	Final concentration
NaCl	16.4 g	280 mM
HEPES ($C_8H_{18}N_2O_4S$)	11.9 g	50 mM
$Na_2HPO_4 \cdot 7H_2O$	0.38 g	1.42 mM
H_2O	to 1 L	

Adjust pH to 7.05 with 10 M NaOH. Sterilize by passing through a 0.22-μm filter and store at room temperature.

REFERENCES

Frecha C, Costa C, Negre D, Gauthier E, Russell SJ, Cosset FL, Verhoeyen E. 2008. Stable transduction of quiescent T cells without induction of cycle progression by a novel lentiviral vector pseudotyped with measles virus glycoproteins. *Blood* 112: 4843–4852.

Kutner RH, Zhang XY, Reiser J. 2009. Production, concentration and titration of pseudotyped HIV-1-based lentiviral vectors. *Nature Protocols* 4: 495–505.

Zhang XY, La Russa VF, Reiser J. 2004. Transduction of bone-marrow-derived mesenchymal stem cells by using lentivirus vectors pseudotyped with modified RD114 envelope glycoproteins. *J Virol* 78: 1219–1229.

Titration of Lentivirus Vectors

The titer of a lentivirus vector is often expressed in transducing units per milliliter. This is a functional titer that reflects the lentivirus' ability to transduce a particular cell line under specific conditions. Transduction of other cell lines is likely to be different and will require optimization. The vast majority of lentivirus vectors are produced with the VSV-G envelope protein, which is compatible with most commonly used cell lines, such as 293, 293T, HT-1080, and HeLa cells. In addition to VSV-G, lentivirus vectors have been pseudotyped (carrying an envelope protein other than the native HIV-1 envelope) with a variety of envelope proteins derived from other viruses, which equips the resulting lentivirus pseudotypes with an ability to transduce particular sets of target cells (Cockrell and Kafri 2007). However, titration of some of these vectors by transduction of reference cell lines can result in inaccurately low titers because the cell-surface receptors for the particular envelope protein may be expressed at only low levels or not at all. In this case, the titer can be calculated by measuring the level of p24 using an enzyme-linked immunosorbent assay (ELISA). However, it is important to bear in mind that this is not a functional titer.

293T cells are used for production of lentivirus stocks, and they can be easily transduced with VSV-G pseudotyped lentivirus vectors. Consequently, this cell line is commonly used to determine the functional titer of lentivirus vector stocks produced with this envelope. For lentivirus vectors encoding fluorescent proteins under the control of promoters functional in these cells, titration can be performed using the limiting dilution method or a flow-cytometry-based method. For lentivirus vectors lacking a fluorescent marker, or for those carrying promoters that may not be functional in 293T cells, titer can be determined either by real-time PCR quantification of viral genomes in genomic DNA from transduced cells or a p24 ELISA–based assay.

Biosafety Considerations

National Institutes of Health (NIH) Guidelines for Research Involving Recombinant DNA Molecules do not directly address work with lentivirus vectors. Therefore, the Recombinant DNA Advisory Committee (RAC) has issued guidelines on how to conduct a risk assessment for lentivirus vector research. The major risks associated with lentivirus vector work are the potential for generation of RCLs and oncogenesis. In the laboratory setting, BL2 or enhanced BL2 containment procedures are adequate for most applications not involving transgenes with high oncogenic potential. A risk assessment by the local Institutional Biosafety Committee (IBC) based on the lentivirus vector design and packaging system, nature of transgene insert, vector titer and total amount of vector produced, inherent biological containment of the animal host, and negative RCL testing will determine the appropriate procedures. Review and approval by the local IBC is mandatory before initiating any work with lentivirus vectors.

MATERIALS

It is essential that you consult the appropriate Material Safety Data Sheets and your institution's Environmental Health and Safety Office for proper handling of equipment and hazardous materials used in this protocol.

Recipes for reagents specific to this protocol, marked <R>, are provided at the end of the protocol. See Appendix 1 for recipes for commonly used stock solutions, buffers, and reagents, marked <A>. Dilute stock solutions to the appropriate concentrations.

Reagents

Complete growth medium <R>
DMEM
Fetal bovine serum (FBS)
Lentivirus vector stock
Paraformaldehyde (4%) (optional; see Step 15)
Penicillin/streptomycin (P/S) solution (100×)
Phosphate-buffered saline (PBS)
Polybrene (hexadimethrine bromide) (8 mg/mL) <R>
QuickTiter Lentivirus Titer Kit (lentivirus-associated HIV p24) (Cell Biolabs)
293T cells
Trypsin–EDTA (0.05%)
UltraRapid Lentiviral Titer Kit (System Biosciences) (optional; see Step 23)

Equipment

Culture dishes, 12 well
Flow cytometer
Hemocytometer
Humidified cell culture incubator (37°C, 5% CO_2)
Inverted fluorescence microscope with appropriate filters for fluorescent protein
Micropipettes
Microplate reader capable of reading at 450 nm
Pipette barrier tips, sterile (20, 200, and 1000 μL)
Tubes, sterile (1.5 mL)

Titer Determination by Limiting Dilution

This protocol applies only to vectors expressing fluorescent proteins under promoters functional in 293T cells.

1. The day before titration, plate 3×10^5 293T cells per well in a 12-well dish in 1 mL of Complete growth medium. Ensure uniform distribution of cells in the wells by gently rocking the plates back and forth and then sideways. Prepare one dish for each lentivirus vector stock to titer. Incubate the cells in a cell culture incubator overnight at 37°C in a 5% CO_2 atmosphere.

2. On day 1, replace the Complete growth medium in all wells with 0.5 mL of Complete growth medium supplemented with 8 μg/mL polybrene (1:1000 dilution of 8 mg/mL polybrene stock solution), prewarmed to 37°C.

3. Prepare serial dilutions of lentivirus vector stocks in Complete growth medium. For nonconcentrated stocks, prepare 1:10, $1:10^2$, and $1:10^3$ dilutions. For concentrated stocks, prepare $1:10^2$, $1:10^3$, $1:10^4$, and $1:10^5$ dilutions. Add 5 μL of vector dilution per well in triplicate (three wells per dilution). For nonconcentrated stocks, include a set of wells receiving 5 μL of nondiluted stock. Return the dishes to the 37°C incubator.

4. On day 2, add 0.5 mL of Complete growth medium prewarmed to 37°C. Incubate the dishes for 2 more days.

5. On day 4, carefully remove the Complete growth medium from the wells and replace it with 1 mL of PBS.

6. Using an inverted fluorescence microscope equipped with appropriate filters to visualize the fluorescent protein, count the total number of colonies per field of view using a 10× objective. Choose a vector dilution that gives an average of one to 20 clones per field of view. Count

five random fields per well for that particular dilution (total of 3 wells × 5 fields/well = 15 fields of view).

Calculate the average number of clones/field of view. Calculate the titer using

$$\text{Titer}(\text{TU/mL}) = (N \times C \times D)/0.005,$$

where N is the average number of clones/field of view, C is the well area divided by the field of view area, and D is the fold dilution of vector stock.

To calculate the area of the field of view, determine the field of view diameter ($\text{FOV}_\emptyset$) by dividing the ocular lens field number (this is the number written on the side of the ocular piece; e.g., 22 corresponds to 22 mm) by the objective magnification (e.g., 10) and use the formula area $= \pi \times (\text{FOV}_\emptyset/2)^2$.

Titer Determination by Flow Cytometry

7. Prepare 293T cells for titration as in Step 1. Prepare the necessary number of wells in 12-well dishes, plus three additional wells.

8. On day 1, count the total number of cells in three wells. Remove the Complete growth medium, wash the cells with 1 mL of PBS, add 0.5 mL of 0.05% trypsin–EDTA, and incubate for 5 min at room temperature. Add 0.5 mL of Complete growth medium and mix. Transfer the cell suspension to a 1.5-mL tube and determine the concentration of cells using a hemocytometer. Calculate the average total number of cells per well.

9. Proceed with the titration procedure as in Steps 2–4.

10. On day 4, remove the Complete growth medium and wash the cells with 1 mL of PBS. Add 0.5 mL of 0.05% trypsin–EDTA per well and incubate for 5 min at room temperature.

11. Add 0.5 mL of Complete growth medium to each well, mix, and transfer the cell suspension to 1.5-mL tubes.

12. Pellet the cells by centrifugation at 500g for 5 min at room temperature. Aspirate the Complete growth medium and resuspend the cell pellet in 1 mL of PBS.

13. Pellet the cells again by centrifugation at 500g for 5 min at room temperature. Remove the supernatant and resuspend the cells in 1–2 mL of PBS.

14. Determine the percentage of fluorescence-positive cells using a flow cytometer.

    If the analysis is not performed within 1 h, fix the cells in 4% paraformaldehyde in PBS for 30 min, wash with PBS, and store at 4°C in PBS.

15. Calculate the titer (TU/mL) using

$$\text{Titer}(\text{TU/mL}) = (N \times F \times D)/0.005,$$

where N is the average number of cells per well on day 1, F is the percentage of fluorescent cells, and D is the fold dilution of vector stock.

The accuracy of the titer depends on the amount of vector used falling within the linear range between vector input and percentage of fluorescent cells. If this percentage is >40%, repeat the assay with additional dilutions.

Titer Determination by PCR Quantification of Vector Genome Genomic DNA of Transduced Cells

16. Perform a titration assay as in Steps 1–4.

    Some protocols include a treatment with DNase I on day 2 to eliminate any plasmid DNA that may be carried over in the lentivirus vector supernatants (Kutner et al. 2009).

17. On day 4, remove the Complete growth medium and wash the cells with 1 mL of PBS. Add 0.5 mL of 0.05% trypsin–EDTA per well and incubate for 5 min at room temperature.

18. Add 0.5 mL of Complete growth medium to each well, mix, and transfer the cell suspension to 1.5-mL tubes.

19. Pellet the cells by centrifugation at 500*g* for 5 min at room temperature. Aspirate the Complete growth medium and resuspend the cell pellet in 1 mL of PBS.

20. Pellet the cells again by centrifugation at 500*g* for 5 min at room temperature.

21. Isolate the genomic DNA from transduced cells using a commercially available kit (e.g., DNeasy Tissue Kit, QIAGEN) and measure the concentration of genomic DNA.

22. Determine the lentivirus vector copy number per target cell diploid genome using real-time PCR quantification as described in Kutner et al. (2009). Alternatively, use the UltraRapid Lentiviral Titer Kit (System Biosciences, Mountain View, CA), also based on real-time PCR. Follow the manufacturer's instructions and bypass Steps 18–24.

23. Calculate the lentivirus vector titer using

$$\mathrm{Titer(TU/mL)} = (N \times C \times D)/0.005,$$

where N is the average number of cells per well on day 1, C is the lentivirus vector copy number per diploid genome, and D is the fold dilution of vector stock.

Titration by p24 ELISA

An alternative to transduction-based assays is to determine the concentration of HIV-1 p24 levels in lentivirus vector stocks using an ELISA. The limitation with this approach is the contamination of lentivirus vector supernatants with free p24 generated by transfected cells during production. Most ELISA Kits quantify the total amount of p24 antigen in the supernatant. The QuickTiter Lentivirus Titer ELISA Kit (Cell Biolabs) quantifies p24 levels associated with the lentivirus virions using a proprietary reagent to separate lentivirus virions from free p24.

24. For nonconcentrated lentivirus vector stocks, prepare 1:10 and $1:10^2$ dilutions of vector stock in OptiMEM and measure in triplicate for each dilution. For concentrated lentivirus vector stocks, prepare $1:10^2$, $1:10^3$, and $1:10^4$ dilutions of vector stock in OptiMEM and measure in triplicate.

25. Use the QuickTiter Lentivirus Titer ELISA Kit to determine the concentration of virion-associated p24 levels in lentivirus vector stocks according to the manufacturer's instructions.

    According to the manufacturer, there are 2000 molecules of p24 per lentivirus particle (LP), and 1 ng of p24 corresponds to 1.25×10^7 LPs.

DISCUSSION

The most accurate, but also the most time-consuming, method to measure the titer of the lentivirus preparation is real-time PCR quantification of vector genomes in target cells. This method is independent of promoter functionality in the cells used to determine the titer. However, this approach can work only in cells expressing the necessary cell-surface receptors for alternative envelope glycoproteins used to pseudotype lentivirus vectors. In these cases, two alternative approaches are available. One is to determine titers based on lentivirus particle-associated p24 levels, and the other is to generate a cell line engineered to overexpress the appropriate cell-surface receptor. It is important to consider that lentiviral titer is a functional measure (transducing units per milliliter) determined using a particular cell line under very specific conditions. Changes in target cell type and transduction conditions have a dramatic effect on transduction efficiency. Consequently, a lentivirus vector titer should be used as a guideline to set up empirical transduction experiments for new cell lines, primary cells, and in vivo gene delivery.

RECIPES

It is essential that you consult the appropriate Material Safety Data Sheets and your institution's Environmental Health and Safety Office for proper handling of equipment and hazardous materials used in this protocol.

Complete Growth Medium

Reagent	Quantity (for 1 L)	Final concentration
DMEM	890 mL	
FBS	100 mL	10%
P/S solution (100×)	10 mL	1×

Polybrene (Hexadimethrine Bromide) (8 mg/mL)
Dissolve 100 mg of polybrene (hexadimethrine bromide) in 12.5 mL of H_2O. Sterilize the solution by passing through a 0.22-μM filter and store at 4°C.

REFERENCES

Cockrell AS, Kafri T. 2007. Gene delivery by lentivirus vectors. *Mol Biotechnol* **36:** 184–204.

Kutner RH, Zhang XY, Reiser J. 2009. Production, concentration and titration of pseudotyped HIV-1-based lentiviral vectors. *Nature Protocols* **4:** 495–505.

Monitoring Lentivirus Vector Stocks for Replication-Competent Viruses

The potential emergence of RCLs during vector production and the significant biosafety risk that this represents has led to the development of lentivirus vector production systems to minimize the risk of generating RCLs. Second- and third-generation lentivirus vector production systems appear to be safe, because there are no reports of RCL generation in either system. Screening of lentivirus vector stocks for RCLs involves serial passaging of transduced cells for 30 d with weekly monitoring of p24 levels in supernatants using an ELISA Kit.

Biosafety Considerations

National Institutes of Health (NIH) Guidelines for Research Involving Recombinant DNA Molecules do not directly address work with lentivirus vectors. Therefore, the Recombinant DNA Advisory Committee (RAC) has issued guidelines on how to conduct a risk assessment for lentivirus vector research. The major risks associated with lentivirus vector work are the potential for generation of RCL and oncogenesis. In the laboratory setting, BL2 or enhanced BL2 containment procedures are adequate for most applications not involving transgenes with high oncogenic potential. A risk assessment by the local Institutional Biosafety Committee (IBC) based on the lentivirus vector design and packaging system, nature of transgene insert, vector titer and total amount of vector produced, inherent biological containment of the animal host, and negative RCL testing will determine the appropriate procedures. Review and approval by the local IBC is mandatory before initiating any work with lentivirus vectors.

MATERIALS

It is essential that you consult the appropriate Material Safety Data Sheets and your institution's Environmental Health and Safety Office for proper handling of equipment and hazardous materials used in this protocol.

Recipes for reagents specific to this protocol, marked <R>, are provided at the end of the protocol. See Appendix 1 for recipes for commonly used stock solutions, buffers, and reagents, marked <A>. Dilute stock solutions to the appropriate concentrations.

Reagents

Complete growth medium <R>
DMEM
Fetal bovine serum (FBS)
Lentivirus vector stock
OptiMEM-I
Penicillin/streptomycin (P/S) solution (100×)
Phosphate-buffered saline (PBS)
Polybrene (hexadimethrine bromide) (8 mg/mL) <R>
QuickTiter Lentivirus Titer Kit (lentivirus-associated HIV p24) (Cell Biolabs)
293T cells
Trypsin–EDTA (0.05%)

Equipment

Culture dishes (12 well)
Humidified cell culture incubator (37°C, 5% CO_2)
Hemocytometer
Micropipettes
Microplate reader capable of reading at 450 nm
Pipette barrier tips, sterile (20, 200, and 1000 μL)
PVDF membrane syringe filters (0.45 μm)
T-25 culture flasks
Tubes, sterile (1.5 mL)

METHOD

1. One day before transduction, plate 3×10^5 293T cells per well in a 12-well dish in 1 mL of Complete growth medium. Incubate in a cell culture incubator overnight set to 37°C and 5% CO_2 atmosphere.

2. On day 1, mix 250 μL of unconcentrated vector stock with 250 μL of Complete growth medium supplemented with 16 μg/mL of polybrene (1:500 dilution of 8 mg/mL polybrene stock). For concentrated lentivirus vector stocks, add 5 μL of vector stock to 0.5 mL of Complete growth medium supplemented with 8 μg/mL of polybrene. Prepare in triplicate. Prepare mock-transduction control tubes containing equal amounts of OptiMEM-I instead of lentivirus vector stock.

3. Remove the Complete growth medium from cells in one well and replace it with 0.5 mL of vector dilution with a final concentration of 8 μg/mL of polybrene. Perform in triplicate for the vector stock/s and mock-transduction control. Incubate the mixtures in a cell culture incubator overnight set to 37°C and 5% CO_2 atmosphere.

4. The following day, transfer the supernatants from the transduced wells (vector stock and mock transduction) to sterile 1.5-mL tubes and replenish the medium in the culture dish wells with 1 mL of Complete growth medium prewarmed to 37°C.

5. Centrifuge the day-1 supernatant at $500g$ for 10 min at room temperature. Transfer the supernatant to a new tube and label it "day 1" and include on the label whether it is vector or mock transduction. Store at −80°C.

6. On day 4, remove the supernatants from lentivirus- and mock-transduced wells and wash the cells with 1 mL of PBS. Remove the PBS, add 0.5 mL of 0.05% tryspin−EDTA, and incubate for 5 min at room temperature. Add 0.5 mL of Complete growth medium and resuspend the cells by repeated pipetting. Transfer all cells from each well into an individual T-25 culture flask (three flasks per vector stock and three per mock transduction) with 5 mL of Complete growth medium. Place the flasks in the cell culture incubator.

7. On day 7, remove the supernatants and filter them through a 0.45-μm syringe filter into new 1.5-mL tubes labeled "day 7" and vector or mock transduction. Store the tubes at −80°C. Split the cells 1:10 twice per week.

8. Repeat Step 7 on days 15, 21, and 30.

9. Assay all of the supernatants for p24 levels using the QuickTiter Lentivirus Titer ELISA Kit. As controls, include vector stock and supernatant from naïve 293T cells. The day-1 supernatant should contain p24 derived from the lentivirus vector stock, but it should gradually disappear and become undetectable in later assay points comparable to mock-transduction supernatants.

RECIPES

It is essential that you consult the appropriate Material Safety Data Sheets and your institution's Environmental Health and Safety Office for proper handling of equipment and hazardous materials used in this protocol.

Complete Growth Medium

Reagent	Quantity (for 1 L)	Final concentration
DMEM	890 mL	
FBS	100 mL	10%
P/S solution (100×)	10 mL	1×

Polybrene (Hexadimethrine Bromide) (8 mg/mL)
Dissolve 100 mg of polybrene (hexadimethrine bromide) in 12.5 mL of H_2O. Sterilize the solution by passing through a 0.22-μM filter and store at 4°C.

ADENOVIRUS VECTORS

Adenoviruses belong to the Adenoviridae family of viruses, which consists of two genera: *Aviadenovirus* (bird adenoviruses) and *Mastadenovirus* (human, simian, bovine, equine, porcine, ovine, canine, and opossum adenoviruses). All of these viruses carry a linear dsDNA genome inside an icosahedral protein capsid, 70–100 mm in diameter. The ~36-kb genome is organized into four early transcription units (E1, E2, E3, and E4) to provide regulatory and replication functions, two delayed early units (IX and IVa2), and one major late transcription unit, which directs synthesis of capsid proteins, among others (Wold and Horwitz 2007). Clinical sequelae resulting from adenovirus infection in humans are primarily acute febrile respiratory syndromes (Wold and Horwitz 2007).

Adenovirus vectors based on human adenovirus serotypes 2 and 5 are the most widely used in the laboratory because of their broad cell-type and tissue tropism and excellent gene-transfer efficiency in culture and in vivo. In fact, these are the most efficient gene-transfer vehicles among dsDNA viral vectors (Amalfitano 2004). Most adenovirus vectors are rendered replication defective by deletion of the E1 region, to accommodate the transgene expression cassette. These vectors are routinely packaged in 293 cells, which carry a stably integrated E1 region of adenovirus serotype 5 that trans-complements the E1 functions necessary for replication and packaging of vector genomes. Adenovirus vector genomes reside in most target cells as nonreplicating episomes (Wold and Horwitz 2007).

The major drawback of using adenovirus vectors for somatic gene transfer in vivo is their ability to induce strong cellular immune responses directed against viral proteins and transgene products (Jooss and Chirmule 2003). Administration of an adenovirus vector elicits robust immune responses in the host on two levels: innate immunity directed to the viral capsid and adaptive T-cell immunity toward de novo expression of the remaining viral genes and transgene in the vector genome. Stimulation of innate immunity by viral capsid protein results in transient elevation of inflammatory cytokines such as IL-6 and IL-10, whereas adaptive T-cell immunity leads to clearance of transduced cells by cytotoxic T lymphocytes specific for viral and transgene proteins (Jooss and Chirmule 2003).

During the past decade, significant progress has been made in the development of new generations of adenovirus vectors with diminished T-cell immunogenicity by deletion of additional early genes in the vector genome. The most advanced generation of adenovirus vector is the gutless vector, in which all viral genes are void. The gutless adenovirus vector not only eliminates the cellular immune responses against de novo–synthesized viral proteins but also significantly expands its capacity for the transgene cassette. Because production of a gutless vector requires helper functions from co-infected E1-deleted adenovirus, it is critical to remove the helper virus efficiently from the gutless vector preparations (Altaras et al. 2005). It is worthwhile to note that there has been little progress in overcoming the innate immune response elicited by the viral capsid after adenovirus vector–mediated in vivo gene transfer.

REFERENCES

Altaras NE, Aunins JG, Evans RK, Kamen A, Konz JO, Wolf JJ. 2005. Production and formulation of adenovirus vectors. *Adv Biochem Eng Biotechnol* **99:** 193–260.

Amalfitano A. 2004. Utilization of adenovirus vectors for multiple gene transfer applications. *Methods* **33:** 173–178.

Jooss K, Chirmule N. 2003. Immunity to adenovirus and adeno-associated viral vectors: Implications for gene therapy. *Gene Ther* **10:** 955–963.

Wold WSM, Horwitz MS. 2007. Adenoviruses. In *Fields virology* (ed Knipe D, et al.), pp. 2395–2436. Wolters Kluwer Health/Lippincott Williams & Wilkins, Philadelphia.

AAV VECTORS

Adeno-associated virus, the first and only ssDNA virus that has been engineered for the purpose of gene delivery, is a nonenveloped, icosahedral particle, 20–26 nm in diameter, encapsidating a linear ssDNA genome of 4.7 kb. Adeno-associated virus is one of the smallest mammalian viruses known, named after its initial discovery as a viral contaminant in adenovirus preparations. It is naturally replication defective in the absence of co-infection with helper viruses such as adenovirus and has not been associated with any disease in humans (Daya and Berns 2008).

Adeno-associated virus was first genetically engineered as a gene-delivery vector in the late 1980s. Since then, its vectors have become the most promising gene-delivery vehicle for effective and safe clinical application in gene therapy for chronic diseases (Daya and Berns 2008). Extensive studies have shown that AAV vectors transduce at exceptionally high efficiency dividing and nondividing/quiescent cells in a number of somatic tissues, including muscle, liver, heart, retina, and CNS. Importantly, adeno-associated virus–mediated gene delivery in vivo appears to be devoid of histopathological alterations or vector-related toxicity (Daya and Berns 2008).

Unlike adenoviral vectors, in vivo administration of AAV vectors in small and large animal models usually does not elicit host immune responses against transduced cells, which leads to long-term in vivo expression of the transgenes (therapeutic genes) (Jooss and Chirmule 2003). AAV vectors mediate stable gene expression by forming circular monomers and concatamers for episomal persistence in host cells (McCarty et al. 2004). A major drawback of adeno-associated virus is that it can accommodate vector genomes only up to ~5 kb (Daya and Berns 2008).

Recent investigations have revealed a diverse family of more than 120 novel primate adeno-associated viruses with unique tissue and cell-type tropisms and efficient gene-transfer capability (Gao et al. 2002, 2003, 2004, 2005). This critical advance in AAV technology has dramatically broadened the potential applications of AAV vectors. Another critical development in vector genome design is the incorporation of one wild-type ITR element and one modified ITR element to allow packaging of double-stranded or self-complementary genomes (McCarty 2008). These self-complementary AAV (scAAV) vectors are capable of initiating transgene expression immediately after capsid uncoating in the nucleus of targeted cells. This elegant vector genome design bypasses the rate-limiting step of AAV vector genome processing in the transduction process—the second-strand synthesis (SSS). This design enhances dramatically the transduction efficiency of AAV vectors both in culture and in vivo. However, the scAAV vector genome design reduces the transgene capacity to 2.5 kb, restricting its potential applications. Recently, Zhong and colleagues documented that mutations of critical surface-exposed tyrosine residues on AAV2 capsids circumvents the ubiquitination step, thereby avoiding proteasome-mediated degradation. This simple modification results in high-efficiency transduction by these vectors in human cells in vitro and murine hepatocytes in vivo (Zhong et al. 2008).

REFERENCES

Daya S, Berns KI. 2008. Gene therapy using adeno-associated virus vectors. *Clin Microbiol Rev* **21:** 583–593.

Gao GP, Alvira MR, Wang L, Calcedo R, Johnston J, Wilson JM. 2002. Novel adeno-associated viruses from rhesus monkeys as vectors for human gene therapy. *Proc Natl Acad Sci* **99:** 11854–11859.

Gao G, Zhou X, Alvira MR, Tran P, Marsh J, Lynd K, Xiao W, Wilson JM. 2003. High throughput creation of recombinant adenovirus vectors=by direct cloning, green-white selection and I-Sce I-mediated rescue of circular adenovirus plasmids in 293 cells. *Gene Ther* **10:** 1926–1930.

Gao G, Vandenberghe LH, Alvira MR, Lu Y, Calcedo R, Zhou X, Wilson JM. 2004. Clades of adeno-associated viruses are widely disseminated in human tissues. *J Virol* **78:** 6381–6388.

Gao G, Vandenberghe LH, Wilson JM. 2005. New recombinant serotypes of AAV vectors. *Curr Gene Ther* **5:** 285–297.

Jooss K, Chirmule N. 2003. Immunity to adenovirus and adeno-associated viral vectors: Implications for gene therapy. *Gene Ther* **10:** 955–963.

McCarty DM. 2008. Self-complementary AAV vectors: Advances and applications. *Mol Ther* **16:** 1648–1656.

McCarty DM, Young SM Jr, Samulski RJ. 2004. Integration of adeno-associated virus (AAV) and recombinant AAV vectors. *Ann Rev Genet* **38:** 819–845.

Zhong L, Li B, Mah CS, Govindasamy L, Agbandje-McKenna M, Cooper M, Herzog RW, Zolotukhin I, Warrington KH Jr, Weigel-Van Aken KA, et al. 2008. Next generation of adeno-associated virus 2 vectors: Point mutations in tyrosines lead to high-efficiency transduction at lower doses. *Proc Natl Acad Sci* **105:** 7827–7832.

LENTIVIRUS VECTORS

Lentiviruses belong to the Retroviridae family, and as all viruses in this family, the virions carry an ssRNA molecule that must undergo reverse transcription and integration into the host cell genome (provirus) for a productive infection. The distinguishing feature of lentiviruses in this family is the complexity of their genome and the cylindrical or conical shape of the nucleocapsid carrying the virus genome. The nucleocapsid is surrounded by a membrane bilayer carrying the envelope glycoprotein responsible for interaction with cell-surface receptors and entry into host cells. Retroviruses are major pathogens in almost all vertebrates. The human immunodeficiency virus type 1 (HIV-1) is responsible for acquired immunodeficiency disease syndrome (AIDS) in humans.

Lentivirus vectors based on HIV-1 were the first to be engineered as gene-delivery vehicles and shown to be highly effective for transduction of dividing and nondividing cells (Cockrell and Kafri 2007). For obvious reasons, HIV-1–derived lentivirus vector packaging systems (Fig. 1) have evolved over the years to minimize the number of HIV-1 genes present in the helper plasmids and thus decrease the probability of generating replication-competent lentiviruses (RCLs) during production. The first-generation packaging system was based on a single expression plasmid carrying an HIV-1 genome where the packaging signal (Ψ) and *env* gene were deleted, and the LTR elements were replaced by the CMV immediate-early promoter and a polyadenylation signal (Cockrell and Kafri 2007). Second-generation systems carry four HIV-1 genes, *gag* and *pol* (structural) and *tat* and *rev* (regulatory), in the complementing plasmid, whereas third-generation systems carry *gag*, *pol*, and *rev* split between two plasmids. Third-generation systems can only be used for production of lentivirus vectors carrying chimeric 5′-LTR, where the HIV-1 promoter is replaced by a CMV or RSV promoter, which renders production of vector genomic RNA necessary for packaging, independent of Tat (absent in third-generation systems). Most lentivirus vectors currently in use carry chimeric 5′-LTR elements (Cockrell and Kafri 2007).

The fact that retrovirus and lentivirus vectors integrate into the host cell genome is an advantage in achieving permanent genetic modification of target cells; however, it is also a source of safety concerns regarding their oncogenic potential through insertional mutagenesis. This issue became apparent in

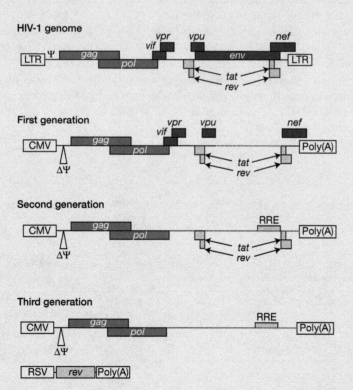

FIGURE 1. **HIV-1–based lentivirus vector packaging systems.** (Top two structures reproduced from Dull et al. 1998. Bottom two structures reproduced from Zufferey et al. 1997.)

clinical trials for severe combined immunodeficiency (SCID), in which a number of patients receiving retrovirus vector–modified hematopoietic stem/progenitor cells developed leukemia. Apparently, MLV-based retrovirus vectors tend to integrate into promoter regions, leading to transactivation of oncogenes. Interestingly, the integration pattern of lentivirus vectors does not show this bias. Recent studies have shown that oncogenesis and abnormal transcription activity resulting from ex vivo modification of hematopoietic stem/progenitor cells with retrovirus or lentivirus vectors is related to the presence of transcriptionally active LTR elements in the vector provirus (i.e., vector integrated in the host genome). Lentivirus vectors carrying transcriptionally active LTR elements are nonetheless 10-fold less likely to induce tumors than comparable MLV-based retrovirus vectors. Self-inactivating retrovirus or lentivirus vectors, where the LTR elements are inactivated after integration into the host genome, have a markedly reduced/absent oncogenic potential (Montini et al. 2009).

REFERENCES

Cockrell AS, Kafri T. 2007. Gene delivery by lentivirus vectors. *Mol Biotechnol* 36: 184–204.

Dull T, Zufferey R, Kelly M, Mandel RJ, Nguyen M, Trono D, Naldini L. 1998. A third-generation lentivirus vector with a conditional packaging system. *J Virol* 72: 8463–8471.

Montini E, Cesana D, Schmidt M, Sanvito F, Bartholomae CC, Ranzani M, Benedicenti F, Sergi LS, Ambrosi A, Ponzoni M, et al. 2009. The genotoxic potential of retroviral vectors is strongly modulated by vector design and integration site selection in a mouse model of HSC gene therapy. *J Clin Invest* 119: 964–975.

Zufferey R, Nagy D, Mandel RJ, Naldini L, Trono D. 1997. Multiply attenuated lentiviral vector achieves efficient gene delivery in vivo. *Nat Biotechnol* 15: 871–875.

BASIC ELEMENTS IN VIRAL VECTORS

Retrovirus Vectors

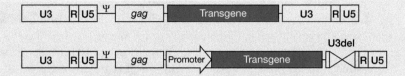

The most commonly used retrovirus vectors are based on the Mo-MLV. The first generation of vectors carried the transgene flanked by two intact LTRs composed of the enhancer (U3), R (defined as the start of transcription and present at both ends of the vector genome in the virion), and U5 region. Packaging of vector genome in virions is mediated by the packaging signal sequence (Ψ), but the adjacent *gag* sequence enhances its efficiency quite considerably. All retrovirus vectors carry an extended packaging signal that includes the Ψ sequence and part of the *gag* sequence, which has been modified to eliminate the start codon. In the first generation of these vectors, the LTR promoter was used to drive transgene expression (top vector in figure above). One of the issues with these vectors was promoter shutdown over time in vivo, and this spurred development of a new generation of SIN retrovirus vectors carrying a 3′-LTR with a nearly complete deletion of the U3 enhancer sequences and an internal mammalian promoter to drive transgene expression. During reverse transcription, the deleted U3 region at the 3′ end of the vector genome is duplicated to the 5′-LTR and thus generates an inactive LTR promoter.

Lentivirus Vectors

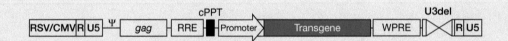

The vast majority of lentivirus vectors currently used in the laboratory and for gene therapy applications in humans are derived from HIV-1, and as such, their design features reflect safety considerations. These vectors carry an extended packaging signal (Ψ+*gag*) as in retrovirus vectors, are self-inactivating (U3-deleted 3′-LTR), and carry a chimeric 5′-LTR with enhancer sequences derived from either RSV or CMV immediate-early gene promoters. This type of design was also previously implemented in retrovirus vectors because it appeared to increase titers considerably. In addition, HIV-1 lentivirus vectors require the presence of the RRE, which is necessary for efficient transport of the vector genome to the cytoplasm and for efficient packaging. Similar to retrovirus vectors, lentivirus vectors also carry a polypurine tract (PPT) near the 3′-LTR, a site necessary to initiate plus-strand DNA synthesis during reverse transcription. One of the puzzling findings with the first generation of lentivirus vectors carrying a single PPT was the observation that transduction of postmitotic neurons in the adult brain was quite effective but not in adult nondividing hepatocytes (Park et al. 2000). Soon thereafter, it became apparent that a second PPT in a central location in the vector genome was key for efficient nuclear translocation of the preintegration complex in postmitotic cells (Follenzi et al. 2000; Zennou et al. 2000). Incorporation of the central PPT (cPPT), or central DNA flap, resulted in lentivirus vectors capable of transducing dividing and nondividing cells at high efficiency. Another element that is now part of most lentivirus vectors is the hepatitis virus woodchuck posttranscriptional regulatory element (WPRE), which has been shown to increase transgene expression fivefold to eightfold when placed downstream from the transgene in the sense orientation (Zufferey et al. 1999).

AAV Vectors

The majority of AAV vectors carry a transgene expression cassette flanked by AAV2 ITRs. No other genetic elements from the wild-type virus are present in the vectors. ITR elements from other AAV serotypes have been used as well (Desmaris et al. 2004; Hewitt et al. 2009). The major limitation to transduction by ssAAV vectors is their conversion to transcriptionally active dsDNA genomes in transduced cells. Self-complementary or dsAAV vectors carry an ITR with a deletion of the terminal resolution site that results in the packaging of double-stranded genomes that rapidly mediate transgene expression after transduction. The total packaging capacity of dsAAV vectors is ~2.4 kb, compared to 4.7–4.8 kb for traditional ssAAV vectors. The strength of these vectors remains their remarkable efficacy and stability of gene expression in vivo, which is not reproduced in cultured cells where other vectors are considerably more effective.

Adenovirus Vectors

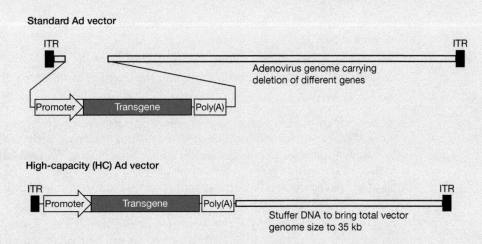

Most adenovirus recombinants are based on Ad5 and carry a transgene expression cassette inserted in place of the E1 early transcriptional unit in the context of a mostly wild-type genome with one or more deletions of the other early transcriptional units (see below). High-capacity adenovirus (HC-Ad) vectors carry only the adenovirus ITRs, a transgene expression cassette, and stuffer DNA to bring the total size to ~35 kb. These HC-Ad vectors do not carry any additional genetic elements from the wild-type adenovirus genome, and they do not have the same immunological complications associated with Ad-mediated in vivo gene delivery resulting from an adaptive response. Innate immunity to the capsid is likely to remain an issue.

REFERENCES

Desmaris N, Verot L, Puech JP, Caillaud C, Vanier MT, Heard JM. 2004. Prevention of neuropathology in the mouse model of Hurler syndrome. *Ann Neurol* **56:** 68–76.

Follenzi A, Ailles LE, Bakovic S, Geuna M, Naldini L. 2000. Gene transfer by lentiviral vectors is limited by nuclear translocation and rescued by HIV-1 pol sequences. *Nat Genet* **25:** 217–222.

Hewitt FC, Li C, Gray SJ, Cockrell S, Washburn M, Samulski RJ. 2009. Reducing the risk of adeno-associated virus (AAV) vector mobilization with AAV type 5 vectors. *J Virol* **83:** 3919–3929.

Park F, Ohashi K, Chiu W, Naldini L, Kay MA. 2000. Efficient lentiviral transduction of liver requires cell cycling in vivo. *Nat Genet* **24:** 49–52.

Zennou V, Petit C, Guetard D, Nerhbass U, Montagnier L, Charneau P. 2000. HIV-1 genome nuclear import is mediated by a central DNA flap. *Cell* **101:** 173–185.

Zufferey R, Donello JE, Trono D, Hope TJ. 1999. Woodchuck hepatitis virus posttranscriptional regulatory element enhances expression of transgenes delivered by retroviral vectors. *J Virol* **73:** 2886–2892.

ASSAYS DONE IN TRANSDUCED CELLS

1. Preparation of vector stocks.

2. Transduction of target cells at different MOIs or vector dose per target cell:

   Adenovirus vectors: 100–10,000 virus particles/target cell

   Adeno-associated virus vectors: 10,000–100,000 virus genomes/target cell

   Lentivirus vectors: MOI 5-500

3. Assess transgene expression at 3–5 d posttransduction.

   For *vector-encoded proteins*, a number of different approaches can be used to assess transgene expression in transduced cells and this depends on the type of transgene, subcellular localization, availability of suitable antibodies, or bioassays.

- If the vector encodes a fluorescent marker protein, assess transduction using a fluorescence microscope equipped with appropriate filters for visualization of the specific marker. Proceed to the next step.

- Extract total cellular proteins and analyze transgene expression by western blot or ELISA quantification.

- Analyze transgene expression using fluorescence activated cell sorting (FACS). This method will quantify both the percentage of positive cells and the intensity of transgene expression.

- Analyze expression and subcellular localization using immunofluorescence or immunocytochemistry.

   i. To distinguish from endogenous protein, or for immunoprecipitation experiments in the absence of effective primary antibodies against the protein of interest, include a protein tag (Table 1) at the amino or carboxyl terminus of the vector-encoded protein. Alternatively, a small tetracysteine peptide can be used to perform live cell imaging of protein expression and localization in the presence of biarsenical compounds that become fluorescent after specific binding to proteins carrying this tag (Griffin et al. 1998).

   ii. The potential effect of these tags on protein biochemistry and function should be carefully considered and analyzed. As an alternative, vectors can be used to transduce cells that do not express the particular protein, and thus, experiments are performed in a null background.

TABLE 1. Amino acid sequence of epitope tags

Tag name	Amino acid sequence
HA	YPYDVPDYA
c-MYC	EQKLISEEDL
His$_6$	HHHHHH
FLAG	DYKDDDDK
AU1	DTYRYI
EE (Glu-Glu)	EYMPME
IRS	RYIRS
Tetracysteine	CCPGCC[a]
GFP	—

Inclusion of a few Gly residues between the protein of interest and epitope tags may be important.

HA, hemagglutinin; IRS, insulin receptor substrate; GFP, green fluorescent protein.

[a]This tag allows for specific labeling of recombinant proteins in live cells incubated with biarsenical compounds that fluoresce on binding. Note that in vectors commercialized by Life Technologies insertion of this tag at the carboxyl terminus of a protein is done in the following context: GAGGCCPGCCGGG.

iii. Measure the biological effect of the vector-encoded protein in cultured cells or in vivo (e.g., use transgene-specific enzymatic assays, monitor changes in current regulation in cells transduced with ion channels such as the cystic fibrosis transmembrane conductance regulator).

For *vector-encoded shRNA/miRNA,* the following approaches can be used.

- Extract total cellular proteins and assess effect on target protein expression by western blot, ELISA, FACS or immunostaining of transduced and control cells.

- In the absence of antibodies, extract total RNA and measure target gene mRNA level in naïve and vector-transduced cells by real-time PCR using a housekeeping gene (GAPDH, β-actin, or 18S rRNA) as reference.

- Biological effect of target down-regulation.

REFERENCE

Griffin BA, Adams SR, Tsien RY. 1998. Specific covalent labeling of recombinant protein molecules inside live cells. *Science* **281:** 269–272.

TRANSGENE EXPRESSION CASSETTES

The most commonly used promoters in viral vectors are ubiquitous promoters, cell type–specific promoters, and polyadenylation signals.

Ubiquitous Promoters

- human CMV immediate-early promoter
- murine stem cell virus (MSCV) LTR promoter
- CAG hybrid promoter carrying the CMV enhancer fused to the chicken β-actin promoter (also known as CGA, CBA, or CB; length varies among different versions)
- human EF-1α
- human phosphoglycerate kinase 1 (PGK1) promoter
- ubiquitin promoter

Cell Type–Specific Promoters

- α1 antitrypsin promoter for liver-specific expression
- muscle creatine kinase (mCK) promoter for skeletal muscle-specific expression
- human synapsin-1 (SYN-1) or rat neuronal-specific enolase (NSE) promoter for neuronal-specific expression
- human Clara cell 10-kDa protein (CC10) promoter for lung-specific expression
- human interphotoreceptor retinoid-binding protein (IRBP) promoter for retinal-specific expression

Polyadenylation Signals Most Commonly Used in Viral Vectors

- bovine growth hormone (BGH)
- rabbit β-globin (RBG)
- SV40
- native signals present in LTR elements in retrovirus and lentivirus vectors.

Introns are commonly present in expression cassettes to enhance transgene expression. Artificial and native introns have been used for this purpose (e.g., RBG intron).

Insertion of posttranscriptional regulatory elements has been shown to increase transgene expression. The most commonly used element is derived from the WPRE (Donello et al. 1998).

Incorporation of miRNA target sequences in the 3′ UTR of viral vectors leads to transcriptional detargeting from cells where the particular miRNA is expressed at high levels. This approach has been successfully used to prevent transgene expression in antigen-presenting cells in vivo and thus blunt transgene-specific adaptive immune response.

Insertion of a loxP-stop-loxP cassette allows for cell-type-specific transgene expression in transgenic mice expressing Cre recombinase in specific cell types.

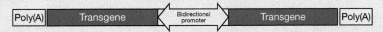

The vector shown immediately above illustrates the expression of two transgenes from the same promoter. The efficiency of transgene expression is not necessarily identical on both sides of the promoter. This may not be the best approach to achieve identical expression of two genes (see other approaches) but is a useful design for studying miRNA regulation of gene expression (Brown and Naldini 2009). Inclusion of the miRNA target in the 3′ UTR of one transgene allows the study of miRNA regulation in the presence of basal transcription from the second transgene, which can be a marker gene (e.g., *GFP.*)]

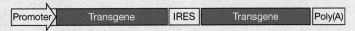

In this vector, transgenes are expressed as separate proteins in their native state from a single transcript. Usually, expression of the transgene downstream from the internal ribosome entry site (IRES) is lower than that of the first transgene, but this aspect depends on the IRES element. The most commonly used IRES element is derived from the encephalomyocarditis virus. Newer IRES elements appear to be considerably more efficient (Chappell et al. 2000; Wang et al. 2005). This cassette design is usually used to coexpress a transgene of interest and a marker gene that confers resistance to a toxic drug, such as puromycin, hygromycin, or neomycin, or a fluorescent protein that allows easy identification of transduced cells. Multimodal proteins that allow fluorescence and bioluminescence imaging have also been developed (Wurdinger et al. 2008; Tannous 2009).

In the example above, a single transgene encodes several proteins separated by 2A sequence/sec. The polyprotein self-processes into individual proteins during translation. Multiple proteins can be expressed from the same polyprotein (e.g., four transcription factors necessary to generate iPS cells [Sommer et al. 2009] and the simultaneous expression of antibody heavy and light chains [Fang et al. 2005]).

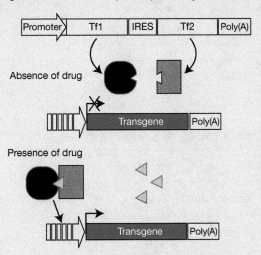

This system uses a bipartite transcription factor complex to regulate gene expression. One plasmid/vector carries a transgene under an inducible promoter composed of 12 ZFHD1-binding sites upstream of a minimal interleukin-2 promoter. Another plasmid/vector is bicistronic and encodes two fusion proteins that function as a transcription activator and a DNA-binding protein. The transcription complex forms in the presence of rapamycin via dimerization of human FKBP12 and FKBP-rapamycin-associated protein (FRAP) domains. The transcription activator (Tf1) consists of the FRB domain of human FRAP fused to an activation domain derived from the p65 subunit of human NF-κB. The DNA-binding fusion protein (Tf2) consists of the ZFHD1 DNA-binding domain fused to three copies of human FKBP12 in tandem. (Ye et al. 1999; Rivera et al. 1999).

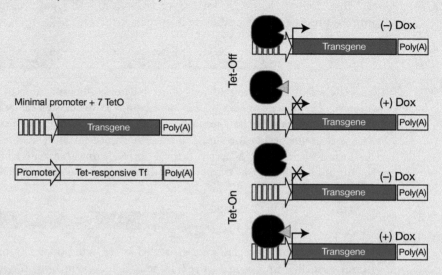

Tetracycline-regulated gene expression was one of the first systems to be developed for application in mammalian cells/organisms and continues to be the most widely used system for drug-regulated expression in biology. It uses a bacterially derived tetracycline-responsive transcription factor, composed of bacterially derived tetracycline binding domain (TetR) fused to the HSV-1 VP16 transcriptional activation domain. Two transcription factors have been developed, tTA and rtTA, that bind the cognate tet operator (tetO) in the absence (Tet-Off system) (Gossen and Bujard 1992) or presence (Tet-On system) (Gossen et al. 1995) of tetracycline, or derivative antibiotics such as doxycycline. Optimized versions of these Tet-responsive transcription factors—namely, rtTA2^S-M2—display lower residual binding to tetO and maximal induction at lower concentrations of doxycycline (Urlinger et al. 2000).

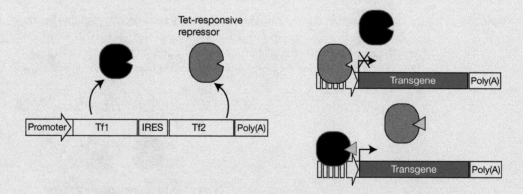

This Tet-regulated system combines a Tet-responsive transcriptional silencer (tTS^{Kid}) with an rtTA transcriptional activator leading to tight control of gene expression over several orders of magnitude (Freundlieb et al. 1999). In the off state, there is active repression of transcription by tTS^{Kid}. This approach is important in the context of viral vectors where other enhancer elements present in the vector may transactivate the Tet-responsive promoter leading to high basal transgene expression. This approach has been successfully used to regulate gene expression in different viral vector systems (Pluta et al. 2005; Candolfi et al. 2007).

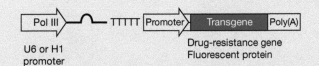

The most common approach for viral vector–based RNA interference (RNAi) is based on using RNA polymerase III (Pol III) promoters such as U6 and H1 (Sibley et al. 2010). The shRNA is followed by a stop sequence composed of a string of five Ts. The shRNA is processed by Dicer, which removes the 6-bp loop, leaving a 21-bp double-stranded molecule with 2-bp overhangs. Incorporation of tetO sequences into H1 or U6 allows for drug-regulated shRNA expression (Pluta et al. 2007).

Expression of shRNA in the context of a miRNA precursor molecule inserted in the 3′ UTR of a transgene expression cassette (Stegmeier et al. 2005) appears to have less toxicity in vivo (McBride et al. 2008; Boudreau et al. 2009). In addition, this design is compatible with RNA Pol II promoters used in the vast majority of transgene expression cassettes to accomplish numerous goals such as tissue-specific expression and developmentally or drug-regulated expression.

REFERENCES

Boudreau RL, Martins I, Davidson BL. 2009. Artificial microRNAs as siRNA shuttles: Improved safety as compared to shRNAs in vitro and in vivo. *Mol Ther* 17: 169–175.

Brown BD, Naldini L. 2009. Exploiting and antagonizing microRNA regulation for therapeutic and experimental applications. *Nat Rev Genet* 10: 578–585.

Candolfi M, Pluhar GE, Kroeger K, Puntel M, Curtin J, Barcia C, Muhammad AK, Xiong W, Liu C, Mondkar S, et al. 2007. Optimization of adenoviral vector-mediated transgene expression in the canine brain in vivo, and in canine glioma cells in vitro. *Neuro Oncol* 9: 245–258.

Chappell SA, Edelman GM, Mauro VP. 2000. A 9-nt segment of a cellular mRNA can function as an internal ribosome entry site (IRES) and when present in linked multiple copies greatly enhances IRES activity. *Proc Natl Acad Sci* 97: 1536–1541.

Donello JE, Loeb JE, Hope TJ. 1998. Woodchuck hepatitis virus contains a tripartite posttranscriptional regulatory element. *J Virol* 72: 5085–5092.

Fang J, Qian JJ, Yi S, Harding TC, Tu GH, VanRoey M, Jooss K. 2005. Stable antibody expression at therapeutic levels using the 2A peptide. *Nat Biotechnol* 23: 584–590.

Freundlieb S, Schirra-Muller C, Bujard H. 1999. A tetracycline controlled activation/repression system with increased potential for gene transfer into mammalian cells. *J Gene Med* 1: 4–12.

Gossen M, Bujard H. 1992. Tight control of gene expression in mammalian cells by tetracycline-responsive promoters. *Proc Natl Acad Sci* 89: 5547–5551.

Gossen M, Freundlieb S, Bender G, Muller G, Hillen W, Bujard H. 1995. Transcriptional activation by tetracyclines in mammalian cells. *Science* 268: 1766–1769.

McBride JL, Boudreau RL, Harper SQ, Staber PD, Monteys AM, Martins I, Gilmore BL, Burstein H, Peluso RW, Polisky B, et al. 2008. Artificial miRNAs mitigate shRNA-mediated toxicity in the brain: Implications for the therapeutic development of RNAi. *Proc Natl Acad Sci* 105: 5868–5873.

Pluta K, Luce MJ, Bao L, Agha-Mohammadi S, Reiser J. 2005. Tight control of transgene expression by lentivirus vectors containing second-generation tetracycline-responsive promoters. *J Gene Med* 7: 803–817.

Pluta K, Diehl W, Zhang XY, Kutner R, Bialkowska A, Reiser J. 2007. Lentiviral vectors encoding tetracycline-dependent repressors and transactivators for reversible knockdown of gene expression: A comparative study. *BMC Biotechnol* 7: 41.

Rivera VM, Ye X, Courage NL, Sachar J, Cerasoli F Jr, Wilson JM, Gilman M. 1999. Long-term regulated expression of growth hormone in mice after intramuscular gene transfer. *Proc Natl Acad Sci* 96: 8657–8662.

Sibley CR, Seow Y, Wood MJ. 2010. Novel RNA-based strategies for therapeutic gene silencing. *Mol Ther* 18: 466–476.

Sommer CA, Stadtfeld M, Murphy GJ, Hochedlinger K, Kotton DN, Mostoslavsky G. 2009. Induced pluripotent stem cell generation using a single lentiviral stem cell cassette. *Stem Cells* 27: 543–549.

Stegmeier F, Hu G, Rickles RJ, Hannon GJ, Elledge SJ. 2005. A lentiviral microRNA-based system for single-copy polymerase II-regulated RNA interference in mammalian cells. *Proc Natl Acad Sci* 102: 13212–13217.

Tannous BA. 2009. Gaussia luciferase reporter assay for monitoring biological expression in culture and in vivo. *Nat Protoc* 4: 582–591.

Urlinger S, Baron U, Thellmann M, Hasan MT, Bujard H, Hillen W. 2000. Exploring the sequence space for tetracycline-dependent transcriptional activators: Novel mutations yield expanded range and sensitivity. *Proc Natl Acad Sci* 97: 7963–7968.

Wang Y, Iyer M, Annala AJ, Chappell S, Mauro V, Gambhir SS. 2005. Noninvasive monitoring of target gene expression by imaging reporter gene expression in living animals using improved bicistronic vectors. *J Nucl Med* 46: 667–674.

Wurdinger T, Badr C, Pike L, de Kleine R, Weissleder R, Breakefield XO, Tannous BA. 2008. A secreted luciferase for ex vivo monitoring of in vivo processes. *Nat Meth* 5: 171–173.

Ye X, Rivera VM, Zoltick P, Cerasoli F Jr, Schnell MA, Gao G, Hughes JV, Gilman M, Wilson JM. 1999. Regulated delivery of therapeutic proteins after in vivo somatic cell gene transfer. *Science* 283: 88–91.

Index

Page references followed by f denote figures; those followed by t denote tables.